# Lecture Notes in Computer Science 3027

*Commenced Publication in 1973*
Founding and Former Series Editors:
Gerhard Goos, Juris Hartmanis, and Jan van Leeuwen

Springer
*Berlin*
*Heidelberg*
*New York*
*Hong Kong*
*London*
*Milan*
*Paris*
*Tokyo*

Christian Cachin Jan Camenisch (Eds.)

# Advances in Cryptology - EUROCRYPT 2004

International Conference on the Theory
and Applications of Cryptographic Techniques
Interlaken, Switzerland, May 2-6, 2004
Proceedings

Volume Editors

Christian Cachin
Jan Camenisch
IBM Zurich Research Laboratory
Säumerstrasse 4, CH-8803 Rüschlikon, Switzerland
E-mail: {cca,jca}@zurich.ibm.com

Library of Congress Control Number: Applied for

CR Subject Classification (1998): E.3, F.2.1-2, G.2.1, D.4.6, K.6.5, C.2, J.1

ISSN 0302-9743
ISBN 3-540-21935-8 Springer-Verlag Berlin Heidelberg New York



Springer-Verlag is a part of Springer Science+Business Media

springeronline.com


Printed in Germany

Typesetting: Camera-ready by author, data conversion by Boller Mediendesign
Printed on acid-free paper SPIN: 10999516 06/3142 5 4 3 2 1 0

# Preface

These are the proceedings of Eurocrypt 2004, the 23rd Annual Eurocrypt Conference. The conference was organized by members of the IBM Zurich Research Laboratory in cooperation with IACR, the International Association for Cryptologic Research.

The conference received a record number of 206 submissions, out of which the program committee selected 36 for presentation at the conference (three papers were withdrawn by the authors shortly after submission). These proceedings contain revised versions of the accepted papers. These revisions have not been checked for correctness, and the authors bear full responsibility for the contents of their papers.

The conference program also featured two invited talks. The first one was the 2004 IACR Distinguished Lecture given by Whitfield Diffie. The second invited talk was by Ivan Damgård who presented "Paradigms for Multiparty Computation." The traditional rump session with short informal talks on recent results was chaired by Arjen Lenstra.

The reviewing process was a challenging task, and many good submissions had to be rejected. Each paper was reviewed independently by at least three members of the program committee, and papers co-authored by a member of the program committee were reviewed by at least six (other) members. The individual reviewing phase was followed by profound and sometimes lively discussions about the papers, which contributed a lot to the quality of the final selection. Extensive comments were sent to the authors in most cases. At the end, the comments and electronic discussion notes filled more than 32,000 lines of text! We would like to thank the members of the program committee for their hard work over the course of several months; it was a pleasure for us to work with them and to benefit from their knowledge and insight. We are also very grateful to the external reviewers who contributed with their expertise to the selection process. Their work is highly appreciated.

The submission of all papers was done using the electronic submission software written by Chanathip Namprempre with modifications by Andre Adelsbach. During the review process, the program committee was mainly communicating using the Web-based review software developed by Bart Preneel, Wim Moreau, and Joris Claessens. We would like to thank Roger Zimmermann for his help with installing and running the software locally, and for solving many other problems, not the least of which was the assembly of these proceedings. The final decisions were made at a meeting in Rüschlikon at the IBM Zurich Research Laboratory. Helga Steimann helped us with the organization and also made sure there was enough coffee and food available so that we could concentrate on the papers and were not distracted by empty stomachs. Thanks a lot!

We are grateful to Endre Bangerter, Martin Hirt, Reto Strobl, and Roger Zimmermann for their help with the local arrangements of the conference.

Eurocrypt 2004 was supported by the IBM Zurich Research Laboratory, Crypto AG, Omnisec, MediaCrypt, HP, Microsoft Research, and Swiss International Air Lines.

Our most important thanks go to our families for bearing with us through this busy period, for their support, and for their love.

Last but not least, we thank all the authors from all over the world who submitted papers. It is due to them and their work that the conference took place.

February 2004 Christian Cachin and Jan Camenisch

# EUROCRYPT 2004

May 2–6, 2004, Interlaken, Switzerland

Sponsored by the
*International Association for Cryptologic Research (IACR)*

in cooperation with the
*IBM Zurich Research Laboratory, Switzerland*

## Program and General Chairs

Christian Cachin and Jan Camenisch
IBM Zurich Research Laboratory, Switzerland

## Program Committee

Alex Biryukov ...................... Katholieke Universiteit Leuven, Belgium
John Black ......................... University of Colorado at Boulder, USA
Christian Cachin ............. IBM Zurich Research Laboratory, Switzerland
Jan Camenisch ............... IBM Zurich Research Laboratory, Switzerland
Jean-Sébastien Coron .................. Gemplus Card International, France
Claude Crépeau ................................ McGill University, Canada
Ivan Damgård ................................ Aarhus University, Denmark
Juan Garay ......................... Bell Labs - Lucent Technologies, USA
Rosario Gennaro ................... IBM T.J. Watson Research Center, USA
Alain Hiltgen ............................................ UBS, Switzerland
Thomas Johansson ............................... Lund University, Sweden
Antoine Joux .................................. DCSSI Crypto Lab, France
Joe Kilian .............................. NEC Laboratories America, USA
Arjen Lenstra ............ Citibank, USA & TU Eindhoven, The Netherlands
Yehuda Lindell .................... IBM T.J. Watson Research Center, USA
Anna Lysyanskaya ................................ Brown University, USA
Daniele Micciancio .................................. UC San Diego, USA
Omer Reingold ....................... Weizmann Institute of Science, Israel
Vincent Rijmen ......................... Cryptomathic and IAIK, Belgium
Phillip Rogaway ......... UC Davis, USA & Chiang Mai University, Thailand
Igor Shparlinski ........................... Macquarie University, Australia
Edlyn Teske ............................... University of Waterloo, Canada
Rebecca Wright ...................... Stevens Institute of Technology, USA

## External Reviewers

Adi Akavia
Joy Algesheimer
Jee Hea An
Siddhartha Annapureddy
Giuseppe Ateniese
Endre Bangerter
Lejla Batina
Amos Beimel
Mihir Bellare
Siddika Berna Ors
Simon Blackburn
Carlo Blundo
Alexandra Boldyreva
Dan Boneh
Colin Boyd
Xavier Boyen
An Braeken
Thomas Brochman
Ran Canetti
Scott Contini
Don Coppersmith
Nora Dabbous
Christophe De Cannière
Alex Dent
Giovanni Di Crescenzo
Christophe Doche
Yevgeniy Dodis
Patrik Ekdahl
Nelly Fazio
Serge Fehr
Marc Fischlin
Matthias Fitzi
Scott Fluhrer
Matt Franklin
Martin Gagne
Steven Galbraith
M. I. Gonzáles Vasco
Jens Groth
Jaime Gutierrez
Stuart Haber
Shai Halevi
Helena Handschuh
Darrel Hankerson
Danny Harnik
Jonathan Herzog
Florian Hess
Alejandro Hevia
Jason Hinek
Susan Hohenberger
Nicholas Hopper
Nick Howgrave-Graham
Jim Hughes
Yuval Ishai
Markus Jakobsson
Stas Jarecki
Eliane Jaulmes
Fredrik Jönsson
Marc Joye
Yael Tauman Kalai
Aggelos Kiayias
Neal Koblitz
David Kohel
Yoshi Kohno
Hugo Krawczyk
Ted Krovetz
Sébastien Kunz-Jacques
John Langford
Joseph Lano
Moses Liskov
Benjamin Lynn
Philip MacKenzie
Chip Martel
Alex May
Dominic Mayers
Ralph C. Merkle
Sara Miner
Ilya Mironov
Siguna Müller
Frédéric Muller
Sean Murphy
Chanathip Namprempre
Moni Naor
Mats Näslund
Phong Nguyen
Antonio Nicolosi
Svetla Nikova
Kobbi Nissim
Luke O'Connor
Roberto Oliveira
Pascal Paillier
Adriana Palacio
Kenneth Paterson
Souradyuti Paul
Thomas Pedersen
Chris Peikert
Erez Petrank
Birgit Pfitzmann
Benny Pinkas
David Pointcheval
Jonathan Poritz
John Proos
Michael Quisquater
Tal Rabin
Zulfikar Ramzan
Leonid Reyzin
Pierre-Michel Ricordel
Alon Rosen
Amit Sahai
Louis Salvail
Palash Sarkar
Jasper Scholten
Hovav Shacham
Taizo Shirai
Thomas Shrimpton
Alice Silverberg
Adam Smith
Patrick Solè
Jessica Staddon
Markus Stadler
Martijn Stam
Andreas Stein
Ron Steinfeld
Reto Strobl
Frédéric Valette
Bart Van Rompay
Luis von Ahn
Shabsi Walfish
Huaxiong Wang
Bogdan Warinschi
John Watrous
Christopher Wolf
Ke Yang

# Table of Contents

## Private Computation



## Signatures I



## Unconditional Security



## Distributed Cryptography

## Foundations I



## Identity-Based Encryption



## Elliptic Curves



## Signatures II



## Public-Key Cryptography

## Foundations II



## Multiparty Computation



## Cryptanalysis



## New Applications



## Algorithms and Implementation

## Anonymity

# Efficient Private Matching and Set Intersection

Michael J. Freedman[1*], Kobbi Nissim[2**], and Benny Pinkas[3]

[1] New York University
(mfreed@cs.nyu.edu)
[2] Microsoft Research SVC
(kobbi@microsoft.com)
[3] HP Labs
(benny.pinkas@hp.com)

**Abstract.** We consider the problem of computing the intersection of private datasets of two parties, where the datasets contain lists of elements taken from a large domain. This problem has many applications for online collaboration. We present protocols, based on the use of homomorphic encryption and balanced hashing, for both semi-honest and malicious environments. For lists of length $k$, we obtain $O(k)$ communication overhead and $O(k \ln \ln k)$ computation. The protocol for the semi-honest environment is secure in the standard model, while the protocol for the malicious environment is secure in the random oracle model. We also consider the problem of approximating the size of the intersection, show a linear lower-bound for the communication overhead of solving this problem, and provide a suitable secure protocol. Lastly, we investigate other variants of the matching problem, including extending the protocol to the multi-party setting as well as considering the problem of approximate matching.

## 1 Introduction

This work considers several two-party set-intersection problems and presents corresponding secure protocols. Our protocols enable two parties that each hold a set of inputs – drawn from a *large* domain – to jointly calculate the intersection of their inputs, without leaking any additional information. The set-intersection primitive is quite useful as it is extensively used in computations over databases, *e.g.*, for data mining where the data is vertically partitioned between parties (namely, each party has different attributes referring to the same subjects).

One could envision the usage of efficient set-intersection protocols for online recommendation services, online dating services, medical databases, and many other applications. We are already beginning to see the deployment of such applications using either trusted third parties or plain insecure communication.

* Research partially done while the author was visiting HP Labs.
** Research done while the author was at NEC Labs.

C. Cachin and J. Camenisch (Eds.): EUROCRYPT 2004, LNCS 3027, pp. 1–19, 2004.

**Contributions.** We study private two-party computation of set intersection, which we also denote as *private matching* (PM):

- Protocols for computing private matching, based on homomorphic encryption and balanced allocations: (i) a protocol secure against semi-honest adversaries; and (ii) a protocol, in the random oracle model, secure against malicious adversaries.[4] Their overhead for input lists of length $k$ is $O(k)$ communication and $O(k \ln \ln k)$ computation, with small constant factors. These protocols are more efficient than previous solutions to this problem.
- Variants of the private matching protocol that (i) compute the intersection *size*, (ii) decide whether the intersection size is greater than a threshold, or (iii) compute some other function of the intersection set.
- We consider private approximation protocols for the intersection size (similar to the private approximation of the Hamming distance by [10]). A simple reduction from the communication lower-bound on disjointness shows that this problem cannot have a sublinear *worst-case* communication overhead. We show a sampling-based private approximation protocol that achieves instance-optimal communication.
- We extend the protocol for set intersection to a multi-party setting.
- We introduce the problem of secure approximate (or "fuzzy") matching and search, and we present protocols for several simple instances.

## 2 Background and Related Work

**Private equality tests (PET).** A simpler form of private matching is where each of the two datasets has a single element from a domain of size $N$. A circuit computing this function has $O(\log N)$ gates, and therefore can be securely evaluated with this overhead. Specialized protocols for this function were also suggested in [9, 18, 17], and they essentially have the same overhead. A solution in [3] provides fairness in addition to security.

A circuit-based solution for computing PM of datasets of length $k$ requires $O(k^2 \log N)$ communication and $O(k \log N)$ oblivious transfers. Another trivial construction compares all combinations of items from the two datasets using $k^2$ instantiations of a PET protocol (which itself has $O(\log N)$ overhead). The computation of this comparison can be reduced to $O(k \log N)$, while retaining the $O(k^2 \log N)$ communication overhead [18]. There are additional constructions that solve the private matching problem at the cost of only $O(k)$ exponentiations [12, 8]. However, these constructions were only analyzed in the random oracle model, against semi-honest parties.

**Disjointness and set intersection.** Protocols for computing (or deciding) the intersection of two sets have been researched both in the general context of communication complexity and in the context of secure protocols. Much attention has been given to evaluating the communication complexity of the disjointness problem, where the two parties in the protocol hold subsets $a$ and $b$ of

[4] For malicious clients, we present a protocol that is secure in the standard model.

$\{1, \ldots, N\}$. The disjointness function $\textsc{Disj}(a, b)$ is defined to be 1 if the sets $a, b$ have an empty intersection. It is well known that $R_\epsilon(\textsc{Disj}) = \Theta(N)$ [14, 22]. An immediate implication is that computing $|a \cap b|$ requires $\Theta(N)$ communication. Therefore, even without taking privacy into consideration, the communication complexity of private matching is at least proportional to the input size.

One may try and get around the high communication complexity of computing the intersection size by approximating it. In the context of secure protocols, this may lead to a sublinear *private approximation* protocol for intersection size.[5] If one settles for an approximation up to additive error $\epsilon N$ (for constant $\epsilon$), it is easy to see that very efficient protocols exist, namely $O(\log N)$ bits in the private randomness model [16, Example 5.5]. However, if we require *multiplicative* error (*e.g.*, an $(\epsilon, \delta)$-approximation), we show a simple reduction from disjointness that proves that a lower-bound of $\Omega(N)$ communication bits is necessary for any such approximation protocol. See Section 6 for details.

## 3 Preliminaries

### 3.1 Private Matching (PM)

A **private matching** (PM) scheme is a two-party protocol between a client (chooser) $\mathcal{C}$ and a server (sender) $\mathcal{S}$. $\mathcal{C}$'s input is a set of inputs of size $k_\mathcal{C}$, drawn from some domain of size $N$; $\mathcal{S}$'s input is a set of size $k_\mathcal{S}$ drawn from the same domain. At the conclusion of the protocol, $\mathcal{C}$ learns which specific inputs are shared by both $\mathcal{C}$ and $\mathcal{S}$. That is, if $\mathcal{C}$ inputs $X = \{x_1, \ldots, x_{k_\mathcal{C}}\}$ and $\mathcal{S}$ inputs $Y = \{y_1, \ldots, y_{k_\mathcal{S}}\}$, $\mathcal{C}$ learns $X \cap Y$: $\{x_u | \exists v, x_u = y_v\} \leftarrow \mathsf{PM}(X, Y)$.

**PM Variants.** Some variants of the private matching protocol include the following. (i) Private cardinality matching ($\mathsf{PM}_C$) allows $\mathcal{C}$ to learn *how many* inputs it shares with $\mathcal{S}$. That is, $\mathcal{C}$ learns $|X \cap Y|$: $|\mathsf{PM}| \leftarrow \mathsf{PM}_C(X, Y)$. (ii) Private threshold matching ($\mathsf{PM}_t$) provides $\mathcal{C}$ with the answer to the *decisional problem* whether $|X \cap Y|$ is greater than some pre-specified threshold $t$. That is, $1 \leftarrow \mathsf{PM}_t(X, Y)$ if $\mathsf{PM}_C > t$ and 0 otherwise. (iii) Generalizing $\mathsf{PM}_C$ and $\mathsf{PM}_t$, one could define arbitrary private-matching protocols that are simple functions of the intersection set, *i.e.*, based on the output of $\mathsf{PM}$ or $\mathsf{PM}_C$.

**Private Matching and Oblivious Transfer.** We show a simple reduction from oblivious transfer (OT) to private matching. The OT protocol we design is a 1-out-of-2 bit-transfer protocol in the semi-honest case. The sender's input contains two bits $b_0, b_1$. The chooser's input is a bit $\sigma$. At the end of the protocol the chooser learns $b_\sigma$ and nothing else, while the sender learns nothing.

First, the parties generate their respective $\mathsf{PM}$ inputs: The sender generates a list of two strings, $\{0|b_0, 1|b_1\}$, and the chooser generates the list $\{\sigma|0, \sigma|1\}$. Then, they run the $\mathsf{PM}$ protocol, at the end of which the chooser learns $\sigma|b_\sigma$. It follows by the results of Impagliazzo and Rudich [13] that there is no black-box reduction of private matching from one-way functions.

---

[5] Informally, a private approximation is an approximation that does not leak information that is not computable given the exact value. See the definition in [10].

Since the reduction is used to show an impossibility result, it is sufficient to show it for the simplest form of OT, as we did above. We note that if one actually wants to build an OT protocol from a PM primitive, it is possible to directly construct a 1-out-of-$N$ bit transfer protocol. In addition, the `PM-Semi-Honest` protocol we describe supports OT of strings.

### 3.2 Adversary Models

This paper considers both semi-honest and malicious adversaries. Due to space constraints, we only provide the intuition and informal definitions of these models. The reader is referred to [11] for the full definitions.

**Semi-honest adversaries.** In this model, both parties are assumed to act according to their prescribed actions in the protocol. The security definition is straightforward, particularly as in our case where only one party ($\mathcal{C}$) learns an output. We follow [18] and divide the requirements into (i) protecting the client and (ii) protecting the sender.
***The client's security – indistinguishability:*** Given that the server $\mathcal{S}$ gets no output from the protocol, the definition of $\mathcal{C}$'s privacy requires simply that the server cannot distinguish between cases in which the client has different inputs.
***The server's security – comparison to the ideal model:*** The definition ensures that the client does not get more or different information than the output of the function. This is formalized by considering an ideal implementation where a trusted third party (TTP) gets the inputs of the two parties and outputs the defined function. We require that in the real implementation of the protocol—that is, one without a TTP—the client $\mathcal{C}$ does not learn different information than in the ideal implementation.

**Malicious adversaries.** In this model, an adversary may behave arbitrarily. In particular, we cannot hope to avoid parties (i) refusing to participate in the protocol, (ii) substituting an input with an arbitrary value, and (iii) prematurely aborting the protocol. The standard security definition (see, *e.g.*, [11]) captures both the correctness and privacy issues of the protocol and is limited to the case in which only one party obtains an output. Informally, the definition is based on a comparison to the ideal model with a TTP, where a corrupt party may give arbitrary input to the TTP. The definition also is limited to the case where at least one of the parties is honest: if $\mathcal{C}$ (resp. $\mathcal{S}$) is honest, then for any strategy that $\mathcal{S}$ (resp. $\mathcal{C}$) can play in the real execution, there is a strategy that it could play in the ideal model, such that the real execution is computationally indistinguishable from execution in the ideal model. We note that main challenge in ensuring security is enforcing the protocol's correctness, rather than its privacy.

### 3.3 Cryptographic Primitives – Homomorphic Encryption Schemes

Our constructions use a semantically-secure public-key encryption scheme that preserves the group homomorphism of addition and allows multiplication by a constant. This property is obtained by Paillier's cryptosystem [20] and subsequent constructions [21, 7]. That is, it supports the following operations that can be performed without knowledge of the private key: (i) Given two encryptions $\mathsf{Enc}(m_1)$ and $\mathsf{Enc}(m_2)$, we can efficiently compute $\mathsf{Enc}(m_1+m_2)$. (ii) Given some constant $c$ belonging to the same group, we can compute $\mathsf{Enc}(cm)$. We will use the following corollary of these two properties: Given encryptions of the coefficients $a_0,\ldots,a_k$ of a polynomial $P$ of degree $k$, and knowledge of a plaintext value $y$, it is possible to compute an encryption of $P(y)$.[6]

# 4 The Semi-Honest Case

## 4.1 Private Matching for Set Intersection (PM)

The protocol follows the following basic structure. $\mathcal{C}$ defines a polynomial $P$ whose roots are her inputs:

$$P(y) = (x_1 - y)(x_2 - y)\ldots(x_{k_C} - y) = \sum_{u=0}^{k_C} \alpha_u y^u$$

She sends to $\mathcal{S}$ homomorphic encryptions of the coefficients of this polynomial. $\mathcal{S}$ uses the homomorphic properties of the encryption system to evaluate the polynomial at each of his inputs. He then multiplies each result by a fresh random number $r$ to get an intermediate result, and he adds to it an encryption of the value of his input, *i.e.*, $\mathcal{S}$ computes $\mathsf{Enc}(r \cdot P(y) + y)$. Therefore, for each of the elements in the intersection of the two parties' inputs, the result of this computation is the value of the corresponding element, whereas for all other values the result is random.[7] See Protocol `PM-Semi-Honest`.[8]

## 4.2 Efficiently Evaluating the Polynomial

As the computational overhead of exponentiations dominates that of other operations, we evaluate the computational overhead of the protocol by counting exponentiations. Equivalently, we count the number of *multiplications* of the

[6] We neglect technicalities that are needed to make sure the resulting ciphertext hides the sequence of homomorphic operations that led to it. This may be achieved, *e.g.*, by multiplying the result by a random encryption of 1.

[7] This construction can be considered a generalization of the oblivious transfer protocols of [19, 1, 17]. In those, a client retrieving item $i$ sends to the server a predicate which is 0 if and only if $i = j$ where $j \in [N]$.

[8] It is sufficient for Step 3 of the protocol that $\mathcal{C}$ is able to decide whether some ciphertext corresponds to $x \in X$ (*i.e.*, decryption is not necessary). This weaker property is of use if, for example, one uses the El Gamal encryption scheme and encodes an element $x$ by $g^x$ (to allow the homomorphic properties under addition). This may prevent $rP(y) + y$ from being recovered in the decryption process, yet it is easy for $\mathcal{C}$ to decide whether $rP(y) + y = x$. The Paillier [20] homomorphic encryption scheme recovers $rP(y) + y$.

Protocol PM-Semi-Honest

INPUT: $\mathcal{C}$'s input is a set $X = \{x_1, \ldots, x_{k_C}\}$, $\mathcal{S}$'s input is a set $Y = \{y_1, \ldots, y_{k_S}\}$. The elements in the input sets are taken from a domain of size $N$.

1. $\mathcal{C}$ performs the following:
   (a) She chooses the secret-key parameters for a semantically-secure homomorphic encryption scheme, and publishes its public keys and parameters. The plaintexts are in a field that contains representations of the $N$ elements of the input domain, but is exponentially larger.
   (b) She uses interpolation to compute the coefficients of the polynomial $P(y) = \Sigma_{u=0}^{k_C} \alpha_u y^u$ of degree $k_C$ with roots $\{x_1, \ldots, x_{k_C}\}$.
   (c) She encrypts each of the $(k_C + 1)$ coefficients by the semantically-secure homomorphic encryption scheme and sends to $\mathcal{S}$ the resulting set of ciphertexts, $\{\mathsf{Enc}(\alpha_0), \ldots, \mathsf{Enc}(\alpha_{k_C})\}$.
2. $\mathcal{S}$ performs the following for every $y \in Y$,
   (a) He uses the homomorphic properties to evaluate the encrypted polynomial at $y$. That is, he computes $\mathsf{Enc}(P(y)) = \mathsf{Enc}(\Sigma_{u=0}^{k_C} \alpha_u y^u)$. See Section 4.2.
   (b) He chooses a random value $r$ and computes $\mathsf{Enc}(rP(y) + y)$. (One can also encrypt some additional payload data $p_y$ by computing $\mathsf{Enc}(rP(y) + (y|p_y))$. $\mathcal{C}$ obtains $p_y$ iff $y$ is in the intersection.)

   He randomly permutes this set of $k_S$ ciphertexts and sends the result back to the client $\mathcal{C}$.
3. $\mathcal{C}$ decrypts all $k_S$ ciphertexts received. She locally outputs all values $x \in X$ for which there is a corresponding decrypted value .

homomorphically-encrypted values by constants (in Step 2(a)), as these multiplications are actually implemented as exponentiations.

Given the encrypted coefficients $\mathsf{Enc}(\alpha_u)$ of a polynomial $P$, a naive computation of $\mathsf{Enc}(P(y))$ as $\mathsf{Enc}(\sum_{u=0}^{k_C} y^u \alpha_u)$ results in an overhead of $O(k_C)$ exponentiations, and hence in a total of $O(k_C k_S)$ exponentiations for PM-Semi-Honest.

The computational overhead can be reduced by noting that the input domain is typically much smaller than the modulus used by the encryption scheme. Hence one may encode the values $x$, $y$ as numbers in the smaller domain. In addition, Horner's rule can be used to evaluate the polynomial more efficiently by eliminating large exponents. This yields a significant (large constant factor) reduction in the overhead.

We achieve a more significant reduction of the overhead by allowing the client to use multiple low-degree polynomials and then allocating input values to polynomials by hashing. This results in reducing the computational overhead to $O(k_C + k_S \ln \ln k_C)$ exponentiations. Details follow.

**Exponents from a small domain.** Let $\lambda$ be the security parameter of the encryption scheme (*e.g.*, $\lambda$ is the modulus size). A typical choice is $\lambda = 1024$ or larger. Yet, the input sets are usually of size $\ll 2^\lambda$ and may be mapped into a small domain—of length $n \approx 2\log(\max(k_C, k_S))$ bits—using pairwise-independent hashing, which induces only a small collision probability. The server should compute $\mathsf{Enc}(P(y))$, where $y$ is $n$ bits long.

**Using Horner's rule.** We get our first overhead reduction by applying Horner's rule: $P(y) = \alpha_0 + \alpha_1 y + \alpha_2 y^2 + \cdots + \alpha_{k_C} y^{k_C}$ is evaluated "from the inside out" as $\alpha_0 + y(\alpha_1 + y(\alpha_2 + y(\alpha_3 + \cdots y(\alpha_{k_C - 1} + y\alpha_{k_C})\cdots)))$. One multiplies each intermediate result by a *short* $y$, compared with $y^i$ in the naive evaluation, which results in $k_C$ *short* exponentiations.

When using the "text book" algorithm for computing exponentiations, the computational overhead is linear in the length of the exponent. Therefore, Horner's rule improves this overhead by a factor of $\lambda/n$ (which is about 50 for $k_C, k_S \approx 1000$). The gain is substantial even when fine-tunes exponentiation algorithms—such as Montgomery's method or Karatsuba's technique—are used.

**Using hashing for bucket allocation.** The protocol's main computational overhead results from the server computing polynomials of degree $k_C$. We now reduce the degree of these polynomials. For that, we define a process that throws the client's elements into $B$ bins, such that each bin contains at most $M$ elements.

The client now defines a polynomial of degree $M$ for each bin: All items mapped to the bin by some function $h$ are defined to be roots of the polynomial. In addition, the client adds the root $x = 0$ to the polynomial, with multiplicity which sets the total degree of the polynomial to $M$. That is, if $h$ maps $\ell$ items to the bin, the client first defines a polynomial whose roots are these $\ell$ items, and then multiplies it by $x^{M-\ell}$. (We assume that 0 is not a valid input.) The process results in $B$ polynomials, all of them of degree $M$, that have a total of $k_C$ non-zero roots.

$\mathcal{C}$ sends to $\mathcal{S}$ the encrypted coefficients of the polynomials, and the mapping from elements to bins.[9] For every $y \in Y$, $\mathcal{S}$ finds the bins into which $y$ could be mapped and evaluates the polynomial of those bins. He proceeds as before and responds to $\mathcal{C}$ with the encryptions $rP(y) + y$ for every possible bin allocation for all $y$.

**Throwing elements into bins – balanced allocations.** We take the mapping from elements to bins to be a random hash function $h$ with a range of size $B$, chosen by the client. Our goal is to reduce $M$, the upper bound on the number of items in a bin. It is well known that if the hash function $h$ maps each item to a random bin, then with high probability (over the selection of $h$), each bin contains at most $k_C/B + O(\sqrt{(k_C/B)\log B} + \log B)$ elements. A better allocation is obtained using the balanced allocation hashing by Azar et al. [2]. The function $h$ now chooses *two* distinct bins for each item, and the item is mapped into the bin which is *less occupied* at the time of placement. In the resulting protocol, the server uses $h$ to locate the two bins into which $y$ might have been mapped, evaluates both polynomials, and returns the two answers to $\mathcal{C}$.

Theorem 1.1 of [2] shows that the maximum load of a bin is now exponentially smaller: with $1 - o(1)$ probability, the maximum number of items mapped to a bin is $M = (1 + o(1)) \ln\ln B / \ln 2 + \Theta(k_C/B)$. Setting $B = k_C / \ln\ln k_C$, we get $M = O(\ln\ln k_C)$.

[9] For our purposes, it is sufficient that the mapping is selected pseudo-randomly, either jointly or by either party.

**A note on correctness and on constants.** One may worry about the case that $C$ is unlucky in her choice of $h$ such that more than $M$ items are mapped to some bin. The bound of [2] only guarantees that this happens with probability $o(1)$. However, Broder and Mitzenmacher [4] have shown that asymptotically, when we map $n$ items into $n$ bins, the number of bins with $i$ or more items falls approximately like $2^{-2.6^i}$. This means that a bound of $M = 5$ suffices with probability $10^{-58}$. Furthermore, if the hashing searches for the emptiest in three bins, then $M = 3$ suffices with probability of about $10^{-33}$. The authors also provide experimental results that confirm the asymptotic bound for the case of $n = 32,000$. We conclude that we can bound $\ln \ln k_C$ by a small constant in our estimates of the overhead. Simple experimentation can provide finer bounds.

**Efficiency.** The communication overhead, and the computation overhead of the client, are equal to the total number of coefficients of the polynomials. This number, given by $B \cdot M$, is $O(k_C)$ if $B = k_C / \ln \ln k_C$. If $k \leq 2^{24}$, then using $B = k_C$ bins implies that the communication overhead is at most 4 times that of the protocol that does not use hashing.

The server computes, for each item in his input, $M$ exponentiations with a small exponent, and one exponentiation with a full-length exponent (for computing $r \cdot P(y)$). Expressing this overhead in terms of full-length exponentiations yields an overhead of $O(k_S + k_S \frac{\ln \ln k_C \cdot n}{\lambda})$ for $B = k_C / \ln \ln k_C$. In practice, the overhead of the exponentiations with a small exponent has little effect on the total overhead, which is dominated by $k_S$ full-length exponentiations.

### 4.3 Security of `PM-Semi-Honest`

We state the claims of security for PM in the semi-honest model.

**Lemma 1 (Correctness).** *Protocol* `PM-Semi-Honest` *evaluates the* PM *function with high probability.*

(The proof is based on the fact that the client receives an encryption of $y$ for $y \in X \cap Y$, and an encryption of a random value otherwise.)

**Lemma 2 ($C$'s privacy is preserved).** *If the encryption scheme is semantically secure, then the views of $S$ for any two inputs of $C$ are indistinguishable.*

(The proof uses the fact that the only information that $S$ receives consists of semantically-secure encryptions.)

**Lemma 3 ($S$'s privacy is preserved).** *For every client $C^*$ that operates in the real model, there is a client $C$ operating in the ideal model, such that for every input $Y$ of $S$, the views of the parties $C, S$ in the ideal model is indistinguishable from the views of $C^*, S$ in the real model.*

(The proof defines a polynomial whose coefficients are the plaintexts of the encryptions sent by $C^*$ to $S$. The $k_C$ roots of this polynomial are the inputs that $C$ sends to the trusted third party in the ideal implementation.)

**Security of the hashing-based protocol.** Informally, the hashing-based protocol preserves $\mathcal{C}$'s privacy since (i) $\mathcal{S}$ still receives semantically-secure encryptions, and (ii) the key is chosen independently of $\mathcal{C}$'s input. Thus, neither the key nor $h$ reveal any information about $X$ to $\mathcal{S}$. The protocol preserves $\mathcal{S}$'s privacy since the total number of non-zero roots of the polynomials is $k_{\mathcal{C}}$.

### 4.4 Variant: Private Matching for Set Cardinality ($\mathsf{PM}_C$)

In a protocol for private cardinality matching, $\mathcal{C}$ should learn the cardinality of $X \cap Y$, but not the actual elements of this set. $\mathcal{S}$ needs only slightly change his behavior from that in Protocol `PM-Semi-Honest` to enable this functionality. Instead of encoding $y$ in Step 2(b), $\mathcal{S}$ now only encodes some "special" string, such as a string of 0's, *i.e.*, $\mathcal{S}$ computes $\mathsf{Enc}(rP(y) + 0^+)$. In Step 3 of the protocol, $\mathcal{C}$ counts the number of ciphertexts received from $\mathcal{S}$ that decrypt to the string $0^+$ and locally outputs this number $c$. The proof of security for this protocol trivially follows from that of `PM-Semi-Honest`.

### 4.5 Variants: Private Matching for Cardinality Threshold ($\mathsf{PM}_t$) and Other Functions

In a protocol for private threshold matching, $\mathcal{C}$ should only learn whether $c = |X \cap Y| > t$. To enable this functionality, we change `PM-Semi-Honest` as follows. (i) In Step 2(b), $\mathcal{S}$ encodes random numbers instead of $y$ in PM (or $0^+$ in $\mathsf{PM}_C$). That is, he computes $\mathsf{Enc}(rP(y)+r_y)$, for random $r_y$. (ii) Following the basic PM protocol, $\mathcal{C}$ and $\mathcal{S}$ engage in a secure circuit evaluation protocol. The circuit takes as input $k_{\mathcal{S}}$ values from each party: $\mathcal{C}$'s input is the ordered set of plaintexts she recovers in Step 3 of the PM protocol. $\mathcal{S}$'s input is the list of random payloads he chooses in Step 2(b), in the same order he sends them. The circuit first computes the equality of these inputs bit-by-bit, which requires $k_{\mathcal{S}}\lambda'$ gates, where $\lambda'$ is a statistical security parameter. Then, the circuit computes a threshold function on the results of the $k_{\mathcal{S}}$ comparisons.

Hence, the threshold protocol has the initial overhead of a PM protocol plus the overhead of a secure circuit evaluation protocol. Note, however, that the overhead of circuit evaluation is not based on the input domain of size $N$. Rather, it first needs to compute equality on the input set of size $k_{\mathcal{S}}$, then compute some simple function of the size of the *intersection set*. In fact, this protocol can be used to compute any function of the intersection set, *e.g.*, check if $c$ within some range, not merely the threshold problem.

## 5 Security against Malicious Parties

We describe modifications to our PM protocol in order to provide security in the malicious adversary model. Our protocols are based on protocol `PM-Semi-Honest`, optimized with the balanced allocation hashing.

We first deal with malicious clients and then with malicious servers. Finally, we combine these two protocols to achieve a protocol in which either party may behave adversarially. We take this non-standard approach as: (i) It provides conceptual clarity as to the security concerns for each party; (ii) These protocols may prove useful in varying trust situations, *e.g.*, one might trust a server but not the myriad clients; and (iii) The client protocol is secure in the standard model, while the server protocol is analyzed in the random oracle model.

Protocol `PM-Malicious-Client`

INPUT: $\mathcal{C}$ has input $X$ of size $k_{\mathcal{C}}$, and $\mathcal{S}$ has input $Y$ of size $k_{\mathcal{S}}$, as before.

1. $\mathcal{C}$ performs the following:
   (a) She chooses a key for a pseudo-random function that realizes the balanced allocation hash function $h$, and she sends it to $\mathcal{S}$.
   (b) She chooses a key $s$ for a pseudo-random function $F$ and gives each item $x$ in her input $X$ a new pseudo-identity, $F_s(G(x))$, where $G$ is a collision-resistant hash function.
   (c) For each of her polynomials, $\mathcal{C}$ first sets roots to the pseudo-identities of such inputs that were mapped to the corresponding bin. Then, she adds a sufficient number of 0 roots to set the polynomial's degree to $M$.
   (d) She repeats Steps (b),(c) for $L$ times to generate $L$ copies, using a different key $s$ for $F$ in each iteration.
2. $\mathcal{S}$ asks $\mathcal{C}$ to open $L/2$ of the copies.
3. $\mathcal{C}$ opens the encryptions of the coefficients of the polynomials for these $L/2$ copies to $\mathcal{S}$, but does not reveal the associated keys $s$. Additionally, $\mathcal{C}$ sends the keys $s$ used in the unopened $L/2$ copies.
4. $\mathcal{S}$ verifies that the each opened copy contains $k_{\mathcal{C}}$ roots. If this verification fails, $\mathcal{S}$ halts. Otherwise, $\mathcal{S}$ uses the additional $L/2$ keys he receives, along with the hash function $G$, to generate the pseudo-identities of his inputs. He runs the protocol for each of the polynomials. However, for an input $y$, rather than encoding $y$ as the payload for each polynomial, he encodes $L/2$ random values whose exclusive-or is $y$.
5. $\mathcal{C}$ receives the results, organized as a list of $k_{\mathcal{S}}$ sets of size $L/2$. She decrypts them, computes the exclusive-or of each set, and compares it to her input.

### 5.1 Malicious Clients

To ensure security against a malicious client $\mathcal{C}$, it must be shown that for any possible client behavior in the real model, there is an input of size $k_{\mathcal{C}}$ that the client provides to the TTP in the ideal model, such that his view in the real protocol is efficiently simulatable from his view in the ideal model.

We first describe a simple solution for the implementation that does not use hashing. We showed in Lemma 3 that if a value $y$ is not a root of the polynomial sent by the client, the client cannot distinguish whether this item is in the server's input. Thus, we have to take care of the possibility that $\mathcal{C}$ sends the encryption of a polynomial with more than $k_{\mathcal{C}}$ roots. This can only happen if all the encrypted coefficients are zero ($P$'s degree is indeterminate). We therefore modify the protocol to require that at least one coefficient is non-zero – in Step 1(b) of Protocol `PM-Semi-Honest`, $\mathcal{C}$ generates the coefficients of $P$ with $\alpha_0$ set to 1, then sends encryptions of the other coefficients to $\mathcal{S}$.

In the protocol that uses hashing, $\mathcal{C}$ sends encryptions of the coefficients of $B$ polynomials (one per bin), each of degree $M$. The server must ensure that the *total* number of roots (different than 0) of these polynomials is $k_{\mathcal{C}}$. For that we use a cut-and-choose method, as shown in Protocol `PM-Malicious-Client`. With overhead $L$ times that of the original protocol, we get error probability that is exponentially small in $L$.

*Proof.* (sketch) In our given cut-and-choose protocol, note that $\mathcal{C}$ learns about an item iff it is a root of all the $L/2$ copies evaluated by $\mathcal{S}$. Therefore, to learn about more than $k_{\mathcal{C}}$ items, she must have $L/2$ copies such that each has more than $k_{\mathcal{C}}$ roots. The probability that all such polynomials are not checked by $\mathcal{S}$ is exponentially small in $L$. This argument can be used to show that, for every adversarial $\mathcal{C}*$ whose success probability is not exponentially small, there is a corresponding $\mathcal{C}$ in the ideal model whose input contains at most $k_{\mathcal{C}}$ items.[10]

### 5.2 Malicious Servers

Protocol `PM-Semi-Honest` of Section 4 enables a malicious server to attack the protocol *correctness*.[11] He can play tricks like encrypting the value $r \cdot (P(y) + P(y')) + y''$ in Step 2(b), so that $\mathcal{C}$ concludes that $y''$ is in the intersection set iff both $y$ and $y'$ are $X$. This behavior does not correspond to the definition of PM in the ideal model. Intuitively, this problem arises from $\mathcal{S}$ using two "inputs" in the protocol execution for input $y$—a value for the polynomial evaluation, and a value used as a payload—whereas $\mathcal{S}$ has a single input in the ideal model.[12]

We show how to modify Protocol `PM-Semi-Honest` to gain security against malicious servers. The protocol based on balanced allocations may be modified similarly. Intuitively, we force the server to run according to its prescribed procedure. Our construction, `PM-Malicious-Server`, is in the random oracle model.

The server's privacy is preserved as in `PM-Semi-Honest`: The pair $(e, h)$ is indistinguishable from random whenever $P(y) \neq 0$. The following lemma shows that the client security is preserved under malicious server behavior.

---

[10] In the proof, the pseudo-random function $F$ hides from $\mathcal{S}$ the identities of the values corresponding to the roots of the opened polynomials. The collision-resistant hash function $G$ prevents $\mathcal{C}$ from setting a root to which $\mathcal{S}$ maps two probable inputs.

[11] He cannot affect $\mathcal{C}$'s privacy as all the information $\mathcal{C}$ sends is encrypted via a semantically-secure encryption scheme.

[12] Actually, the number of "inputs" is much higher, as $\mathcal{S}$ needs to be consistent in using the same $y$ for all the steps of the polynomial-evaluation procedure.

Protocol `PM-Malicious-Server`

INPUT: $\mathcal{C}$ has input $X$ of size $k_{\mathcal{C}}$, and $\mathcal{S}$ has input $Y$ of size $k_{\mathcal{S}}$, as before.
RANDOM ORACLES: $H_1, H_2$.

1. $\mathcal{C}$ performs the following:
   (a) She chooses a secret-key/public-key pair for the homomorphic encryption scheme, and sends the public-key to $\mathcal{S}$.
   (b) She generates the coefficients of a degree $k_{\mathcal{C}}$ polynomial $P$ whose roots are the values in $X$. She sends to $\mathcal{S}$ the encrypted coefficients of $P$.
2. $\mathcal{S}$ performs the following for every $y \in Y$,
   (a) He chooses a random $s$ and computes $r = H_1(s)$. We use $r$ to "derandomize" the rest of $\mathcal{S}$'s computation for $y$, and we assume that it is of sufficient length.
   (b) He uses the homomorphic properties of the encryption scheme to compute $(e, h) \leftarrow (\mathsf{Enc}(r' \cdot P(y) + s), H_2(r'', y))$. In this computation, $r$ is parsed to supply $r', r''$ and all the randomness needed in the computation.

   $\mathcal{S}$ randomly permutes this set of $k_{\mathcal{S}}$ pairs and sends it to $\mathcal{C}$.
3. $\mathcal{C}$ decrypts all the $k_{\mathcal{S}}$ pairs she received. She performs the following operations for every pair $(e, h)$,
   (a) She decrypts $e$ to get $\hat{s}$ and computes $\hat{r} = H_1(\hat{s})$.
   (b) She checks whether, for some $x \in X$, the pair $(e, h)$ is consistent with $x$ and $\hat{s}$. That is, whether the server yields $(e, h)$ using her encrypted coefficients on $y \leftarrow x$ and randomness $\hat{r}$. If so, she puts $x$ in the intersection set.

**Lemma 4 (Security for the client).** *For every server $\mathcal{S}^*$ that operates in the real model, there is a server $\mathcal{S}$ operating in the ideal model, such that the views of the parties $\mathcal{C}, \mathcal{S}$ in the ideal model is computationally indistinguishable from the views of $\mathcal{C}, \mathcal{S}^*$ in the real model.*

*Proof.* (sketch) We describe how $\mathcal{S}$ works.

1. $\mathcal{S}$ generates a secret-key/public-key pair for the homomorphic encryption scheme, chooses a random polynomial $P(y)$ of degree $k_{\mathcal{C}}$ and gives $\mathcal{S}^*$ his encrypted coefficients. Note that $\mathcal{S}^*$ does not distinguish the encryption of $P(y)$ from the encryption of any other degree $k_{\mathcal{C}}$ polynomial.
2. $\mathcal{S}$ records all the calls $\mathcal{S}^*$ makes to the random oracles $H_1, H_2$. Let $\hat{S}$ be the set of input values to $H_1$ and $\hat{Y}$ be the set of $y$ input values to $H_2$.
3. For every output pair $(e, h)$ of $\mathcal{S}^*$, $\mathcal{S}$ checks whether it agrees with some $\hat{s} \in \hat{S}$ and $\hat{y} \in \hat{Y}$. We call such a pair a consistent pair. That is, $\mathcal{S}$ checks that (i) $e$ is a ciphertext resulting from applying the server's prescribed computation using the encrypted coefficients, the value $\hat{y}$, and randomness $r'$; and (ii) $h = H_2(r'', \hat{y})$, where $r', r''$ and the randomness in the computation are determined by $H_1(\hat{s})$. If such consistency does occur, $\mathcal{S}$ sets $y = \hat{y}$, otherwise it sets $y = \perp$.
4. $\mathcal{S}$ sends the values $y$ it computed to the TTP, and $\mathcal{S}$ outputs the same output as $\mathcal{S}^*$ in the real model.

It is easy, given the view of $\mathcal{S}^*$, to decide whether a pair is consistent. As $\mathcal{S}^*$ cannot distinguish the input fed to it by $\mathcal{S}$ from the input it receives from $\mathcal{C}$ in the real execution, we get that $\mathcal{S}^*$'s distributions on consistent and inconsistent pairs, when run by the simulator and in the real execution, are indistinguishable.

Whenever $(e, h)$ forms an inconsistent pair, giving an invalid symbol $\perp$ as input to the TTP does not affect its outcome. Let $(e, h)$ be a consistent pair, and let $y$ be the value that is used in its construction. In the real execution, $y \in X$ would result in adding $y$ to the intersection set, and this similarly would happen in the simulation. The event that, in the real execution, an element $x \neq y$ would be added to the intersection set occurs with negligible probability.

We get that the views of the parties $\mathcal{C}, \mathcal{S}$ in the ideal model is computationally indistinguishable from the views of $\mathcal{C}, \mathcal{S}^*$ in the real model, as required.

### 5.3 Handling Both Malicious Clients and Servers

We briefly describe how to *combine* these two schemes yield a PM protocol fully secure in the malicious model. We leave the detailed description to the full version of this paper.

$\mathcal{C}$ generates $B$ bins as before; for each bin $B_i$, she generates a polynomial of degree $M$ with $P(z) = 0$, where $z \in B_i$ if it is (1) mapped to $B_i$ by our hashing scheme (for $z = F_s(G(x))$ for $x \in X$) or (2) added as needed to yield $M$ items. The latter should be set outside the range of $F_s$. For each polynomial, $\mathcal{C}$ prepares $L$ copies and sends their commitments to $\mathcal{S}$.

Next, $\mathcal{S}$ opens the encryptions of $L/2$ copies and verifies them. If verification succeeds, $\mathcal{S}$ opens the $F_s$ used in the other $L/2$ copies. He chooses a random $s$, splits it into $L/2$ shares, and then acts as in `PM-Malicious-Server`, albeit using the random shares as payload, $H_1(s)$ as randomness, and appending $H_2(r'', y)$.

Finally, $\mathcal{C}$ receives a list of the unopened $L/2$ copies. For each, she computes candidates for $s$'s shares and recovers $s$ from them. She uses a procedure similar to `PM-Malicious-Server` to check the consistency of the these $L/2$ shares.

## 6 Approximating Intersection

In this section, we focus on a problem related to private matching: set intersection and its approximation. Assume $\mathcal{C}$ and $\mathcal{S}$ hold strings $X$ and $Y$ respectively, where $|X| = |Y| = N$. Define $\text{INTERSECT}(X, Y) = |\{i : X_i = Y_i\}|$. Equivalently, $\text{INTERSECT}(X, Y)$ is the scalar product of $X, Y$. Let $0 < \epsilon, \delta$ be constants. An $(\epsilon, \delta)$-approximation protocol for intersection yields, on inputs $X, Y$, a value $\hat{\alpha}$ such that $\Pr[(1-\epsilon)\alpha < \hat{\alpha} < (1+\epsilon)\alpha] \geq 1-\delta$ where $\alpha = |X \cap Y|$. The probability is taken over the randomness used in the protocol.

**A lower bound.** Let $0 < \eta \leq N$. It is easy to see that an $(\epsilon, \delta)$-approximation may be used for distinguishing the cases $|X \cap Y| \leq \eta$ and $|X \cap Y| \geq \eta(1+\epsilon)^2$, as (with probability $1 - \delta$) its output is less than $\eta(1+\epsilon)$ in the former case and greater than $\eta(1+\epsilon)$ in the latter.

Protocol `Private-Sample-`$B$

INPUT: $\mathcal{C}$ and $\mathcal{S}$ hold $N$-bit strings $X, Y$, respectively.

1. $\mathcal{C}$ picks a random mask $m_C \in_R \{0,1\}$ and shift amount $r_C \in_R [N]$. She computes the $N$-bit string $X' = (X \lll r_C) \oplus m_C$ (*i.e.*, she shifts $X$ cyclicly $r_C$ positions and XORs every location in the resulting string with $m_C$). Similarly, $\mathcal{S}$ picks $m_S, r_S$ and computes $Y' = (Y << r_S) \oplus m_S$.
2. $\mathcal{C}$ and $\mathcal{S}$ invoke two $\binom{N}{1}$-OT protocols where $\mathcal{C}$ retrieves $s_C = Y'_{r_C}$ and $\mathcal{S}$ retrieves $s_S = X'_{r_S}$.
3. $\mathcal{C}$ computes $T_{00} = B(m_C, s_S), T_{01} = B(m_C, s_S \oplus 1), T_{10} = B(m_C \oplus 1, s_S), T_{11} = B(m_C \oplus 1, s_S \oplus 1)$.
4. $\mathcal{C}$ and $\mathcal{S}$ invoke a $\binom{4}{1}$-OT protocol where $\mathcal{S}$ retrieves $T_{m_S,s_S}$. $\mathcal{S}$ sends $T_{m_S,s_S}$ back to $\mathcal{C}$.

A protocol that distinguishes $|X \cap Y| \leq \eta$ and $|X \cap Y| \geq \eta(1+\epsilon)$ may be used for deciding disjointness, as defined in Section 2. Given inputs $a, b$ of length $m$ for DISJ, $\mathcal{C}$ sets her input to be $X = 1^\eta | a^{(2\epsilon+\epsilon^2)\eta}$ (*i.e.*, $\eta$ ones followed by $(2\epsilon+\epsilon^2)\eta$ copies of $a$). Similarly, $\mathcal{S}$ sets $Y = 1^\eta | b^{(2\epsilon+\epsilon^2)\eta}$. The length of these new inputs is $N = |X| = |Y| = \eta + (2\epsilon + \epsilon^2)\eta m$ bits. Note that if $a, b$ are disjoint, then $|X \cap Y| = \eta$; otherwise, $|X \cap Y| \geq \eta(1+\epsilon)^2$. Hence, for constant $\epsilon$, it follows that the randomized communication complexity of distinguishing the two cases is at least $\Omega(m) = \Omega(N/\eta)$. By setting $\eta$ to a constant, we get that the randomized communication complexity of an $(\epsilon, \delta)$ approximation for INTERSECT is $\Theta(N)$.

**A private approximation protocol for intersection.** We describe a protocol for the semi-honest case. Informally, a protocol realizes a private approximation to a function $f(X,Y)$ if it computes an approximation to $f(X,Y)$ and does not leak any information that is not efficiently computable from $f(X,Y)$. This is formulated by the requirement that each party should be able to simulate her view given her input and $f(X,Y)$. We refer the reader to [10] for the formal definition.

Our building block – protocol `Private-Sample-`$B$ – is a simple generalization of the private sampler of [10]. `Private-Sample-`$B$ samples a random location $\ell$ and checks if a predicate $B$ holds on $(X_\ell, Y_\ell)$. The location $\ell$ is shared by $\mathcal{C}$ and $\mathcal{S}$ as $\ell = r_C + r_S \pmod N$, with each party holding one of the random shares $r_C, r_S$ at the end of Step 1. Step 2 results in $\mathcal{C}$ and $\mathcal{S}$ holding random shares of $X_\ell = m_C \oplus s_S$ and $Y_\ell = m_S \oplus s_C$. Finally, both parties learn $B(m_C \oplus s_C, m_S \oplus s_S) = B(X_\ell, Y_\ell)$.

It is easy to see that the views of $\mathcal{C}$ and $\mathcal{S}$ in Protocol `Private-Sample-`$B$ are simulatable given $v = |\{i : B(X_i, Y_y)\}|$. It follows that any approximation based on the outcome of the protocol is a *private approximation* for $v$.

The communication costs of `Private-Sample-`$B$ are dominated by the cost of the $\binom{N}{1}$-OT protocol in use. Naor and Pinkas [19] showed how to combine a $\binom{N}{1}$-OT protocol with any computational PIR scheme, under the DDH assumption. Combining this result with PIR scheme of Cachin et al. [5] (or of Kiayias and Yung [15]) results in $\lambda\,\mathsf{polylog}(N)$ communication, for security parameter $\lambda$.

Our protocol `Intersect-Approx` repeatedly invokes `Private-Sample-`$B$ with $B(\alpha, \beta) = \alpha \wedge \beta$, for a maximum of $M$ invocations. We call an invocation *positive* if it concludes with $B$ evaluated as 1. If $T$ invocations occur in $t < M$ rounds, the protocol outputs $T/t$ and halts. Otherwise (after $M$ invocations) the protocol outputs 0 and halts.

The random variable $t$ is the sum of $T$ independent geometric random variables. Hence, $\mathbf{E}[t] = T/p$ and $\mathbf{Var}[t] = T(1-p)/p^2$, where $p = v/N$. Using the Chebyshev Inequality, we get that $\Pr\left[|t - T/p| \geq \beta T/p\right] \leq \left(T\frac{1-p}{p^2}\right) / \left((\beta\frac{T}{p})^2\right) \leq \frac{1}{\beta^2 T}$. Let $\beta = \frac{\epsilon}{1+\epsilon}$, taking $T = \frac{2}{\beta^2 \delta}$ ensures that, if $T$ positive invocations occur, then the protocol's output is within $(1-\epsilon)\frac{v}{N}$ and $(1+\epsilon)\frac{v}{N}$, except for $\delta/2$ probability. To complete the protocol, we set $M = N(\ln \delta + 1)$ so that if $v \neq 0$, the probability of not having $T$ positive invocations is at most $\delta/2$.

Note that the number of rounds in protocol `Intersect-Approx` is not fixed, and depends on the exact intersection size $v$. The protocol is optimal in the sense that it matches the lower-bound for distinguishing inputs with intersection size $k$ from inputs with intersection size $k(1+\epsilon)$ in an expected $O(N/k)$ invocations of `Private-Sample-`$B$.

**Caveat.** As the number of rounds in our protocol is a function of its outcome, an observer that only counts the number of rounds in the protocol, or the time it takes to run it, may estimate its outcome. The problem is inherent in our security definitions—both for semi-honest and malicious parties—as they only take into account the parties that "formally" participate in the protocol (unlike, *e.g.*, in universal composability [6]). In particular, these definitions allow for any information that is learned by all the participating parties to be sent in the clear. While it may be that creating secure channels for the protocol (*e.g.*, using encryption) prevents this leakage in many cases, this is not a sufficient measure in general nor specifically for our protocol (as one must hide the communication length of `Intersect-Approx`).

# 7 The Multi-party Case

We briefly discuss computing the intersection in a multi-party environment. Assume that there are $n$ parties, $P_1, \ldots, P_n$, with corresponding lists of inputs $X_1, \ldots, X_n$; w.l.o.g., we assume each list contains $k$ inputs. The parties compute the intersection of *all* $n$ lists. We only sketch a protocol for semi-honest parties, starting with a basic protocol that is secure with respect to client parties $P_1, \ldots, P_{n-1}$ and then modifying it get security with respect to all parties.

**A strawman protocol.** Let client parties $P_1, \ldots, P_{n-1}$ each generate a polynomial encoding their input, as for Protocol `PM-Semi-Honest` in the two-party case. Each client uses her own public key and sends the encrypted polynomials to $P_n$, which we refer to as the *leader*. This naming of parties as *clients* and the *leader* is done for conceptual clarity.

For each item $y$ in his list, leader $P_n$ prepares $(n-1)$ random shares that XOR to $y$. He then evaluates the $(n-1)$ polynomials he received, encoding the

$l$th share of $y$ as the payload of the evaluation of the $l$th polynomial. Finally, he publishes a shuffled list of $(n-1)$-tuples. Each tuple contains the encryptions that the leader obtained while evaluating the polynomials on input $y$, for every $y$ in his input set. Note that every tuple contains exactly one entry encrypted with the key of client $P_l$, for $1 \leq l \leq n-1$.

To obtain the outcome, each client $P_l$ decrypts the entries that are encrypted with her public key and publishes them. If XOR-ing the decrypted values results in $y$, then $y$ is in the intersection.

**Achieving security with respect to semi-honest parties.** This strawman approach is flawed. The leader $P_n$ generates the shares that the clients decrypt. Hence, he may recognize, for values $y$ in his set but not in the intersection, which clients also hold $y$: these clients, and only these clients, would publish the right shares. We can fix this problem by letting each client generate $k$ sets of random shares that XOR to zero (one set for each of the leader's inputs). Then, each client encrypts one share from each set to every other client. Finally, the clients publish the XOR of the original share from the leader with the new shares from other clients. If $y$ is in the intersection set, then the XOR of all published values for each of the leader's $k$ inputs is still $y$, otherwise it looks random to any coalition. More concretely, the protocol for semi-honest parties is as follows.

1. A client party $P_i$, for $1 \leq i \leq n-1$, operates as in the two-party case. She generates a polynomial $Q_i$ of degree $k$ encoding her inputs, and generates homomorphic encryptions of the coefficients (with her own public key). $P_i$ also chooses $k$ sets of $n-1$ random numbers, call these $\{s^i_{j,1}, \ldots, s^i_{j,n-1}\}_{j=1}^{k}$. We can view this as a matrix with $k$ rows and $(n-1)$ columns: Each column corresponds to the values given to party $P_l$; each row corresponds to the random numbers generated for one of the leader's inputs. This matrix is chosen such that the XOR of each row sums to zero, *i.e.*, for $j=1\ldots k$, $\bigoplus_{l=1}^{n-1} s^i_{j,l} = 0$. For each column $l$, she encrypts the corresponding shares using the public key of client $P_l$. She sends all her encrypted data to a public bulletin board (or just to the leader who acts in such a capacity).
2. For each item $y$ in his list $X_n$ (the rows), leader $P_n$ prepares $(n-1)$ random shares $\sigma_{y,l}$ (one for each column), where $\bigoplus_{l=1}^{n-1} \sigma_{y,l} = y$. Then, for each of the $k$ elements of the matrix column representing client $P_l$, he computes the encryption of $(r_{y,l} \cdot Q_l(y) + \sigma_{y,l})$ using $P_l$'s public key and a fresh random number $r_{y,l}$. In total, the leader generates $k$ tuples of $(n-1)$ items each. He randomly permutes the order of the tuples and publishes the resulting data.
3. Each client $P_l$ decrypts the $n$ entries that are encrypted with her public key: namely, the $l$th column generated by $P_n$ (of $k$ elements) and the $(n-1)$ $l$th columns generated by clients (each also of $k$ elements). $P_l$ computes the XOR of each row in the resulting matrix: $(\bigoplus_{i=1}^{n-1} s^i_{j,l}) \oplus \sigma_{j,l}$. She publishes these $k$ results.
4. Each $P_i$ checks if the XOR of the $(n-1)$ published results for each row is equal to a value $y$ in her input: If this is the case, $\bigoplus_{l=1}^{n-1} \left( (\bigoplus_{i=1}^{n-1} s^i_{j,l}) \oplus \sigma_{j,l} \right) = y$, and she concludes that $y$ is in the intersection.

Intuitively, the values output by each client (Step 3) appear random to the leader, so he cannot differentiate between the output from clients with $y$ in their input and those without, as he could in the strawman proposal.

Note that the communication involves two rounds in which $P_1, \ldots P_{n-1}$ submit data, and a round where $P_n$ submits data. This is preferable to protocols consisting of many rounds with $n^2$ communication. The computation overhead of $P_n$ can be improved by using the hashing-to-bins method of Section 4.2.

## 8 Fuzzy Matching and Fuzzy Search

In many applications, database entries are not always accurate or full (*e.g.*, due to errors, omissions, or inconsistent spellings of names). In these cases, it would be useful to have a private matching algorithm that reports a match even if two entries are only *similar*.

We let each database entry be a list of $T$ attributes, and consider $X = (x_1, \ldots, x_T)$ and $Y = (y_1, \ldots, y_T)$ similar if they agree on (at least) $t < T$ attributes. One variant is fuzzy search, where the client specifies a list of attributes and asks for all the database entries that agree with at least $t$ of the attributes. This may be achieved by a simple modification of our basic `PM-Semi-Honest` protocol, by letting the server reply with the encryptions of $r_i \cdot P_i(y_i) + s_i$, where $t$ shares of $s_1, \ldots, s_T$ are necessary and sufficient for recovering $Y$. This fuzzy search scheme may be used to compare two "databases" each containing just one element comprised of many attributes.

The protocol may be modified to privately compute fuzzy matching in larger databases, *e.g.*, when a match is announced if entries agree on $t$ out of $T$ attributes. In this section, we present a scheme, in the semi-honest model, that considers a simple form of this fuzzy private matching problem.

**A 2-out-of-3 fuzzy matching protocol** A client $\mathcal{C}$ has $k_{\mathcal{C}}$ 3-tuples $X_1, \ldots, X_{k_{\mathcal{C}}}$. Let $P_1, P_2, P_3$ be polynomials, such that $P_j$ is used to encode the $j$th element of the three tuple, $X_i^j$, for $1 \le i \le k_{\mathcal{C}}$. For all $i$, let $\mathcal{C}$ choose a new random value $R_i$ and set $R_i = P_1(X_i^1) = P_2(X_i^2) = P_2(X_i^3)$. In general, the degree of each such polynomial is $k_{\mathcal{C}}$, and therefore, two non-equal polynomials can match in at most $k_{\mathcal{C}}$ positions. $\mathcal{C}$ sends $(P_1, P_2, P_3)$ to $\mathcal{S}$ as encrypted coefficients, as earlier. The server $\mathcal{S}$, for every three-tuple $Y_i$ in his database of size $k_{\mathcal{S}}$, responds to $\mathcal{C}$ in a manner similar to Protocol `PM-Semi-Honest`: He computes the encrypted values $r(P_1(Y_i^1) - P_2(Y_i^2)) + Y_i$, $r'(P_1(Y_i^2) - P_3(Y_i^3)) + Y_i$, and $r''(P_1(Y_i^1) - P_3(Y_i^3)) + Y_i$. If two elements in $Y_i$ are the same as those in $X_i$, the client receives $Y_i$ in one of the entries.

We leave as open problems the design of more efficient fuzzy matching protocols (without incurring a $\binom{T}{t}$ factor in the communication complexity) and of protocols secure in the malicious model.

## References

[1] Bill Aiello, Yuval Ishai, and Omer Reingold. Priced oblivious transfer: How to sell digital goods. In *Advances in Cryptology—EUROCRYPT 2001*, Innsbruck, Austria, May 2001.

[2] Yossi Azar, Andrei Z. Broder, Anna R. Karlin, and Eli Upfal. Balanced allocations. *SIAM Journal on Computing*, 29(1):180–200, 1999.

[3] Fabrice Boudot, Berry Schoenmakers, and Jacques Traore. A fair and efficient solution to the socialist millionaires' problem. *Discrete Applied Mathematics*, 111(1-2):23–036, 2001.

[4] Andrei Z. Broder and Michael Mitzenmacher. Using multiple hash functions to improve ip lookups. In *IEEE INFOCOM'01*, pages 1454–1463, Anchorage, Alaska, April 2001.

[5] Christian Cachin, Silvio Micali, and Markus Stadler. Computationally private information retrieval with polylogarithmic communication. In *Advances in Cryptology—EUROCRYPT '99*, pages 402–414, Prague, Czech Republic, May 1999.

[6] Ran Canetti. Universally composable security: A new paradigm for cryptographic protocols. In *42nd Annual Symposium on Foundations of Computer Science*, pages 136–145, Las Vegas, Nevada, October 2001.

[7] Ivan Damgård and Mads Jurik. A generalisation, a simplification and some applications of Paillier's probabilistic public-key system. In *4th International Workshop on Practice and Theory in Public Key Cryptosystems (PKC 2001)*, pages 13–15, Cheju Island, Korea, February 2001.

[8] Alexandre Evfimievski, Johannes Gehrke, and Ramakrishnan Srikant. Limiting privacy breaches in privacy preserving data mining. In *Proc. 22nd ACM Symposium on Principles of Database Systems (PODS 2003)*, pages 211–222, San Diego, CA, June 2003.

[9] Ronald Fagin, Moni Naor, and Peter Winkler. Comparing information without leaking it. *Communications of the ACM*, 39(5):77–85, 1996.

[10] Joan Feigenbaum, Yuval Ishai, Tal Malkin, Kobbi Nissim, Martin Strauss, and Rebecca N. Wright. Secure multiparty computation of approximations. In *Automata Languages and Programming: 27th International Colloquim (ICALP 2001)*, pages 927–938, Crete, Greece, July 2001.

[11] Oded Goldreich. Secure multi-party computation. In *Available at Theory of Cryptography Library,* `http://philby.ucsb.edu/cryptolib/BOOKS`, 1999.

[12] Bernardo A. Huberman, Matt Franklin, and Tad Hogg. Enhancing privacy and trust in electronic communities. In *Proc. ACM Conference on Electronic Commerce*, pages 78–86, Denver, Colorado, November 1999.

[13] Russell Impagliazzo and Steven Rudich. Limits on the provable consequences of one-way permutations. In *Proc. 21st Annual ACM Symposium on Theory of Computing*, pages 44–61, Seattle, Washington, May 1989.

[14] Bala Kalyanasundaram and Georg Schnitger. The probabilistic communication complexity of set intersection. *SIAM J. Discrete Mathematics*, 5(4):545–557, 1992.

[15] Aggelos Kiayias and Moti Yung. Secure games with polynomial expressions. In *Automata Languages and Programming: 27th International Colloquim (ICALP 2001)*, pages 939–950, Crete, Greece, July 2001.

[16] Eyal Kushilevitz and Noam Nisan. *Communication Complexity.* Cambridge University Press, Cambridge, 1997.

[17] Helger Lipmaa. Verifiable homomorphic oblivious transfer and private equality test. In *Advances in Cryptology—ASIACRYPT 2003*, pages 416–433, Taipei, Taiwan, November 2003.
[18] Moni Naor and Benny Pinkas. Oblivious transfer and polynomial evaluation. In *Proc. 31st Annual ACM Symposium on Theory of Computing*, pages 245–254, Atlanta, Georgia, May 1999.
[19] Moni Naor and Benny Pinkas. Efficient oblivious transfer protocols. In *SIAM Symposium on Discrete Algorithms (SODA)*, pages 448–457, Washington, D.C., January 2001.
[20] Pascal Paillier. Public-key cryptosystems based on composite degree residuosity classes. In *Advances in Cryptology—EUROCRYPT '99*, pages 223–238, Prague, Czech Republic, May 1999.
[21] Pascal Paillier. Trapdooring discrete logarithms on elliptic curves over rings. In *Advances in Cryptology—ASIACRYPT 2000*, pages 573–584, Kyoto, Japan, 2000.
[22] Alexander A. Razborov. Application of matrix methods to the theory of lower bounds in computational complexity. *Combinatorica*, 10(1):81–93, 1990.

# Positive Results and Techniques for Obfuscation

Benjamin Lynn[1], Manoj Prabhakaran[2], and Amit Sahai[2]

[1] Stanford University, USA
blynn@theory.stanford.edu
[2] Princeton University, USA
{mp, sahai}@cs.princeton.edu

**Abstract.** Informally, an *obfuscator* $\mathcal{O}$ is an efficient, probabilistic "compiler" that transforms a program $P$ into a new program $\mathcal{O}(P)$ with the same functionality as $P$, but such that $\mathcal{O}(P)$ protects any secrets that may be built into and used by $P$. Program obfuscation, if possible, would have numerous important cryptographic applications, including: (1) "Intellectual property" protection of secret algorithms and keys in software, (2) Solving the long-standing open problem of homomorphic public-key encryption, (3) Controlled delegation of authority and access, (4) Transforming Private-Key Encryption into Public-Key Encryption, and (5) Access Control Systems. Unfortunately however, program obfuscators that work on arbitrary programs *cannot* exist [1]. No positive results for program obfuscation were known prior to this work.

In this paper, we provide the first *positive* results in program obfuscation. We focus on the goal of access control, and give several provable obfuscations for complex access control functionalities, in the random oracle model. Our results are obtained through non-trivial compositions of obfuscations; we note that general composition of obfuscations is impossible, and so developing techniques for composing obfuscations is an important goal. Our work can also be seen as making initial progress toward the goal of obfuscating finite automata or regular expressions, an important general class of machines which are not ruled out by the impossibility results of [1]. We also note that our work provides the *first* formal proof techniques for obfuscation, which we expect to be useful in future work in this area.

## 1 Introduction

Software Obfuscation is an important cryptographic concept with wide applications. However until recently there was little theoretical investigation of obfuscation, despite the great success theoretical cryptography has had in tackling other challenging notions of security.

Roughly speaking, the goal of (program) obfuscation is to hide the secrets inside a program while preserving its functionality. Ideally, an obfuscated program should be a "virtual black box," in the sense that anything one can compute from it could also be computed from the input-output behavior of the program. To be clear (but still informal), an *obfuscator* $\mathcal{O}$ is an efficient, probabilistic "compiler" that transforms a program $P$ into a new program $\mathcal{O}(P)$ such that:

C. Cachin and J. Camenisch (Eds.): EUROCRYPT 2004, LNCS 3027, pp. 20–39, 2004.

- **(Functionality Preservation.)** The input/output behavior of $\mathcal{O}(P)$ is the same as $P$.
- **(Secrecy.)** "Anything that can be efficiently computed from $\mathcal{O}(P)$ can be efficiently computed given oracle access to $P$."

This second property seeks to formalize the notion that all aspects of $P$ which are not obvious from its input/output behavior should be hidden by $\mathcal{O}(P)$. By considering the problem of obfuscation restricted to specific classes of interesting programs, one can further specify exactly what needs to be hidden by the obfuscation, and what doesn't need to be[3].

Program obfuscation, if possible, would have numerous important cryptographic applications, including: (1) "Intellectual property" protection of secret algorithms and keys in software, (2) Solving the long-standing open problem of homomorphic public-key encryption, (3) Controlled delegation of authority and access, and (4) Transforming Private-Key Encryption into Public-Key Encryption. (See [1] for more discussion.) We discuss another important application, access control, in more detail below.

Barak, Goldreich, Impagliazzo, Rudich, Sahai, Vadhan, and Yang [1] initiated the formal cryptographic study of obfuscation, and established several important impossibility results (which we discuss further below). There have been many ad-hoc approaches to program obfuscation (see *e.g.* [3]); Many of these have been broken (e.g. [4] broken by [7]), and none of these have proofs of their security properties. Proven results are known only in models where the adversary has only partial access to the obfuscated program or circuit [5,6].

In this paper, we provide the first *positive* results in program obfuscation. We focus on the goal of access control, and give several provable obfuscations for complex access control functionalities, in the random oracle model. Our results are obtained through non-trivial compositions of obfuscations; we note that general composition of obfuscations is impossible, and so developing techniques for composing obfuscations is an important goal. Our work can also be seen as making initial progress toward the goal of obfuscating finite automata or regular expressions, an important general class of machines which are not ruled out by the impossibility results of [1]. We also note that our work provides the *first* formal proof techniques for obfuscation, which we expect to be useful in future work in this area.

**Context for our work.** In order to understand the challenge of program obfuscation, we first recall the impossibility results of [1]. Their central construction demonstrates the existence of a particular family $\mathcal{F}$ of programs, for which no obfuscator can exist. More precisely, every function in $\mathcal{F}$ has an associated secret key such that: (1) no efficient algorithm can extract the secret key given the input/output functionality of a random function from $\mathcal{F}$; (2) however, there exists an adversary which can *always* extract the

3 In general, one can define a class of programs parametrized by the secrets which are meant to be protected by the obfuscation. For instance, for a program $P$ which sorts the input and then signs it using a secret signature key $sk$, one can define a program class $\mathcal{F} = \{P_{sk} :$ $P$ using key $sk\}$. An obfuscator for $\mathcal{F}$ would then only be required to protect the secret key; it would not be required, for example, to protect the exact nature of the sorting algorithm, since this is the same for all programs in $\mathcal{F}$.

secret key given *any* program which implements a function in $\mathscr{F}$. There are several important observations to be made:

- The program family $\mathscr{F}$ consists of programs which have inputs and outputs of bounded length. Under a widely believed complexity assumption (factoring Blum integers is hard), $\mathscr{F}$ can implemented by constant-depth polynomial-size threshold circuits (*i.e.* $\mathscr{F} \subset \mathbf{TC^0}$). Furthermore, $\mathscr{F}$ can be embedded into specific constructions of most cryptographic primitives, thus ruling out obfuscators that work on, say, any signature scheme.
- If the obfuscated program runs in time $T$, the adversary which extracts the secret key runs in time roughly only $\tilde{O}(T^2)$. Note also that the adversary's probability of success is $1$.
- The impossibility result (with all the properties above) extends to the random oracle model.

The above properties highlight the difficulty of obtaining any *general* methods for obfuscation: Because the adversary runs quickly and always succeeds in extracting the secret key (and the impossibility result holds in the random oracle model), there seems little hope to relax our security requirement: General purpose obfuscation under any meaningful relaxed secrecy definition[4] would seem to find a counterexample in $\mathscr{F}$.

This has consequences for the techniques we can hope to develop to build and prove obfuscations. One of the most useful techniques we could hope for is composition. However, note that any single logic gate is trivially obfuscatable; indeed even a depth 1 threshold circuit ($\mathbf{TC^0_1}$) is trivially obfuscatable since it is learnable with oracle queries. Obviously, an arbitrary circuit can be built from a composition of logic gates; and any $\mathbf{TC_0}$ circuit can be built from just a *constant* number of compositions of $\mathbf{TC^0_1}$ circuits. Thus, no general theorem showing how to compose even a constant number of obfuscations is possible (under reasonable complexity assumptions).

**Our Results.** We now describe our results in more detail. The starting point for our work is the simple observation that a commonly used practice for hiding passwords can be viewed as a provably secure obfuscation of a "point function" under the random oracle model. That is, consider the family of functions $\{f_\alpha\}$ where $f_\alpha(x) = 1$ if $x = \alpha$, and $f_\alpha(x) = 0$ otherwise. If $\mathcal{R}$ is a random oracle[5] (with a large enough range), then the program which stores $\rho = \mathcal{R}(\alpha)$, and on input $x$ outputs 1 iff $\mathcal{R}(x) = \rho$

[4] There is one intriguing, if limited, possibility that we can imagine: There is nothing known to rule out a general purpose obfuscator that takes circuits of size $s$, and outputs circuits of size, say, $O(sk)$, such that no adversary running in time $\Omega(sk^2)$ could obtain meaningful information. If $k$ were large enough, this could conceivably provide enough of a slowdown to be useful in some cases. No such transformation is known to exist.

[5] The work of [2] on "perfectly one-way hash functions" can be seen as a way to implement the random oracle within this obfuscation in certain models. By considering an extension of such models, it is possible to apply the techniques of [2] to remove the random oracles from all our constructions. However, these models are not satisfactory, because in general [2] cannot deal with partial information being available to the adversary, which is an important part of the obfuscation model we consider. Extending [2] to deal with partial information is an important open problem. Progress there would lead to progress toward removing the random oracle in our constructions. However, since we seek to give the *first* positive results regarding obfuscation,

is an obfuscation of $f_\alpha$ with high probability over $\mathcal{R}$. Starting with this most basic of access control functionalities, we give a number of novel reduction and composition techniques for obfuscation, and use these to build obfuscations of much more complex access control functionalities.

We show how to obfuscate a functionality we call an *Access Automaton*. Consider a large organization (such as a government) that wishes to implement a complex hierarchical access control system for a large collection of private information. In such a system, a single piece of information may need to accessible by persons with a variety of different credentials (*e.g.* the co-chair of one subcommittee and the secretary of an unrelated working group may need access to the same piece of secret information). In our setting, we allow for an *exponential* number of sets of credentials to give access to a common piece of information. We model this framework as an arbitrary directed graph, where each edge is labeled with a password/credential, and each node is attached to a secret. At the start, the structure of the graph is completely unknown to a user, but by supplying passwords/credentials, the user can explore and learn as much of the graph as she has access to, given the set of passwords/credentials she has. We show how to provably obfuscate this functionality in the random oracle model. We also show that our obfuscation can be dynamically updated, such that secrecy is preserved even if the adversary observes the entire history of obfuscated programs.

A potential drawback of the above functionality concerns *weak* passwords. Suppose there is a document which is accessible by giving a sequence of 5 passwords, but the adversary has partial information allowing him to narrow each password to a (different) set of $10^4$ possibilities. The adversary could efficiently "break" each password one by one, and access the document, even though the document itself had $\log(10^{20})$ "bits" of security. We show how to address this problem: Suppose we have a public regular expression over hidden strings (*e.g.* the expression "$x_1(x_1|x_4)^*(x_2|x_3)x_3x_4$", where $x_1, x_2, x_3, x_4$ are unknown strings). Then we show how to essentially obfuscate this expression in a way that preserves the natural security inherent in the expression. In the example above, the adversary would not gain any partial information even if he knew that $x_3$ was one of only two possibilities – without knowing $x_1$ and $x_4$, he cannot resolve his uncertainty about $x_3$. The main difference between this case and the Access Automaton is that the overall structure of the regular expression is not hidden by the obfuscation. We also give another obfuscation for public regular expressions over "black boxes" – this does not have the security property above, but can be seen as providing a nontrivial obfuscation of a composition of individually obfuscatable functions. We also show how to go beyond just "equality checking" by giving an obfuscation for *proximity checking* in tree metrics.

We believe that the proof techniques we introduce are as important as the results we obtain. In particular, we give a new notion of reduction between classes of functions which implies that if one is obfuscatable, then so is the other. The significance of this is that this allows obfuscations of complex functions to be built using obfuscations of simpler functions. The latter may be implemented in anyway, possibly in the hardware. From a theoretical perspective, this is important because obfuscations built this way

we do not concern ourselves with removing the random oracle in this work. We stress that it is indeed an important problem to address in the future.

need not be based on the random-oracle model, but can be in a model where the simpler obfuscations are available as primitives. We also make many observations about the possibility of putting together multiple obfuscations. We believe our techniques and observations will be of further use in the nascent field of program obfuscation.

## 2 Preliminaries

Following Barak et al. [1] we define obfuscation of a family of functions $\mathscr{F}$ as follows.

**Definition 1.** *A family of functions $\mathscr{F}$ is* obfuscatable *if there exists an algorithm $\mathcal{O}$ which takes a Turing Machine (or circuit) that computes $F \in \mathscr{F}$ and outputs a Turing Machine (circuit, respectively) such that the following conditions hold (the TM or circuit is also denoted by $F$).*

1. *(Functionality) For all $F \in \mathscr{F}$ and all inputs $x \in \{0,1\}^*$ we have $\mathcal{O}(F)(x) = F(x)$*
2. *(Polynomial Slowdown) There exists a polynomial $p$ such that for all $F \in \mathscr{F}$ we have $|\mathcal{O}(F)| \leq p(|F|)$ and (in the case of Turing Machines) if $F$ takes $t$ time steps on an input $x \in \{0,1\}^*$, $\mathcal{O}(F)$ takes at most $p(t)$ time steps.*
3. *(Virtual Blackbox) For all PPT $\mathcal{A}$, there exists a PPT $\mathcal{S}$ and a negligible function $\nu$ such that for all $F \in \mathscr{F}$ we have*

$$|\mathbf{Pr}[\mathcal{A}(\mathcal{O}(F))) = 1] - \mathbf{Pr}[\mathcal{S}^F(1^{|F|}) = 1]| \leq \nu(|M|).$$

*Here the probabilities are taken over the randomness of $\mathcal{A}$ and $\mathcal{S}$ (and $\mathcal{O}$ and $F$ if they are randomized).*

*$\mathcal{O}$ is called an* obfuscator *for $\mathscr{F}$, and $\mathcal{O}(F)$ an obfuscation of $F$. $\mathcal{O}$ is said to be* efficient *if it runs in polynomial time, in which case we say $\mathscr{F}$ is* efficiently obfuscatable.

Now we extend this definition so that random oracles are taken into account.

We consider a parameter $k$ associated with the family $\mathscr{F}_k$ of functions being obfuscated. The size of $F \in \mathscr{F}_k$ is polynomial in $k$, and the random oracle that can be used in the obfuscation will be a random member of $\mathscr{R}_k$, the set of all functions from $\{0,1\}^*$ to $\{0,1\}^{\ell(k)}$ for some polynomial $\ell$. We shall refer to $k$ as the *feasibility parameter*.

**Definition 2. (Obfuscation in the Random Oracle Model)** *An oracle algorithm $\mathcal{O}$ which takes as input a Turing Machine (or circuit) and produces an oracle Turing Machine (or oracle circuit) is said to be an obfuscator of the family $\mathscr{F} = \cup_k \mathscr{F}_k$ if we have that*

1′. *(Approximate Functionality) There exists a negligible function $\nu$ such that, for all $k$, for all $F \in \mathscr{F}_k$ we have $\mathbf{Pr}[\exists x \in \{0,1\}^* : \mathcal{O}^{\mathcal{R}}(F)(x) \neq F(x)] \leq \nu(k)$.*[6]

2′. *(Polynomial Slowdown) There exists a polynomial $p$ such that for all $k$, for all $F \in \mathscr{F}_k$ we have $|\mathcal{O}(F)| \leq p(k)$ and (in the case of Turing Machines) if $F$ takes $t$ time steps on an input $x \in \{0,1\}^*$, $\mathcal{O}(F)$ takes at most $p(t)$ time steps.*

[6] A weaker requirement would be that for all $F \in \mathscr{F}_k$ and $x \in \{0,1\}^*$, we have $\mathbf{Pr}[\mathcal{O}^{\mathcal{R}}(F)(x) \neq F(x)] \leq \nu(k)$.

3$'$. *(Virtual Blackbox) For all PPT $\mathcal{A}$, there exists a PPT $\mathcal{S}$ and a negligible function $\nu$ such that for all $k$, for all $F \in \mathscr{F}_k$ we have*

$$|\,\mathbf{Pr}\,[\mathcal{A}^{\mathcal{R}}(\mathcal{O}^{\mathcal{R}}(F))) = 1] - \mathbf{Pr}\,[\mathcal{S}^F(1^k) = 1]| \leq \nu(k)$$

*Here the probabilities are taken over $\mathcal{R} \in \mathscr{R}_k$ as well as the randomness of $\mathcal{A}$ and $\mathcal{S}$ (and $\mathcal{O}$ if it is randomized).*

*$\mathcal{O}$ is called an* obfuscator *for $\mathscr{F}$, and $\mathcal{O}(F)$ an obfuscation of $F$. $\mathcal{O}$ is said to be* efficient *if it runs in polynomial time, in which case we say $\mathscr{F}$ is* efficiently obfuscatable.

In the sequel, all our results will apply to the definition presented here (in the random oracle model). For notational convenience we shall often abbreviate $\mathcal{O}^{\mathcal{R}}, \mathcal{A}^{\mathcal{R}}$ etc. to simply $\mathcal{O}, \mathcal{A}$ etc.

## 3 Reductions and Composition

### 3.1 Reductions

**Definition 3.** *A class of Turing Machines (or circuits) $\mathscr{F}$ is said to be* polynomial-time black-box implementable relative to *$\mathscr{G}$ (denoted $\mathscr{F} \ll \mathscr{G}$) if there exist polynomial time TMs (circuits) $M$ and $N$ such that for every $F \in \mathscr{F}$ there is a $G \in \mathscr{G}$, such that $M^G$ computes the same function as $F$, and $N^F$ computes the same function as $G$.*

So, if $\mathscr{F} \ll \mathscr{G}$, for every $F \in \mathscr{F}, \mathscr{G}$ contains a function $G$ which is "equivalent" to $F$ in some extended sense. Now we give the main tool which lets us reuse results on obfuscatability.

**Lemma 1.** *If $\mathscr{F} \ll \mathscr{G}$ and $\mathscr{G}$ is obfuscatable (when every $G \in \mathscr{G}$ is given as $N^F$ for an $F \in \mathscr{F}$),[7] then so is $\mathscr{F}$. Further if $\mathscr{G}$ is efficiently obfuscatable, then $\mathscr{F}$ is efficiently obfuscatable too.*

*Proof:* Given $F \in \mathscr{F}$, let $G \in \mathscr{G}$ be such that $M^G \equiv F$ and $G \equiv N^F$. Since $\mathscr{G}$ is obfuscatable, let $\mathcal{O}'$ be an obfuscator for $\mathscr{G}$. We claim that $\mathcal{O}(F) = M^{\mathcal{O}'(G)}$ (i.e., the code of $M$ and the code $\mathcal{O}'(G)$) is an obfuscation of $F$.

Clearly, conditions 1$'$ and 2$'$ of Definition 2 are satisfied. To prove condition 3$'$, consider any adversary $\mathcal{A}$ which accepts the code $\mathcal{O}(F) = M^{\mathcal{O}'(G)}$. We need to demonstrate a PPT $\mathcal{S}$ as required by condition 3$'$. First, we build an adversary $\mathcal{A}'$ which accepts the code $\mathcal{O}'(G)$, adds the code of $M$ to it to get $\mathcal{O}(F)$, passes it on to an internally simulated copy of $\mathcal{A}$, and outputs whatever $\mathcal{A}$ outputs. Now, since $\mathcal{O}'(G)$ is an obfuscation of $G$, there exists a simulator $\mathcal{S}'$ such that

$$|\,\mathbf{Pr}\,[\mathcal{S}'^G(|\mathcal{O}'(G)|) = 1] - \mathbf{Pr}\,[\mathcal{A}'(\mathcal{O}'(G)) = 1]| \leq \epsilon \qquad (1)$$

for some negligible function $\epsilon(|\mathcal{O}'(G)|)$.

---

[7] If $G \in \mathscr{G}$ is obfuscatable only when represented in some other format, still this Lemma holds, but now the obfuscator for $\mathscr{F}$ takes $F$ as $M^G$ with $G$ specified in that obfuscatable format.

We use $\mathcal{S}'$ to build $\mathcal{S}$, as follows. Note that $\mathcal{S}$ gets oracle access to $F$ and receives $|\mathcal{O}(F)|$ as input. $\mathcal{S}^F$ can implement an oracle equivalent to $G$ as $N^F$, using its oracle access to $F$. It runs $\mathcal{S}'$ with oracle access to $G$ implemented in this way, and input $|\mathcal{O}'(G)|$ calculated from $|\mathcal{O}(F)|$ (by subtracting the size of $M$). $\mathcal{S}$ outputs whatever $\mathcal{S}'$ outputs.

Clearly, by construction,

$$\mathbf{Pr}\left[\mathcal{A}(\mathcal{O}(F)) = 1\right] = \mathbf{Pr}\left[\mathcal{A}'(\mathcal{O}'(G)) = 1\right]$$
$$\mathbf{Pr}\left[\mathcal{S}^F(|\mathcal{O}(F)|) = 1\right] = \mathbf{Pr}\left[\mathcal{S}'^G(|\mathcal{O}'(G)|) = 1\right]$$

and so by Equation (1), $|\,\mathbf{Pr}\left[\mathcal{S}^F(|\mathcal{O}(F)|) = 1\right] - \mathbf{Pr}\left[\mathcal{A}(\mathcal{O}(F)) = 1\right]| \leq \epsilon$. Finally $|\mathcal{O}(F)| \geq |\mathcal{O}'(G)|$, so that $\epsilon$ is still negligible when considered a function of $\mathcal{O}(F)$, completing the proof.

Note that in building $\mathcal{O}(F) = M^{\mathcal{O}'(G)}$, the obfuscator $\mathcal{O}$ needs to obtain $\mathcal{O}'(G)$, given $F$. Since $G$ can be specified as $N^F$ to $\mathcal{O}'$, if $\mathcal{O}'$ is efficient so is $\mathcal{O}$. □

### 3.2 Extending Lemma 1

We extend Definition 3, and Lemma 1 to allow reductions to probabilistic families of functions. We do this for proving Theorem 3. In fact, somewhat more general extensions are possible. But for the sake of simplicity we restrict ourselves more or less to the minimum extensions we will need. The reader may skip this section, and return to it while reading Section 5. The other results in this paper do not need these extensions.

**Definition 4.** *Suppose $\widetilde{\mathcal{G}}$ is a family of probabilistic Turing Machines (or circuits), and $\mathcal{F}$ a family of deterministic TMs (circuits). We say $\mathcal{F} \lll \widetilde{\mathcal{G}}$ if there exist probabilistic polynomial time TMs (circuits) $M$ and $N$ such that for every $F \in \mathcal{F}$ there is a $G \in \widetilde{\mathcal{G}}$, such that the distributions of outputs of $M^G$ and $F$ are computationally indistinguishable, and those of $N^F$ and $G$ are computationally indistinguishable.*

Note that unlike Definition 3, the above definition is *not* information theoretic. It involves the notion of computational indistinguishability, and hence inherently all the results which use the following lemma requires the adversary ($\mathcal{A}$ and $\mathcal{S}$) to be PPT machines or circuits. The proof of the lemma closely follows that of Lemma 1. It is given in the extended version [8].

**Lemma 2.** *Suppose $\mathcal{F} \lll \widetilde{\mathcal{G}}$. Let $\mathcal{G}$ be the family of deterministic TMs (circuits) obtained by fixing in all possible ways the random-tapes of the TMs (circuits) in $\widetilde{\mathcal{G}}$. Then, if $\mathcal{G}$ is obfuscatable, so is $\mathcal{F}$.*

### 3.3 Composition of Obfuscations

An obfuscated program can be idealized as oracle access to the corresponding function. We ask if obfuscations compose: can we put together different obfuscations and expect them to behave ideally as the corresponding collection of oracles. Note that here we use the term *compose* in the same way as one refers to composition of cryptographic protocols- to ask whether having multiple instances in the system breaks the security or not. It does not necessarily refer to composition of functions in the usual mathematical sense, something which we will address later in this section. We make the following definition to define a simple composition of obfuscations, where there is no interaction between the different instances.

**Definition 5.** *An* array *of t functions* $F_1, \ldots, F_t$ *is defined as follows:*

$$[\![F_1, \ldots, F_t]\!](i, x) = F_i(x) \quad \textit{if } i \in \{1, \ldots, t\}\textit{; else } \perp$$

Let $[\![\mathcal{O}(F), \mathcal{O}(G)]\!]$, by abuse of notation stands for the code which consists of the codes $\mathcal{O}(F)$ and $\mathcal{O}(G)$ as modules, and a small driving unit which directs the calls to one of the modules as appropriate.

**Definition 6. (Simply Composing Obfuscations)** *An obfuscator $\mathcal{O}$ for a family $\mathscr{F}$ is said to produce* simply $t$-self-composing obfuscations *if*

$$\mathcal{O}^*([\![F_1, \ldots, F_t]\!]) = [\![\mathcal{O}(F_1), \ldots, \mathcal{O}(F_t)]\!]$$

*is an obfuscation of the family* $\{[\![F_1, \ldots, F_t]\!] | F_i \in \mathscr{F}\}$.[8]

*This can be extended to multiple families of obfuscatable functions to define a set of* simply composing obfuscations.

In fact, in the random oracle model we have the following claim (which we conjecture to extend to the plain model too):

**Claim 1.** *There exists a class of functions $\mathscr{F}$, and an obfuscator $\mathcal{O}$ for $\mathscr{F}$ in the random oracle model, such that obfuscations produced by $\mathcal{O}$ are* not *simply 2-self-composing.*

*Proof:* We consider the class of point functions $\mathscr{P}$ (defined later, in Section 4). By Lemma 4, this class is obfuscatable in the random oracle model. Note that when $F$ and $G$ are identical (randomly chosen) functions, oracle access to the function $[\![F, G]\!]$ does not reveal the fact that they are identical, to a PPT machine. On the other hand the obfuscation given in Lemma 4 does reveal this. (Of course, it is easy to modify the obfuscation, in order to avoid this problem.) Thus no simulator can simulate the behaviour of an adversary $\mathcal{A}$ (which has access to these obfuscations) which outputs 1 if $F = G$ and 0 otherwise. □

**Conjecture 1.** *If there are non-trivial obfuscations in the plain model, Claim 1 holds in the plain model too. Indeed, in that case, we conjecture that there exists an obfuscatable family $\mathscr{F}$, such that $\mathscr{A} = \{[\![F, G]\!] : F, G \in \mathscr{F}\}$ is* unobfuscatable.

The difficulty in attempting to prove this conjecture is that it requires a non-trivial obfuscatable family $\mathscr{F}$, and we have virtually nothing known beyond what is being presented in this work (which is in the random oracle model).

On the other hand, an obfuscatable function composes with any *trivially obfuscatable* function (defined below).

[8] We can have $t$ constant, or polynomial in the feasibility parameter $k$.

**Definition 7.** *A family of functions $\mathscr{F}$ is* learnable as polynomial time circuits *if there exists an oracle circuit $P$ such that for all $F \in \mathscr{F}$, $P^F$ outputs a polynomial sized circuit $C_F$ which computes $F$.*

If $\mathscr{F}$ is learnable it is obfuscatable: the obfuscator $\mathcal{O}$ takes a circuit for $F$ and runs $P$ with oracle access to that circuit; it outputs $C_F$ produced by $P$ as $\mathcal{O}(F)$. This is clearly an obfuscation, because for every adversary $\mathcal{A}$, a simulator $\mathcal{S}$ simply runs $P$ with the oracle for $F$, obtains $C_F$ and runs $\mathcal{A}$ on it.

**Definition 8.** *A family of learnable functions is called a* family of trivially obfuscatable functions. *The obfuscation obtained via learning the function is called the* trivial obfuscation of the function.

Simple as the following lemma is, it is interesting that its intuitive extension from *trivially* obfuscatable family to *any* obfuscatable family is an open problem.

**Lemma 3.** *Let $\mathscr{F}$ be a trivially obfuscatable family of functions. Then, $\mathscr{G}$ is obfuscatable, if and only if the family of functions $\mathscr{A} = \{[\![F, G]\!] \ : \ F \in \mathscr{F}, G \in \mathscr{G}\}$ is obfuscatable.*

*Proof:* First, we show that $\mathscr{G} \ll \mathscr{A}$. Then it follows from Lemma 1 that $\mathscr{G}$ is obfuscatable if $\mathscr{A}$ is.

To see that $\mathscr{G} \ll \mathscr{A}$, for each $G \in \mathscr{G}$ we choose $A = [\![F, G]\!] \in \mathscr{A}$, where $F \in \mathscr{F}$ is a fixed function for all $G$. Then a machine $M$ which internally implements $F$ can implement $A$ with access to only $G$. On the other hand a machine $N$ which has access to $A$ can clearly implement $G$.

Now we show that $\mathscr{A}$ is obfuscatable if $\mathscr{G}$ is. Intuitively, an obfuscation of $\mathscr{A}$ does not "hide" the $\mathscr{F}$ component (which is easily learnable). So it is sufficient if we are able to obfuscate the $\mathscr{G}$ part. Formally, we show that for $A = [\![F, G]\!] \in \mathscr{A}$, the following is a valid obfuscation: $\mathcal{O}(A) = [\![\mathcal{O}'(F), \mathcal{O}'(G)]\!]$, where $\mathcal{O}'(F)$ is the trivial obfuscation of $F$ and $\mathcal{O}'(G)$ is the obfuscation of $G$ given by the assumption that $\mathscr{G}$ is obfuscatable. As earlier the notation $[\![\mathcal{O}'(F), \mathcal{O}'(G)]\!]$ refers to the code which has $\mathcal{O}'(F)$ and $\mathcal{O}'(G)$ as internal modules, plus a small control module to activate the appropriate one depending on the input.

To show that $\mathcal{O}(A)$ is a valid obfuscation, for every adversary $\mathcal{A}$ which accepts $\mathcal{O}(A)$, we show a simulator $\mathcal{S}$ such that $|\,\mathbf{Pr}[\mathcal{S}^A(|\mathcal{O}(A)|) = 1] - \mathbf{Pr}[\mathcal{A}(\mathcal{O}(A)) = 1]|$ is negligible. The structure of the argument is similar to that in the proof of Lemma 1.

From $\mathcal{A}$, we first build an adversary $\mathcal{A}'$ which takes as input $\mathcal{O}'(G)$, uses it to build the code $\mathcal{O}(A) = [\![\mathcal{O}'(F), \mathcal{O}'(G)]\!]$, passes it on to an internally simulated copy of $\mathcal{A}$, and outputs whatever $\mathcal{A}'$ outputs. Using the fact that $\mathcal{O}'(G)$ is an obfuscation of $G$, there exists a simulator $\mathcal{S}'$ such that

$$|\,\mathbf{Pr}[\mathcal{S}'^G(|\mathcal{O}'(G)|) = 1] - \mathbf{Pr}[\mathcal{A}'(\mathcal{O}'(G)) = 1]| \leq \epsilon \tag{2}$$

for some negligible function $\epsilon(|\mathcal{O}'(G)|)$.

We use $\mathcal{S}'$ to build a simulator $\mathcal{S}$ as follows. Note that $\mathcal{S}$ gets oracle access to $A$ and receives $|\mathcal{O}(A)|$ as input. Oracle access to $A$ in particular gives oracle access to $F$.

Since $F$ is trivially obfuscatable, it is possible to obtain the trivial obfuscation $\mathcal{O}'(F)$ just using this oracle access to $F$. So $\mathcal{S}$ first computes $\mathcal{O}'(F)$. Next, note that given oracle access to $A$, oracle access to $G$ can also be implemented. So $\mathcal{S}$ runs $\mathcal{S}'$ with oracle access to $G$ implemented in this way, and input $|\mathcal{O}'(G)|$ calculated from $|\mathcal{O}(A)|$ (by subtracting the size of $\mathcal{O}'(F)$). $\mathcal{S}$ outputs whatever $\mathcal{S}'$ outputs.

By construction,

$$\mathbf{Pr}\left[\mathcal{A}(\mathcal{O}(A)) = 1\right] = \mathbf{Pr}\left[\mathcal{A}'(\mathcal{O}'(G)) = 1\right]$$
$$\mathbf{Pr}\left[\mathcal{S}^{A}(|\mathcal{O}(A)|) = 1\right] = \mathbf{Pr}\left[\mathcal{S}'^{G}(|\mathcal{O}'(G)|) = 1\right]$$

and so by Equation (2), $|\,\mathbf{Pr}\left[\mathcal{S}^{F}(|\mathcal{O}(F)|) = 1\right] - \mathbf{Pr}\left[\mathcal{A}(\mathcal{O}(F)) = 1\right]| \leq \epsilon$. Finally to complete the proof, we note that $|\mathcal{O}(A)| \geq |\mathcal{O}'(G)|$ and so $\epsilon$ is still negligible when considered a function of $\mathcal{O}(A)$. □

Now we consider the question of more complex composition of obfuscations. We ask if obfuscations of composed functions can be obtained by using obfuscations of the component functions. In particular we look at function compositions (in the usual mathematical sense, of one function invoking another).

**Conjecture 2.** *Conjecture on Obfuscatability of Function Compositions: Given two classes $\mathscr{F}$ and $\mathscr{G}$ of obfuscatable programs, the family $\mathscr{A} = \{A(x) = F(G(x)) : F \in \mathscr{F}, G \in \mathscr{G}\}$ is obfuscatable.*

**Theorem 1.** *The Conjecture on Obfuscatability of Function Compositions is false, if factoring Blum integers is hard or the DDH assumption is true.*

*Proof Sketch:* The Conjecture on Obfuscatability of Function Compositions, if true, could be applied any constant number of times: if $\mathscr{F}$ is obfuscatable, then $\cup_t\{A(x) = F_1(F_2(\cdots(F_t(x))\cdots))|F_i \in \mathscr{F}\}$ is obfuscatable. However, it is known that if the assumptions of the theorem hold, then there exists a family of functions $\mathscr{A} \subset \mathbf{TC^0}$ that is unobfuscatable. On the other hand it is not hard to see that $\mathscr{F} = \mathbf{TC^0_1}$, the family of depth 1 threshold circuits, is trivially obfuscatable, because they can be easily learned from input/output queries. Noting that $\mathscr{A}$ is obtained by a constant number of compositions of functions from $\mathscr{F}$ completes the contradiction, and the proof. □

## 4 Point Functions and Extensions

In this section we define a few basic functions which can be obfuscated under the random oracle model. The proofs are easy and we include a couple of them.

**Definition 9. (Class of Point Functions)** *A point function $P_\alpha : \{0,1\}^k \to \{0,1\}$ is defined by $P_\alpha(x) = 1$ if $x = \alpha$ and 0 otherwise. Define $\mathscr{P}_k = \{P_\alpha : \alpha \in \{0,1\}^k\}$ and $\mathscr{P} = \cup_k \mathscr{P}_k$.*

We observe that the following simple obfuscation heuristic is indeed an obfuscation in the random oracle model (Definition 2).

**Lemma 4.** *For random oracles $\mathcal{R} : \{0,1\}^* \to \{0,1\}^{2k}$, let $\mathcal{O}^{\mathcal{R}}(P_\alpha)$ be a program which stores $r = \mathcal{R}(\alpha)$, and on input $x \in \{0,1\}^k$, checks if $\mathcal{R}(x) = r$; if so it outputs 1, else 0.*

*Then, $\mathcal{O}$ is an obfuscator of $\mathcal{P}$ as defined in Definition 2.*

*Proof:* Polynomial Slowdown is evident (by convention oracle queries are answered in one time step). The Approximate Functionality condition is true since

$$\begin{aligned} &\mathbf{Pr}_{\mathcal{R}}[\exists x \in \{0,1\}^k \backslash \{\alpha\} : \mathcal{R}(x) = \mathcal{R}(\alpha)] \\ \leq &\sum_{x \in \{0,1\}^k \backslash \{\alpha\}} \mathbf{Pr}_{\mathcal{R}}[\mathcal{R}(x) = \mathcal{R}(\alpha)] = (2^k - 1)/2^{2k} \end{aligned}$$

which is negligible in $k$.

To show the Virtual Black-Box property (3′), for any adversary $\mathcal{A}$, define the simulator $\mathcal{S}$ (with oracle access to $P_\alpha$ which does the following. Pick a random string $r \leftarrow \{0,1\}^{2k}$, prepare a purported obfuscation of $P_\alpha$ with this $r$ and hand it to an internally simulated copy of $\mathcal{A}$. Recall that $\mathcal{A}$ can make queries to a random oracle, which in this case will be simulated by $\mathcal{S}$. W.l.o.g we assume $\mathcal{A}$'s queries to the oracle are distinct, since oracle replies can be cached. When $\mathcal{A}$ makes a query $q$ to the random oracle, $\mathcal{S}$ queries the $P_\alpha$ oracle with $q$. If $P_\alpha$ answers 1, it answers $\mathcal{A}$'s query with $r$. Else it picks a random string in $\{0,1\}^{2k}$ and sends it to $\mathcal{A}$. Finally $\mathcal{S}$ outputs whatever $\mathcal{A}$ outputs. It is easy to see that the view of this internally simulated $\mathcal{A}$ is *identical* to that of an $\mathcal{A}$ which receives the obfuscation and access to the random oracle. Thus the Virtual Black-box requirement is satisfied (with $\nu(k) = 0$). □

Though we defined the point function as $P_\alpha : \{0,1\}^k \to \{0,1\}$ with $\alpha \in \{0,1\}^k$, it is easy to see that it can be modified to $P_\alpha : \cup_{i=0}^k \{0,1\}^i \to \{0,1\}$ with $\alpha \in \cup_{i=0}^k \{0,1\}^i$

### 4.1 Composable Obfuscations of Point Functions with General Output

**Definition 10. (Class of Point Functions with General Output)** *A point function with general output $Q_{(\alpha,\beta)} : \{0,1\}^k \to \{0,1\}^{s(k)}$ is defined by $Q_{\alpha,\beta}(x) = \beta$ if $x = \alpha$ and $\perp$ otherwise. Define $\mathcal{Q}_k = \{P_\alpha \ : \ \alpha \in \{0,1\}^k\}$ and $\mathcal{Q} = \cup_k \mathcal{Q}_k$.*

We omit the proof of the following theorem, as it is similar to the proof of Lemma 4.

**Theorem 2.** *For random oracles $\mathcal{R} : \{0,1\}^* \to \{0,1\}^{2k+s(k)}$, let $\mathcal{O}^{\mathcal{R}}(P_{\alpha,\beta})$ be a program as follows: Let $\mathcal{R}_1(\cdot)$ denote the first $2k$ bits of $\mathcal{R}(\cdot)$, and $\mathcal{R}_2(\cdot)$ denote the remaining bits. Choose $\psi$ at random from $\{0,1\}^k$. Let $a = \mathcal{R}_1(\psi, \alpha)$ and $b = \mathcal{R}_2(\psi, \alpha)$. The program stores $\psi$, $a$ and $c = \beta \oplus b$. On input $x \in \{0,1\}^k$, it computes $a' = \mathcal{R}_1(\psi, x)$ and $b' = \mathcal{R}_2(\psi, x)$; if $a' = a$ it outputs $b' \oplus c$; else it outputs $\perp$.*

*Then, $\mathcal{O}$ is an obfuscator of $\mathcal{P}$ as defined in Definition 2.*

We further observe that the above obfuscation self-composes according to Definition 6. As long as there only polynomially many (polynomial in $k$) obfuscations in the system, the probability that two of the obfuscations will have the same value of $\psi$ is negligible. Conditioned on this (negligible probability) event not happening, a simulator with black-box access to all the (polynomially many) $Q_{\alpha,\beta}$ functions can perfectly simulate the behavior of an adversary with access to the obfuscations. Note that here the obfuscator is a randomized algorithm.

### 4.2 Multi-point Functions with General Output

Finally, we define a multi-point function *with general output* as follows.

**Definition 11. (Class of Multi-Point Functions with General Output)** *A multi-point function* $Q_{(\alpha_1,\beta_1)\dots,(\alpha_t,\beta_t)} : \{0,1\}^k \to \left(\{0,1\}^{s(k)}\right)^t$ *is defined as follows: On input* $x$, *output* $b \in \left(\{0,1\}^{s(k)}\right)^t$ *where* $b_i = \beta_i$ *if* $x = \alpha_i$, *and else* $b_i = \perp$. *Define* $\mathscr{Q}^t_k = \{Q_{(\alpha_1,\dots,\alpha_{t(k)})} : \alpha_i \in \{0,1\}^k\}$ *and* $\mathscr{Q}^t = \cup_k \mathscr{Q}^t_k$. *Define* $\mathscr{Q}^* = \cup_{\text{polynomials } t} \mathscr{Q}^t$.

Since from last section we have a self-composable obfuscation for the single point function with general output, we simply put together the $t$ programs $\mathcal{O}(Q_{\alpha_i,\beta_i})$, $i = 1,\dots,t$ to obtain an obfuscation for $Q_{(\alpha_1,\beta_1)\dots,(\alpha_t,\beta_t)}$.

**Lemma 5.** *The family of functions* $\mathscr{Q}^*$ *is efficiently obfuscatable in the random oracle model, in a self-composable manner.*

*Proof Sketch:* It is easy to see that $\mathscr{Q}^t \ll \{[\![F_1,\dots,F_t]\!] : F_i \in \mathscr{Q}\}$. Since the obfuscation in Theorem 2 is self-composable, $\{[\![F_1,\dots,F_t]\!] : F_i \in \mathscr{Q}\}$ is obfuscatable, and by Lemma 1, so is $\mathscr{Q}^t$ (and hence $\mathscr{Q}^*$). To see that this composition is self-composable, note that the obfuscation of an array of functions from $\mathscr{Q}^*$ is identical to the obfuscation of a (much larger) array of functions from $\mathscr{Q}$. □

## 5 Obfuscating a Complex Access Control Mechanism

Consider the following (interactive) access control task. There are multiple access points to various functions or secrets. There is an underlying directed multi-graph (possibly with multiple edges between nodes, and self-loops), with each node representing an access point. The user starts at a predefined access point, or "start node" and proceeds to establish her access privileges which allows her to move from one access point to another, through the edges of the graph. The access control task is the following:

- The user can reach an access point only by presenting credentials that can take her from the start node to that point.
- The user gains complete access to a function or secret available at an access point if and only if the user has reached that access point.
- The user does not learn anything about the structure of the graph, except what is revealed by the secrets at the access points she reached and the edges she traversed.

We specify this task as access to a black-box with which the user interacts, giving her credentials at various points and receiving the secrets; the black-box internally maintains the current access point of the user. But we would like to implement this task as a program which we then hand over to the user. To maintain the security of the task, we need to obfuscate this program.

In this section we explore this obfuscation problem. We show that in the random oracle model this access control mechanism can indeed be obfuscated. We model the interactive task as a non-interactive function (formulated below) which takes the "history" of interaction and gives a response to the last query.

**Definition 12.** *A graph-based access control problem $X_G$ with parameters $k$ and $d$ is defined by the following:*

1. *Directed multi-graph $G$ on $k$ vertices. Each node $u \in k$ has at most $d$ ordered neighbors $\mu_u^{(1)}, \ldots, \mu_u^{(d)}$. Let $E = \{(u, v, i) : v = \mu_u^{(i)}$ for some $i \in [d]\ \}$ be the set of all edges ($i$ is used to differentiate between the multiple edges possible between the same pair of nodes).*
2. *A set of passwords on the edges $\{\pi_e | e \in E\}$, and*
3. *A set of secrets at the nodes $\{\sigma_v | v \in [k]\}$.*

*Then,*

$$X_G((i_1, x_1), \ldots, (i_n, x_n)) = \begin{cases} (v_n, \sigma_{v_n}) & \text{if } \exists v_0, \ldots, v_n, \in [k] \text{ and } e_0, \ldots, e_{n-1} \in E \\ & \text{such that } v_0 = 1, e_j = (v_j, v_{j+1}, i_j), \text{ and} \\ & x_j = \pi_{e_j} \\ \perp & \text{otherwise.} \end{cases}$$

*We define the family of functions $\mathscr{X}$ as the set of all $X_G$ with parameters $(k, d)$ over all multi-graphs G, sets of edge-passwords and sets of node-secrets.*

Above, $(i, x)$ is a query in which the user provides a purported password $x$ for the $i$-th edge going out of the "current" node. For later notational convenience we shall assume that there is no secret available at node 1: i.e., $\sigma_1 = \perp$.

We are interested in cases where the inputs to $X_G$ are of size polynomial in $k$ and $d$. We point out that there may be exponentially many *valid* inputs for which $X_G$ outputs a secret (though the number of distinct secrets is only $k$). So it is not possible to obfuscate $X_G$ directly using Lemma 5.

Instead we proceed in the following manner: each node is represented by the tuple $(v, \sigma_v, e_1, \ldots, e_d, \pi_{e_1}, \ldots, \pi_{e_d})$ where $e_i \in E$ (if there are less than $d$ outgoing edges pick dummy values for the remaining edges). For each node $1 < u \leq k$ pick a random "key" $\kappa_u$ from $\{0,1\}^\ell$; let $\kappa_1 = 0^\ell$ (recall that 1 is the start node). Define the function $W_G^{\bar{\kappa}}$ as follows:

$$W_G^{\bar{\kappa}}(u, z, i, x) = \begin{cases} (v, \sigma_v, \kappa_v) & \text{if } z = \kappa_u \text{ and} \\ & \exists v \in [k] \text{ such that } \pi_{u,v,i} = x \\ \perp & \text{otherwise.} \end{cases}$$

The obfuscation consists of an obfuscation of $W_G^{\bar{\kappa}}$ (which is a multi-point function with at most $kd$ input points where the output is not $\perp$, and hence can be obfuscated).

Intuitively, this is a good obfuscation because the adversary cannot find the randomly chosen key of a node $\kappa_v$, unless it was given out by the (obfuscated) function $W_G^{\bar{\kappa}}$. But the only way to obtain that is to give $\pi_e$ for an edge leading to $v$ from a node $u$ to which the adversary already has the key. Since, to start with, the only key the adversary knows is $\kappa_1$, it must indeed traverse a path from 1 to $v$ by providing the all the edge-passwords in order to get to $v$.

Formally, we first define a probabilistic program $\widetilde{W}_G$ which picks the random keys above to get a particular deterministic function $W_G^{\bar{\kappa}}$. Then we show that the family $\mathscr{X} \lll \widetilde{\mathscr{W}}$, where $\widetilde{\mathscr{W}}$ is the family of all $\widetilde{W}_G$ as above.

**Definition 13.** *Define the randomized algorithm $\widetilde{W}_G$ as follows: for $v \in [k]$, pick random keys $\kappa_v \leftarrow \{0,1\}^k$. On input $(u, z, i, x)$ return $W_G^{\bar{\kappa}}(u, z, i, x)$.*

*We define the family of functions $\widetilde{\mathscr{W}}$ as the set of all $\widetilde{W}_G$ (with parameters $(k, d)$) over all multi-graphs G, sets of edge-passwords and sets of node-secrets.*

**Lemma 6.** $\mathscr{X} \lll \widetilde{\mathscr{W}}$.

*Proof:* For $X_G \in \mathscr{X}$ we pick $\widetilde{W}_G \in \widetilde{\mathscr{W}}$ and demonstrate $M$ and $N$ as required by the definition of the relation $\lll$.

*M such that* $M_G^{\widetilde{W}} \equiv X_G$ : On input $(i_1, x_1), \ldots, (i_n, x_n)$ query $\widetilde{W}_G$ with $(1, 0^\ell, i_1, x_1)$; if $\widetilde{W}_G$ returns $(v_2, \sigma_{v_2})$, query it with $(v_2, \sigma_{v_2}, i_2, x_2)$ and so on, until it either returns $\perp$ or we reach the end of the input and receive $(v_n, \sigma_{v_n})$. In either case output this value.

*N such that* $N^{X_G} \approx \widetilde{W}_G$ : $N$ internally maintains two tables: one table is for keys $\kappa_i$, and one for *paths* to each node $v$ from node 1, with edge passwords for each edge appearing on the edge. Initially it sets $\kappa_1 = 0^k$ and all other keys as $\perp$, and does not have any paths recorded for any node. On input $(u, z, i, x)$ $N$ checks if $z = \kappa_u \neq \perp$. If not it returns $\perp$. Else it will have recorded a path $(v_1 = 1, v_2, i_1, x_1), \ldots, (v_t, v_{t+1} = u, i_t, x_t)$ such that $x_j = \pi_{(v_j, v_{j+1}, i_j)}$. It makes a query $(i_1, x_1), \ldots, (i_t, x_t), (i, x)$ to $X_G$. If $X_G$ responds with $\perp$, $N$ outputs $\perp$. Else, it receives $(v, \sigma_v)$ from $X_G$. It checks if a key has been already assigned to $v$; if not it picks a random key and assigns that to $v$. Then it returns $(v, \sigma_v, \kappa_v)$.

It is not hard to see that for any PPT $S'$ interacting with $\widetilde{W}_G$ or $N^{X_G}$, the output distribution of $N^{X_G}$ is the same as that of $\widetilde{W}_G$, but both distributions conditioned on the event that $S'$ never makes a query with a valid key which it did not receive as answer to a previous query. But that event is of negligible probability, and so $N^{X_G} \approx \widetilde{W}_G$. □

Note that $\widetilde{\mathscr{W}}$ is a family of probabilistic machines, such that if we consider the family obtained by fixing the random-tapes of machines in $\widetilde{\mathscr{W}}$ in all possible ways, we get a sub-family of $\mathscr{Q}^*$ (Definition 11). This sub-family is obfuscatable (because $\mathscr{Q}^*$ is obfuscatable, by Lemma 5). Then, from the above lemma and Lemma 2, we conclude the following.

**Theorem 3.** *The family $\mathscr{X}$ is efficiently obfuscatable in the random oracle model.*

## 6 Regular Expressions and Obfuscations

Let $\Sigma$ be an alphabet (of constant size). We consider regular expressions over $\Sigma \cup \{\zeta^{L_1}, \ldots, \zeta^{L_t}\}$, where $\zeta^{L_i}$ are formal symbols corresponding to languages $L_i$. We define whether or not a string $s \in \Sigma^*$ *matches* such a regular expression $\rho(L_1, \ldots, L_t)$ as follows: $s$ matches a symbol $\zeta^{L_i}$ if $s \in L_i$. The rest of the rules are the usual ones: a single character $a \in \Sigma$ matches itself; $s \in \Sigma^*$ matches $\rho_1|\rho_2$ if it matches either $\rho_1$ or $\rho_2$; $s$ matches $\rho_1 \cdot \rho_2$ if $s = s_1 \cdot s_2$ such that $s_1$ matches $\rho_1$ and $s_2$ matches $\rho_2$; finally $s$ matches $\rho^*$ if $s$ is the null-string, or $s = s_1 \cdot s_2 \cdots s_k$ where each $s_i$ matches $\rho$. If $s$ matches a regular expression $\rho$, we write $s \sim \rho$. Below $\mathcal{L}_{\rho(L_1,\ldots,L_t)}$ stands for the language defined as the set of all strings matching $\rho(L_1, \ldots, L_t)$.

### 6.1 Obfuscating $\mathcal{L}_{\rho(P_{\alpha_1},\ldots,P_{\alpha_t})}$

Consider the case when the languages $L_i$ above are the point functions $P_{\alpha_i}$. In this section we consider a family of functions $\mathscr{U}_\rho = \cup_k \mathscr{U}_{\rho k}$ where for all $k$ and all $U_\rho^{\alpha_1,\ldots,\alpha_t} \in \mathscr{U}_{\rho k}$ there is a single fixed regular expression $\rho$. However, for each $k$, the point functions $P_{\alpha_i}$ belong to the $\mathscr{P}_k$, the family of point functions on $\cup_{j=0}^k \{0,1\}^j$. For brevity we denote $\mathcal{L}_{\rho(P_{\alpha_1},\ldots,P_{\alpha_t})}$ by $\mathcal{L}_{\rho(\alpha_1,\ldots,\alpha_t)}$.

**Definition 14.** *Define the function $U_\rho^{\alpha_1,\ldots,\alpha_t}$ as follows: on input $x \in \{0,1\}^*$, check if $x \in \mathcal{L}_{\rho(\alpha_1,\ldots,\alpha_t)}$. If so return $\alpha_1, \ldots, \alpha_t$; else return $\perp$. Let $\mathscr{U}_{\rho k} = \{U_\rho^{\alpha_1,\ldots,\alpha_t} : \alpha_i \in \cup_{j=0}^k \{0,1\}^j\}$, and $\mathscr{U}_\rho = \cup_k \mathscr{U}_{\rho k}$.*

Unless a string in the language $\mathcal{L}_{\rho(\alpha_1,\ldots,\alpha_t)}$ is given as input $U_\rho^{\alpha_1,\ldots,\alpha_t}$ reveals nothing beyond the fact that the string is not in the language. We show that this function can be completely obfuscated.

**Theorem 4.** *For any regular expression $\rho$, the family $\mathscr{U}_\rho$ is efficiently obfuscatable in the random oracle model.*

To prove this, we introduce another family of functions $\mathscr{V}_\rho$, and show that $\mathscr{U}_\rho \ll \mathscr{V}_\rho$. Then, we show that $\mathscr{V}_\rho$ can be obfuscated (in the random oracle model).

Recall that $\rho$ is a regular expression over the symbols $\Sigma \cup \{\zeta^{\alpha_1}, \ldots, \zeta^{\alpha_t}\}$. We can convert this to a deterministic finite-state automaton (DFA), with some of the edges labeled with $\zeta^{\alpha_i}$. Define a set $\mathcal{Z}_\rho \subseteq 2^{[t]}$ of subsets of $[t]$ as follows. If there is a path in the above DFA from the start state to some accept state, in which the set of non-$\Sigma$ symbols appearing are $\{\zeta^{\alpha_i} : i \in Z \subseteq [t]\}$, then $Z \in \mathcal{Z}_\rho$. In other words, $\mathcal{Z}_\rho$ is the set of all subsets of $\alpha_i$'s, such that knowing $\alpha_i$'s in any of these subsets will enable one to construct a string in $\mathcal{L}_{\rho(\alpha_1,\ldots,\alpha_t)}$. Note that $\mathcal{Z}_\rho$ can be constructed from $\rho$, independent of $\alpha_1, \ldots, \alpha_t$.

**Definition 15.** *Define the function $V_\rho^{\alpha_1,\ldots,\alpha_t}$ as follows: on input $(\beta_1, \ldots, \beta_t)$, $\beta_i \in \{0,1\}^*$, check if $\exists Z \in \mathcal{Z}_\rho$ such that $\forall i \in Z$, $\beta_i = \alpha_i$. If so return $\alpha_1, \ldots, \alpha_t$; else return $\perp$. Let $\mathscr{V}_{\rho k} = \{V_\rho^{\alpha_1,\ldots,\alpha_t} : \alpha_i \in \cup_{j=0}^k \{0,1\}^j\}$, and $\mathscr{V}_\rho = \cup_k \mathscr{V}_{\rho k}$.*

**Lemma 7.** *$\mathscr{U}_\rho \ll \mathscr{V}_\rho$ for all regular expressions $\rho$.*

*Proof:* Corresponding to $U_\rho^{\alpha_1,\ldots,\alpha_t} \in \mathscr{U}_\rho$ we pick $V_\rho^{\alpha_1,\ldots,\alpha_t} \in \mathscr{V}_\rho$.

*Constructing $M$ such that $M^{V_\rho^{\alpha_1,\ldots,\alpha_t}} \equiv U_\rho^{\alpha_1,\ldots,\alpha_t}$:* As input $M^{V_\rho^{\alpha_1,\ldots,\alpha_t}}$ receives a string $x \in \{0,1\}^*$. It needs to check if $x \in \mathcal{L}_{\rho(\alpha_1,\ldots,\alpha_t)}$. $M$ chooses $t$ substrings of $x$ as guesses for $\alpha_1,\ldots,\alpha_t$. If $|x| = n$ there are $O(n^{2t}$ such choices. But by our convention, since $\rho$ is fixed, $t$ is a constant and $n^{2t}$ is still polynomial in $n$, the size of input to $M$. For each such guess $(\beta_1,\ldots,\beta_t)$, $M$ queries $V_\rho^{\alpha_1,\ldots,\alpha_t}$ on $(\beta_1,\ldots,\beta_t)$. If $V_\rho^{\alpha_1,\ldots,\alpha_t}$ returns $\perp$ for all choices, $M$ also outputs $\perp$. If $V_\rho^{\alpha_1,\ldots,\alpha_t}$ returns $(\alpha_1,\ldots,\alpha_t)$ for any choice of $(\beta_1,\ldots,\beta_t)$, then $M$ constructs the complete DFA (replacing the variables $\zeta^{\alpha_i}$ with $\alpha_i$) and checks if $x$ is accepted by the DFA. If so, $M$ outputs $\alpha_1,\ldots,\alpha_t$; if not it outputs $\perp$.

If $x \in \mathcal{L}_{\rho(\alpha_1,\ldots,\alpha_t)}$, then there is some path in the DFA for $\rho$ which accepts $x$. Let $Z$ be the set of all $i$ such that $\zeta^{\alpha_i}$ appears on this accepting path. By the way $\mathcal{Z}_\rho$ was constructed, $Z \in \mathcal{Z}_\rho$. Further all these $\zeta^{\alpha_i}$ appear as part of $x$. Thus, for some guess $\beta_1,\ldots,\beta_t$, it will be the case that for all of $i \in Z$ $\beta_i = \alpha_i$. Thus if $x \in \mathcal{L}_{\rho(\alpha_1,\ldots,\alpha_t)}$, $M$ will obtain all of $\alpha_1,\ldots,\alpha_t$ from $V_\rho^{\alpha_1,\ldots,\alpha_t}$, and will be able to verify that $x \in \mathcal{L}_{\rho(\alpha_1,\ldots,\alpha_t)}$. On the other hand if $x \notin \mathcal{L}_{\rho(\alpha_1,\ldots,\alpha_t)}$ either $\alpha_1,\ldots,\alpha_t$ are not revealed to $M$, or they are and $M$ will discover that $x \notin \mathcal{L}_{\rho(\alpha_1,\ldots,\alpha_t)}$. In either case $M$ will output $\perp$, as required.

*Constructing $N$ such that $N^{U_\rho^{\alpha_1,\ldots,\alpha_t}} \equiv V_\rho^{\alpha_1,\ldots,\alpha_t}$:* As input $N^{U_\rho^{\alpha_1,\ldots,\alpha_t}}$ receives $t$ strings $(\beta_1,\ldots,\beta_t)$. It needs to check if there is any $Z \in \mathcal{Z}_\rho$ such that $\forall i \in Z\ \alpha_i = \beta_i$. Associated with each $Z$ is a path from the start state to an accept state in which the variable $\zeta^{\alpha_i}$ appear for exactly those $i \in Z$. $N$ chooses for each $Z$ such a path, and constructs a string $x_Z$ corresponding to that path, substituting $\beta_i$ for $\zeta^{L_i}$. It then submits $x_Z$ to $U_\rho^{\alpha_1,\ldots,\alpha_t}$ (to which it has oracle access). If $U_\rho^{\alpha_1,\ldots,\alpha_t}$ responds with $\perp$ for all $x_Z$, $Z \in \mathcal{Z}_\rho$ then $N$ outputs $\perp$. If $U_\rho^{\alpha_1,\ldots,\alpha_t}$ responds with $\alpha_1,\ldots,\alpha_t$ for any $x_Z$, then $N$ then checks if $\exists Z \in \mathcal{Z}_\rho\ \forall i \in Z\ \alpha_i = \beta_i$, and responds accordingly. It can be easily verified that $N^{U_\rho^{\alpha_1,\ldots,\alpha_t}} \equiv V_\rho^{\alpha_1,\ldots,\alpha_t}$. □

Next we observe that $\mathscr{V}_\rho \ll \mathscr{Q}^*$, where $\mathscr{Q}^*$ is the class of multi-point functions with general output (Definition 11).

**Lemma 8.** $\mathscr{V}_\rho \ll \mathscr{Q}^*$

*Proof:* Let $\mathcal{Z}_\rho = \{Z_1,\ldots,Z_\ell\}$, and for each $Z_i \in \mathcal{Z}_\rho$, let the string $\gamma_i$ be $(\gamma_i^1,\ldots,\gamma_i^t)$ where if $j \in Z_i$, $\gamma_i^j = \alpha_j$ and else $\gamma_i^j = 0$.

For every $V_\rho^{\alpha_1,\ldots,\alpha_t} \in \mathscr{V}_\rho$, consider $Q = Q_{(\gamma_1,\Delta),\ldots,(\gamma_\ell,\Delta)} \in \mathscr{Q}^*$ where $\Delta = (\alpha_1,\ldots,\alpha_t)$ (i.e., if $Q$ is given one of the strings $\gamma_1,\ldots,\gamma_\ell$, it outputs $\Delta$. It is easy to verify that the following machines $M$ and $N$ are as required by Definition 4.

$M^Q$, on input $(\beta_1,\ldots,\beta_t)$ does the following: for each $Z_i \in \mathcal{Z}_\rho$ it constructs a string $\delta_i = (\delta_i^1,\ldots,\delta_i^t)$ where if $j \in Z_i$, $\delta_i^j = \beta_j$ and else $\delta_i^j = 0$; then it queries $Q$ with $\delta_i$; if for any $i$ it receives $\Delta$ from $Q$ it outputs that and else $\perp$.

$N^{V_\rho^{\alpha_1,\ldots,\alpha_t}}$ on input $\delta = (\delta^1,\ldots,\delta^t)$, queries $V_\rho^{\alpha_1,\ldots,\alpha_t}$ with $\delta$. If it receives $\perp$ as an answer, it also outputs $\perp$. Else it receives $\Delta$, and can then can compute $Q(\delta)$, which it outputs. □

By Lemma 5, $\mathcal{Q}^*$ is obfuscatable, thereby completing the proof of $\mathcal{V}_\rho$ being obfuscatable. To complete the proof of Theorem 4, we appeal to Lemma 1, along with Lemma 7 and the above fact that $\mathcal{V}_\rho$ is obfuscatable.

We remark that the construction above can easily be extended to also produce an arbitrary secret output if the input matches the regular expression.

### 6.2 Obfuscating a Function Related to $\rho(L_1, \ldots, L_t)$

In this section we allow $\rho$ to be part of the function (and therefore can have size polynomial in $k$). We are interested in matching a given string against $\rho(L_1, \ldots, L_t)$ without compromising the black-box nature of $[\![L_1, \ldots, L_t]\!]$. The family of functions we are interested in is $\mathcal{F}_C$ below.

**Definition 16.** *Define $G_\rho^{L_1,\ldots,L_t}$ and $F_\rho^{L_1,\ldots,L_t}$ as follows:*

$$G_\rho^{L_1,\ldots,L_t}(a,x) = \begin{cases} \rho & \text{if } a = 1 \\ L_{a-1}(x) & \text{if } a \in \{2,\ldots,t+1\} \\ \perp & \text{otherwise} \end{cases}$$

$$F_\rho^{L_1,\ldots,L_t}(a,x) = \begin{cases} 1 & \text{if } a = 0 \text{ and } x \text{ matches } \rho(L_1,\ldots,L_t) \\ 0 & \text{if } a = 0 \text{ and } x \text{ does not match } \rho(L_1,\ldots,L_t) \\ G_\rho^{L_1,\ldots,L_t}(a,x) & \text{otherwise} \end{cases}$$

$$\mathcal{G}_C = \{G_\rho^{L_1,\ldots,L_t} \ : \ \rho \text{ a regular expression and } L_i \in C\}$$

$$\mathcal{F}_C = \{F_\rho^{L_1,\ldots,L_t} \ : \ \rho \text{ a regular expression and } L_i \in C\}$$

In other words, both $G_\rho^{L_1,\ldots,L_t}$ and $F_\rho^{L_1,\ldots,L_t}$ provide access to the languages $L_i$ and to (the description of) the regular expression $\rho$. In addition, $F_\rho^{L_1,\ldots,L_t}$ gives access to the language defined by the regular expression $\rho(L_1, \ldots, L_t)$.

**Theorem 5.** *$\mathcal{F}_C$ is obfuscatable if and only if $\{[\![L_1, \ldots, L_t]\!] \ : \ L_i \in C\}$ is. Further this statement holds restricted to efficient obfuscations too.*

First we prove the following lemma, which is the heart of the proof. It shows how to evaluate the regular expressions involving $L_i$'s just with access to $\mathcal{G}_C$.

**Lemma 9.** *$\mathcal{F}_C \ll \mathcal{G}_C$ and $\mathcal{G}_C \ll \mathcal{F}_C$, for all families $C$.*

*Proof:* It is easy to see that $\mathcal{G}_C \ll \mathcal{F}_C$. For the other direction, we have to demonstrate the polynomial time oracle machines $M$ and $N$ as in Definition 3. But $N$ is trivial, and so is $M$'s behaviour when on input $(a, x)$, it sees $a \neq 0$. The non-trivial case is when $a = 0$: $M$ should match the input $x$ with the regular expression $\rho$ with only black-box access to $L_i$. We give a fairly efficient algorithm using dynamic programming to achieve this.

First $M$ obtains the regular expression $\rho$ from $G$ (by giving input $(1, \epsilon)$. It constructs a tree corresponding to $\rho$ with leaf nodes corresponding to symbols from $\Sigma \cup \{\zeta^{L_1}, \ldots, \zeta^{L_n}\}$. Each internal node corresponds to one of the three operators $|$, $\cdot$ and

*; in the first two cases the node will have two children and in the last case a single child. The root node corresponds to the whole regular expression $\rho$. The algorithm will consider the set $S$ of all substrings of the input string $x = x_1 \ldots x_n$; i.e., $S = \{x_i^j : 1 \leq i \leq j \leq n\} \cup \{\epsilon\}$. For each node it will try to find out all the strings in $S$ which match the regular expression at that node. This is done bottom-up in the tree. To obtain this information at the leaf nodes, $M$ makes $O(n^2)$ queries to each $L_i$.

Given this information for the children of a node, the information for that node itself can be obtained. In the case of a (|)-node (denoted by $\mathcal{Q} = \mathcal{Q}_1|\mathcal{Q}_2$) this is simple: for each string $s \in S$ check if $s \sim \mathcal{Q}_1$ or $s \sim \mathcal{Q}_2$. If either case holds record that $s \sim \mathcal{Q}$. For ($\cdot$)-node $\mathcal{Q} = \mathcal{Q}_1 \cdot \mathcal{Q}_2$ we do the following:

```
for each s ∈ S do
  for i = 0 to |s| do
    if s_1^i ~ Q_1 AND s_{i+1}^{|s|} ~ Q_2 then
      record s ~ Q
```

The checks $s_1^i \sim \mathcal{Q}_1$ and $s_{i+1}^{|s|} \sim \mathcal{Q}_2$ are done by checking if those matchings have already been recorded. The (*)-nodes require a little more work. At a node $\mathcal{Q} = \mathcal{Q}_1^*$ we do the following:

```
Let Q_1^1 denote Q_1
for k = 2 to n do
  for each s ∈ S\{ε} do
    for i = 0 to |s| do
      if s_1^i ~ Q_1^{k-1} AND s_{i+1}^{|s|} ~ Q_1 then
        record s ~ Q_1^k
record ε ~ Q
for each s ∈ S\{ε} do
  if s ~ Q_1^k for some k ∈ {1,...,n} then
    record s ~ Q
```

It is not hard to see that at each node the algorithm correctly records all $s \in S$ which match the node. Finally, it checks if $x \sim \rho$ by checking if it is recorded at the root node. □

*Proof:* **(of Theorem 5)** By the above Lemma and Lemma 1, we can obfuscate $\mathscr{F}_C$, if and only if we can obfuscate $\mathscr{G}_C$. We can view $G \in \mathscr{G}_C$ as $[\![\langle\rho\rangle, [\![L_1, \ldots, L_n]\!]]\!]$, where $\langle\rho\rangle$ stands for the constant (and hence trivially obfuscatable) function which outputs $\rho$. Then by Lemma 3, $\mathscr{G}_C$ is obfuscatable if and only if $\{[\![L_1, \ldots, L_n]\!] : L_i \in C\}$ is obfuscatable. □

## 7 Obfuscating Neighborhoods in Tree Metrics

Point functions are identity checks- they check if the input is identical to a particular value. A natural relaxation thereof is a neighborhood check. Consider some metric space from which the inputs are drawn. We would like to have a program which checks if the input is "near" a hidden point.

We work in a restricted metric space- the space of "tree metrics," where the the points are nodes in a (rooted, undirected) tree, and the distance between two points is the length of the (unique) path between them. (We can allow a metric space that can be decomposed as a collection of a *constant* number of tree metrics, but for simplicity we stick to a single tree-metric.)

Let $\mathcal{M}$ stand for the metric space as well as (by abuse of notation) the tree defining it. Let $\mathrm{d}_{\mathcal{M}}(\cdot,\cdot)$ be the distance function in $\mathcal{M}$.

**Definition 17.** *Define the function $T_\alpha^{\mathcal{M}} : \mathcal{M} \to \mathcal{M} \cup \{\perp\}$ as follows:*

$$T_\alpha^{\mathcal{M}}(x) = \begin{cases} \alpha & \mathrm{d}_{\mathcal{M}}(\alpha, x) \leq \delta \\ \perp & \mathrm{d}_{\mathcal{M}}(\alpha, x) > \delta \end{cases}$$

$\mathscr{T}_k = \{T_\alpha^{\mathcal{M}} : \mathcal{M}$ *a tree-metric* $, |\mathcal{M}| = 2^{O(k)}, \alpha \in \mathcal{M}\}$ *and* $\mathscr{T} = \cup_k \mathscr{T}_k$.

Obfuscating $\delta$-neighborhoods in general metric spaces (beyond what can be achieved by exhaustively searching the entire $\delta$-neighborhood of a point) is a challenging problem. But we show that for tree metrics this problem can be satisfactorily solved using a simple technique. To obfuscate $T_\alpha^{\mathcal{M}}$, traverse the tree $\mathcal{M}$, starting at the node $\alpha$, towards the root of the tree, for a distance $\delta$, and pick the node at which we finish. (If we reach the root before $\delta$ steps pick the root.) Call this node $\beta$. We show that obfuscating $T_\alpha^{\mathcal{M}}$ is essentially the same as obfuscating the point function on $\beta$ with output $\alpha$ (which as we have shown, can be efficiently obfuscated in the random oracle model).

**Lemma 10.** $\mathscr{T} \ll \mathscr{Q}$ *(where $\mathscr{Q}$ is the point function with general output, as in Definition 10).*

*Proof:* For $T^{\mathcal{M}} \in \mathscr{T}$ we pick $Q_{\beta,\alpha} \in \mathscr{Q}$. $Q_{\beta,\alpha}$ is the function which outputs $\alpha$ on input $\beta$ and $\perp$ everywhere else.

$N^{T_\alpha^{\mathcal{M}}}$ *works as follows* : On input $x \in \mathcal{M}$ query $T_\alpha^{\mathcal{M}}$ with $x$. If $x$ were indeed equal to $\beta$ then $T_\alpha^{\mathcal{M}}$ would respond with $\alpha$. So if $T_\alpha^{\mathcal{M}}$ gives $\perp$ return $\perp$. If it gives $\alpha$, locate $\beta$ by traversing $\mathcal{M}$, and check if the $x$ is indeed $\beta$ or not and answer accordingly.

$M^{Q_{\beta,\alpha}}$ *works as follows* : on input $x \in \mathcal{M}$, check the first $2\delta$ ancestors of $x$ for being identical to $\beta$ (using $Q_{\beta,\alpha}$). If $Q_{\beta,\alpha}$ returns $\alpha$ on some query, check $\mathrm{d}_{\mathcal{M}}(x,\alpha)$ and answer appropriately. If it returns $\perp$ in all $2\delta$ queries, then it is easy to see that the distance $\mathrm{d}_{\mathcal{M}}(x,\alpha) > \delta$. In this case, output $\perp$. □

By Lemma 1 and Theorem 2, we get:

**Theorem 6.** $\mathscr{T}$ *is obfuscatable in the random oracle model.*

## 8 Conclusions and Open Problems

We have given the first positive results and techniques for program obfuscation, but many important open problems remain. We are hopeful our reduction and composition

parties can choose to give an arbitrary input to the trusted party, and to terminate the protocol prematurely, even at a stage where they have received their output and the other parties have not. We limit it to the case where both parties compute the same function $f : \{0,1\}^* \times \{0,1\}^* \rightarrow \{0,1\}^*$.

**Definition 1 (The Ideal Model).** *A strategy for party A in the ideal model is a pair of PPT (probabilistic polynomial time) algorithms, $A_I(X, r)$ that uses the input $X$ and a sequence of coin flips $r$ to generate an input that A sends to the trusted party, and $A_O(X, r, Z)$ which takes as an additional input the value $Z$ that A receives from the TTP, and outputs A's final output. If A is honest then $A_I(X, r) = X$ and $A_O(X, r, Z) = Z$. A strategy for party B is similarly defined using functions $B_I(Y, r)$ and $B_O(Y, r, Z)$.*

*The definition is limited to the case where at least one of the parties is honest. We call an adversary that corrupts only one of the parties an* admissible adversary. *The joint execution of A and B in the ideal model, denoted* $\mathsf{IDEAL}_{A,B}(X, Y)$, *is defined to be*

- *If B is honest,*
  - $\mathsf{IDEAL}_{A,B}(X, Y)$ *equals* $(A_O(X, r, f(X', Y)), f(X', Y))$, *where* $X' = A_I(X, r)$ *(in the case that A did not abort the protocol),*
  - *or,* $\mathsf{IDEAL}_{A,B}(X, Y)$ *equals* $(A_O(X, r, f(X', Y)), -)$, *where* $X' = A_I(X, r)$ *(if A terminated the protocol prematurely).*
- *If A is honest*
  - $\mathsf{IDEAL}_{A,B}(X, Y)$ *equals* $(f(X, Y'), B_O(Y, r, f(X, Y')))$, *where* $Y' = B_I(Y, r)$,
  - *or,* $\mathsf{IDEAL}_{A,B}(X, Y)$ *equals* $(-, B_O(Y, r, f(X, Y')))$, *where* $Y' = B_I(Y, r)$.

In the real execution a malicious party could follow any strategy that can be implemented by a PPT algorithm. The strategy is an algorithm mapping a partial execution history to the next message sent by the party in the protocol.

**Definition 2 (The Real Model (for semi-honest and malicious adversaries)).** *Let $f$ be as in Definition 1, and $\Pi$ be a two-party protocol for computing $f$. Let $(A', B')$ be a pair of PPT algorithms representing the parties' strategies. This pair is admissible w.r.t. $\Pi$ if at least one of $(A', B')$ is the strategy specified by $\Pi$ for the corresponding party. In the* semi-honest case *the other party could have an arbitrary output function. In the* malicious *case, the other party can behave arbitrarily throughout that protocol.*

*The joint execution of $\Pi$ in the real model, denoted* $\mathsf{REAL}_{\Pi,A',B'}(X, Y)$ *is defined as the output pair resulting from the interaction between $A'(X)$ and $B'(Y)$.*

The definition of security states that an execution of a secure real model protocol under any admissible adversary can be simulated by an admissible adversary in the ideal model.

**Definition 3 (Security (for both the semi-honest case and the malicious case)).** *Let $f$ and $\Pi$ be as in Definition 2. Protocol $\Pi$* **securely computes** *$f$ if for every PPT pair $(A', B')$ that is admissible in the real model (of Definition 2) there is a PPT pair $(A, B)$ that is admissible in the ideal model (of Definition 1), such that* $\mathsf{REAL}_{\Pi,A',B'}(X, Y)$ *is computationally indistinguishable from* $\mathsf{IDEAL}_{A,B}(X, Y)$.

**Reactive Computations** A reactive computation consists of steps in which parties provide inputs and receive outputs. Each step generates a *state* which is used by the following step. The input that a party provides at step $i$ can depend on the outputs that it received in previous steps. (We limit ourselves to synchronous communication, and to an environment in which there are secure channels between the parties.) The protocols that we design for the malicious case implement reactive computation. Security definitions and constructions for reactive computation were discussed in [3,5] (in particular, they enable parties to abort the protocol at arbitrary stages). We will not describe these definitions in this extended abstract, due to their length and detail.

**A Composition Theorem** Our protocols implement the computation of the $k^{th}$-ranked element by running many invocations of secure computation of simpler functionalities. Such constructions are covered by theorems of secure composition [2,3]. Loosely speaking, consider a *hybrid model* where the protocol uses a trusted party that computes the functionalities $f_1, \ldots, f_\ell$. The secure composition theorem states that if we consider security in terms of comparing the real computation to the ideal model, then if a protocol is secure in the hybrid model, and we replace the calls to the trusted party by calls to secure protocols computing $f_1, \ldots, f_\ell$, then the resulting protocol is secure. A secure composition theorem applies to reactive computation, too [3,5].

## 2 Two-Party Computation of the $k^{th}$ Element

This section describes protocols for secure two-party computation of the $k^{th}$-ranked element of the union of two databases. The protocols are based on the observation that a natural algorithm for computing the $k^{th}$-ranked element discloses very little information that cannot be computed from the value of the $k^{th}$-ranked element itself. Some modification to that protocol can limit the information that is leaked by the execution to information that can be computed from the output alone.

To simplify the description of the basic, insecure, protocol, we describe it for the case of two parties, A and B, each of which has an input of size $n/2$, that wish to compute the value of the median, i.e. $(n/2)^{th}$-ranked element, of the union of their two inputs sorted in increasing order of their values. This protocol is a modification of the algorithm given in [17,13]. Assume for simplicity that all input values are different. The protocol operates in rounds. In each round, each party computes the median value of his or her input, and then the two parties compare their two median values. If A's median value is smaller than B's then A adjusts her input by removing the values which are less than or equal to her median, and B removes his input items which are greater than his median. Otherwise, A removes her items which are greater than her median and B removes his items which are less than or equal to his median. The protocol continues until the inputs are of length 1 (thus the number of rounds is logarithmic in the number of input items). The protocol is correct since when A's median is smaller than B's median, each of the items that A removes is smaller than A's median, which is smaller than at least $n/4$ inputs of A and $n/4$ inputs of B. Therefore the removed item cannot be the median. Also, the protocol removes $n/4$ items which are smaller than the median and $n/4$ which are greater than it, and therefore the median of the new data is the same as that of the original input. Other cases follow similarly.

Suppose now that the comparison is done privately, i.e. that the parties only learn whose median value is greater, and do not learn any other information about each others median value. We show below that in this case the protocol is secure. Intuitively this is true since, e.g., if party A knows the median value of her input and the median of the union of the two inputs, and observes that her median is smaller than the median of the union, then she can deduce that her median value is smaller than that of B. This means that given the final output of the protocol party A can simulate the results of the comparisons. Consequently, we have a reduction from securely computing the median of the union to securely computing comparisons.

**Secure comparison:** The main cryptographic primitive that is used by the protocol is a two-party protocol for secure comparison. The protocol involves two parties, where party A has an input $x$ and party B has an input $y$. The output is 0 if $x \geq y$ and 1 otherwise. The protocol (which essentially computes a solution to Yao's millionaires problem) can be implemented by encoding the comparison function as a binary circuit which compares the bits encoding the two inputs, and applying to it Yao's protocol for secure two-party computation. The overhead is $|x|$ oblivious transfers, and $O(|x| + |y|)$ applications of a pseudo-random function, as well as $O(|x|+|y|)$ communication. More efficient, non-interactive comparison protocols also exist (see e.g. [7]).

## 2.1 A Protocol for Semi-Honest and Malicious Parties

Following is a description of a protocol that finds the $k^{th}$-ranked element in the union of two databases and is secure against semi-honest parties. The computation of the median is a specific case where $k$ is set to be the sum of the two inputs divided by two. The protocol reduces the general problem of computing the $k^{th}$-ranked element of arbitrary size inputs, to the problem of computing the median of two inputs of equal size, which is also a power of 2. To simplify the exposition, we assume that all the inputs are *distinct*. This issue is further discussed later.

**Security against a malicious adversary.** The protocol for the semi-honest case can be amended to be secure against malicious adversaries. The main change is that the protocol must now verify that the parties provide consistent inputs to the different invocations of the secure computation of the comparisons. For example, if party A gave an input of value 100 to a secure comparison computation, and the result was that A must delete all its input items which are smaller than 100, then $A$ cannot provide an input which is smaller than 100 to any subsequent comparison. We provide a proof that given this enforcement, the protocol is secure against malicious behavior. For this protocol, we do not force the input elements to be integers. However, if such an enforcement is required (e.g. if the input consists of rounded salary data), then the protocol for the malicious case verifies that there is room for sufficiently many distinct integers between the reported values of different elements of the input. This is made more precise later.

In protocol FIND-RANKED-ELEMENT that we describe here, we specify the additional *functionality* that is required in order to ensure security against malicious parties. Then in Section 2.3 we describe how to implement this functionality, and prove that given this functionality the protocol is secure against malicious adversaries. Of course, to obtain a protocol which is only secure against semi-honest adversaries, one should ignore the additional highlighted steps that provide security in the malicious case.

**Protocol** FIND-RANKED-ELEMENT
**Input:** $D_A$ known to A, and $D_B$ known to B. Public parameter $k$ (for now, we assume that the numerical value of the rank of the element is known). All items in $D_A \cup D_B$ are distinct.
**Output:** The $k^{th}$-ranked element in $D_A \cup D_B$.

1. Party A (resp., B) initializes $S_A$ (resp., $S_B$) to be the sorted sequence of its $k$ smallest elements in $D_A$ (resp., $D_B$).
2. If $|S_A| < k$ then Party A pads $(k - |S_A|)$ values of "$+\infty$" to its sequence $S_A$. Party B does the same: if $|S_B| < k$ then it pads $(k - |S_B|)$ values of "$+\infty$" to its sequence $S_B$.
3. Let $2^j$ be the smallest power of 2 greater than or equal to $k$. Party A pre-pads $S_A$ with $(2^j - k)$ values of "$-\infty$" and Party B pads $S_B$ with $(2^j - k)$ values of "$+\infty$". (The result is two input sets of size $2^j$ each, whose median is the $k^{th}$-ranked element in $D_A \cup D_B$ .)
   **In the malicious case:** The protocol sets bounds $l_A = l_B = -\infty$ and $u_A = u_B = \infty$.
4. For $i = (j-1), \ldots, 0$ :
   A. A computes the $(2^i)^{th}$ element of $S_A$, denoted $m_A$, and B computes the $(2^i)^{th}$ element of $S_B$, $m_B$. (I.e., they compute the respective medians of their sets.)
   B. A and B engage in a *secure computation* which outputs 0 if $m_A \geq m_B$, and 1 if $m_A < m_B$.
   **In the malicious case:** The secure computation first checks that $l_A < m_A < u_B$ and $l_B < m_B < u_B$. If we want to force the input to be integral, then we check that $l_A + 2^i < m_A \leq u_A - 2^i$ and $l_B + 2^i < m_B \leq u_B - 2^i$. If these conditions are not satisfied, then the protocol is aborted. Otherwise, if $m_A \geq m_B$, the protocol sets $u_A$ to be $m_A$ and $l_B$ to be $m_B$. Otherwise it updates $l_A$ to $m_A$ and $u_B$ to $m_B$. Note that the lower and upper bounds are not revealed to either party.
   C. If $m_A < m_B$, then A removes all elements ranked $2^i$ or less from $S_A$, while B removes all elements ranked greater than $2^i$ from $S_B$. On the other hand, if $m_A \geq m_B$, then A removes all elements ranked higher than $2^i$ from $S_A$, while B removes all elements ranked $2^i$ or lower from $S_B$.
5. (Here every party has an input set of size 1.) Party $A$ and party $B$ output the result of a *secure computation* of the minimum value of their respective elements.

   **In the malicious case:** The secure computation checks that the inputs given in this step are consistent with the inputs given earlier. Specifically, for any item other than item $2^j$ of the original set of A (respectively B), this means that the value must be equal to $u_A$ (respectively $u_B$). For item $2^k$ of step A (respectively B), it is verified that its value is greater than $l_A$ (respectively $l_B$).

**Overhead:** Since the value $j$ is at most $\log 2k$ and the number of rounds of communication is $(j+1)$, the total number of rounds of communication is $\log(2k)$. In each round, the protocol performs at most one secure computation, which requires a comparison of $(\log M)$ bit integers. Thus the total communication cost is $O(\log M \cdot \log k)$ times the security parameter.

**Proof of Correctness** Regardless of security issues, we first have to show that the protocol indeed computes the $k^{th}$-ranked item. We need to show that (a) The preprocessing performed in Steps 1-3 does not eliminate the $k^{th}$-ranked value and (b) The $(2^{i+1})^{st}$ value of $S_A^i \cup S_B^i$ is the $k^{th}$-ranked value in $D_A \cup D_B$ for each $i = j-1, \ldots, 0$ (where $S_A^i, S_B^i$ are the sorted sequences maintained by parties $A$, $B$, respectively, during iteration $i$). These two properties are shown in Lemma 1.

**Lemma 1.** *In Protocol* FIND-RANKED-ELEMENT*, the* $(2^{i+1})^{st}$*-ranked element of* $S_A \cup S_B$ *in round* $i$ *of Step 4 (i.e., the median) is equal to the* $k^{th}$*-ranked element in* $D_A \cup D_B$*, for* $i = (j-1), \ldots, 0$.

*Proof.* Note that in the preprocessing (Step 1) we do not eliminate the $k^{th}$-ranked element since the $k^{th}$-ranked element cannot appear in position $(k+1)$ or higher in the sorted version of $D_A$ or $D_B$. Step 2 ensures that both sequences have size exactly $k$ without affecting the $k^{th}$-ranked element (since padding is performed at the end of the sequences). And, Step 3 not only ensures that the length of both sequences is a power of 2, but also pads $S_A$ and $S_B$ so that the $(2^j)^{th}$ element of the union of the two sequences is the $k^{th}$-ranked element of $D_A \cup D_B$. This establishes the Lemma for the case where $i = (j-1)$.

The remaining cases of $i$ follow by induction. We have essentially transformed the original problem to that of computing the median between two sets of equal size $2^{i+1}$. Note that neither party actually removes the median of $S_A \cup S_B$: if $m_A < m_B$ then there are $2 \cdot 2^i$ points in $S_A$ and $S_B$ that are larger than $m_A$ and $2 \cdot 2^i$ points in $S_A$ and $S_B$ that are smaller than $m_B$, thus no point in $S_A$ that is less than or equal to $m_A$ can be the median, nor can any point in $S_B$ greater than $m_B$. A similar argument follows in the case that $m_A > m_B$. Furthermore, the modifications made to $S_A$ and $S_B$ maintain the median of $S_A \cup S_B$ since at each iteration an equal number of elements are removed from above and below the median (exactly half of the points of each party are removed). The lemma follows.

## 2.2 Security for the Semi-Honest Case

In the semi-honest case, the security definition in the ideal model is identical to the definition which is based on simulation (which we haven't explicitly described). Thus, it is sufficient to show that, assuming that the number of elements held by each party is public information, party A (and similarly party B), given its own input and the value of the $k^{th}$-ranked element, can simulate the execution of the protocol in the hybrid model, where the comparisons are done by a trusted party (the proof follows by the composition theorem). We describe the proof detail for the case of party A simulating the execution.

Let $x$ be the $k^{th}$-ranked element which the protocol is supposed to find. Then, party A simulates the protocol as follows:

**Algorithm** SIMULATE-FIND-RANK
**Input:** $D_A$ and $x$ known to A. Public parameter $k$. All items in $D_A \cup D_B$ are distinct.
**Output:** Simulation of running the protocol for finding the $k^{th}$-ranked element in $D_A \cup D_B$.

1. Party A initializes $S_A$ to be the sorted sequence of its $k$ smallest elements in $D_A$
2. If $|S_A| < k$ then Party A pads $(k - |S_A|)$ values of "+∞" to its sequence $S_A$.
3. Let $2^j$ be the smallest power of 2 larger than $k$. Party A pre-pads $S_A$ with $(2^j - k)$ values of "-∞".
4. For $i = (j-1), \ldots, 0$:
   A. A computes the $(2^i)^{th}$ element of $S_A$, $m_A$
   B. If $m_A < x$, then the secure computation is made to output 1, i.e., $m_A < m_B$, else it outputs 0.
   C. If $m_A < x$, then A removes all elements ranked $2^i$ or less from $S_A$. On the other hand, if $x \leq m_A$, then A removes all elements ranked higher than $2^i$ from $S_A$.
5. The final secure computation outputs 1 if $m_A < x$ and 0 otherwise (in this case $m_A = x$ is the median).

**Lemma 2.** *The transcript generated by Algorithm* SIMULATE-FIND-RANK *is the same as the transcript generated by Protocol* FIND-RANKED-ELEMENT. *In addition, the state information that Party A has after each iteration of Step 4, namely ($S_A$, k), correctly reflects the state of Protocol* FIND-RANKED-ELEMENT *after the same iteration.*

*Proof.* We prove the lemma by induction on the number of iterations. Assume that the lemma is true at the beginning of an iteration of Step 4, i.e. Algorithm SIMULATE-FIND-RANK has been correctly simulating Protocol FIND-RANKED-ELEMENT and its state correctly reflects the state of Protocol FIND-RANKED-ELEMENT at the beginning of the iteration. We show that $m_A < x$ if and only if $m_A < m_B$. If $m_A < x$ then the number of points in $S_A^i$ smaller than $x$ is at least $2^i$. If by way of contradiction $m_B \leq m_A$, then $m_B < x$, implying that the number of points in $S_B^i$ smaller than $x$ is at least $2^i$. Thus the total number of points in $S_A^i \cup S_B^i$ smaller than $x$ would be at least $2^{i+1}$, contradicting that $x$ is the median. So, $m_A < m_B$. On the other hand, if $m_A < m_B$, and by way of contradiction, $m_A \geq x$, then $x \leq m_A < m_B$. Thus the number of points in $S_B^i$ greater than $x$ is strictly more than $2^i$. Also, at least $2^i$ points in $S_A^i$ are greater than $x$. Thus, the number of points in $S_A^i \cup S_B^i$ greater than $x$ is strictly more than $2^{i+1}$, again contradicting that $x$ is the median. So, $m_A < x$. Thus, the secure computations in Step 4 of Algorithm Simulate-Find-Rank return the same outputs as in Protocol FIND-RANKED-ELEMENT.

**Duplicate Items** Protocol FIND-RANKED-ELEMENT preserves privacy as long as no two input elements are identical (this restriction must be met for each party's input, and also for the union of the two inputs). The reason for this restriction is that the execution of the protocol reveals to each party the exact number of elements in the other party's input which are smaller than the $k^{th}$ item of the union of the two inputs. If all elements are distinct then given the $k^{th}$-ranked value each party can compute the number of elements in its own input that are smaller than it, and therefore also the number of such elements in the other party's input. This information is sufficient for simulating the execution of the protocol. However, if the input contains identical elements then given the $k^{th}$-ranked value it is impossible to compute the exact number of elements in the other party's input which are smaller than it and to simulate the protocol. (For example, if several items in A's input are equal to the $k^{th}$-ranked element then the protocol could have ended with a comparison involving any one of them. Therefore A does not know which of the possible executions took place.)

**Handling duplicate items** Protocol FIND-RANKED-ELEMENT-MULTIPARTY in Section 3 can securely computed the $k^{th}$-ranked item even if the inputs contain duplicate elements, and can be applied to the two-party case (although with $\log M$ rounds, instead of $\log k$). Also, protocol FIND-RANKED-ELEMENT can be applied to inputs that might contain identical elements, if they are transformed into inputs containing distinct elements. This can be done, for example, in the following way: Let the total number of elements in each party's input be $n$. Add $\lceil \log n \rceil + 1$ bits to every input element, in the least significant positions. For every element in A's input let these bits be a "0" followed by the rank of the element in a sorted list of A's input values. Apply the same procedure to B's inputs using a "1" instead of a "0". Now run the original protocol using the new inputs, but ensure that the output does not include the new least significant bits of the $k^{th}$ item. The protocol is privacy preserving with regard to the new inputs (which are all distinct). Also, this protocol does not reveal to party A more information than running the original protocol with the original inputs and in addition providing A with the number of items in B's input which are smaller than the $k^{th}$ value (the same holds of course w.r.t. B). This property can be verified by observing that if A is given the $k^{th}$-ranked element of the union of the two inputs, as well as the number of elements in $B$'s input which are smaller than this value, it can simulate the operation of the new protocol with the transformed input elements.

**Hiding the Size of the Inputs** Assume that the parties wish to hide from each other the size of their inputs. Note that if $k$ is public then the protocol that we described indeed hides the sizes of the inputs, since each party transforms its input to one of size $k$. This solution in insufficient, though, if $k$ discloses information about the input sizes. For example, if the protocol computes the median, then $k$ is equal to half the sum of the sizes of the two inputs. We next show now how to hide the size of the inputs when the two parties wish to compute the value of the $p^{th}$ percentile, which includes the case of computing the median (which is the $50^{th}$ percentile).

**$p^{th}$-Percentile** The $p^{th}$ percentile is the element with rank $\lceil \frac{p}{100} \cdot (|D_A| + |D_B|) \rceil$. We assume that an upper bound $U$ on the number of elements held by each party is known. Both parties first pad their inputs to get $U$ elements each, in a way that keeps the value of the $p^{th}$ percentile. For this, if a party A needs to add $X = U - |D_A|$ elements to its input, it adds $\frac{p}{100}X$ elements with value $-\infty$ and $\frac{(100-p)}{100}X$ elements with value $+\infty$ (to simplify the exposition, we assume that $\frac{p}{100}X$ is an integer). Party B acts in a similar way. Then the parties engage in a secure computation of the $p^{th}$ percentile, which is the $(\frac{p}{100} \cdot 2U)^{th}$-ranked element of the new inputs, using the protocol we described above.

### 2.3 Security for the Malicious Case

We assume that the comparison protocol is secure against malicious parties. We then show that although the malicious party can choose its input values adaptively during the execution of the protocol, it could as well have constructed an input apriori and given it to a trusted third party to get the same output. In other words, although the adversary can define the values of its input points depending on whether that input point needs to be compared or not in our protocol, this does not give it any more power. The proof is composed of two parts. First, we show that the functionality provided by the protocol definition provides the required security. Second, we show how to implement this functionality efficiently.

**Lemma 3.** *For every adversary $A'$ in the real model there is an adversary $A''$ in the ideal model, such that the outputs generated by $A'$ and $A''$ are computationally indistinguishable.*

**Proof Sketch:** Based on the composition theorem, we can consider only a protocol in the hybrid model where we assume that the comparisons are done securely by a trusted party. (We actually need here a composition theorem for a reactive scenario. We refer the reader to [3,5] for a treatment of this issue.)

Visualize the operation of $A'$ as a binary tree. The root is the first comparison it performs in the protocol. Its left child is the comparison which is done if the answer to the first comparison is 0, and the right child is the comparison that happens if the first answer is 1. The tree is constructed recursively following this structure, where every node corresponds to a comparison done at Step 4(B). We add leaves corresponding to the secure computation of Step 5 of the protocol following the sequence of comparisons that lead to a leaf.

Fix the random input used by the adversary $A'$. We also limit ourselves to adversaries that provide inputs that correspond to the bounds maintained by the protocol (otherwise the protocol aborts as in early termination, and since this is legitimate in the ideal model, we are done). We must generate an input to the trusted party that corresponds to the operation of $A'$. Let us run $A'$ where we provide it with the output of the comparisons. We go over all execution paths (i.e. paths in the tree) by stopping and rewinding the operation. (This is possible since the tree is of logarithmic depth.) Note that each of the *internal* nodes corresponds to a comparison involving a *different* location in the sorted list that $A'$ is required to generate according to the protocol. Associate with each node the value that $A'$ provides to the corresponding comparison.

Observe the following facts:

- For any three internal nodes $L, A, R$ where $L$ and $R$ are the left and right children of $A$, the bounds checked by the protocol enforce that the value of $L$ is smaller than that of $A$, which is smaller than that of $R$. Furthermore, an inorder traversal of the internal nodes of the tree results in a list of distinct values appearing in ascending order.

- When the computation reaches a leaf (Step 5), $A'$ provides a single value to a comparison. For the rightmost leaf, the value is larger than any value seen till now, while for each of the remaining leaves, the value is the same as the value on the rightmost internal node on the path from the root to the leaf (this is enforced by checking that the value is the same as $u_A$ or $u_B$ respectively).
- Each item in the input of $A'$ is used in at most a single internal node, and exactly a single leaf of the tree.

Consequently, the values associated with the leaves are sorted, and agree with all the values that $A'$ provides to comparisons in the protocol. We therefore use these values as the input to the trusted third party in the ideal model. When we receive the output from the trusted party we simulate the route that the execution takes in the tree, provide outputs to $A'$ and $B$, and perform any additional operation that $A'$ might apply to its view in the protocol.[6] □

**Implementing the Functionality of the Malicious Case Protocol** The functionality that is required for the malicious case consists of using the results of the first $i$ comparisons in order to impose bounds on the possible inputs to the following comparison. This is a *reactive secure computation*, which consists of several steps, where each step operates based on input from the parties and state information that is delivered from the previous step. This scenario, as well as appropriate security definitions and constructions, was described in [3,5]. (We are interested, however, in a simpler synchronous environment with secure channels.)

In order to implement secure reactive computation, each step should output shares of a state-information string, which are then input to the following step. The shares must be encrypted and authenticated by the secure computation, and be verified and decrypted by the secure computation of the following step. This functionality can be generically added to the secure computation

## 3 Multi-party Computation of the $k^{th}$ Ranked Element

We now describe a protocol that outputs the exact value of the $k^{th}$-ranked element of the union of multiple databases. For this protocol we assume that the elements of the sets are integer-valued, but they need not be distinct. Let $[\alpha, \beta]$ be the (publicly-known) range of input values, and let $M = \beta - \alpha + 1$. The protocol runs a series of rounds in which it (1) suggests a value for the $k^{th}$-ranked element, (2) performs a secure computation to which each party reports the number of its inputs which are smaller than this suggested value, adds these numbers and compares the result to $k$, and (3) updates the guess. The number of rounds of the protocol is logarithmic in $M$.

[6] Note that we are assuming that the inputs can be arbitrary Real numbers. If, on the other hand, there is some restriction on the form of the inputs, the protocol must verify that $A'$ provides values which are consistent with this restriction. For example, if the inputs are integers then the protocol must verify that the distance between the reported median and the bounds is at least half the number of items in the party's input (otherwise the input items cannot be distinct).

**Malicious adversaries.** We describe a protocol which is secure against semi-honest adversaries. Again, the protocol can be amended to be secure against malicious adversaries by verifying that the parties are providing it with consistent inputs. We specify in the protocol the additional functionality that should be implemented in order to provide security against malicious adversaries.

**Protocol** FIND-RANKED-ELEMENT-MULTIPARTY
**Input:** Party $P_i$, $1 \leq i \leq s$, has database $D_i$. The sizes of the databases are public, as is the value $k$. The range $[\alpha, \beta]$ is also public.
**Output:** The $k^{th}$-ranked element in $D_1 \cup \cdots \cup D_s$.

1. Each party ranks its elements in ascending order. Initialize the current range $[a, b]$ to $[\alpha, \beta]$ and set $n = \sum |D_i|$.
   **In the malicious case:** Set for each party $i$ bounds $(l)^i = 0,\ (g)^i = 0$. These values are used to bound the inputs that party $i$ reports in the protocol. $(l)^i$ reflects the number of inputs of party $i$ strictly smaller than the current range, while $(g)^i$ reflects the number of inputs of party $i$ strictly greater than the current range.
2. Repeat until "done"
   (a) Set $m = \lceil (a + b)/2 \rceil$ and announce it.
   (b) Each party computes the number of elements in its database which are strictly smaller than $m$, and the number of elements strictly greater than $m$. Let $l_i$ and $g_i$ be these values for party $i$.
   (c) The parties engage in the following secure computation:
   **In the malicious case:** Verify for every party $i$ that $l_i + g_i \leq |D_i|$, $l_i \geq (l)^i$, and $g_i \geq (g)^i$. In addition, if $m = \alpha$, then we check that $l_i = 0$; or if $m = \beta$, we verify that $g_i = 0$.
      - Output "done" if $\sum l_i \leq k - 1$ and $\sum g_i \leq n - k$. (This means that $m$ is the $k^{th}$-ranked item.)
      - Output "0" if $\sum l_i \geq k$. In this case set $b = m - 1$. (This means that the $k^{th}$-ranked element is smaller than $m$.)
        **In the malicious case:** Set $(g)^i = |D_i| - l_i$. (Since the left end-point of the range remains the same, $(l)^i$ remains unchanged.)
      - Output "1" if $\sum g_i \geq n - k + 1$. In this case set $a = m + 1$. (This means that the $k^{th}$-ranked element is larger than $m$.)
        **In the malicious case:** Set $(l)^i = |D_i| - g_i$.

**Correctness:** The correctness of this algorithm follows from observing that if $m$ is the $k^{th}$-ranked element then the first condition will be met and the algorithm will output it. In the other two cases, the $k^{th}$-ranked element is in the reduced range that the algorithm retains.

**Overhead:** The number of rounds is $\log M$. Each round requires a secure multi-party computation that computes two summations and performs two comparisons. The size of the circuit implementing this computation is $O(s \log M)$, which is also the number of input bits. The secure evaluation can be implemented using the protocols of [11,1,8].

**Security for the semi-honest case:** We provide a sketch of a proof for the security of the protocol. Assume that the multi-party computation in step 2(c) is done by a trusted party. Denote this scenario as the hybrid model. We show that in this case the protocol is secure against an adversary that controls up to $s-1$ of the parties. Now, implement the multi-party computation by a protocol which is secure against an adversary that controls up to $t$ parties, e.g. using [11,1,8]. (Of course, $t < s-1$ in the actual implementation, since the protocols computing the "simple" functionalities used in the hybrid model are not secure against $s-1$ parties, but rather against, say, any coalition of less than $s/3$ corrupt parties.) It follows from the composition theorem that the resulting protocol is secure against this adversary.

In the hybrid model, the adversary can simulate its view of the execution of the protocol, given the output of the protocol (and without even using its input). Indeed, knowing the range $[a, b]$ that is used at the beginning of a round, the adversary can compute the target value $m$ used in that round. If $m$ is the same as the output, it concludes that the protocol must have ended in this round with $m$ as the output (if the real execution did not output $m$ at this stage, $m$ would have been removed from the range and could not have been output). Otherwise, it simply updates the range to that side of $m$ which contains the output (if the real execution had not done the same, the output would have gone out of the active range and could not have been the output). Along with the knowledge of the initial range, this shows that the adversary can simulate the execution of the protocol.

**Security for the malicious case:** We show that the protocol is secure given a secure implementation of the functionality that is described in Step 3 of algorithm FIND-RANKED-ELEMENT-MULTIPARTY. Since this is a multi-party reactive system we refer the reader to [3,5] for a description of such a secure implementation. (The idea is that the parties run a secure computation of each step using, e.g., the protocol of [1]. The output contains encrypted and authenticated *shares* of the current state, which are then input to the computation of the following step, and checked by it.)

For every adversary that corrupts up to $s-1$ parties in the computation in the hybrid model, there is an adversary with the same power in the ideal model. We limit the analysis to adversaries that provide inputs that agree with all the boundary checks in the algorithm (otherwise the protocol aborts, and this is a legitimate outcome in the ideal model).

Imagine a tree of size $M$ corresponding to the comparisons done in the protocol (i.e. the root being the comparison for $m = (\beta - \alpha)/2$, etc.). Consider also the range $[\alpha, \beta]$ where each element is associated with the single node $u$ in the tree in which $m$ is set to the value of this element. Fix the random values (coin flips) used by the adversary in its operation. Run the adversary, with rewinding, checking the values that are given by each of the parties it controls to each of the comparisons. The values that party $i$ provides to the comparison of node $u$ define, for the corresponding element in the range, the number of items in the input of party $i$ which are smaller than, larger than, and equal to that value.

Assume that we first examine the adversary's behavior for the root node, then for the two children of the root, and continue layer by layer in the tree. Then the boundary checks ensure that the nodes are consistent. Let $l_u, e_u, g_u$ denote the number of items that are specified by the adversary to be less than, equal to, and greater than $u$, respectively. Then, for any three nodes $L, A, R$ that appear in this order in an inorder traversal of the tree, the boundary checks ensure that $l_L + e_L \leq l_A$ and $g_R + e_R \leq g_A$. Since $l_A + e_A + g_A = l_R + e_R + g_R$, the second inequality implies that $l_A + e_A \leq l_R$. Thus, for any two nodes $u$ and $v$ with $u < v$, we have $l_u + e_u \leq l_v$. In particular, for $i = \alpha, \ldots, \beta - 1$, we have $l_i + e_i \leq l_{i+1}$, which implies that $\Sigma_{i=\alpha}^{\beta} e_i \leq l_\beta + e_\beta - l_\alpha$. Since $l_\alpha = 0$ and $g_\beta = 0$ (enforced by our checks), we know that $l_\beta + e_\beta - l_\alpha = D_i$. Thus, $e_i = l_{i+1} - l_i$ for $\alpha \leq i < \beta$.

We use the result of this examination to define the input that each corrupt party provides to the trusted party in the ideal model. We set the input to contain $e_u$ items of value $u$, for every $u \in [\alpha, \beta]$. The trusted party computes the $k^{th}$ value (say, using the same algorithms as in the protocol). Since in the protocol itself the values provided by each party depend only on the results of previous comparisons (i.e. path in the tree) the output of the trusted party is the same as in the protocol.

**Acknowledgements**

We would like to thank Yehuda Lindell and Kobbi Nissim for stimulating discussions on this work.

## References

1. D. Beaver, S. Micali, and P. Rogaway. The round complexity of secure protocols. In *Proc. 22nd Annual ACM Symposium on the Theory of Computing*, pages 503–513, 1990.
2. R. Canetti. Security and composition of multiparty cryptographic protocols. *Journal of Cryptology*, 13(1):143–202, 2000.
3. R. Canetti. Universally composable security: A new paradigm for cryptographic protocols. In *Proc. 42nd IEEE Symposium on Foundations of Computer Science*, pages 136–145, 2001.
4. R. Canetti, Y. Ishai, R. Kumar, M. Reiter, R. Rubinfeld, and R. Wright. Selective private function evaluation with applications to private statistics. In *Proc. of 20th ACM Symposium on Principles of Distributed Computing*, pages 293–304, 2001.
5. R. Canetti, Y. Lindell, R. Ostrovsky, and A. Sahai. Universally composable two party computation. In *Proc. 34th ACM Symp. on the Theory of Computing*, pages 494–503, 2002.
6. J. Feigenbaum, Y. Ishai, T. Malkin, K. Nissim, M. Strauss, and R. Wright. Secure multiparty computation of approximations. In *Proc. 28th ICALP*, pages 927–938, 2001.
7. M. Fischlin. A cost-effective pay-per-multiplication comparison method for millionaires. In *CT-RSA 2001: The Cryptographers' Track at RSA Conference*, pages 457–472, 2001.
8. M. Franklin and M. Yung. Communication complexity of secure computation (extended abstract). In *Proc. 24th ACM Symp. on the Theory of Computing*, pages 699–710, 1992.
9. P. Gibbons, Y. Matias, and V. Poosala. Fast incremental maintenance of approximate histograms. In *Proc. 23rd Int. Conf. Very Large Data Bases*, pages 466–475, 1997.
10. O. Goldreich. Secure multi-party computation. In *Theory of Cryptography Library*, 1998.
11. O. Goldreich, S. Micali, and A. Wigderson. How to play any mental game or A completeness theorem for protocols with honest majority. In *Proc. 19th Annual ACM Symposium on Theory of Computing*, pages 218–229, 1987.
12. H. Jagadish, N. Koudas, S. Muthukrishnan, V. Poosala, K. Sevcik, and T. Suel. Optimal histograms with quality guarantees. In *Proc. 24th Int. Conf. Very Large Data Bases*, pages 275–286, 1998.

13. E. Kushilevitz and N. Nisan. *Communication Complexity*. Cambridge University Press, 1997.
14. Y. Lindell and B. Pinkas. Privacy preserving data mining. *Journal of Cryptology*, 15(3):177–206, 2002.
15. M. Naor and K. Nissim. Communication preserving protocols for secure function evaluation. In *Proc. 33rd Annual ACM Symposium on Theory of Computing*, pages 590–599, 2001.
16. V. Poosala, V. Ganti, and Y. Ioannidis. Approximate query answering using histograms. *IEEE Data Engineering Bulletin*, 22(4):5–14, 1999.
17. M. Rodeh. Finding the median distributively. *Journal of Computer and Systems Sciences*, 24:162–166, 1982.
18. A. Yao. How to generate and exchange secrets. In *Proc. 27th IEEE Symposium on Foundations of Computer Science*, pages 162–167, 1986.

# Short Signatures Without Random Oracles

Dan Boneh[1*] and Xavier Boyen[2]

[1] Computer Science Department, Stanford University, Stanford CA 94305-9045
dabo@cs.stanford.edu
[2] Voltage Security, Palo Alto, California
xb@boyen.org

**Abstract.** We describe a short signature scheme which is existentially unforgeable under a chosen message attack without using random oracles. The security of our scheme depends on a new complexity assumption we call the *Strong Diffie-Hellman* assumption. This assumption has similar properties to the Strong RSA assumption, hence the name. Strong RSA was previously used to construct signature schemes without random oracles. However, signatures generated by our scheme are much shorter and simpler than signatures from schemes based on Strong RSA. Furthermore, our scheme provides a limited form of message recovery.

## 1 Introduction

Boneh, Lynn, and Shacham (BLS) [BLS01] recently proposed a short digital signature scheme where signatures are about half the size of DSA signatures with the same level of security. Security is based on the Computational Diffie-Hellman (CDH) assumption on certain elliptic curves. The scheme is shown to be existentially unforgeable under a chosen message attack in the random oracle model.

In this paper we describe a signature scheme where signatures are almost as short as BLS signatures, but whose security does not require random oracles. We prove security of our scheme using a complexity assumption we call the Strong Diffie-Hellman assumption, or SDH for short. Roughly speaking, the $q$-SDH assumption in a group $\mathbb{G}$ of prime order $p$ states that the following problem is intractable: given $g, g^x, g^{(x^2)}, \ldots, g^{(x^q)} \in \mathbb{G}$ as input, output a pair $(c,\ g^{1/(x+c)})$ where $c \in \mathbb{Z}_p^*$. Precise definitions are given in Section 2.3. Using this assumption we construct a signature scheme that is existentially unforgeable under a chosen message attack *without using random oracles*.

Currently, the most practical signature schemes secure without random oracles [GHR99, CS00] are based on the Strong RSA assumption (given an RSA modulus $N$ and $s \in \mathbb{Z}_N^*$ it is difficult to construct a non-trivial pair $(c, s^{1/c})$ where $c \in \mathbb{Z}$). Roughly speaking, what makes Strong RSA so useful for constructing secure signature schemes is the following property: given a Strong RSA problem instance $(N, s)$ it is possible to construct a new instance $(N, s')$ with $q$ known

* Supported by NSF and the Packard Foundation.

C. Cachin and J. Camenisch (Eds.): EUROCRYPT 2004, LNCS 3027, pp. 56–73, 2004.

solutions $(c_i, (s')^{1/c_i})$, where the construction of any other solution $(c, (s')^{1/c})$ makes it possible to solve the original problem instance. This property provides a way to prove security against a chosen message attack. In Section 3.1 we show that the $q$-SDH problem has a similar property. Hence, $q$-SDH may be viewed as a discrete logarithm analogue of the Strong RSA assumption. We believe that the properties of $q$-SDH make it a useful tool for constructing cryptographic systems and we expect to see many other systems based on it.

To gain some confidence in the $q$-SDH assumption we provide in Section 5 a lower bound on the computational complexity of solving the $q$-SDH problem in a generic group model. This shows that no generic attack on $q$-SDH is possible. Mitsunari, Sakai, and Kasahara [MSK02] previously used a weaker variant of the $q$-SDH assumption to construct a traitor tracing scheme. The ideas in their paper are nice, and we use some of them here. Unfortunately, their application to tracing traitors is insecure [TSNZ03].

We present our secure signature scheme in Section 3 and prove its security against existential forgery under chosen message attack. The resulting signatures are as short as DSA signatures, but are provably secure in the absence of random oracles. Our signatures also support limited message recovery, which makes it possible to further reduce the total length of a message/signature pair. In Section 4 we show that with random oracles the $q$-SDH assumption gives even shorter signatures. A related system using random oracles was recently described by Zhang et al. [ZSNS04].

We refer to [BLS01] for applications of short signatures. We only mention that short digital signatures are needed in environments with stringent bandwidth constraints, such as bar-coded digital signatures on postage stamps [NS00, PV00]. We also note that Patarin et al. [PCG01, CDF03] construct short signatures whose security depends on the Hidden Field Equation (HFE) problem.

## 2 Preliminaries

Before presenting our results we briefly review two notions of security for signature schemes, review the definition for groups equipped with a bilinear map, and precisely state the $q$-SDH assumption.

### 2.1 Secure Signature Schemes

A signature scheme is made up of three algorithms, *KeyGen*, *Sign*, and *Verify*, for generating keys, signing, and verifying signatures, respectively.

#### Strong Existential Unforgeability

The standard notion of security for a signature scheme is called existential unforgeability under a chosen message attack [GMR88]. We consider a slightly stronger notion of security, called strong existential unforgeability [ADR02], which is defined using the following game between a challenger and an adversary $\mathcal{A}$:

**Setup:** The challenger runs algorithm *KeyGen* to obtain a public key *PK* and a private key *SK*. The adversary $\mathcal{A}$ is given *PK*.

**Queries:** Proceeding adaptively, $\mathcal{A}$ requests signatures on at most $q_S$ messages of his choice $M_1, \ldots, M_{q_s} \in \{0,1\}^*$, under *PK*. The challenger responds to each query with a signature $\sigma_i = Sign(SK, M_i)$.

**Output:** Eventually, $\mathcal{A}$ outputs a pair $(M, \sigma)$ and wins the game if
(1) $(M, \sigma)$ is not any of $(M_1, \sigma_1), \ldots, (M_{q_s}, \sigma_{q_s})$, and
(2) $Verify(PK, M, \sigma) = \texttt{valid}$.

We define $\mathsf{Adv\,Sig}_{\mathcal{A}}$ to be the probability that $\mathcal{A}$ wins in the above game, taken over the coin tosses made by $\mathcal{A}$ and the challenger.

**Definition 1.** *A forger $\mathcal{A}$ $(t, q_S, \epsilon)$-breaks a signature scheme if $\mathcal{A}$ runs in time at most $t$, $\mathcal{A}$ makes at most $q_S$ signature queries, and $\mathsf{Adv\,Sig}_{\mathcal{A}}$ is at least $\epsilon$. A signature scheme is $(t, q_S, \epsilon)$-existentially unforgeable under an adaptive chosen message attack if no forger $(t, q_S, \epsilon)$-breaks it.*

When proving security in the random oracle model we add a fourth parameter $q_H$ denoting an upper bound on the number of queries that the adversary $\mathcal{A}$ makes to the random oracle.

We note that the definition above captures a stronger version of existential unforgeability than the standard one: we require that the adversary cannot even generate a new signature on a previously signed message. This property is required for some applications [ADR02, Sah99, CHK04]. All our signature schemes satisfy this stronger security notion.

### Weak Chosen Message Attacks

We will also use a weaker notion of security which we call existential unforgeability under a weak chosen message attack. Here we require that the adversary submit all signature queries before seeing the public key. This notion is defined using the following game between a challenger and an adversary $\mathcal{A}$:

**Query:** $\mathcal{A}$ sends the challenger a list of $q_S$ messages $M_1, \ldots, M_{q_s} \in \{0,1\}^*$.

**Response:** The challenger runs algorithm *KeyGen* to generate a public key *PK* and private key *SK*. Next, the challenger generates signatures $\sigma_i = Sign(SK, M_i)$ for $i = 1, \ldots, q_S$. The challenger then gives $\mathcal{A}$ the public key *PK* and the $q_S$ signatures $\sigma_1, \ldots, \sigma_{q_s}$.

**Output:** Algorithm $\mathcal{A}$ outputs a pair $(M, \sigma)$ and wins the game if
(1) $M$ is not any of $M_1, \ldots, M_{q_s}$, and
(2) $Verify(PK, M, \sigma) = \texttt{valid}$.

We define $\mathsf{Adv\,W\text{-}Sig}_{\mathcal{A}}$ to be the probability that $\mathcal{A}$ wins in the above game, taken over the coin tosses of $\mathcal{A}$ and the challenger.

**Definition 2.** *A forger $\mathcal{A}$ $(t, q_S, \epsilon)$-weakly breaks a signature scheme if $\mathcal{A}$ runs in time at most $t$, $\mathcal{A}$ makes at most $q_S$ signature queries, and $\mathsf{Adv\,W\text{-}Sig}_{\mathcal{A}}$ is at least $\epsilon$. A signature scheme is $(t, q_S, \epsilon)$-existentially unforgeable under a weak chosen message attack if no forger $(t, q_S, \epsilon)$-weakly breaks it.*

## 2.2 Bilinear Groups

Signature verification in our scheme requires a bilinear map. We briefly review the necessary facts about bilinear maps and bilinear map groups. We follow the notation in [BLS01]:

1. $\mathbb{G}_1$ and $\mathbb{G}_2$ are two (multiplicative) cyclic groups of prime order $p$;
2. $g_1$ is a generator of $\mathbb{G}_1$ and $g_2$ is a generator of $\mathbb{G}_2$;
3. $\psi$ is an isomorphism from $\mathbb{G}_2$ to $\mathbb{G}_1$, with $\psi(g_2) = g_1$; and
4. $e$ is a bilinear map $e : \mathbb{G}_1 \times \mathbb{G}_2 \to \mathbb{G}_T$.

For simplicity one can set $\mathbb{G}_1 = \mathbb{G}_2$. However, as in [BLS01], we allow for the more general case where $\mathbb{G}_1 \neq \mathbb{G}_2$ so that we can take advantage of certain families of elliptic curves to obtain short signatures. Specifically, elements of $\mathbb{G}_1$ have a short representation whereas elements of $\mathbb{G}_2$ may not. The proofs of security require an efficiently computable isomorphism $\psi : \mathbb{G}_2 \to \mathbb{G}_1$. When $\mathbb{G}_1 = \mathbb{G}_2$ and $g_1 = g_2$ one could take $\psi$ to be the identity map. On elliptic curves we can use the trace map as $\psi$.

Let thus $\mathbb{G}_1$ and $\mathbb{G}_2$ be two groups as above, with an additional group $\mathbb{G}_T$ such that $|\mathbb{G}_1| = |\mathbb{G}_2| = |\mathbb{G}_T|$. A bilinear map is a map $e : \mathbb{G}_1 \times \mathbb{G}_2 \to \mathbb{G}_T$ with the following properties:

1. Bilinear: for all $u \in \mathbb{G}_1, v \in \mathbb{G}_2$ and $a, b \in \mathbb{Z}$, $e(u^a, v^b) = e(u, v)^{ab}$.
2. Non-degenerate: $e(g_1, g_2) \neq 1$.

We say that $(\mathbb{G}_1, \mathbb{G}_2)$ are bilinear groups if there exists a group $\mathbb{G}_T$, an isomorphism $\psi : \mathbb{G}_2 \to \mathbb{G}_1$, and a bilinear map $e : \mathbb{G}_1 \times \mathbb{G}_2 \to \mathbb{G}_T$ as above, and $e$, $\psi$, and the group action in $\mathbb{G}_1$, $\mathbb{G}_2$, and $\mathbb{G}_T$ can be computed efficiently.

Joux and Nguyen [JN01] showed that an efficiently computable bilinear map $e$ provides an algorithm for solving the Decision Diffie-Hellman problem (DDH). Our results can be stated using a generic algorithm for DDH. Nevertheless, for the sake of concreteness we instead describe our results by directly referring to the bilinear map.

## 2.3 The Strong Diffie-Hellman Assumption

Before describing the new signature schemes, we first state precisely the hardness assumption on which they are based. Let $\mathbb{G}_1, \mathbb{G}_2$ be two cyclic groups of prime order $p$, where possibly $\mathbb{G}_1 = \mathbb{G}_2$. Let $g_1$ be a generator of $\mathbb{G}_1$ and $g_2$ a generator of $\mathbb{G}_2$.

*q-Strong Diffie-Hellman Problem.* The $q$-SDH problem in $(\mathbb{G}_1, \mathbb{G}_2)$ is defined as follows: given a $(q+2)$-tuple $(g_1, g_2, g_2^x, g_2^{(x^2)}, \ldots, g_2^{(x^q)})$ as input, output a pair $(c, g_1^{1/(x+c)})$ where $c \in \mathbb{Z}_p^*$. An algorithm $\mathcal{A}$ has advantage $\epsilon$ in solving $q$-SDH in $(\mathbb{G}_1, \mathbb{G}_2)$ if

$$\Pr\left[\mathcal{A}(g_1, g_2, g_2^x, \ldots, g_2^{(x^q)}) = (c, g_1^{\frac{1}{x+c}})\right] \geq \epsilon$$

where the probability is over the random choice of $x$ in $\mathbb{Z}_p^*$ and the random bits consumed by $\mathcal{A}$.

**Definition 3.** *We say that the $(q,t,\epsilon)$-SDH assumption holds in $(\mathbb{G}_1,\mathbb{G}_2)$ if no $t$-time algorithm has advantage at least $\epsilon$ in solving the $q$-SDH problem in $(\mathbb{G}_1,\mathbb{G}_2)$.*

Occasionally we drop the $t$ and $\epsilon$ and refer to the $q$-SDH assumption rather than the $(q,t,\epsilon)$-SDH assumption. As we will see in the next section the $q$-SDH assumption has similar properties to the Strong RSA problem and we therefore view $q$-SDH as a discrete logarithm analogue of the Strong RSA assumption.

To provide some confidence in the $q$-SDH assumption, we prove in Section 5 a lower bound on the complexity of solving the $q$-SDH problem in a generic group. Furthermore, we note that the Strong Diffie-Hellman problem has a simple random self-reduction in $(\mathbb{G}_1,\mathbb{G}_2)$.

A weaker version of the $q$-SDH assumption was previously used by Mitsunari, Sakai, and Kasahara [MSK02] to construct a traitor tracing system (see [TSNZ03] for an analysis). Using our notation, their version of the assumption requires Algorithm $\mathcal{A}$ to output $g_1^{1/(x+c)}$ for a *given input value* $c$. In the assumption above we allow $\mathcal{A}$ to choose $c$. When $c$ is pre-specified the $q$-SDH problem is equivalent to the following problem: given $(g_1, g_2, g_2^x, g_2^{x^2}, \ldots, g_2^{x^q})$ output $g_1^{1/x}$. We note that when $\mathcal{A}$ is allowed to choose $c$ no such equivalence is known. The weaker variant of the assumption was recently used to construct an efficient selective identity secure identity based encryption (IBE) system without random oracles [BB04a].

## 3 Short Signatures Without Random Oracles

We now construct a fully secure short signature scheme in the standard model using the $q$-SDH assumption. We consider this to be the main result of the paper.

Let $(\mathbb{G}_1,\mathbb{G}_2)$ be bilinear groups where $|\mathbb{G}_1| = |\mathbb{G}_2| = p$ for some prime $p$. As usual, $g_1$ is a generator of $\mathbb{G}_1$ and $g_2$ a generator of $\mathbb{G}_2$. For the moment we assume that the messages $m$ to be signed are elements in $\mathbb{Z}_p^*$, but as we mention in Section 3.5, the domain can be extended to all of $\{0,1\}^*$ using a collision resistant hash function $H : \{0,1\}^* \to \mathbb{Z}_p^*$.

**Key generation:** Pick random $x, y \xleftarrow{R} \mathbb{Z}_p^*$, and compute $u \leftarrow g_2^x \in \mathbb{G}_2$ and $v \leftarrow g_2^y \in \mathbb{G}_2$. The public key is $(g_1, g_2, u, v)$. The secret key is $(x, y)$.

**Signing:** Given a secret key $x, y \in \mathbb{Z}_p^*$ and a message $m \in \mathbb{Z}_p^*$, pick a random $r \in \mathbb{Z}_p^*$ and compute $\sigma \leftarrow g_1^{1/(x+m+yr)} \in \mathbb{G}_1$. Here $1/(x+m+yr)$ is computed modulo $p$. In the unlikely event that $x+m+yr = 0$ we try again with a different random $r$. The signature is $(\sigma, r)$.

**Verification:** Given a public key $(g_1, g_2, u, v)$, a message $m \in \mathbb{Z}_p^*$, and a signature $(\sigma, r)$, verify that

$$e(\sigma,\ u \cdot g_2^m \cdot v^r) = e(g_1, g_2)$$

If the equality holds the result is `valid`; otherwise the result is `invalid`.

*Signature length.* A signature contains two elements $(\sigma, r)$, each of length approximately $\log_2(p)$ bits, therefore the total signature length is approximately $2\log_2(p)$. When using the elliptic curves described in [BLS01] we obtain a signature whose length is approximately the same as a DSA signature with the same security, but which is provably existentially unforgeable under a chosen message attack without the random oracle model.

*Performance.* Key and signature generation times are comparable to BLS signatures. Verification time is faster since verification requires only one pairing and one multi-exponentiation. The value $e(g_1, g_2)$ only needs to be computed at initialization time and cached. In comparison, BLS signature verification requires two pairing computations. Since exponentiation tends to be significantly faster than pairing, signature verification is faster than in the BLS system.

*Security.* The following theorem shows that the scheme above is existentially unforgeable in the strong sense under chosen message attacks, provided that the $q$-SDH assumption holds in $(\mathbb{G}_1, \mathbb{G}_2)$.

**Theorem 1.** *Suppose the $(q, t', \epsilon')$-SDH assumption holds in $(\mathbb{G}_1, \mathbb{G}_2)$. Then the signature scheme above is $(t, q_S, \epsilon)$-secure against existential forgery under a chosen message attack provided that*

$$t \le t' - o(t')\ , \qquad q_S < q \qquad \text{and} \qquad \epsilon \ge 2(\epsilon' + q_S/p) \approx 2\epsilon'$$

*Proof.* We prove the theorem using two lemmas. In Lemma 1, we first describe a simplified signature scheme and prove its existential unforgeability against *weak* chosen message attacks under the $q$-SDH assumption. In Lemma 2, we then show that the security of the weak scheme implies the security of the full scheme. From these results (Lemmas 1 and 2), Theorem 1 follows easily. We present the proof in two steps since the construction used to prove Lemma 1 will be used later on in the paper. □

### 3.1 A Weakly Secure Short Signature Scheme

We first show how the $q$-SDH assumption can be used to construct an existentially unforgeable scheme under a *weak* chosen message attack. This construction demonstrates the main properties of the $q$-SDH assumption. In the next section we show that the security of this weak scheme implies the security of the full scheme above.

The weakly secure short signature scheme is as follows. As before, let $(\mathbb{G}_1, \mathbb{G}_2)$ be bilinear groups where $|\mathbb{G}_1| = |\mathbb{G}_2| = p$ for some prime $p$. As usual, $g_1$ is a generator of $\mathbb{G}_1$ and $g_2$ a generator of $\mathbb{G}_2$. For the moment we assume that the messages $m$ to be signed are elements in $\mathbb{Z}_p^*$.

**Key generation:** Pick random $x \stackrel{R}{\leftarrow} \mathbb{Z}_p^*$, and compute $v \leftarrow g_2^x \in \mathbb{G}_2$. The public key is $(g_1, g_2, v)$. The secret key is $x$.

**Signing:** Given a secret key $x \in \mathbb{Z}_p^*$ and a message $m \in \mathbb{Z}_p^*$, output the signature $\sigma \leftarrow g_1^{1/(x+m)} \in \mathbb{G}_1$. Here $1/(x+m)$ is computed modulo $p$. By convention in this context we define $1/0$ to be 0 so that in the unlikely event that $x+m=0$ we have $\sigma \leftarrow 1$.

**Verification:** Given a public key $(g_1, g_2, v)$, a message $m \in \mathbb{Z}_p^*$, and a signature $\sigma \in \mathbb{G}_1$, verify that

$$e(\sigma, v \cdot g_2^m) = e(g_1, g_2)$$

If equality holds output `valid`. If $\sigma = 1$ and $v \cdot g_2^m = 1$ output `valid`. Otherwise, output `invalid`.

We show that the basic signature scheme above is existentially unforgeable under a *weak* chosen message attack. The proof of the following lemma uses a similar method to the proof of Theorem 3.5 of Mitsunari et al. [MSK02].

**Lemma 1.** *Suppose the $(q, t', \epsilon)$-SDH assumption holds in $(\mathbb{G}_1, \mathbb{G}_2)$. Then the basic signature scheme above is $(t, q_S, \epsilon)$-secure against existential forgery under a weak chosen message attack provided that*

$$t \le t' - O(q^2) \qquad \text{and} \qquad q_S < q$$

*Proof.* Assume $\mathcal{A}$ is a forger that $(t, q_S, \epsilon)$-breaks the signature scheme. We construct an algorithm $\mathcal{B}$ that, by interacting with $\mathcal{A}$, solves the $q$-SDH problem in time $t'$ with advantage $\epsilon$. Algorithm $\mathcal{B}$ is given an instance $(g_1, g_2, A_1, \ldots, A_q)$ of the $q$-SDH problem, where $A_i = g_2^{(x^i)} \in \mathbb{G}_2$ for $i = 1, \ldots, q$ and for some unknown $x \in \mathbb{Z}_p^*$. For convenience we set $A_0 = g_2$. Algorithm $\mathcal{B}$'s goal is to produce a pair $(c, g_1^{1/(x+c)})$ for some $c \in \mathbb{Z}_p^*$. Algorithm $\mathcal{B}$ does so by interacting with the forger $\mathcal{A}$ as follows:

**Query:** Algorithm $\mathcal{A}$ outputs a list of distinct $q_S$ messages $m_1, \ldots, m_{q_s} \in \mathbb{Z}_p^*$, where $q_S < q$. Since $\mathcal{A}$ must reveal its queries up front, we may assume that $\mathcal{A}$ outputs exactly $q-1$ messages to be signed (if the actual number is less, we can always virtually reduce the value of $q$ so that $q = q_S + 1$).

**Response:** $\mathcal{B}$ must respond with a public key and signatures on the $q-1$ messages from $\mathcal{A}$. Let $f(y)$ be the polynomial $f(y) = \prod_{i=1}^{q-1}(y+m_i)$. Expand $f(y)$ and write $f(y) = \sum_{i=0}^{q-1} \alpha_i y^i$ where $\alpha_0, \ldots, \alpha_{q-1} \in \mathbb{Z}_p$ are the coefficients of the polynomial $f(y)$. Compute:

$$g_2' \leftarrow \prod_{i=0}^{q-1} (A_i)^{\alpha_i} = g_2^{f(x)} \quad \text{and} \quad h \leftarrow \prod_{i=1}^{q} A_i^{\alpha_{i-1}} = g_2^{xf(x)} = (g_2')^x$$

Also, let $g_1' = \psi(g_2')$. The public key given to $\mathcal{A}$ is $(g_1', g_2', h)$. Next, for each $i = 1, \ldots q-1$, Algorithm $\mathcal{B}$ must generate a signature $\sigma_i$ on $m_i$. To do so, let $f_i(y)$ be the polynomial $f_i(y) = f(y)/(y+m_i) = \prod_{j=1, j\ne i}^{q-1}(y+m_j)$. As before, we expand $f_i$ and write $f_i(y) = \sum_{j=0}^{q-2} \beta_j y^j$. Compute

$$S_i \leftarrow \prod_{j=0}^{q-2} A_j^{\beta_j} = g_2^{f_i(x)} = (g_2')^{1/(x+m_i)} \in \mathbb{G}_2$$

Observe that $\sigma_i = \psi(S_i) \in \mathbb{G}_1$ is a valid signature on $m$ under the public key $(g_1', g_2', h)$. Algorithm $\mathcal{B}$ gives $\mathcal{A}$ the $q-1$ signatures $\sigma_1, \ldots, \sigma_{q-1}$.

**Output:** Algorithm $\mathcal{A}$ returns a forgery $(m_*, \sigma_*)$ such that $\sigma_* \in \mathbb{G}_1$ is a valid signature on $m_* \in \mathbb{Z}_p^*$ and $m_* \notin \{m_1 \ldots, m_{q-1}\}$ since there is only one valid signature per message. In other words, $e(\sigma_*, h \cdot (g_2')^{m_*}) = e(g_1', g_2')$. Since $h = (g_2')^x$ we have that $e(\sigma_*, (g_2')^{x+m_*}) = e(g_1', g_2')$ and therefore

$$\sigma_* = (g_1')^{1/(x+m_*)} = (g_1)^{f(x)/(x+m_*)} \tag{1}$$

Using long division we write the polynomial $f$ as $f(y) = \gamma(y)(y+m_*) + \gamma_{-1}$ for some polynomial $\gamma(y) = \sum_{i=0}^{q-2} \gamma_i y^i$ and some $\gamma_{-1} \in \mathbb{Z}_p$. Then the rational fraction $f(y)/(y+m_*)$ in the exponent on the right side of Equation (1) can be written as

$$f(y)/(y+m_*) = \frac{\gamma_{-1}}{y+m_*} + \sum_{i=0}^{q-2} \gamma_i y^i$$

Note that $\gamma_{-1} \neq 0$, since $f(y) = \prod_{i=1}^{q-1}(y+m_i)$ and $m_* \notin \{m_1, \ldots, m_{q-1}\}$, as thus $(y+m_*)$ does not divide $f(y)$. Then algorithm $\mathcal{B}$ computes

$$w \leftarrow \left(\sigma_* \cdot \prod_{i=0}^{q-1} \psi(A_i)^{-\gamma_i}\right)^{1/\gamma_{-1}} = g_1^{1/(x+m_*)}$$

and returns $(m_*, w)$ as the solution to the $q$-SDH instance.

The claimed bounds are obvious by construction of the reduction. □

## 3.2 From Weak Security to Full Security

We now present a reduction from the security of the basic scheme of Lemma 1 to the security of the full signature scheme described at the onset of Section 3. This will complete the proof of Theorem 1.

**Lemma 2.** *Suppose that the basic signature scheme of Lemma 1 is $(t', q_S, \epsilon')$-weakly secure. Then the full signature scheme is $(t, q_S, \epsilon)$-secure against existential forgery under a chosen message attack provided that*

$$t \leq t' - O(q_S) \qquad \textit{and} \qquad \epsilon \geq 2(\epsilon' + q_S/p) \approx 2\epsilon'$$

*Proof.* Assume $\mathcal{A}$ is a forger that $(t, q_S, \epsilon)$-breaks the full signature scheme. We construct an algorithm $\mathcal{B}$ that $(t+O(q_S), q_S, \epsilon/2 - q_S/p)$-weakly breaks the basic signature scheme of Lemma 1.

Before describing Algorithm $\mathcal{B}$ we distinguish between two types of forgers that $\mathcal{A}$ can emulate. Let $(h_1, h_2, u, v)$ be the public key given to forger $\mathcal{A}$ where $u = g_2^x$ and $v = g_2^y$. Suppose $\mathcal{A}$ asks for signatures on messages $m_1, \ldots, m_{q_s} \in \mathbb{Z}_p^*$ and is given signatures $(\sigma_i, r_i)$ for $i = 1, \ldots, q_S$ on these messages. Let $w_i = m_i + y r_i$ and let $(m_*, \sigma_*, r_*)$ be the forgery produced by $\mathcal{A}$. We distinguish between two types of forgers:

**Type-1 forger:** a forger that either (i) makes a signature query for the message $m = -x$, or (ii) outputs a forgery where $m_* + yr_* \notin \{w_1, \ldots, w_{q_s}\}$.

**Type-2 forger:** a forger that (i) never makes a signature query for the message $m = -x$, and (ii) outputs a forgery where $m_* + yr_* = w_i$ for some $i \in \{1, \ldots, q_S\}$.

We show that either forger can be used to forge signatures for the weak signature scheme of Lemma 1. However, the reduction works differently for each forger type. Therefore, initially $\mathcal{B}$ will choose a random bit $c_{\text{mode}} \in \{1,2\}$ that indicates its guess for the type of forger that $\mathcal{A}$ will emulate. The simulation proceeds differently for each mode.

We are now ready to describe Algorithm $\mathcal{B}$. It produces a forgery for the signature scheme of Lemma 1 as follows:

**Setup:** Algorithm $\mathcal{B}$ first picks a random bit $c_{\text{mode}} \in \{1,2\}$. Next, $\mathcal{B}$ sends to its own challenger a list of $q_S$ random messages $w_1, \ldots, w_{q_s} \in \mathbb{Z}_p^*$ for which it requests a signature. The challenger responds with a public key $(g_1, g_2, u)$ and signatures $\sigma_1, \ldots, \sigma_{q_s} \in \mathbb{G}_1$ on these messages. We know that $e(\sigma_i, g_2^{w_i} u) = e(g_1, g_2)$ for all $i = 1, \ldots, q_S$. Then:

- (If $c_{\text{mode}} = 1$). $\mathcal{B}$ picks a random $y \in \mathbb{Z}_p^*$ and gives $\mathcal{A}$ the public key $PK_1 = (g_1, g_2, u, g_2^y)$.
- (If $c_{\text{mode}} = 2$). $\mathcal{B}$ picks a random $x \in \mathbb{Z}_p^*$ and gives $\mathcal{A}$ the public key $PK_2 = (g_1, g_2, g_2^x, u)$.

In either case, we note that $\mathcal{B}$ provides the adversary $\mathcal{A}$ with a valid public key $(g_1, g_2, U, V)$.

**Signature queries:** The forger $\mathcal{A}$ can issue up to $q_S$ signature queries in an adaptive fashion. In order to respond, $\mathcal{B}$ maintains a list $H$-list of tuples $(m_i, r_i, W_i)$ and a query counter $\ell$ which is initially set to 0. Upon receiving a signature query for $m$, Algorithm $\mathcal{B}$ increments $\ell$ by one. Then:

- (If $c_{\text{mode}} = 1$). Check if $g_2^{-m} = u$. If so, then $\mathcal{B}$ just obtained the private key for the public key $(g_1, g_2, u)$ it was given, which allows it to forge the signature on any message of its choice. At this point $\mathcal{B}$ successfully terminates the simulation.
  Otherwise, set $r_\ell = (w_\ell - m)/y \in \mathbb{Z}_p^*$. In the very unlikely event that $r_\ell = 0$, Algorithm $\mathcal{B}$ reports failure and aborts. Otherwise, Algorithm $\mathcal{B}$ gives $\mathcal{A}$ the signature $(\sigma_\ell, r_\ell)$. This is a valid signature on $m$ under $PK_1$ since $r_\ell$ is uniform in $\mathbb{Z}_p^*$ and

$$e(\sigma_\ell,\ U \cdot g_2^m \cdot V^{r_\ell}) = e(\sigma_\ell,\ u \cdot g_2^m \cdot g_2^{yr_\ell}) = e(\sigma_\ell, u \cdot g_2^{w_\ell}) = e(g_1, g_2)$$

- (If $c_{\text{mode}} = 2$). Set $r_\ell = (x + m)/w_\ell \in \mathbb{Z}_p^*$. If $r_\ell = 0$, Algorithm $\mathcal{B}$ reports failure and aborts. Otherwise, give $\mathcal{A}$ the signature $(\sigma_\ell^{1/r_\ell}, r_\ell)$. This is a valid signature on $m$ under $PK_2$ since $r_\ell$ is uniform in $\mathbb{Z}_p^*$ and

$$e(\sigma_\ell^{1/r_\ell},\ U \cdot g_2^m \cdot V^{r_\ell}) = e(\sigma_\ell^{1/r_\ell},\ g_2^x \cdot g_2^m \cdot u^{r_\ell}) = e(\sigma_\ell,\ g_2^{w_\ell} u) = e(g_1, g_2)$$

In either case if $\mathcal{B}$ does not stop it responds with a valid signature on $m$. In either case Algorithm $\mathcal{B}$ adds the tuple $(m, r_\ell, g_2^m V^{r_\ell})$ to the $H$-list.

**Output:** Eventually, $\mathcal{A}$ returns a forgery $(m_*, \sigma_*, r_*)$, where $(\sigma_*, r_*)$ is a valid forgery distinct from any previously given signature on message $m_*$. Note that by adding dummy queries as necessary, we may assume that $\mathcal{A}$ made exactly $q_S$ signature queries. Let $W_* \leftarrow g_2^{m_*} V^{r_*}$. Algorithm $\mathcal{B}$ searches the $H$-list for a tuple whose rightmost component is equal to $W_*$. There are two possibilities:

**Type-1 forgery:** No tuple of the form $(\cdot, \cdot, W_*)$ appears on the $H$-list.

**Type-2 forgery:** The $H$-list contains at least one tuple $(m_j, r_j, W_j)$ such that $W_j = W_*$.

Let $b_{\text{type}} \leftarrow 1$ if $\mathcal{A}$ produced a type-1 forgery, or $\mathcal{A}$ made a signature query for a message $m$ such that $g_2^{-m} = U$. In all other cases, set $b_{\text{type}} \leftarrow 2$. If $b_{\text{type}} \neq c_{\text{mode}}$ then $\mathcal{B}$ reports failure and aborts. Otherwise, $\mathcal{B}$ outputs an existential forgery on the basic signature scheme as follows:

- (If $c_{\text{mode}} = b_{\text{type}} = 1$). If $\mathcal{A}$ made a signature query for a message $m$ such that $g_2^{-m} = U$ then $\mathcal{B}$ is already done. Therefore, we assume $\mathcal{A}$ produced a type-1 forgery. Since the forgery is valid, we have

$$e(g_1, g_2) = e(\sigma_*,\ U \cdot g_2^{m_*} \cdot V^{r_*}) = e(\sigma_*,\ u \cdot g_2^{m_* + y r_*})$$

Let $w_* = m_* + y r_*$. It follows that $(w_*, \sigma_*)$ is a valid message/signature pair in the basic signature scheme. Furthermore, it is a valid existential forgery for the basic scheme since in a type-1 forgery Algorithm $\mathcal{B}$ did not request a signature on the message $w_* \in \mathbb{Z}_p^*$. Indeed, $\mathcal{B}$ only requested signatures on messages $w_j = m_j + y r_j$ where $(m_j, r_j, g_2^{w_j})$ is a tuple in the $H$-list, but $g_2^{w_*}$ is not equal to any $g_2^{w_j}$ on the $H$-list. Algorithm $\mathcal{B}$ outputs $(w_*, \sigma_*)$ as the required existential forgery.

- (If $c_{\text{mode}} = b_{\text{type}} = 2$). Let $(m_j, r_j, W_j)$ be a tuple on the $H$-list where $W_j = W_*$. Since $V = u$ we know that $g_2^{m_j} u^{r_j} = g_2^{m_*} u^{r_*}$. Write $u = g_2^z$ for some $z \in \mathbb{Z}_p^*$ so that $m_j + z r_j = m_* + z r_*$. We know that $(m_j, r_j) \neq (m_*, r_*)$, otherwise the forgery would be identical to a previously given signature on the query message $m_j$. Since $g_2^{m_j} u^{r_j} = g_2^{m_*} u^{r_*}$ it follows that $m_j \neq m_*$ and $r_j \neq r_*$. Therefore, $z = (m_* - m_j)/(r_j - r_*) \in \mathbb{Z}_p^*$. Hence, $\mathcal{B}$ just recovered the private key for the public key $(g_1, g_2, u)$ it was given. Algorithm $\mathcal{B}$ can now forge a signature on any message of its choice.

This completes the description of Algorithm $\mathcal{B}$.

A standard argument shows that if $\mathcal{B}$ does not abort, then, from the viewpoint of $\mathcal{A}$, the simulation provided by $\mathcal{B}$ is indistinguishable from a real attack scenario. In particular, (i) the view from $\mathcal{A}$ is independent of the value of $c_{\text{mode}}$, (ii) the public keys are uniformly distributed, and (iii) the signatures are correct. Therefore, $\mathcal{A}$ produces a valid forgery in time $t$ with probability at least $\epsilon$.

It remains to bound the probability that $\mathcal{B}$ does not abort. We argue as follows:

- Conditionally on the event $c_{\text{mode}} = b_{\text{type}} = 1$, Algorithm $\mathcal{B}$ aborts if $\mathcal{A}$ issued a signature query $m_\ell = w_\ell$. This happens with probability at most $q_S/p$.
- Conditionally on the event $c_{\text{mode}} = b_{\text{type}} = 2$, Algorithm $\mathcal{B}$ does not abort.

Since $c_{\text{mode}}$ is independent of $b_{\text{type}}$ we have that $\Pr[c_{\text{mode}} = b_{\text{type}}] = 1/2$. It now follows that $\mathcal{B}$ produces a valid forgery with probability at least $\epsilon/2 - q_S/p$, as required. □

Since in the full scheme a single message has many valid signatures, it is worth repeating that the full signature scheme is existentially unforgeable in the strong sense: the adversary cannot make any forgery, even on messages which are already signed.

### 3.3 Relation to Chameleon Hash Signatures

It is instructive to consider the relation between the full signature scheme above and a signature construction based on the Strong RSA assumption due to Gennaro, Halevi, and Rabin (GHR) [GHR99]. GHR signatures are pairs $(r, s^{1/H(m,r)})$ where $H$ is a Chameleon hash [KR00], $r$ is random in some range, and arithmetic is done modulo an RSA modulus $N$. Looking closely, one can see some parallels between the proof of security in Lemma 2 above and the proof of security in [GHR99]. There are three interesting points to make:

- The $m + yr$ component in our signature scheme provides us with the functionality of a Chameleon hash: given $m$, we can choose $r$ so that $m + yr$ maps to some predefined value of our choice. This makes it possible to handle the chosen message attack. Embedding the hash $m + yr$ directly in the signature scheme results in a much more efficient construction than using an explicit Chameleon hash (which requires additional exponentiations). This is not known to be possible with Strong RSA signatures.
- One difficulty with GHR signatures is that given a solution $(6, s^{1/6})$ to the Strong RSA problem one can deduce another solution, e.g. $(3, s^{1/3})$. Thus, given a GHR signature on one message it possible to deduce a GHR signature on another message (see [GHR99, CN00] for details). Gennaro et al. solve this problem by ensuring that $H(m, r)$ always maps to a prime; However, that makes it difficult to compute the hash (a different solution is given in [CS00]). This issue does not come up at all in our signature scheme above.
- We obtain short signatures since, unlike Strong RSA, the $q$-SDH assumption applies to groups with a short representation.

Thus, we see that Strong Diffie-Hellman leads to signatures that are simpler, more efficient, and shorter than their Strong RSA counterparts.

### 3.4 Limited Message Recovery

We now describe another useful property of the signature schemes whereby the total size of signed messages can be further reduced at the cost of increasing the verification time. The technique applies equally well to the fully secure signature scheme as to the weakly secure one.

A standard technique for shortening the total length of message/signature pairs is to encode a part of the message in the signature [MVV97]. Signatures based on trapdoor permutations support very efficient message recovery.

At the other end of the spectrum, a trivial signature compression mechanism that applies to any signature scheme is as follows: Rather than transmit a message/signature pair $(M, \sigma)$, the sender transmits $(\hat{M}, \sigma)$ where $\hat{M}$ is the same as $M$ except that the last $t$ bits are truncated. In other words, $\hat{M}$ is $t$ bits shorter than $M$. To verify $(\hat{M}, \sigma)$ the verifier tries all $2^t$ possible values for the truncated bits and accepts the signature if one of them verifies. To reconstruct the original signed message $M$, the verifier appends to $\hat{M}$ the $t$ bits for which the signature verified.

This trivial method shows that the pair $(M, \sigma)$ can be shortened by $t$-bits at the cost of increasing verification time by a factor of $2^t$. For our signature scheme we obtain a better tradeoff: the pair $(M, \sigma)$ can be shortened by $t$ bits at the cost of increasing verification time by a factor of $2^{t/2}$ only. We refer to this property as limited message recovery.

*Limited Message Recovery.* Limited message recovery applies to both the full signature scheme and the weakly secure signature scheme of Lemma 1. For simplicity, we only show how limited message recovery applies to the full signature scheme. Assume messages are $k$-bit strings represented as integers in $\mathbb{Z}_p^*$. Let $(g_1, g_2, u, v)$ be a public key in the full scheme. Suppose we are given the signed message $(\hat{m}, \sigma, r)$ where $\hat{m}$ is a truncation of the last $t$ bits of $m \in \mathbb{Z}_p^*$. Thus $m = \hat{m} \cdot 2^t + \delta$ for some integer $0 \leq \delta < 2^t$. Our goal is to verify the signed message $(\hat{m}, \sigma, r)$ and to reconstruct the missing bits $\delta$ in time $2^{t/2}$. To do so, we first rewrite the verification equation $e(\sigma, \, u \cdot v^r \cdot g_2^m) = e(g_1, g_2)$ as

$$e(\sigma, g_2)^m = \frac{e(g_1, g_2)}{e(\sigma, u \cdot v^r)}$$

Substituting $m = \hat{m} \cdot 2^t + \delta$ we obtain

$$e(\sigma, g_2)^\delta = \frac{e(g_1, g_2)}{e(\sigma, \, u \cdot v^r \cdot g_2^{\hat{m}2^t})} \qquad (2)$$

Now, we say that $(\hat{m}, \sigma, r)$ is valid if there exists an integer $\delta \in [0, 2^t)$ satisfying equation (2). Finding such a $\delta$ takes time approximately $2^{t/2}$ using Pollard's Lambda method [MVV97, p.128] for computing discrete logarithms. Thus, we can verify the signature and recover the $t$ missing message bits in time $2^{t/2}$, as required.

*Ultra Short Weakly Secure Signatures.* Obvious applications of limited message recovery are situations where bandwidth is extremely limited, such as when the signature is an authenticator that is to be typed-in by a human. The messages in such applications are typically chosen and signed by a central authority, so that adaptive chosen message attacks are typically not a concern. It is safe in those cases to use the weakly secure signature scheme of Lemma 1, and apply limited message recovery to further shrink the already compact signatures it produces. Specifically, using $t$-bit truncation as above we obtain a total signature overhead of $(160 - t)$ bits for common security parameters, at the cost of requiring $2^{t/2}$ pairing computations for signature verification. We emphasize that the security of this system does not rely on random oracles.

### 3.5 Arbitrary Message Signing

We can extend our signature schemes to sign arbitrary messages in $\{0,1\}^*$, as opposed to merely messages in $\mathbb{Z}_p^*$, by first hashing the message using a collision-resistant hash function $H : \{0,1\}^* \to \mathbb{Z}_p^*$ prior to both signing and verifying. A standard argument shows that if the scheme above is secure against existential forgery under a chosen message attack (in the strong sense) then so is the scheme with the hash. The result is a signature scheme for arbitrary messages in $\{0,1\}^*$. We note that there is no need for a full domain hash into $\mathbb{Z}_p^*$; a collision resistant hash function $H : \{0,1\}^* \to \{1,\ldots,2^b\}$ for $2^b < p$ is sufficient for the security proof. This transformation applies to both the fully and the weakly secure signature schemes described above.

## 4 Shorter Signatures with Random Oracles

For completeness we show that the weakly secure signature scheme of Lemma 1 gives rise to very efficient and fully secure short signatures in the random oracle model. To do so, we show a general transformation from any existentially unforgeable signature scheme under a *weak* chosen message attack into an existentially unforgeable signature scheme under a *standard* chosen message attack (in the strong sense), in the random oracle model. This gives a very efficient short signature scheme based on $q$-SDH in the random oracle model. We analyze our construction using a method of Katz and Wang [KW03] which gives a very tight reduction to the security of the underlying signature. We note that a closely related system with a weaker security analysis was independently discovered by Zhang et al. [ZSNS04].

Let (*KeyGen*, *Sign*, *Verify*) be an existentially unforgeable signature under a *weak* chosen message attack. We assume that the scheme signs messages in some finite set $\Sigma$ and that the private keys are in some set $\Pi$. We need two hash functions $H_1 : \Pi \times \{0,1\}^* \to \{0,1\}$ and $H_2 : \{0,1\} \times \{0,1\}^* \to \Sigma$ that will be viewed as random oracles in the security analysis. The hash-signature scheme is as follows:

**Key generation:** Same as *KeyGen*. The public key is *PK*; The secret key is $SK \in \Pi$.

**Signing:** Given a secret key *SK*, and given a message $M \in \{0,1\}^*$, compute $b \leftarrow H_1(SK, M) \in \{0,1\}$ and $m \leftarrow H_2(b, M) \in \Sigma$. Output the signature

$(b, Sign(m))$. Note that signatures are one bit longer than in the underlying signature scheme.

**Verification:** Given a public key $PK$, a message $M \in \{0,1\}^*$, and a signature $(b, \sigma)$, output `valid` if $Verify(PK, H_2(b, M), \sigma) =$ `valid`.

Theorem 2 below proves security of the scheme. Note that the security reduction in Theorem 2 is tight, namely, an attacker on the hash-signature scheme with success probability $\epsilon$ is converted to an attacker on the underlying signature with success probability approximately $\epsilon/2$. Proofs of signature schemes in the random oracle model are often far less tight. The proof is given in the full version of the paper [BB04b].

**Theorem 2.** *Suppose (KeyGen, Sign, Verify) is $(t', q'_S, \epsilon')$-existentially unforgeable under a weak chosen message attack. Then the corresponding hash-signature scheme is $(t, q_S, q_H, \epsilon)$-secure against existential forgery under an adaptive chosen message attack, in the random oracle model, whenever $q_S + q_H < q'_S$, and for all all $t$ and $\epsilon$ satisfying*

$$t \leq t' - o(t') \qquad \textit{and} \qquad \epsilon \geq 2\epsilon'/(1 - \frac{q'_S}{|\Sigma|}) \approx 2\epsilon'$$

Applying Theorem 2 to the weakly secure scheme of Lemma 1 gives an efficient short signature existentially unforgeable under a *standard* chosen message attack in the random oracle model assuming $(q_S + q_H + 1)$-SDH. For a public key $(g_1, g_2, v = g_2^x)$ and a hash function $H : \{0,1\}^* \to \mathbb{Z}_p^*$ a signature on a message $m$ is defined as the value $\sigma \leftarrow g_1^{1/(x+H(b,m))} \in \mathbb{G}_1$ concatenated with the bit $b \in \{0,1\}$. To verify the signature, check that $e(\sigma, v \cdot g_2^{H(b,m)}) = e(g_1, g_2)$. We see that signature length is essentially the same as in BLS signatures, but verification time is approximately half that of BLS. During verification, exponentiation is always base $g_2$ which enables a further speed-up by pre-computing certain powers of $g_2$.

*Full Domain Hash.* Another method for converting a signature scheme secure under a weak chosen message attack into a scheme secure under a standard chosen message attack is to simply apply *Sign* and *Verify* to $H(M)$ rather than $M$. In other words, we hash $M \in \{0,1\}^*$ using a full domain hash $H$ prior to signing and verifying. Security in the random oracle model is shown using a similar argument to Coron's analysis of the Full Domain Hash [Cor00]. However, the resulting reduction is not tight: an attacker on this hash-then-sign signature with success probability $\epsilon$ yields an attacker on the underlying signature with success probability approximately $\epsilon/q_S$. We note, however, that these proofs are set in the random oracle model and therefore it is not clear whether the efficiency of the security reduction is relevant to actual security in the real world. Therefore, since this full domain hash signature scheme is slightly simpler that the system in Theorem 2 it might be preferable to use it rather than the system of Theorem 2. When we apply the full domain hash to the weakly secure scheme of Lemma 1, we obtain a secure signature under a standard chosen message attack assuming

$(q_S+q_H+1)$-SDH. A signature is one element, namely $\sigma \leftarrow g_1^{1/(x+H(m))} \in \mathbb{G}_1$. As before, signature verification is twice as fast as in BLS signatures. As mentioned above, a similar scheme was independently proposed by Zhang et al. [ZSNS04]. We also note that, in the random oracle model, security of this full domain hash scheme can be proven under a slightly weaker complexity assumption than $q$-SDH, namely that the value $c$ in the $q$-SDH assumption is pre-specified rather than chosen by the adversary. However, the resulting security reduction is far less efficient.

## 5 Generic Security of the $q$-SDH Assumption

To provide more confidence in the $q$-SDH assumption we prove a lower bound on the computational complexity of the $q$-SDH problem for generic groups in the sense of Shoup [Sho97].

In the generic group model, elements of $\mathbb{G}_1$, $\mathbb{G}_2$, and $\mathbb{G}_T$ appear to be encoded as unique random strings, so that no property other than equality can be directly tested by the adversary. Five oracles are assumed to perform operations between group elements, such as computing the group action in each of the three groups $\mathbb{G}_1$, $\mathbb{G}_2$, $\mathbb{G}_T$, as well as the isomorphism $\psi : \mathbb{G}_2 \to \mathbb{G}_1$, and the bilinear pairing $e : \mathbb{G}_1 \times \mathbb{G}_2 \to \mathbb{G}_T$. The opaque encoding of the elements of $\mathbb{G}_1$ is modeled as an injective function $\xi_1 : \mathbb{Z}_p \to \Xi_1$, where $\Xi_1 \subset \{0,1\}^*$, which maps all $a \in \mathbb{Z}_p$ to the string representation $\xi_1(g^a)$ of $g^a \in \mathbb{G}_1$. We similarly define $\xi_2 : \mathbb{Z}_p \to \Xi_2$ for $\mathbb{G}_2$ and $\xi_T : \mathbb{Z}_p \to \Xi_T$ for $\mathbb{G}_T$. The attacker $\mathcal{A}$ communicates with the oracles using the $\xi$-representations of the group elements only.

**Theorem 3.** *Let $\mathcal{A}$ be an algorithm that solves the q-SDH problem in the generic group model, making a total of at most $q_G$ queries to the oracles computing the group action in $\mathbb{G}_1, \mathbb{G}_2, \mathbb{G}_T$, the oracle computing the isomorphism $\psi$, and the oracle computing the bilinear pairing e. If $x \in \mathbb{Z}_p^*$ and $\xi_1$, $\xi_2$, $\xi_T$ are chosen at random, then the probability $\epsilon$ that $\mathcal{A}(p, \xi_1(1), \xi_2(1), \xi_2(x), \ldots, \xi_2(x^q))$ outputs $(c, \xi_1(\frac{1}{x+c}))$ with $c \in \mathbb{Z}_p^*$, is bounded by*

$$\epsilon \leq \frac{(q_G+q+2)^2 q}{p} = O\left(\frac{(q_G)^2 q + q^3}{p}\right)$$

*Proof.* Consider an algorithm $\mathcal{B}$ that plays the following game with $\mathcal{A}$.

$\mathcal{B}$ maintains three lists of pairs $L_1 = \{(F_{1,i}, \xi_{1,i}) : i = 0, \ldots, \tau_1 - 1\}$, $L_2 = \{(F_{2,i}, \xi_{2,i}) : i = 0, \ldots, \tau_2 - 1\}$, $L_T = \{(F_{T,i}, \xi_{T,i}) : i = 0, \ldots, \tau_T - 1\}$, such that, at step $\tau$ in the game, $\tau_1 + \tau_2 + \tau_T = \tau + q + 2$. The $F_{1,i}$ and $F_{2,i}$ are polynomials of degree $\leq q$ in $\mathbb{Z}_p[x]$, and the $F_{T,i}$ are polynomials of degree $\leq 2q$ in $\mathbb{Z}_p[x]$. The $\xi_{1,i}$, $\xi_{2,i}$, $\xi_{T,i}$ are strings in $\{0,1\}^*$. The lists are initialized at step $\tau = 0$ by taking $\tau_1 = 1$, $\tau_2 = q+1$, $\tau_T = 0$, and posing $F_{1,0} = 1$, and $F_{2,i} = x^i$ for $i \in \{0, \ldots, q\}$. The corresponding $\xi_{1,0}$ and $\xi_{2,i}$ are set to arbitrary distinct strings in $\{0,1\}^*$.

We may assume that $\mathcal{A}$ only makes oracle queries on strings previously obtained form $\mathcal{B}$, since $\mathcal{B}$ can make them arbitrarily hard to guess. We note that $\mathcal{B}$

can determine the index $i$ of any given string $\xi_{1,i}$ in $L_1$ (resp. $\xi_{2,i}$ in $L_2$, or $\xi_{T,i}$ in $L_T$), breaking ties between multiple matches arbitrarily.

$\mathcal{B}$ starts the game by providing $\mathcal{A}$ with the $q+2$ strings $\xi_{1,0}, \xi_{2,0}, \ldots, \xi_{2,q}$. Queries go as follows.

**Group action:** Given a multiply/divide selection bit and two operands $\xi_{1,i}$, $\xi_{1,j}$ with $0 \le i, j < \tau_1$, we compute $F_{1,\tau_1} \leftarrow F_{1,i} \pm F_{1,j} \in \mathbb{Z}_p[x]$ depending on whether a multiplication or a division is requested. If $F_{1,\tau_1} = F_{1,l}$ for some $l < \tau_1$, we set $\xi_{1,\tau_1} \leftarrow \xi_{1,l}$; otherwise, we set $\xi_{1,\tau_1}$ to a string in $\{0,1\}^*$ distinct from $\xi_{1,0}, \ldots, \xi_{1,\tau_1-1}$. We add $(F_{1,\tau_1}, \xi_{1,\tau_1})$ to $L_1$ and give $\xi_{1,\tau_1}$ to $\mathcal{A}$, then increment $\tau_1$ by one. Group action queries in $\mathbb{G}_2$ and $\mathbb{G}_T$ are treated similarly.

**Isomorphism:** Given a string $\xi_{2,i}$ with $0 \le i < \tau_2$, we let $F_{1,\tau_1} \leftarrow F_{2,i} \in \mathbb{Z}_p[x]$. If $F_{1,\tau_1} = F_{1,l}$ for some $l < \tau_1$, we set $\xi_{1,\tau_1} \leftarrow \xi_{1,l}$; otherwise, we set $\xi_{1,\tau_1}$ to a string in $\{0,1\}^* \setminus \{\xi_{1,0}, \ldots, \xi_{1,\tau_1-1}\}$. We add $(F_{1,\tau_1}, \xi_{1,\tau_1})$ to $L_1$, give $\xi_{1,\tau_1}$ to $\mathcal{A}$, and increment $\tau_1$ by one.

**Pairing:** Given two operands $\xi_{1,i}$ and $\xi_{2,j}$ with $0 \le i < \tau_1$ and $0 \le j < \tau_2$, we compute the product $F_{T,\tau_T} \leftarrow F_{1,i}F_{2,j} \in \mathbb{Z}_p[x]$. If $F_{T,\tau_T} = F_{T,l}$ for some $l < \tau_T$, we set $\xi_{T,\tau_T} \leftarrow \xi_{T,l}$; otherwise, we set $\xi_{T,\tau_T}$ to a string in $\{0,1\}^* \setminus \{\xi_{T,0}, \ldots, \xi_{T,\tau_T-1}\}$. We add $(F_{T,\tau_T}, \xi_{T,\tau_T})$ to $L_T$, give $\xi_{T,\tau_T}$ to $\mathcal{A}$, and increment $\tau_T$ by one.

$\mathcal{A}$ terminates and returns a pair $(c, \xi_{2,\ell})$ where $0 \le \ell < \tau_2$. At this point $\mathcal{B}$ chooses a random $x^* \in \mathbb{Z}_p$. The simulation provided by $\mathcal{B}$ is perfect unless the choice of $x$ creates an equality relation between the simulated group elements that was not revealed to $\mathcal{A}$. Thus, the success probability of $\mathcal{A}$ is bounded by the probability that any of the following holds:

1. $F_{1,i}(x^*) - F_{1,j}(x^*) = 0$ for some $i, j$ such that $F_{1,i} \ne F_{1,j}$,
2. $F_{2,i}(x^*) - F_{2,j}(x^*) = 0$ for some $i, j$ such that $F_{2,i} \ne F_{2,j}$,
3. $F_{T,i}(x^*) - F_{T,j}(x^*) = 0$ for some $i, j$ such that $F_{T,i} \ne F_{T,j}$,
4. $(x^* + c)\, F_{2,\ell}(x^*) = 0$.

Since $F_{1,i} - F_{1,j}$ for fixed $i$ and $j$ is a polynomial of degree at most $q$, it vanishes at a random $x^* \in \mathbb{Z}_p$ with probability at most $q/p$. Similarly, for fixed $i$ and $j$, the second case occurs with probability $\le q/p$, the third with probability $\le 2q/p$ (since $F_{T,i} - F_{T,j}$ has degree at most $2q$), and the fourth with probability $\le (q+1)/p$. By summing over all valid pairs $(i, j)$ in each case, we find that $\mathcal{A}$ wins the game with probability $\epsilon \le \binom{\tau_1}{2}\frac{q}{p} + \binom{\tau_2}{2}\frac{q}{p} + \binom{\tau_T}{2}\frac{2q}{p} + \frac{q+1}{p}$. Since $\tau_1 + \tau_2 + \tau_T \le q_G + q + 2$, the required bound follows: $\epsilon \le (q_G + q + 2)^2(q/p) = O((q_G)^2(q/p) + q^3/p)$. □

**Corollary 1.** *Any adversary that solves the q-SDH problem with constant probability $\epsilon > 0$ in generic groups of order $p$ such that $q < o(\sqrt[3]{p})$ requires $\Omega(\sqrt{\epsilon p/q})$ generic group operations.*

## 6 Conclusions

We presented a number of short signature schemes based on the $q$-SDH assumption. Our main result is a short signature which is fully secure without using the random oracle model. The signature is as short as DSA signatures, but is provably secure in the standard model. We also showed that the scheme supports limited message recovery, for even greater compactness.

These constructions are possible thanks to properties of the $q$-SDH assumption. The assumption can be viewed as a discrete logarithm analogue of the Strong RSA assumption. We believe the $q$-SDH assumption is a useful tool for constructing cryptographic systems and we expect to see many other schemes based on it. For example, we mention a new group signature scheme of Boneh et al. [BBS04].

## References

[ADR02] J.H. An, Y. Dodis, and T. Rabin. On the security of joint signature and encryption. In *Proceedings of Eurocrypt 2002*, volume 2332 of *LNCS*. Springer-Verlag, 2002.

[BB04a] Dan Boneh and Xavier Boyen. Efficient selective-ID secure identity based encryption without random oracles. In *Proceedings of Eurocrypt '04*, 2004.

[BB04b] Dan Boneh and Xavier Boyen. Short signatures without random oracles. In *Proceedings of Eurocrypt 2004*, 2004. Full version at: http://crypto.stanford.edu/~dabo/abstracts/sigssdh.html.

[BBS04] Dan Boneh, Xavier Boyen, and Hovav Shacham. Short group signatures using strong Diffie-Hellman. Manuscript, 2004.

[BLS01] Dan Boneh, Ben Lynn, and Hovav Shacham. Short signatures from the Weil pairing. In *Proceedings of Asiacrypt 2001*, volume 2248 of *LNCS*, pages 514–32. Springer-Verlag, 2001.

[CDF03] Nicolas Courtois, Magnus Daum, and Patrick Felke. On the security of HFE, HFEv- and Quartz. In *Proceedings of PKC 2003*, volume 2567 of *LNCS*, pages 337–50. Springer-Verlag, 2003.

[CHK04] Ran Canetti, Shai Halevi, and Jonathan Katz. Chosen-ciphertext security from identity-based encryption. In *Proceeings of Eurocrypt 2004*, LNCS. Springer-Verlag, 2004. http://eprint.iacr.org/2003/182/.

[CN00] J. Coron and D. Naccache. Security analysis of the Gennaro-Halevi-Rabin signature scheme. In *Proceedings of Eurocrypt 2000*, pages 91–101, 2000.

[Cor00] Jean-Sébastien Coron. On the exact security of full domain hash. In *Proceedings of Crypto 2000*, volume 1880 of *LNCS*, pages 229–35. Springer-Verlag, 2000.

[CS00] Ronald Cramer and Victor Shoup. Signature schemes based on the strong RSA assumption. *ACM TISSEC*, 3(3):161–185, 2000. Extended abstract in Proc. 6th ACM CCS, 1999.

[GHR99] Rosario Gennaro, Shai Halevi, and Tal Rabin. Secure hash-and-sign signatures without the random oracle. In *Proceedings of Eurocrypt 1999*, LNCS, pages 123–139. Springer-Verlag, 1999.

[GMR88] Shafi Goldwasser, Silvio Micali, and Ron Rivest. A digital signature scheme secure against adaptive chosen-message attacks. *SIAM J. Computing*, 17(2):281–308, 1988.

[JN01] Antoine Joux and Kim Nguyen. Separating decision Diffie-Hellman from Diffie-Hellman in cryptographic groups. Cryptology ePrint Archive, Report 2001/003, 2001. http://eprint.iacr.org/2001/003/.

[KR00] Hugo Krawczyk and Tal Rabin. Chameleon signatures. In *Proceedings of NDSS 2000*. Internet Society, 2000. http://eprint.iacr.org/1998/010/.

[KW03] Jonathan Katz and Nan Wang. Efficiency improvements for signature schemes with tight security reductions. In *Proceedings of ACM CCS*, 2003.

[MSK02] Shigeo Mitsunari, Ryuichi Sakai, and Masao Kasahara. A new traitor tracing. *IEICE Trans. Fundamentals*, E85-A(2):481–484, 2002.

[MVV97] Alfred J. Menezes, Paul C. Van Oorschot, and Scott A. Vanstone. *Handbook of Applied Cryptography*. CRC Press, 1997.

[NS00] David Naccache and Jacques Stern. Signing on a postcard. In *Proceedings of Financial Cryptography 2000*, 2000.

[PCG01] Jacques Patarin, Nicolas Courtois, and Louis Goubin. QUARTZ, 128-bit long digital signatures. In *Proceedings of RSA 2001*, volume 2020 of *LNCS*, pages 282–97. Springer-Verlag, 2001.

[PV00] Leon Pintsov and Scott Vanstone. Postal revenue collection in the digital age. In *Proceedings of Financial Cryptography 2000*, 2000.

[Sah99] Amit Sahai. Non-malleable non-interactive zero knowledge and adaptive chosen-ciphertext security. In *Proceeings 40 IEEE Symp. on Foundations of Computer Science*, 1999.

[Sho97] Victor Shoup. Lower bounds for discrete logarithms and related problems. In *Proceedings of Eurocrypt 1997*. Springer-Verlag, 1997.

[TSNZ03] V. To, R. Safavi-Naini, and F. Zhang. New traitor tracing schemes using bilinear map. In *Proceedings of 2003 DRM Workshop*, 2003.

[ZSNS04] Fangguo Zhang, Reihaneh Safavi-Naini, and Willy Susilo. An efficient signature scheme from bilinear pairings and its applications. In *Proceedings of PKC 2004*, 2004.

# Sequential Aggregate Signatures from Trapdoor Permutations

Anna Lysyanskaya[1], Silvio Micali[2], Leonid Reyzin[3], and Hovav Shacham[4]

[1] Brown University
anna@cs.brown.edu
[2] Massachusetts Institute of Technology
[3] Boston University
reyzin@cs.bu.edu
[4] Stanford University
hovav@cs.stanford.edu

**Abstract.** An aggregate signature scheme (recently proposed by Boneh, Gentry, Lynn, and Shacham) is a method for combining $n$ signatures from $n$ different signers on $n$ different messages into one signature of unit length. We propose *sequential aggregate signatures*, in which the set of signers is ordered. The aggregate signature is computed by having each signer, in turn, add his signature to it. We show how to realize this in such a way that the size of the aggregate signature is independent of $n$. This makes sequential aggregate signatures a natural primitive for certificate chains, whose length can be reduced by aggregating all signatures in a chain. We give a construction in the random oracle model based on families of certified trapdoor permutations, and show how to instantiate our scheme based on RSA.

## 1 Introduction

Authentication constitutes one of the core problems in cryptography. Much modern research focuses on constructing authentication schemes that are: (1) as secure as possible, i.e., provably secure under the most general assumptions; and (2) as efficient as possible, i.e., communication- and computation-efficient. For cryptographic schemes to be adopted in practice, efficiency is crucial. Moreover, communication and storage efficiency – namely, the size of the authentication data, for example the size of a signature – plays an even greater role than computation: While computational power of modern computers has experienced rapid growth over the last several decades, the growth in bandwidth of communication networks seems to have more constraints.

As much as we wish to reduce the size of a stand-alone signature, its length is lower-bounded by the security parameter. The problem becomes more interesting, however, once we have $n$ different signers with public keys $PK_1, \ldots, PK_n$, and each of them wants to sign her own message, $M_1, \ldots, M_n$, respectively. Suppose that the public keys and the messages are known to the signature recipient ahead of time, or clear from context. We want, in some way, to combine the

C. Cachin and J. Camenisch (Eds.): EUROCRYPT 2004, LNCS 3027, pp. 74–90, 2004.

authenticating information associated with this set of signers and messages into one short signature, whose length is independent of $n$.

This problem actually arises in practice. For example, in a Public Key Infrastructure (PKI) of depth $n$, a certificate on a user's public key consists of a chain of certificates issued by a hierarchy of certification authorities (CAs): the CA at depth $i$ certifies the CA at depth $i+1$. Which CAs were responsible for certifying a given user is usually clear from the context, and the public keys of these CAs may be available to the recipient off-line. The user's certificate, however, needs to be included in all of his communications, and therefore it is highly desirable to make its length independent of the length of the certification chain. Even if the entire certificate chain must be transmitted, significant space savings can be realized. In a typical X.509 certificate, 15% of the length is due to the signature.

Recently, Boneh et al. [5] introduced and realized *aggregate signatures.* An aggregate signature scheme is a signature scheme which, in addition to the usual setup, signing, and verification algorithms, admits an efficient algorithm for *aggregating* $n$ signatures under $n$ different public keys into one signature of unit length. Namely, suppose each of $n$ users has a public-private key pair $(PK_i, SK_i)$; each wishes to attest to a message $M_i$. Each user first signs her message $M_i$, obtaining a signature $\sigma_i$; the $n$ signatures can then be combined by an unrelated party into an aggregate $\sigma$. An aggregate signature scheme also includes an extra verification algorithm that verifies such an aggregate signature. An aggregate signature provides non-repudiation simultaneously on message $M_1$ for User 1, message $M_2$ for User 2, and so forth. Crucially, such repudiation holds for each user *regardless* of whether other users are malicious. Boneh et al. construct an aggregate signature scheme in the random oracle model under the bilinear Diffie-Hellman assumption (see, for example, Boneh and Franklin [4] and references therein).

For applications such as certificate chains, the ability to combine preexisting individual signatures into an aggregate is unnecessary. Each user, when producing a signature, is aware of the signatures above his in the chain. Thus aggregation for certificate chains should be performed incrementally and sequentially, so that User $i$, given an aggregate on messages $M_1, \ldots, M_{i-1}$ under keys $PK_1, \ldots, PK_{i-1}$, outputs an aggregate on messages $M_1, \ldots, M_{i-1}, M_i$ under keys $PK_1, \ldots, PK_{i-1}, PK_i$. We call such a procedure *sequential aggregation*, and a signature scheme supporting it, a *sequential aggregate signature scheme.*

In this paper, we begin by giving a formal definition of sequential aggregate signatures. We then show how to realize such signatures from a family of certified[5] trapdoor permutations (TDPs) over the same domain, as long as the domain is a group under some operation. We prove security (with exact security analysis) of our construction in the random oracle model; we give tighter security guarantees for the special cases of *homomorphic* and *claw-free* TDPs. As compared to the scheme of Boneh et al. [5], our scheme place more restrictions

[5] A TDP is *certified* [2] if one can verify from the public key that it is actually a permutation.

on the signers because of the sequentiality requirement, but relies on a more accepted, more general assumption.

Finally, we show how to instantiate our construction with the RSA trapdoor permutation. This instantiation turns out to be more difficult than may be expected, because of the possibility of maliciously generated RSA keys: We need to provide security for User $i$ regardless of whether other users are honest. There are essentially four problems. The first is that our scheme assumes multiple trapdoor permutations over the same domain, which RSA does not provide. The second is that RSA is not a *certified* trapdoor permutation: for a maliciously generated public-key, it can indeed be very far from a permutation. The third is that the domain of RSA is not the convenient $\mathbb{Z}_N$, but rather $\mathbb{Z}_N^*$, which can be much smaller for maliciously generated $N$. Finally, the natural group operation on $\mathbb{Z}_N^*$ (multiplication) is not a group operation on $\mathbb{Z}_N$. We overcome these problems with techniques that may be of independent interest. In particular, we turn RSA into a *certified* trapdoor permutation over *all* of $\mathbb{Z}_N$.

*Other Related Work.* Aggregate signatures are related to multisignatures [14, 16, 15, 3]. In particular, our aggregate signature scheme has similarities with the multisignature scheme of Okamoto [16] (though the latter has no security proof and, indeed, is missing important details that would make the security proof possible, as shown by Micali et al. [13]). Also of interest are threshold signatures, in particular the non-interactive threshold signature scheme due to Shoup [18], where we have a set of $n$ signers, and a threshold $t$, such that signature shares from any $t < k \leq n$ signers can be combined into one signature. They are different from aggregate signatures in several crucial aspects: threshold signatures require an expensive (or trusted) setup procedure; pieces of a threshold signature do not constitute a stand-alone signature; pieces of a threshold signature can only be combined into one once there are enough of them; and a threshold signature looks the same no matter which of the signers contributed pieces to it.

## 2 Preliminaries

We recall the definitions of trapdoor permutations and ordinary digital signatures, and the full-domain hash signatures based on trapdoor permutations. We also define certified trapdoor permutations, which are needed for building sequential aggregate signatures. In addition, we define claw-free permutations, and homomorphic trapdoor permutations, whose properties are used to achieve a better security reduction.

### 2.1 Trapdoor One-Way Permutations

Let $D$ be a group over some operation $\odot$. For simplicity, we assume that choosing an element of $D$ at random, computing $\odot$, and inverting $\odot$ each take unit time.

A trapdoor permutation family $\Pi$ over $D$ is defined as a triple of algorithms: *Generate*, *Evaluate*, and *Invert*. The randomized generation algorithm *Generate* outputs the description $s$ of a permutation along with the corresponding trapdoor $t$. The evaluation algorithm *Evaluate*, given the permutation description $s$ and a value $x \in D$, outputs $a \in D$, the image of $x$ under the permutation. The inversion algorithm *Invert*, given the permutation description $s$, the trapdoor $t$, and a value $a \in D$, outputs the preimage of $a$ under the permutation.

We require that $\mathit{Evaluate}(s, \cdot)$ be a permutation of $D$ for all $(s,t) \xleftarrow{\mathrm{R}} \mathit{Generate}$, and that $\mathit{Invert}(s, t, \mathit{Evaluate}(s, x)) = x$ hold for all $(s,t) \xleftarrow{\mathrm{R}} \mathit{Generate}$ and for all $x \in D$. The algorithms *Generate*, *Evaluate*, and *Invert* are assumed to take unit time for simplicity.

**Definition 1.** *The advantage of an algorithm $\mathcal{A}$ in inverting a trapdoor permutation family is*

$$\mathsf{Adv\,Invert}_{\mathcal{A}} \stackrel{\mathrm{def}}{=} \Pr\left[x = \mathcal{A}(s, \mathit{Evaluate}(s, x)) : (s,t) \xleftarrow{\mathrm{R}} \mathit{Generate}, x \xleftarrow{\mathrm{R}} D\right] .$$

*The probability is taken over the coin tosses of Generate and of $\mathcal{A}$. An algorithm $\mathcal{A}$ $(t, \epsilon)$-inverts a trapdoor permutation family if $\mathcal{A}$ runs in time at most $t$ and $\mathsf{Adv\,Invert}_{\mathcal{A}}$ is at least $\epsilon$. A trapdoor permutation family is $(t, \epsilon)$-one-way if no algorithm $(t, \epsilon)$-inverts the trapdoor permutation family.*

Note that this definition of a trapdoor permutation family requires that there exist multiple trapdoor permutations over the same domain $D$. We avoid the use of an infinite sequence of domains $D$, one for each security parameter, by simply fixing the security parameter and considering concrete security.

When it engenders no ambiguity, we consider the output of the generation algorithm *Generate* as a probability distribution $\Pi$ on permutations, and write $(\pi, \pi^{-1}) \xleftarrow{\mathrm{R}} \Pi$; here $\pi$ is the permutation $\mathit{Evaluate}(s, \cdot)$, and $\pi^{-1}$ is the inverse permutation $\mathit{Invert}(s, t, \cdot)$.

## 2.2 Certified Trapdoor Permutations

The trapdoor permutation families used in sequential aggregation must be certified trapdoor permutation families [2]. A certified trapdoor permutation family is one such that, for any string $s$, it is easy to determine whether $s$ can have been output by *Generate*, and thereby ensure that $\mathit{Evaluate}(s, \cdot)$ is a permutation. This is important when permutation descriptions $s$ can be generated by malicious parties.

Applying the definitions above to the RSA permutation family requires some care. RSA gives permutations over domains $\mathbb{Z}_N^*$, where each user has a distinct modulus $N$. Moreover, given just a public key $(N, e)$, certifying that the key describes a permutation is difficult. We consider this further in Sect. 5.

## 2.3 Claw-Free Permutations, Homomorphic Trapdoor Permutations

We now describe two variants of trapdoor permutations: claw-free permutations and homomorphic trapdoor permutations. The features these variants provide are not needed in the description of the sequential aggregate signature scheme, but allow a more efficient security reduction in Theorem 4.

A *claw-free* permutation family $\Pi$ [11] is a trapdoor permutation family with an additional permutation $g : D \to D$, evaluated by algorithm $EvaluateG(s, \cdot)$. More generally, $g$ can map any domain $E$ onto $D$ as long as the uniform distribution on $E$ induces the uniform distribution on $g(E)$. We assume that algorithm *EvaluateG* runs in unit time, and choosing an element of $E$ at random also takes unit time, just as above.

**Definition 2.** *The advantage of an algorithm $\mathcal{A}$ in finding a claw in a claw-free permutation family is*

$$\mathsf{Adv\,Claw}_{\mathcal{A}} \stackrel{\text{def}}{=} \Pr\left[\begin{array}{r} Evaluate(s,x) = EvaluateG(s,y)\ : \qquad\qquad\qquad \\ (s,t) \stackrel{\text{R}}{\leftarrow} Generate, (x,y) \stackrel{\text{R}}{\leftarrow} \mathcal{A}(s) \end{array}\right] .$$

*The probability is taken over the coin tosses of Generate and of $\mathcal{A}$. An algorithm $\mathcal{A}$ $(t, \epsilon)$-breaks a claw-free permutation family if $\mathcal{A}$ runs in time at most $t$ and* $\mathsf{Adv\,Claw}_{\mathcal{A}}$ *is at least $\epsilon$. A permutation family is $(t, \epsilon)$-claw-free if no algorithm $(t, \epsilon)$-breaks the claw-free permutation family.*

When it engenders no ambiguity, we abbreviate $EvaluateG(s, \cdot)$ as $g(\cdot)$, and write $(\pi, \pi^{-1}, g) \stackrel{\text{R}}{\leftarrow} \Pi$. In this compact notation, a claw is a pair $(x, y)$ such that $\pi(x) = g(y)$.

One obtains from every claw-free permutation family a trapdoor permutation family, simply by ignoring *EvaluateG* [11]. The proof is straightforward. Suppose there exists an algorithm $\mathcal{A}$ that inverts $\pi$ with nonnegligible probability. One selects $y \stackrel{\text{R}}{\leftarrow} E$, and provides $\mathcal{A}$ with $z = g(y)$, which is uniformly distributed in $D$. If $\mathcal{A}$ outputs $x$ such that $x = \pi^{-1}(z)$, then it has uncovered a claw $\pi(x) = g(y)$.

A trapdoor permutation family is *homomorphic* if $D$ is a group with some operation $*$ and if, for all $(s, t)$ generated by *Generate*, the permutation $\pi : D \to D$ induced by $Evaluate(s, \cdot)$ is an automorphism on $D$ with $*$. That is, if $a = \pi(x)$ and $b = \pi(y)$, then $a * b = \pi(x * y)$. The group action $*$ is assumed to be computable in unit time. The operation $*$ can be different from the operation $\odot$ given above; we do not require any particular relationship (e.g., distributivity) between $\odot$ and $*$.

One obtains from every homomorphic trapdoor permutation family a claw-free permutation family [10]. Pick some $z \neq 1 \in D$, and define $g(x) = z * \pi(x)$. In this case, $E = D$. Then a claw $\pi(x) = g(y) = z * \pi(y)$ reveals $\pi^{-1}(z) = x * (1/y)$ (where the inverse is with respect to $*$).

### 2.4 Digital Signatures

We review the well-known definition of security for ordinary digital signatures.

Existential unforgeability under a chosen message attack [11] in the random oracle model [1] for a signature scheme (*KeyGen*, *Sign*, and *Verify*) with a random oracle $H$ is defined using the following game between a challenger and an adversary $\mathcal{A}$:

**Setup.** The challenger runs algorithm *KeyGen* to obtain a public key *PK* and private key *SK*. The adversary $\mathcal{A}$ is given *PK*.

**Queries.** Proceeding adaptively, $\mathcal{A}$ requests signatures with *PK* on at most $q_S$ messages of his choice $M_1, \ldots, M_{q_s} \in \{0,1\}^*$. The challenger responds to each query with a signature $\sigma_i = \mathit{Sign}(SK, M_i)$. Algorithm $\mathcal{A}$ also adaptively asks for at most $q_H$ queries of the random oracle $H$.

**Output.** Eventually, $\mathcal{A}$ outputs a pair $(M, \sigma)$ and wins the game if (1) $M$ is not any of $M_1, \ldots, M_{q_s}$, and (2) $\mathit{Verify}(PK, M, \sigma) = \texttt{valid}$.

We define $\mathsf{Adv\,Sig}_{\mathcal{A}}$ to be the probability that $\mathcal{A}$ wins in the above game, taken over the coin tosses of *KeyGen* and of $\mathcal{A}$.

**Definition 3.** *A forger $\mathcal{A}$ $(t, q_H, q_S, \epsilon)$-breaks a signature scheme if $\mathcal{A}$ runs in time at most $t$; $\mathcal{A}$ makes at most $q_S$ signature queries and at most $q_H$ queries to the random oracle; and $\mathsf{Adv\,Sig}_{\mathcal{A}}$ is at least $\epsilon$. A signature scheme is $(t, q_H, q_S, \epsilon)$-existentially unforgeable under an adaptive chosen-message attack if no forger $(t, q_H, q_S, \epsilon)$-breaks it.*

### 2.5 Full-Domain Signatures

We review the full-domain hash signature scheme. The scheme, introduced by Bellare and Rogaway [1], works in any trapdoor one-way permutation family. The more efficient security reduction given by Coron [8] additionally requires that the permutation family be homomorphic. Dodis and Reyzin show that Coron's analysis can be applied for any claw-free permutation family [10]. The scheme makes use of a hash function $H : \{0,1\}^* \to D$, which is modeled as a random oracle. The signature scheme comprises three algorithms: *KeyGen*, *Sign*, and *Verify*.

**Key Generation.** For a particular user, pick random $(s,t) \xleftarrow{R} \mathit{Generate}$. The user's public key *PK* is $s$. The user's private key *SK* is $(s,t)$.

**Signing.** For a particular user, given the private key $(s,t)$ and a message $M \in \{0,1\}^*$, compute $h \leftarrow H(M)$, where $h \in D$, and $\sigma \leftarrow \mathit{Invert}(s,t,h)$. The signature is $\sigma \in D$.

**Verification.** Given a user's public key $s$, a message $M$, and a signature $\sigma$, compute $h \leftarrow H(M)$; accept if $h = \mathit{Evaluate}(s, \sigma)$ holds.

The following theorem, due to Coron, shows the security of full-domain signatures under the adaptive chosen message attack in the random oracle model. The terms given in the exact analysis of $\epsilon$ and $t$ have been adapted to agree with the accounting employed by Boneh et al. [6].

**Theorem 1.** *Let $\Pi$ be a $(t', \epsilon')$-one-way homomorphic trapdoor permutation family. Then the full-domain hash signature scheme on $\Pi$ is $(t, q_H, q_S, \epsilon)$-secure against existential forgery under an adaptive chosen-message attack (in the random oracle model) for all $t$ and $\epsilon$ satisfying*

$$\epsilon \geq e(q_S + 1) \cdot \epsilon' \qquad \textit{and} \qquad t \leq t' - 2(q_H + 2q_S) \ .$$

*Here $e$ is the base of the natural logarithm.*

## 3 Sequential Aggregate Signatures

We introduce sequential aggregate signatures and present a security model for them.

### 3.1 Aggregate and Sequential Aggregate Signatures

Boneh et al. [5] present a new signature primitive, aggregate signatures. Aggregate signatures are a generalization of multisignatures [14, 16, 15, 3] wherein signatures by several users on several distinct messages may be combined into an aggregate whose length is the same as that of a single signature. Using an aggregate signature in place of several individual signatures in a protocol yields useful space savings. In an aggregate signature, signatures are first individually generated and then combined into an aggregate.

Sequential aggregate signatures are different. Each would-be signer transforms a sequential aggregate into another that includes a signature on a message of his choice. Signing and aggregation are a single operation; sequential aggregates are built in layers, like an onion; the first signature in the aggregate is the inmost. As with non-sequential aggregate signatures, the resulting sequential aggregate is the same length as an ordinary signature. This behavior closely mirrors the sequential nature of certificate chains in a PKI.

Let us restate the intuition given above more formally. Key generation is a randomized algorithm that outputs a public-private keypair $(PK, SK)$.

Aggregation and signing is a combined operation. The operation takes as input a private key $SK$, a message $M_i$ to sign, and a sequential aggregate $\sigma'$ on messages $M_1, \ldots, M_{i-1}$ under respective public keys $PK_1, \ldots, PK_{i-1}$, where $M_1$ is the inmost message. All of $M_1, \ldots, M_{i-1}$ and $PK_1, \ldots, PK_{i-1}$ must be provided as inputs. If $i$ is 1, the aggregate $\sigma$ is taken to be empty. It adds a signature on $M_i$ under $SK$ to the aggregate, outputting a sequential aggregate $\sigma$ on all $i$ messages $M_1, \ldots, M_i$.

The aggregate verification algorithm is given a sequential aggregate signature $\sigma$, messages $M_1, \ldots, M_i$, and public keys $PK_1, \ldots, PK_i$, and verifies that $\sigma$ is a valid sequential aggregate (with $M_1$ inmost) on the given messages under the given keys.

### 3.2 Sequential Aggregate Signature Security

The security of sequential aggregate signature schemes is defined as the non-existence of an adversary capable, within the confines of a certain game, of existentially forging a sequential aggregate signature. Existential forgery here means that the adversary attempts to forge a sequential aggregate signature, on messages of his choice, by some set of users not all of whose private keys are known to the forger.

We formalize this intuition as the sequential aggregate chosen-key security model. In this model, the adversary $\mathcal{A}$ is given a single public key. His goal is the existential forgery of a sequential aggregate signature. We give the adversary power to choose all public keys except the challenge public key. The adversary is also given access to a sequential aggregate signing oracle on the challenge key. His advantage, $\mathsf{Adv\,AggSig}_{\mathcal{A}}$, is defined to be his probability of success in the following game.

**Setup.** The aggregate forger $\mathcal{A}$ is provided with a public key $PK$, generated at random.

**Queries.** Proceeding adaptively, $\mathcal{A}$ requests sequential aggregate signatures with $PK$ on messages of his choice. For each query, he supplies a sequential aggregate signature $\sigma$ on some messages $M_1, \ldots, M_{i-1}$ under distinct keys $PK_1, \ldots, PK_{i-1}$, and an additional message $M_i$ to be signed by the oracle under key $PK$ (where $i$ is at most $n$, a game parameter).

**Response.** Finally, $\mathcal{A}$ outputs $i$ distinct public keys $PK_1, \ldots, PK_i$. Here $i$ is at most $n$, and need not equal the lengths (also denoted $i$) of $\mathcal{A}$'s requests in the query phase above. One of these keys must equal $PK$, the challenge key. Algorithm $\mathcal{A}$ also outputs messages $M_1, \ldots, M_i$, and a sequential aggregate signature $\sigma$ by the $i$ users, each on his corresponding message, with $PK_1$ inmost.

The forger wins if the sequential aggregate signature $\sigma$ is a valid sequential aggregate signature on messages $M_1, \ldots, M_i$ under keys $PK_1, \ldots, PK_i$, and $\sigma$ is nontrivial, i.e., $\mathcal{A}$ did not request a sequential aggregate signature on messages $M_1, \ldots, M_{i^*}$ under keys $PK_1, \ldots, PK_{i^*}$, where $i^*$ is the index of the challenge key $PK$ in the forgery. Note that $i^*$ need not equal $i$: the forgery can be made in the middle of $\sigma$. The probability is over the coin tosses of the key-generation algorithm and of $\mathcal{A}$.

**Definition 4.** *A sequential aggregate forger $\mathcal{A}$ $(t, q_H, q_S, n, \epsilon)$-breaks an $n$-user aggregate signature scheme in the sequential aggregate chosen-key model if: $\mathcal{A}$ runs in time at most $t$; $\mathcal{A}$ makes at most $q_H$ queries to the hash function and at most $q_S$ queries to the aggregate signing oracle;* $\mathsf{Adv\,AggSig}_{\mathcal{A}}$ *is at least $\epsilon$; and the forged sequential aggregate signature is by at most $n$ users. A sequential aggregate signature scheme is $(t, q_H, q_S, n, \epsilon)$-secure against existential forgery in the sequential aggregate chosen-key model if no forger $(t, q_H, q_S, n, \epsilon)$-breaks it.*

## 4 Sequential Aggregates from Trapdoor Permutations

We describe a sequential aggregate signature scheme arising from any family of trapdoor permutations, and prove the security of the scheme.

We first introduce some notation for vectors. We write a vector as $\boldsymbol{x}$, its length as $|\boldsymbol{x}|$, and its elements as $\boldsymbol{x}_1, \boldsymbol{x}_2, \ldots, \boldsymbol{x}_{|\boldsymbol{x}|}$. We denote concatenating vectors as $\boldsymbol{x} \| \boldsymbol{y}$ and appending an element to a vector as $\boldsymbol{x} \| z$. For a vector $\boldsymbol{x}$, $\boldsymbol{x}|_a^b$ is the sub-vector containing elements $\boldsymbol{x}_a, \boldsymbol{x}_{a+1}, \ldots, \boldsymbol{x}_b$. It is necessarily the case that $1 \le a \le b \le |\boldsymbol{x}|$.

### 4.1 The Scheme

We now describe three algorithms, *KeyGen*, *AggregateSign*, and *AggregateVerify*, for our sequential aggregate signature scheme. The scheme employs a full-domain hash function $H : \{0,1\}^* \to D$, viewed as a random oracle, and resembles full-domain hash described in Sect. 2.5. The trick to aggregation is to incorporate the sequential aggregate signature of previous users by multiplying it (via the group operation $\odot$) together with the hash of the message. Actually, the hash now needs to include not only the signer's message, but also her public key and the prior messages and keys.[6]

**Key Generation.** For a particular user, pick random $(s,t) \xleftarrow{\text{R}} \mathit{Generate}$. The user's public key *PK* is $s$. The user's private key *SK* is $(s,t)$.

**Aggregate Signing.** The input is a private key $(s,t)$, a message $M \in \{0,1\}^*$ to be signed, and a sequential aggregate $\sigma'$ on messages $\boldsymbol{M}$ under public keys $\boldsymbol{s}$. Verify that $\sigma'$ is a valid signature on $\boldsymbol{M}$ under $\boldsymbol{s}$ using the verification algorithm below; if not, output $\star$, indicating error. Otherwise, compute $h \leftarrow H(\boldsymbol{s}\|s, \boldsymbol{M}\|M)$, where $h \in D$, and $\sigma \leftarrow \mathit{Invert}(s,t,h \odot \sigma')$. The sequential aggregate signature is $\sigma \in D$.

**Aggregate Verification.** The input is a sequential aggregate $\sigma$ on messages $\boldsymbol{M}$ under public keys $\boldsymbol{s}$. If any key appears twice in $\boldsymbol{s}$, if any element of $\boldsymbol{s}$ does not describe a valid permutation, or if $|\boldsymbol{M}|$ and $|\boldsymbol{s}|$ differ, reject. Otherwise, let $i$ equal $|\boldsymbol{M}| = |\boldsymbol{s}|$. Set $\sigma_i \leftarrow \sigma$. Then, for $j = i, \ldots, 1$, set $\sigma_{j-1} \leftarrow \mathit{Evaluate}(\boldsymbol{s}_j, \sigma_j) \odot H(\,\boldsymbol{s}|_1^j\,, \boldsymbol{M}|_1^j)^{-1}$. Accept if $\sigma_0$ equals 1, the unit of $D$ with respect to $\odot$.

Written using $\pi$-notation, a sequential aggregate signature is of the form

$$\pi_i^{-1}(h_i \odot \pi_{i-1}^{-1}(h_{i-1} \odot \pi_{i-2}^{-1}(\cdots \pi_2^{-1}(h_2 \odot \pi_1^{-1}(h_1))\cdots))) \enspace ,$$

where $h_j = H(\,\boldsymbol{s}|_1^j\,, \boldsymbol{M}|_1^j)$. Verification evaluates the permutations in the forward direction, peeling layers away until the center is reached.

### 4.2 Security

The following theorem demonstrates that our scheme is secure when instantiated on any certified trapdoor permutation family.

**Theorem 2.** *Let $\Pi$ be a certified $(t', \epsilon')$-trapdoor permutation family. Then our sequential aggregate signature scheme on $\Pi$ is $(t, q_H, q_S, n, \epsilon)$-secure against existential forgery under an adaptive sequential aggregate chosen-message attack (in the random oracle model) for all $t$ and $\epsilon$ satisfying*

$$\epsilon \geq (q_H + q_S + 1) \cdot \epsilon' \qquad \text{and} \qquad t \leq t' - (4nq_H + 4nq_S + 7n - 1) \enspace .$$

[6] This is done not merely because we do not know how to prove the scheme secure otherwise. Micali et al. [14] pointed out that if the signature does not include the public key, then an adversary may attack the scheme by deciding on the public key after the signature is issued. Our approach is the same as that of Boneh et al. [5, Sect. 3.2].

Following Coron's work [8], a better security reduction is obtained if the trapdoor permutations are, additionally, homomorphic under some operation $*$. (The operation $*$ need not be the same as the operation $\odot$ used in the description of the signature scheme in Sect. 4.)

**Theorem 3.** *Let $\Pi$ be a certified homomorphic $(t', \epsilon')$-trapdoor permutation family. Then our sequential aggregate signature scheme on $\Pi$ is $(t, q_H, q_S, n, \epsilon)$-secure against existential forgery under an adaptive sequential aggregate chosen-message attack (in the random oracle model) for all $t$ and $\epsilon$ satisfying*

$$\epsilon \geq e(q_S + 1) \cdot \epsilon' \qquad \text{and} \qquad t \leq t' - ((4n+1)q_H + (4n+1)q_S + 7n + 3) \ .$$

*Here $e$ is the base of the natural logarithm.*

Finally, following the work of Dodis and Reyzin [10], the homomorphic property is not really necessary, and can be replaced with the more general claw-free property:

**Theorem 4.** *Let $\Pi$ be a certified $(t', \epsilon')$-claw-free permutation family. Then the sequential aggregate signature scheme on $\Pi$ is $(t, q_H, q_S, n, \epsilon)$-secure against existential forgery under an adaptive sequential aggregate chosen-message attack (in the random oracle model) for all $t$ and $\epsilon$ satisfying*

$$\epsilon \geq e(q_S + 1) \cdot \epsilon' \qquad \text{and} \qquad t \leq t' - (4nq_H + 4nq_S + 7n) \ .$$

*Here $e$ is the base of the natural logarithm.*

The proofs of these theorems are very similar (in fact, Theorem 3 is just a corollary of Theorem 4, because, as we already saw, homomorphic trapdoor permutations are claw-free). We will prove all three at once.

*Proofs.* Suppose there exists a forger $\mathcal{A}$ that breaks the security of our sequential aggregate signature scheme. We describe three algorithms that use $\mathcal{A}$ to break one of the three possible security assumptions (trapdoor one-wayness, homomorphic one-wayness, and claw-freeness). In fact, the algorithms are quite similar regardless of the assumption. Therefore, we present only one of them: $\mathcal{B}$ that uses $\mathcal{A}$ to find a claw in a (supposedly) claw-free permutation family $\Pi$. We will point out later the changes needed to make the reduction to ordinary and homomorphic trapdoor permutations.

Suppose $\mathcal{A}$ is a forger algorithm that $(t, q_H, q_S, n, \epsilon)$-breaks the sequential aggregate signature scheme. We construct an algorithm $\mathcal{B}$ that finds a claw in $\Pi$.

Crucial in our construction is the following fact about our signature scheme: once the function $H$ is fixed on $i$ input values $(\boldsymbol{s}|_1^j, \boldsymbol{M}|_1^j)$, $1 \leq j \leq i$, there exists only one valid aggregate signature on $\boldsymbol{M}$ using keys $\boldsymbol{s}$. Thus, by answering hash queries properly, $\mathcal{B}$ can prepare for answering signature queries and for taking advantage of the eventual forgery.

Algorithm $\mathcal{B}$ is given the description $s$ of an element of $\Pi$, and must find values $x \in D$ and $y \in E$ such that $\mathit{Evaluate}(s, x) = \mathit{EvaluateG}(s, y)$. Algorithm $\mathcal{B}$

supplies $\mathcal{A}$ with the public key $s$. It then runs $\mathcal{A}$ and answers its oracle queries as follows.

**Hash Queries.** Algorithm $\mathcal{B}$ maintains a list, to which we refer as the $H$-list, of tuples $\langle \boldsymbol{s}^{(j)}, \boldsymbol{M}^{(j)}, w^{(j)}, r^{(j)}, c^{(j)} \rangle$. The list is initially empty. When $\mathcal{A}$ queries the oracle $H$ at a point $(\boldsymbol{s}, \boldsymbol{M})$, algorithm $\mathcal{B}$ responds as follows.

First we consider the easy cases.

- If some tuple $\langle \boldsymbol{s}, \boldsymbol{M}, w, r, c \rangle$ on the $H$-list already contains the query $(\boldsymbol{s}, \boldsymbol{M})$, then algorithm $\mathcal{B}$ answers the query as $H(\boldsymbol{s}, \boldsymbol{M}) = w \in D$.
- If $|\boldsymbol{M}|$ and $|\boldsymbol{s}|$ differ, if $|\boldsymbol{s}|$ exceeds $n$, if some key is repeated in $\boldsymbol{s}$, or if any key in $\boldsymbol{s}$ does not describe a valid permutation, then $(\boldsymbol{s}, \boldsymbol{M})$ can never be part of a sequential aggregate signature. Algorithm $\mathcal{B}$ picks $w \xleftarrow{R} D$, and sets $r \leftarrow \star$ and $c \leftarrow \star$, both placeholder values. It adds $\langle \boldsymbol{s}, \boldsymbol{M}, w, r, c \rangle$ to the $H$-list and responds to the query as $H(\boldsymbol{s}, \boldsymbol{M}) = w \in D$.

Now for the more complicated cases. Set $i = |\boldsymbol{s}| = |\boldsymbol{M}|$. If $i$ is greater than 1, $\mathcal{B}$ runs the hashing algorithm on input $(\,\boldsymbol{s}|_1^{i-1}, \boldsymbol{M}|_1^{i-1})$, obtaining the corresponding entry on the $H$-list, $\left\langle \boldsymbol{s}|_1^{i-1}, \boldsymbol{M}|_1^{i-1}, w', r', c' \right\rangle$. If $i$ equals 1, $\mathcal{B}$ sets $r' \leftarrow 1$. Algorithm $\mathcal{B}$ must now choose elements $r$,$w$, and $c$ to include, along with $\boldsymbol{s}$ and $\boldsymbol{M}$, in a new entry on the $H$-list. There are three cases to consider.

- If the challenge key $s$ does not appear at any index of $\boldsymbol{s}$, $\mathcal{B}$ chooses $r \xleftarrow{R} D$ at random, sets $c \leftarrow \star$, a placeholder value, and computes

$$w \leftarrow \mathit{Evaluate}(\boldsymbol{s}_i, r) \odot (r')^{-1} \ .$$

- If the challenge key $s$ appears in $\boldsymbol{s}$ at index $i^* = i$, Algorithm $\mathcal{B}$ generates a random coin $c \in \{0, 1\}$ such that $\Pr[c = 0] = 1/(q_S + 1)$. If $c = 1$, $\mathcal{B}$ chooses $r \xleftarrow{R} D$ at random and sets

$$w \leftarrow \mathit{Evaluate}(s, r) \odot (r')^{-1} \ .$$

(In this case, $w$ is uniform in $D$ and independent of all other queries because $r$ has been chosen uniformly and independently at random from $D$, and *Evaluate* and combining with $(r')^{-1}$ are both permutations.) If $c = 0$, $\mathcal{B}$ chooses $r \xleftarrow{R} E$ at random and sets

$$w \leftarrow \mathit{EvaluateG}(s, r) \odot (r')^{-1} \ .$$

(In this case, $w$ is uniform in $D$ and independent of all other queries because $r$ has been chosen uniformly and independently at random from $E$, *EvaluateG* maps uniformly onto $D$, and combining with $(r')^{-1}$ is a permutation.)

- If the challenge key $s$ appears in $\boldsymbol{s}$ at index $i^* \lneq i$, algorithm $\mathcal{B}$ picks $w \xleftarrow{R} D$ at random, and sets $r \leftarrow \star$ and $c \leftarrow \star$, both placeholder values.

Finally, $\mathcal{B}$ adds $\langle s, \boldsymbol{M}, w, r, c\rangle$ to the $H$-list, and responds to the query as $H(\boldsymbol{s}, \boldsymbol{M}) = w$.

In all cases, $\mathcal{B}$'s response, $w$, is uniform in $D$ and independent of $\mathcal{A}$'s current view, as required.

**Aggregate Signature Queries.** Algorithm $\mathcal{A}$ requests a sequential aggregate signature, under key $s$, on messages $\boldsymbol{M}$ under keys $\boldsymbol{s}$.

If $|\boldsymbol{s}|$ and $|\boldsymbol{M}|$ differ, if $|\boldsymbol{s}|$ exceeds $n$, if any key appears more than once in $\boldsymbol{s}$, or if any key in $\boldsymbol{s}$ does not describe a valid permutation, $(\boldsymbol{s}, \boldsymbol{M})$ is not a valid aggregate, and $\mathcal{B}$ responds to $\mathcal{A}$ with $\star$, indicating error. Let $i = |\boldsymbol{s}| = |\boldsymbol{M}|$. If $\boldsymbol{s}_i$ differs from $s$, $(\boldsymbol{s}, \boldsymbol{M})$ is not a valid query to the aggregate signing oracle, and $\mathcal{B}$ again responds with $\star$.

Algorithm $\mathcal{A}$ also supplies a purported sequential aggregate signature $\sigma'$ on messages $\boldsymbol{M}|_1^{i-1}$ under keys $\boldsymbol{s}|_1^{i-1}$. If $i$ equals 1, $\mathcal{B}$ verifies that $\sigma'$ equals 1. Otherwise, $\mathcal{B}$ uses *AggregateVerify* to ensure that $\sigma'$ is the correct sequential aggregate signature on $(\boldsymbol{s}|_1^{i-1}, \boldsymbol{M}|_1^{i-1})$. If $\sigma'$ is incorrect, $\mathcal{B}$ again responds with $\star$.

Otherwise, $\mathcal{B}$ runs the hash algorithm on $(\boldsymbol{s}, \boldsymbol{M})$, obtaining the corresponding entry on the $H$-list, $\langle \boldsymbol{s}, \boldsymbol{M}, w, r, c\rangle$. Since $\boldsymbol{s}_i$ equals $s$, $c$ must be 0 or 1. If $c = 0$ holds, $\mathcal{B}$ reports failure and terminates. Otherwise, $\mathcal{B}$ responds to the query with $\sigma \leftarrow r$.

**Output.** Eventually algorithm $\mathcal{A}$ halts, outputting a message vector $\boldsymbol{M}$, a public-key vector $\boldsymbol{s}$, and a corresponding sequential aggregate signature forgery $\sigma$. The forgery must be valid: No key may occur more than once in $\boldsymbol{s}$, each key in $\boldsymbol{s}$ must describe a valid permutation, the two vectors $\boldsymbol{s}$ and $\boldsymbol{M}$ must have the same length $i$, which is at most $n$. The forgery must also be nontrivial: The challenge key $s$ must occur in $\boldsymbol{s}$, at some location $i^*$, and $\mathcal{A}$ must not have asked for a sequential aggregate signature on messages $\boldsymbol{M}|_1^{i^*}$ under keys $\boldsymbol{s}|_1^{i^*}$. If $\mathcal{A}$ fails to output a valid and nontrivial forgery, $\mathcal{B}$ reports failure and terminates.

Algorithm $\mathcal{B}$ begins by checking the hashes included in $\sigma$. For each $j$, $1 \le j \le i$, $\mathcal{B}$ runs its hash algorithm on $(\boldsymbol{s}|_1^j, \boldsymbol{M}|_1^j)$, obtaining a series of tuples $\left\langle \boldsymbol{s}|_1^j, \boldsymbol{M}|_1^j, w^{(j)}, r^{(j)}, c^{(j)} \right\rangle$. Note that $\mathcal{B}$ always returns $w$ as the answer to a hash query, so, for each $j$, $H(\boldsymbol{s}|_1^j, \boldsymbol{M}|_1^j) = w^{(j)}$.

Algorithm $\mathcal{B}$ then examines $c^{(i^*)}$. Since $s^{(i^*)}$ equals $s$, $c^{(i^*)}$ must be 0 or 1. If $c^{(i^*)} = 1$ holds, $\mathcal{B}$ reports failure and terminates. Then $\mathcal{B}$ applies the aggregate signature verification algorithm to $\sigma$. It sets $\sigma^{(i)} \leftarrow \sigma$. For $j = i, \ldots, 1$, it sets $\sigma^{(j-1)} \leftarrow \mathit{Evaluate}(s^{(j)}, \sigma^{(j)}) \odot (w^{(j)})^{-1}$.

If $\sigma^{(0)}$ does not equal 0, $\sigma$ is not a valid aggregate signature, and $\mathcal{B}$ reports failure and terminates. Otherwise, $\sigma$ is valid and, moreover, each $\sigma^{(j)}$ computed by $\mathcal{B}$ is the (unique) valid aggregate signature on messages $\boldsymbol{M}|_1^j$ under keys $\boldsymbol{s}|_1^j$.

Finally, $\mathcal{B}$ sets $x \leftarrow \sigma^{(i^*)}$ and $y \leftarrow r^{(i^*)}$.

This completes the description of algorithm $\mathcal{B}$.

It is easy to modify this algorithm for homomorphic trapdoor permutations. Now the algorithm's goal is not to find a claw, but to invert the permutation given by $s$ on a given input $z$. Simply replace, when answering hash queries for $c = 0$, invocation of $EvaluateG(s, r)$ with $z * Evaluate(s, r)$. The a claw $(x, y)$ allows $\mathcal{B}$ to recover the inverse of $z$ under the permutation by computing $z = x * (1/y)$, where $1/y$ is the inverse of $y$ under $*$.

Finally, it is also easy to modify this algorithm for ordinary trapdoor permutations:

- In answering hash queries where the challenge key $s$ is outmost in $\boldsymbol{s}$, instead of letting $c = 0$ with probability $1/(q_S + 1)$, set $c = 0$ for exactly one query, chosen at random. There can be at most $q_H + q_S + 1$ such queries.
- For the $c = 0$ query, set $w \leftarrow z \odot (r')^{-1}$. Then $w$ is random given $\mathcal{A}$'s view.
- If Algorithm $\mathcal{A}$'s forgery is such that $c^{(i^*)} = 0$, $\mathcal{B}''$ outputs $x \leftarrow \sigma^{(i^*)}$.

In the full version of this paper [12], we show that $\mathcal{B}$ correctly simulates $\mathcal{A}$'s environment, and analyze its running time and success probability. □

## 5 Aggregating with RSA

Here we consider the details of instantiating the sequential aggregate signature scheme presented in Sect. 4 using the RSA permutation family.

The RSA function was introduced by Rivest, Shamir, and Adleman [17]. If $N = pq$ is the product of two large primes and $ed = 1 \bmod \phi(N)$, then $\pi(x) = x^e \bmod N$ is a permutation on $\mathbb{Z}_N^*$, and $\pi^{-1}(x) = x^d \bmod N$ is its inverse. Setting $s = (N, e)$ and $t = (d)$ gives a one-way trapdoor permutation that is multiplicatively homomorphic.

A few difficulties arise when we try to instantiate the above scheme with RSA. We tackle them individually.

The first problem is that RSA is not a *certified* trapdoor permutation. Raising to the power $e$ may not be a permutation over $\mathbb{Z}_N^*$ if $e$ is not relatively prime with $\phi(N)$. Moreover, even if it is a permutation of $\mathbb{Z}_N^*$, it may not be a permutation of all of $\mathbb{Z}_N$ if $N$ is maliciously generated (in particular, if $N$ is not square-free). Note that, for maliciously generated $N$, the difference between $\mathbb{Z}_N^*$ and $\mathbb{Z}_N$ may be considerable. The traditional argument used to dismiss this issue (that if one finds $x$ outside $\mathbb{Z}_N^*$, one factors $N$) has no relevance here: $N$ may be generated by the adversary, and our ability to factor it has no impact on the security of the scheme for the honest signer who is using a different modulus. Our security proof substantially relied on the fact that even the adversarial public keys define permutations, for uniqueness of signatures and proper distribution of hash query answers. Indeed, this is not just a "proof problem," but a demonstrable security concern: If the adversary is able to precede the honest user's key $(N_i, e_i)$ with multiple keys $(N_1, e_1), \ldots, (N_{i-1}, e_{i-1})$, each of which defines a collision-prone function rather than a permutation, then it is quite possible that no matter value one takes for $\sigma_i$, it will be likely to verify correctly: for example, there will be two valid $\sigma_1$ values, four valid $\sigma_2$ values, eight valid $\sigma_3$ values, ..., $2^i$ valid $\sigma_i$ values.

One way to resolve this problem is to make sure that *every* key participating in an aggregate signature has been verified to be of correct form. This could be accomplished by having a trusted certification authority check that $N$ is a product of two large primes and $e$ is relatively prime to $\phi(N)$ before issuing a certificate. This check, however, requires one to place more trust in the authority than usual: the authority must be trusted not just to verify the identity of a key's purported owner, but also to perform verification of some complicated properties of the key. Moreover, the security of an honest signer can be compromised *without the signer's knowledge or participation* by dishonest signers whose keys are of incorrect form, when the dishonest signers form an aggregate signature that verifies with the honest signer's public key. The only way to prevent this is to trust that the verifier of the aggregate signature only accepts certificates from certification authorities who verify the correctness of the key.

In the case when it is best to avoid assuming such complex trust relationships, we propose to tackle this problem in the same way as Micali et al. [13], though at the expense of longer verification time. First, we require $e$ to be a prime larger than $N$ (this idea also appeared in Cachin et al. [7]). Then it is guaranteed to be relatively prime with $\phi(N)$, and therefore provide a permutation over $\mathbb{Z}_N^*$. To extend to a permutation over $\mathbb{Z}_N$, we define *Evaluate*$((N, e), x)$ as follows: if $\gcd(x, N) = 1$, output $x^e \bmod N$; else output $x$.

The second problem is that the natural choice for the group operation $\odot$, multiplication, is not actually a group operation over $\mathbb{Z}_N$. Thus, signature verification, which requires computation of an inverse under $\odot$, may be unable to proceed. Moreover, our security proof, which relies on the fact that $\odot$ is a group operation for uniqueness of signatures and proper distribution of hash query answers, will no longer hold. This difficulty is simple to overcome: Use addition modulo $N$ as the group operation $\odot$. Recall that no properties were required of $\odot$ beyond being a group operation on the domain.

The third problem is that two users cannot share the same modulus $N$. Thus the domains of the one-way permutations belonging to the aggregating users differ, making it difficult to treat RSA as a family of trapdoor permutations. We give two approaches that allow us to create sequential aggregates from RSA nonetheless.

The first approach is to require the users' moduli to be arranged in increasing order: $N_1 < N_2 \ldots < N_n$. At verification, it is important to check that the $i$-th signature $\sigma_i$ is actually less than $N_i$, to ensure that correct signatures are unique if $H$ is fixed. As long as $\log N_1 - \log N_n$ is constant, and the range of $H$ is a subset of $\mathbb{Z}_{N_1}$ whose size is a constant fraction of $N_1$, the scheme will be secure. The same security proof still goes through, with the following minor modification for answering hash queries. Whenever a hash query answer $w$ is computed by first choosing a random $r$ in $\mathbb{Z}_{N_i}$, there is a chance that $w$ will be outside of the range of $H$. In this case, simply repeat with a fresh random $r$ until $w$ falls in the right range (the expected number of repetitions is constant). Note that because we insisted on *Evaluate* being a permutation and $\odot$ being a group operation, the resulting distribution of $w$ is uniform on the range of $H$. Therefore, the distribution of answers to hash queries is uniform. Since signatures are uniquely determined by answers to hash queries, the adversary's whole view is correct, and the proof works without other modifications. (This technique is related to Coron's partial-domain hash analysis [9], though Coron deals with the more complicated case when the partial domain is exponentially smaller than the full domain.)

Our second approach allows for more general moduli: We do not require them to be in increasing order. However, we do require them to be of the same length $l$ (constant differences in the lengths will also work, but we do not address them here for simplicity of exposition). The signature will expand by $n$ bits $b_1 \dots b_n$, where $n$ is the total number of users. Namely, during signing, if $\sigma_i \geq N_{i+1}$, let $b_i = 1$; else, let $b_i = 0$. During verification, if $b_i = 1$, add $N_{i+1}$ to $\sigma_i$ before proceeding with the verification of $\sigma_i$. Always check that $\sigma_i$ is in the correct range $0 \leq \sigma_i < N_i$ (to ensure, again, uniqueness of signatures). The security proof requires no major modifications.[7]

To summarize, the resulting RSA aggregate signature schemes for $n$ users with moduli of length $l$ are as follows. Let $H : \{0,1\}^* \to \{0,1\}^{l-1}$ be a hash function.

*Restricted Moduli.* We first present the scheme where the moduli must be ordered.

**Key Generation.** Each user $i$ generates an RSA public key $(N_i, e_i)$ and secret key $(N_i, d_i)$, ensuring that $2^{l-1}(1+(i-1)/n) \leq N_i < 2^{l-1}(1+i/n)$ and that $e_i > N_i$ is a prime.

**Signing.** User $i$ is given an aggregate signature $\sigma'$, the messages $M_1, \dots, M_{i-1}$, and the corresponding keys $(N_1, e_1), \dots, (N_{i-1}, e_{i-1})$. User $i$ first verifies $\sigma'$, using the verification procedure below. If the verification succeeds, user $i$ computes $h_i = H((M_1, \dots, M_i), ((N_1, e_1), \dots, (N_i, e_i)))$, $y = h_i + \sigma'$ and outputs $\sigma = y^{d_i} \bmod N_i$. The user may first check that $\gcd(y, N) = 1$ and, if not, output $y$; however, the chances that the check will fail are negligible, because the user is honest.

**Verifying.** The verifier is given as input an aggregate signature $\sigma$, the messages $M_1, \dots, M_i$, and the corresponding keys $(N_1, e_1), \dots, (N_i, e_i)$, and proceeds as follows. Check that no key appears twice, that $e_i > N_i$ is a prime and that $N_i$ is of length $l$ (this needs to be checked only once per key, and need not be done with every signature verification). and that $0 \leq \sigma < N_i$. If $\gcd(\sigma, N_i) = 1$, let $y \leftarrow \sigma^{e_i} \bmod N_i$. Else let $y \leftarrow \sigma$ (this

[7] We need to argue that correct signatures are unique given the hash answers. At first glance it may seem that the adversary may have choice on whether to use $b_i = 0$ or $b_i = 1$. However, this will result in two values $\sigma_{i-1}$ that are guaranteed to be different: one will be less than $N_i$ and the other at least $N_i$. Hence uniqueness of $\sigma_{i-1}$ implies uniqueness of $b_i$ and, therefore, $\sigma_i$. Thus, by induction, signatures are still unique. In particular, there is no need to include $b_i$ in the hash function input.

check is crucial, because we do not know if user $i$ is honest). Compute $h_i \leftarrow H((M_1,\ldots,M_i),((N_1,e_1),\ldots,(N_i,e_i)))$ and $\sigma' \leftarrow y - h_i \bmod N_i$. Verify $\sigma'$ recursively. The base case for recursion is $i = 0$, in which case simply check that $\sigma = 0$.

*Unrestricted Moduli.* We present the scheme for unordered moduli by simply demonstrating the required modifications. First, the range of $N_i$ is now $2^{l-1} < N_i < 2^l$. Second, to sign, upon verifying $\sigma'$, check if $\sigma' \geq N_i$. If so, replace $\sigma'$ with $\sigma' - N_i$ and set $b_i = 1$; else, set $b_i = 0$. Finally, to verify, replace $\sigma'$ with $\sigma' + b_i N_i$ before proceeding with the recursive step.

*Security.* Because RSA over $\mathbb{Z}_N^*$ is homomorphic with respect to multiplication, it is claw-free (not just over $\mathbb{Z}_N^*$, but over entire $\mathbb{Z}_N$, because finding a claw outside of $\mathbb{Z}_N^*$ implies factoring $N$ and hence being able to invert RSA). Therefore, the conclusions of Theorem 4 apply to this scheme.

## Acknowledgments

The authors thank Dan Boneh, Stanisław Jarecki, and Craig Gentry for helpful discussions about this work, Eu-Jin Goh for his detailed and helpful comments on the manuscript, and the anonymous referees for valuable feedback.

## References

[1] M. Bellare and P. Rogaway. Random oracles are practical: A paradigm for designing efficient protocols. In D. Denning, R. Pyle, R. Ganesan, R. Sandhu, and V. Ashby, editors, *Proceedings of CCS 1993*, pages 62–73. ACM Press, 1993.

[2] M. Bellare and M. Yung. Certifying permutations: Non-interactive zero-knowledge based on any trapdoor permutation. *J. Cryptology*, 9(1):149–66, 1996.

[3] A. Boldyreva. Efficient threshold signature, multisignature and blind signature schemes based on the gap-Diffie-Hellman-group signature scheme. In Y. Desmedt, editor, *Proceedings of PKC 2003*, volume 2567 of *LNCS*, pages 31–46. Springer-Verlag, 2003.

[4] D. Boneh and M. Franklin. Identity-based encryption from the Weil pairing. *SIAM J. Computing*, 32(3):586–615, 2003. Extended abstract in *Proceedings of Crypto 2001*.

[5] D. Boneh, C. Gentry, B. Lynn, and H. Shacham. Aggregate and verifiably encrypted signatures from bilinear maps. In E. Biham, editor, *Proceedings of Eurocrypt 2003*, volume 2656 of *LNCS*, pages 416–32. Springer-Verlag, 2003.

[6] D. Boneh, B. Lynn, and H. Shacham. Short signatures from the Weil pairing. In *Proceedings of Asiacrypt 2001*, volume 2248 of *LNCS*, pages 514–32. Springer-Verlag, 2001. Full paper: http://crypto.stanford.edu/~dabo/pubs.html.

[7] C. Cachin, S. Micali, and M. Stadler. Computationally private information retrieval with polylogarithmic communication. In J. Stern, editor, *Proceedings of Eurocrypt 1999*, volume 1592 of *LNCS*, pages 402–414. Springer-Verlag, 1999.

[8] J.-S. Coron. On the exact security of full domain hash. In M. Bellare, editor, *Proceedings of Crypto 2000*, volume 1880 of *LNCS*, pages 229–35. Springer-Verlag, 2000.

[9] J.-S. Coron. Security proof for partial-domain hash signature schemes. In M. Yung, editor, *Proceedings of Crypto 2002*, volume 2442 of *LNCS*, pages 613–26. Springer-Verlag, 2002.

[10] Y. Dodis and L. Reyzin. On the power of claw-free permutations. In S. Cimato, C. Galdi, and G. Persiano, editors, *Proceedings of SCN 2002*, volume 2576 of *LNCS*, pages 55–73. Springer-Verlag, 2002.

[11] S. Goldwasser, S. Micali, and R. Rivest. A digital signature scheme secure against adaptive chosen-message attacks. *SIAM J. Computing*, 17(2):281–308, 1988.

[12] A. Lysyanskaya, S. Micali, L. Reyzin, and H. Shacham. Sequential aggregate signatures from trapdoor permutations. Cryptology ePrint Archive, Report 2003/091, 2003. http://eprint.iacr.org/.

[13] S. Micali, K. Ohta, and L. Reyzin. Provable-subgroup signatures. Unpublished manuscript, 1999.

[14] S. Micali, K. Ohta, and L. Reyzin. Accountable-subgroup multisignatures (extended abstract). In *Proceedings of CCS 2001*, pages 245–54. ACM Press, 2001.

[15] K. Ohta and T. Okamoto. Multisignature schemes secure against active insider attacks. *IEICE Trans. Fundamentals*, E82-A(1):21–31, 1999.

[16] T. Okamoto. A digital multisignature scheme using bijective public-key cryptosystems. *ACM Trans. Computer Systems*, 6(4):432–41, November 1988.

[17] R. Rivest, A. Shamir, and L. Adleman. A method for obtaining digital signatures and public key cryptosystems. *Commun. ACM*, 21:120–126, 1978.

[18] V. Shoup. Practical threshold signatures. In B. Preneel, editor, *Proceedings of Eurocrypt 2000*, volume 1807 of *LNCS*, pages 207–20. Springer Verlag, 2000.

# On the Key-Uncertainty of Quantum Ciphers and the Computational Security of One-Way Quantum Transmission

Ivan Damgård, Thomas Pedersen*, and Louis Salvail*

BRICS**, FICS***, Dept. of Computer Science, University of Århus, {ivan|pede|salvail}@brics.dk

**Abstract.** We consider the scenario where Alice wants to send a secret (classical) $n$-bit message to Bob using a classical key, and where only one-way transmission from Alice to Bob is possible. In this case, quantum communication cannot help to obtain perfect secrecy with key length smaller then $n$. We study the question of whether there might still be fundamental differences between the case where quantum as opposed to classical communication is used. In this direction, we show that there exist ciphers with perfect security producing quantum ciphertext where, even if an adversary knows the plaintext and applies an optimal measurement on the ciphertext, his Shannon uncertainty about the key used is almost maximal. This is in contrast to the classical case where the adversary always learns $n$ bits of information on the key in a known plaintext attack. We also show that there is a limit to how different the classical and quantum cases can be: the most probable key, given matching plain- and ciphertexts, has the same probability in both the quantum and the classical cases. We suggest an application of our results in the case where only a short secret key is available and the message is much longer. Namely, one can use a pseudorandom generator to produce from the short key a stream of keys for a quantum cipher, using each of them to encrypt an $n$-bit block of the message. Our results suggest that an adversary with bounded resources in a known plaintext attack may potentially be in a much harder situation against quantum stream-ciphers than against any classical stream-cipher with the same parameters.

## 1 Introduction

In this paper, we consider the scenario where Alice wants to send a secret (classical) $n$-bit message to Bob using an $m$-bit classical shared key, and where only one-way transmission from Alice to Bob is possible (or at least where interaction is only available with a prohibitively long delay). If interaction had been available, we could have achieved (almost) perfect secrecy using standard quantum

* Part of this research was funded by European project PROSECCO.

** Funded by the Danish National Research Foundation.

*** FICS, Foundations in Cryptography and Security, funded by the Danish Natural Sciences Research Council.

C. Cachin and J. Camenisch (Eds.): EUROCRYPT 2004, LNCS 3027, pp. 91–108, 2004.

key exchange, even if $m < n$. But with only one-way communication, we need $m \geq n$ even with quantum communication [1].

We study the question of whether there might still be some fundamental differences between the case where quantum as opposed to classical communication is used. In this direction, we present two examples of cryptosystems with perfect security producing $n$-bit quantum ciphertexts, and with key length $m = n + 1$, respectively $m = 2n$. We show that given plaintext and ciphertext, and even when applying an optimal measurement to the ciphertext, the adversary can learn no more than $n/2$, respectively 1 bit of Shannon information on the key. This should be compared to the fact that for a classical cipher with perfect security, the adversary always learns $n$ bits of information on the key. While proving these results, we develop a method which may be of independent interest, for estimating the maximal amount of Shannon information that a measurement can extract from a mixture. We note that the first example can be implemented without quantum memory, it only requires technology similar to what is needed for quantum key exchange, and is therefore within reach of current technology. The second example can be implemented with a circuit of $O(n^3)$ gates out of which only $O(n^2)$ are elementary quantum gates.

We also discuss the composition of ciphers, i.e., what happens to the uncertainty of keys when the same quantum cipher is used to encrypt several blocks of data using independent keys. This requires some care, it is well known that cryptographic constructions do not always compose nicely in the quantum case. For composition of our ciphers, however, a rather simple argument shows that the adversary's uncertainty about the keys grows with the number of blocks encrypted exactly as one would expect classically.

On the other hand, we show that there is a limit to how different the quantum and classical cases can be. Namely, the most probable key (i.e. the min-entropy of the key), given matching plain- and ciphertexts, has the same probability in both cases.

On the technical side, a main observation underlying our results on Shannon key-uncertainty is that our method for estimating the optimal measurement w.r.t. Shannon entropy can be combined with known results on so called entropic uncertainty relations [5,3,7] and mutually unbiased bases [8]. We note that somewhat related techniques are used in concurrent independent work by DiVincenzo et al. [2] to handle a different, non-cryptographic scenario.

While we believe the above results are interesting, and perhaps even somewhat surprising from an information theoretic point of view, they have limited practical significance if perfect security is the goal: a key must never be reused, and so we do not really have to care whether the adversary learns information about it when it is used.

However, there is a different potential application of our results to the case where only a short secret key is available, and where no upper bound on the message length is known a priori. In such a case, only computational security is possible and the standard classical way to encrypt is to use a stream-cipher: using a pseudorandom generator, we expand the key into a long random looking keystream, which is then combined with the plaintext to form the ciphertext. The simplest way of doing such a combination is to take the bit-wise XOR of key and plaintext streams. In a known plaintext attack, an adversary will then be able to learn full information on a part of the keystream and can try to analyze it to find the key or guess other parts of the keystream better than at random. In general, any cipher with perfect secrecy, $n$-bit plain- and ciphertext and $m$-bit keys can be used: we simply take the next $m$ bits from the keystream and use these as key in the cipher to encrypt the next $n$ bits of the plaintext. It is easy to see that for any classical cipher, if the adversary knows some $n$-bit block of plaintext and also the matching ciphertext, then he learns $n$ bit of Shannon information on the keystream.

If instead we use quantum communication and one of our quantum ciphers mentioned above, intuition suggests that an adversary with limited resources is in a more difficult situation when doing a known plaintext attack: if measuring the state representing the ciphertext only reveals a small amount of information on the corresponding part of the keystream, then the adversary will need much more known plaintext than in the classical case before being able to cryptanalyze the keystream.

Care has to be taken in making this statement more precise: our results on key uncertainty tell us what happens when keys are random, whereas in this application they are pseudorandom. It is conceivable that the adversary could design a measurement revealing more information by exploiting the fact that the keystream is not truly random. This, however, is equivalent to cryptanalyzing the generator using a *quantum* computation, and is likely to be technologically much harder than implementing the quantum ciphers. In particular, unless the generator is very poorly designed, it will require keeping a coherent state much larger than what is required for encryption and decryption – simply because one will need to involve many bits from the keystream simultaneously in order to distinguish it efficiently from random. Thus, an adversary limited to measurements involving only a *small* number of qubits will simply have to make many such measurements, hoping to gather enough classical information on the keystream to cryptanalyze it. Our results apply to this situation: first, since the adversary makes many measurements, we should worry about what he learns on average, so Shannon information is the appropriate measure. Second, even though the keystream is only pseudorandom, it may be genuinely random when considering only a small part of it (see Maurer and Massey [4]).

In Sect. 9, we prove a lower bound on the amount of known plaintext the adversary would need in order to obtain a given amount of information on the keystream, for a particular type of keystream generator and assuming the size of coherent states the adversary can handle is limited. We believe that quantum communication helps even for more general adversaries and generators. However, quantifying this advantage is an open problem. We stress that our main goal here is merely to point out the potential for improved security against a bounded adversary.

## 2 Preliminaries

We assume the reader is familiar with the standard notions of Shannon entropy $H(\cdot)$ of a probability distribution, conditional entropy, etc. A related notion that also measures "how uniform" a distribution is, is the so called *min-entropy*. Given a probability distribution $\{p_1, ..., p_n\}$, the min-entropy is defined as

$$H_\infty(p_1, ..., p_n) = -\log_2(max\{p_1, ..., p_n\}) \quad (1)$$

As usual, $H_\infty(X)$ for random variable $X$ is the min-entropy of its distribution. Min-entropy is directly related to the "best guess" probability: if we want to guess which value random variable $X$ will take, the best strategy is to guess at a value with maximal probability, and then we will be correct with probability $2^{-H_\infty(X)}$. Given the value of another random variable $Y$, we can define $H_\infty(X|Y = y)$ simply as the min-entropy of the distribution of $X$ given that $Y = y$, and similarly to Shannon entropy, we can define $H_\infty(X|Y) = \sum_y Pr(Y = y) \cdot H_\infty(X|Y = y)$.

The min-entropy can be thought of as a worst-case measure, which is more relevant when you have access to only one sample of some random experiment, whereas Shannon entropy measures what happens on average over several experiments. To illustrate the difference, consider the two distributions $(1/2, 1/2)$ and $(1/2, 1/4, 1/4)$. They both have min-entropy 1, even though it intuitively seems there should be more uncertainty in the second case, indeed the Shannon entropies are 1 and 1.5. In fact, we always have $H(X) \geq H_\infty(X)$, with equality if $X$ is uniformly distributed.

## 3 Classical Ciphers

Consider a classical cryptosystem with $n$-bit plain and ciphertexts, $m$-bit keys and perfect secrecy (assuming, of course, that keys are used only once). We identify the cryptosystem with its encryption function $E(\cdot, \cdot)$. We call this an $(m, n)$-cipher for short.

**Definition 1.** *Consider an $(m, n)$-cipher $E$. We define the* Shannon key-uncertainty *of $E$ to be the amount of Shannon entropy that remains on an $m$-bit key given $n$-bit blocks of plain- and ciphertexts, i.e. $H(K|P, C)$, where $K, P, C$ are random variables corresponding to the random choices of key, plaintext and ciphertext blocks for $E$, and where the key is uniformly chosen. The* min-entropy key-uncertainty *of $E$ is defined similarly, but w.r.t. min-entropy, as $H_\infty(K|P, C)$.*

From the definition, it may seem that the key uncertainties depend on the distribution of the plaintext. Fortunately, this is not the case. The key-uncertainty in the classical case is easy to compute, using the following slight generalization of the classical perfect security result by Shannon:

**Proposition 1.** *Let $E$ be a cipher with perfect security, and with plaintext, ciphertext and keyspace $\mathcal{P}, \mathcal{C}, \mathcal{K}$, where $|\mathcal{P}| = |\mathcal{C}|$. Furthermore, assume that keys* *are chosen uniformly. For any such cipher, it holds that the distribution of the key, given any pair of matching ciphertext and plaintext is uniform over a set of* $|\mathcal{K}|/|\mathcal{P}|$ *keys.*

*Proof.* By perfect security, we must have $|\mathcal{K}| \geq |\mathcal{P}|$. Now, let us represent the cipher in a table as follows: we index rows by keys and columns by plaintexts, and we fill each entry in the table with the ciphertext resulting from the key and plaintext on the relevant row and column. Then, since correct decryption must be possible and $|\mathcal{P}| = |\mathcal{C}|$, each ciphertext appears exactly once in each row. Fix any ciphertext $c$, and let $t_c$ be the number of times $c$ appears in, say, the first column. Since the probability distribution of the ciphertext must be the same no matter the plaintext, $c$ must appear $t_c$ times in every column. Since it also appears in every row, it follows that the length of a column satisfies $|\mathcal{K}| = t_c|\mathcal{P}|$. So $t_c = |\mathcal{K}|/|\mathcal{P}|$ is the same for every $c$. If we know a matching plaintext/ciphertext pair, we are given some $c$ and a column, and all we know is that the key corresponds to one of the $t_c$ possible rows. The proposition follows. □

**Corollary 1.** *For any classical* $(m,n)$*-cipher, both the Shannon- and min-entropy key-uncertainty is* $m-n$ *bits.*

This result shows that there is no room for improvement in classical schemes: the natural constraints on $(m,n)$-ciphers imply that the key-uncertainty is always the same, once we fix $m$ and $n$. As we shall see, this is not true for quantum ciphers. Although they cannot do better in terms of min-entropy key uncertainty, they can when it comes to Shannon key-uncertainty.

## 4 Quantum Ciphers and Min-Entropy Key-Uncertainty

In this section, we consider quantum ciphers which encrypt classical messages using classical keys and produce quantum ciphers.

We model both the encryption and decryption processes by unitary operations on the plaintext possibly together with an ancilla. This is the same model as used in [1], with the restriction that we only encrypt classical messages.

**Definition 2 ((m, n)-quantum cipher).** *A* general $(m,n)$-quantum cipher *is a tuple* $(\mathcal{P},\mathcal{E})$*, such that*

- $\mathcal{P} \subseteq \mathcal{H}$ *is a finite set of orthonormal pure-states (plaintexts) in the Hilbert space* $\mathcal{H}$*, and* $\|\mathcal{P}\| = N$ *and* $N = 2^n$.
- $\mathcal{E} = \{\mathsf{E}_k : \mathcal{H} \to \mathcal{H} |\ k = 1,\ldots,M\}$ *is a set of unitary operators (encryptions), and* $M = 2^m$*. Decryption using key* $k$ *is performed using* $\mathsf{E}_k^\dagger$.

*And the following properties hold:*

- *Key hiding:* $(\forall k,k' \in \{1,\ldots,M\})$,

$$\sum_{a\in\mathcal{P}} \frac{1}{N}\mathsf{E}_k|a\rangle|0\rangle\langle 0|\langle a|\mathsf{E}_k^\dagger = \sum_{a\in\mathcal{P}} \frac{1}{N}\mathsf{E}_{k'}|a\rangle|0\rangle\langle 0|\langle a|\mathsf{E}_{k'}^\dagger. \tag{2}$$

– *Data hiding:* $(\forall |a\rangle, |b\rangle \in \mathcal{P})$,

$$\sum_{k=1}^{M} \frac{1}{M} \mathsf{E}_k |a\rangle|0\rangle\langle 0|\langle a| \mathsf{E}_k^\dagger = \sum_{k=1}^{M} \frac{1}{M} \mathsf{E}_k |b\rangle|0\rangle\langle 0|\langle b| \mathsf{E}_k^\dagger. \quad (3)$$

The key and data hiding properties guarantee that an adversary cannot gain any information about the key and message respectively when an arbitrary ciphertext is seen. In [1], it was shown that data hiding implies that $m \geq n$.

The key hiding property states that an adversary with no information on the message encrypted expects to see the same ensemble no matter what key was used. We denote this ensemble

$$\rho = \sum_{a \in \mathcal{P}} \frac{1}{N} \mathsf{E}_k |a\rangle|0\rangle\langle 0|\langle a| \mathsf{E}_k^\dagger, \quad (4)$$

for any $k \in \{1, 2, \ldots, M\}$. As motivation for the key-hiding property, we mention that it is always satisfied if ciphertexts are as short as possible ($dim(\mathcal{H}) = 2^n$). On the other hand, if the key-hiding property does not hold then the cipher-state on its own reveals information about the secret-key. This is certainly an unnecessary weakness that one should avoid when designing ciphers.

The data hiding property states that the adversary expects to see the same ensemble no matter what message was encrypted. We denote this ensemble

$$\sigma = \sum_{k=1}^{M} \frac{1}{M} \mathsf{E}_k |a\rangle|0\rangle\langle 0|\langle a| \mathsf{E}_k^\dagger, \quad (5)$$

for any $a \in \mathcal{P}$. We first prove that $\rho = \sigma$.

**Lemma 1.** $\rho = \sigma$.

*Proof.* Define the state

$$\xi = \sum_{k=1}^{M} \sum_{a \in \mathcal{P}} \frac{1}{MN} \mathsf{E}_k |a\rangle|0\rangle\langle 0|\langle a| \mathsf{E}_k^\dagger. \quad (6)$$

Observe that

$$\xi = \sum_{k=1}^{M} \sum_{a \in \mathcal{P}} \frac{1}{MN} \mathsf{E}_k |a\rangle|0\rangle\langle 0|\langle a| \mathsf{E}_k^\dagger = \sum_{k=1}^{M} \frac{1}{M} \rho = \rho. \quad (7)$$

Similarly, when switching the sums in (6), we get $\xi = \sigma$. We conclude that $\rho = \sigma$. □

We are now ready to prove that for any $(m, n)$-quantum cipher there exists a measurement that returns the secret key with probability $2^{n-m}$ given any plaintext and its associated cipher-state. In other words and similarly to the classical case, the min-entropy key-uncertainty of any $(m, n)$-quantum cipher is at most $m - n$.

**Theorem 1 (Min-entropy key uncertainty).** *Let $(\mathcal{P}, \mathcal{E})$ be an $(m,n)$-quantum cipher, encoding the set $\mathcal{P}$. Then*

$$(\forall a \in \mathcal{P})(\exists \text{ POVM } \{M_i\}_{i=1}^{M})(\forall k \in \{1, \dots, M\})[\mathrm{tr}(M_k \mathcal{E}_k(|a\rangle\langle a|)) = 2^{n-m}]. \quad (8)$$

*Proof.* Let $|a\rangle \in \mathcal{P}$ be given. Consider the set $\mathcal{M} = \{M_k = \frac{N}{M} \mathsf{E}_k |a\rangle|0\rangle\langle 0|\langle a| \mathsf{E}_k^\dagger \mid k = 1, \dots, M\}$. Lemma 1 gives

$$\sum_{k=1}^{M} M_k = \sum_{k=1}^{M} \frac{N}{M} \mathsf{E}_k |a\rangle|0\rangle\langle 0|\langle a| \mathsf{E}_k^\dagger = N\sigma = N\rho. \quad (9)$$

Since the plaintexts are orthogonal quantum states, and since unitary operators preserve angles, we have that $N \sum_{a \in \mathcal{P}} \frac{1}{N} \mathsf{E}_k |a\rangle|0\rangle\langle 0|\langle a| \mathsf{E}_k^\dagger$ is the eigen decomposition of $N\rho$, and that 1 is the only eigenvalue. Therefore there exists a positive operator $P$ such that $N\rho + P = \mathbb{I}$, and thus

$$\sum_{k=1}^{M} M_k + P = N\rho + P = \mathbb{I}, \quad (10)$$

and $\mathcal{M} \cup \{P\}$ (and therefore also $\mathcal{M}$) is a valid POVM.

The probability of identifying the key with the measurement $\mathcal{M}$ is

$$\begin{aligned} \mathrm{tr}(M_k \mathsf{E}_k |a\rangle|0\rangle\langle 0|\langle a| \mathsf{E}_k^\dagger) &= \mathrm{tr}(\frac{N}{M} \mathsf{E}_k |a\rangle|0\rangle\langle 0|\langle a| \mathsf{E}_k^\dagger \mathsf{E}_k |a\rangle|0\rangle\langle 0|\langle a| \mathsf{E}_k^\dagger) \\ &= \frac{N}{M} \mathrm{tr}(\mathsf{E}_k |a\rangle|0\rangle\langle 0|\langle a| \mathsf{E}_k^\dagger) \\ &= 2^{n-m}, \end{aligned} \quad (11)$$

which proves the theorem. □

## 5 Some Example Quantum Ciphers

In this section, we suggest a general method for designing quantum ciphers that can do better in terms of Shannon key-uncertainty than any classical cipher with the same parameters. The properties of our ciphers are analyzed in the next section.

The first example is extremely simple:

**Definition 3.** *The $H_n$ cipher is an $(n+1, n)$-quantum cipher. Given message $b_1, b_2, \dots, b_n$ and key $c, k_1, \dots, k_n$, it outputs the following $n$ q-bit state as ciphertext:*

$$(H^{\otimes n})^c (X^{k_1} \otimes X^{k_2} \otimes \dots \otimes X^{k_n} |b_1 b_2 \dots b_n\rangle), \quad (12)$$

*where $X$ is the bit-flip operator and $H$ is the Hadamard transform. That is, we use the last $n$ bits of key as a one-time pad, and the first key bit determines whether or not we do a Hadamard transform on all $n$ resulting q-bits.*

Decryption is trivial by observing that the operator $(X^{k_1} \otimes X^{k_2} \otimes \cdots \otimes X^{k_n})(H^{\otimes n})^c$ is the inverse of the encryption operator. It is also easy to see that the data hiding property is satisfied: if $c, k_1, \ldots, k_n$ are uniformly random, then the encryption of any message produces the complete mixture (in fact this would be the case, already if only $k_1, \ldots, k_n$ were uniformly random).

This cipher can be described from a more general point of view: let $\mathcal{B} = \{B_0, \ldots, B_{2^t-1}\}$ be a set of $2^t$ orthonormal bases for the Hilbert space of dimension $2^n$. We require that the bases do not overlap, i.e., no unit vector occurs in more than one basis. For instance $\mathcal{B}$ could consist of the computational basis and the diagonal basis (i.e. $\{H^{\otimes n}|x\rangle | x \in \{0,1\}^n\}$). Let $U_i$ be the unitary operator that performs a basis shift from the computational basis to the basis $B_i$. Finally, let $[k_1, \ldots, k_t]$ be the number with binary representation $k_1, \ldots, k_t$. Then we can define an $(n+t, n)$-cipher $C_{\mathcal{B}}$ which on input a key $c_1, \ldots, c_t, k_1, \ldots, k_n$ and a plaintext $b_1, \ldots, b_n$ outputs

$$U_{[c_1,\ldots,c_t]}(X^{k_1} \otimes X^{k_2} \otimes \ldots \otimes X^{k_n}|b_1 b_2 \ldots b_n\rangle). \tag{13}$$

The $H_n$-cipher above is a special case with $U_0 = Id, U_1 = H^{\otimes n}$. Using arguments similar to the above, it is easy to see that

**Lemma 2.** *For any set of orthonormal non-overlapping bases $\mathcal{B}$, $C_{\mathcal{B}}$ is a quantum cipher satisfying the data hiding and unique decryption properties.*

The lemma holds even if $\mathcal{B}$ contains only the computational basis, in which case $C_{\mathcal{B}}$ is equivalent to the classical one-time pad. The point of having several bases is that if they are well chosen, this may create additional confusion for the adversary, so that he will not learn full information on the key, even knowing the plaintext. We shall see this below.

For now, we note that Wootters and Fields have shown that in a Hilbert space of dimension $2^n$, there exists $2^n + 1$ orthonormal bases that are *mutually unbiased*, i.e., the inner product between any pair of vectors from different bases has norm $2^{-n/2}$. Using, say, the first $2^n$ of these bases, we get immediately from the construction above a $(2n, n)$ cipher:

**Definition 4.** *The $W_n$-cipher is the cipher $C_{\mathcal{B}}$ obtained from the above construction when $\mathcal{B}$ is the set of $2^n$ mutually unbiased bases obtained from [8].*

### 5.1 Efficient Encoding/Decoding

In this section we look at how to implement $W_n$ efficiently. In [8], a construction for $2^n + 1$ mutually unbiased bases in the space of $n$ qubits is given. In the following, we denote by $v_{\boldsymbol{s}}^{(\boldsymbol{r})}$ with $\boldsymbol{s}, \boldsymbol{r} \in \{0,1\}^n$ the $\boldsymbol{s}$-th vector in the $\boldsymbol{r}$-th mutually unbiased basis. We write $v_{\boldsymbol{s}}^{(\boldsymbol{r})}$ in the computational basis as,

$$|v_{\boldsymbol{s}}^{(\boldsymbol{r})}\rangle = \sum_{\boldsymbol{l} \in \{0,1\}^n} \left(v_{\boldsymbol{s}}^{(\boldsymbol{r})}\right)_{\boldsymbol{l}} |\boldsymbol{l}\rangle, \tag{14}$$

where $\sum_{\boldsymbol{l}} |(v_{\boldsymbol{s}}^{(\boldsymbol{r})})_{\boldsymbol{l}}|^2 = 1$. Wootters and Field[8] have shown that $2^n$ mutually unbiased bases are obtained whenever

$$\left(v_{\boldsymbol{s}}^{(\boldsymbol{r})}\right)_{\boldsymbol{l}} = \frac{1}{\sqrt{2^n}} i^{\boldsymbol{l}^T(\boldsymbol{r}\cdot\boldsymbol{\alpha})\boldsymbol{l}}(-1)^{\boldsymbol{s}\cdot\boldsymbol{l}}, \tag{15}$$

for $\boldsymbol{\alpha}$ a vector of $n$ matrices each of dimensions $n \times n$ with elements in $\{0,1\}$. The arithmetic in the exponent of $i$ should be carried out over the integers (or equivalently mod 4). The elements of $\boldsymbol{\alpha}$ are defined by

$$f_i f_j = \sum_{m=1}^{n} \boldsymbol{\alpha}_{i,j}^{(m)} f_m, \tag{16}$$

where $\{f_i\}_{i=1}^n$ is a basis for $GF(2^n)$ when seen as a vector space. Therefore, $\boldsymbol{\alpha}$ can be computed on a classical computer (and on a quantum one) in $O(n^3)$.

Let $c = c_1, \ldots, c_n$ and $k = k_1, \ldots, k_n$ be the $2n$ bits of key with $c$ defining one out of $2^n$ mutually unbiased basis and $k$ defining the key for the one-time-pad encoding. The circuit for encrypting classical message $a$ starts by computing:

$$|\psi_a^k\rangle = H^{\otimes n} X^{\otimes k} |a\rangle = H^{\otimes n} |a \oplus k\rangle = 2^{-n/2} \sum_{\boldsymbol{l}} (-1)^{(a\oplus k)\cdot \boldsymbol{l}} |\boldsymbol{l}\rangle. \tag{17}$$

The state (17) differs from (14) only with respect to the phase factor $i^{\boldsymbol{l}^T(\boldsymbol{r}\cdot\boldsymbol{\alpha})\boldsymbol{l}}$ in front of each $|\boldsymbol{l}\rangle$ with $\boldsymbol{r} = c$. Transforming (17) into (14) (i.e. that is transforming $|\psi_a^k\rangle \mapsto |v_{k\oplus a}^{(c)}\rangle$) can easily be achieved using a few controlled operations as described in App. A. The complexity of the quantum encryption circuit is $O(n^3)$ out of which only $O(n^2)$ are quantum gates. The decryption circuit is the same as for the encryption except that it is run in reverse order. A similar encryption/decryption circuit can easily be implemented for any $C_{\mathcal{B}}$-cipher where $\mathcal{B}$ is a set of mutually unbiased bases.

## 6 Optimal Measurements w.r.t. Shannon Entropy

Our ultimate goal is to estimate the Shannon key-uncertainty of an $(m,n)$-quantum cipher, i.e., the amount of entropy that remains on the key after making an optimal measurement on a ciphertext where the plaintext is given. But actually, this scenario is quite general and not tied to the cryptographic application: what we want to answer is: given a (pure) state chosen uniformly from a given set of states, how much Shannon entropy must (at least) remain on the choice of state after having made a measurement that is optimal w.r.t. minimizing the entropy?

So what we should consider is the following experiment: choose a key $k \in \mathcal{K}$ uniformly. Encrypt a given plaintext $p$ under key $k$ to get state $|c_k\rangle$ (we assume here for simplicity that this is a pure state). Perform some measurement (that may depend on $p$) and get outcome $u$. Letting random variables $K, U$ correspond to the choices of key and outcome, we want to estimate

$$H(K|U) = \sum_u Pr(U = u)H(K|U = u). \tag{18}$$

Now, $H(K|U = u)$ is simply the Shannon entropy of the probability distribution $\{Pr(K = k|U = u)|k \in \mathcal{K}\}$. By the standard formula for conditional probabilities, we have

$$Pr(K = k|U = u) = \frac{Pr(U = u|K = k)Pr(K = k)}{Pr(U = u)}. \tag{19}$$

Note that neither $Pr(U = u)$, nor $Pr(K = k)$ depend on the particular value of $k$ (since keys are chosen uniformly).

The measurement in question can be modeled as a POVM, which without loss of generality can be assumed to contain only elements of the form $a_u|u\rangle\langle u|$, i.e., a constant times a projection determined by a unit vector $|u\rangle$. This is because the elements of any POVM can be split in a sum of scaled projections, leading to a measurement with more outcomes which cannot yield less information than the original one. It follows immediately that

$$Pr(U = u|K = k) = |a_u|^2|\langle u|c_k\rangle|^2. \tag{20}$$

Note that also the factor $|a_u|^2$ does not depend on $k$. Then by (19) and (20), we get

$$1 = \sum_{l \in \mathcal{K}} Pr(K = l|U = u) = \frac{|a_u|^2 Pr(K = l)}{Pr(U = u)} \sum_{l \in \mathcal{K}} |\langle u|c_l\rangle|^2. \tag{21}$$

Which means that we have

$$Pr(K = k|U = u) = \frac{|\langle u|c_k\rangle|^2}{\sum_{l \in \mathcal{K}} |\langle u|c_l\rangle|^2}. \tag{22}$$

In other words, $H(K|U = u)$ can be computed as follows: compute the set of values $\{|\langle u|c_k\rangle|^2|k \in \mathcal{K}\}$, multiply by a normalization factor so that the resulting probabilities sum to 1, and compute the entropy of the distribution obtained. We call the resulting entropy $H[|u\rangle, S_K]$, where $S_K$ is the set of states that may occur $\{|c_k\rangle|k \in \mathcal{K}\}$. This is to emphasize that $H[|u\rangle, S_K]$ can be computed only from $|u\rangle$ and $S_K$, we do not need any information about other elements in the measurement. From (18) and $H(K|U = u) = H[|u\rangle, S_K]$ follows immediately

**Lemma 3.** *With notation as above, we have:*

$$H(K|U) \geq min_{|u\rangle}\{H[|u\rangle, S_K]\}, \tag{23}$$

*where $|u\rangle$ runs over all unit vectors in the space we work in.*

This bound is not necessarily tight, but it will be, exactly if it is possible to construct a POVM consisting only of (scaled) projections $a_u|u\rangle\langle u|$, that minimize $H[|u\rangle, S_K]$. In general, it may not be easy to solve the minimization problem suggested by the lemma, particularly if $S_K$ is large and lives in many dimensions. But in some cases, the problem is tractable, as we shall see.

## 7 The Shannon Key-Uncertainty of Quantum Ciphers

In this section we study the cipher $C_{\mathcal{B}}$ that we constructed earlier based on a set of $2^t$ orthonormal bases $\mathcal{B}$. For this, we first need a detour: each basis in our set defines a projective measurement. Measuring a state $|u\rangle$ in basis $B_i \in \mathcal{B}$ produces a result, whose probability distribution depends on $|u\rangle$ and $B_i$. Let $H[|u\rangle, B_i]$ be the entropy of this distribution. We define the Minimal Entropy Sum (MES) of $\mathcal{B}$ as follows:

$$MES(\mathcal{B}) = min_{|u\rangle}\{\sum_{i=0}^{2^t-1} H[|u\rangle, B_i]\}, \tag{24}$$

where $|u\rangle$ runs over all unit vectors in our space. Lower bounds on the minimal entropy sum for particular choices of $\mathcal{B}$ have been studied in several papers, under the name of entropic uncertainty relations [5,7,3]. This is motivated by the fact that if the sum is large, then it is impossible to simultaneously have small entropy on the results of all involved measurements. One can think of this as a "modern" version of Heisenberg's uncertainty relations. It turns out that the key uncertainty of $C_{\mathcal{B}}$ is directly linked to $MES(\mathcal{B})$:

**Lemma 4.** *The Shannon key uncertainty of the cipher $C_{\mathcal{B}}$ (with $2^t$ bases) is at least $MES(\mathcal{B})/2^t + t$.*

*Proof.* We may use Lemma 3, where the set of states $S_K$ in our case consists of all basis states belonging to any of the bases in $\mathcal{B}$. To compute $H[|u\rangle, S_K]$, we need to consider the inner products of unit vector $|u\rangle$ with all vectors in $S_K$. In our case, this is simply the coordinates of $|u\rangle$ in each of the $2^t$ bases, so clearly the norm squares of the inner products sum to $2^t$. Let $z_{ij}$ be the $i$'th vector in the $j$'th basis from $\mathcal{B}$. We compute as follows:

$$\begin{aligned}
H[|u\rangle, S_K] &= \sum_{j=0}^{2^t-1}\sum_{i=0}^{2^n-1} \frac{1}{2^t}|\langle u|z_{ij}\rangle|^2 \log(2^t|\langle u|z_{ij}\rangle|^{-2}) \\
&= \sum_{j=0}^{2^t-1}\sum_{i=0}^{2^n-1} \frac{1}{2^t}|\langle u|z_{ij}\rangle|^2 \log(|\langle u|z_{ij}\rangle|^{-2}) + \sum_{j=0}^{2^t-1}\sum_{i=0}^{2^n-1} \frac{1}{2^t}|\langle u|z_{ij}\rangle|^2 \log(2^t) \\
&= \frac{1}{2^t}\sum_{j=0}^{2^t-1}\sum_{i=0}^{2^n-1} |\langle u|z_{ij}\rangle|^2 \log(|\langle u|z_{ij}\rangle|^{-2}) + t\frac{1}{2^t}\sum_{j=0}^{2^t-1}\sum_{i=0}^{2^n-1} |\langle u|z_{ij}\rangle|^2 \\
&= \frac{1}{2^t}\sum_{j=0}^{2^t-1} H[|u\rangle, B_j] + t \geq \frac{1}{2^t}MES(\mathcal{B}) + t.
\end{aligned} \tag{25}$$

The lemma follows. □

We warn the reader against confusion about the role of $|u\rangle$ and $\mathcal{B}$ at this point. When we estimate the key uncertainty of $C_{\mathcal{B}}$, we are analyzing a POVM, where $|u\rangle$ is one of the unit vectors defining the POVM. But when we do the proof of the above lemma and use the entities $H[|u\rangle, B_j]$, we think instead of $|u\rangle$ as the vector being measured according to basis $B_j$. There is no contradiction, however, since what matters in both cases is the inner products of $|u\rangle$ with the vectors in the bases in $\mathcal{B}$. We are now in a position to give results for our two concrete ciphers $H_n$ and $W_n$ defined earlier.

**Theorem 2.** *The $H_n$-cipher has Shannon key-uncertainty $n/2+1$ bits.*

*Proof.* The main result of [5] states that when $\mathcal{B}$ is a set of two mutually unbiased bases in a Hilbert space of dimension $2^n$ then $MES(\mathcal{B}) \geq n$. Using Lemma 4, it follows that $H_n$ has Shannon key-uncertainty at least $n/2+1$. Moreover, there exists measurements (i.e. for example the Von Neumann measurement in either the rectilinear or Hadamard basis) achieving $n/2+1$ bit of Shannon key-uncertainty. The result follows. □

For the case of $W_n$, we can use a result by Larsen[3]. He considers the probability distributions induced by measuring a state $|u\rangle$ in $N+1$ mutually unbiased bases, for a space of dimension $N$. Let the set of bases be $B_1, \ldots, B_{N+1}$, and let $\pi_{|u\rangle,i}$ be the collision probability for the $i$'th distribution, i.e., the sum of the squares of all probabilities in the distribution. Then Larsen's result (actually a special case of it) says that

$$\sum_{i=1}^{N+1} \pi_{|u\rangle,i} = 2 \tag{26}$$

In our case, $N = 2^n$. However, to apply this to our cipher $W_n$, we would like to look at a set of only $2^n$ bases and we want a bound on the sum of the entropies $H[|u\rangle, B_i]$ and not the sum of the collision probabilities. This can be solved following a line of arguments from Sánchez-Ruiz[7]. Using Jensen's inequality, we can compute as follows:

$$\begin{aligned}
\sum_{i=1}^{N} H[|u\rangle, B_i] &\geq -\sum_{i=1}^{N} \log \pi_{|u\rangle,i} \\
&\geq -N \log\left(\frac{1}{N}\sum_{i=1}^{N} \pi_{|u\rangle,i}\right) \\
&= -N \log\left(\frac{1}{N}\left(-\pi_{|u\rangle,N+1} + \sum_{i=1}^{N+1} \pi_{|u\rangle,i}\right)\right) \\
&= N \log\left(\frac{N}{2-\pi_{|u\rangle,N+1}}\right) \geq N \log\left(\frac{N}{2-1/N}\right).
\end{aligned} \tag{27}$$

Together with Lemma 4 we get:

**Theorem 3.** *The $W_n$-cipher has Shannon key-uncertainty at least $2n-1$ bits.*

The result stated here is only a lower bound on the key uncertainty for $W_n$. Given what we know, however, it is very plausible to conjecture that the minimal entropy sum of a set of $2^t$ mutually unbiased bases in a space of dimension $2^n$ is $(2^t - 1)n$ bits, or at least very close to this value. This is true for $t = 1$ and we know that for $t = n$, the sum is at least $2^n(n-1)$.

*Conjecture 1.* Let $\mathcal{B}$ be a set of $2^n$ mutually unbiased bases of a space of dimension $2^n$. Then $MES(\mathcal{B}) = (2^n - 1)n$ bits.

Under this conjecture, cipher $W_n$ has almost full Shannon key-uncertainty:

**Lemma 5.** *Under Conjecture 1, $W_n$ has Shannon key-uncertainty $2n - n2^{-n}$ bits.*

The $H_n$ and $W_n$-ciphers represent two extremes, using the minimal non-trivial number of bases, respectively as many of the known mutually unbiased bases as we can address with an integral number of key bits. It is not hard to define example ciphers that are "in between" and prove results on their key-uncertainty using the same techniques as for $W_n$. However, what can be derived from Larsen's result using the above line of argument becomes weaker as one considers a smaller number of bases.

## 8 Composing Ciphers

What happens to the key uncertainty if we use a quantum cipher twice to encrypt two plaintext blocks, using independently chosen keys? Intuition based on classical behavior suggests that the key uncertainty should now be twice that of a single application of the cipher, since the keys are independent. But in the quantum case, this requires proof: the adversary will be measuring a product state composed of the state of the two ciphertext blocks, and may make a coherent measurement involving both blocks simultaneously. This may in general produce results different from those obtained by measuring blocks individually. In our case, however, a simple information theoretic argument shows that the key uncertainty will still be what one might expect:

For any quantum $(m, n)$-cipher $C$, let $C^v$ be the cipher composed $v$ times with itself, i.e., we encrypt $v$ blocks of plaintext using $v$ independently and randomly chosen keys. We have

**Theorem 4.** *Let $C$ be a quantum $(m, n)$-cipher with Shannon key-uncertainty $h$. The Shannon key-uncertainty of $C^v$ is at least $v \cdot h$.*

*Proof.* Consider the following random experiment: choose $v$ keys for $C$ independently, represented by random variables $K_1, ..., K_v$. Encrypt an arbitrary $v$-block plaintext $p$ using $C^v$ and $K_1, ..., K_v$ as key. Perform some measurement $\mathcal{M}$ on the ciphertext, where $\mathcal{M}$ may be chosen as a function of $p$. Let the measurement result be represented by random variable $Y_{\mathcal{M}}$. We will assume that given $p$, we choose a measurement $\mathcal{M}_p$ so that the Shannon key-uncertainty on $K_1, \ldots, K_v$

provided by $Y_{\mathcal{M}_p}$(i.e. $H(K_1,\ldots,K_v|Y_{\mathcal{M}_p})$) is minimum. Since the $K_i$'s are independent, we have

$$\min_p (H(K_1,\ldots,K_v|Y_{\mathcal{M}_p})) = \min_p \left( \sum_{i=1}^{v} H(K_i|Y_{\mathcal{M}_p}) \right). \tag{28}$$

We can then estimate each term $H(K_i|Y_{\mathcal{M}_p})$ for fixed $p$ and $i$ by considering a different experiment: suppose we are given a *single* ciphertext block $c_i$ produced by $C$ from a plaintext block $p_i$ where $p_i$ is the $i$'th block of $p$. Then, choose $v-1$ keys $\{K_j|j=1,..,v,j\neq i\}$ at random, encrypt the $j$'th block of $p$ (except $p_i$) under $K_j$ to get ciphertext block $c_j$, and perform measurement $\mathcal{M}_p$ on the concatenation of $c_1,\ldots,c_v$ (including the block $c_i$ we received as input). Clearly, the state measured is the same mixture as in the first experiment, and so the result will have the same distribution as $Y_{\mathcal{M}_p}$. On other hand, this second experiment is equivalent to performing a measurement with ancilla on $c_i$. It follows that by assumption on $C$, $H(K_i|Y_{\mathcal{M}_p}) \geq h$. Combining this with (28) gives the desired result. □

## 9 Application to Stream-Ciphers

We can use the quantum ciphers we just described to build a (computationally secure) quantum stream-cipher using a short key $K$ of length independent from the message length. In fact, any $(m,n)$-cipher and classical pseudorandom generator can be used: we seed the generator with key $K$, and use its output as a keystream. To encrypt, we simply take the next $m$ bits from the keystream and use these as key in the cipher to encrypt the next $n$ bits of the plaintext.

Since an $(m,n)$-cipher has perfect security, this construction would have perfect security as well if the keystream was genuinely random. By a standard reduction, this implies that breaking it is at least as hard as distinguishing the output of the generator from a truly random string.

All this is true whether we use a classical or an $(m,n)$-quantum cipher. However, by our results on Shannon key-uncertainty, the adversary is in a potentially much harder situation in the quantum case. For intuition on this, we refer to the discussion in the introduction. As a more concrete illustration, we consider the following scenario:

1. We have a pseudorandom generator $G$, expanding a $k$-bit seed $K$ into an $N$-bit sequence $G(K)$. Furthermore, any subset containing at most $e$ bits of $G(K)$ is uniformly random. Finally, no polynomial time (in $k$) classical algorithm can with non-negligible advantage distinguish $G(K)$ from a truly random sequence when given any piece of data that is generated from $G(K)$ and contains at most $t$ bits of Shannon information on $G(K)$. Both $e$ and $t$ are assumed to be polynomial in $k$.
2. Coherent measurements simultaneously involving $\mu$ qubits or more are not possible to implement in practice. However, technology has advanced so that the $W_n$-cipher can be implemented for some $n << \mu$.

3. We will consider an adversary that first obtains some amount of known plaintext. Given the plaintext, he decides on a number of complete measurements that he executes on parts of the ciphertext (under the constraints of assumption 2). For simplicity we assume that each measurement involves an integral number of $n$-bit ciphertext blocks.[1] Finally he executes any polynomial time classical algorithm to analyze the results.

The first assumption can be justified using a result by Maurer and Massey [4] on locally random pseudorandom generators. Their result asserts that there exists pseudorandom generators satisfying the assumption that any $e$ bits are genuinely random, provided $e \leq k/\log_2 N$. Their generators may not behave well against attacks having access to more than $e$ bits of the sequence, but one can always xor the output from their generator with the output of a more conventional one using an independent key. This will preserve the local randomness. Note that the size of $k$ does not influence the size of the quantum computer required for the honest party to encrypt or decrypt. The third assumption essentially says that we do not expect that results of (incomplete) measurements obtained on one part of the ciphertext will help significantly in designing measurements on other parts. This is justified, as long as not too many measurements are performed: as long as results from previous measurements contain less than $t$ bits of information on the keystream, then by assumption 1, these results might (from the adversary's point of view) as well have been generated from measuring a random source, and so they do not help in designing the next measurement. This assumption can therefore be dropped in a more careful analysis since it esssentially follows from assumptions 1 and 2. For simplicity, we choose to make it explicit.

**Lemma 6.** *Assume we apply the $W_n$-cipher for stream encryption using a pseudorandom generator and with an adversary as defined by assumptions 1,2, and 1 above. Suppose we choose $e = 2\mu$ and $k \geq 2\mu \log_2 N$, then the adversary will need to obtain more than $tn$ bits of known plaintext in order to distinguish the case of a real encryption from the case where the keystream is random. Assuming Conjecture 1, this number becomes $t2^n$ bits of known plaintext.*

*Proof.* Assume the PRG satisfies assumption 1 which is possible since $k \geq e \log_2 N$. By assumption 2, any attack that measures several blocks of ciphertext in one coherent measurement can handle at most $\mu = e/2$ qubits at any one time. By construction, this ciphertext was created using less than $e$ bits of the keystream, which is random by assumption 1. Therefore, the measurement will give the same result as when attacking the composition $W_n^{v/n}$ since the measurement involves $v \leq \mu$ qubits (since different blocks of the keystream are independent if the stream is truly random) and by assumption 1. Hence, by Theorem 4 and 3 the adversary can learn at most $v/n$ bits of information from each measurement, and under Conjecture 1, at most $v2^{-n}$ bits. Hence, if the

[1] This assumption can be dropped so that we can still prove Lemma 6 using a more complicated argument and provided the local randomness of the generator is expanded from $e$ to $n^2 e$

adversary has $T$ bits of known plaintext, and hence measures $T$ ciphertext bits, he needs to have $T/n > t$ in order for the classical distinguisher to work, by assumption 1 (or $T2^{-n} > t$ under the conjecture). The lemma follows. □

This lemma essentially says that for a generator with the right properties, and for an adversary constrained as we have assumed, quantum communication allows using the generator securely to encrypt $tn$ ($t2^n$) bits, rather than the $t$ bits we would have in the classical case. A similar result can be shown for the $H_n$ cipher, saying that by using $H_n$, we gain essentially a factor 2 in plaintext size over the classical case.

Of course, these result do not allow to handle adversaries as general as we would like, in particular our constraints are different from just assuming the adversary is quantum polynomial time. Nevertheless, we believe that the scenario we have described can be reasonable with technology available in the foreseeable future. Moreover, it seems to us that quantum communication should help even for more general adversaries and generators. Quantifying this advantage is an open problem.

## 10 Conclusion and Open Problems

We have seen that, despite the fact that quantum communication cannot help to provide perfect security with shorter keys when only one-way communication is used, there are fundamental differences between classical and quantum ciphers with perfect security, in particular the Shannon key uncertainty can be much larger in the quantum case. However, the min-entropy key-uncertainty is the same in the two cases. It is an open question whether encryption performed by general quantum operations allows for quantum ciphers to have more min-entropy key-uncertainty than classical ones.

We have also seen an application of the results on Shannon key uncertainty to some example quantum ciphers that could be used to construct a quantum stream-cipher where, under a known plaintext attack, a resource-bounded adversary would be in a potentially much worse situation than with any classical stream-cipher with the same parameters.

For the ciphers we presented, the Shannon key-uncertainty is known exactly for the $H_n$-cipher but not for the $W_n$-cipher. It is an interesting open question to determine it. More generally, is Conjecture 1 true?

## References

1. A. Ambainis, M. Mosca, A. Tapp and R. de Wolf, *Private Quantum Channels*, Proceedings of the 41st Annual Symposium on Foundations of Computer Science, 2000, pp. 547–553.
2. D. DiVincenzo, M. Horodecki, D. Leung, J. Smolin and B. Terhal, *Locking Classical Correlation in Quantum States*, Phys. Rev. Letters,vol. 92, 067902, 2004.
3. U. Larsen, *Superspace Geometry: the exact uncertainty relationship between complementary aspects*, J.Phys. A: Math. Gen. 23 (1990), pp. 1041–1061.

4. U. Maurer and J. Massey, *Local Randomness in Pseudorandom Sequences*, Journal of Cryptology, vol. 4, 1991, pp. 135–149.
5. H. Maassen and J. B. M. Uffink, *Generalized Entropic Uncertainty Relations*, Phys. Rev. Letters, vol. 60, 1988, pp. 1103–1106.
6. M. Nielsen and I. Chuang, *Quantum Computation and Quantum Information*, Cambridge University Press, 2000.
7. J. Sánchez-Ruiz, *Improved bounds in the entropic uncertainty and certainty relations for complementary observables*, Physics Letters A 201, 1995, pp. 125–131.
8. W.K. Wootters and B.D. Fields, *Optimal state-determination by mutually unbiased measurements*, Annals of Physics 191, pp. 363–381.

# A Encryption Circuit for the $W_n$-Cipher

The circuit depicted in Fig. 2 implements the encryption of any plaintext $a = a_1, \ldots, a_n \in \{0,1\}^n$ according the secret key $(c,k) \in \{0,1\}^{2n}$. It uses three sub-circuits (1), (2), and (3) as defined in Fig. 1.

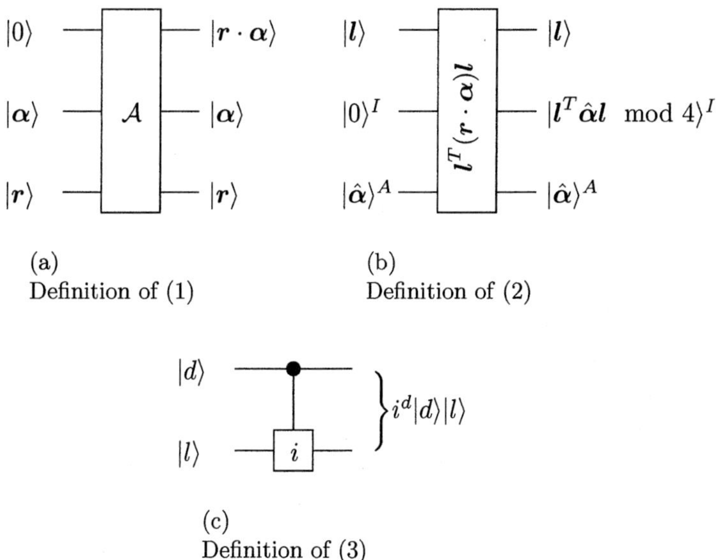


**Fig. 1.** Sub-circuits to the encryption circuit of Fig. 2.

$\mathcal{A}$, given $c$ and $\boldsymbol{\alpha}$, produces the matrix $c\cdot\boldsymbol{\alpha}$ in the register denoted $A$. Notice that circuit $\mathcal{A}$ is a classical circuit. It can be implemented with $O(n^3)$ classical gates. The sub-circuit (2) accepts as input $\hat{\boldsymbol{\alpha}} = c\cdot\boldsymbol{\alpha}$ together with $\boldsymbol{l}$, computes $d = \boldsymbol{l}^T\hat{\boldsymbol{\alpha}}\boldsymbol{l} \in [0,\ldots,3]$, and stores the result in a 2-qubit register $I$. In (3), an overall phase factor $i^d$ is computed in front of the computational basis element $|\boldsymbol{l}\rangle$.

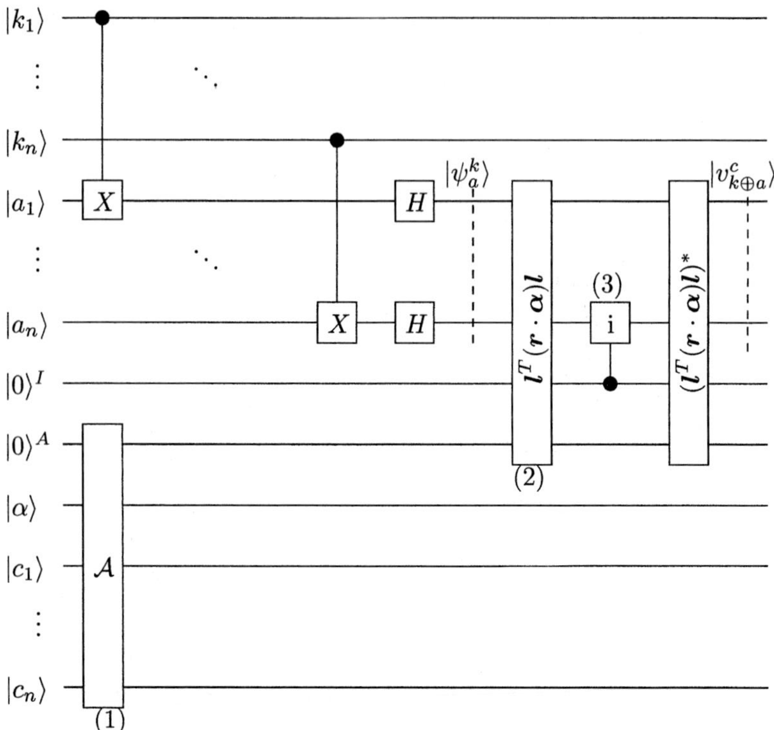


**Fig. 2.** Encoding circuit for cipher $W_n$.

The last gates allow to reset registers $I$ and $A$ making sure registers containing the encrypted data are separable from the other registers. It is straightforward to verify that registers initially in state $|a_1\rangle \otimes \ldots \otimes |a_n\rangle$ ends up in state $|v_{k\oplus a}^{(c)}\rangle$ as required. The overall complexity is $O(n^2)$ quantum gates since (3) requires only $O(n^2)$ CNOT's which is of the same complexity as super-gate (2). In conclusion, the total numbers of gates is $O(n^3)$ out of which $O(n^2)$ are quantum.

# The Exact Price for Unconditionally Secure Asymmetric Cryptography

Renato Renner[1] and Stefan Wolf[2]

[1] Department of Computer Science, ETH Zürich, Switzerland
renner@inf.ethz.ch

[2] Département d'Informatique et R.O., Université de Montréal, Canada
wolf@iro.umontreal.ca

**Abstract.** A completely insecure communication channel can only be transformed into an unconditionally secure channel if some information-theoretic primitive is given to start from. All previous approaches to realizing such authenticity and privacy from weak primitives were symmetric in the sense that security for both parties was achieved. We show that asymmetric information-theoretic security can, however, be obtained at a substantially lower price than two-way security—like in the computational-security setting, as the example of public-key cryptography demonstrates. In addition to this, we show that also an unconditionally secure bidirectional channel can be obtained under weaker conditions than previously known. One consequence of these results is that the assumption usually made in the context of quantum key distribution that the two parties share a short key initially is unnecessarily strong.

**Keywords.** Information-theoretic security, authentication, information reconciliation, privacy amplification, quantum key agreement, reductions of information-theoretic primitives.

## 1 Motivation and Main Results

### 1.1 Realizing Unconditional Security from Other Primitives

There are mainly two types of cryptographic security, namely *computational* and *information-theoretic* security. Systems of the first type can in principle be broken by adversaries with sufficient computing power; their security is based on the hardness of certain computational tasks—such as factoring large integers or computing discrete logarithms. However, no proofs can be given up to date for the security of such schemes. To make things even worse, the realization of a *quantum computer* would allow for breaking many presently-used systems efficiently. These facts serve as a strong motivation for the study of *information-theoretically secure cryptography.* Systems of this type are provably unbreakable even by computationally unlimited adversaries. Clearly, this is the most desirable type of security—but it has its price [21], the exact determination of which

C. Cachin and J. Camenisch (Eds.): EUROCRYPT 2004, LNCS 3027, pp. 109–125, 2004.

has been an open problem and subject to intensive study. Most generally speaking, this price is some *information-theoretic primitive* [15] $I$, such as shared keys that are fully [20], [8], [11] or partially random [9], [18] and secret [19], authenticated and/or noisy classical [22], [6] or quantum [1] communication channels, or correlated pieces of information [13].

In order to describe these previous—and our new—results on a conceptual level, we use the following "channel calculus" introduced in [16]. Here, $A \longrightarrow B$ denotes an insecure communication channel from Alice to Bob, $A \bullet\!\!\longrightarrow B$ is an authentic channel from Alice to Bob (i.e., the *exclusivity*—represented by "$\bullet$"—sits on the sender's side, whereas the actual *security* is on the receiver's side: according to *his* view and knowledge, the message comes indeed from the legitimate sender), $A \longrightarrow\!\!\bullet B$ is a confidential channel (in the sender's view, the channel's output is accessible exclusively by the legitimate receiver), and the channel $A \bullet\!\!\longrightarrow\!\!\bullet B$ offering both authenticity *and* confidentiality is called a secure channel. The bidirectional channel $A \bullet\!\!\!\longleftrightarrow B$, for instance, is authentic from Alice to Bob and confidential in the opposite direction.

A number of previous results showed when and how an unconditionally secure channel can be obtained from completely insecure and from authentic but public channels, respectively. In [22], [6], [17], [19], examples of information-theoretic primitives $I$ are given that allow for obtaining an unconditionally secure channel from completely insecure communication, i.e., for realizing the transformation

$$\left.\begin{array}{c} A \longleftrightarrow B \\ I \end{array}\right\} \; A \bullet\!\!\!\longleftrightarrow\!\!\!\bullet B \; ,$$

whereas it was shown in [13], for instance, that the required primitive $I'$ can generally be much weaker if the communication channel is assumed to be *authentic* initially:

$$\left.\begin{array}{c} A \bullet\!\!\longrightarrow B \\ A \longleftarrow\!\!\bullet B \\ I' \end{array}\right\} \; A \bullet\!\!\!\longleftrightarrow\!\!\!\bullet B \; .$$

Note that in the context of *computational* security, this latter channel transformation is possible *without* any additional primitive $I'$ (e.g., by using the Diffie-Hellman protocol [7]). In sharp contrast to this, unconditional authenticity *alone* is not sufficient for realizing unconditional confidentiality [13], [14], [17].

Clearly, a typical example of a primitive $I$ which works in both of the above cases is a shared secret key of sufficient length. The question is whether much weaker primitives can be sufficient as well. More specifically, some of the open questions are the following.

– All known examples of protocols achieving the above transformations do this via the generation of a shared private key. The generated (unconditional) security then sits on both sides of the channel (as shown in the diagrams above). Is it possible to realize unconditional security *on only one end of the channel* under weaker assumptions? In other words, what is the price for

realizing *asymmetric*[3] *unconditional security*? What is the minimal price for an unconditional "•"?
- Unconditional secret-key agreement protocols consist of different phases (such as interactive error correction, called information reconciliation, or privacy amplification). The assumption is normally made that the public communication channel over which these protocols are carried out is authentic. Which of these protocol steps *do* require authentic channels, and which do not?
- If authentic channels *are* indeed necessary (such as in quantum key distribution), what is the minimal price (i.e., the weakest possible primitive) for obtaining them?

In the present paper, we give answers to all three questions. First, we describe a class of information-theoretic primitives $I''$ that allow for obtaining unconditional asymmetric security, i.e., for realizing the transformation

$$\left.\begin{array}{c} A \longleftrightarrow B \\ I'' \end{array}\right\} A \bullet\!\!\longleftrightarrow B .$$

We show that such a primitive $I''$ is generally *not* sufficient for obtaining a *two-way* secure channel, and that our class of primitives is optimal in the sense that weaker primitives do normally not allow for obtaining *any information-theoretic security at all* in the setting of completely insecure communication. Because of these two optimality results, one can say that we give the exact price for unconditional security, i.e., for realizing an unconditional "•", which can be seen as an "atom of information-theoretic security".

Among the protocols used to achieve these results are methods for so-called *information reconciliation* (i.e., interactive error correction) not requiring authentic channels. Together with a similar result for privacy amplification [19], this implies that in many cases, information-theoretically secure key agreement protocols exist which do not require authentic channels *at all.*

If, on the other hand, such authenticity *is* required for a protocol, it can be achieved under much weaker assumptions than previously believed. For instance, it has been a standard assumption in quantum key distribution that the processing of the key requires a short secret key to start with—therefore, quantum key agreement is sometimes said to be key *expansion.* We show that neither a short secret key [11] nor a partially secret *common* string [19] are required for quantum key distribution, but that much weaker assumptions are in fact sufficient.

### 1.2 Main Results

We now give the main results of this paper. We first introduce the entropy measures required to formulate them. (For an introduction to information theory, see, for instance, [5].)

[3] Note that the term *asymmetric* is used here with respect to the high-level functionality and not—as usual—with respect to the keys held by the parties. In spite of this difference, it is fair to say that we try to realize the functionality of public-key authentication and encryption, but in the setting of unconditional security.

**Definition 1** Let $X$ and $Y$ be two random variables (with ranges $\mathcal{X}$ and $\mathcal{Y}$). The *min-entropy* $H_\infty(Y)$ of $Y$ is[4] $H_\infty(Y) := -\log(\max_{y\in\mathcal{Y}}(P_Y(y)))$. The *0-entropy* $H_0(Y)$ is defined as $H_0(Y) := \log|\{y \in \mathcal{Y} \mid P_Y(y) > 0\}|$, and let

$$H_0^{\max}(Y|X) := \max_{x\in\mathcal{X}}(H_0(Y|X = x)) .$$

It has been shown in [19] that a common key $S$ an arbitrarily large fraction of which (in terms of min-entropy) is known to the adversary is sufficient for obtaining two-way unconditional security.

**Previous Result.** [19] Let Alice and Bob be connected by a completely insecure bidirectional channel and share a binary string $S$, whereas an adversary Eve knows a random variable $U$ such that[5]

$$H_\infty(S|U = u) = \Omega(\text{len}(S))$$

holds (where $u \in \mathcal{U}$ is the particular value known to Eve). Then Alice and Bob can obtain an unconditionally authentic and confidential bidirectional channel between each other.[6]

In this paper, we show that unconditional security on only one side of the channel can be achieved at a substantially lower price; in particular, the parties are not required to share any common string initially. The following result and its tightness are shown in Sections 2 and 3.

**Asymmetric Result.** Assume that Alice and Bob—who are connected by a completely insecure bidirectional channel—, and an adversary Eve know random variables $X$, $Y$, and $U$, respectively, such that

$$H_\infty(Y|U = u) - H_0^{\max}(Y|X) = \Omega(\log|\mathcal{Y}|) \tag{1}$$

holds. Then Alice and Bob can obtain an unconditionally authentic channel from Alice to Bob and an unconditionally confidential channel from Bob to Alice.

The length of the message which can be sent in a confidential way is (asymptotically) equal to the expression on the left hand side of (1). It is shown in Section 3.2 that this is optimal.

We also give a symmetric result which improves on the previous result above: Even a completely secure bidirectional channel can be obtained by parties not sharing a common string to start with. This is shown in Section 4.

---

4 All logarithms in this paper are with respect to the base 2.

5 It is only for simplicity that we give asymptotic formulations of the previous and new results here. The involved hidden constants are small, and the protocols are useful already for relatively small values of $n$.

6 More precisely, the length of a message that can be sent in an almost-perfectly secret way, for instance, is $(1 - o(1))H_\infty(S|U = u)$.

**Symmetric Result.** Assume that Alice and Bob—who are connected by a completely insecure bidirectional channel—, and an adversary Eve know random variables $X$, $Y$, and $U$, respectively, such that

$$\max(H_\infty(X|U=u), H_\infty(Y|U=u)) - H_0^{\max}(Y|X) - H_0^{\max}(X|Y) = \Omega(\max(\log|\mathcal{X}|, \log|\mathcal{Y}|))$$

holds. Then Alice and Bob can obtain an unconditionally authentic and confidential bidirectional channel between each other.

In contrast to many previous secret-key agreement protocols, our protocols are not restricted to specific probability distributions but are universal in the sense that they work for *any* element in the class of distributions characterized by the given entropy conditions, where Alice and Bob do not have to know what the actual distribution is. Of course, such a condition is just *one* possible way of defining classes of distributions; it is a natural one, however, since a direct connection can be made to, for instance, an adversary's memory space. In Section 3 it is shown that our protocols are—in their universality—optimal.

Note that we have conditioned the involved random variables on an adversary's knowledge $U = u$. Alternatively, our results can be interpreted as to concern the model of unconditional security from keys generated by *correlated weak random sources* (other examples of such results are given in [9] and [18]).

If, on the other hand, $Y$ is a *a priori* uniformly distributed key and $U$ is Eve's information, then inequality (1) can be replaced by the—somewhat stronger—assumption

$$H_0(U) + H_0^{\max}(Y|X) = (1 - \Omega(1)) \log |\mathcal{Y}| \qquad (2)$$

because of Lemma 2 below. Condition (2) is directly comparable to related bounds and results in *quantum* cryptography since all the involved quantities now have natural "translations" on the quantum side: The entropy of the involved random variables can simply be replaced by the entropy of the corresponding quantum states. Bounds on these quantities naturally arise from bounds on the size of an adversary's (quantum) memory [12], for instance.

## 2 Asymmetric Unconditional Security from Minimal Primitives

### 2.1 Authentication Between Parties NOT Sharing a Common String

The first ingredient for our protocols is an unconditional authentication method that is secure even between parties not sharing the same string; furthermore, none of the two parties' initial strings has to be secret, the only condition being that a non-vanishing fraction of the receiver's string is unknown to the adversary (in terms of min-entropy). More precisely, we show that the interactive authentication method presented in [19]—there in the context of parties sharing a partially secret key—has the following property: Under the *sole condition* that an adversary Eve is not fully aware of the receiver Bob's string, the latter can receive authenticated messages from Alice: He will (almost) never accept if the message was *not* the one sent by Alice (whatever her string and Eve's knowledge about it is). In other words, the protocol is secure also if Alice and Bob do not share the same key. More precisely, whereas they will only accept if their initial strings *are* identical—a fact that they enforce by interactive error correction—, Eve is unable to mount a successful active attack *even if they are not.*

We review Protocol AUTH of [19]—using identical keys $s$ there; here, we will later replace $s$ by two not necessarily equal strings $y$ and $y'$. For parameters $k \cdot l = n$, let $s = s_0||s_1|| \cdots ||s_{k-1}$ be the decomposition of the $n$-bit string $s$ into $l$-bit substrings, interpreted as elements of $GF(2^l)$, and let, for $x \in GF(2^l)$,

$$p_s(x) := \sum_{i=0}^{k-1} s_i \cdot x^i \tag{3}$$

be the evaluation in $x$ of the polynomial represented by $s$. Then the protocol consists of repeating the following three rounds: First, Alice—the sender of the message to be authenticated—sends a random challenge $c' \in \{0,1\}^l$ to Bob which he replies to by sending back the pair $(p_s(c'), c)$, where $c \in \{0,1\}^l$ is another random challenge. Alice (after having checked the correctness of Bob's message—if it is incorrect, she rejects and aborts the protocol) then sends a message bit and, if this bit is 1, the value $p_s(c)$ to confirm. Under the assumption that an encoding of messages is used such that any insertion of a 0-bit (something Eve obviously can do) as well as any bit flip from 1 to 0 can be detected—because the resulting string is not a valid codeword—, this protocol was proven secure in [19]; more precisely, it was shown to be hard for Eve (having non-vanishing uncertainty in terms of min-entropy about $S = s$) to respond to a challenge, made by one party, without being able to use the other as an oracle, and that this fact implies the security of the protocol. Furthermore, it was shown that an encoding of $m$-bit messages with the mentioned properties exists with code word length $M = (1 + o(1))m$.

Below, we will show the security of this protocol—from the receiver's point of view (like in one-way authentication)—even when the parties do *not* share the same string and under the only assumption that Eve has some uncertainty about Bob's string ($y$). The main technical ingredient of this is Lemma 1, which implies, roughly speaking, that under the given conditions, Eve can, with overwhelming probability, either not respond to Alice's challenges ($c'$) or not to Bob's ($c$)—even when given Alice's string ($y'$). The intuitive reason for this is that it is *either useless or impossible* for Eve to (impersonate Bob and) talk to Alice—depending on whether her uncertainty about Alice's string is small or not. Without loss of generality, we state and prove Lemma 1 with respect to *deterministic* adversarial strategies (given by the functions $f$ and $g$).

**Lemma 1** *Let $Y'$ and $Y$ be two random variables with joint distribution $P_{Y'Y}$ and ranges $\mathcal{Y}' = \mathcal{Y} = \{0,1\}^l$. Let $f : \{0,1\}^l \to \{0,1\}^l$ and $g : \{0,1\}^l \times \{0,1\}^n \to \{0,1\}^l$ be two functions and, for uniformly—and independently of $Y'Y$—distributed random variables $C'$ and $C$ with ranges $\{0,1\}^l$, let*

$$\alpha := \text{Prob}_{Y'YC'C}[p_{Y'}(C') = f(C') \text{ and } p_Y(C) = g(C, Y')] ,$$

*where $p_{\cdot}(\cdot)$ is the polynomial function (3). Then there exists $y \in \mathcal{Y}$ with*

$$P_Y(y) \geq \left(\alpha - \frac{2k}{2^l}\right)^k .$$

*Proof.* Let for every particular value $y' \in \mathcal{Y}'$

$$r_{y'} := \text{Prob}_{C'}[p_{y'}(C') = f(C')] = \frac{|\{c' \in \{0,1\}^l \mid p_{y'}(c') = f(c')\}|}{2^l} ,$$

and for every pair $(y, y') \in \mathcal{Y} \times \mathcal{Y}'$

$$r_{y|y'} := \text{Prob}_C[p_y(C) = g(C, y')] = \frac{|\{c \in \{0,1\}^l \mid p_y(c) = g(c, y')\}|}{2^l} .$$

Then we have

$$\alpha = \text{E}_{Y'Y}[r_{Y'} \cdot r_{Y|Y'}] . \tag{4}$$

Let us now consider the random experiment defined by

$$P_{Y'YC'_1\cdots C'_kC_1\cdots C_k} := P_{Y'Y} \cdot P_{C'_1\cdots C'_kC_1\cdots C_k} ,$$

where $P_{C'_1\cdots C'_kC_1\cdots C_k}$ is the uniform distribution over the subset of $(\{0,1\}^l)^{2k}$ satisfying that all the $C'_i$ and all the $C_i$ are distinct among each other. We then have

$$\begin{aligned}
&\text{Prob}\,[p_{Y'}(C'_i) = f(C'_i) \text{ for } i = 1, \ldots, k \text{ and } p_Y(C_i) = g(C_i, Y') \text{ for } i = 1, \ldots, k] \\
&\quad \geq \text{E}_{Y'Y}\left[r_{Y'} \cdot \left(r_{Y'} - \frac{1}{2^l}\right) \cdots \left(r_{Y'} - \frac{k-1}{2^l}\right) \cdot r_{Y|Y'} \cdots \left(r_{Y|Y'} - \frac{k-1}{2^l}\right)\right] \\
&\quad \geq \text{E}_{Y'Y}\left[\left(r_{Y'} - \frac{k-1}{2^l}\right)^k \left(r_{Y|Y'} - \frac{k-1}{2^l}\right)^k\right] \\
&\quad \geq \text{E}_{Y'Y}\left[\left(r_{Y'} \cdot r_{Y|Y'} - (r_{Y'} + r_{Y|Y'}) \cdot \frac{k-1}{2^l}\right)^k\right] \geq \left(\alpha - \frac{2k}{2^l}\right)^k .
\end{aligned} \tag{5}$$

The last inequality in (5) follows from the fact that $x \mapsto x^k$ is a convex function and Jensen's inequality [5], from (4), and from $r_{Y'}, r_{Y|Y'} \leq 1$.

Let $\mathcal{A}_k$ be the event the probability of which is bounded in (5). Since, for $x \in \{0,1\}^n$, $k$ values $p_x(c)$ (for $k$ distinct $c$'s) uniquely determine $x$, we have, given that $\mathcal{A}_k$ occurs, that $Y'$ is uniquely determined and $Y$ is uniquely determined given $Y'$; together, we get that there exist $y' \in \mathcal{Y}'$ and $y \in \mathcal{Y}$ such that

$P_{Y'Y|\mathcal{A}_k}(y', y) = 1$, hence $P_Y(y) \geq \text{Prob}[\mathcal{A}_k]$ for this particular value $y$. □

We will now state and prove the described property of the interactive authentication protocol AUTH (Theorem 3). This and other proofs in the paper make use of Lemma 2 (see [4], [17], [19]), which implies that when $d$ (physical) bits of side information about a random variable are leaked, then its conditional min-entropy is not reduced by much more than $d$ except with small probability.

**Lemma 2** [4], [17], [19] *Let $S$, $V$, and $W$ be random variables such that $S$ and $V$ are independent, and let $b \geq 0$. Then*

$$\text{Prob}_{VW}[H_\infty(S|V = v, W = w) \geq H_\infty(S) - \log|\mathcal{W}| - b] \geq 1 - 2^{-b} .$$

**Theorem 3** *Assume that two parties Alice and Bob know $n$-bit strings $Y'$ and $Y$, respectively. Given that $H_\infty(Y|U = u) \geq tn$ holds for some constant $0 < t \leq 1$, where $U = u$ summarizes an adversary Eve's entire knowledge, Alice can use Protocol AUTH to send authenticated messages of length $m$ of order at most $O(tn/(\log n)^2)$ to Bob by communication over a completely insecure channel. The probability of a successful active attack, which is the event that Bob accepts although the message he received is not the correct one (or although Alice rejects) is of order $2^{-\Omega(tn/m)}$. If, on the other hand, Eve is passive and $Y' = Y$ holds, then Alice and Bob accept with certainty and Bob receives the correct message.*

*Proof.* Let $m$ be the length of the message Alice wants to send to Bob; the number of executions of the three-round step in Protocol AUTH is then $M = (1+o(1))m$.

Since each party responds to at most $M$ challenges during the protocol execution (and would then reject and abort), the min-entropy of $Y$, from Eve's viewpoint, at *any* point of the protocol, given all the communication $C = c$ she has seen, is, according to Lemma 2 (applied $2M$ times), at least

$$H_\infty(Y|U = u, C = c) \geq tn - 2Ml - 2Ma$$

with probability at least $1 - 2M2^{-a}$. We conclude that there exist choices of the protocol parameters of order $l = \Theta(n/M)$ and $k = \Theta(M)$—and a suitable choice of the auxiliary parameter $a$—such that we get the following:

$$\text{There exists } f(n) = \Omega(n) \text{ with } \text{Prob}[H_\infty(Y|U = u, C = c) \leq f(n)] \leq 2^{-f(n)} . \tag{6}$$

As described above, a successful attack of the protocol implies that Eve has been able to answer a challenge generated by one of the parties without help from the other party (i.e., without receiving any message from the other party between receiving the challenge and sending the corresponding response). The first possibility is that a challenge of *Alice* is responded without Bob's help; here, it is necessary for Eve to also answer at least one of Bob's challenges successfully (an attack is successful only if Bob is fooled)—possibly with Alice's "help", however. Let therefore $\mathcal{A}$ be the event that Eve correctly responds to one of the at most $M$ challenges by Alice, and to one of Bob's at most $M$ challenges *given Alice's string* $Y'$. According to Lemma 1, and because of the union bound, we have

$$\text{Prob}\left[\mathcal{A}\right] \leq M^2 \cdot \left(2^{-H_\infty(Y|U=u,C=c)/k} + 2k/2^l\right) .$$

Hence, because of (6), the success probability of this attack is at most

$$M^2 \cdot \left(2^{-\Omega(n/M)} + \frac{\Theta(M)}{2^l}\right) + 2^{-\Omega(n)} = 2^{-\Omega(n/M)}$$

(note that $M/2^l = 2^{-\Omega(n/M)}$ and $M^2 2^{-\Omega(n/M)} = 2^{-\Omega(n/M)}$ hold since $M = O(tn/(\log n)^2)$). The second possibility of an attack is that a challenge of *Bob* is responded without Alice's help. The probability of this is, because of (6) and by a similar but simpler reasoning as the one used above, of order $2^{-\Omega(n/M)}$. The application of the union bound concludes the proof. □

## 2.2 Information Reconciliation over Unauthenticated Channels

We will now use the described authentication protocol, and its new property established in the previous section, for the construction of a protocol for *information reconciliation* by completely insecure communication. Information reconciliation is interactive error correction: Two parties, knowing strings $X$ and $Y$, respectively, should share a common string at the end (e.g., one of the initial strings). The idea is to use Protocol AUTH in such a way that the parties can detect active attacks at any point in the protocol.

According to Lemma 4, the error correction itself can be done by exchanging redundancy, where the latter is generated by applying universal hashing[7] to the input strings; this is efficient with respect to the required communication, but computationally inefficient for one of the parties (Alice in our case). In the special but typical scenario where $X$ and $Y$ are bitstrings which differ in a certain limited number of positions, more efficient methods, based on *concatenated codes* [10], can be used instead in Protocol IR below.

**Lemma 4** *Let $X$ and $Y$ be distributed according to $P_{XY}$ such that $H_0^{\max}(Y|X) \leq r$ holds. Let, for some integer $s \geq 0$, $\mathcal{H}$ be a universal class of functions $h : \mathcal{Y} \rightarrow \{0,1\}^{r+s}$, and let $H$ be the random variable corresponding to the random choice, independently of $X$ and $Y$, of a function in $\mathcal{H}$ according to the uniform distribution. Then*

$$\text{Prob}\left[\text{there exists } \overline{Y} \neq Y \text{ with } H(\overline{Y}) = H(Y) \text{ and } P_{Y|X}(Y,X) > 0\right] \leq 2^{-s} .$$

[7] A class $\mathcal{H}$ of functions $h : \mathcal{A} \rightarrow \mathcal{B}$ is *2-universal*—or *universal* for short—if, for all $a, a' \in \mathcal{A}$, $a \neq a'$, we have $|\{h \mid h(a) = h(a')\}| = |\mathcal{H}|/|\mathcal{B}|$.

*Proof.* For $x \in \mathcal{X}$, let $\mathcal{Y}_x := \{y \in \mathcal{Y} \,|\, P_{Y|X}(y,x) > 0\}$. We have $|\mathcal{Y}_x| \leq 2^r$. Since for any $y, \overline{y} \in \mathcal{Y}_x$, $\overline{y} \neq y$, and random $H \in \mathcal{H}$ the probability that $H(y) = H(\overline{y})$ holds is at most $1/2^{r+s}$, we have

$$\begin{aligned} &\text{Prob}_{YH}[\text{there exists } \overline{Y} \in \mathcal{Y}_x,\ \overline{Y} \neq Y, \text{ such that } H(\overline{Y}) = H(Y)] \\ &\quad \leq |\mathcal{Y}_x| \cdot \text{Prob}\,[H(\overline{y}) = H(y) \text{ for some } \overline{y} \neq y] \ \leq\ 2^r \cdot \frac{1}{2^{r+s}} \ =\ \frac{1}{2^s} \end{aligned}$$

by the union bound. The statement then follows when the expectation over $X$ is taken. □

In Protocol IR, $D$ and $T$ are parameters to be determined below, and $\mathcal{H}$ is a universal class of functions from $\{0,1\}^n$ to $\{0,1\}^D$. Furthermore, $\text{AUTH}_{Y',Y}(M)$ means that the message $M$ is sent using Protocol AUTH, where the "keys" used by the sender (Alice) and the receiver (Bob) are $Y'$ and $Y$, respectively.

**Protocol IR (Information Reconciliation)**

| Alice | | Bob |
|---|---|---|
| $X \in \{0,1\}^n$ | | $Y \in \{0,1\}^n$ |
| | | $H \in_r \mathcal{H}$, $H : \{0,1\}^n \to \{0,1\}^D$ |
| | $\xleftarrow{\quad H,\ H(Y) \quad}$ | |
| $Y' \in \mathcal{Y}_X$ with $H(Y') = H(Y)$<br>$R \in_r \{0,1\}^T$<br>compute $p_{Y'}(R)$ | | |
| | $\xrightarrow{\ \text{AUTH}_{Y',Y}((R, p_{Y'}(R)))\ }$ | |
| accept, $Y'$ | | if $p_{Y'}(R) = p_Y(R)$:<br>accept, $Y$<br>otherwise: reject. |

The content of the second message serves as a verification of whether the string $Y'$ computed by Alice is correct. Clearly, it has to be authenticated because of possible substitution attacks. It is an interesting point that because of this authentication, Alice can choose the "challenge" string $R$ herself: If the authentication is successful, Bob knows that $R$ is indeed the challenge generated by Alice, and hence random.

Note that although applied in a—symmetric—context where two parties want to generate a common secret key, Protocol IR is secure (for Bob) in the same—asymmetric—sense as the authentication protocol: Either everything goes well or Bob will know it did not (with high probability).

**Theorem 5** *Assume that two parties Alice and Bob know the value of a random variable $X$ and an $n$-bit string $Y$, respectively, and that*

$$H_\infty(Y|U=u) - H_0^{\max}(Y|X) \geq tn \tag{7}$$

*holds for some constant $0 < t \leq 1$, where $U = u$ summarizes an adversary's entire knowledge. Then Protocol IR (with suitable parameter choices)—carried out over a completely insecure channel—achieves the following. If Eve is passive, then Alice and Bob both accept and the string $Y'$ computed by Alice is equal to $Y$ except with probability $2^{-\Omega(n/\log n)}$. In general, it is true except with probability $2^{-\Omega(\sqrt{n}/\log n)}$ that either Bob rejects or both accept and $Y' = Y$ holds. Furthermore, the remaining conditional min-entropy of $Y$ given Eve's initial information and the protocol communication is of order $(1-o(1))tn$ with probability $1 - 2^{-\Omega(n/\log n)}$.*

*Proof.* Let us assume that Eve is passive. Let the parameter $D$ be of order $D = H_0^{\max}(Y|X) + \Theta(n/\log n)$. Then we have, according to Lemma 4, that Alice's guess $Y'$—from $X$ and $H(Y)$—is uniquely determined and hence correct except with probability $2^{-\Omega(n/\log n)}$.

Let us now consider the general case where Eve is possibly an active adversary. We first analyze the properties of the authentication of the confirmation message sent from Alice to Bob. Let the parameter $T$ be of order $T = \Theta(\sqrt{n})$. We will argue that with high probability, either Bob rejects or Alice and Bob both accept and the values $(R, p_{Y'}(R))$ as received by Bob are the ones sent by Alice and, finally, that this implies that $Y' = Y$ holds, i.e., that Alice and Bob share the same string, the min-entropy of which, from Eve's viewpoint, is still $(1 - o(1))tn$.

First, we get, using Lemma 2 with the parameter choice $b = \Theta(n/\log n)$, that there exist functions $f(n) = (1 - o(1))tn$ and $g(n) = \Omega(n/\log n)$ such that

$$\text{Prob}\left[H_\infty(Y|U=u, H=h, H(Y)=h(y)) \geq f(n)\right] \geq 1 - 2^{-g(n)} .$$

Because of this, Theorem 3 implies that the authentication works—even if, for instance, Eve had modified the error-correction information sent in the first message and knows $Y'$ perfectly. The length of the message to be authenticated with Protocol AUTH is of order $\Theta(\sqrt{n})$, and we choose the protocol parameter $l$ to be $l = \Theta(\sqrt{n}/\log n)$ to make sure that the remaining min-entropy, given all the communication, is still an arbitrarily large fraction of $tn$. The success probability of the protocol is then, according to the proof of Theorem 3, $1 - 2^{-\Omega(\sqrt{n}/\log n)}$.

Let us hence assume now that Bob actually received the correct message $(R, p_{Y'}(R))$ as sent by Alice. Since $R$ are the truly random bits (in particular, independent of $Y'$) chosen by Alice, and since $p_y(r) = p_{y'}(r)$ can hold for at most $\deg(p_y) = n/T - 1 = \Theta(\sqrt{n})$ different values of $r$ for any $y' \neq y$, we have that with probability $1 - 2^{-\Omega(\sqrt{n})}$ either Alice has the correct string, or Bob realizes that she does not.

Finally, the remaining min-entropy is still roughly the same with high probability since the total number of bits sent is of order $\Theta(n/\log n) = o(n)$. From

Lemma 2, we get that there exist $f(n) = (1 - o(1))tn$ and $g(n) = \Omega(n/\log n)$ such that we have $\text{Prob}\,[H_\infty(Y|U = u, C = c) \geq f(n)] \geq 1 - 2^{-g(n)}$, where $C = c$ is the entire protocol communication. This concludes the proof. □

**Remark.** In Theorem 5—as well as in Theorems 6, 7, and 8 and Corollary 9 below—the assumed entropy bounds can be conditioned on an event $\mathcal{A}$ if at the same time the claimed protocol failure probabilities are increased by $1 - \text{Prob}\,[\mathcal{A}]$. An example for which this can lead to substantially stronger statements is when the random variables $X = (X_1, \ldots, X_n)$, $Y = (Y_1, \ldots, Y_n)$, and $U = (U_1, \ldots, U_n)$ arise from $n$ independent repetitions of a certain random experiment $P_{X_i Y_i U_i}$. In this case, $\mathcal{A}$ can be the event that the actual outcome sequences are *typical* (see [5]). This is a good choice because $\mathcal{A}$ occurs except with exponentially (in $n$) small probability, and because

$$H_\infty(Y|U = u, \mathcal{A}) \approx H(Y_i|U_i) \cdot n \gg H_\infty(Y|U = u)$$

and

$$H_0^{\max}(Y|X, \mathcal{A}) \approx H(Y_i|X_i) \cdot n \ll H_0^{\max}(Y|X)$$

can hold. (See also Example 1 below.)

### 2.3 The Price for One-Sided Authenticity and Confidentiality

In [19], Protocol PA, allowing for *privacy amplification* over a completely insecure channel, was presented. Privacy amplification [3], [2] means to generate, from an only weakly secret shared string, a shorter but highly secret key. Protocol PA—which uses Protocol AUTH as well as *extractors* as its main ingredients—has been shown to extract virtually all the min-entropy of an arbitrarily weakly secret string.

**Theorem 6 [19]** *Assume that Alice and Bob both know the same $n$-bit string $Y$ satisfying $H_\infty(Y|U = u) \geq tn$ for some constant $0 < t \leq 1$, where $U = u$ summarizes Eve's entire information about $Y$. Then Protocol PA, using two-way communication over a completely insecure channel, has the following properties. Both Alice and Bob either reject or accept and compute strings $S_A$ and $S_B$, respectively, such that if Eve is passive, then Alice and Bob accept and there exists a $(1 - o(1))tn$-bit string $S$ that is uniformly distributed from Eve's viewpoint and such that $S_A = S_B = S$ holds except with probability $2^{-\Omega(n/(\log n)^2)}$. In general (i.e., if Eve is possibly active), either both parties reject or there exists a string $S$ with the above properties, except with probability $2^{-\Omega(\sqrt{n}/\log n)}$.*

Putting everything together, we can now conclude that the combination of Protocols IR and PA achieves what we had stated initially, namely asymmetric unconditional security for Bob from a very weak initial primitive. Given that Bob accepts at the end of the protocol, he shares a secret key with Alice. He can then send unconditionally confidential messages *to* her and receive authenticated messages *from* her.

**Theorem 7** *Assume that two parties Alice and Bob know a random variable $X$ and an $n$-bit string $Y$, respectively, and that $H_\infty(Y|U=u) - H_0^{\max}(Y|X) \geq tn$ holds for some constant $0 < t \leq 1$, where $U = u$ summarizes an adversary's entire knowledge. Then the combination of Protocols IR and PA, carried out over a completely insecure channel, achieves the following. Alice and Bob both either reject or accept and compute strings $S_A$ and $S_B$, respectively, such that if Eve is passive, then Alice and Bob accept and there exists a $(1-o(1))tn$-bit string $S$ that is uniformly distributed from Eve's viewpoint and such that $S_A = S_B = S$ holds except with probability $2^{-\Omega(n/(\log n)^2)}$. In general, either Bob rejects or Alice and Bob accept, and the above holds, except with probability $2^{-\Omega(\sqrt{n}/\log n)}$.*

*Proof.* Follows from Theorems 5 and 6. □

## 3 Impossibility Results and Lower Bounds

### 3.1 Two-Sided Security Requires Stronger Conditions

All protocols presented in Section 2 are asymmetric in the sense that the generated security is on Bob's side only. (Alice, for instance, could be talking to Eve instead of Bob without realizing this.) Example 1 shows that security for Alice simply *cannot* be achieved under assumptions as weak as that. This implies that the price for unconditional security on one side is strictly lower than for such security on both sides. The same is already well-known in the computational-security model, as the example of public-key cryptography demonstrates.

**Example 1.** Let $X = (X_1, \ldots, X_n)$ be a uniformly distributed $n$-bit string, and let $Y = (Y_1, \ldots, Y_n)$ and $U = (U_1, \ldots, U_n)$ be $n$-bit strings jointly distributed with $X$ according to[8]

$$P_{YU|X}((y_1, \ldots, y_n), (u_1, \ldots, u_n), (x_1, \ldots, x_n)) = \prod_{i=1}^{n} |\delta_{y_i x_i} - \varepsilon| \cdot |\delta_{u_i x_i} - \varepsilon| \quad (8)$$

for some $0 < \varepsilon < 1/2$. Equation (8) means that the $i$-th bits of $Y$ and $U$ are generated by sending $X_i$ over two independent binary symmetric channels with error probability $\varepsilon$.

Let now $\mathcal{A}$ be the event—which occurs except with exponentially (in $n$) small probability—that all the involved strings are typical sequences. Then we have, roughly,[9] $H_\infty(Y|U=u,\mathcal{A}) \approx h(2\varepsilon - 2\varepsilon^2)n$ and $H_0^{\max}(Y|X,\mathcal{A}) \approx h(\varepsilon)n$. Because of $2\varepsilon - 2\varepsilon^2 > \varepsilon$, the condition of Theorem 7 is satisfied. On the other hand, Bob has no advantage over Eve from Alice's viewpoint since Eve is able to *simulate* [17] Bob towards Alice: She can generate a random variable from $U$—in fact, she can use $U$ itself—which has the same joint distribution with $X$ as $Y$ does—$P_{XU} = P_{XY}$. Hence Alice will never be able to tell Bob and Eve apart.

---

[8] Here, $\delta_{ij}$ is the Kronecker symbol, i.e., $\delta_{ij} = 1$ if $i = j$ and $\delta_{ij} = 0$ otherwise.

[9] We denote by $h(p)$ the binary entropy function $h(p) = -(p\log p + (1-p)\log(1-p))$.

### 3.2 Optimality of the Achieved Secret-Key Length

The protocols we have presented in Section 2 are universal and work for all possible specific probability distributions under the only assumption that the entropy condition (7) is satisfied. In other words, our protocols work for large *classes* of probability distributions, where Alice and Bob do not have to know the nature of Eve's information, i.e., the particular distribution, but only that the corresponding entropy bound is satisfied. In this sense, our protocols are optimal: In many situations, no protocol can extract a longer secret key—*even when the communication channel is assumed authentic.* (It should be noted, however, that there are *specific* settings in which key agreement by *authenticated* public communication is possible even though the expression in (7) is negative [13].)

This can be illustrated with the setting where Bob's random variable $Y$ is uniformly distributed (also from Eve's viewpoint) and Alice's $X$ provides her uniformly with *deterministic* information about $Y$: For every value $x$ it can take, $P_{Y|X=x}$ is the uniform distribution over the set $|\mathcal{Y}_x|$ of size $|\mathcal{Y}|/|\mathcal{X}|$ (and these sets are disjoint for different values of $X$). After the execution of a key-agreement protocol, Alice has to know (with overwhelming probability) the key $S$ generated by Bob. Eve, on the other hand, should be (almost) completely ignorant about it. Clearly, this can be satisfied only if there are at least as many possible values Alice can initially have as possible keys. Therefore, we always have, roughly, $|S| \leq |\mathcal{X}| = |\mathcal{Y}|/|\mathcal{Y}_x|$, and hence

$$\text{len}(S) \approx \log|S| \leq \log|\mathcal{Y}| - \log|\mathcal{Y}_x| = H_\infty(Y) - H_0^{\max}(Y|X) .$$

## 4 Two-Way Security Under New and Weaker Assumptions

In this section we determine the price for achieving unconditional security for *both* Alice and Bob. The conditions we will find are weaker than the ones known previously (such as, for instance, a highly insecure but *common* string [19]).

We first give Protocol IR+, an extension of Protocol IR offering security also for Alice. After the first two protocol steps, which are the same as in Protocol IR, Alice sends error correction information $H'(X)$ about her initial string $X$ (here, $H'$ is from a universal class $\mathcal{H}'$ with suitable parameters) to Bob, who then uses his "estimate" $X'$ of $X$ as the authentication key for sending a challenge-response pair for $Y$. If Alice receives this correctly, and if it corresponds to the value $p_{Y'}(R')$ she can compute herself, she can be convinced that $Y' = Y$ holds. The crucial observation for proving Theorem 8 is that the given entropy condition on $Y$ *also* implies that Eve, having seen all the error-correction information and other messages, still has $\Omega(n)$ of min-entropy about $X$—because the same holds for $Y$. The reason is that given all the protocol communication, $Y$ can—with overwhelming probability—be computed from $X$, and vice versa.

**Protocol IR+ (Two-Secure Information Reconciliation)**

| Alice | | Bob |
|---|---|---|
| $X \in \{0,1\}^n$ | | $Y \in \{0,1\}^n$ |
| | | $H \in_r \mathcal{H}$,<br>$H : \{0,1\}^n \to \{0,1\}^D$ |
| | $\xleftarrow{H,\ H(Y)}$ | |
| $Y' \in \mathcal{Y}_X$<br>with $H(Y') = H(Y)$<br>$R \in_r \{0,1\}^T$<br>compute $p_{Y'}(R)$ | | |
| | $\xrightarrow{\mathrm{AUTH}_{Y',Y}((R, p_{Y'}(R)))}$ | |
| $H' \in_r \mathcal{H}'$,<br>$H' : \{0,1\}^n \to \{0,1\}^{D'}$ | | if $p_{Y'}(R) \neq p_Y(R)$: reject |
| | $\xrightarrow{H',\ H'(X)}$ | |
| | | $X' \in \mathcal{X}_Y$<br>with $H'(X') = H'(X)$<br>$R' \in_r \{0,1\}^T$ |
| | $\xleftarrow{\mathrm{AUTH}_{X',X}((R', p_Y(R')))}$ | |
| if $p_Y(R') \neq p_{Y'}(R')$: reject<br>otherwise: accept, $Y'$ | | accept, $Y$. |

**Theorem 8** *Assume that two parties Alice and Bob know n-bit strings X and Y, respectively, and that*

$$H_\infty(Y|U=u) - H_0^{\max}(Y|X) - H_0^{\max}(X|Y) \geq tn$$

*holds for some constant $0 < t \leq 1$, where $U = u$ summarizes an adversary's entire knowledge. Then Protocol IR+ (for suitable parameter choices)—carried out over a completely insecure channel—achieves the following. If Eve is passive, then Alice and Bob both accept and the string $Y'$ computed by Alice is equal to $Y$ except with probability $2^{-\Omega(n/\log n)}$. In general, it is true except with probability $2^{-\Omega(\sqrt{n}/\log n)}$ that either both parties reject or $Y' = Y$ holds. Furthermore, the remaining min-entropy of $Y$ given Eve's initial information and the protocol communication is of order $(1 - o(1))tn$ with probability $1 - 2^{-\Omega(n/\log n)}$.*

*Proof.* Follows from Theorem 5, Lemma 4, and Theorem 3. □

**Corollary 9** *Assume that two parties Alice and Bob know n-bit strings X and Y, respectively, and that*

$$H_\infty(Y|U=u) - H_0^{\max}(Y|X) - H_0^{\max}(X|Y) \geq tn$$

*holds for some constant* $0 < t \leq 1$*, where* $U = u$ *summarizes an adversary's entire knowledge. Then the combination of Protocols IR+ and PA, carried out over a completely insecure channel, achieves the following. Alice and Bob both either reject or accept and compute strings* $S_A$ *and* $S_B$*, respectively, such that if Eve is passive, then Alice and Bob both accept and there exists a* $(1 - o(1))tn$*-bit string* $S$ *that is uniformly distributed from Eve's viewpoint and such that* $S_A = S_B = S$ *holds except with probability* $2^{-\Omega(n/(\log n)^2)}$*. In general, either both parties reject or there exists a string* $S$ *with the above properties, except with probability* $2^{-\Omega(\sqrt{n}/\log n)}$*.*

*Proof.* Follows from Theorems 8 and 6. □

## 5 Concluding Remarks

In this paper we have determined, so to speak, a minimal price for unconditional security. For two parties connected by a completely insecure bidirectional communication channel, we have described the weakest possible information-theoretic primitive necessary for obtaining security on *one* end of the channel—i.e., guaranteed *exclusivity* of read and write access to the channel on its other end. Roughly speaking, we found that whenever Eve's uncertainty about the information of the party at one end of the channel exceeds the uncertainty about the same information as seen by the party at the channel's other end, then the entire entropy difference can be transformed into a key which is secret for the former party. This asymmetric notion of security for one party means that *either* the two parties share a secret key, *or* this—designated—party knows that they do not.

One of the consequences of our protocols is that the required conditions for the possibility of secret-key agreement in general, and quantum key distribution in particular, can be relaxed substantially: Quantum key agreement has sometimes been perceived to be rather key *extension* than actual key *generation* in view of the usually-made assumption that the two parties share a short unconditionally secret key already initially, from which they can then produce a longer key (where the initial key is required for authenticating the public communication exchanged for processing the raw key). Our results show that this condition is unnecessary and can be replaced by a much weaker assumption.

## References

1. C. H. Bennett and G. Brassard, Quantum cryptography: public key distribution and coin tossing, *Proceedings of the IEEE International Conference on Computers, Systems and Signal Processing*, pp. 175–179, 1984.
2. C. H. Bennett, G. Brassard, C. Crépeau, and U. M. Maurer, Generalized privacy amplification, *IEEE Trans. on Information Theory*, Vol. 41, No. 6, pp. 1915–1923, 1995.
3. C. H. Bennett, G. Brassard, and J.-M. Robert, Privacy amplification by public discussion, *SIAM Journal on Computing*, Vol. 17, pp. 210–229, 1988.

4. C. Cachin, *Entropy measures and unconditional security in cryptography*, Ph. D. Thesis, ETH Zürich, Hartung-Gorre Verlag, Konstanz, 1997.
5. T. M. Cover and J. A. Thomas, *Elements of information theory*, Wiley Series in Telecommunications, 1992.
6. I. Csiszár and J. Körner, Broadcast channels with confidential messages, *IEEE Trans. on Information Theory*, Vol. 24, pp. 339–348, 1978.
7. W. Diffie and M. E. Hellman, New directions in cryptography, *IEEE Trans. on Information Theory*, Vol. 22, No. 6, pp. 644–654, 1976.
8. Y. Dodis and A. Smith, Fooling an unbounded adversary with a short key: a notion of indistinguishability, relations to extractors, and lower bounds, manuscript, 2003.
9. Y. Dodis and J. Spencer, On the (non)universality of the one-time pad, *Proceedings of FOCS 2002*, pp. 376–385, 2002.
10. G. D. Forney Jr., *Concatenated codes*, Massachusetts Institute of Technology, Cambridge, Massachusetts, 1966.
11. P. Gemmell and M. Naor, Codes for interactive authentication, *Advances in Cryptology - CRYPTO '93*, LNCS, Vol. 773, pp. 355–367, Springer-Verlag, 1993.
12. R. König, U. M. Maurer, and R. Renner, On the power of quantum memory, available on www.arxiv.org, quant-ph/0305154, 2003.
13. U. M. Maurer, Secret key agreement by public discussion from common information, *IEEE Trans. on Information Theory*, Vol. 39, No. 3, pp. 733–742, 1993.
14. U. M. Maurer, Information-theoretically secure secret-key agreement by NOT authenticated public discussion, *Advances in Cryptology - EUROCRYPT '97*, LNCS, Vol. 1233, pp. 209–225, Springer-Verlag, 1997.
15. U. M. Maurer, Information-theoretic cryptography, *Advances in Cryptology - CRYPTO '99*, LNCS, Vol. 1666, pp. 47–64, Springer-Verlag, 1999.
16. U. M. Maurer and P. Schmid, A calculus for security bootstrapping in distributed systems, *Journal of Computer Security*, Vol. 4, No. 1, pp. 55–80, 1996.
17. U. M. Maurer and S. Wolf, Secret-key agreement over unauthenticated public channels – Parts I–III, *IEEE Trans. on Information Theory*, Vol. 49, No. 4, pp. 822–851, 2003.
18. J. L. McInnes and B. Pinkas, On the impossibility of private key cryptography with weakly random keys, *Advances in Cryptology - CRYPTO '90*, LNCS, Vol. 537, pp. 421–436, Springer-Verlag, 1990.
19. R. Renner and S. Wolf, Unconditional authenticity and privacy from an arbitrarily weak secret and completely insecure communication, *Advances in Cryptology - CRYPTO 2003*, LNCS, Vol. 2729, pp. 78–95, Springer-Verlag, 2003.
20. A. Russell and H. Wang, How to fool an unbounded adversary with a short key, *Advances in Cryptology - EUROCRYPT 2002*, LNCS, Vol. 2332, pp. 133–148, Springer-Verlag, 2002.
21. C. E. Shannon, Communication theory of secrecy systems, *Bell System Technical Journal*, Vol. 28, pp. 656–715, 1949.
22. A. D. Wyner, The wire-tap channel, *Bell System Technical Journal*, Vol. 54, No. 8, pp. 1355–1387, 1975.

# On Generating the Initial Key in the Bounded-Storage Model

Stefan Dziembowski[1*] and Ueli Maurer[2]

[1] Institute of Informatics, Warsaw University
Banacha 2, PL-02-097 Warsaw, Poland
std@mimuw.edu.pl

[2] Department of Computer Science, ETH Zurich
CH-8092 Zurich, Switzerland
maurer@inf.ethz.ch

**Abstract.** In the bounded-storage model (BSM) for information-theoretically secure encryption and key-agreement one uses a random string $R$ whose length $t$ is greater than the assumed bound $s$ on the adversary Eve's storage capacity. The legitimate parties Alice and Bob share a short initial secret key $K$ which they use to select and combine certain bits of $R$ to obtain a derived key $X$ which is much longer than $K$. Eve can be proved to obtain essentially no information about $X$ even if she has infinite computing power and even if she learns $K$ after having performed the storage operation and lost access to $R$.

This paper addresses the problem of generating the initial key $K$ and makes two contributions. First, we prove that without such a key, secret key agreement in the BSM is impossible unless Alice and Bob have themselves very high storage capacity, thus proving the optimality of a scheme proposed by Cachin and Maurer. Second, we investigate the hybrid model where $K$ is generated by a computationally secure key agreement protocol. The motivation for the hybrid model is to achieve provable security under the sole assumption that Eve cannot break the key agreement scheme *during* the storage phase, even if afterwards she may gain infinite computing power (or at least be able to break the key agreement scheme). In earlier work on the BSM, it was suggested that such a hybrid scheme is secure because if Eve has no information about $K$ during the storage phase, then she has missed any opportunity to know anything about $X$, even when later learning $K$. We show that this very intuitive and apparently correct reasoning is false by giving an example of a secure (according to the standard definition) computational key-agreement scheme for which the BSM-scheme is nevertheless completely insecure. One of the surprising consequences of this example is that existing definitions for the computational security of key-agreement and encryption are still too weak and therefore new, stronger definitions are needed.

* Part of this work was done while the first author was a Post-Doc at ETH Zurich, Switzerland. Supported in part by the Polish KBN grant no. 4 T11C 042 25, by the European Communitiy Research Training Network ("GAMES" contract HPRN-CT-2002-00283), and by the Foundation for Polish Science (FNP).

C. Cachin and J. Camenisch (Eds.): EUROCRYPT 2004, LNCS 3027, pp. 126–137, 2004.

# 1 Introduction

In the bounded-storage model (BSM) for information-theoretically secure encryption and key-agreement one can prove the security of a scheme based on the sole assumption that the adversary's storage capacity is bounded, say by $s$ bits, even if her computing power is unlimited. Assume that a random $t$-bit string $R$ is either temporarily available to the public (e.g. the signal of a deep space radio source) or broadcast by one of the legitimate parties. If $s < t$, then the adversary can store only partial information about $R$. The legitimate parties Alice and Bob, sharing a short secret key $K$ initially, can therefore potentially generate a very long $n$-bit one-time pad $X$ with $n \gg |K|$ about which the adversary has essentially no information.

## 1.1 Definition of the Bounded-Storage Model

We define the bounded-storage model for key-expansion (and encryption) more formally. Alice and Bob share a short secret *initial key* $K$, selected uniformly at random from a key space $\mathcal{K}$, and they wish to generate a much longer $n$-bit *expanded key* $X = (X_1, \ldots, X_n)$ (i.e. $n \gg \log_2 |\mathcal{K}|$).

In a first phase, a $t$-bit random string $R$ is available to all parties, i.e., the randomizer space is $\mathcal{R} = \{0,1\}^t$. For instance, $R$ is sent from Alice to Bob or broadcast by a satellite. In fact, $R$ need not be uniformly random, it suffices to know a lower bound on the min-entropy $H_\infty(R)$ of $R$. Alice and Bob apply a known *key-expansion function*

$$f : \mathcal{R} \times \mathcal{K} \to \{0,1\}^n$$

to compute the expanded (or derived) key as $X = f(R, K)$. Of course, the function $f$ must be efficiently computable and based on only a very small portion of the bits of $R$ such that Alice and Bob need not read the entire string $R$.

Eve can store arbitrary $s$ bits of information about $R$, i.e., she can apply an arbitrary storage function

$$h : \mathcal{R} \to \mathcal{U}$$

for some $\mathcal{U}$ with the only restriction that $|\mathcal{U}| \leq 2^s$.[3] The memory size during the evaluation of $h$ need not be bounded. The value stored by Eve is $U = h(R)$. After storing $U$, Eve loses the ability to access $R$. (This is also referred to as the second phase.) All she knows about $R$ is $U$. In order to prove as strong a result as possible, one assumes that Eve can now even learn $K$, although in a practical system one would of course keep $K$ secret. This strong security property will be of special importance in this paper.

A key-expansion function $f$ is secure in the bounded-storage model if, with overwhelming probability, Eve, knowing $U$ and $K$, has essentially no information about $X$. More precisely, the conditional probability distribution $P_{X|U=u,K=k}$ is

[3] Since for every probabilistic strategy there is a best choice of the randomness, we can without loss of generality consider only deterministic adversary strategies.

very close to the uniform distribution over the $n$-bit strings, with overwhelming probability over values $u$ and $k$. Hence $X$ can be used as a secure one-time pad. Of course the security of $f$ depends on Eve's memory size $s$.

### 1.2 The Subject of this Paper and Previous Results

The bounded-storage model was proposed initially in 1992 [15], but the really strong (and essentially optimal) security results were proved only recently in a sequence of papers [2,1,10,11,14,17]. The first security proof for general storage functions $h$ was obtained by Aumann and Rabin [2], but only for $n = 1$ (i.e., for a scheme in which the derived key $X$ is much shorter than the initial key) or for $s \ll t$ (i.e., when the size of the memory of the adversary is much smaller than the length of the randomizer). The first fully general security proof was given in [11]. Lu [14] and Vadhan [17] showed that a special type of randomness extractor can be used to construct secure schemes, also improving on the size of the initial key $K$.

In all these papers one assumes that Alice and Bob initially share a secret key $K$, usually without considering how such a key $K$ is obtained by Alice and Bob. In this paper we address the problem of generating this key $K$ and investigate how this key generation process relates to the security proof of the BSM. We discuss the two most natural approaches to generating $K$, in a setting where Alice and Bob are connected only by an authenticated communication channel, without a trusted third party that initially distributes the key $K$.

The first approach is to generate $K$ within the context of the BSM itself or, equivalently, to perform key agreement in the BSM without sharing any secret key $K$ initially. This approach was discussed by Cachin and Maurer in [3] where a scheme was proposed in which both Alice and Bob need storage on the order of $\sqrt{t}$. More precisely, they each store a random subset (with pairwise independent indices) of the bits of $R$ and, after $R$ has disappeared for all parties, publicly agree on which bits they have both stored. With very high probability, Eve has only partial information about these bits, and therefore Alice and Bob can apply privacy amplification (i.e., randomness extraction using a strong extractor with a public extractor parameter) to distill an essentially perfect key $X$, which they can then use as a one-time pad. We show (Section 3) that the protocol of [3] is essentially optimal (in terms of the ratio between the storage size of the honest parties and the adversary) if $s$ is on the order of $t$. Since the storage requirement of $\sqrt{t}$ (which is also on the order of $\sqrt{s}$) bits for Alice and Bob may be too high for a reasonable storage bound $s$ for Eve, the practicality of this approach is questionable.

The second approach is to generate $K$ by a computationally secure key-agreement protocol, for instance based on the Diffie-Hellman protocol [7]. At first, this approach may appear to be completely useless since the provable information-theoretic security of the BSM-scheme would be lost: A computationally unbounded adversary could break the computational key-agreement protocol and then play the role of either Alice or Bob, with the same (small) storage requirements. However, at second sight, this approach is quite attractive as it allows to preserve the security of the key agreement protocol, which is only computational, even if the adversary can later break it and even if she gains infinite computing power.

It was claimed in [1] (Section IV B) (also, less formally in [10] (p. 5), [9] (p. 11) and [14] (p. 2)) that this implies the security of the hybrid scheme, for the following reason. Let $T$ be the transcript of the key agreement protocol. The adversary has (computationally) no information about $K$, given $T$, when performing the storage operation. More precisely, she could not distinguish $K$ from a truly random and independently generated key (as in the pure BSM). Therefore she could just as well forget $T$ and generate a random key $K$ himself, in which case she obviously has no advantage over the pure BSM setting. Since in this setting Eve learns $K$ anyway after finishing the storage operation, it does not hurt in the computational setting if Eve can now break the key-agreement scheme and compute $K$ (from $T$). Note that all the remaining aspects of the security proof are entirely information-theoretic.

It may come as a surprise that this reasoning is false, as is proved in Section 4. More specifically, we give an example of a computationally secure (according to the standard definition) key-agreement scheme which, when used to generate $K$ in the BSM context, renders the latter completely insecure. This shows that security arguments in a mixed computational/information-theoretic context can be very subtle. More interestingly, it demonstrates that existing definitions for the computational security of key-agreement and encryption are still too weak. Therefore new, stronger definitions are needed.

## 2 Preliminaries

The treatment in this paper is intentionally quite informal, but it is obvious how all aspects could be formalized in the traditional manner used in cryptography. The computation of a party (or algorithm) can be modelled by a probabilistic Turing machine, a protocol for two parties can be modelled as two interactive Turing machines, cryptographic primitives (such as key agreement) can be modelled as an asymptotic family with a security parameter, efficient can be defined as polynomial time, and negligible can also be defined in the traditional manner.

### 2.1 Secure Key-Agreement

A key-agreement scheme is a protocol between two parties Alice and Bob, at the end of which each party computes the same key $K \in \mathcal{K}$ (with overwhelming probability), for some key space $\mathcal{K}$. Let $T$ be the transcript of the protocol, i.e., the entire list of exchanged messages. Throughout the paper, we consider security against *passive* attacks, i.e., we assume that Alice and Bob can communicate over an authenticated channel. This is sufficient to illustrate our point, but note that security definitions for key-agreement are much more subtle in a setting with an active adversary who can tamper with messages exchanged between Alice and Bob.

A key-agreement scheme is *computationally secure* if no efficient distinguisher, when given $T$, can distinguish $K$ from a key $K'$ chosen independently and uniformly at random from the key space $\mathcal{K}$, with non-negligible advantage. For example, the computational security of the Diffie-Hellman key-agreement protocol [7] is equivalent to the so-called Decision-Diffie-Hellman assumption.

A computationally secure key-agreement scheme can also be obtained from any semantically secure public-key encryption scheme: Alice selects a random key $K \in \mathcal{K}$ and sends it to Bob, encrypted with Bob's public key.

### 2.2 Private Information Retrieval

The idea of private information retrieval (PIR) was introduced in [4]. A PIR scheme is a protocol for two parties, a user $U$ and a database $D$, allowing the user to access database entries in a way that $D$ cannot learn which information $U$ requested. More precisely, the database content can be modelled as a string $x = (x_1, \ldots, x_l) \in \{0,1\}^l$, and $U$ wants to access the $i$th bit $x_i$ of $x$, for some $i \in \{1, \ldots, l\}$, such that $D$ does not learn $i$. It is not relevant whether $U$ learns more than $x_i$.

A trivial solution to this problem is that $D$ sends all bits $x_1, \ldots, x_l$ to $U$, allowing $U$ to pick the bits he wants. The purpose of PIR protocols is to reduce the required communication. Depending on the protocol, the secrecy of $i$ can be computational or information-theoretic. In this paper we consider computationally secure PIR protocols [13].

A typical PIR protocol proceeds in three stages. First, $U$ sends a query, depending on $i$. Let $\mathcal{Q}(i)$ denote the query for index $i$. Second, $D$ computes the reply $\mathcal{R}(\mathcal{Q}(i), x)$ and sends it to $U$. Third, $U$ extracts $x_i$ from $\mathcal{R}(\mathcal{Q}(i), x)$. The scheme is computationally private if no efficient distinguisher can distinguish $\mathcal{Q}(i)$ from $\mathcal{Q}(i')$, for any $i, i' \in \{1, \ldots, l\}$.

In this paper we need an additional property of the PIR scheme, namely that $x_i$ is determined by $i$, $Q(i)$, and $\mathcal{R}(\mathcal{Q}(i), x)$ (even if it cannot be efficiently computed). Note that in a PIR scheme, $U$ typically holds a secret key which allows to extract $x_i$ efficiently from $i$ and $\mathcal{R}(\mathcal{Q}(i), x)$.

A well-known PIR scheme proposed in [13] makes use of the homomorphic property of the quadratic residues and the computational difficulty of the quadratic residuosity problem. More precisely, $U$ generates an RSA modulus $n = pq$. The string $(x_1, \ldots, x_l)$ is divided into $v = \lceil l/t \rceil$ blocks of length $t$, for some $t$. Let $1 \leq j \leq t$ be the index of $x_i$ within its block. The query $\mathcal{Q}(i)$ consists of a list $(y_1, \ldots, y_t)$ of $t$ elements of $Z_n^*$, all (independent) random quadratic residues, except for $y_j$ which is a random quadratic non-residue with Jacobi symbol 1. The database's reply consists of $v$ elements in $Z_n^*$, one for each of the $v$ blocks, where for each block $D$ computes the product of all the $y_m$ corresponding to 1's in the block. More precisely, $\mathcal{R}(\mathcal{Q}(i), x)$ consists of one element of $Z_n^*$ for each block, where for the first block $(x_1, \ldots, x_t)$ the value is

$$\prod_{k=1}^{t} y_k^{x_k},$$

for the second block it is $\prod_{k=1}^{t} y_k^{x_{t+k}}$, and similarly for the other blocks. Let $m \in \{1, \ldots, v\}$ be the index of the block to which $x_i$ belongs. It is easy to see that $x_i = 0$ if and only if the reply for the $m$th block is a quadratic residue. Clearly this can be efficiently checked by the user $U$ (who knows $p$ and $q$). Note that the user ignores all other received values not relevant for obtaining $x_i$. The communication complexity of this scheme is as follows: The query consists of $t$ elements of $Z_n^*$ and the reply consists of $v$ elements of $Z_n^*$. A reasonable trade-off is to let $t \approx \sqrt{l}$.

## 3 Limitations of Key-Agreement in the BSM

### 3.1 The Setting

In this section we consider the BSM without an initially shared secret key between Alice and Bob. In this setting, in the first phase when $R$ is available to all parties, Alice and Bob may use a randomized strategy (where the random strings of Alice and Bob are independent and denoted as $R_A$ and $R_B$, respectively) to execute a protocol resulting in transcript $T$, and to each store some information about $R$. Alice stores $M_A = f_A(R, T, R_A)$, and Bob stores $M_B = f_B(R, T, R_B)$, for some functions $f_A$ and $f_B$. Eve also stores some information $M_E = f_E(R, T, R_E)$ about $R$, for some random string $R_E$.

In the second phase, when $R$ has disappeared, Alice and Bob execute a second (probabilistic) protocol based on the stored values $M_A$ and $M_B$, resulting in a second transcript $T'$ and in both computing the key $K$.[4] The security requirement is that Eve must have only a negligible amount of information about $K$, i.e., $I(K; M_E T') \approx 0$. In fact, for the sake of simplicity, we assume here that Eve should obtain zero information about $K$, i.e.,

$$I(K; M_E T') = 0,$$

but the analysis can easily be generalized to a setting where Eve is allowed to obtain some minimal amount of information about $K$. The lower bound result changes only marginally.

We prove the following result, which shows that the practicality of such an approach without shared initial key is inherently limited. Alice or Bob must have storage capacity $\sqrt{s}$. The proof is given in Section 3.3.

**Theorem 1.** *For any key-agreement protocol secure in the BSM with no initial key for which $I(K; M_E T') = 0$, the entropy $H(K)$ of the generated secret key $K$ is upper bounded by*

$$H(K) \leq \frac{s_A s_B}{s},$$

*where $s_A$ and $s_B$ are the storage requirements for Alice and Bob, respectively, and $s$ is the assumed storage bound for Eve.*

[4] Here we assume that Alice and Bob generate the same key $K$, but this is of course a requirement of the scheme. The results can easily be generalized to a setting where the two key values must agree only with high probability.

We note that this bound also implies a bound on the memory of the adversary in the protocol for the oblivious transfer in the bounded-storage model.[5] Namely, if the memory of the honest parties is $s_A$ then the memory of a cheating party has to be much smaller than $s_A^2$. This shows that the protocol of [8] is essentially optimal and answers the question posted in [8,9].

### 3.2 The Cachin-Maurer Scheme

Indeed, as shown in [3], key agreement can be possible in such a BSM setting where Alice and Bob share no secret initial key . In this scheme, Alice and Bob each stores an (independent) random subset (with pairwise independent indices) of the bits of $R$. After $R$ has disappeared for all parties, they publicly check which bits they have stored. Eve has only partial information about these bits (with overwhelming probability), no matter what she has stored. Therefore Alice and Bob can use privacy amplification using an extractor to distill an essentially perfect key $K$.

In this scheme, due to the birthday paradox, the number of bits stored by Alice and Bob must be greater than $\sqrt{t}$ since otherwise the number of bits known to both Alice and Bob would be very small with high probability. This also shows that for $s$ on the order of $t$, the scheme of [3] has parameters close to the lower bound given by Theorem 1.[6]

### 3.3 Proof of Theorem 1

We first need the following information-theoretic lemma.

**Lemma 1.** *Consider a random experiment with random variables $Y, Z, Z_1, \ldots, Z_n$ such that conditioned on $Y$, the variables $Z, Z_1, \ldots, Z_n$ are independent and identically distributed, i.e.,*

$$P_{ZZ_1,\ldots,Z_n|Y}(z, z_1, \ldots, z_n, y) = P_{Z|Y}(z, y) \prod_{i=1}^{n} P_{Z|Y}(z_i, y)$$

*for all $y, z, z_1, \ldots, z_n$ and for some conditional probability distribution $P_{Z|Y}$.*[7] *Then*

$$I(Y; Z|Z_1 \cdots Z_n) \le \frac{H(Y)}{n+1}.$$

---

[5] This is because there exists a black-box reduction of the key-agreement problem to the oblivious transfer problem [12]. (It is easy to see that the reduction of [12] works in the bounded-storage model.)

[6] It should be mentioned that the security analysis given in [3] is quite weak, but this could potentially be improved by a better scheme or a tighter security analysis.

[7] In other words, $Z, Z_1, \ldots, Z_n$ can be considered as being generated from $Y$ by sending $Y$ over $n+1$ independent channels specified by $P_{Z|Y}$, i.e.,

$$P_{Z|Y}(z, y) = P_{Z_1|Y}(z, y) = \cdots = P_{Z_n|Y}(z, y)$$

for all $y$ and $z$.

*Proof.* The random experiment has a strong symmetry between the random variables $Z, Z_1, \ldots, Z_n$, both when considered without $Y$, and also when considered conditioned on $Y$. Any information-theoretic quantity involving some of the random variables $Z, Z_1, \ldots, Z_n$ (and possibly $Y$) depends only on how many of these random variables occur, but not which ones. Let therefore $H(u)$ denote the entropy of (any) $u$ of the random variables $Z, Z_1, \ldots, Z_n$. Similarly, we can define $H(u|v)$ as the conditional entropy of $u$ of them, given any other $v$ of them. The quantities $H(Y,u|v)$, $H(u|Y,v)$, and $I(Y;u|v)$ can be defined analogously. We refer to [5] for an introduction to information theory.

In this notation, the lemma states that

$$I(Y;1|n) \le \frac{H(Y)}{n+1}.$$

The chain rule for conditional information[8] implies that

$$I(Y;n+1) = \sum_{i=0}^{n} I(Y;1|i). \tag{1}$$

We next show that

$$I(Y;1|i) \le I(Y;1|i-1). \tag{2}$$

This can be seen as follows:

$$\begin{aligned} I(Y;1|i-1) - I(Y;1|i) &= H(Y,i-1) + H(i) - H(Y,i) - H(i-1) \\ &\quad -(H(Y,i) + H(i+1) - H(Y,i+1) - H(i)) \\ &= \underbrace{2H(i) - H(i-1) - H(i+1)}_{=I(1;1|i-1)} \\ &\quad - \underbrace{(2H(Y,i) - H(Y,i-1) - H(Y,i+1))}_{I(1;1|Y,i-1)=0} \\ &\ge 0 \end{aligned}$$

The first step follows from

$$I(U;V|W) = H(UW) + H(VW) - H(UVW) - H(W),$$

the second step by rearranging terms, and the last step since $I(1;1|i-1) \ge 0$ but $I(1;1|Y,i-1) = 0$. This last fact follows since when given $Y$, any disjoint sets of $Z$-variables are independent.

Now using $I(Y;n+1) \le H(Y)$ and combining (1) and (2) completes the proof since the right side of (2) is the sum of $n+1$ terms, the smallest of which is $I(Y;1|n)$. □

[8] Recall that the chain rule for information (see eg. [5], Theorem 2.5.2) states that for arbitrary random variables $V_1, \ldots, V_n$, and $U$ we have

$$I(U;V_1,\ldots,V_n) = \sum_{i=1}^{n} I(U;V_i|V_{i-1},\ldots,V_1)$$

To prove Theorem 1, recall that $s_A$, $s_B$ and $s$, are the storage capacities of Alice, Bob, and Eve, respectively. We have to specify a strategy for Eve to store information (i.e., the function $f_E$). Such an admissible strategy is the following: For the fixed observed randomizer $R = r$ and transcript $T = t$, Eve generates $\lfloor s/s_B \rfloor$ independent copies of what Bob stores, according to the distribution $P_{M_B|R=r,T=t}$. In other words, Eve plays, independently, $\lfloor s/s_B \rfloor$ times the role of Bob. We denote Eve's stored information (consisting of $\beta$ parts) as $M_E$. The above lemma implies that

$$I(M_A; M_B|M_E) \le \frac{H(M_A)}{\left\lfloor \frac{s}{s_B} \right\rfloor + 1} \le \frac{H(M_A)}{\frac{s}{s_B}} \le \frac{s_A s_B}{s}.$$

The last step follows from $H(M_A) \le s_A$. Now we can apply Theorem 3 in [16] which considers exactly this setting, where Alice, Bob, and Eve have some random variables $M_A$, $M_B$, and $M_E$, respectively, jointly distributed according to some distribution $P_{M_A M_B M_E}$. The theorem states that the entropy of a secret key $K$ that can be generated by public discussion is upper bounded as

$$H(K) \le \min\left(I(M_A; M_B), I(M_A; M_B|M_E)\right),$$

i.e., in particular by $I(M_A; M_B|M_E)$. This concludes the proof. □

## 4 The Hybrid Model

As described in Section 1.2, some authors have suggested that one can securely combine the BSM with a (computationally secure) public-key agreement scheme KA used to generate the initial key $K$. We call this model the *hybrid model*. The motivation for the hybrid model is to achieve provable security under the sole assumption that Eve cannot break the key agreement scheme *during* the storage phase, even if afterwards she may gain infinite computing power, or for some other reason might be able to break the key agreement scheme. The BSM can hence potentially be used to preserve the security of a computationally secure scheme for eternity. The reason is that because if Eve has no information about $K$ during the storage phase, she has missed any opportunity to know anything about the derived key $X$, even if she later learns $K$. Note that in the standard BSM Eve learns $K$ by definition, but in the hybrid scheme she may at a later stage learn it because she can possibly break the key agreement scheme based on the stored transcript.

We show that this very intuitive and apparently correct reasoning is false by giving an example of a secure (according to the standard definition) computational key-agreement scheme for which the BSM-scheme is nevertheless completely insecure.

The hybrid model can be formalized as follows. During the first phase, Eve is computationally bounded (typically a polynomial bound), and Alice and Bob carry out a key agreement protocol, resulting in transcript $T$. Eve performs an efficient computation on $R$ and $T$ (instead of performing an unbounded computation on $R$ alone), and stores the result $U$ of the computation (which is again bounded to be at most $s$ bits). Then she loses access to $R$ and obtains infinite computing power. Without much loss of generality we can assume that she stored $T$ as part of $U$ and hence she can now compute $K$.

**Theorem 2.** *Assume that a computationally secure PIR scheme*[9] *exists, and assume its communication complexity is at most $l^{2/3}$, where $l$ is the size of the database. Then there exists a key-expansion function $f$ secure in the standard BSM but insecure in the hybrid model, for the same bound on Eve's storage capacity.*

Clearly, the scheme of [13] (described in Section 2.2) satisfies the requirements stated in the theorem. The key-expansion function $f$ (whose existence we claim in the theorem) can be basically any key-expansion function proven secure in the literature.

*Proof (of Theorem 2).* We are going to construct a (rather artificial) key agreement scheme KA such that $f$ is not secure in the hybrid model. To construct KA we will use an arbitrary computationally secure key agreement scheme KA$'$. In [12,6] it was shown that the existence of computationally secure PIR schemes implies the existence of a key agreement scheme. Therefore we can assume that such a scheme exists (since we assume a secure PIR scheme). It is also reasonable to assume that the communication complexity of this scheme is small when compared to the size of the randomizer. One can also have in mind a concrete key-agreement scheme, for instance the Diffie-Hellman protocol, in which case the transcript consists of $g^x$ and $g^y$ (for $x$ and $y$ chosen secretly be Alice and Bob, respectively) and the resulting shared key is $g^{xy}$. This protocol is secure under the so-called decision Diffie-Hellman assumption.

Let us fix some PIR scheme. For the key-expansion we will use an arbitrary function $f$ (secure in the BSM) with the property that the number $m$ of bits accessed by the honest parties is much smaller than the total length $t$ of the randomizer, say $m \leq t^{1/4}$ (without essential loss of generality as any practical scheme satisfies this). We assume that $f$ is secure in the BSM against an adversary who can store at most $s = t/2$ bits. An example of a function satisfying these requirements is the function $f$ of [11] (for a specific choice of the parameters).

In our scenario Eve will be able (at the end of the second phase) to reconstruct each bit accessed by the honest parties. The basic idea is to execute $m$ times independently and in parallel the PIR query protocol. More precisely the protocol KA is defined as follows:

1. Alice and Bob invoke the given key-agreement scheme KA$'$. Let $K$ be the agreed key and let $T'$ be the transcript of the key agreement scheme.

[9] as defined in Section 2.2

2. Let $\kappa_1, \ldots, \kappa_m$ be the indices of the bits in the randomizer that are accessed by the parties (for the given initial key $K$ and the BSM scheme $f$). Alice sends to Bob a sequence $\mathcal{Q}(\kappa_1), \ldots, \mathcal{Q}(\kappa_m)$ of $m$ PIR queries, where each query is generated independently (with fresh random coins).
3. Alice and Bob (locally) output $K$ as their secret key.

It is not hard to see that the security of KA$'$ and the privacy of PIR imply the security of KA. Step 2 is an artificial extension of KA$'$ needed to make the resulting scheme insecure in the hybrid BSM, i.e., to encode into the transcript some useful information that can be used by Eve. Her strategy (for a given transcript of KA and a randomizer $R$) is as follows. In the first phase she computes the answers to the queries („acting" as a database). She does not send them anywhere, but simply stores them in her memory. She also stores the queries and the transcript $T'$ of the key-agreement scheme KA$'$. In other words:

$$U := ((\mathcal{Q}(\kappa_1), \mathcal{R}(\mathcal{Q}(\kappa_1), R)), \ldots, (\mathcal{Q}(\kappa_m), \mathcal{R}(\mathcal{Q}(\kappa_m), R), T).$$

(where $R$ denotes the randomizer). Since the PIR is efficient, so is Eve's computation. Because of the communication efficiency of the PIR scheme and of the key-agreement protocol KA$'$, the length of $U$ is at most $m \cdot t^{2/3} + |T|$, which is at most $t^{1/4+2/3} + |T|$. Since $|T|$ is much smaller than $t$, this value has to be smaller than $s = \frac{1}{2} \cdot t$ for sufficiently large $t$.

In the second phase the adversary can easily compute (from $T'$) the value of $K$ and therefore she can obtain $\kappa_1, \ldots, \kappa_m$. For every $i$ she also knows $(\mathcal{Q}(\kappa_i, R))$, thus she can (using the unlimited computing power) compute the bit of $R$ at position $\kappa_i$.[10] Therefore she can compute the value of $f(K, R)$. □

## 5 Discussion

One of the surprising consequences of Theorem 2 is that existing definitions for the computational security of key-agreement and encryption are too weak to cover settings in which the adversary's computing power may change over time, as is the case in real life. We thus need a new security definition of a key-agreement scheme and a public-key cryptosystem which implies, for example, the security in the BSM, as discussed above. It is quite possible that existing schemes such as the Diffie-Hellman protocol satisfy such a stronger security definition, and that only artificial schemes as the one described in the proof of Theorem 2 fail to be secure according to the stronger definition. It is an interesting problem to formalize in a more general context how and when security can be preserved even though a scheme gets broken a a certain point in time.

[10] Recall that in the definition of a PIR scheme (Section 2.2) we assumed that the values $i, \mathcal{Q}(i), \mathcal{R}(\mathcal{Q}(i), x)$ determine the value of $x_i$.

## References


1. Y. Aumann, Y. Z. Ding, and M. O. Rabin. Everlasting security in the bounded storage model. *IEEE Transactions on Information Theory*, 48(6):1668–1680, 2002.
2. Y. Aumann and M. O. Rabin. Information theoretically secure communication in the limited storage space model. In *Advances in Cryptology - CRYPTO '99*, pages 65–79, 1999.
3. C. Cachin and U. Maurer. Unconditional security against memory-bounded adversaries. In *Advances in Cryptology - CRYPTO '97*, pages 292–306, 1997.
4. B. Chor, O. Goldreich, E. Kushilevitz, and M. Sudan. Private information retrieval. *Journal of the ACM*, 45(6):965–981, 1998.
5. T. M. Cover and J. A. Thomas. *Elements of Information Theory.* John Wiley and Sons, Inc., 1991.
6. G. Di Crescenzo, T. Malkin, and R. Ostrovsky. Single database private information retrieval implies oblivious transfer. In *Advances in Cryptology - EUROCRYPT 2000*, pages 122–138, 2000.
7. W. Diffie and M.E. Hellman. New directions in cryptography. *IEEE Transactions on Information Theory*, 22(6):644–654, 1976.
8. Y. Z. Ding. Oblivious transfer in the bounded storage model. In *Advances in Cryptology - CRYPTO 2001*, pages 155–170, 2001.
9. Y. Z. Ding. *Provable Everlasting Security in the Bounded Storage Model.* PhD thesis, Harvard University, 2001.
10. Y. Z. Ding and M. O. Rabin. Hyper-encryption and everlasting security. In *STACS 2002, 19th Annual Symposium on Theoretical Aspects of Computer Science*, pages 1–26, 2002.
11. S. Dziembowski and U. Maurer. Tight security proofs for the bounded-storage model. In *Proceedings of the 34th Annual ACM Symposium on Theory of Computing*, pages 341–350, 2002.
12. Y. Gertner, S. Kannan, T. Malkin, O. Reingold, and M. Viswanathan. Relationship between public key encryption and oblivious transfer. In *41st Annual Symposium on Foundations of Computer Science*, pages 325–339, 2000.
13. E. Kushilevitz and R. Ostrovsky. Replication is not needed: Single database, computationally-private information retrieval. In *38th Annual Symposium on Foundations of Computer Science*, pages 364–373, 1997.
14. C. Lu. Hyper-encryption against space-bounded adversaries from on-line strong extractors. In *Advances in Cryptology - CRYPTO 2002*, pages 257–271, 2002.
15. U. Maurer. Conditionally-perfect secrecy and a provably-secure randomized cipher. *Journal of Cryptology*, 5(1):53–66, 1992.
16. U. Maurer. Secret key agreement by public discussion. *IEEE Transactions on Information Theory*, 39(3):733–742, 1993.
17. S. Vadhan. On constructing locally computable extractors and cryptosystems in the bounded storage model. In *Advances in Cryptology - CRYPTO 2003*, pages 61–77, 2003.

# Practical Large-Scale Distributed Key Generation

John Canny and Stephen Sorkin

University of California, Berkeley, CA 94720 USA
{jfc,ssorkin}@cs.berkeley.edu

**Abstract.** Generating a distributed key, where a constant fraction of the players can reconstruct the key, is an essential component of many large-scale distributed computing tasks such as fully peer-to-peer computation and voting schemes. Previous solutions relied on a dedicated broadcast channel and had at least quadratic cost per player to handle a constant fraction of adversaries, which is not practical for extremely large sets of participants. We present a new distributed key generation algorithm, sparse matrix DKG, for discrete-log based cryptosystems that requires only polylogarithmic communication and computation per player and no global broadcast. This algorithm has nearly the same optimal threshold as previous ones, allowing up to a $\frac{1}{2} - \epsilon$ fraction of adversaries, but is probabilistic and has an arbitrarily small failure probability. In addition, this algorithm admits a rigorous proof of security. We also introduce the notion of matrix evaluated DKG, which encompasses both the new sparse matrix algorithm and the familiar polynomial based ones.

**Keywords:** Threshold Cryptography. Distributed Key Generation. Discrete Logarithm. Random Walk. Linear Algebra.

## 1 Introduction

Distributed key generation (DKG) is an essential component of fully-distributed threshold cryptosystems. In many contexts, it is impractical or impossible to assume that a trusted third party is present to generate and distribute key shares to users in the system. In essence, DKG allows a set of players to collectively generate a public/private key pair with the "shares" of the private key spread over the players so that any sufficiently large subset can reveal or use the key. The generated key pair is then used in a discrete-log based threshold cryptosystem. Commonly the security parameter of such a system is called the threshold, $t$. This is the number of players that can be corrupted without the key being compromised.

Most distributed key generation schemes in the literature do not carefully consider the communication and computation cost required of each server. Specifically, most schemes require $O(nt)$ computation and communication per player, where $n$ is the number of players participating in the scheme. In this paper, we present a randomized algorithm called sparse matrix DKG that reduces this

C. Cachin and J. Camenisch (Eds.): EUROCRYPT 2004, LNCS 3027, pp. 138–152, 2004.

cost to $O(\log^3 n)$ per player, both in terms of computation and communication, in the presence of $\Omega(n)$ malicious parties. For large systems, this difference is quite significant. The cost of this gain is a slight chance that the algorithm fails. We formalize this cost in the definition of a probabilistic threshold distributed key generation algorithm. We also show how sparse matrix DKG is a specific instance of a more broad family of DKG algorithms.

## 2 Basic System Model

The systems we describe involve a group of $n$ players. The players are modeled by probabilistic polynomial-time Turing machines. They are connected with secure point-to-point channels, but without any broadcast channel. We feel that these assumptions are realistic for practical situations: true broadcast is only achievable in practice using Byzantine agreement but secure channels can be achieved through a public key infrastructure. We also assume that the players have access to a common source of randomness. The adversaries in this system are static and are assumed to be fixed when the algorithm begins. This is a reasonable assumption since, in practice, this algorithm is sufficiently fast that successful dynamic adversaries are unlikely.

For simplicity, we assume that there is an honest, but not trusted, dealer present to aid in the initialization of the algorithm. The first task of the dealer is to establish the set of $n$ players that will participate in the algorithm. This task makes the dealer resemble a directory server for the players. The users who wish to generate a distributed key are assumed to know the identity of the dealer. These users will then contact the dealer in order to secure a unique identifier for the algorithm. This dealer may also assist in the creation of a public key infrastructure so that users can transmit messages securely through encryption or authenticate messages with digital signatures. The dealer then decides when enough users are present for the algorithm to begin. This is decided by either having a fixed number of users known up front or requiring that users who wish to participate send a message to the dealer within a fixed time limit. In either case, this process determines the value of $n$, the number of users participating in the algorithm. Part of our initial assumption is that the dealer has access to a random oracle [1] so some randomness may be distributed to the players. Based on the value of $n$, the dealer will decide on a set of random bits and distribute a subset of them to each player.

In practice, the dealer need not be a single party, and could be implemented through a logarithmic size group of users or a hierarchy of users.

## 3 Distributed Key Generation Protocols

### 3.1 Previous Work

Existing literature on DKG algorithms is quite broad. One main line of approach began with the polynomial-based algorithm presented by Pedersen [8]. This algorithm was presented by other authors in varied forms as the basis of various threshold cryptosystems. This basic algorithm and these modifications, however, were vulnerable to a variety of attacks that would allow an adversary to bias the distribution of the private and public keys. This flaw was remedied by Gennaro, et al. [4] in a protocol that operates in two phases and uses Pedersen's verifiable secret sharing algorithm [9] to protect the bit commitments of the public key against static adversaries. All of these approaches require $O(t)$ broadcasts and $O(n)$ point-to-point messages for each player. Gennaro, et al. followed up their paper [5] with an explanation of how Pedersen's original algorithm is secure when used for Schnorr signatures. One main advantage for using the basic Pedersen algorithm is saving one broadcast round.

### 3.2 Definitions

The following definitions apply to discrete-log based cryptosystems. The globally known constants are $p$, a large prime; $q$, a large prime that divides $p-1$ and $g$ an element of order $q$ in $Z_p$. The first three criteria of the following definition have been used widely to define DKG protocols. The fourth was added by Gennaro, et al. in order to quantify the secrecy of an algorithm's key against malicious participants in the generation phase.

**Definition 1.** *A $t$-secure distributed key generation algorithm satisfies the following requirements, assuming that fewer than $t$ players are controlled by the adversary:*

- *(C1) All subsets of $t+1$ shares provided by honest players define the same unique secret key $x$.*
- *(C2) All honest parties have the same value of the public key $y = g^x \mod p$, where $x$ is the unique secret guaranteed by (C1).*
- *(C3) $x$ is uniformly distributed in $Z_q$ (and hence $y$ is uniformly distributed in the subgroup generated by $g$).*

- *(S1) The adversary can learn no information about $x$ except for what is implied by the value $y = g^x \mod p$.*

We propose the following modification to allow a DKG algorithm to fail with small probability. This will allow for algorithms that are considerably more efficient with arbitrarily small impacts to security.

**Definition 2.** *A probabilistic threshold $(\alpha, \beta, \delta)$ distributed key generation algorithm satisfies requirements (C2), (C3) and (S1) above as well as (C1'), (C4') below (which replace (C1)), all with probability $1-\delta$, assuming that the set of players controlled by the adversary is less than $\alpha n$.*

- *(C1') The shares of any subset of honest players define the same key, $x$, or no key at all.*

- *(C4') Any subset of at least $\beta n$ honest players can recover the key with probability $1-\delta$.*

By these definitions, a $t$-secure DKG algorithm is a probabilistic threshold $\left(\frac{t}{n}, \frac{t+1}{n}, 0\right)$ DKG algorithm as well.

### 3.3 Matrix Evaluated DKG

The sparse matrix distributed key generation algorithm we propose for very large sets of players is a specific instance in a family of protocols which we call matrix evaluated DKG. In section 3.4 we show that this family includes the familiar DKG algorithm introduced by Gennaro, et al. This general technique consists of three primary phases. In the first phase, all the players create their secrets and share them with the others. After it is decided which players have correctly shared their secrets, the public key is recovered. Finally, after the generation is complete, the algorithm provides a method for recovering the secret with only a subset of the shareholders. The use of a matrix to codify the relation between the key and its shares is similar to the technique Blakley proposed for secret sharing [2].

#### Master Algorithm

1. Start with a dealing phase so that all players know $E, v, g, h, p, q$ where $p$ is a large prime, $q$ is a large prime that divides $p-1$, $g$ and $h$ are elements of order $q$ in $Z_p$, $E$ is an $m \times n$ matrix over $Z_q$ and $v$ is an $m$ element vector over $Z_q$.
2. **Generate** $x$:
   (a) Each player $i$ chooses two row vectors, $a_i, a'_i \in Z_q^m$.
   (b) Player $i$ then calculates $s_i \triangleq a_i E, s'_i \triangleq a'_i E$. Define the *checking group*: $Q_i = \{j | s_{ij} \neq 0 \vee s'_{ij} \neq 0\}$. Player $i$ sends the $j$th element of each, $s_{ij}, s'_{ij}$ to player $j \in Q_i$ and broadcasts[1] $C_i \triangleq g^{a_i} h^{a'_i} \mod p$ (where both the exponentiation and multiplication are element-wise) to all $j \in Q_i$.
   (c) Each player $j$ verifies the shares received from other players. For each $i$ such that $j \in Q_i$, player $j$ checks if:

$$g^{s_{ij}} h^{s'_{ij}} = \prod_{k=1}^{m} (C_{ik})^{E_{kj}} \mod p \tag{1}$$

   If this check fails for $i$, player $j$ broadcasts a complaint against player $i$ to $Q_i$.
   (d) Every player $i$ who was complained against by player $j$ will broadcast $s_{ij}, s'_{ij}$ to $Q_i$.
   (e) The other players in $Q_i$ will check the broadcast $s_{ij}, s'_{ij}$ and mark as invalid each $i$ for which Eq. 1 does not hold.
3. Each player $i$ builds the set $V$ of all players who were not marked invalid and sets their share of the secret as $x_i \triangleq \sum_{j \in V} s_{ji}$. Note that $x_i$ is the $i$th element of the vector $\left(\sum_{j \in V} a_j\right) E$. The secret key is defined as $x \triangleq \left(\sum_{i \in V} a_i\right) v$.

[1] All the following broadcasts are to $Q_i$ only. In the sparse matrix algorithm, $|Q_i| = O(\log n)$.

4. **Reveal** $y = g^x \mod p$
   (a) Each valid player $i \in V$ broadcasts the vector $A_i \triangleq g^{a_i} \mod p$ to $Q_i$.
   (b) Each player $j$ verifies that each $A_i$ is valid by checking for each $i \in V$ such that $j \in Q_i$ if:

$$g^{s_{ij}} = \prod_{k=0}^{m} (A_{ik})^{E_{kj}} \mod p \tag{2}$$

   If this check fails for $i$, player $j$ broadcasts a complaint against player $i$ as well as $s_{ij}, s'_{ij}$ to $Q_i$.
   (c) If at least one valid complaint is filed against player $i$ (Eq. 1 holds for $s_{ij}, s'_{ij}$ but Eq. 2 does not), then all players in $Q_i$ will reconstruct $a_i$ in public by solving a set of linear equations ($s_i = a_i E$), for the valid values of $s_i$.
   (d) Each of the honest members of $Q_i$ knows both whether player $i$ is valid and, if so, the correct value $A_i$. To find $y$, it suffices to ask each member of each $Q_i$, find the set $V$, and then take $y \triangleq \prod_{i \in V} \prod_j A_{ij}^{v_j}$.

**Sharing the Secret** The basis of this algorithm is what we call an *evaluation matrix* $E$, which has $m$ rows and $n$ columns, where $n$ is the total number of players in the system. Each player $i$ picks an internal secret $a_i$, which is a row vector with $m$ elements and, from that, creates an external secret $s_i \triangleq a_i E \mod q$, another row vector, now with $n$ elements. Player $i$ then reveals the $j$th column of the external secret $s_{ij}$ to player $j$. If the $a_i$ and $E$ are structured, it is possible that fewer than $n$ users are assigned non-zero shares by player $i$.

In order to demonstrate that player $i$ is creating consistent shares of the external secret, she must broadcast a committed version of her internal secret. For this we employ the VSS scheme introduced by Pedersen [9]. Player $i$ creates another, random internal secret $a'_i$ and broadcasts a commitment vector $C_i = g^{a_i} h^{a'_i} \mod p$ (the multiplication here is element-wise) to the *checking group* $Q_i$ of players who receive the non-trivial shares. Each member of $Q_i$ can then verify that her share of $s_i$ was created from the same $a_i$ as the others. This is achieved by each player $j$ checking that Eq. 1 holds for each $i$, since all honest players are assumed to agree on $C_i$

This equality will certainly hold if $s_{ij} = \sum_{k=1}^{m} a_{ik} E_{kj}$ as specified above. If it does not, then the secret share is invalid and player $j$ broadcasts a complaint against player $i$. In response, player $i$ will broadcast $s_{ij}, s'_{ij}$ to demonstrate that Eq. 1 is satisfied. If player $i$ broadcasts a pair of secret shares that did not satisfy Eq. 1, then the other players will mark $i$ as invalid, since the external secret of player $i$ is not consistent with any internal secret, and hence it will be impossible to recover the secret key if it includes this share.

At this point, $V$ is well defined and each checking group $Q_i$ knows whether $i \in V$, since the decision is based on broadcast information. At this point both the global private key $x$ and public key $y$ are well defined by the following equations, where $T$ is a linear function (e.g., an inner product with a fixed vector, $v$):

$$
\begin{aligned}
a &\triangleq \sum_{i \in V} a_i \\
x &\triangleq \sum_{i \in V} T(a_i) \\
&= T(a) \\
y &\triangleq g^x \mod p
\end{aligned}
$$

**Revealing the Public Key** Note that the public key is not yet publicly known. This is necessary to ensure that an attacker does not skew its distribution by manipulating the set $V$. After $V$ is established, the public key may be safely extracted. Every player $i \in V$ broadcasts their share of the public key, the vector $A_i = g^{a_i} \mod p$. The other players will check if Eq. 2 holds.

Like the previous check, if the rules of the algorithm are followed, the equality holds. If it does not and player $j$ broadcasts a complaint, the allegation of cheating is treated with more caution than before. Specifically, player $j$ must prove that her complaint is valid by broadcasting $s_{ij}, s'_{ij}$. The other players will check that Eq. 1 holds for these secret shares. If it does, then it is very likely that these shares are indeed from player $i$. They will also check Eq. 2 to validate the complaint. If it is valid, all honest users will reconstruct $a_i$ by broadcasting their shares. This is the straightforward process of solving a set of linear equations that specify the internal secret for that player. The number of shares required for reconstruction depends on the structure of $E$ and $a_i$.

**Using the Private Key** Based on the definition of $x$, it suffices to find $a$. We can find $a$ directly through the relation $aE = (x_1, x_2, \ldots, x_n)$ if we know a sufficient fraction of the $x_i$. This fraction is a function of the structure of $E$. It is desirable, however, to never reveal $a$. Note that since $a$ is a linear function of the $x_i$ we can write $a = (x_1, x_2, \ldots, x_n)\Sigma$, where $\Sigma$ is the pseudo-inverse of $E$. If $T(a) = av$ then $T(a) = (x_1, x_2, \ldots, x_n)\Sigma v$. If the signature or encryption is of the form $g^x$, then each player can sign $g^{x_i}$ and the signatures or encryptions are combined as $\prod g^{x_i L_i}$, where $L_i$ is the $i$th row of $\Sigma v$. Of course, if only a subset of the $x_i$ are available, a different $\Sigma$ must be used.

### 3.4 GJKR DKG

Before diving into sparse matrix DKG, we will show how the algorithm introduced by Gennaro, Jarecki, Krawczyk, and Rabin fits into the more generic matrix evaluated DKG framework. This protocol was introduced to eliminate a security flaw in the simpler Joint-Feldman protocol proposed by Pedersen where an adversary could bias the distribution over public keys.

Let $E$ be the Vandermonde matrix with $m \triangleq t + 1$ rows and $n$ columns, where $t$ is the security threshold of the system:

$$E_{ij} = j^{i-1}$$

Each $a_i$ is a random $m$ element column vector so that $|Q_i| = n$. The matrix makes the external secret, $s_i$, the same as evaluations of a polynomial (with coefficients defined by the internal secret) at the points $1, 2, \ldots, n$.

The private key is just a function applied to the sum of the internal secrets of the valid players. The function in this case is $T(b_1, b_2, \ldots, b_m) = b_1$. Each player's share of this secret is the sum of the shares they hold from the other valid players. Equivalently, each player's share is the evaluation at a certain point of the polynomial created by summing the polynomials of the valid players. Since all the polynomials are of degree $t$, any $t+1$ of the $\sum_{i \in V} s_{ij}$ will allow interpolation to reconstruct $\sum_{i \in V} a_i$. Hence with $t+1$ valid shares, the algorithm can succeed in revealing the public or private key. Of course, since the private key should never be revealed, it is possible to use the key using just the evaluations of the polynomial through Lagrange interpolation coefficients.

Assuming that each server behaves properly, this algorithm incurs communication cost per player of $O(t)$ broadcast bits and $O(n)$ messages, assuming no faulty players. Achieving the broadcast through Byzantine agreement is very expensive. In the the two phases of the algorithm, each server performs $O(tn)$ operations. This occurs because each server must check every other server, and checking each server is an $O(t)$ operation.

## 4 Sparse Matrix DKG

To reduce both communication cost and computational complexity, we introduce a technique that that relies on a sparse evaluation matrix. This produces a $(\gamma - \epsilon, \gamma + \epsilon, \delta)$ probabilistic threshold DKG algorithm for $0 < \gamma < \frac{1}{2}, \epsilon > 0, \delta > 0$.

### 4.1 Intuition

The basic algorithm only requires a user to communicate with her checking group (to share her secret) and with all the checking groups to which she belongs (to determine whether those users behaved as the protocol dictates). To reduce the communication complexity of the algorithm it is sufficient to make these checking groups small. However, they must be sufficiently large so that a majority of the group is honest. If the groups are logarithmic in size, Hoeffding bounds give an inverse-polynomial probability of a group failing this criteria, assuming that the fraction of honest players is greater than a half. For recovery to be possible, the internal secret of a player must have fewer degrees of freedom than the number of honest recipients of her shares. A sparse, structured evaluation matrix is the tool that we use to map small secrets onto small checking groups.

### 4.2 Sparse Evaluation Matrix

By imposing constraints on the number of non-zero entries in each user's internal secret and having a very structured evaluation matrix, we reduce the cost of the problem from quadratic to polylogarithmic in the number of users. Figure 1

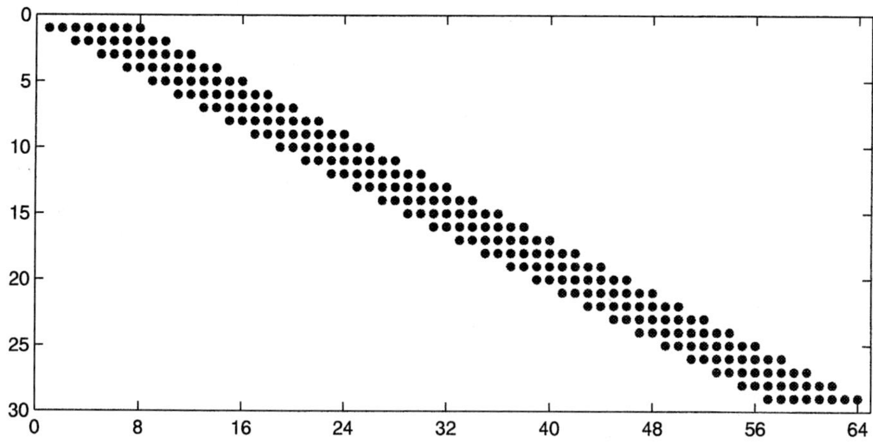


**Fig. 1.** An illustration of an evaluation matrix for $\gamma = \frac{1}{2}$. Here $n = 64, \ell = 8$. Each dot indicates a non-zero position in the matrix.

illustrates the evaluation matrix used for an $\left(\frac{1}{2} - \epsilon, \frac{1}{2} + \epsilon, \delta\right)$ instance of the algorithm. Each row of the evaluation matrix has $\ell$ consecutive, random entries offset by two columns from the previous row. Since the matrix must have $n$ columns, it must also have $m \triangleq (n - \ell)/2 + 1$ rows. We will show that for $\ell = O(\log n)$, the algorithm can recover the key with a $\frac{1}{2} + \epsilon$ fraction of honest shares with high probability. In general, for $\gamma \leq \frac{1}{2}$, the evaluation matrix $E$ has the band in each row offset by $\gamma^{-1}$ columns (on average) from the previous row and hence there are $m \approx \gamma(n - \ell)$ rows in $E$. Intuitively, this offset allows most sets of $\gamma n$ columns of the matrix to have full rank. The rest of the description of the algorithm and the accompanying proofs assume $\gamma = \frac{1}{2}$.

### 4.3 Dealing Phase

The dealer must generate $O(n \log n)$ independent random bits for the analysis of this algorithm to succeed. This randomness is used to determine two things. First, it establishes the evaluation matrix $E$. The second use is to create a permutation of the users. All aspects of the algorithm require the adversary and honest players to be distributed at random. Any large cluster of dishonest players leaves part of the private key vulnerable.

Unfortunately, it isn't practical to distribute that many bits to all the players. It is reasonable, however, for the dealer to send each player his identifier as determined by the permutation. Then player $i$ only needs to know the columns $Q_i$ and $\bigcup_{j | i \in Q_j} Q_j$ of $E$. The first group is needed to produce the shares to send to other players and the second, larger group is needed to check and possibly reconstruct other players' shares. As we will see, this is a logarithmic number of columns, and hence a polylogarithmic number of random bits. All players will also query the dealer for the addresses of their peers. Since the dealer is assumed to be honest, these bits need not be "broadcast"; a simple point-to-point distribution suffices.

### 4.4 Making and Sharing the Secret

Player $i$ creates a (sparse) internal secret vector, $a_i$ that contains only $u = \ell/2\epsilon^2$ non-zero, consecutive elements. The position of these elements is chosen to equally distribute the checking load among all the players. Specifically, player $i$ will have her terms start at $\lceil i\frac{m-u}{n}\rceil$. The structure of both the internal secret vectors and the evaluation matrix mean that not all players are given non-zero shares by player $i$. This defines the checking group, $Q_i$. The only recipients of broadcasts regarding player $i$ are members of her checking group. In this setting of $\gamma$, there will be $\ell + 2(u-1)$ (non-zero) shares, so that an adversary who controls any $1/2 - \epsilon/2$ fraction of these still has fewer than $u$ shares.

To prove that these shares were constructed from the same internal secret vector, it is necessary for player $i$ to broadcast the $C_{ij}$ as defined above to $Q_i$. Since our system model uses Byzantine agreement to achieve broadcast, the relatively small number of non-zero shares reduces the number of participants in this operation. In this algorithm, since the set of players outside of $Q_i$ has no evidence as to whether player $i$ is or is not valid, the other players will take a majority vote of the checking group to determine that. We are guaranteed that all the honest players in the checking group will agree whether player $i$ is valid since they are privy to the same information. We must then show that a majority of $Q_i$ is honest. This is a straightforward application of Hoeffding bounds since the expected fraction of honest players in any group is $\gamma + \epsilon$:

$$\begin{aligned}\Pr\left(\left|\{x \in Q_i : x \text{ is dishonest}\}\right| > \left(\gamma - \frac{\epsilon}{2}\right)|Q_i|\right) &\le 2\exp\left(-\frac{(|Q_i|\epsilon/2)^2}{2|Q_i|}\right)\\ &= 2\exp\left(-\frac{|Q_i|\epsilon^2}{8}\right)\end{aligned}$$

Hence, since $|Q_i| > \frac{\ell}{\epsilon^2}$, if $\ell > O((1+\kappa)\log n)$, then the probability is bounded by $n^{-1-\kappa}$ and the probability that all checking groups have a majority honest players is $1 - n^{-\kappa}$ by a union bound.

### 4.5 Extracting the Public Key

Revealing the public key occurs just as in the more general description. In the proof of correctness, we show that it happens with high probability. A third party who wishes to discover the public key that the algorithm has generated needs to ask each checking group for the checked user's share and take the majority (the checking group may also respond that the player is invalid). This is an $O(n \log n)$ operation. It is reasonable to assume that the dealer poll each player about the other players that they have checked, tally these votes for each player, multiply the public key shares for the valid players and then send the public key to every player.

It may be desirable to do this operation in a more distributed fashion. It is certainly not practical for each player to poll each checking group. Instead, a message passing algorithm to tally the public key is feasible. Here we set up a binary partition of all the players. At the bottom of a binary tree are small[2], non-overlapping groups of users for which the public key share of each (valid) member is known to all in the group. As one moves up the tree, these shares are multiplied to establish the shares of larger groups. The global public key is found at the top of the tree. A logarithmic sized group is responsible for each aggregation point within the binary tree. The public key, once found, may be passed back down the tree to each user. By restricting communication to the edges of the tree, each player need only communicate with $O(\log n)$ other players.

### 4.6 Using the Global Secret

As described above, the global secret is maintained as shares kept by each of the valid users. However, to be useful in threshold cryptosystems, it must be possible to apply this key for a signature without having all shareholders act. The structure of the evaluation matrix assists us in this regard. Discovering the value of the private key requires the solving for $a$ (implicitly). If we can find the inversion matrix $\Sigma$, we will be able to recover the private key. This is possible if the submatrix of the evaluation matrix corresponding to the valid users has rank $m$.

### 4.7 Recoverability

We assume that there are $\beta n$ players ($\beta \triangleq \gamma + \epsilon$) that will contribute their shares to sign (or decrypt) a message. Let $S$ be the set of these players. Our goal is to show that $E|_S$, the $m \times |S|$ submatrix of $E$ consisting of the columns corresponding to good shares has rank $m$. We first state a lemma that concerns biased random walks. In the appendix, we prove this lemma for $\gamma = \frac{1}{2}$. It is straightforward to extend this to arbitrary $\gamma$.

**Lemma 1.** *An $n$ step random walk that reflects at zero, with $(\gamma + \epsilon)n$ steps of $1 - \gamma^{-1}$ and $(1 - \gamma - \epsilon)n$ steps of $+1$, will reach $O(\kappa \log n)$ with probability less than $n^{-\kappa}$, for fixed $\epsilon$.*

We can now prove our theorem concerning the rank of $E|_S$, in the case that $\gamma = \frac{1}{2}$.

**Theorem 1.** *The matrix $E|_S$ formed by randomly deleting $\alpha n$ columns from $E$, $\alpha = \frac{1}{2} - \epsilon$, will have rank $m$ with probability $1 - n^{-\kappa}$ if each row of $E$ has $\ell = -\frac{(2+\kappa)}{\log(1-2\epsilon)} \log n$ consecutive, non-zero, random entries.*

[2] It is important that these groups be large enough that there be a low probability of a corrupt majority. It suffices for them to be of size $\Omega(\log n)$.

*Proof.* The theorem is proved by showing that a subset of $m$ columns of $E|_S$, taken as a matrix $E|'_S$, has random elements along its main diagonal. Consider the process adding columns from $E$ to $E|_S$. Each row of $E$ has $\ell$ non-zero, consecutive entries, called the band. We consider the incremental process of examining columns in order. If a column is present, and the column is not to the left of the band for the current row being considered (i.e., the row $r$ for which we want a random value at $(E'_S)_{rr}$), that column is added to $E|'_S$, and the next row is considered. Let $X_i$ denote the offset of column $i$ in the non-zero entries of the currently considered row, $R_i$ (where $i$ is a column index for the full matrix, $E$). For example, in Fig. 1, if $i = 8$ for row 3 the offset is 4. In general, $X_i = i - 2(R_i - 1)$. If $X_i$ ever exceeds $\ell$, the process fails since none of the later columns will be able to contribute a random element for the diagonal entry at the current row. Define $X_1 = 1, R_1 = 1$. Now consider the state $X_i$. If column $i$ is missing, we stay at the current row, $R_{i+1} = R_i$, but step forward in relation to the beginning of the band. Hence $X_{i+1} = X_i + 1$. Now if column $i$ is present, there are two possibilities. If $X_i \geq 1$, the column is added and $R_{i+1} = R_i + 1$, so that column $i + 1$ would be one step forward in the same row, but one step behind in the next row (since the rows are offset by two relative to each other), and $X_{i+1} = X_i - 1$. Now if $X_i = 0$ a present column does not help since we are in front of the band, so $X_{i+1} = 0, R_{i+1} = R_i$.

Observing just the $X_i$, the process is identical to the reflecting random walk process defined above. Applying the random walk lemma, we may bound the probability that the walk passes $t$ after $n$ steps, one for each column of $A$. This probability is just:

$$P(\text{process fails}) = P(\text{walk passes } \ell \text{ in } n \text{ or fewer steps}) < n^{-\kappa}$$

### 4.8 Correctness

**Theorem 2.** *The sparse matrix DKG algorithm described above is a probabilistic threshold* $\left(\frac{1}{2} - \epsilon, \frac{1}{2} + \epsilon, 2n^{-\kappa}\right)$ *DKG algorithm for* $\ell = O(\kappa \log n)$, *the width of the band in the constraint matrix.*

*Proof.* (C1) The honest players within checking group $Q_i$ always agree whether player $i$ is in $V$, since that decision is made based on broadcast information. Honest players outside the checking group will rely on a majority vote of the checking group to determine if player $i$ is included in $V$. Then the set $V$ that is established in step 3 of the algorithm is unique if each checking group has a majority of honest players, which happens with probability $1 - n^{-\kappa}$. Assuming the adversary is not able to compute $\text{dlog}_g h$, the check in Eq. 1 implies that the player's shares are all consistent with the same $a_i$. The reconstruction theorem implies that the honest players for any set of rows will have full rank, so they are consistent with a unique $a_i$, with probability $1 - n^{-\kappa}$. Since all honest players have shares that satisfy Eq. 1, any set of them are able to recover $x$ or not recover anything (in the case that the submatrix that they define is not invertible).

(C2) In the case where a valid complaint is filed against player $i$, this $a_i$ is reconstructed in public using the reconstruction theorem and $A_i = g^{a_i} \bmod p$.

Now we consider the case where no valid complaints are filed. Since $a_i$ could have been reconstructed with the shares from the honest players, there are at least $u$ linearly independent equations and $u$ unknowns, so that there must be a unique solution. Since the broadcast from player $i$ agrees with these equations, through Eq. 2, the broadcast $A_i$ is exactly $g^{a_i}$. Hence all honest players within each checking group have the correct value of $A_i, i \in V$, and since there are a majority of honest players in each checking group, all honest players have the same value for $A_i, i \in V$ and also $y = g^x \mod p$.

(C3) The secret is defined as $x = \left(\sum_{i \in V} a_i\right) v$. Since $v$ is random, any random term of some $a_i$ that is independent from the other $a_j$ will cause $x$ to be uniform. In the proof of secrecy, we show that the adversary cannot determine any $a_i$ completely for an honest player $i$, so the dishonest players' $a_j$ is independent of a term of every honest $a_i$. Also honest players choose all their terms independently, so the $a_i$ from any honest player will satisfy this requirement.

(C4) This is a direct application of the reconstruction theorem.

### 4.9 Secrecy

To show that the adversary is not able to learn any information about the private key $x$ other than the fact that it is the discrete log of the public key $y$, we create a simulator. Formally, a *simulator* is a probabilistic polynomial-time algorithm that given $y \in Z_p$, such that $y = g^x \mod p$ for some $x$, can produce a distribution of messages that is indistinguishable from a normal run of the protocol where the players controlled by the simulator are controlled instead by honest players. This is the familiar technique used to show that zero-knowledge proofs do not reveal any private information. However, since our algorithm relies on a random distribution of adversarially controlled players, our simulator will only have a high probability of success.

We assume that the adversary controls no more than a $\frac{1}{2} - \epsilon$ fraction of the players, chosen before the start of the algorithm. Recall that in section 4.4 we required that the adversary control no more than a $\frac{1-\epsilon}{2}$ fraction of any checking group. Hence each checking group contains less than $\frac{1-\epsilon}{2}\left(\ell + 2\left(\frac{\ell}{2\epsilon^2} - 1\right)\right) < \frac{\ell}{2\epsilon^2} < u$ adversarially controlled players and the adversary cannot learn an honest player's entire internal secret.

The input to the simulator is a $y$ that could have been established at the end of a normal run of the protocol. Assume that the adversary controls the set $\mathcal{B} = \{B_1, B_2, \ldots, B_t\}$ and that the honest parties (controlled by the simulator) are $\mathcal{G} = \{G_1, G_2, \ldots, G_{n-t}\}$. In a non-simulated run of the protocol that ends with the public key $y$, the $a_i'$ are uniform random vectors, subject to the sparseness constraint, and the $a_i$ are random vectors subject to both $\prod_{i \in V} \prod_j g^{a_{ij} v_j} = y$ and the sparseness constraint.

Consider the following algorithm for the simulator, SIM:

1. Each honest player $i \in \mathcal{G}$ performs steps 2 and 3 of the sparse matrix DKG protocol. At this point:

- The set $V$ is well defined and $\mathcal{G} \subseteq V$.
- The adversary $\mathcal{B}$ has seen $a_i, a'_i$ for $i \in \mathcal{B}$, $s_{ij}, s'_{ij}$ for $i \in V, j \in \mathcal{B}$ and $C_{ij}$ for $i \in V$.
- SIM knows $a_i, a'_i$ for all $i \in V$ (including those in $V \cap \mathcal{B}$, the internal secrets for the consistent adversary players).

2. Perform the following calculations:
   - Compute $A_i \triangleq g^{a_i} \mod p$ for $i \in V \setminus \{G_1\}$.
   - Let $S$ be a subset of $Q_{G_1}$, the checking group for $G_1$, that contains all of $Q_{G_1} \cap \mathcal{B}$ and enough of $Q_{G_1} \cap \mathcal{G}$ so that the rank of $E|_S$, the columns of $E$ corresponding to $S$, has rank $u-1$. Let $r$ be some $i$ such that the columns $S \cup \{r\}$ have rank $u$.
   - Assign $s^*_{G_1j} \triangleq s_{G_1j}$ for $j \in S$.
   - Let $S'$ be a subset of $u-1$ elements of $S$ such that the columns $S'$ of $E$ have rank $u-1$.
   - Compute $\Sigma$, the inverse of the submatrix of $E$ corresponding to the columns $S' \cup \{r\}$ and the rows $1, \ldots, u$ (i.e., assume wlog that $G_1 = 1$).
   - Note that $a^*_{G_1} = (s^*_{G_1S'_1}, \ldots, s^*_{G_1S'_{u-1}}, s^*_{G_1r})\Sigma$, but $s^*_{G_1r}$ has not yet been fixed. Similarly $A^*_{G_1} = (g^{\alpha_1+\beta_1 s^*_{G_1r}}, \ldots, g^{\alpha_u+\beta_u s^*_{G_1r}})$, where the $\alpha_i, \beta_i$ are functions of $\Sigma$ and $s^*_{G_1j}, j \in S'$.
   - For our construction to succeed, $\prod_j A^{v_j}_{G_ij} = y\left(\prod_{i\in V\setminus\{G_1\}} \prod_j A^{v_j}_{ij}\right)^{-1}$. The right hand side is known and of the form $g^\gamma$, and the left hand side is of the form $g^{\alpha+\beta s^*_{G_1r}}$. Hence we can solve to find $g^{s^*_{G_1r}}$ and then evaluate the expression for $A^*_{G_1}$.
3. Broadcast $A_i$ for $i \in \mathcal{G} \setminus G_1$ and $A^*_{G_1}$.
4. Perform the checks of step 4.(b) of the algorithm for each player $i \in \mathcal{G}$ on the $A_j, j \in \mathcal{B}$ broadcast by the adversary's players and broadcast any necessary complaints.
5. The adversary cannot file a valid complaint against any honest player since all the messages followed the protocol. However, the simulator must recover from the adversary's actions. The simulator will follow the protocol of 4.(c) to recover the $a_i$ for players who did not share their $A_i$ properly.

The simulator will result in an identical distribution of the observed $a_i, a'_i$ as would be expected from a non-simulated run that generated the public key $y$. This is because the simulator uses the original, random $a_i, a'_i$ for $i \in \mathcal{G} \setminus G_1$. Also $a^*_{G_1}$ is chosen to be random subject to the constraint above. It remains to be shown that $a'^*_{G_1}$ is also random. The relationship between $a^*_{G_1}, a'^*_i$ is established through the public $C_{G_1} = g^{a^*_{G_1}} h^{a'^*_{G_1}}$. This implies $a'^*_{G_1} = \mathrm{dlog}_g(h)^{-1} \cdot (a_{G_1} - a^*_{G_1}) + a'_{G_1}$, which inherits its randomness from $a'_{G_1}$.

### 4.10 Communication Complexity and Running Time

The primary objective of the matrix-based constraint algorithm is to reduce both the communication complexity and running time for the users participating in the key generation without significantly affecting the chance that the algorithm fails to properly generate a key. As described above, a fraction of each user's secret as well as a bit-committed version of the secret will be sent to $\ell/2\epsilon^2 = O(\log n)$ other users. Hence this communication accounts for $O(\ell^2)$ messages since the secret is of length $O(\ell)$, using Byzantine agreement for the broadcast. If $\gamma > \frac{1}{3}$, then we must use authenticated Byzantine agreement [6]. Also, we will incur greater costs in the dealing phase since this operation cannot be simply composed [7]. If $\gamma < \frac{1}{3}$ we can reduce the dealing cost by using the technique proposed by Cachin, Kursawe and Shoup [3]. In the presence of dishonest players, this cost grows by a factor of $O(\ell)$ since each dishonest player can cause two more broadcasts to occur within the checking group. In the second phase of the protocol, without any adversaries the cost is again $O(\ell^2)$ messages. With adversaries, this cost again increases by a factor of $O(\ell)$ since each member of the checking group must broadcast.

The running time for this algorithm is also much shorter than that of previous solutions to this problem. Each player is a member of $O(\ell)$ checking groups and must check one equation for each. Each equation is a product of $O(\ell)$ modular exponentiations, so the cost is $O(\ell^2)$ exponentiations.

The constants in these asymptotic expressions are very reasonable. For an $n^{-2}$ chance of failure, if $\epsilon = 1/10$, then a suitable setting for $\ell$ is $17 \log n$. A linear increase of $\ell$ results in either an exponentially smaller failure probability or a linear decrease in $\epsilon$. Since the checking groups are of size $\frac{\ell}{2\epsilon^2}$, they are more sensitive to the value of $\epsilon$. In practice, it is reasonable for the gap between the the fraction of dishonest parties and the fraction of shares required for reconstruction to be a fixed constant, so the size of the $Q_i$ is logarithmic with a small leading constant.

## References

[1] Bellare, M., Rogaway, P.: Random oracles are practical: A paradigm for designing efficient protocols. In: Proceedings of the First ACM Conference on Computer and Communications Security, ACM Press (1993) 62–73

[2] Blakley, G.R.: Safeguarding cryptographic keys. **48** (1979) 313–317

[3] Cachin, C., Kursawe, K., Shoup, V.: Random oracles in constantipole: Practical asynchronous byzantine agreement using cryptography (extended abstract). In: Proceedings of the Nineteenth Annual ACM Symposium on Principles of Distributed Computing, ACM Press (2000) 123–132

[4] Gennaro, R., Jarecki, S., Krawczyk, H., Rabin, T.: Secure distributed key generation for discrete-log based cryptosystems. Lecture Notes in Computer Science **1592** (1999) 295+

[5] Gennaro, R., Jarecki, S., Krawczyk, H., Rabin, T.: Revisiting the distributed key generation for discrete-log based cryptosystems. (2003)

[6] Lamport, L., Shostak, R., Pease, M.: The byzantine generals problem. ACM Transactions on Programming Languages and Systems (TOPLAS) **4** (1982) 382–401

[7] Lindell, Y., Lysyanskaya, A., Rabin, T.: On the composition of authenticated byzantine agreement. In: Proceedings of the Thirty-fourth Annual ACM Symposium on Theory of Computing, ACM Press (2002) 514–523

[8] Pedersen, T.P.: A threshold cryptosystem without a trusted party. Lecture Notes in Computer Science **547** (1991) 522–526
[9] Pedersen, T.P.: Non-interactive and information-theoretic secure verifiable secret sharing. Lecture Notes in Computer Science **576** (1991) 129–140

## 5 Appendix

**Lemma 2.** *Consider a (reflecting) random walk $X_i$ defined in terms of a sequence of differences $D_i$:*

$$X_0 = 0$$
$$X_{i+1} = \max(0, X_i + D_{i+1})$$

*The sequence of differences $D_1, D_2, \ldots, D_n$, is generated at random and satisfies:*

$$|\{i|D_i = 1\}| = \alpha n \triangleq r, |\{i|D_i = -1\}| = (1-\alpha)n \triangleq s$$

*Let $\alpha \triangleq 1/2-\epsilon$. Then the probability that the walk has reached $\ell \triangleq -\frac{(2+\kappa)}{\log(1-2\epsilon)} \log n$ in $n$ or fewer steps is $P(X_j = \ell, j \leq n) < n^{-\kappa}$.*

*Proof.* Let $B_{i,j}$ be the event that $X_{i+j} = \ell, X_i = 0, X_k \neq 0, i < k < i+j$. That is the event where $i$ is the last time that the walk is at 0 and the walk is at $\ell$ after $j$ more steps. Of those $j$ steps, such a walk will have exactly $\ell$ more steps to the right than steps to the left. Hence there are $d \triangleq \frac{j-\ell}{2}$ left steps and $d+\ell$ right steps. There are fewer than $\binom{j}{d}$ ways to choose the order of the steps. Condition the sequence of differences on those $j$ steps:

$$P(B_{i,j}) < \binom{j}{d} \frac{\left(\prod_{k=0}^{d-1}(s-k)\right)\left(\prod_{k=0}^{d+\ell-1}(r-k)\right)}{\prod_{k=0}^{j-1}(n-k)}$$
$$= \binom{j}{d} \prod_{k=0}^{d-1} \frac{(s-k)(r-k)}{(n-2k)(n-2k-1)} \prod_{k=0}^{\ell-1} \frac{r-d-k}{n-2d-k}$$

Observe that if $a+b=c$ and $b < \frac{c}{2}-1$ then $\frac{ab}{c(c-1)} < \frac{1}{4}$. Also $\frac{r-d-k}{n-2d-k} < \frac{r}{n} = \frac{1}{2}-\epsilon$, for arbitrary $d, k$ since $r < \frac{n}{2}$ so that:

$$P(B_{i,j}) < \binom{j}{d} \left(\frac{1}{4}\right)^{\frac{j-\ell}{2}} \left(\frac{1}{2}-\epsilon\right)^{\ell}$$
$$= \binom{j}{d} \left(\frac{1}{2}\right)^{j} (1-2\epsilon)^{\ell}$$
$$< (1-2\epsilon)^{\ell}$$

So $P(B_{i,j}) < (1-2\epsilon)^{\ell} = n^{-(2+\kappa)}$. Now there are fewer than $n$ choices for either $i$ or $j$ so that $P(\cup_{i,j} B_{i,j}) < n^2 n^{-(2+\kappa)} = n^{-\kappa}$.

# Optimal Communication Complexity of Generic Multicast Key Distribution*

Daniele Micciancio and Saurabh Panjwani

University of California, San Diego
9500 Gilman Drive, Mail Code 0114,
La Jolla, CA 92093, USA
{daniele, panjwani}@cs.ucsd.edu

**Abstract.** We prove a tight lower bound for generic protocols for secure multicast key distribution where the messages sent by the group manager for rekeying the group are obtained by arbitrarily nested application of a symmetric-key encryption scheme, with random or pseudorandom keys. Our lower bound shows that the amortized cost of updating the group key for a secure multicast protocol (measured as the number of messages transmitted per membership change) is $\log_2(n) + o(1)$. This lower bound matches (up to a small additive constant) the upper bound of Canetti, Garay, Itkis, Micciancio, Naor and Pinkas (Infocomm 1999), and is essentially optimal.

**Keywords:** Multicast, Key Distribution, Lower Bounds.

## 1 Introduction

Broadcast and multicast are communication primitives of fundamental importance for many emerging internet (or more generally, network) applications, like teleconferencing, pay TV, on-line gaming, electronic news delivery, etc. Roughly speaking, broadcast allows data to be (simultaneously) delivered to all nodes in a network at a much smaller cost (in terms of network resources) than transmitting it individually to each intended recipient, and it is essential for the scalability of the applications to groups of medium and large size. Multicast achieves a similar goal, but with an arbitrary (and, often, dynamically changing) set of recipients that does not necessarily include all the nodes in the network. From a security point of view, broadcast and multicast raise many new and challenging issues that are not directly addressed by conventional (point-to-point) cryptographic techniques. (See [3] for a survey.)

* This material is based upon work supported by the National Science Foundation under Grant CCR-0313241 and a Sloan Research Fellowship. Any opinions, findings, and conclusions or recommendations expressed in this material are those of the author(s) and do not necessarily reflect the views of the National Science Foundation.

C. Cachin and J. Camenisch (Eds.): EUROCRYPT 2004, LNCS 3027, pp. 153–170, 2004.

*Security properties.* As in point-to-point communication (unicast), the two main security concerns are *secrecy* and *authenticity*. In this paper we concentrate on the secrecy property, i.e., making sure that only group members can receive the transmitted data (See Sect. 3 for a precise definition of our communication and security model). Two distinct models have been considered within the cryptographic community to study secrecy properties in broadcast and multicast scenarios. One, called *broadcast encryption*, is motivated mostly by pay TV and similar applications where an information provider communicates with a large (and highly dynamic) set of low-end receivers (e.g., set-top boxes). The other, usually called *multicast encryption* or *multicast key distribution*, is more closely related to internet applications, where a dynamically changing, but relatively stable, group of users wants to broadcast messages within the group, while keeping the content of the messages hidden from users that do not currently belong to the group. This is the model we study in this paper, and we refer the reader to Sect. 2 for a brief discussion of related work, including broadcast encryption as well as other security properties like authenticity.

In unicast, secrecy is easily achieved by establishing a secret key between the two communicating parties, who, in turn, use the key to encrypt all communication using a conventional (symmetric-key) encryption scheme. A similar approach may be used for multicast as well: once a secret key (common to all group members) is established, secrecy can be achieved by encrypting all communication under the common key. However, in the presence of a dynamically changing group, establishing a common secret key can be quite an onerous task: each time a user leaves the group (voluntarily or not), a new group key needs to be established in order to protect future communication. We consider a setting where a single (physical or logical) entity (called the group center) has authority over deciding group membership and is in charge of group key distribution. The problem is how the group center can securely communicate a new key to all the remaining group members, after one of them leaves the group, in such a way that the evicted user cannot recover the new key. Since the old group key can no longer be used to secure communication, communicating a new key seemingly requires unicasting the new key individually to all group members (e.g., as advocated in [10,9]), but this is clearly not a scalable solution as it would lose essentially all the potential efficiency benefits of using a multicast channel. The now standard approach to this problem, suggested in [16,15], is to maintain not only a group key, known to all group members, but also a collection of auxiliary keys known to selected subsets of members that can be used to efficiently communicate to subsets of the group when the group membership changes. The solution described in [16,15] requires the transmission of $2\log_2 n$ messages[1] each time a user leaves and another one joins the group, where $n$ is the size of the

[1] We measure the communication complexity in basic messages, where each message is a fixed size packet of sufficiently large size to allow for the transmission of a single (possibly encrypted) key.

group[2]. Although this is exponentially more efficient than the trivial solution requiring $n$ (unicast) transmissions, it would be desirable to have even more efficient solutions with smaller communication complexity. In [3] an improved solution is given, where the number of transmissions is reduced by a factor of 2, but remains logarithmic in the number of group members.

*Previous lower bounds.* Our inability to find even better solutions to the multicast key distribution problem has prompted many researchers to explore lower bounds, showing that no such improvement is indeed possible, under reasonable assumptions about the protocol. The first non-trivial communication lower bound for multicast security was proved in [4] for a restricted class of protocols, namely protocols where the group members have a bounded amount of memory, or the key distribution scheme has some special "structure preserving" property. A different, and seemingly optimal, lower bound, for a more general class of protocols without memory or structure restrictions was subsequently proved in [14], where it was shown that any secure multicast key distribution protocol (within a certain class) can be forced to transmit at least $3 \log_3 n$ messages for every group update operation (averaged over a long sequence of update operations). [14] also suggested a simple variant of the protocol of [16,15] (basically, replacing binary trees with ternary ones) meeting their lower bound. This apparently closed the gap between upper and lower bounds for multicast key distribution protocols, putting a final word to our search of an optimal solution. The class of protocols considered in [14] restricts the group center to transmit messages of the form $E_{k_1}(k_2)$ consisting of a key $k_2$ encrypted with another key $k_1$. Although not explicitly stated in [14], it is important to note that more general protocols are indeed possible and have also been considered in practice. For example, two relatively standard, and eminently practical, techniques very common in cryptography are the following:

- The use of a pseudorandom generator, say $G$, to expand a single key $k_0$ into two or more (seemingly random and independent) keys $(k_1, k_2, \ldots, k_m) = G(k_0)$. In principle, this allows to transmit multiple keys at the price of one, by sending the seed $k_0$, instead of transmitting the pseudorandom keys individually.
- The use of double (or multiply iterated) encryption, where more encryption functions are applied in a sequence to the same message before transmission. For example, consider a group of four users $u_1, u_2, u_3, u_4$, where each user $u_i$ knows a private key $k_i$. Assume two auxiliary keys $k$ and $k'$ are known to groups $u_1, u_2, u_3$ and $u_2, u_3, u_4$ respectively. Then a new key $k''$ can be sent to users $u_2$ and $u_3$ by transmitting a single (doubly encrypted) message $E_k(E_{k'}(k''))$. Notice that using single encryption, as in the model considered

[2] For simplicity, we consider groups of fixed size in analyzing multicast key distribution i.e. we assume that each time a user leaves, another one is immediately added. So the size of the group is always equal to $n$. We refer to each leave/join operation as a "group update operation". See Sect. 6 for a discussion on variable-sized groups with separate leave and join operations.

in [4,14], communicating to the same group of users requires the transmission of two messages $E_{k_2}(k'')$ and $E_{k_3}(k'')$.

The inadequacy of the model used by previous lower bounds [4,14] is clearly demonstrated by known (and practical) protocols that "beat" the lower bound proved in [14]. For example, [3] uses pseudorandom generators to improve the communication complexity of [16,15] by a factor of 2, resulting in a secure key distribution protocol where all update operations can be performed by transmitting $\log_2(n)$ messages, which is strictly smaller than the $3\log_3(n) \approx 1.89\log_2(n)$ lower bound proved in [14]. This observation opens up again the possibility of further improving the communication complexity of multicast key distribution, or proving more satisfactory lower bounds for more general classes of protocols.

*Our contribution.* In this paper, we consider generic protocols for multicast key distribution that make arbitrary use of pseudorandom generators and encryption algorithms, where both techniques can be mixed and iteratively applied multiple times in arbitrary ways. In our model, keys can be either freshly generated (i.e. are purely random) or produced by applying a pseudorandom generator (polynomially many times) on freshly generated keys. Messages sent out by the group center for rekeying the group are composed by encrypting keys *iteratively* using different (random or pseudorandom) keys for encrytion at each iteration (See Sect. 3 for a complete description of our model).

The lower bound we prove in this paper on multicast key distribution protocols matches the upper bound of [3] up to a small *additive* term. We demonstrate that in any protocol where the group center broadcasts arbitrary expressions built according to our formal logic for messages, the center must transmit $\log_2(n) + o(1)$ messages per group update operation in the worst case (here, $n$ is the size of the group and the number of messages per update operation is measured by amortizing over an infinite sequence of such operations). In other words, we demonstrate that the use of pseudorandom generators suggested in [3] is essentially optimal, and that even a combined use of iterated encryption does not substantially help to improve the worst-case communication complexity below $\log_2(n)$ messages per update.

*Organization.* In Sect. 2 we briefly review related work. In Sect. 3 we give a detailed description of the model used to prove our lower bound. The actual lower bound is proved in Sects. 4 and 5. Section 6 concludes the paper with a discussion on possible extensions to our model.

## 2 Related Work

Previous work on secure communication in broadcast and multicast scenarios is based on two distinct formulations. The first one, often referred to as *broadcast encryption*, has received much attention from the cryptographic community (e.g., [6,11].) In this model, as originally introduced by Fiat and Naor [6], receivers are stateless, in the sense that they receive a set of keys at the very beginning of the protocol, and they never update their state during protocol execution. However, broadcast encryption schemes are typically secure only against coalitions of bounded size. An essentially optimal lower bound on the communication complexity of broadcast encryption (as a function of the amount of key storage allowed per user) was given by Luby and Staddon in [11],

In this paper we consider a different scenario more closely related to internet applications, where the users maintain state, the group of recipients changes over time, and all users in the group may broadcast information to the other group members. As discussed in the following section, this problem is equivalent to the key distribution problem, where a common secret key is established among all current group members, and updated over time as the group membership changes. This problem, usually called *multicast encryption* or *multicast key distribution*, is the one studied for example in [16,15,4,14] already discussed in the introduction.

Besides secrecy, other important security issues are *authenticity*, i.e., making sure that only authorized users can transmit messages and these messages cannot be altered during transmission, *independence*, i.e., emulating a synchronous network where all players transmit and receive messages at the same time (e.g., see [7]), and *availability*, e.g., protecting the network against denial of service attacks. These are all different security concerns that can be addressed separately using the appropriate cryptographic techniques. Here we briefly discuss *authenticity*. As discussed in [3], one can distinguish different kinds of authenticity. The simplest kind only ensures that the sender of the information is one of the current group members. This can be achieved using the same techniques studied in this paper (e.g., establishing a common secret key and using it within a message authentication protocol.) Individual authentication is a much harder problem, and it has been shown that it is actually equivalent to using public key digital signatures [2].

## 3 The Model

We consider a scenario in which an information provider wishes to communicate to a selected (and dynamically changing) set of users over a broadcast channel. At any point in time, all users may receive the information sent over the broadcast channel, and we want to ensure that only current group members can decipher the transmitted information and recover the original messages sent by the information provider. A centralized trusted authority, called the group *center*, governs access to the group[3]. The problem of secure multicast communication is easily seen to be equivalent to the problem of establishing a common secret key, known to all and only the current group members: on the one hand, given a secret key shared among all current group members, the information provider can securely communicate with all group members by encrypting its messages

[3] We remark that such a group center is only a logical abstraction, and does not necessarily correspond to any single physical entity, like the information provider or any of the group members.

using a secure symmetric key encryption scheme. On the other hand, given a secure multicast protocol, the center can immediately establish a common secret key among all current group members by picking a new key at random and securely transmitting it to all group members using the secure multicast protocol. Therefore, in the rest of the paper we identify secure multicast encryption with the group key distribution problem. We remark that a common secret key allows all group members to act as information providers and securely transmit information encrypted under the common secret key. The common secret key can also be used to achieve additional security goals besides secrecy (for eg., message integrity against non-members).

### 3.1 Protocol Initialization

We assume users come from a fixed, but potentially infinite, set $\mathcal{U}$ and that they communicate with the group center using a reliable and authenticated broadcast channel. At every time instant $t$ a finite set of users, $\mathcal{M}_t \subset \mathcal{U}$, referred to as members, holds a shared secret key which is supposed to be known only to the users in this set. All users and the center have black-box access to three functions, $E$, $D$ and $G$, where the functions $(E, D)$ model an encryption/decryption pair and $G$ models a pseudorandom generator. We think of these three functions as abstract operations satisfying the following conditions :

- $E$ takes as input two expressions, $K$ (a key) and $\gamma$ (a message), and outputs another expression, $\beta$ (a ciphertext). $D$ takes two expressions, $K'$ and $\beta'$, as input and outputs a third expression $\gamma'$. These operations satisfy the obvious correctness condition : $D(K, E(K, \gamma)) = \gamma$. We write $E_K(\gamma)$ for $E(K, \gamma)$.
- $G$ takes as input a key $K$ and outputs two keys, denoted $G_0(K)$ and $G_1(K)$. In other words, the function $G$ models a length-doubling pseudorandom generator. We remark that our choice of using a length-doubling generator (and not a more general one) is only for the purpose of simplifying the analysis and it does not impact our lower bound in any way[4].

Every user $u_i \in \mathcal{U}$ also has a secret key $K_i$ (referred to as the *unique* key of that user) that is known only to him and the group center $C$ from the beginning of protocol execution (Such a key may be established using different techniques in a setup phase using, say, unicast and public key cryptography).

### 3.2 Rekey Messages

Changes in the group membership (i.e. the set $\mathcal{M}_t$) over time are modeled using an adversary who adaptively chooses to add and delete members from the group.

[4] Indeed, our lower bound can be shown to hold even if we replace $G$ with a function that takes as input a single key and outputs arbitrarily many pseudorandom keys. An intuitive reason for this is that any pseudorandom generator with arbitrary expansion-factor can be easily built using only a length-doubling generator. The proof of Lemma 2 makes this clearer.

At every point in time $t$, our adversary examines the history and current state of the protocol and issues one of the following three commands:

- $JOIN(u_i)$: set $\mathcal{M}_{t+1} = \mathcal{M}_t \cup \{u_i\}$,
- $LEAVE(u_i)$: set $\mathcal{M}_{t+1} = \mathcal{M}_t \setminus \{u_i\}$,
- $REPLACE(u_i, u_j)$: set $\mathcal{M}_{t+1} = \mathcal{M}_t \setminus \{u_i\} \cup \{u_j\}$,

In response to a membership change request, the group center transmits a set of messages $S_t = (\gamma_1, \ldots, \gamma_{|S_t|})$, known as *rekey messages*, over the broadcast channel where each rekey message, $\gamma_i$, is a symbolic expression derived using the following grammar :

$$\begin{aligned} \mathbf{M} &\rightarrow E_{\mathbf{K}}(\mathbf{M}) \mid \mathbf{K} \\ \mathbf{K} &\rightarrow K \mid G_0(\mathbf{K}) \mid G_1(\mathbf{K}) \end{aligned} \tag{1}$$

Here, the symbol $\mathbf{M}$ represents *messages* while the symbol $\mathbf{K}$ represents *keys*. The expression $K$ models any basic (i.e. freshly generated) key, including unique keys of users. Messages can be built from keys by iterated application of the encryption function, $E$, with basic keys or derived keys (obtained using the pseudorandom generator)[5].

*Communication Complexity.* The communication complexity of a group key distribution protocol is defined in terms of the *number* of rekey messages transmitted by the center per update operation performed on the group. The cost of transmitting a set of messages $S_t$ equals the number of basic messages in the set (i.e. $|S_t|$). The *amortized cost* of a group key distribution protocol in the course of a sequence of such adversarial operations is the ratio of the total number of messages transmitted by the center in that period to the total number of operations performed. This is expressed in terms of the *size* of the group, which is the maximum number of members in the group at any stage in that sequence of operations (As we will see, in our lower bound analysis, the number of members is kept constant across time). The *amortized communication complexity* of the protocol is the maximum amortized cost it has to incur in the course of *any* sequence of adversarial operations. We are interested in a lower bound on the amortized communication complexity for any group key distribution protocol satisfying certain constraints. We next describe what these constraints are.

### 3.3 Security Definition

We analyze the security of key distribution protocols with respect to the abstract cryptographic operations $E$, $D$ and $G$. This approach is similar to that taken in

[5] Note that we do not allow the use of expressions of the form $E_{\mathbf{K}}(\mathbf{M})$ (i.e. ciphertexts) either as keys or as inputs to the pseudorandom generator because ciphertexts do not necessarily have the (pseudo)randomness properties necessary to prove that such an application would be secure. For example, given any (provably secure) encryption function, $E$, it is possible to build another (provably secure) encryption function, $E'$, such that one can easily recover a message, $\gamma$, from a corresponding ciphertext $E'_{E'_{K_0}(K_1)}(\gamma)$ even without knowing any of the keys $K_0$ and $K_1$.

previous lower bounds for this problem[4,14], except that we also allow for the use of pseudorandomness and arbitrarily nested encryption as dictated by our grammar.

**Definition 1.** *For any set, $S$, of messages obtained using grammar 1, we define the set of keys that can be derived from $S$ as the smallest set, $\mathbf{Keys}(S)$, which satisfies the following three conditions :*

- *If $K_0 \in S$, then $K_0 \in \mathbf{Keys}(S)$.*
- *If $K_0 \in \mathbf{Keys}(S)$, then $G_0(K_0) \in \mathbf{Keys}(S)$ and $G_1(K_0) \in \mathbf{Keys}(S)$.*
- *If $E_{K_1}(E_{K_2}(\cdots(E_{K_l}(K_0)))) \in S$ and $K_1, \ldots, K_l \in \mathbf{Keys}(S)$, then $K_0 \in \mathbf{Keys}(S)$.*

This definition corresponds to the intuitive idea that given $E_K(M)$ one can compute $M$ if and only if $K$ is known, and given $K$ everybody can compute $G_0(K)$ and $G_1(K)$ applying the pseudorandom generator to $K$. However, since pseudorandom generators are one-way, given $G_0(K)$ or $G_1(K)$ (or both) one cannot recover $K$, or even tell if $G_0(K), G_1(K)$ is in the range of the pseudorandom generator. This is essentially a straightforward generalization of the Dolev-Yao [5] model of encryption, extended with pseudorandom generation. Analyzing security of protocols with respect to this formal cryptographic model is motivated by the fact that we would like the protocols to be secure independently of the specific instantiation of the underlying cryptographic building blocks. The formal analysis can be made precise and interpreted in standard complexity-theoretic terms, by extending known soundness and completeness results of [1,12].

**Definition 2.** *We say that a group key distribution protocol is* **secure** *if for any sequence of adverserial operations, and for every time instant $t$, there exists a key, $K$, such that*

- *$K \in \mathbf{Keys}(S_1 \cup \cdots \cup S_t \cup \{K_i\})$ for all $u_i \in \mathcal{M}_t$, i.e., key $K$ can be computed by all current group members at time $t$;*
- *$K \notin \mathbf{Keys}(S_1 \cup \cdots \cup S_t \cup \{K_i : u_i \notin \mathcal{M}_t\})$, i.e., the users that do not belong to the group at time $t$ cannot compute $K$ even if they collude and pool together all the information available to them.*

The first clause in the definition is a correctness criterion while the second clause is the main security condition. Note that our definition of security is a bit restrictive in that it requires non-members not to be able to obtain the shared secret key at any instant of time based only on the information obtained *at or before* that instant. Intuitively, this captures the idea that if a user leaves the group (i.e. becomes a non-member) at some point, then he should not be able to decrypt any future communication (even if he colludes with other non-members to do so), unless, of course, he is added back to the group. This kind of security is often referred to as *forward secrecy.* However, one could also require that the shared secret key at any instant be such that the non-members (at that instant) not to be able to compute it even *later on* i.e. even if some of them become members in the future. Such a security requirement is more stringent and it captures the notion that any new entrant to the group should not be able to compute the shared key for any past instant when he was not a member (a requirement often referred to as *backward secrecy*). In order for a protocol to satisfy both forward and backward secrecy, we must strengthen the security condition above so that $K \notin \mathbf{Keys}(S_1 \cup \cdots \cup S_{t'} \cup \{k_i : u_i \notin \mathcal{M}_t\})$ *for all* $t' \geq t$. We remark that backward secrecy is usually considered a less important property than forward secrecy, as in many multicast applications (e.g., stock quotes) information looses value over time. The lower bound proved in this paper only requires forward secrecy and is, thus, applicable to protocols satisfying the more stringent definition, too.

Another important remark is the following. Since most networking protocols do not provide any form of security, it is a good practice to assume that an adversary attacking the network has access to all transmitted data, which needs to be properly protected using appropriate cryptographic techniques. Moreover, this allows for the development of security solutions that are independent of the underlying networking technology. In the above definition, the security criterion models the fact that the adversary has complete knowledge of all past communication. The assumption of infinite memory is less reasonable in the case of group members in our correctness criterion, but giving all past broadcast messages to all users makes our security definition less stringent, and consequently it only makes our lower bound stronger. We refer the interested reader to Sect. 6 for a discussion on possible extensions to our model.

## 4 The Multicast Game

For the actual lower bound analysis, it is useful to view every secure group key distribution protocol as an abstract game, which we call the *multicast game*, played between the group center, $C$, and the adversary, $A$. In this game, keys are modelled as nodes in an infinite hypergraph. Each node corresponds to either a basic key (recall that basic keys include unique keys of users as well) or a derived key obtained by applying the pseudorandom generator to some other key. Messages transmitted by the group center are modeled as directed hyperedges, so that the cost incurred by the center equals the number of hyperedges in the graph. For any user, the set of keys known to him at any time is defined as the set of nodes that can be "reached" from the node representing his unique key following the hyperedges. Details follow.

### 4.1 Game Configurations

The playing board for the multicast game is an infinite collection of rooted binary trees, $T = \{T_1, T_2, \cdots\}$ each containing an infinite number of nodes. The entire set of nodes in these trees is denoted $\mathcal{V}$. The edges are directed edges and every tree in $T$ has one root node which has zero in-degree while all other nodes have in-degree equal to 1. The out-degree of all nodes, including the root, is equal to 2. Every node in this playing board represents a key $K$. The roots of the trees are associated to the basic keys, while the internal nodes are pseudorandom keys.

The two children of a node represent keys $G_0(K)$ and $G_1(K)$, the keys that can be obtained by applying the pseudorandom generator to the key, $K$, of the parent node.

The root nodes of some (but not all) trees in $T$ correspond to the unique keys, $K_i$, of all users. We refer to these special trees as *user trees* and denote the entire set of user trees by $U$. At any given point in time during the game, the root of every tree in $U$ has one of two labels associated with it – *member* or *non-member*. We refer to the edges in the trees in $T$ as *tree-edges* or simply $t$-edges and the entire set of $t$-edges in all the trees is denoted $\mathcal{T}$. A $t$-edge from a node $v_1$ to a node $v_2$ is denoted $v_1 \xrightarrow{\text{t}} v_2$.

Rekey messages sent by the group center are modeled as hyperedges as follows. A *directed hyper-edge*, or simply an $h$-*edge*, over nodes in $\mathcal{V}$ is a pair $\{V, v\}$, denoted $V \xrightarrow{\text{h}} v$, where $V$ is a finite subset of $\mathcal{V}$ and $v$ is a single node. The $h$-edge $V \xrightarrow{\text{h}} v$ is said to be incident on $v$. The hyperedge, $\{K_1, \cdots, K_d\} \xrightarrow{\text{h}} K$ models a rekey message of the form $E_{K_1}(E_{K_2}(\cdots E_{K_d}(K)\cdots))$. Here, $K_1, \ldots, K_d, K$ can be either basic or derived keys (i.e., keys associated to either root or internal nodes), and the encryptions can be performed in any order.

A *configuration*, $\mathcal{C}$, of the multicast game is defined as a triple $\mathcal{C} = (\mathcal{M}, \mathcal{N}, \mathcal{H})$, where $\mathcal{M}$ is the set of all member nodes, $\mathcal{N}$ is the set of all non-member nodes and $\mathcal{H}$ is a (finite) set of $h$-edges over nodes in $\mathcal{V}$. The union $\mathcal{M} \cup \mathcal{N}$ is always equal to the set of roots of the user trees in $U$. A configuration of the game at time $t$ corresponds to the state of the group key distribution protocol at time $t$ with $\mathcal{M}$ representing the set of members, $\mathcal{N}$ the set of non-members and $\mathcal{H}$ the set $S_0 \cup S_1 \cup S_2 \cup \cdots \cup S_t$ of rekey messages transmitted by $C$ in response to the first $t$ group update operations (plus an optional set $S_0$ corresponding to the initial configuration of the game).

### 4.2 Defining Moves of Players

Each move by player $C$ in our game involves adding zero or more $h$-edges and each move by player $A$ involves changing the label on a node labelled *member* to *non-member* or vice versa, or swapping a member with a non-member. Formally, if the game is in a configuration $\mathcal{C} = (\mathcal{M}, \mathcal{N}, \mathcal{H})$, then

- a move by player $C$ changes the configuration of the game to $\mathcal{C}' = (\mathcal{M}, \mathcal{N}, \mathcal{H}')$ where $\mathcal{H}' = \mathcal{H} \bigcup \mathcal{H}_a$ and $\mathcal{H}_a$ is a finite (possibly empty) set of $h$-edges over nodes in $\mathcal{V}$;
- a move by player $A$ changes the configuration to $\mathcal{C}' = (\mathcal{M}', \mathcal{N}', \mathcal{H})$ where either
  - $\mathcal{M}' = \mathcal{M} \setminus \{v_m\}$ and $\mathcal{N}' = \mathcal{N} \bigcup \{v_m\}$ for some $v_m \in \mathcal{M}$ (we call this a `delete` move and we say that the node $v_m$ gets deleted from $\mathcal{M}$); or
  - $\mathcal{M}' = \mathcal{M} \bigcup \{v_n\}$ and $\mathcal{N}' = \mathcal{N} \setminus \{v_n\}$ for some $v_n \in \mathcal{N}$ (we call this an `add` move and we say that the node $v_n$ gets added to $\mathcal{M}$).
  - $\mathcal{M}' = \mathcal{M} \bigcup \{v_n\} \setminus \{v_m\}$ and $\mathcal{N}' = \mathcal{N} \setminus \{v_n\} \cup \{v_m\}$ for some $v_n \in \mathcal{N}$ and $v_m \in \mathcal{M}$ (we call this a `replace` move and we say that the node

$v_m$ gets replaced by $v_n$). This corresponds to a simultaneous execution of an `add` move and a `delete` move, and it leaves the size, $|\mathcal{M}|$, of the group unchanged.

At any time instant $t$, a pair of moves is played, the first move being played by $A$, followed by a response by $C$. Associated with each player's move is a cost function. The cost of a move by player $C$ is the number of $h$-edges added by him i.e. if a move by player $C$ takes the game from $\mathcal{C} = (\mathcal{M}, \mathcal{N}, \mathcal{H})$ to $\mathcal{C}' = (\mathcal{M}, \mathcal{N}, \mathcal{H}')$, then the cost of the move is $|\mathcal{H}'| - |\mathcal{H}|$. The cost of any move by player $A$ is 1. For simplicity, we concentrate on `replace` operations that leave the size of the group unchanged (since we are interested in proving a lower bound, considering only `replace` operations only makes our result stronger).

### 4.3 Defining Goals of Players

The security notion described in Sect. 3 is easily modeled in terms of reachability between nodes in the hypergraph corresponding to the current configuration.

**Definition 3.** *A node, $v \in \mathcal{V}$, is called h-reachable from a set of nodes, $V \subseteq \mathcal{V}$, under a configuration $\mathcal{C} = (\mathcal{M}, \mathcal{N}, \mathcal{H})$ if any of the following conditions hold:*

- *$v \in V$.*
- *There exists a t-edge from some node $v'$ to $v$ and $v'$ is h-reachable from $V$.*
- *For some $m > 0$ and a set of nodes, $V = \{v_1, v_2 \cdots, v_m\} \subseteq \mathcal{V}$, there exists an h-edge $V \xrightarrow{\text{h}} v$ in $\mathcal{H}$ and each of the nodes, $v_1, \cdots, v_m$ is h-reachable from $V$.*

We write $V \Rightarrow_{\mathcal{C}} v$ to denote that $v$ is $h$-reachable from $V$ under $\mathcal{C}$ and $V \not\Rightarrow_{\mathcal{C}} v$ to denote the converse. We say that $v$ is $h$-reachable from a *node* $v'$ under $\mathcal{C}$ if $\{v'\} \Rightarrow_{\mathcal{C}} v$ holds; this is denoted simply by $v' \Rightarrow_{\mathcal{C}} v$ (similarly, $v' \not\Rightarrow_{\mathcal{C}} v$ denotes that $v$ is not $h$-reachable from $v'$ under $\mathcal{C}$). If $\mathcal{S}$ is the set of rekey messages represented by $\mathcal{H}$, the set of $h$-edges in $\mathcal{C}$, and $\mathcal{K}$ the set of keys represented by $V$, then the set of nodes $h$-reachable from $V$ under $\mathcal{C}$ corresponds exactly to the set of keys **Keys**$(\mathcal{K} \cup \mathcal{S})$ that can be computed from $\mathcal{K}$ and $\mathcal{S}$ according to the Dolev-Yao model of abstract encryption described in Sect. 3.

A configuration which satisfies the security constraint for group key distribution is called a secure configuration:

**Definition 4. (Secure Configuration)** *A configuration $\mathcal{C} = \{\mathcal{M}, \mathcal{N}, \mathcal{H}\}$ is called a secure configuration if there exists a node, $v_s \in \mathcal{V}$, such that*

- *$v_s$ is h-reachable from every node in $\mathcal{M}$ under $\mathcal{C}$ i.e. $\forall v \in \mathcal{M}, v \Rightarrow_{\mathcal{C}} v_s$*
- *$v_s$ is not h-reachable from $\mathcal{N}$ under $\mathcal{C}$ i.e. $\mathcal{N} \not\Rightarrow_{\mathcal{C}} v_s$*

*A node, $v_s$, which satisfies this property is called a secret node for the corresponding secure configuration.*

Clearly, the shared secret key at any instant of time, $t$, in the protocol, must be (represented by) one of the secret nodes for the game configuration corresponding to time $t$.

Goals of the players can now be defined in terms of secure configurations. The goal of player $C$ is that at the end of *each* of his moves, the game be in a secure configuration. The goal of player $A$ is the converse of this i.e. at the end of *at least one* of player $C$'s moves, the configuration of the game is not secure. Our aim here is to determine the minimum cost that *every* player $C$ needs to pay, relative to the cost paid by player $A$, in order to be able to attain his goal in the game against *any* player $A$.

## 5 The Lower Bound Proof

In this section we present our main technical result on multicast games which directly implies the lower bound for secure group key distribution protocols.

### 5.1 Usefulness of $h$-edges and Canonical Graphs

Let us fix a configuration, $\mathcal{C} = (\mathcal{M}, \mathcal{N}, \mathcal{H})$ in the multicast game for this entire subsection. An $h$-edge, $V \xrightarrow{\text{h}} v$, in $\mathcal{H}$ is said to be *useless* under $\mathcal{C}$ if $\mathcal{N} \Rightarrow_{\mathcal{C}} v$. An $h$-edge which is not useless under $\mathcal{C}$ is called *useful* under it. By the definition of $h$-reachability, for every useful $h$-edge, $V \xrightarrow{\text{h}} v$, in $\mathcal{H}$, there must exist at least one node in $V$ which is not $h$-reachable from $\mathcal{N}$ under $\mathcal{C}$. We assume an arbitrary total order on the set $\mathcal{V}$ of all nodes. For any useful $h$-edge $V \xrightarrow{\text{h}} v$, the first node (according to the total ordering) in $V$ which is not $h$-reachable from $\mathcal{N}$ (under $\mathcal{C}$) is referred to as the *canonical node* of that $h$-edge. Canonical nodes are defined only for useful $h$-edges.

A *canonical edge*, or *c-edge*, corresponding to a useful $h$-edge, $V \xrightarrow{\text{h}} v$, is a simple directed edge from the canonical node, $v_c$, of that $h$-edge to $v$ and is denoted $v_c \xrightarrow{\text{c}} v$. The definitions of canonical nodes and edges are both specific to the configuration $\mathcal{C}$.

**Definition 5.** *Let $\mathcal{C} = (\mathcal{M}, \mathcal{N}, \mathcal{H})$ be a configuration of the multicast game. A canonical path or a c-path from a node $v_1$ to another node $v_2$ ($v_1, v_2 \in \mathcal{V}$) under $\mathcal{C}$, denoted $v_1 \leadsto_{\mathcal{C}} v_2$, is a path consisting of zero or more t-edges and c-edges such that all nodes on this path are h-reachable from $v_1$.*

At this point it is not clear whether a canonical path must exist from any node $v_1$ to any other node $v_2$. Indeed, this does not hold for every pair $(v_1, v_2)$. The following lemma characterizes the existence of canonical paths for certain pairs of nodes - a canonical path from $v_1$ to $v_2$ must exist if $v_2$ is $h$-reachable from $v_1$ but is not $h$-reachable from the set $\mathcal{N}$.

**Lemma 1.** *For any configuration $\mathcal{C} = (\mathcal{M}, \mathcal{N}, \mathcal{H})$ and any two nodes $v_1, v_2 \in \mathcal{V}$, if $\{v_1\} \Rightarrow_{\mathcal{C}} v_2$ and $\mathcal{N} \not\Rightarrow_{\mathcal{C}} v_2$, then there exists a c-path from $v_1$ to $v_2$ under $\mathcal{C}$.*

*Proof.* Let $R(v_1) \subseteq \mathcal{V}$ denote the set of all nodes which are $h$-reachable from $v_1$ (here, and everywhere else in the proof, $h$-reachable means $h$-reachable under $\mathcal{C}$). Let $B \subseteq R(v_1)$ be the set of *bad* nodes such that for all $v_2 \in B$, $v_2$ is not $h$-reachable from $\mathcal{N}$ and yet, there exists no $c$-path from $v_1$ to $v_2$. Let $G = R(v_1) \backslash B$ (the set of *good* nodes). We claim that either the set of bad nodes is empty or (if not so) $v_1$ is in it (i.e. $B = \phi$ or $v_1 \in B$).

Suppose this is not the case i.e. suppose that $B$ is non-empty and it still doesn't contain $v_1$. Then for all nodes in $B$ to be $h$-reachable from $v_1$, there exists some node $v_2 \in B$ such that one of the following conditions hold *(i)* For some $v \in G$, $v \xrightarrow{t} v_2 \in \mathcal{T}$; *(ii)* For some $V \subseteq G$, $V \xrightarrow{h} v \in \mathcal{H}$. Since any $v_2 \in B$ is not $h$-reachable from $\mathcal{N}$, an $h$-edge incident on it must be useful and thus, must have a $c$-edge corresponding to it. So, if $B$ is non-empty and doesn't contain $v_1$ there must exist a $t$-edge or a $c$-edge from some node $v \in G$ to some node $v_2 \in B$. By the definition of $B$ there exists no $c$-path from $v_1$ to such a $v_2$. Which means there must not be a $c$-path from $v_1$ to $v$ as well (else joining such a path with the edge between $v$ and $v_2$ would give us a $c$-path from $v_1$ to $v_2$). At the same time $v$ must not be $h$-reachable from $\mathcal{N}$ for that would imply $\mathcal{N} \Rightarrow_{\mathcal{C}} v_2$. Both these two conditions qualify $v$ to be a member of $B$, which it is not. We, thus, conclude that the set $B$ is either empty or contains the node $v_1$. If $B$ is an empty set, we're done. If it isn't and it contains $v_1$, then the definition of $B$ is defied since there exists a trivial $c$-path (with 0 edges) from $v_1$ to itself. Thus, the set $B$ must be empty and the lemma holds. ∎

**Canonical Graphs** We focus our attention on secure configurations from now on. Let $\mathcal{C} = (\mathcal{M}, \mathcal{N}, \mathcal{H})$ be a secure configuration with secret node $v_s$. By the definition of a secret node and by Lemma 1, for every $v_m \in \mathcal{M}$ there must exist a canonical path from $v_m$ to $v_s$. For every $v_m \in \mathcal{M}$ select a $c$-path $P_m \equiv v_m \rightsquigarrow_{\mathcal{C}} v_s$. The canonical graph for $\mathcal{C}$, denoted $G(\mathcal{C})$, is defined as the graph formed by superimposing the $c$-paths $P_m$ associated to the member nodes $v_m \in \mathcal{M}$. While superimposing paths, if there is more than one $c$-paths containing an edge between the same two nodes and if at least one of these edges is a $c$-edges then we insert a single $c$-edge between the nodes in $G(\mathcal{C})$, and if all these edges are $t$-edges, then we insert a single $t$-edge between the nodes. If there is no edge (a $c$-edge or a $t$-edge) between any two nodes then there is no edge between them in $G(\mathcal{C})$ also. Note that in the graph $G(\mathcal{C})$, there may be more than one paths from any member node $v_m$ to $v_s$, but only one of them corresponds (modulo replacement of $t$-edges by $c$-edges) to the canonical path $P_m$ associated to $v_m$. Figure 1(a) shows a toy example of a canonical graph for a configuration with three members nodes $\{v_1, v_2, v_3\}$ and secret node $v_s$.

For each member node, $v_m$, in $\mathcal{M}$, we define the *incidence weight* of $v_m$ in the graph $G(\mathcal{C})$ as the number of $c$-edges in this graph incident on any node along the $c$-path $P_m \equiv v_m \rightsquigarrow_{\mathcal{C}} v_s$. This is at least equal to the number of $c$-edges on the $c$-path itself. The *maximum incidence weight* of the graph $G(\mathcal{C})$ is the maximum among the incidence weights of all member nodes in it. A useful property on

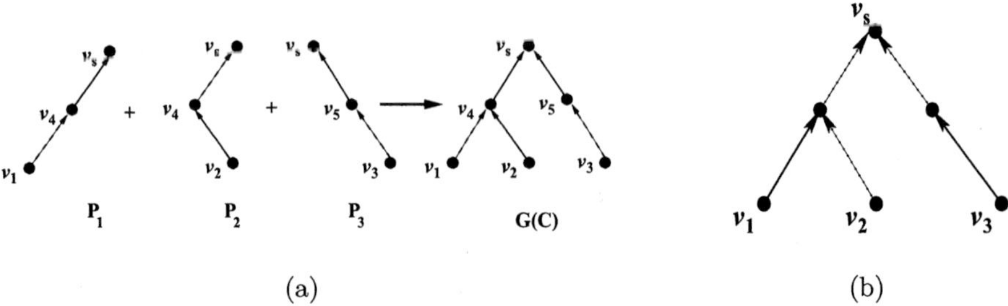


(a) (b)

**Fig. 1. Canonical Graphs :** Figure (a) shows the construction of a canonical graph for a configuration with three member nodes $\{v_1, v_2, v_3\}$ and secret node $v_s$. $c$-edges are shown by dark lines while $t$-edges are shown by dotted ones. Path $P_i$ goes from member $i$ to $v_s$. Note that there is a $c$-edge between $v_4$ and $v_s$ in $P_1$ and a $t$-edge between the same two nodes in $P_2$; the final graph has a $c$-edge between these nodes because of the higher precedence given to $c$-edges. In this graph, $v_1, v_2$ and $v_3$ have incidence weights 3,3 and 2 respectively. Figure (b) shows an example of a graph that cannot be a canonical graph since the topmost node, $v_s$, has two $t$-edges enterring it. This restriction on $t$-edges will be crucial in proving Lemma 2.

the maximum incidence weight of any canonical graph is given by the following lemma.

**Lemma 2.** *Let $\mathcal{C} = (\mathcal{M}, \mathcal{N}, \mathcal{H})$ be a secure configuration such that $|\mathcal{M}| = n$. Then, any canonical graph for $\mathcal{C}$ has maximum incidence weight at least $\lceil \log_2 n \rceil$.*

*Proof.* We shall prove something stronger than what the lemma states. We say that a node $v \in \mathcal{V}$ is *$\mathcal{M}'$-secure* under $\mathcal{C}$ (for some $\mathcal{M}' \subseteq \mathcal{M}$) if $\mathcal{N} \not\to_{\mathcal{C}} v$ and for all nodes $u \in \mathcal{M}'$, $u \not\to_{\mathcal{C}} v$. For a set $\mathcal{M}' \subseteq \mathcal{M}$ and a node $v$ which is $\mathcal{M}'$-secure under $\mathcal{C}$, we define a *sub-canonical graph* for $\mathcal{C}$ *over $\mathcal{M}'$ and $v$*, denoted $G_{\mathcal{C}}(\mathcal{M}', v)$, as a graph formed by superimposing $c$-paths from nodes in $\mathcal{M}'$ to $v$, one $c$-path being selected for every node in $\mathcal{M}'$. The set of $c$-paths from nodes in $\mathcal{M}'$ to $v$ used for constructing $G_{\mathcal{C}}(\mathcal{M}')$ is denoted $P(G_{\mathcal{C}}(\mathcal{M}', v))$. As a special case, observe that any canonical graph for $\mathcal{C}$ is a sub-canonical graph over $\mathcal{M}$ and $v_s$.

We hypothesize that for all $i > 0$, if there exists a pair $(\mathcal{M}', v')$ where $\mathcal{M}' \subseteq \mathcal{M}$, $|\mathcal{M}'| = i$ and $v'$ is $\mathcal{M}'$-secure then for every such pair $(\mathcal{M}', v')$ the maximum incidence weight of any graph $G_{\mathcal{C}}(\mathcal{M}', v')$ is at least $\lceil \log_2 i \rceil$. This hypothesis clearly implies the above lemma.

The proof uses an inductive argument on $i$. For the base case observe that a single node in the set $\mathcal{M}$ is a trivial sub-canonical graph with $\lceil \log_2 1 \rceil = 0$ $c$-edges. Suppose that for some $j > 1$ and for all $i < j$, the maximum incidence weight of any graph, $G_{\mathcal{C}}(\mathcal{M}_i, v_i)$, with $\mathcal{M}_i \subseteq \mathcal{M}$ and $|\mathcal{M}_i| = i$, if there exists such a graph, is at least $\lceil \log_2 i \rceil$. Suppose there exists a pair $(\mathcal{M}_j, v_j)$ such that, $\mathcal{M}_j \subseteq \mathcal{M}$, $|\mathcal{M}_j| = j$ and $v_j$ is $\mathcal{M}_j$-secure under $\mathcal{C}$. Consider the graph $G_{\mathcal{C}}(\mathcal{M}_j, v_j)$ and let $v_r$ be the (unique) node in this graph ($v_r$ may be the same as $v_j$) such that $v_r$ has in-degree greater than 1, say $m$, and all nodes on the path from $v_r$ to $v_j$ have in-degree exactly 1 (By in-degree of a node we mean the number of $c$-edges *and* $t$-edges in $G_{\mathcal{C}}(\mathcal{M}_j, v_j)$ incident on it). Since $v_j$ is not $h$-reachable from $\mathcal{N}$ and since a $c$-edge is defined only for a pair of nodes both of which are not $h$-reachable from $\mathcal{N}$, $v_r$ must also not be $h$-reachable from $\mathcal{N}$. Let $v_1, v_2, \cdots, v_m$ be the nodes which point to $v_r$ and $\mathcal{M}_1, \mathcal{M}_2 \cdots \mathcal{M}_m$ be the sets of member nodes in $\mathcal{M}_j$ for which the canonical paths to $v_j$ go through $v_1, v_2 \cdots v_m$ respectively. Since $v_r$ is not $h$-reachable from $\mathcal{N}$ under $\mathcal{C}$, none of the nodes $v_1, v_2, \cdots, v_m$ must be so, too. It is not hard to see that, for any $l \in \{1, \cdots, m\}$, the graph formed by superimposing the portion of the $c$-paths from nodes in $\mathcal{M}_l$ upto the node $v_l$ is also a sub-canonical graph for $\mathcal{C}$ (over $\mathcal{M}_l$ and $v_l$). Furthermore, there exists some $l' \in [m]$ such that $|\mathcal{M}_{l'}| \geq \lceil j/m \rceil$. From the induction hypothesis, the maximum incidence weight the graph $G_{\mathcal{C}}(\mathcal{M}_{l'}, v_{l'})$ is at least $\lceil \log_2 \left(\frac{j}{m}\right) \rceil$. Finally, the maximum incidence weight of $G_{\mathcal{C}}(\mathcal{M}_j, v_j)$ must be at least equal to the maximum incidence weight of $G_{\mathcal{C}}(\mathcal{M}_{l'}, v_{l'})$ plus the number of $c$-edges incident on $v_r$ in $G_{\mathcal{C}}(\mathcal{M}_j, v_j)$. A crucial observation is that there can be at most one $t$-edge incident on $v_r$, which means at least $m-1$ out of the $m$ edges incident on it must be $c$-edges. Thus, the maximum incidence weight of $G_{\mathcal{C}}(\mathcal{M}_j, v_j)$ is at least $\min_{m \in [j]} \lceil \log_2 \left(\frac{j}{m}\right) \rceil + m - 1$ which is not less than $\lceil \log_2 j \rceil$. ∎

### 5.2 The Main Theorem

We consider multicast games in which the group center, $C$, always maintains the game in a secure configuration. The following theorem establishes a logarithmic lower bound on the amortized cost of the moves performed by $C$, when the moves of $A$ are adversarially chosen. The lower bound holds for any initial configuration, and even if $A$ only issues `replace` operations that do not affect the size of the group. This lower bound directly implies a $\lceil \log_2 n \rceil$ lower bound on the amortized communication complexity of any secure group key distribution protocol.

**Theorem 1.** *For every strategy of player $C$ and initial configuration $(\mathcal{M}_0, \mathcal{N}_0, \mathcal{H}_0)$, there exists a strategy of player $A$ consisting of* `replace` *operations only, such that for any $t \geq 1$, the amortized cost, $\tilde{c}_t$, of the first $t$ moves of $C$ is at least $\lceil \log_2 n \rceil - |\mathcal{H}_0|/t$, where $n = |\mathcal{M}_0|$ is the size of the group. In particular, the asymptotic amortized cost of the moves of $C$ is*

$$\lim_{t \to \infty} \tilde{c}_t \geq \lceil \log_2 n \rceil .$$

*Proof.* Let $(\mathcal{M}_i, \mathcal{N}_i, \mathcal{H}_i)$ be the sequence of configurations following each move by $C$. We know by assumption that all configurations are secure. Notice that for all $i$, $\mathcal{H}_{i-1} \subseteq \mathcal{H}_i$, and the cost of each move by $C$ equals $c_i = |\mathcal{H}_i| - |\mathcal{H}_{i-1}|$. The moves of $A$ are chosen as follows. All moves are *replace* moves that substitute one of the current member nodes with a non-member node. In particular, the size of the group is always equal to $n = |\mathcal{M}_0| = |\mathcal{M}_i|$. By Lemma 2, the maximum incidence weight of the canonical graph, $G(\mathcal{C}_i)$, for any configuration $\mathcal{C}_i = (\mathcal{M}_i, \mathcal{N}_i, \mathcal{H}_i)$ is at least $\lceil \log_2 n \rceil$. Let $v_i$ be a member node achieving the maximum incidence weight in $G(\mathcal{C}_i)$. In his $i$th move, player $A$ replaces the member node $v_i$ with a new node from $\mathcal{N}_i$ that never was a member node before.

For the configuration, $\mathcal{C}_i$, consider the graph, $G(\mathcal{C}_i)$, and the $c$-path $v_i \rightsquigarrow_{\mathcal{C}_i} v_s$ in this graph (here, $v_s$ is a secret node for $\mathcal{C}_i$). Let $I_{\mathcal{C}_i}(v_i)$ be the set of $c$-edges in $G(\mathcal{C}_i)$ which are incident on the nodes in $v_i \rightsquigarrow_{\mathcal{C}_i} v_s$ and let $H_{\mathcal{C}_i}(v_i)$ be the set of (useful) $h$-edges corresponding (uniquely) to the $c$-edges in $I_{\mathcal{C}_i}(v_i)$. The key observation is that once $v_i$ gets labeled as a non-member node (and its label doesn't change after that), the nodes on the $c$-path $v_i \rightsquigarrow_{\mathcal{C}_i} v_s$ become $h$-reachable from the set of non-member nodes under *any* configuration of the game following $\mathcal{C}_i$. This implies that all $h$-edges in $H_{\mathcal{C}_i}(v_i)$ become (and remain) useless for all configurations from time $i$ onwards, since they are incident on $v_i \rightsquigarrow_{\mathcal{C}_i} v_s$.

Since $v_i$ is a node with maximum incidence weight in $G(\mathcal{C}_i)$, there are at least $\lceil \log_2 n \rceil$ $c$-edges in $I_{\mathcal{C}_i}(v_i)$ and an equal number of $h$-edges in $H_{\mathcal{C}_i}(v_i)$. So, each time $A$ performs a move, the number of useless $h$-edges increases by $\lceil \log_2 n \rceil$, and after $t$ move there are at least $t \cdot \lceil \log_2 n \rceil$ useless $h$-edges in $\mathcal{C}_t$. Clearly, the number of useless $h$-edges cannot be greater than the number of $h$-edges in the final configuration, i.e.,

$$\begin{aligned} t \cdot \lceil \log_2 n \rceil &\leq |\mathcal{H}_t| \\ &= |\mathcal{H}_0| + \sum_{i=1}^{t} |\mathcal{H}_i \setminus \mathcal{H}_{i-1}| \\ &= |\mathcal{H}_0| + \sum_{i=1}^{t} c_i \end{aligned}$$

where $c_i$ is the cost of the $i$th move performed by $C$. From this, we immediately get the desired bound on the amortized cost of $C$'s moves : $\frac{\sum_{i=1}^{t} c_i}{t} \geq \lceil \log_2 n \rceil - \frac{|\mathcal{H}_0|}{t}$ ∎

## 6 Extensions to Our Model

In this section, we address some of the possible extensions and modifications one could make to our model for secure group key distribution described in Sect. 3. Some of these extensions yield models that are equivalent to the model we have already described while others lead to interesting open problems for the group key distribution problem.

**1. Allowing Message Pairs:** We have proved a lower bound for protocols where the rekey messages sent by the group center consist of a single key encrypted with multiple other keys. It is easy to see that our lower bound also applies to more general protocols where every rekey message can also consist of "pairs" of other rekey messages (i.e. protocols in which the grammar for messages also includes a rule $\mathbf{M} \rightarrow (\mathbf{M}, \mathbf{M})$). Allowing messages pairs does not affect communication complexity in any way.

**2. Groups without Simultaneous Leave and Join:** Our lower bound for group key distribution is proved using a sequence of simultaneous `join` and `leave` operations (which we refer to as a `replace` operation) performed on the group by an adaptive adversary. One reason for having `replace` operations is that they simplify our analysis considerably (by helping us keep the group size constant over time). In groups where `replace` operations are not allowed, the bound that we get using our technique is $\log_2(n)/2$. We remark that it is possible to construct a practical protocol (without replace operations) in which every individual `join` and `leave` can be performed at the cost of $\log_2(n)/2$ multicast messages and $\log_2(n)/2$ *unicast* messages (this can be done by combining the protocol of [3] with ideas from [13]). The interesting question is whether our bound can be extended so that it is tight even when the $\log_2(n)/2$ unicast cost is included in computing communication complexity or whether one can come up with better protocols that involve no unicast at all. We are unable to resolve this question at the moment and leave it open for future work.

**3. Other Cryptographic Primitives:** Our model for secure group key distribution allows the usage of iterated encryption and pseudorandom generation for the center's rekey messages and the best known protocols for this problem also use just these cryptographic primitives. It would be interesting to find out if better protocols can be constructed using other cryptographic primitives (for eg., pseudorandom functions, secret sharing) or whether our lower bound can be extended to even more general classes of protocols that allow the usage of such primitives[6].

*Analyzing Upper Bounds.* The model for secure multicast key distribution we study in this paper can also be used to analyze upper bounds but in doing so, one must take care of some efficiency issues which we ignore in our framework (note that ignoring such issues only helps to *strengthen* our lower bound). For example, in our model, the group members can compute the shared secret key at any instant by looking at the rekey messages sent out in the entire history of the protocol. Practical protocols should require that members be able to get the key using just the rekey messages sent since they joined the group. Also, we do not address the issue of storage limitations of the users or the group center. In practice, the key update should be made possible not only with minimal communication overhead but also with minimal storage requirements for the users.

## References

1. M. Abadi and P. Rogaway. Reconciling two views of cryptography (the computational soundness of formal encryption). *Journal of Cryptology*, 15(2):103–127, 2002.

[6] We note that some known protocols do make use of pseudorandom functions (e.g., [13]), but not in a substantial way, meaning that whatever they do can be easily achieved using pseudorandom generators instead.

2. D. Boneh, G. Durfee, and M. Franklin. Lower bounds for multicast message authentication. In B. Pfitzmann, editor, *Advances in Cryptology - EUROCRYPT 2001, Proceedings of the Internarional Conference on the Theory and Application of Cryptographic Techniques*, volume 2045 of *Lecture Notes in Computer Science*, pages 437–452, Innsbruck, Austria, May 2001. Springer-Verlag.
3. R. Canetti, J. Garay, G. Itkis, D. Micciancio, M. Naor, and B. Pinkas. Multicast security: A taxonomy and some efficient constructions. In *INFOCOM 1999. Proceedings of the Eighteenth Annual Joint Conference of the IEEE Computer and Communications Societies*, volume 2, pages 708–716, New York, NY, Mar. 1999. IEEE.
4. R. Canetti, T. Malkin, and K. Nissim. Efficient communication-storage tradeoffs for multicast encryption. In J. Stern, editor, *Advances in Cryptology - EUROCRYPT '99, Proceedings of the International Conference on the Theory and Application of Cryptographic Techniques*, volume 1592 of *Lecture Notes in Computer Science*, Prague, Czech Republic, May 1999. Springer-Verlag.
5. D. Dolev and A. Yao. On the security of public key protocols. *IEEE Transactions on Information Theory*, 29(2):198–208, 1983.
6. A. Fiat and M. Naor. Broadcast encryption. In D. R. Stinson, editor, *Advances in Cryptology - CRYPTO '93, Proceedings of the 13th annual international Cryptology conference*, volume 773 of *Lecture Notes in Computer Science*, pages 480–491, Santa Barbara, California, USA, Aug. 1993. Springer-Verlag.
7. R. Gennaro. A protocol to achieve independence in constant rounds. *IEEE Transactions on Parallel and Distributed Systems*, 11(7):636–647, 2000.
8. O. Goldreich, S. Goldwasser, and S. Micali. How to construct random functions. *Journal of the ACM*, 33:792–807, 1986.
9. H. Harney and C. Muckenhirn. Group key management protocol (GKMP) architecture. Request for Comments 2094, Internet Engineering Task Force, July 1997.
10. H. Harney and C. Muckenhirn. Group key management protocol (GKMP) specification. Request for Comments 2093, Internet Engineering Task Force, July 1997.
11. M. Luby and J. Staddon. Combinatorial Bounds for Broadcast Encryption. In K. Nyberg, editor, *Advances in Cryptology - EUROCRYPT '98, Proceedings of the International Conference on the Theory and Application of Cryptographic Techniques*, volume 1403 of *Lecture Notes in Computer Science*, pages 512–526, Espoo, Finland, May 1998. Springer-Verlag.
12. D. Micciancio and B. Warinschi. Completeness theorems for the abadi-rogaway logic of encrypted expressions. *Journal of Computer Security*, 12(1):99–129, 2004. Preliminary version in WITS 2002.
13. A. Perrig, D. X. Song, and J. D. Tygar. ELK, A New Protocol for Efficient Large-Group Key Distribution. In *IEEE Symposium on Security and Privacy*, pages 247–262, Oakland, CA, USA, May 2001. IEEE.
14. J. Snoeyink, S. Suri, and G. Varghese. A lower bound for multicast key distribution. In *INFOCOM 2001. Proceedings of the Twentieth Annual Joint Conference of the IEEE Computer and Communications Societies*, volume 1, pages 422–431, New York, NY, Apr. 2001. IEEE.
15. D. M. Wallner, E. G. Harder, and R. C. Agee. Key management for multicast: issues and architecture. Request for Comments 2627, Internet Engineering Task Force, June 1999.
16. C. K. Wong, M. Gouda, and S. S. Lam. Secure group communications using key graphs. *IEEE/ACM Transactions on Networking*, 8(1):16–30, Feb. 2000. Preliminary version in SIGCOMM 1998.

# An Uninstantiable Random-Oracle-Model Scheme for a Hybrid-Encryption Problem

Mihir Bellare, Alexandra Boldyreva, and Adriana Palacio

Dept. of Computer Science & Engineering, University of California, San Diego
9500 Gilman Drive, La Jolla, CA 92093-0114, USA
{mihir,aboldyre,apalacio}@cs.ucsd.edu
http://www.cse.ucsd.edu/users/{mihir,aboldyre,apalacio}

**Abstract.** We present a simple, natural random-oracle (RO) model scheme, for a practical goal, that is uninstantiable, meaning is proven in the RO model to meet its goal yet admits *no* standard-model instantiation that meets this goal. The goal in question is *IND-CCA-preserving asymmetric encryption* which formally captures security of the most common practical usage of asymmetric encryption, namely to transport a symmetric key in such a way that symmetric encryption under the latter remains secure. The scheme is an ElGamal variant, called Hash ElGamal, that resembles numerous existing RO-model schemes, and on the surface shows no evidence of its anomalous properties. These results extend our understanding of the gap between the standard and RO models, and bring concerns raised by previous work closer to practice by indicating that the problem of RO-model schemes admitting no secure instantiation can arise in domains where RO schemes are commonly designed.

## 1 Introduction

A random-oracle (RO) model scheme is one whose algorithms have oracle access to a random function. Its security is evaluated with respect to an adversary with oracle access to the same function. An "instantiation" of such a scheme is the standard-model scheme obtained by replacing this function with a member of a polynomial-time computable family of functions, described by a short key. The security of the scheme is evaluated with respect to an adversary given the same key. In the random-oracle paradigm, as enunciated by Bellare and Rogaway [6], one first designs and proves secure a scheme in the RO model, and then instantiates it to get a (hopefully still secure) standard-model scheme.

The RO model has proven quite popular and there are now numerous practical schemes designed and proven secure in this model. But the important issue of whether such schemes can be securely instantiated, and, if so, how, remains less clear. This paper adds to existing concerns in this regard. Let us begin by reviewing previous work and then explain our results.

C. Cachin and J. Camenisch (Eds.): EUROCRYPT 2004, LNCS 3027, pp. 171–188, 2004.

### 1.1 Previous Work

Let us call a RO-model scheme *uninstantiable*, with respect to some underlying cryptographic goal, if the scheme can be proven to meet this goal in the random-oracle model, but *no* instantiation of the scheme meets the goal in question.

Canetti, Goldreich and Halevi [8] provided the first examples of uninstantiable schemes, the goals in question being IND-CPA-secure asymmetric encryption and digital signatures secure against chosen-message attacks. Further examples followed: Nielsen [19] presented an uninstantiable RO-model scheme for the goal of non-interactive, non-committing encryption [7], and Goldwasser and Taumann [17] showed the existence of a 3-move protocol which, when collapsed via a RO as per the Fiat-Shamir heuristic [14], yields an uninstantiable RO-model signature scheme.

The results of [8] indicate that it is possible for the RO paradigm to fail to yield secure "real-world" schemes. The example schemes provided by [8], however, are complex and contrived ones that do not resemble the kinds of RO schemes typically being designed. (Their schemes are designed to return the secret key depending on the result of some test applied to an output of the oracle, and they use diagonalization and CS proofs [18].) The same is true of the scheme of [17]. In contrast, the scheme of [19] is simple, but the goal, namely non-interactive, non-committing encryption, is somewhat distant from ones that are common practical targets of RO-model designs. Accordingly, based on existing work, one might be tempted to think that "in practice," or when confined to "natural" schemes for practical problems commonly being targeted by RO-scheme designers, the RO paradigm is sound.

This paper suggests that even this might not always be true. For a practical cryptographic goal, we present an uninstantiable RO-model scheme that is simple and natural, closely resembling the types of schemes being designed in this domain. We begin below by discussing the goal, which we call IND-CCA-preserving asymmetric encryption and which arises in the domain of hybrid encryption.

### 1.2 IND-CCA-Preserving Asymmetric Encryption

In practice, the most common usage of asymmetric encryption is to transport a symmetric key that is later used for symmetric encryption of the actual data. The notion of an asymmetric encryption scheme $\mathsf{AS}$ being IND-CCA-preserving, that we introduce, captures the security attribute that $\mathsf{AS}$ must possess in order to render this usage of $\mathsf{AS}$ secure. We now elaborate.

Encryption, in practice, largely employs the "hybrid" paradigm. The version of this paradigm that we consider here is quite general. In a first phase, the sender picks at random a "session" key $K$ for a symmetric encryption scheme, encrypts $K$ asymmetrically under the receiver's public key to get a ciphertext $C_a$, and transfers $C_a$ to the receiver. In a second phase, it can encrypt messages of its choice symmetrically under $K$ and transfer the corresponding ciphertexts to the receiver. We call this multi-message (mm) hybrid encryption.[1]

A choice of an asymmetric encryption scheme $\mathsf{AS}$ and a symmetric encryption scheme $\mathsf{SS}$ gives rise to a particular mm-hybrid scheme. We introduce in Section 2 a definition of the IND-CCA security of this mm-hybrid scheme which captures the privacy of the encrypted messages even in the presence of an adversary allowed chosen-ciphertext attacks on both component schemes and allowed to choose the messages to be encrypted adaptively and as a function of the asymmetric ciphertext, denoted $C_a$ above, that transports the symmetric key.

Now let us say that an asymmetric encryption scheme $\mathsf{AS}$ is *IND-CCA preserving* if the mm-hybrid associated to $\mathsf{AS}$ and symmetric encryption scheme $\mathsf{SS}$ is IND-CCA secure for *every* IND-CCA secure $\mathsf{SS}$. This notion of security for an asymmetric encryption scheme captures the security attribute of its being able to securely transport a session key for the purpose of mm-hybrid encryption. The goal we consider is IND-CCA-preserving asymmetric encryption.

It is easy to see that any IND-CCA-secure asymmetric encryption scheme is IND-CCA preserving. (For completeness, this is proved in the full version of this paper [3].) IND-CCA preservation, however, is actually a weaker requirement on an asymmetric encryption scheme than IND-CCA security itself. In fact, since the messages to be encrypted using the asymmetric scheme are randomly-chosen symmetric keys, the encryption itself need not even be randomized. Hence there might be IND-CCA-preserving asymmetric encryption schemes that are simpler and more efficient than IND-CCA-secure ones. In particular, it is natural to seek an efficient IND-CCA-preserving scheme in the RO model along the lines of existing hybrid encryption schemes such as those of [9,10,15,20].

### 1.3 The Hash ElGamal Scheme and Its Security

It is easy to see that the ElGamal encryption scheme [13] is not IND-CCA preserving. An effort to strengthen it to be IND-CCA preserving lead us to a variant that we call the Hash ElGamal scheme. It uses the idea underlying the Fujisaki-Okamoto [15] transformation, namely to encrypt under the original (ElGamal) scheme using coins obtained by applying a random oracle $H$ to the message. Specifically, encryption of a message $K$ under public key $(q, g, X)$ in the Hash ElGamal scheme is given by

$$\mathsf{AE}^{G,H}((q,g,X),K) \;=\; (g^{H(K)}\,,\,G(X^{H(K)}) \oplus K)\;, \tag{1}$$

[1] The term multi-message refers to the fact that multiple messages may be encrypted, in the second phase, under the same session key. The main reason for using such a hybrid paradigm, as opposed to directly encrypting the data asymmetrically under the receiver's public key, is that the number-theoretic operations underlying popular asymmetric encryption schemes are computationally more expensive than the block-cipher operations underlying symmetric encryption schemes, so hybrid encryption brings significant performance gains.

where $G, H$ are random oracles, $q, 2q+1$ are primes, $g$ is a generator of the order $q$ cyclic subgroup of $\mathbb{Z}_{2q+1}^*$, and the secret key is $(q, g, x)$ where $g^x = X$. Decryption is performed in the natural way as detailed in Figure 1.

The Hash ElGamal scheme is very much like practical RO-model schemes presented in the literature. In fact, it is a particular case of an asymmetric encryption scheme proposed by Baek, Lee and Kim [2,4].

We note that the Hash ElGamal asymmetric encryption scheme is not IND-CCA secure, or even IND-CPA secure, in particular because the encryption algorithm is deterministic. But Theorem 1 guarantees that the Hash ElGamal asymmetric encryption scheme is IND-CCA-preserving in the RO model, if the Computational Diffie-Hellman (CDH) problem is hard in the underlying group.

We follow this with Theorem 2, however, which says that the Hash ElGamal scheme is uninstantiable. In other words, the standard-model asymmetric encryption scheme obtained by instantiating the RO-model Hash ElGamal scheme is not IND-CCA preserving, regardless of the choice of instantiating functions.[2] (We allow these to be drawn from any family of polynomial-time computable functions.)

### 1.4 A Closer Look

As noted above, we show that no instantiation of the Hash ElGamal scheme is IND-CCA-preserving. The way we establish this is the following. We let AS be some (any) instantiation of the Hash ElGamal scheme. Then, we construct a particular IND-CCA-secure symmetric encryption scheme SS such that the mm-hybrid associated to AS and SS is not IND-CCA secure. The latter is proven by presenting an explicit attack on the mm-hybrid. We clarify that the symmetric scheme SS constructed in this proof is not a natural one. It is contrived, but not particularly complex. We do not view this as subtracting much from the value of our result, which lies rather in the nature of the Hash ElGamal scheme itself and the practicality of the underlying goal.

What we suggest is interesting about the result is that the Hash ElGamal scheme, on the surface, seems innocuous enough. It does not seem to be making any "peculiar" use of its random oracle that would lead us to think it is "wrong." (Indeed, it uses random oracles in ways they have been used previously, in particular by [15,2,4].) The scheme is simple, efficient, and similar to other RO-model schemes out there. In addition, we contend that the definition of IND-CCA-preserving asymmetric encryption is natural and captures a practical requirement. The fact that the Hash ElGamal scheme is uninstantiable thus points to the difficulty of being able to distinguish uninstantiable RO-model schemes from ones that at least *may* be securely instantiable, even in the context of natural and practical goals.

[2] This result is based on the assumption that one-way functions exist (equivalently, IND-CCA-secure symmetric encryption schemes exist), since, otherwise, by default, *any* asymmetric encryption scheme is IND-CCA preserving, and, indeed, the entire mm-hybrid encryption problem we are considering is vacuous. This assumption is made implicitly in all results in this paper.

### 1.5 Generalizations

In the full version of the paper [3] we provide some results that generalize the above. We consider the class of IND-CCA-preserving asymmetric encryption schemes that possess a pair of properties that we call *key verifiability* and *ciphertext verifiability*. Key verifiability means there is a way to recognize valid public keys in polynomial time. Ciphertext verifiability means there is a polynomial-time procedure to determine whether a given ciphertext is an encryption of a given message under a given valid public key. Note that ciphertext verifiability contradicts IND-CPA security, but it need not prevent a scheme from being IND-CCA preserving, since the latter notion considers the use of the asymmetric scheme only for the encryption of messages that are chosen at random.

In [3] we prove that the goal of key-verifiable, ciphertext-verifiable IND-CCA-preserving asymmetric encryption is achievable in the RO model, by the Hash El Gamal scheme in particular, assuming the CDH problem is hard in the underlying group. However, as we also prove in [3], this goal is not achievable in the standard model. In other words, there exist RO-model schemes meeting this goal, but there exist no standard-model schemes meeting it. This generalizes Theorem 2 because any instantiation of the Hash ElGamal scheme is key-verifiable and ciphertext-verifiable, and hence cannot be IND-CCA-preserving.

In [3] we lift our results from being about a particular scheme to being about a primitive, or class of schemes. The generalization also helps better understand what aspects of the Hash ElGamal scheme lead to its admitting no IND-CCA-preserving instantiation. In particular, we see that this is not due to some "peculiar" use of random oracles but rather due to some simply stated properties of the resulting asymmetric encryption scheme itself.

### 1.6 Related Work

In the cryptographic community, the term "hybrid encryption" seems to be used quite broadly, to refer to a variety of goals or methods in which symmetric and asymmetric primitives are combined to achieve privacy. We have considered one goal in this domain, namely mm-hybrid encryption. We now discuss related work that has considered other goals or problems in this domain.

Works such as [9,10,15,20,12,21] provide designs of IND-CCA-secure asymmetric encryption schemes that are referred to as "hybrid encryption schemes" because they combine the use of asymmetric and symmetric primitives. (Possible goals of such designs include gaining efficiency, increasing the size of the message space, or reducing the assumptions that must be made on the asymmetric component in order to guarantee the IND-CCA security of the construction.) The schemes of [9,10,15,20] are in the RO model and, although addressing a different goal, form an important backdrop for our work because the Hash ElGamal scheme is based on similar techniques and usage of random oracles. We stress, however, that we have no reason to believe that any of these schemes, or that of [2,4] of which Hash ElGamal is a special case, are uninstantiable.

## 2 Definitions

NOTATION AND CONVENTIONS. If $S$ is a randomized algorithm, then $[S(x, y, \ldots)]$ denotes the set of all points having positive probability of being output by $S$ on inputs $x, y, \ldots$. If $x$ is a binary string, then $|x|$ denotes its length, and if $n \geq 1$ is an integer, then $|n|$ denotes the length of its binary encoding, meaning the unique integer $\ell$ such that $2^{\ell-1} \leq n < 2^{\ell}$. The string-concatenation operator is denoted "$\|$".

Formal definitions in the RO model provide as an oracle, to the algorithms and the adversary, a single random function $R$ mapping $\{0,1\}^*$ to $\{0,1\}$. Schemes might, however, use and refer to multiple random functions of different domains and ranges. These can be derived from $R$ via standard means [6].

SYMMETRIC ENCRYPTION. A symmetric encryption scheme $\mathsf{SS} = (\mathsf{SK}, \mathsf{SE}, \mathsf{SD})$ is specified by three polynomial-time algorithms: via $K \stackrel{\$}{\leftarrow} \mathsf{SK}(1^k)$ one can generate a key; via $C \stackrel{\$}{\leftarrow} \mathsf{SE}(K, M)$ one can encrypt a message $M \in \{0,1\}^*$; and via $M \leftarrow \mathsf{SD}(K, C)$ one can decrypt a ciphertext $C$. It is required that $\mathsf{SD}(K, \mathsf{SE}(K, M)) = M$ for all $K \in [\mathsf{SK}(1^k)]$ and all $M \in \{0,1\}^*$. We assume (without loss of generality) that $[\mathsf{SK}(1^k)] \subseteq \{0,1\}^k$. In the RO model, all algorithms have access to the RO.

We define security following [5] and addressing the possibility of the symmetric scheme being in the RO model. Let $\mathrm{LR}(M_0, M_1, b) = M_b$ if $M_0, M_1$ are strings of equal length, and $\perp$ otherwise. Associate to $\mathsf{SS}$, an adversary $\boldsymbol{S}$, and $k \in \mathbb{N}$, the following experiment.

Experiment $\mathbf{Exp}^{\text{ind-cca}}_{\mathsf{SS},\boldsymbol{S}}(k)$
 Randomly choose RO $R_s \colon \{0,1\}^* \to \{0,1\}$
 $K \stackrel{\$}{\leftarrow} \mathsf{SK}^{R_s}(1^k)$ ; $b \stackrel{\$}{\leftarrow} \{0,1\}$
 Run $\boldsymbol{S}$ with input $1^k$ and oracles $\mathsf{SE}^{R_s}(K, \mathrm{LR}(\cdot,\cdot,b))$, $\mathsf{SD}^{R_s}(K,\cdot)$, $R_s$
 Let $d$ denote the output of $\boldsymbol{S}$
 If $d = b$ then return 1 else return 0.

We say that adversary $\boldsymbol{S}$ is legitimate if it never queries $\mathsf{SD}^{R_s}(K,\cdot)$ with a ciphertext previously returned by $\mathsf{SE}^{R_s}(K, \mathrm{LR}(\cdot,\cdot,b))$. Symmetric encryption scheme $\mathsf{SS}$ is said to be IND-CCA secure if the function

$$\mathsf{Adv}^{\text{ind-cca}}_{\mathsf{SS},\boldsymbol{S}}(k) \;=\; 2 \cdot \Pr\left[\, \mathbf{Exp}^{\text{ind-cca}}_{\mathsf{SS},\boldsymbol{S}}(k) = 1 \,\right] - 1$$

is negligible for all legitimate polynomial-time adversaries $\boldsymbol{S}$.

ASYMMETRIC ENCRYPTION. An asymmetric encryption scheme $\mathsf{AS} = (\mathsf{AK}, \mathsf{AE}, \mathsf{AD})$ is specified by three polynomial-time algorithms: via $(pk, sk) \stackrel{\$}{\leftarrow} \mathsf{AK}(1^k)$ one can generate keys; via $C \stackrel{\$}{\leftarrow} \mathsf{AE}(pk, K)$ one can encrypt a message $K \in \{0,1\}^k$; and via $K \leftarrow \mathsf{AD}(sk, C)$ one can decrypt a ciphertext $C$. (We denote the message by $K$ because we will set it to a key for a symmetric encryption scheme.) It is required that $\mathsf{AD}(sk, \mathsf{AE}(pk, K)) = K$ for all $(pk, sk) \in [\mathsf{AK}(1^k)]$ and all $K \in \{0,1\}^k$. In the RO model, all algorithms have access to the RO.

Discussions and peripheral results in this paper sometimes refer to standard notions of security for such schemes like IND-CPA and IND-CCA, but these are not required for the main results and, accordingly, are not defined here but recalled in [3].

IND-CCA-PRESERVING ASYMMETRIC ENCRYPTION. We provide the formal definitions first and explanations later. A *multi-message hybrid (mm-hybrid) encryption scheme* is simply a pair $(\mathsf{AS}, \mathsf{SS})$ consisting of an asymmetric encryption scheme $\mathsf{AS} = (\mathsf{AK}, \mathsf{AE}, \mathsf{AD})$ and a symmetric encryption scheme $\mathsf{SS} = (\mathsf{SK}, \mathsf{SE}, \mathsf{SD})$. We associate to $(\mathsf{AS}, \mathsf{SS})$, a *hybrid adversary* $\boldsymbol{H}$, and $k \in \mathbb{N}$, the following experiment.

Experiment $\mathbf{Exp}^{\text{ind-cca}}_{\mathsf{AS},\mathsf{SS},\boldsymbol{H}}(k)$
Randomly choose RO $R$: $\{0,1\}^* \to \{0,1\}$
Define ROs $R_s(\cdot) = R(0\|\cdot)$ and $R_a(\cdot) = R(1\|\cdot)$
$(pk, sk) \stackrel{\$}{\leftarrow} \mathsf{AK}^{R_a}(1^k)$ ; $K \stackrel{\$}{\leftarrow} \mathsf{SK}^{R_s}(1^k)$ ; $b \stackrel{\$}{\leftarrow} \{0,1\}$
$C_a \stackrel{\$}{\leftarrow} \mathsf{AE}^{R_a}(pk, K)$
Run $\boldsymbol{H}$ with inputs $pk, C_a$ and
oracles $\mathsf{SE}^{R_s}(K, \mathrm{LR}(\cdot,\cdot,b))$, $\mathsf{SD}^{R_s}(K,\cdot)$, $\mathsf{AD}^{R_a}(sk,\cdot)$, $R$
Let $d$ denote the output of $\boldsymbol{H}$
If $d = b$ then return 1 else return 0.

We say that adversary $\boldsymbol{H}$ is legitimate if it does not query $\mathsf{SD}^{R_s}(K,\cdot)$ on a ciphertext previously returned by $\mathsf{SE}^{R_s}(K, \mathrm{LR}(\cdot,\cdot,b))$, and it does not query $\mathsf{AD}^{R_a}(sk,\cdot)$ on $C_a$. Mm-hybrid encryption scheme $(\mathsf{AS}, \mathsf{SS})$ is said to be IND-CCA secure if the function

$$\mathsf{Adv}^{\text{ind-cca}}_{\mathsf{AS},\mathsf{SS},\boldsymbol{H}}(k) = 2 \cdot \Pr\left[\mathbf{Exp}^{\text{ind-cca}}_{\mathsf{AS},\mathsf{SS},\boldsymbol{H}}(k) = 1\right] - 1$$

is negligible for all legitimate polynomial-time adversaries $\boldsymbol{H}$.

Finally, we say that an asymmetric encryption scheme $\mathsf{AS}$ is *IND-CCA preserving* if the mm-hybrid encryption scheme $(\mathsf{AS}, \mathsf{SS})$ is IND-CCA secure for *all* IND-CCA-secure symmetric encryption schemes $\mathsf{SS}$. Here, the set of symmetric encryption schemes over which we quantify includes RO-model ones if $\mathsf{AS}$ is a RO-model scheme, and includes only standard-model ones if $\mathsf{AS}$ is a standard-model scheme.

Let us now explain the ideas behind these formalisms. Recall that we are modelling the security of the following two-phase scenario: in phase one, the sender picks a key $K$ for symmetric encryption, asymmetrically encrypts it under the receiver's public key to get a ciphertext $C_a$, and sends $C_a$ to the receiver; in phase two, the sender symmetrically encrypts messages of its choice under $K$ and transmits the resulting ciphertexts to the receiver. The definition above captures the requirement of privacy of the symmetrically encrypted data under a chosen-ciphertext attack. Privacy is formalized in terms of indistinguishability via left-or-right oracles, and the chosen-ciphertext attack is formalized via the adversary's access to decryption oracles for *both* the symmetric and asymmetric schemes. The legitimacy requirement, as usual, disallows decryption queries on challenge ciphertexts since they would lead to trivial adversary victory. The experiment reflects the possibility that SS and AS are RO-model schemes by picking random oracles for their encryption and decryption algorithms. The standard model is the special case where the algorithms of the schemes do not refer to any oracles, and thus the definition above covers security in both models. The notion of AS being IND-CCA preserving reflects a valuable pragmatic requirement, namely that one may use, in conjunction with AS, any symmetric encryption scheme and be guaranteed security of the mm-hybrid under the minimal assumption that the symmetric scheme itself is secure.

*Remark 1.* Suppose we have two RO-model schemes, and are composing them, or executing them in a common context. (Above, this is happening with the asymmetric encryption scheme and the symmetric encryption scheme.) We claim that, in this case, the ROs of the two schemes should be chosen independently. (This does not mean that we need to assume two RO oracles are given. The formal model always provides just one RO. But one can easily derive several independent ROs from a single one, as we did above.) The correctness of this principle of independent instantiation of ROs in a common context can be seen in many ways. First, it is easy to come up with an example of a pair of secure RO-model schemes that, when composed, yield an insecure one if the ROs in the two schemes are defined to be the same. Second, one can reason by analogy with the way we need to choose keys in composing primitives. For example, suppose we have a MAC and symmetric encryption scheme, each individually secure. If we use them to construct an authenticated-encryption scheme, we should use different keys for the MAC and the symmetric encryption scheme. (There is no reason to think otherwise that the composition will be secure.) The principle, for ROs, is exactly the same. They are just like keys provided to primitives.

The existence of IND-CCA-preserving asymmetric encryption schemes is easy to establish since, as we show in [3], any IND-CCA-secure asymmetric encryption scheme is IND-CCA preserving. The interesting question is to find IND-CCA-preserving asymmetric encryption schemes that are more efficient than existing IND-CCA-secure asymmetric encryption schemes. Hash El Gamal is one such scheme.

## 3 The HEG Scheme and Its Security in the RO Model

In this section we introduce a variant of the ElGamal encryption scheme [13] that, although not IND-CCA secure, is IND-CCA preserving in the RO model under a standard assumption. In Section 4, we will show that this scheme admits no IND-CCA-preserving instantiation.

PRELIMINARIES. A *cyclic-group generator* is a randomized, polynomial-time algorithm CG which on input $1^k$ outputs a pair $(q, g)$, where $q$ is a prime such that $p = 2q + 1$ is also a prime, $g$ is a generator of the cyclic, order $q$ subgroup $\langle g \rangle$ of $\mathbb{Z}_p^*$, and $|p| = k$. Recall that the Computational Diffie-Hellman (CDH) problem is said to be hard for CG if the function

| $\mathsf{AK}(1^k)$ | $\mathsf{AE}^{G,H}((q,g,X),K)$ | $\mathsf{AD}^{G,H}((q,g,x),(Y,W))$ |
|---|---|---|
| $(q,g) \xleftarrow{\$} \mathsf{CG}(1^k)$ | $y \leftarrow H(K)$ | $T \leftarrow G(Y^x)$ |
| $x \xleftarrow{\$} \mathbb{Z}_q$ | $Y \leftarrow g^y$ | $K \leftarrow T \oplus W$ |
| $X \leftarrow g^x$ | $T \leftarrow G(X^y)$ | If $g^{H(K)} = Y$ then |
| Return $((q,g,X),(q,g,x))$ | $W \leftarrow T \oplus K$ | Return $K$ |
| | Return $(Y,W)$ | else Return $\perp$ EndIf |

**Fig. 1.** Algorithms of the RO-model asymmetric encryption scheme $\mathsf{HEG}[\mathsf{CG}] = (\mathsf{AK},\mathsf{AE},\mathsf{AD})$ associated to cyclic-group generator CG. Here $G\colon \langle g\rangle \to \{0,1\}^k$ and $H\colon \{0,1\}^k \to \mathbb{Z}_q$ are random oracles.

$$\mathsf{Adv}^{\mathrm{cdh}}_{\mathsf{CG},C}(k) \;=\; \Pr\Big[\,(q,g) \xleftarrow{\$} \mathsf{CG}(1^k)\,;\, x,y \xleftarrow{\$} \mathbb{Z}_q \;:\; C(q,g,g^x,g^y) = g^{xy}\,\Big]$$

is negligible for all polynomial-time *cdh adversaries* $C$.

SCHEME AND RESULT STATEMENT. To any cyclic-group generator CG we associate the RO-model asymmetric encryption scheme $\mathsf{HEG}[\mathsf{CG}] = (\mathsf{AK},\mathsf{AE},\mathsf{AD})$ whose constituent algorithms are depicted in Figure 1. (The scheme makes reference to two ROs, namely $G\colon \langle g\rangle \to \{0,1\}^k$ and $H\colon \{0,1\}^k \to \mathbb{Z}_q$, while the formal definition of an asymmetric encryption scheme provides a single RO $R\colon \{0,1\}^* \to \{0,1\}$, but $G, H$ may be implemented via $R$ in standard ways [6].) We call this variant of the ElGamal encryption scheme the *Hash ElGamal* encryption scheme associated to CG. Our result about its security in the RO model is the following.

**Theorem 1.** *If the CDH problem is hard for cyclic-group generator* CG, *then the associated Hash ElGamal asymmetric encryption scheme* HEG[CG] *is IND-CCA preserving in the RO model.*

For the definition of what it means to be IND-CCA preserving, we refer the reader to Section 2.

REMARKS. We note that the encryption algorithm AE of HEG[CG] is deterministic. For this reason alone, HEG[CG] is not an IND-CCA secure, or even IND-CPA secure, asymmetric encryption scheme. Nonetheless, Theorem 1 says that it is IND-CCA preserving as long as the CDH problem is hard for CG. This is not a contradiction. Very roughly, the reason HEG[CG] can preserve IND-CCA while not itself being even IND-CPA is that the former notion considers the use of the scheme only for the encryption of messages that are symmetric keys, which (as long as the associated symmetric encryption scheme is secure) have relatively high entropy, and the entropy in these messages compensates for the lack of any introduced by AE. We add that previous work [9,10,15,20] has shown that in the RO model, relatively weak asymmetric components suffice to ensure strong security properties of the hybrid based on them. Thus, it is not surprising that, although HEG[CG] is not secure with respect to standard measures like IND-CPA and IND-CCA, it is secure enough to permit its use for transport of a symmetric encryption key as indicated by Theorem 1.

The full proof of Theorem 1 is in [3]. Below we provide an intuitive overview that highlights the main areas of novelty.

PROOF SETUP. Let $\mathsf{AS} = \mathsf{HEG}[\mathsf{CG}]$ and let $\mathsf{AK}, \mathsf{AE}, \mathsf{AD}$ denote its constituent algorithms. Let $\mathsf{SS} = (\mathsf{SK}, \mathsf{SE}, \mathsf{SD})$ be any IND-CCA-secure symmetric encryption scheme. We need to show that $(\mathsf{AS}, \mathsf{SS})$ is an IND-CCA-secure mm-hybrid encryption scheme.

Let $\boldsymbol{H}$ be a polynomial-time hybrid adversary attacking $(\mathsf{AS}, \mathsf{SS})$. We will construct polynomial-time adversaries $\boldsymbol{S}$ and $\boldsymbol{C}$ such that

$$\mathsf{Adv}^{\text{ind-cca}}_{\mathsf{AS},\mathsf{SS},\boldsymbol{H}}(k) \leq \text{poly}(k) \cdot \text{poly}\left( \mathsf{Adv}^{\text{ind-cca}}_{\mathsf{SS},\boldsymbol{S}}(k) \, , \, \mathsf{Adv}^{\text{cdh}}_{\mathsf{CG},\boldsymbol{C}}(k) \right) + \frac{\text{poly}(k)}{2^k} \, . \quad (2)$$

Since $\mathsf{SS}$ is assumed IND-CCA secure and the CDH problem is hard for $\mathsf{CG}$, the advantage functions related to $\boldsymbol{S}$ and $\boldsymbol{C}$ above are negligible, and thus so is the advantage function related to $\boldsymbol{H}$. To complete the proof, we need to specify adversaries $\boldsymbol{S}, \boldsymbol{C}$ for which Equation (2) is true.

Consider $\mathbf{Exp}^{\text{ind-cca}}_{\mathsf{AS},\mathsf{SS},\boldsymbol{H}}(k)$. Let $(q, g, X)$ be the public key and $(q, g, x)$ the secret key chosen, where $X = g^x$. Let $C_a = (Y, W)$ where $Y = g^y$. Let $K$ denote the symmetric encryption key chosen. Let $\mathsf{GH}$ be the event that there is a time at which $g^{xy}$ is queried to $G$ but $K$ has not been queried to $H$; $\mathsf{HG}$ the event that there is a time at which $K$ is queried to $H$ but $g^{xy}$ has not been queried to $G$; and $\mathsf{Succ}(\boldsymbol{H})$ the event that $\boldsymbol{H}$ is successful at guessing the value of its challenge bit $b$. We will construct $\boldsymbol{C}$ so that

$$\Pr[\, \mathsf{GH} \,] \leq \text{poly}(k) \cdot \mathsf{Adv}^{\text{cdh}}_{\mathsf{CG},\boldsymbol{C}}(k) + \frac{\text{poly}(k)}{2^k} \, ,$$

and we will construct $\boldsymbol{S}$ so that

$$\Pr[\, \mathsf{HG} \vee (\mathsf{Succ}(\boldsymbol{H}) \wedge \neg\mathsf{GH} \wedge \neg\mathsf{HG}) \,] \leq \mathsf{Adv}^{\text{ind-cca}}_{\mathsf{SS},\boldsymbol{S}}(k) + \frac{\text{poly}(k)}{2^k} \, . \quad (3)$$

Equation (2) follows.

THE ADVERSARIES. The design of $\boldsymbol{C}$ relies mostly on standard techniques, and so we leave it to [3]. We turn to $\boldsymbol{S}$. The latter gets input $1^k$ and oracles $\mathsf{SE}^{R_s}(K, \text{LR}(\cdot,\cdot,b))$, $\mathsf{SD}^{R_s}(K,\cdot)$, $R_s$, and begins with the initializations

$$\begin{aligned} &((q,g,X),(q,g,x)) \stackrel{\$}{\leftarrow} \mathsf{AK}(1^k) \, ; \\ &y \stackrel{\$}{\leftarrow} \mathbb{Z}_q \, ; \, Y \leftarrow g^y \, ; \, W \stackrel{\$}{\leftarrow} \{0,1\}^k \, ; \, C_a \leftarrow (Y, W) \, . \end{aligned} \quad (4)$$

It then runs $\boldsymbol{H}$ on inputs $(q, g, X), C_a$, itself responding to the oracle queries of the latter. Its aim is to do this in such a way that the key $K$ underlying $\boldsymbol{S}$'s oracles plays the role of the quantity of the same name for $\boldsymbol{H}$. Eventually, it will output what $\boldsymbol{H}$ outputs. The difficulty faced by this adversary is that $\boldsymbol{H}$ might query $K$ to $H$. (Other oracle queries are dealt with in standard ways.) In that case, $\boldsymbol{H}$ expects to be returned $y$. (And it cannot be fooled since, knowing $Y = g^y$, it can verify whether or not the value returned is $y$.) The difficulty for

$\boldsymbol{S}$ is not that it does not know the right answer —via Equation (4), it actually knows $y$— but rather that it is not clear how it would know that a query being made to $H$ equals the key $K$ underlying its oracles, so that it would know *when* to return $y$ as the answer to a query to $H$.

In order to "detect" when query $K$ is made, we would, ideally, like a test that can be performed on a value $L$, accepting if $L = K$ and rejecting otherwise. However, it is not hard to see that, in general, such a test does not exist.[3] Instead, we introduce a test that has a weaker property and show that it suffices for us.

Our test $\mathsf{KeyTest}$ takes input $L$ and has access to $\boldsymbol{S}$'s $\mathsf{SE}^{R_s}(K, \mathrm{LR}(\cdot,\cdot,b))$ oracle. It returns a pair $(\mathrm{dec}, \mathrm{gs})$ such that: (1) If $L = K$ then $(\mathrm{dec}, \mathrm{gs}) = (1, b)$, meaning in this case it correctly computes the challenge bit $b$, and (2) If $L \neq K$ then, with overwhelming probability, either $\mathrm{dec} = 0$ (the test is saying $L \neq K$) or $(\mathrm{dec}, \mathrm{gs}) = (1, b)$ (the test is saying it does not know whether or not $L = K$, but it has successfully calculated the challenge bit anyway). With $\mathsf{KeyTest}$ in hand, $\boldsymbol{S}$ can answer a query $L$ made to $H$ as follows. It runs $(\mathrm{dec}, \mathrm{gs}) \xleftarrow{\$} \mathsf{KeyTest}(L)$. If $\mathrm{dec} = 0$, it can safely assume $L \neq K$ and return a random answer, while if $\mathrm{dec} = 1$, it can output gs as its guess to challenge bit $b$ and halt.

A precise description and analysis of $\mathsf{KeyTest}$ are in [3], but we briefly sketch the ideas here. The algorithm has two phases. In the first phase, it repeatedly tests whether or not

$$
\begin{aligned}
\mathsf{SD}^{R_s}(L, \mathsf{SE}^{R_s}(K, \mathrm{LR}(T_0, T_0, b))) &= T_0 \text{ and} \\
\mathsf{SD}^{R_s}(L, \mathsf{SE}^{R_s}(K, \mathrm{LR}(T_1, T_1, b))) &= T_1 ,
\end{aligned}
$$

where $T_0, T_1$ are some distinct "test" messages. If any of these checks fails, it knows that $L \neq K$ and returns $(0, 0)$. (However, the checks can succeed with high probability even if $L \neq K$.) In the next phase, it repeatedly computes $\mathsf{SD}^{R_s}(L, \mathsf{SE}^{R_s}(K, \mathrm{LR}(T_0, T_1, b)))$ and, if *all* these computations yield $T_{\mathrm{gs}}$ for some bit gs, it returns $(1, \mathrm{gs})$. The analysis shows that, conditional on the first phase not returning $(0, 0)$, the bit gs from the second stage equals $b$ with overwhelming probability.

A subtle point arises with relation to the test. Recall that $\boldsymbol{H}$ is making queries to $\mathsf{SD}^{R_s}(K, \cdot)$. $\boldsymbol{S}$ will answer these via its own oracle of the same name. Now, consider the event that $\boldsymbol{H}$ queries to $\mathsf{SD}^{R_s}(K, \cdot)$ a ciphertext $C$ generated in some execution of $\mathsf{KeyTest}$. If $\boldsymbol{S}$ calls $\mathsf{SD}^{R_s}(K, C)$ to obtain the answer, it would immediately become an illegitimate adversary and thus forgo its advantage, since $C$ is a result of a call to $\mathsf{SE}^{R_s}(K, \mathrm{LR}(\cdot,\cdot,b))$ made by $\boldsymbol{S}$ via subroutine $\mathsf{KeyTest}$. There are a few ways around this, and the one we use is to choose the initial "test" messages randomly so that $\boldsymbol{H}$ has low probability of being able to query a ciphertext $C$ generated in some execution of $\mathsf{KeyTest}$.

[3] Suppose, for example, that algorithms $\mathsf{SE}, \mathsf{SD}$ only depend on the first half of the bits of their $k$-bit key. This is consistent with their being IND-CCA secure (in the sense that, if there exists an IND-CCA-secure symmetric encryption scheme, there also exists one with this property), but now, any test has probability at most $2^{-k/2}$ of being able to differentiate between $K$ and a key $L \neq K$ that agrees with $K$ in its first half.

| $\overline{\mathsf{AK}}(1^k)$ | $\overline{\mathsf{AE}}(\overline{pk}, K)$ | $\overline{\mathsf{AD}}(\overline{sk}, C)$ |
|---|---|---|
| $fk \xleftarrow{\$} \{0,1\}^{\text{FKL}(k)}$ | Parse $\overline{pk}$ as $(pk, fk)$ | Parse $\overline{sk}$ as $(sk, fk)$ |
| $(pk, sk) \xleftarrow{\$} \mathsf{AK}^{\overline{F}^k(fk,\cdot)}(1^k)$ | $C \xleftarrow{\$} \mathsf{AE}^{\overline{F}^k(fk,\cdot)}(pk, K)$ | $K \leftarrow \mathsf{AD}^{\overline{F}^k(fk,\cdot)}(sk, C)$ |
| Return $((pk, fk), (sk, fk))$ | Return $C$ | Return $K$ |

**Fig. 2.** Algorithms of the standard-model asymmetric encryption scheme $\overline{\mathsf{AS}} = (\overline{\mathsf{AK}}, \overline{\mathsf{AE}}, \overline{\mathsf{AD}})$ obtained by instantiating RO-model asymmetric encryption scheme $\mathsf{AS} = (\mathsf{AK}, \mathsf{AE}, \mathsf{AD})$ via poly-time family of functions $\overline{F}$.

We note that one might consider an alternative solution to $\boldsymbol{S}$'s problem of wanting to "detect" query $K$ to $H$. Namely, reply to queries to $H$ at random, then, after $\boldsymbol{H}$ terminates, pick one such query $L$ at random, decrypt a challenge ciphertext via $L$, and use that to predict the challenge bit. Unfortunately, even though $L = K$ with probability $1/\operatorname{poly}(k)$, the advantage over one-half obtained by $\boldsymbol{S}$ via the strategy just outlined could be negligible because the wrong answers from the wrong random choices could overwhelm the right answer that arises when $K$ is chosen.

We provide all the details and justify Equation (2) in [3].

## 4 Uninstantiability of the Hash ElGamal Scheme

In this section we show (cf. Theorem 2) that the RO-model Hash ElGamal scheme admits no IND-CCA-preserving instantiation. Below we begin by detailing what we mean by instantiation of a RO-model asymmetric encryption scheme. This will refer to a RO-model scheme which, as per the formal definitions in Section 2, uses a single random oracle mapping $\{0,1\}^*$ to $\{0,1\}$.

INSTANTIATING RO-MODEL ASYMMETRIC ENCRYPTION SCHEMES. A *poly-time family of functions* $\overline{F}$ associates to security parameter $k \in \mathbb{N}$ and key $fk \in \{0,1\}^{\text{FKL}(k)}$ a map $\overline{F}^k(fk, \cdot)\colon \{0,1\}^* \to \{0,1\}$. The *key length* FKL of the family of functions is a polynomial in $k$. We require that there exist a polynomial $t$ such that $\overline{F}^k(fk, x)$ is computable in $t(k + |x|)$ time for all $k \in \mathbb{N}$, $fk \in \{0,1\}^{\text{FKL}(k)}$ and $x \in \{0,1\}^*$.

An *instantiation* of a RO-model asymmetric encryption scheme $\mathsf{AS} = (\mathsf{AK}, \mathsf{AE}, \mathsf{AD})$ via family $\overline{F}$ is the standard-model asymmetric encryption scheme $\overline{\mathsf{AS}} = (\overline{\mathsf{AK}}, \overline{\mathsf{AE}}, \overline{\mathsf{AD}})$ whose constituent algorithms are illustrated in Figure 2. As these indicate, the public and secret keys of the original scheme are enhanced to also include a key $fk$ specifying the function $\overline{F}^k(fk, \cdot)$, and calls to the random oracle are then replaced by evaluations of this function in all algorithms.

THE UNINSTANTIABILITY RESULT. The formal statement of the result is the following.

**Theorem 2.** *Let* $\mathsf{HEG}[\mathsf{CG}] = (\mathsf{AK}, \mathsf{AE}, \mathsf{AD})$ *be the RO-model Hash ElGamal scheme associated to a cyclic-group generator* $\mathsf{CG}$. *Let* $\overline{\mathsf{HEG}}[\mathsf{CG}] = (\overline{\mathsf{AK}}, \overline{\mathsf{AE}}, \overline{\mathsf{AD}})$ *be any instantiation of* $\mathsf{HEG}[\mathsf{CG}]$ *via a poly-time family of functions. Then* $\overline{\mathsf{HEG}}[\mathsf{CG}]$ *is not IND-CCA preserving.*

Proof of Theorem 2. Let $\overline{F}$ be the poly-time family of functions used in $\overline{\mathsf{HEG}}[\mathsf{CG}]$ to replace the random oracle. We will construct an IND-CCA-secure symmetric encryption scheme $\mathsf{SS}$ such that the mm-hybrid encryption scheme $(\overline{\mathsf{HEG}}[\mathsf{CG}], \mathsf{SS})$ is not IND-CCA secure. This proves the theorem.

Let us say that a value $\overline{pk}$ is a $(\overline{\mathsf{HEG}}[\mathsf{CG}], k)$*-valid public key* if there exists a value $\overline{sk}$ such that $(\overline{pk}, \overline{sk}) \in [\overline{\mathsf{AK}}(1^k)]$. We first define two polynomial-time algorithms $\mathsf{VfPK}$ and $\mathsf{VfCtxt}_{\overline{F}}$ which are used by $\mathsf{SS}$.

Algorithm $\mathsf{VfPK}$, which we call a *key verifier*, takes inputs $1^k$ and $\overline{pk}$, and outputs 1 if and only if $\overline{pk}$ is a $(\overline{\mathsf{HEG}}[\mathsf{CG}], k)$-valid public key. The algorithm works by parsing $\overline{pk}$ as $((q, g, X), fk)$, where $fk \in \{0,1\}^{\mathrm{FKL}}$, and then returning 1 if and only if $q$ and $2q+1$ are primes, $g$ is a generator of the order $q$ cyclic subgroup $\langle g \rangle$ of $\mathbb{Z}^*_{2q+1}$, $|2q+1| = k$, and $X \in \langle g \rangle$. This algorithm can be implemented in polynomial-time based on standard facts from computational number theory, and even deterministically, given the existence of polynomial-time primality tests [1]. We omit the details.

Algorithm $\mathsf{VfCtxt}_{\overline{F}}$, which we call a *ciphertext verifier*, takes inputs $1^k, \overline{pk}, K, C$, where $\overline{pk}$ is a $(\overline{\mathsf{HEG}}[\mathsf{CG}], k)$-valid public key and $K \in \{0,1\}^k$. It runs $\overline{\mathsf{AE}}(\overline{pk}, K)$ and outputs 1 if the result is $C$, and 0 otherwise. In other words, $\mathsf{VfCtxt}_{\overline{F}}$ verifies whether $C$ is indeed an encryption of message $K$ under the given public key $\overline{pk}$. This is possible because the encryption algorithm $\mathsf{AE}$ of $\mathsf{HEG}[\mathsf{CG}]$ (cf. Figure 1), and hence the encryption algorithm $\overline{\mathsf{AE}}$ of $\overline{\mathsf{HEG}}[\mathsf{CG}]$, is deterministic.

Let $\mathsf{SS}' = (\mathsf{SK}', \mathsf{SE}', \mathsf{SD}')$ be any standard-model IND-CCA-secure symmetric encryption scheme. (Recall an implicit assumption is that some such scheme exists, since otherwise *all* asymmetric encryptions schemes are by default IND-CCA preserving and the entire problem we are considering is moot.) The construction of $\mathsf{SS}$ is in terms of $\mathsf{SS}'$ and algorithms $\mathsf{VfPK}$ and $\mathsf{VfCtxt}_{\overline{F}}$. We use the notation $\langle(\cdot,\cdot)\rangle$ to denote an injective, polynomial-time computable encoding of pairs of strings as strings such that given $\langle(M_1, M_2)\rangle$, $M_1$ and $M_2$ can be recovered in polynomial time. If $s$ is a string and $a \le b$ are integers then $s[a \ldots b]$ denotes the string consisting of bit positions $a$ through $b$ of $s$. The algorithms constituting $\mathsf{SS} = (\mathsf{SK}, \mathsf{SE}, \mathsf{SD})$ are depicted in Figure 3. To conclude the proof, we need only establish the following propositions.

**Proposition 1.** *Symmetric encryption scheme* $\mathsf{SS}$ *is IND-CCA secure.*

**Proposition 2.** *Multi-message hybrid encryption scheme* $(\overline{\mathsf{HEG}}[\mathsf{CG}], \mathsf{SS})$ *is not IND-CCA secure.*

*Proof (Proposition 1).* Let us first provide some intuition. Note that on input $M$, encryption algorithm $\mathsf{SE}(K_1' \| K_2, \cdot)$ uses the encryption algorithm $\mathsf{SE}'$ of an

| $\mathsf{SK}(1^k)$ | $\mathsf{SE}(K, M)$ | $\mathsf{SD}(K, C)$ |
|---|---|---|
| $K' \xleftarrow{\$} \mathsf{SK}'(1^{\lceil k/2 \rceil})$<br>$K_2 \xleftarrow{\$} \{0,1\}^{\lfloor k/2 \rfloor}$<br>Return $K' \| K_2$ | $k \leftarrow \lvert K \rvert$<br>$K' \leftarrow K[1 \ldots \lceil k/2 \rceil]$<br>$K_2 \leftarrow K[1 + \lceil k/2 \rceil \ldots k]$<br>$C' \leftarrow \mathsf{SE}'(K', M)$<br>Parse $M$ as $\langle (M_1, M_2) \rangle$<br>If the parsing fails then<br>  Return $C' \| 1$ EndIf<br>$p \leftarrow \mathsf{VfPK}(1^k, M_1)$<br>$c \leftarrow \mathsf{VfCtxt}_{\overline{F}}(1^k, M_1, K, M_2)$<br>If ($p = 1$ and $c = 1$) then<br>  Return $C' \| 0$<br>else Return $C' \| 1$ EndIf | $k \leftarrow \lvert K \rvert$<br>$K' \leftarrow K[1 \ldots \lceil k/2 \rceil]$<br>$K_2 \leftarrow K[1 + \lceil k/2 \rceil \ldots k]$<br>Parse $C$ as $C' \| d$,<br>  where $d \in \{0, 1\}$<br>$M' \leftarrow \mathsf{SD}'(K', C')$<br>Parse $M'$ as $\langle (M_1, M_2) \rangle$<br>If the parsing fails then<br>  If $d = 1$ then Return $M'$<br>  else Return $\perp$ EndIf<br>$p \leftarrow \mathsf{VfPK}(1^k, M_1)$<br>$c \leftarrow \mathsf{VfCtxt}_{\overline{F}}(1^k, M_1, K, M_2)$<br>If ($d = 0$ and $p = 1$ and $c = 1$)<br>then Return $M'$ EndIf<br>If ($d = 1$ and ($p \neq 1$ or $c \neq 1$))<br>then Return $M'$ EndIf<br>Return $\perp$ |

**Fig. 3.** Algorithms of the symmetric encryption scheme $\mathsf{SS} = (\mathsf{SK}, \mathsf{SE}, \mathsf{SD})$ for the proof of Theorem 2. Above, $\langle (M_1, M_2) \rangle$ denotes an encoding of the pair of strings $(M_1, M_2)$ as a string.

---

IND-CCA-secure scheme to compute $C' \xleftarrow{\$} \mathsf{SE}'(K_1', M)$ and outputs $C' \| 0$ or $C' \| 1$, depending on whether $M$ has some "special" form or not. The ciphertext ends with 0 if $M$ parses as a pair $(M_1, M_2)$ such that algorithms $\mathsf{VfPK}, \mathsf{VfCtxt}_{\overline{F}}$ indicate that $M_1$ is a $(\overline{\mathsf{HEG}}[\mathsf{CG}], k)$-valid public key and $M_2 \in [\overline{\mathsf{AE}}(M_1, K_1' \| K_2)]$. The decryption algorithm $\mathsf{SD}(K_1' \| K_2, \cdot)$ on input $C' \| d$, where $d$ is a bit, computes $M' \leftarrow \mathsf{SD}'(K_1', C')$ and returns $M'$ only if either $M'$ is of the special form and $d = 0$, or $M'$ is not of this form and $d = 1$. Therefore, an obvious strategy for an adversary against $\mathsf{SS}$ is to query its oracle $\mathsf{SE}(K, \mathrm{LR}(\cdot, \cdot, b))$ on a pair of messages such that one of them is of this special form and the other is not. Using the unique decryptability of $\overline{\mathsf{AE}}$ and the fact that $K_2$ is chosen at random, independently from the adversary's view, we show that it cannot find such queries except with negligible probability. Moreover, we show that any strategy for the adversary can be employed by an attacker against scheme $\mathsf{SS}'$ to win its game. Details follow.

Let $\boldsymbol{S}$ be a legitimate polynomial-time adversary attacking $\mathsf{SS}$. We will construct a legitimate polynomial-time adversary $\boldsymbol{S}'$ such that

$$\mathsf{Adv}^{\text{ind-cca}}_{\mathsf{SS}, \boldsymbol{S}}(k) \leq \mathsf{Adv}^{\text{ind-cca}}_{\mathsf{SS}', \boldsymbol{S}'}(\lceil k/2 \rceil) + \frac{O(Q(k))}{2^{\lfloor k/2 \rfloor}}, \tag{5}$$

where $Q$ is a polynomial upper bounding the total number of queries made by $\boldsymbol{S}$ to its different oracles. Since $\mathsf{SS}'$ is assumed IND-CCA secure, the advantage function associated to $\boldsymbol{S}'$ above is negligible, and thus so is the advantage function associated to $\boldsymbol{S}$. To complete the proof, we need to specify adversary $\boldsymbol{S}'$ and prove Equation (5).

Adversary $\boldsymbol{S}'$ is given input $1^{\lceil k/2 \rceil}$ and has access to oracles $\mathsf{SE}'(K_1', \mathrm{LR}(\cdot,\cdot,b))$ and $\mathsf{SD}'(K_1', \cdot)$. Its goal is to guess the bit $b$. It runs $\boldsymbol{S}$ on input $1^k$. In this process, $\boldsymbol{S}$ will query its two oracles $\mathsf{SE}(K, \mathrm{LR}(\cdot,\cdot,b))$ and $\mathsf{SD}(K,\cdot)$. To answer a query to the first of these oracles, $\boldsymbol{S}'$ forwards the query to its oracle $\mathsf{SE}'(K_1', \mathrm{LR}(\cdot,\cdot,b))$, appends 1 to the oracle's reply and returns the result to $\boldsymbol{S}$. To answer a query to the second oracle, $\boldsymbol{S}'$ checks the last bit of the query. If it is 0, $\boldsymbol{S}'$ returns $\perp$ to $\boldsymbol{S}$. Otherwise, it removes the last bit, forwards the result to its oracle $\mathsf{SD}'(K_1', \cdot)$, and returns the answer to $\boldsymbol{S}$. When $\boldsymbol{S}$ outputs its guess $b'$, $\boldsymbol{S}'$ returns $b'$.

We now analyze $\boldsymbol{S}'$. Consider the experiment in which $\boldsymbol{S}'$ attacks $\mathsf{SS}'$. We define the following events.

| | | |
|---|---|---|
| $\mathsf{Succ}(\boldsymbol{S}')$ | : | $\boldsymbol{S}'$ is successful, meaning its output equals the challenge bit $b$ |
| $\mathsf{BadE}$ | : | $\boldsymbol{S}$ makes a query to oracle $\mathsf{SE}(K, \mathrm{LR}(\cdot,\cdot,b))$ in which one of the messages can be parsed as $\langle (M_1, M_2) \rangle$ such that $M_1$ is a $(\overline{\mathsf{HEG}}[\mathsf{CG}], k)$-valid public key and $M_2 \in [\overline{\mathsf{AE}}(M_1, K)]$ |
| $\mathsf{BadD}$ | : | $\boldsymbol{S}$ makes a query to oracle $\mathsf{SD}(K,\cdot)$ that can be parsed as $C' \| d$, where $d$ is a bit, such that $\mathsf{SD}'(K_1', C') = \langle (M_1, M_2) \rangle$, where $M_1$ is a $(\overline{\mathsf{HEG}}[\mathsf{CG}], k)$-valid public key and $M_2 \in [\overline{\mathsf{AE}}(M_1, K)]$ |

For the experiment in which $\boldsymbol{S}$ attacks $\mathsf{SS}$, we define the following event.

$\mathsf{Succ}(\boldsymbol{S})$ : $\boldsymbol{S}$ is successful, meaning its output equals the challenge bit $b$

We claim that if events $\mathsf{BadE}$ and $\mathsf{BadD}$ do not occur, then $\boldsymbol{S}'$ simulates perfectly the environment provided to $\boldsymbol{S}$ in its attack against $\mathsf{SS}$. First, note that answers to queries to oracle $\mathsf{SE}(K, \mathrm{LR}(\cdot,\cdot,b))$ can only be off by the last bit. In the absence of the "bad" events, each ciphertext returned to $\boldsymbol{S}$ as a reply to a query to oracle $\mathsf{SE}(K, \mathrm{LR}(\cdot,\cdot,b))$ has 1 as the last bit. This is also the case in $\boldsymbol{S}$'s real attack. If $\boldsymbol{S}$ queries $\mathsf{SD}(K,\cdot)$ with a ciphertext $C'\|0$, assuming events $\mathsf{BadE}$ and $\mathsf{BadD}$ do not occur, $\boldsymbol{S}'$ gives $\boldsymbol{S}$ the response it would get in the real attack, namely $\perp$. Since $\boldsymbol{S}$ is legitimate, if it queries oracle $\mathsf{SD}(K,\cdot)$ with a ciphertext $C'\|1$, then $C'$ must not have previously been returned by oracle $\mathsf{SE}'(K_1', \mathrm{LR}(\cdot,\cdot,b))$. Thus $\boldsymbol{S}'$ can legitimately make query $C'$ to its oracle $\mathsf{SD}'(K_1', \cdot)$. If $M$ is the response, then, assuming that events $\mathsf{BadE}$ and $\mathsf{BadD}$ do not occur, the answer $\boldsymbol{S}$ expects is exactly $M$. Therefore,

$$\begin{aligned}\Pr[\,\mathsf{Succ}(\boldsymbol{S}')\,] &\geq \Pr[\,\mathsf{Succ}(\boldsymbol{S}') \mid \neg\mathsf{BadE} \wedge \neg\mathsf{BadD}\,] - \Pr[\,\mathsf{BadE} \vee \mathsf{BadD}\,] \\ &\geq \Pr[\,\mathsf{Succ}(\boldsymbol{S})\,] - \Pr[\,\mathsf{BadE} \vee \mathsf{BadD}\,]\ .\end{aligned}$$

We now provide an upper bound for the probability of event $\mathsf{BadE} \vee \mathsf{BadD}$. Let $q_e(k)$ and $q_d(k)$ be the number of queries $\boldsymbol{S}$ makes to oracles $\mathsf{SE}(K, \mathrm{LR}(\cdot,\cdot,b))$ and $\mathsf{SD}(K,\cdot)$, respectively, on input $1^k$. We observe that if $M_1$ is a $(\overline{\mathsf{HEG}}[\mathsf{CG}], k)$-valid public key, then for any $M_2 \in \{0,1\}^*$, there exists a unique $K' \in [\mathsf{SK}(1^k)]$ such that $M_2 \in [\overline{\mathsf{AE}}(M_1, K')]$. Recall that the key for oracles $\mathsf{SE}(K, \mathrm{LR}(\cdot,\cdot,b))$ and $\mathsf{SD}(K,\cdot)$ is $K = K_1'\|K_2$, where $K_2$ is chosen uniformly at random from

$\{0,1\}^{\lfloor k/2 \rfloor}$ and is independent from $\boldsymbol{S}$'s view. Therefore, for any query made by $\boldsymbol{S}$ to oracle $\mathsf{SE}(K, \mathrm{LR}(\cdot,\cdot,b))$, the probability that one of the messages in the query parses as $\langle (M_1, M_2) \rangle$ such that $M_1$ is a $(\overline{\mathsf{HEG}}[\mathsf{CG}], k)$-valid public key and $M_2 \in [\overline{\mathsf{AE}}(M_1, K)]$ is at most $2/2^{\lfloor k/2 \rfloor}$. Similarly, for any query $C' \| d$, where $d$ is a bit, made by $\boldsymbol{S}$ to oracle $\mathsf{SD}(K, \cdot)$, the probability that $\mathsf{SD}'(K_1', C') = M'$, where $M'$ parses as $\langle (M_1, M_2) \rangle$, $M_1$ is a $(\overline{\mathsf{HEG}}[\mathsf{CG}], k)$-valid public key and $M_2 \in [\overline{\mathsf{AE}}(M_1, K)]$ is at most $1/2^{\lfloor k/2 \rfloor}$. Therefore,

$$\Pr[\,\mathsf{BadE} \vee \mathsf{BadD}\,] \;\leq\; \frac{2q_e(k) + q_d(k)}{2^{\lfloor k/2 \rfloor}} \;\leq\; \frac{2 \cdot Q(k)}{2^{\lfloor k/2 \rfloor}}\,,$$

where $Q(k) = q_e(k) + q_d(k)$. Hence

$$\mathsf{Adv}^{\text{ind-cca}}_{\mathsf{SS}',\boldsymbol{S}'}(\lceil k/2 \rceil) = 2 \cdot \Pr[\,\mathsf{Succ}(\boldsymbol{S}')\,] - 1 \;\geq\; 2\left(\Pr[\,\mathsf{Succ}(\boldsymbol{S})\,] - \frac{O(Q(k))}{2^{\lfloor k/2 \rfloor}}\right) - 1$$

$$= \mathsf{Adv}^{\text{ind-cca}}_{\mathsf{SS},\boldsymbol{S}}(k) - \frac{O(Q(k))}{2^{\lfloor k/2 \rfloor}}\,.$$

Rearranging terms gives Equation (5).

*Proof (Proposition 2).* We define a hybrid adversary $\boldsymbol{H}$ attacking $(\overline{\mathsf{HEG}}[\mathsf{CG}], \mathsf{SS})$. $\boldsymbol{H}$ is given inputs $\overline{pk} = ((q,g,X), \mathit{fk})$ and $C_a$ and has access to oracles $\mathsf{SE}(K, \mathrm{LR}(\cdot,\cdot,b))$, $\mathsf{SD}(K,\cdot)$, and $\overline{\mathsf{AD}}(\overline{sk},\cdot)$, where $\overline{sk} = ((q,g,x), \mathit{fk})$. Its goal is to guess the challenge bit $b$. By the definition of experiment $\mathbf{Exp}^{\text{ind-cca}}_{\overline{\mathsf{HEG}}[\mathsf{CG}],\mathsf{SS},\boldsymbol{H}}(k)$, $\overline{pk}$ is a $(\overline{\mathsf{HEG}}[\mathsf{CG}], k)$-valid public key and $C_a \in [\overline{\mathsf{AE}}(\overline{pk}, K)]$. Therefore, $\langle (\overline{pk}, C_a) \rangle$ is a message which, when encrypted with $\mathsf{SE}(K,\cdot)$, yields a ciphertext that has last bit 0. We observe that for any string $C$ chosen at random from $\{0,1\}^{|C_a|} \setminus \{C_a\}$, the probability that $K = \overline{\mathsf{AD}}(\overline{sk}, C)$ is 0 (since $\overline{\mathsf{AE}}(\overline{pk}, K) = C_a$ and $\overline{\mathsf{AE}}$ is deterministic), i.e., the probability that $C \in [\overline{\mathsf{AE}}(\overline{pk}, K)]$ is 0. Hence $\langle (\overline{pk}, C) \rangle$ is a message which, when encrypted with $\mathsf{SE}(K,\cdot)$, yields a ciphertext that has last bit 1. (If $C \notin [\overline{\mathsf{AE}}(\overline{pk}, K)]$, then the last bit will be 1.) Thus, adversary $\boldsymbol{H}$ can construct two messages for which it can guess with probability 1 the last bit of the corresponding ciphertext. Using this information it can then guess the challenge bit. Details follow.

Adversary $\boldsymbol{H}$ chooses $C$ at random from $\{0,1\}^{|C_a|} \setminus \{C_a\}$, makes a query $\langle (\overline{pk}, C_a) \rangle, \langle (\overline{pk}, C) \rangle$ to oracle $\mathsf{SE}(K, \mathrm{LR}(\cdot,\cdot,b))$, parses the response as $C' \| d$, where $d$ is a bit, and returns $d$. The running time of $\boldsymbol{H}$ is clearly polynomial in $k$. We claim that $\mathsf{Adv}^{\text{ind-cca}}_{\overline{\mathsf{HEG}}[\mathsf{CG}],\mathsf{SS},\boldsymbol{H}}(k) = 1$. To prove this, we consider the event

$\mathsf{Succ}(\boldsymbol{H})$ : $\boldsymbol{H}$ is successful, meaning its output equals the challenge bit $b$

If challenge bit $b$ is 0, then the response to $\boldsymbol{H}$'s query is a ciphertext that has last bit 0. If bit $b$ is 1, then the response is a ciphertext that has last bit 1. Thus

$$\Pr[\,\mathsf{Succ}(\boldsymbol{H})\,] \;=\; \frac{1}{2} + \frac{1}{2} \;=\; 1\,.$$

Hence

$$\mathsf{Adv}^{\text{ind-cca}}_{\overline{\mathsf{HEG}}[\mathsf{CG}],\mathsf{SS},\boldsymbol{H}}(k) \;=\; 2 \cdot \Pr[\,\mathsf{Succ}(\boldsymbol{H})\,] - 1 \;=\; 1\,,$$

as desired.

Notice that the adversary constructed in the proof of Proposition 2 does not make any queries to its oracles $\mathsf{SD}(K, \cdot)$ and $\overline{\mathsf{AD}}(\overline{sk}, \cdot)$.

*Remark 2.* An interesting question at this point may be why the proof of Theorem 2 fails for the RO-model Hash ElGamal scheme $\mathsf{HEG}[\mathsf{CG}]$ associated to a cyclic-group generator $\mathsf{CG}$ —it must, since otherwise Theorem 1 would be contradicted— but succeeds for any instantiation of this scheme. The answer is that symmetric encryption scheme $\mathsf{SS}$, depicted in Figure 3 runs a ciphertext verifier $\mathsf{VfCtxt}_{\overline{F}}$ for the asymmetric encryption scheme in question. In the case of the RO-model scheme $\mathsf{HEG}[\mathsf{CG}]$, any ciphertext verifier must query random oracles $G$ and $H$. But as we clarified in Section 2, $\mathsf{SS}$ does not have access to these oracles (although it might have access to its own, independently chosen oracle $R_s$), and so cannot run such a ciphertext verifier. The adversary of course does have access to $G, H$, but has no way to "pass" these objects to the encryption algorithm of the symmetric encryption scheme. On the other hand, in the instantiated scheme, the keys describing the functions instantiating the random oracles may be passed by the adversary to the encryption algorithm of $\mathsf{SS}$ in the form of a message containing the public key, giving $\mathsf{SS}$ the ability to run the ciphertext verifier. This might lead one to ask why $\mathsf{SS}$ does not have oracle access to $G, H$. This is explained in Remark 1.

As we discussed in Section 1, in [3] we provide a more general impossibility result.

## 5 Acknowledgements

We thank the anonymous referees for their comments. Mihir Bellare and Alexandra Boldyreva were supported in part by NSF grants CCR-0098123, ANR-0129617 and CCR-0208842, and an IBM Faculty Partnership Development Award. Adriana Palacio is supported by an NSF Graduate Research Fellowship.

## References

1. M. Agarwal, N. Saxena and N. Kayal, "PRIMES is in P," *Preprint. Available at* http://www.cse.iitk.ac.in/news/primality.html, August 6, 2002.
2. J. Baek, B. Lee and K. Kim, "Secure length-saving ElGamal encryption under the Computational Diffie-Hellman assumption," *Proceedings of the Fifth Australasian Conference on Information Security and Privacy– ACISP 2000*, LNCS Vol. 1841, E. Dawson, A. Clark and C. Boyd ed., Springer-Verlag, 2000.
3. M. Bellare, A. Boldyreva and A. Palacio, "An Uninstantiable Random-Oracle-Model Scheme for a Hybrid-Encryption Problem," Full version of this paper. Available at http://www-cse.ucsd.edu/users/mihir/.
4. J. Baek, B. Lee and K. Kim, "Provably secure length-saving public-key encryption scheme under the computational Diffie-Hellman assumption," *ETRI Journal*, 22(4), 2000.

5. M. Bellare, A. Desai, E. Jokipii and P. Rogaway, "A concrete security treatment of symmetric encryption: Analysis of the DES modes of operation," *Proceedings of the 38th Symposium on Foundations of Computer Science*, IEEE, 1997.
6. M. Bellare and P. Rogaway, Random oracles are practical: a paradigm for designing efficient protocols, *First ACM Conference on Computer and Communications Security*, ACM, 1993.
7. R. Canetti, U. Feige, O. Goldreich and M. Naor, "Adaptively secure multiparty computation," *Proceedings of the 28th Annual Symposium on the Theory of Computing*, ACM, 1996.
8. R. Canetti, O. Goldreich, S. Halevi, "The random oracle methodology, revisited," *Proceedings of the 30th Annual Symposium on the Theory of Computing*, ACM, 1998.
9. J.-S. Coron, H. Handschuh, M. Joye, P. Paillier, D. Pointcheval, C. Tymen, "GEM: A Generic Chosen-Ciphertext Secure Encryption Method", *Topics in Cryptology – CT-RSA 2002*, LNCS Vol. 2271 , B. Preneel ed., Springer-Verlag, 2002.
10. J.-S. Coron, H. Handschuh, M. Joye, P. Paillier, D. Pointcheval, C. Tymen, "Optimal Chosen-Ciphertext Secure Encryption of Arbitrary-Length Messages," *Proceedings of the Fifth International workshop on practice and theory in Public Key Cryptography– PKC 2002*, LNCS Vol. 1431, D. Naccache and P. Paillier eds., Springer-Verlag, 2002.
11. R. Cramer and V. Shoup, "A practical public key cryptosystem provably secure against adaptive chosen ciphertext attack," *Advances in Cryptology – CRYPTO 1998*, LNCS Vol. 1462, H. Krawczyk ed., Springer-Verlag, 1998.
12. R. Cramer and V. Shoup, "Design and analysis of practical public-key encryption schemes secure against adaptive chosen ciphertext attack," IACR ePrint archive, record 2001/108, 2001, http://eprint.iacr.org/.
13. T. ElGamal, "A public key cryptosystem and signature scheme based on discrete logarithms," *IEEE Transactions on Information Theory*, Vol. 31, 1985.
14. A. Fiat and A. Shamir, "How to prove yourself: practical solutions to identification and signature problems," *Advances in Cryptology – CRYPTO 1986*, LNCS Vol. 263, A. Odlyzko ed., Springer-Verlag, 1986.
15. E. Fujisaki, T. Okamoto, "Secure Integration of Asymmetric and Symmetric Encryption Schemes," *Advances in Cryptology – CRYPTO 1999*, LNCS Vol. 1666, M. Wiener ed., Springer-Verlag, 1999.
16. S. Goldwasser and S. Micali, "Probabilistic encryption," *Journal of Computer and System Science*, Vol. 28, 1984, pp. 270–299.
17. S. Goldwasser and Y. Taumann, "On the (in)security of the Fiat-Shamir paradigm," *Proceedings of the 44th Symposium on Foundations of Computer Science*, IEEE, 2003.
18. S. Micali, "Computationally sound proofs," *SIAM Journal on Computing*, Vol. 30, No. 4, 2000, pp. 1253-1298.
19. J. B. Nielsen "Separating Random Oracle Proofs from Complexity Theoretic Proofs: The Non-committing Encryption Case," *Advances in Cryptology – CRYPTO 2002*, LNCS Vol. 2442 , M. Yung ed., Springer-Verlag, 2002.
20. T. Okamoto and D. Pointcheval "REACT: Rapid Enhanced-security Asymmetric Cryptosystem Transform," *Topics in Cryptology – CT-RSA 2001*, LNCS Vol. 2020, D. Naccache ed., Springer-Verlag, 2001.
21. V. Shoup, "A proposal for an ISO standard for public key encryption", IACR ePrint archive, record 2001/112, 2001, http://eprint.iacr.org/.

# Black-Box Composition Does Not Imply Adaptive Security

Steven Myers

Department of Computer Science
University of Toronto, Canada
myers@cs.toronto.edu

**Abstract.** In trying to provide formal evidence that composition has security increasing properties, we ask if the composition of non-adaptively secure permutation generators necessarily produces adaptively secure generators. We show the existence of oracles relative to which there are non-adaptively secure permutation generators, but where the composition of such generators fail to achieve security against adaptive adversaries. Thus, any proof of security for such a construction would need to be non-relativizing. This result can be used to partially justify the lack of formal evidence we have that composition increases security, even though it is a belief shared by many cryptographers.

## 1 Introduction

While there is arguably no strong theory that guides the development of block-ciphers such as DES and AES, there is a definite belief in the community that the composition of functions often results in functions that have stronger security properties than their constituents. This is evident as many ciphers such as DES, AES and MARS have a "round structure" at the heart of their constructions, and a large part of the ciphers' apparent security comes from the composition of these rounds.

In an attempt to understand the security benefits of composition, there have been several papers that have tried to quantify different ways in which the composition of functions increases security properties as compared to the constituent functions [14,1]. A natural question along these lines is to look at functions that are pseudo-random from the perspective of a non-adaptive adversary, but not that of the standard adaptive adversary, and ask if composition of these functions necessarily provides security against adaptive adversaries. It appears that at least some people in the cryptographic community believe this to be true. In fact, recently Maurer and Pietrzak [16] have shown the cascade of two non-adaptively *statistically-secure* permutations results in an adaptively secure construction, where the cascade of two generators is the composition of the first with the inverse of the second. Additionally, they ask if their cascade construction can be proven secure in the computational setting.

In this paper we show that there is no non-relativizing proof that composition of functions provides security against adaptive adversaries. Thus, this

C. Cachin and J. Camenisch (Eds.): EUROCRYPT 2004, LNCS 3027, pp. 189–206, 2004.

work falls into a general research program that demonstrates the limitations of black-box constructions in cryptography. Examples of such research include [12,19,13,5,7,6]. In the final section, we discuss how the techniques used here can be lifted and used on at least one other natural construction: the XOR of function generators.

We note that it is not possible to strictly separate non-adaptively secure function generators from adaptively secure ones in the black-box model, as there are several black-box constructions that construct the stronger object from the weaker one. The first treats the non-adaptively secure generator as a pseudo-random number generator and then uses the construction of Goldreich, Goldwasser and Micali [9] in order to construct a pseudo-random function generator. The second construction treats the non-adaptively secure function generator as a synthesizer and then constructs a function generator as described by Naor and Reingold in [17]. In both cases, we can go from function generators to permutation generators through the well known Luby-Rackoff construction [15]. However, there are several reasons why these constructions are unsatisfying: first, these constructions are not representative of what is done in practice to construct block-ciphers; second, they require $\Omega(\frac{n}{\log n})$ calls to the non-adaptively secure functions generators. Therefore it is natural to ask if the more efficient constructions used in practice can provide adaptive security.

Finally, since it is possible to construct adaptively secure generators from non-adaptively secure generators using black box techniques, this result suggests the possibility that one reason there may be few general theorems championing the general security amplification properties of compositions is that such theorems are not establishable using standard black-box proof techniques.

### 1.1 Black-Box Constructions and Proofs

Since the existence of most modern cryptographic primitives imply $\mathcal{P} \neq \mathcal{NP}$, much of modern cryptography revolves around trying to construct more complex primitives from other simpler primitives that are assumed to exist. That is, if we assume primitives of type $P$ exist, and wish to show that a primitive of type $Q$ exists, then we give a construction $C$, where $C(M_P)$ is an implementation of $Q$ whenever $M_P$ is an implementation of $P$. However, most constructions in modern cryptography are black-box. More specifically, when given a a primitive $P$, we construct a primitive $Q$ by a construction $C^P$, where the primitive $P$ is treated as an oracle. The difference between the two constructions is that in the former case the construction may make use of the machine description, while in the latter it only treats the primitive as an oracle to be queried: it's as if P were inside of a black box.

Observe that it is not immediately clear how to prove that there can be no black-box construction $C^P$ of a primitive $Q$ from an implementation $M_P$ of a primitive $P$, as the implementation $C$ and the proof of its correctness and security could always ignore the presence of the oracle $P$, and independently use the implementation $M_P$ in the construction $C$. The notion of proving black-box separation results was initiated by Baker, Gill and Solovay [2], who were interested in the techniques necessary to answer the $\mathcal{P}$ vs. $\mathcal{NP}$ question. Building on this work, Impagliazzo and Rudich [12] gave a model in which one can prove separations for cryptographic primitives. In their model they note that black-box constructions and proofs work relative to any oracle, that is they relativize, and therefore it is sufficient to provide an oracle $O$ which implements a primitive $P$, but all constructions $C^O$ of primitive $Q$ are not secure relative to $O$. Gertner, Malkin and Reingold [8] have shown that if one's goal is to rule out black-box constructions, then a weaker type of theorem will suffice: for each black-box construction $C^P$ of primitive $Q$, it suffices to demonstrate an oracle $O$ that implements primitive $P$, but for which $C^O$ is insecure. Our result will be of this flavor.

As was stated previously, we cannot separate non-adaptive generators from adaptive ones, as there are black-box constructions of one from the other. However, we show that certain constructions (those which are the composition of permutation generators) cannot provide provable adaptive security using black-box techniques. This is done by constructing an oracle for each construction that provides a natural representation of a non-adaptively secure permutation generator, but where the composition of these generators is not adaptively secure.

Finally, we note that there are several techniques that are used in cryptography, such as Zero-Knowledge in its many incarnations (to name but a few [11,10,4,18,3]) that are often used in cryptographic constructions in such a way that the construction, and not necessarily the technique, is non-black-box.

### 1.2 Our Results

Our main result involves permutation generators. These generators have an associated domain-size parameter $n \in \mathbb{N}$ that fixes the set $\{0,1\}^n$ over which the permutations are defined.

**Theorem 1.** *For every polynomial $m$, there exists a pair of oracles relative to which there exist non-adaptively secure pseudo-random permutation generators $P$ such that the generator $\underbrace{P \circ \ldots \circ P}_{m(n)}$ is not adaptively secure.*

In the theorem $P \circ P'$ denotes the natural composition construction: it is the generator constructed by fixing the security parameter and randomly choosing a $p \in P$ and $p' \in P'$ and computing the permutation $p \circ p'$. The construction $\underbrace{P \circ \ldots \circ P}_{m(n)}$ defines the generator that generates permutations over the set $\{0,1\}^n$ by composing $m(n)$ generators $P$.

### 1.3 Preliminaries & Notations

Let $S$ be a finite set, and let $x \in_{\mathcal{U}} S$ denote the act of choosing an element $x$ uniformly at random from $S$. To describe some of the probabilistic experiments we adopt the notation $\Pr[R_1; \ldots; R_k :: E|C]$ to denote the probability that if random processes $R_1$ through $R_k$ are performed, in order, then, conditioned on event $C$, the event $E$ occurs.

**Notation 1.** *Let $D$ be any finite set, and let $Q_1, Q_2 \subseteq D \times D$ be arbitrary sets of pairs. We define $Q_1 \circ Q_2 = \{(a, c) | \exists a, b, c \in D \text{ s.t. } (a, b) \in Q_1 \wedge (b, c) \in Q_2\}$. Generalizing this notion, for a collection of sets $Q_{k_1}, \ldots, Q_{k_m} \subseteq D \times D$ and for a vector $\boldsymbol{K} = (k_1, ..., k_m)$, let $Q_{\boldsymbol{K}} = Q_{k_1} \circ \ldots \circ Q_{k_m}$.*

**Notation 2.** *For any finite set $D$, we denote by $D^{[k]}$ the set of all $k$-tuples of* distinct *elements from $D$. In a slight abuse of notation, for a $k$-tuple $\mathbf{d} = (d_1, \ldots, d_k) \in D^k$, we say $x \in \mathbf{d}$ if there exists an $i \leq k$ such that $x = d_i$. Additionally, for a function $f$ of the form $D \to D$, we write $f(\mathbf{d})$ to denote $(f(d_1), \ldots, f(d_k))$.*

**Notation 3.** *We denote by $\Pi^n$ the set of all permutations over $\{0,1\}^n$, and we denote by $\mathcal{F}^n$ the set off all functions of the form $\{0,1\}^n \to \{0,1\}^n$.*

### 1.4 Organization

In Section 2 we introduce the standard definitions related to Pseudo-Random Permutation and Function Generators, the difference between adaptive and non-adaptive security, and we discuss how these definitions are lifted into relativized worlds. In Section 3 we present the oracles relative to which we will prove our result. We show that, relative to these oracles, non-adaptively secure permutation generators exist, but that their composition does not provide adaptive security. This is done by showing that non-adaptive adversaries cannot make effective use of one of the oracles that an adaptive adversary can make use of. We demonstrate the oracles' lack of effectiveness to the non-adaptive adversary by demonstrating how the oracles responses could easily be simulated by a non-adaptive adversary. In Section 4 we present the proofs of the combinatorial lemmas behind the simulation just mentioned. We finish in Section 5 by discussing how the techniques presented can be lifted to get similar results for other constructions, such as those based on XOR. Finally, we discuss some directions for future work.

## 2 Standard Definitions

We use the standard, Turing machine based, uniform definitions for pseudo-random function generators and adversaries.

**Definition 1 (Function Ensembles).** *We call $G : \{0,1\}^\kappa \times \{0,1\}^n \to \{0,1\}^n$ a function generator. We say that $k \in \{0,1\}^\kappa$ is a* key *of $G$, write $G(k, \cdot)$ as $g_k(\cdot)$ and say that key $k$ chooses the function $g_k$. Let $g \in_{\mathcal{U}} G$ represent the act of uniformly at random choosing a key $k$ from $\{0,1\}^\kappa$, and then using the key $k$ to choose the function $g_k$.*

*Let $\ell$ be a polynomial, and let $\mathcal{N} \subseteq \mathbb{N}$ be an infinitely large set. For each $n \in \mathcal{N}$, let $G^n : \{0,1\}^{\ell(n)} \times \{0,1\}^n \rightarrow \{0,1\}^n$ be a function generator. We call $G = \{G^n | n \in \mathcal{N}\}$ a* function ensemble. *Given an ensemble $G$, if for every $n \in \mathcal{N}$, the function $g \in_{\mathcal{U}} G^n$ is a permutation, then we say $G$ is a* permutation ensemble. *We say an ensemble $G$ is* efficiently computable *if there exists a Turing machine $M$ and a polynomial $p$ such that for all sufficiently large $n$, for all $x \in \{0,1\}^n$ and $k \in \{0,1\}^{\ell(n)}$ the Turing machine's $M(k,x)$ output is $G^n_k(x)$ and $M(k,x)$ runs in time $p(n)$.*

**Definition 2 ((Non-)Adaptive Adversaries).** *An adversary, $A$, is a probabilistic, polynomial-time Turing machine with oracle access that outputs an element in $\{0,1\}$. We denote an adversary $A$ with access to an oracle $f$ as $A^f$. In order to query an oracle $f$, $A$ writes its query to a special oracle-query-tape, and enters a specified query request state. The response to the query is then written to an oracle-response-tape by the oracle, and $A$ continues its computation. For accounting purposes, we assume that it takes unit time to write the response of the oracle to the tape, once $A$ has entered the query state. An adversary is* adaptive *if it can make multiple queries to the oracle, where future queries can depend on the results of previous queries. A* non-adaptive *adversary may make multiple queries to the oracle, but all queries must be made in parallel at the same time. Formally, the adversary is permitted to write several queries at a time to the oracle-query-tape. When the machine enters the specified query state, the response to all of the queries are written to the response tape.*

**Definition 3 ((Non-)Adaptive Pseudo-Random Function Generator Ensembles).** *Let $m$ and $\ell$ be polynomials. Let $G = \{G^n | n \in \mathbb{N}\}$ be an efficiently computable function generator ensemble such that for each $n$ the generator $G^n$ is of the form $\{0,1\}^{\ell(n)} \times \{0,1\}^n \rightarrow \{0,1\}^{m(n)}$. Define $\mathcal{F} = \{\mathcal{F}^n | n \in \mathbb{N}\}$.*

*We say that $G$ is adaptively (resp. non-adaptively) secure if for all constants $c > 0$, for all adaptive (resp. non-adaptive) polynomial time adversaries $A$ and for all sufficiently large $n$:*

$$\left| \Pr_{\substack{g \in_{\mathcal{U}} G^n \\ r \in_{\mathcal{U}} \{0,1\}^*}} [A^g(1^n) = 1] - \Pr_{\substack{f \in_{\mathcal{U}} \mathcal{F}^n \\ r \in_{\mathcal{U}} \{0,1\}^*}} \left[A^f(1^n) = 1\right] \right| \leq \frac{1}{n^c},$$

*where the $r \in \{0,1\}^*$ represent the random coin-tosses made by $A$.*

In this work we are concerned with the above definitions, but in worlds where a pair of oracles $(O, R)$ exist. We note we use a pair of oracles, as opposed to just one, to simplify the presentation of the proof. We extend the definitions of function ensembles and adaptive/non-adaptive adversaries by allowing Turing machines to have access to the oracles $O$ and $R$. We stress that non-adaptive adversaries *are* permitted to query $O$ and $R$ in an adaptive manner: the non-adaptive restriction on oracle queries in the definition of the adversary (Defn. 2) are only for the oracles $f$ and $g$ specified in the definition of pseudo-random function generator ensembles (Defn. 3).

## 3 The Separating Oracles for Composition

We will construct an oracle that contains an information theoretically secure pseudo-random permutation generator (PRPG). By this we mean that for each $n$ it will include $2^n$ random permutations over $\{0,1\}^n$. Clearly, a non-adaptively secure PRPG $F$ can be immediately constructed from such an oracle, but it is also clear that the same generator will be adaptively secure. Therefore, we add another oracle $R$ that weakens the security of $O$. To help describe $R$, suppose the construction of interest is the composition of two permutations from $O$, and suppose the adversary has access to a function $g$ that is either a function chosen randomly from $\mathcal{F}^{n \to n}$ or $\pi_1 \circ \pi_2$ for $\pi_2, \pi_1 \in_{\mathcal{U}} \Pi^n$. The oracle $R$ iteratively requests the values of $y_i = g(x_i)$ for enough (but still a small number of) randomly chosen values $x_i$ that it should be able to uniquely identify $\pi_1, \pi_2 \in O$, if it is the case that $g = \pi_1 \circ \pi_2$. If $R$ determines that that there exists a $\pi_1, \pi_2 \in O$ such that $y_i = \pi_1 \circ \pi_2(x_i)$ for each $i$, then it will predict a random input/output pair $(x^*, y^*)$, where $y^* = \pi_1 \circ \pi_2(x^*)$. Alternatively, if there is no pair of permutations in $O$ whose composition is consistent with all of the $(x_i, y_i)$ then the oracle rejects and outputs $\perp$.

The oracle $R$ provides a trivial way for an adaptive adversary to break the security of the composed generators: such an adversary can easily supply the $y_i = g(x_i)$ values $R$ requests as responses to its $x_i$ challenges. If $R$ returns a prediction $(x^*, y^*)$ that is consistent with $y^* = g(x^*)$ then almost surely $g$ is a composition of permutations from $O$. In contrast, if the adversary is *non-adaptive* then the oracle $R$ will be of essentially no use to the adversary because of $R$'s iterative nature. Therefore, it is as if $R$ does not exist to the adversary, and therefore the adversary cannot use $R$ to help identify permutations that are in $O$.

### 3.1 Oracle Definitions

**Definition 4 (The Oracle O).** *Let $O^n \stackrel{O}{\leftarrow} \Pi^n$ denote the process of choosing an indexed set of $2^n$ random permutations from $\Pi^n$ with replacement. Let $O_k^n$ denote the kth permutation in $O^n$. Let $O = \{O^n | n \in \mathbb{N}\}$. Where n is clear we write $O_k$ to denote $O_k^n$. For $k_1, ..., k_m \in \{0,1\}^n$ and $\boldsymbol{K} = (k_1, ..., k_m)$, let $O_{\boldsymbol{K}}$ denote $O_{k_m} \circ \ldots \circ O_{k_1}$. Further, for $x_1, ..., x_\ell \in \{0,1\}^n$ and $\boldsymbol{x} = (x_1, ..., x_\ell)$, denote $O_k(\boldsymbol{x}) = \boldsymbol{y} = (O_k(x_1), \ldots, O_k(x_\ell))$.*

**Definition 5 (Composition Construction).** *Let $m : \mathbb{N} \to \mathbb{N}$ be a polynomial where for every $i \in \mathbb{N}$, $m(i) > 2$. For an oracle $O$, for every $n \in \mathbb{N}$ and $k_1, ..., k_{m(n)} \in \{0,1\}^n$ let $F^n_{(k_1,...,k_{m(n)})}(x) = O_{k_1,...k_{m(n)}}(x)$. Let $F = \cup_n \{F^n\}$ be the proposed construction for an adaptively secure PRPG.*

**Definition 6 (The Oracle $R$).** *For an oracle $O$ as described in Definition 4 and a construction $F$ of $m$ compositions as described in Definition 5 we define the oracle $R$ as follows. Define, with foresight, $\ell(n) = m(n) + 1$. Let*

$R = \{(R_1, R_2, R_3)\}$ *be an oracle that for each n is chosen randomly according to the random process* $\Psi^n(O)$, *and the fixed. The process* $\Psi^n(0)$ *is described below:*

$R_1(1^n) \to x_1, \ldots, x_{\ell(n)}$ *where* $(x_1, \ldots, x_{\ell(n)}) \in_{\mathcal{U}} (\{0,1\}^n)^{[\ell(n)]}$.

$R_2(1^n, x_1, \ldots, x_{\ell(n)}, y_1, ..., y_{\ell(n)}) \to x_{\ell(n)+1}, x_{\ell(n)+2}$ *for the* $x_1, \ldots, x_{\ell(n)}$ *output by* $R_1(1^n)$; *any* $y_1, \ldots, y_{\ell(n)} \in \{0,1\}^n$; *and* $(x_{\ell(n)+1}, x_{\ell(n)+2}) \in_{\mathcal{U}} (\{0,1\}^n \setminus \{x_1, \ldots, x_{\ell(n)}\})^{[2]}$.

$R_3(1^n, x_1, \ldots, x_{\ell(n)+2}, y_1, \ldots, y_{\ell(n)+2}) = (x^*, y^*)$ *for the* $(x^{\ell(n)+1}, x^{\ell(n)+2})$ *output by* $R_2(1^n, x_1, \ldots, x_{\ell(n)}, y_1, \ldots, y_{\ell(n)})$; *any* $y_{\ell(n)+1}, y_{\ell(n)+2} \in \{0,1\}^n$; $\kappa \in_{\mathcal{U}} \{\kappa = (k_1, \ldots, k_{m(n)}) \in \{0,1\}^{n \cdot m(n)} | O_\kappa(x_1, \ldots, x_{\ell(n)+2}) = (y_1, \ldots, y_{\ell(n)+2})\}$; $x^* \in_{\mathcal{U}} \{0,1\}^n$; *and* $y^* = O_\kappa(x^*)$.

*On all other inputs to the oracles* $R_1, R_2$ *and* $R_3$ *the result is* $\perp$. *Finally, we denote by* $R \overset{R}{\leftarrow} \Psi(O)$ *the process of randomly choosing* $R$ *given a fixed* $O$, *according to the random process* $\Psi^n(O)$ *described above for each* $n \in \mathbb{N}$.

## 3.2 The Oracle $O$ Provides Adaptive Security

We state, without proof, the following lemma that states that most of the oracles $O$ provide a natural, adaptively secure permutation generator.

**Lemma 1.** *For all probabilistic, polynomial-time, adaptive adversaries* $A$ *and for all sufficiently* $n$*:*

$$\Pr_{O \overset{O}{\leftarrow} \Pi}\left[\left|\Pr_{\substack{f \in_{\mathcal{U}} O^n \\ r \in_{\mathcal{U}} \{0,1\}^*}}\left[A^{f,O}(1^n) = 1\right] - \Pr_{\substack{g \in_{\mathcal{U}} \Pi^n \\ r \in_{\mathcal{U}} \{0,1\}^*}}\left[A^{g,O}(1^n) = 1\right]\right| \le \frac{1}{2^{n/2}}\right] \ge 1 - \frac{1}{2^{n/2}},$$

*where* $r \in_{\mathcal{U}} \{0,1\}^*$ *represents the random coin-tosses of* $A$.

## 3.3 The Oracle $R$ Breaks Adaptive Security

**Lemma 2.** *There exists an efficient adversary, Adv, such that for all oracle pairs* $(O, R)$ *that could possibly be constructed, Adv breaks the adaptive security of* $F$ *relative to* $O$ *and* $R$.

*Proof.* We show that the following adversary has a significant chance of distinguishing between the composition of $m(n)$ functions from $O$ and a random function. Note that this adversary calls $f$ *adaptively.*

$Adv^{f,O,R}(1^n)$
 $\boldsymbol{x}_1 = (x_1, \ldots, x_{\ell(n)}) \leftarrow R_1(1^n)$.
 $\boldsymbol{y}_1 = (y_1, ..., y_{\ell(n)}) \leftarrow f(\boldsymbol{x}_1)$.
 $\boldsymbol{x}_2 = (x_{\ell(n)+1}, x_{\ell(n)+2}) \leftarrow R_2(1^n, \boldsymbol{x}_1, \boldsymbol{y}_1)$.
 $\boldsymbol{y}_2 = (y_{\ell(n)+1}, y_{\ell(n)+2}) \leftarrow f(\boldsymbol{x}_2)$.
 If $\perp = R_3(1^n, \boldsymbol{x}_1, \boldsymbol{x}_2, \boldsymbol{y}_1, \boldsymbol{y}_2)$ output 0.
 Otherwise output 1.

Fix the oracles $O$ and $R$. We show that if $f$ was chosen from $F$ then we output 1 and otherwise (w.h.p.) we output 0. It is an easy observation that if $f \in F$ then, by the construction of *Adv* and $R$, the adversary *necessarily* outputs 1. Alternatively, if $f \in \diamond^n$ then it is easy to see by the following claim that there is not likely to be any key $\boldsymbol{k}$ where $O_{\boldsymbol{k}}(\boldsymbol{x}) = \boldsymbol{y}$ holds, and therefore the oracle $R$ will output $\perp$, and thus (w.h.p.) *Adv* will output 0. We remind the reader of the notation defined in Notn. 2, as it is used in the statement of the claim.

*Claim 1. For all sufficiently large $n$, for $\boldsymbol{x} = (x_1, \ldots, x_{\ell(n)}) \leftarrow R(1^n)$:*

$$\Pr[f \in_{\mathcal{U}} \Pi^n :: \exists \boldsymbol{K} \in \{0,1\}^{n \cdot m(n)} \text{ s.t. } f(\boldsymbol{x}) = O_{\boldsymbol{K}}(\boldsymbol{x})] \leq 2^{-n}.$$

*Proof.* Let $S = \{O_{\boldsymbol{K}}(\boldsymbol{x}) | \boldsymbol{K} \in \{0,1\}^{n \cdot m(n)}\}$. Clearly $|S| \leq 2^{n \cdot m(n)}$. Consider the probability that $f(\boldsymbol{x}) \in S$, and since $f \in_{\mathcal{U}} \Pi^n$ it is easy to see that this probability is bound by $2^{n \cdot m(n)} / \prod_{i=1}^{\ell(n)} (2^n - i) \leq 2^{n \cdot (m(n) - \ell(n) - 1)} < 2^{-n}$, as $\ell(n) = m(n) + 1$. □

### 3.4 Simulating the Oracle *R* for Non-adaptive Adversaries

It needs to be shown that $R$ does not destroy the non-adaptive security of $O$. We show that for every non-adaptive adversary with access to the oracle $R$ we can construct another non-adaptive adversary that is essentially just as successful at breaking $O$, but that has no access to $R$. Since $O$ is a large set of random permutations, it is clear that without $R$ there can be no successful distinguishing adversary, and therefore there must be no successful non-adaptive adversary relative to $R$ either.

We will begin by showing that for every adversary $B$ relative to $R$, there exists an adversary $\widehat{B}$ that distinguishes nearly as well as $B$, but does not make queries to $R_3$. This is done by having $\widehat{B}$ simulate the responses of $R_3$. In this simulation there are two general cases: first, there are queries which are likely to be made, and in these cases it turns out that $\widehat{B}$ can simulate $R_3$'s responses with only access to $O$. Next, there are queries that are unlikely to be made, and we cannot simulate $R_3$'s responses in these cases: we show it is incredibly unlikely that $B$ will make such queries, and thus incorrect answers will not significantly affect the acceptance probability of $\widehat{B}$. Finally, it is then a simple observation that $\widehat{B}$ can easily simulate $R_1$ and $R_2$ perfectly, and thus there is no need for $\widehat{B}$ to query the oracle $R$.

In order to construct $\widehat{B}$ we need $B$ to be in a normal form. First, we assume that an adversary never makes the same oracle query twice. Any adversary that does can be converted to one that does not by storing all of its previous oracle queries and the corresponding responses; it can then look up responses on duplicate queries. Next, for our adversary $B$, with access to a function oracle $f : \{0,1\}^n \to \{0,1\}^n$, we assume without loss of generality that $B^{f,O,R}(1^n)$ always makes exactly $T(n)$ combined queries to $f, O$ and $R$, for some polynomial $T$. Further, we will assume that $B$ records all of its queries and their corresponding responses on its tape in a manner which is efficiently retrievable. In particular, we will assume that for each $k \in \{0,1\}^n$ there is a set $Q_k = \{(q,r)\}$ that contains all of the query/response pairs $O_k(q) \rightarrow r$ that have been made; a set $Q_f = \{(q,r)\}$ that contains all the query/response pairs to the challenge function, where $f(q) \rightarrow r$; and a set $SR_2$ that contains all the query/response pairs to $R_2$.

**Lemma 3.** *Let $T$ be a polynomial. For every oracle adversary $B$ that on input of size $n$ makes $T(n)$ queries, there exists an oracle adversary $\widehat{B}$: $\widehat{B}$ on input of size $n$ makes at most $T(n) \cdot m(n)$ oracle queries; $\widehat{B}$ never queries $R_3$; and for all sufficiently large $n$ it is the case that*

$$\Pr[O \overset{O}{\leftarrow} \Pi; R \overset{R}{\leftarrow} \Psi(O); f \in_{\mathcal{U}} O^n :: \widehat{B}^{O,R,f}(1^n) \neq B^{O,R,f}(1^n)] \leq 5 \cdot T(n)/2^{n/2}$$

*and*

$$\Pr[O \overset{O}{\leftarrow} \Pi; R \overset{R}{\leftarrow} \Psi(O); f \in_{\mathcal{U}} \Pi^n :: \widehat{B}^{O,R,f}(1^n) \neq B^{O,R,f}(1^n)] \leq 5 \cdot T(n)/2^{n/2}.$$

*Proof.* Fix $n$. We now construct an adversary $\widehat{B}$ that doesn't make queries to $R_3$. We note that in the statement of the lemma and its proof we don't concern ourselves with the random coin-tosses of $B$ or $\widehat{B}$. It will be obvious from the proof, that the random coin-tosses do not affect any of the probabilities we discuss, and that we could simply fix the random coin tosses of $B$ and prove the theorem for each such sequence.

We will consider a series of hybrid adversaries. Let $\widehat{A}_i(1^n)$ be an adversary that runs $B(1^n)$ but on the first $i$ oracle queries rather than make an oracle query $q$ it runs the sub-routine $G(q)$ and takes the output of $G(q)$ as the result of the query. Before giving the description of $G$ we remind the reader of the notation defined in Notn. 1 and 2. The sub-routine $G$ is defined below:

$G(q)$

If $q$ is not a query to $R_3$ perform query $q$ and let $a$ be the oracle's response: output $a$

Otherwise $q = R_3(x_1, ..., x_{l(n)+2}, y_1, ..., y_{\ell(n)+2})$.
$\boldsymbol{x}_1 = (x_1, ..., x_{l(n)})$ and let $\boldsymbol{x}_2 = (x_{l(n)+1}, x_{l(n)+2})$.
$\boldsymbol{y}_1 = (y_1, ..., y_{l(n)})$ and let $\boldsymbol{y}_2 = (y_{l(n)+1}, y_{l(n)+2})$.
(6) If $((\boldsymbol{x}_1, \boldsymbol{y}_1), \boldsymbol{x}_2) \notin SR_2$ output $\perp$.
(7) $\mathcal{K} = \{\boldsymbol{K} \in (\{0,1\}^n \cup \{f\})^{m(n)} | ((\boldsymbol{x}_1, \boldsymbol{x}_2), (\boldsymbol{y}_1, \boldsymbol{y}_2)) \in Q_{\boldsymbol{K}}\}$.
(8) If $|\mathcal{K}| \neq 1$ output $\perp$.
(9) $\mathcal{K} = \{\boldsymbol{k}\}$.
(10) If $f \in \boldsymbol{k}$ output $\perp$
(11) Choose $x^* \in_{\mathcal{U}} \{0,1\}^n$ and query $y^* \leftarrow O_{\boldsymbol{k}}(x^*)$.
(12) Output $(x^*, y^*)$.

The intuition behind $G$ is the following: for any query $q$ to an oracle other than $R_3$ it behaves *identically* to that oracle on query $q$; for queries to $R_3$ it will almost surely output the same result as the query to $R_3$. We give a quick outline of the intuition for the latter case. First, in Line 6, when $G$ outputs $\perp$ it is almost surely the correct answer because if $(\boldsymbol{x}_1, \boldsymbol{y}_1)$ has not been queried from $R_2$, then the probability of the adversary guessing $\boldsymbol{x}_2$ correctly is negligible. On Line 8 we really have two cases: first, when $|\mathcal{K}| = 0$, it is unlikely there is a key, $\boldsymbol{K}$, such that $O_{\boldsymbol{K}}(\boldsymbol{x}_1, \boldsymbol{x}_2) = (\boldsymbol{y}_1, \boldsymbol{y}_2)$, and thus the response $\perp$ is almost surely correct; next, when $|\mathcal{K}| \geq 2$, and in this case $G$ will output the incorrect answer, but the probability of this case occurring is negligible. The intuition for Line 10 is really the main point behind the proof. If the adversary manages to find a key where it can substitute the function $f$ for part of the key, then the simulation $G$ will output an incorrect answer. However, because of the iterative nature in which $\boldsymbol{x}_1$ and $\boldsymbol{x}_2$ are exposed and the adversary's limitation of accessing $f$ in a non-adaptive manner, we show that the probability of this event occurring is negligible. Finally, If $(x^*, y^*)$ is output on Line 12 then the output is almost surely correct.

We now look at the cumulative errors that can be made in the hybrid process. We use the following two lemmas that are proven in Section 4.

**Lemma 4.** *For all sufficiently large n:*

$$\Pr[O \stackrel{O}{\leftarrow} \Pi; R \stackrel{R}{\leftarrow} \Psi(O); f \in_{\mathcal{U}} O^n :: \widehat{A}_{i+1}^{O,R,f}(1^n) \neq \widehat{A}_i^{O,R,f}(1^n)] \leq 5/2^{n/2}$$

**Lemma 5.** *For all sufficiently large n:*

$$\Pr[O \stackrel{O}{\leftarrow} \Pi; R \stackrel{R}{\leftarrow} \Psi(O); f \in_{\mathcal{U}} \Pi^n :: \widehat{A}_{i+1}^{O,R,f}(1^n) \neq \widehat{A}_i^{O,R,f}(1^n)] \leq 5/2^{n/2}$$

We note that by the previous two lemmas, the probabilities that $B$ and $\widehat{A}_{T(n)+1}$ have differing outputs in the same experiments is less than $T(n) \cdot 5/2^{n/2}$, and, since $T$ is a polynomial, this is a negligible amount. Let $\widehat{B}$ be the Turing machine $\widehat{A}_{T(n)+1}$. We note that by inspection of $G$, and remembering that the call to $O_{\boldsymbol{k}}(x^*)$ on Line 11 of G can mask $m(n)$ queries, $\widehat{B}(1^n)$ makes $\widehat{T}(n) \cdot m(n)$ queries. Further, the probability that $B^{O,R,f}(1^n)$ and $\widehat{B}^{O,R,f}(1^n)$ have differing outputs for the either experiment defined in Lemma 3 is less than $T(n) \cdot 5/2^{n/2}$. □

The last remaining step is to get rid of the queries to $R_1$ and $R_2$ that are made by $\widehat{B}$. We note that the results of queries to $R_1$ and $R_2$ are independent of $O$ and the challenge function $f$, and since the results of such queries are random bit strings, they are easy to simulate. Specifically, we consider a Turing machine $C$ that executes $\widehat{B}$ faithfully, but before beginning the simulation $C(1^n)$ will randomly select $(x_1, ..., x_{\ell(n)}) \in_{\mathcal{U}} (\{0,1\}^n)^{[\ell(n)]}$. During the simulation of $\widehat{B}(1^n)$, if there is a query to $R_1(1^n)$, it will respond with $(x_1, ..., x_{\ell(n)})$ and if for $y_1, \ldots, y_{\ell(n)} \in \{0,1\}^n$ there is a query to $R_2(1^n, x_1, ..., x_\ell(n), y_1, .., y_\ell(n))$ it responds with $(x'_{\ell(n)+1}, x'_{\ell(n)+2}) \in_{\mathcal{U}} \left(\{0,1\}^n \setminus \{x_1, \ldots, x_{\ell(n)}\}\right)^{[2]}$. Note that this simulation is perfect. We can now prove the final result of this section.

**Lemma 6.** *For every probabilistic, polynomial-time, non-adaptive adversary $A$ and for all sufficiently large $n$:*

$$\Pr_{\substack{O \xleftarrow{\mathcal{Q}} \Pi \\ R \xleftarrow{R} \Psi(O)}} \left[ \left| \Pr_{f \in_{\mathcal{U}} O^n} \left[ A^{O,R,f}(1^n) = 1 \right] - \Pr_{g \in_{\mathcal{U}} \Pi^n} \left[ A^{O,R,g}(1^n) = 1 \right] \right| \leq \frac{1}{2^{n/3}} \right] \geq 1 - \frac{1}{2^{n/2}},$$

*where the $r \in \{0,1\}^*$ represent the random coin-tosses made by $A$.*

*Proof.* Assume for contradiction that there exists a probabilistic, polynomial time adversary $B$ and infinitely many $n$ for which:

$$\Pr_{\substack{O \xleftarrow{\mathcal{Q}} \Pi \\ R \xleftarrow{R} \Psi(O)}} \left[ \left| \Pr_{\substack{f \in_{\mathcal{U}} O^n \\ r \in \{0,1\}^*}} \left[ B^{O,R,f}(1^n) = 1 \right] - \Pr_{\substack{g \in_{\mathcal{U}} \Pi^n \\ r \in \{0,1\}^*}} \left[ B^{O,R,g}(1^n) = 1 \right] \right| > \frac{1}{2^{n/3}} \right] > \frac{1}{2^{n/2}}$$

By Lemmas 4 and 5 and the discussion following them, there exists a probabilistic, polynomial time, non-adaptive adversary $C$ that does not query oracle $R$ and infinitely many $n$ such that:

$$\Pr_{\substack{O \xleftarrow{\mathcal{Q}} \Pi \\ R \xleftarrow{R} \Psi(O)}} \left[ \left| \Pr_{\substack{f \in_{\mathcal{U}} O^n \\ r \in \{0,1\}^*}} \left[ C^{O,f}(1^n) = 1 \right] - \Pr_{\substack{g \in_{\mathcal{U}} \Pi^n \\ r \in \{0,1\}^*}} \left[ C^{O,g}(1^n) = 1 \right] \right| > \frac{1}{2^{n/3-1}} \right] > \frac{1}{2^{n/2}}$$

Observing that the choices over $R \xleftarrow{R} \Psi(O)$ have no effect and can be removed, this result contradicts Lemma 1. □

By using standard counting arguments and the previous lemma, we get the following theorem.

**Theorem 2.** *There exists a pair of oracles $(O, R)$ where $O$ is a non-adaptively secure permutation generator and where $F$ is not an adaptively secure permutation generator.*

## 4 Combinatorial Lemmas

### 4.1 Unique Paths Lemma

An essential point in proving Lemmas 4 & 5 is the following: unless an adversary has already determined by oracle queries to $O$ that for a given key, $\boldsymbol{K}$, of $F$ and $\ell$-tuples, $\boldsymbol{x}$ and $\boldsymbol{y}$, where $O_{\boldsymbol{K}}(\boldsymbol{x}) = \boldsymbol{y}$; then the probability that $O_{\boldsymbol{K}}(\boldsymbol{x}) = \boldsymbol{y}$ holds is negligible. The following lemma and its corollary formalizes this concept.

**Lemma 7 (Unique Paths Lemma).** *Let $T$, $\ell$ and $m$ be polynomials. For all sufficiently large $n \in \mathbb{N}$: let $\boldsymbol{x} = (x_1, ..., x_{\ell(n)}), \boldsymbol{y} = (y_1, ..., y_{\ell(n)}) \in (\{0,1\}^n)^{[\ell(n)]}$; for each $i \in \{0,1\}^n$ there is a set $Q_i \subseteq (\{0,1\}^n)^2$ such that $\sum_{i \in \{0,1\}^n} |Q_i| \leq$*

*$T(n)$; let $\boldsymbol{K} = (k_1,..,k_{m(n)}) \in (\{0,1\}^n)^{m(n)}$ such that there is no $i$ where $(x_i, y_i) \in Q_{\boldsymbol{K}}$, then:*

$$\Pr[O \stackrel{O}{\leftarrow} \Pi :: O_{\boldsymbol{K}}(\boldsymbol{x}) = \boldsymbol{y} | \forall i,\ \forall (a,b) \in Q_i,\ O_i(a) = b] \leq 2^{(n-1)\cdot \ell(n))}.$$

*Proof.* We consider several cases. First, we consider the case that there exists a pair $(a,b) \in Q_{\boldsymbol{K}}$ such that either there exists an $i$ s.t. $x_i = a$ but $y_i \neq b$ or there exists a $j$ s.t. $y_j = b$ but $x_j \neq a$. In this case it is not possible for $O_{\boldsymbol{K}}(\boldsymbol{x}) = \boldsymbol{y}$, so the probability is 0.

Second, we consider the case that there exists a $k_i \in \boldsymbol{K}$ where $Q_{k_i} = \{\}$. A necessary condition for $O_{\boldsymbol{K}}(\boldsymbol{x}) = \boldsymbol{y}$ is that $O_{k_i}\left(O_{k_{i-1},\ldots,k_1}(\boldsymbol{x})\right) = O^{-1}_{k_{i+1},\ldots,k_{m(n)}}(\boldsymbol{y})$. The probability of this event is no more than $\prod_{j=1}^{\ell(n)} (1/(2^n - i)) \leq \frac{1}{2^{(n-1)\cdot \ell(n)}}$ (for sufficiently large $n$).

Thirdly, we consider the case where for every $k_i \in \boldsymbol{K}$ the corresponding set $Q_{k_i}$ is not empty. Because of our conditioning on the probability, for each $x_i$ there exist a value $k_j$ where $\alpha_i = O_{k_1,\ldots,k_{j-1}}(x_i)$ and $\beta_i = O^{-1}_{k_{m(n)},\ldots,k_{j+1}}(y_i)$, but $(\alpha_i, \beta_i) \notin Q_{k_j}$, as otherwise $(x_i, y_i) \in Q_{\boldsymbol{K}}$ which is not permitted by the statement of the lemma. Therefore, the probability that $O_{k_j}(\alpha_i) = \beta_i$ is less than $\frac{1}{2^n - |Q_{k_j}| - \ell(n)}$ (We subtract $\ell(n)$ in the denominator as several $x_i$'s may have this condition occur for the same key $k_j$.). Therefore, the probability that $O_{\boldsymbol{K}}(\boldsymbol{x}) = \boldsymbol{y}$ is less than $\prod_{i=1}^{\ell(n)} \frac{1}{2^n - |Q_i| - \ell(n)} \leq \frac{1}{2^{(n-1)\cdot \ell(n)}}$, for sufficiently large $n$ (remembering $|Q_i| \leq T(n)$). □

**Corollary 1.** *Let $T$, $\ell$ and $m$ be polynomials. For all sufficiently large $n \in \mathbb{N}$: let $\boldsymbol{x} = (x_1, ..., x_{\ell(n)}), \boldsymbol{y} = (y_1, ..., y_{\ell(n)}) \in (\{0,1\}^n)^{[\ell(n)]}$; for each $i \in \{0,1\}^n$ there is a a set $Q_i \subseteq (\{0,1\}^n)^2$ such that $\sum_{i \in \{0,1\}^n} |Q_i| \leq T(n)$; let $KS \subseteq (\{0,1\}^n)^{m(n)}$ such that for each $\boldsymbol{K} \in KS$ there is no $i$ such that $(x_i, y_i) \in Q_{\boldsymbol{K}}$, then:*

$$\Pr[O \stackrel{O}{\leftarrow} \Pi :: \exists \boldsymbol{K} \in KS \text{ s.t. } O_{\boldsymbol{K}}(\boldsymbol{x}) = \boldsymbol{y} | \forall i,\ \forall (a,b) \in Q_i,\ O_i(a) = b] \leq 2^{n\cdot(m(n)-\ell(n))-\ell(n)}.$$

*Proof.* This proof follows directly from Lemma 7 and a union bound over the probabilities of each of the keys $\boldsymbol{K} \in KS$. □

### 4.2 Proof of Lemma 4

For the convenience of the reader, we restate Lemma 4.

**Lemma 8.** *For all sufficiently large $n$:*

$$\Pr[O \stackrel{O}{\leftarrow} \Pi; R \stackrel{R}{\leftarrow} \Psi(O); f \in_{\mathcal{U}} O^n :: \widehat{A}^{O,R,f}_{i+1}(1^n) \neq \widehat{A}^{O,R,f}_{i}(1^n)] \leq \frac{5}{2^{n/2}}$$

*Proof.* We note that in the statement of the lemma and its proof we don't concern ourselves with the random coin-tosses of $A_i$ or $A_{i+1}$. It will be obvious from the proof, that the random coin-tosses do not affect any of the probabilities we discuss, and that we could simply fix the random coin tosses of $B$ and prove the theorem for each such sequence.

We begin by proving upper-bounds on two probabilistic events that we will frequently want to condition on. We will frequently want to bound the probability that $A_{i+1}$ makes any of its first $i$ queries to $O_k$, where $f = O_k$. We will call such an event $\mathbf{F}$.

*Claim 2. For all sufficiently large $n$:* $\Pr[O \overset{O}{\leftarrow} \Pi; R \overset{R}{\leftarrow} \Psi(O); f \in_{\mathcal{U}} O^n ::$ $\widehat{A}_{i+1}^{f,O,R}(1^n)$ *makes a query to* $O_k = f$ *in one of the first* $i$ *calls to* $G] \le 2i/2^n$.

*Proof.* Observe that queries to $R_1$ and $R_2$ are statistically independent of $f$ and $O$. Further, the first $i$ queries are all made by $G$, and therefore there have been no queries to $R_3$. Thus the probability of making a query to $O_k$ corresponding to $f$ is no more than the probability of drawing at random the unique red ball from a vase of $2^n$ balls in $i$ draws without replacement, as there are most $i$ different keys on which $O$ can be queried. Therefore, the probability is bound by $\sum_{j=0}^{i-1} 1/(2^n - j) < 2i/2^n$, for sufficiently large $n$. □

We also frequently want to bound the probability that by $\widehat{A}_{i+1}$'s $i$th call to $G$ two oracle queries to $O$ (or $O$ and $f$) have been made that have the same output. We call such queries *collisions* and we denote such an event by $\mathbf{E}$.

*Claim 3. For all sufficiently large $n$:* $\Pr[O \overset{O}{\leftarrow} \Pi; R \overset{R}{\leftarrow} \Psi(O); f \in_{\mathcal{U}} O^n ::$ *after* $\widehat{A}_{i+1}^{f,O,R}(1^n)$ *makes* $i$ *calls to* $G$ *there exists* $k \ne j \in \{0,1\}^n \cup \{f\}$ *s.t.* $(a,b) \in Q_i \wedge (c,b) \in Q_j] \le 2(i \cdot m(n))^2/2^n$.

*Proof.* We note that since we are only concerned with queries made in the first $i$ calls to $G$, there have been no queries to $R_3$. Next, we condition on $\overline{\mathbf{F}}$ from Claim 2, so query results on $f$ and $O_k$ for $k \in \{0,1\}^n$ are independent of each other. It can easily be observed that to maximize the probability of a collision the adversary should make all of its queries to different functions. The structure of $G$ does not necessarily permit this, but this permits an easy upper-bound on the probability of a collision. Since each call to $G$ makes at most $m(n)$ queries, the probability of $\mathbf{E}$ can be upper-bounded by $\sum_{j=1}^{i \cdot m(n)} \frac{j}{2^n} \le (i \cdot m(n))^2/2^n$ Since for sufficiently large $n$ the probability of event $\mathbf{F}$ is bound by $2i/2^n$, we can bound the probability of the claim by $2(i \cdot m(n))^2/2^n$. □

To prove Lemma 4, we note that any difference in executions between $A_{i+1}^{O,R,f}(1^n)$ and $A_i^{O,R,f}(1^n)$ must occur in $G$. We will consider the places where $G$ could have an output different from that of the actual query to the oracle, and bound this probability. We note that this can only occur on lines 6, 8, 10 and 12, and we bound the probability of error on each of these lines with the following series of claims. In order to prove the lemma we take the union bound of the errors from these claims.

*Claim 4. The probability that G gives incorrect output on line 6 is less than $\frac{1}{2^{2n-1}}$ for all sufficiently large n.*

*Proof.* The response to query $q = R_3(\boldsymbol{x}_1, \boldsymbol{y}_1, \boldsymbol{x}_2, \boldsymbol{y}_2)$ will always be $\perp$ unless $R_2(\boldsymbol{x}_1, \boldsymbol{y}_1) = \boldsymbol{x}_2$. If $R_2$ has not yet been queried, then it is easily seen, by the definition of $R_2$, that the probability of the adversary correctly guessing $\boldsymbol{x}_2$ in its query to $R_3$ is $\frac{1}{(2^n-\ell(n))(2^n-\ell(n)-1)}$. For sufficiently large $n$, this value is upper-bounded by $\frac{1}{2^{2n-1}}$. □

*Claim 5. The probability that G gives incorrect output on line 8 is less than $\frac{1}{2^{n/2}}$ for all sufficiently large n.*

*Proof.* For this claim, we consider two separate cases: first we consider the case in which $|\mathcal{K}| = 0$ and next we consider the case where $|\mathcal{K}| \geq 2$.

**[Case** $|\mathcal{K}| = 0$**]:** We first show that for query $R_3(\boldsymbol{x}_1, \boldsymbol{x}_2, \boldsymbol{y}_1, \boldsymbol{y}_2)$, if we let $\overline{KS} = \{(k_1, ..., k_{m(n)}) | \exists i s.t. (x_i, y_i) \in Q_{k_1,...,k_{m(n)}}\}$, then with high probability $|\overline{KS}| \leq \ell(n)+2$. Next, we show that for each element $\boldsymbol{K} \in \overline{KS}$ that there is a very small chance that $O_{\boldsymbol{K}}(\boldsymbol{x}_1, \boldsymbol{x}_2) = (\boldsymbol{y}_1, \boldsymbol{y}_2)$. We then show that for $\boldsymbol{K} \in (\{0,1\}^n)^{m(n)} \setminus \overline{KS}$ that the chances that $O_{\boldsymbol{K}}(\boldsymbol{x}_1, \boldsymbol{x}_2) = (\boldsymbol{y}_1, \boldsymbol{y}_2)$ holds is very small using the Unique Paths Corollary (Corollary 1).

In order to bound (w.h.p.) the size of $\overline{KS}$ we will condition on event $\overline{\mathbf{E}}$ from Claim 3 (i.e. there are no collisions). Observe that if $\overline{\mathbf{E}}$ holds, then it is not possible for $|\overline{KS}| > \ell(n)+2$, as otherwise by the pigeonhole principle there would be two keys, $\boldsymbol{\kappa} = (\kappa_1, ..., \kappa_{m(n)})$ and $\boldsymbol{\kappa}' = (\kappa'_1, ..., \kappa'_{m(n)})$, where for some $a$ $(1 \leq a \leq \ell(n)+2)$ we would have $y_a = O_{\boldsymbol{\kappa}}(x_a) = O_{\boldsymbol{\kappa}'}(x_a)$, and letting $j$ be the largest index where $\kappa_j \neq \kappa'_j$ this implies $O_{\kappa_j}\left(O_{\kappa_1,..,\kappa_{j-1}}(x_a)\right) = O_{\kappa'_j}\left(O_{\kappa'_1,..,\kappa'_{j-1}}(x_a)\right)$, which is a collision and thus this contradicts our conditioning on event $\overline{\mathbf{E}}$.

Next, we condition on $\overline{\mathbf{F}}$ from Claim 2 to ensure that responses for queries to $O$ are statistically independent of responses for queries to $f$. We now bound the probability that for any specific key $\boldsymbol{K} \in \overline{KS}$ that $O_{\boldsymbol{K}}(\boldsymbol{x}_1, \boldsymbol{x}_2) = (\boldsymbol{y}_1, \boldsymbol{y}_2)$. We wish to consider the probability that for a key $\boldsymbol{K} = (k_1, ..., k_{m(n)}) \in \overline{KS}$ that $(\boldsymbol{y}_1, \boldsymbol{y}_2) = O_{\boldsymbol{K}}(\boldsymbol{x}_1, \boldsymbol{x}_2)$. For each such key $\boldsymbol{K}$ there exists an $i$ s.t. $(x_i, y_i) \notin Q_{\boldsymbol{K}}$ (otherwise $|\mathcal{K}| \geq 1$ contradicting the case we are in) Consider the smallest $j$ such that there exists a $b \in \{0,1\}^n$ where $(x_i, b) \in Q_{k_1...k_{j-1}}$, and such that for every $b' \in \{0,1\}^n$ it is the case that $(x_i, b') \notin Q_{k_1...k_j}$. The probability that $O_{k_j}(b) = O^{-1}_{k_{j+1},...,k_{m(n)}}(y_i)$ is less than $1/(2^n - |Q_{k_j}|) \leq 1/(2^n - i \cdot m(n))$, as at most $i \cdot m(n)$ queries have been made. Therefore, the probability there exists a key $\boldsymbol{K} \in \overline{KS}$ such that $O_{\boldsymbol{K}}(\boldsymbol{x}) = \boldsymbol{y}$ is less than $\frac{\ell(n)+2}{(2^n - i \cdot m(n))}$, as $|\overline{KS}| \leq \ell(n)+2$ by our conditioning on $\overline{\mathbf{E}}$.

For the remaining set of keys $KS = (\{0,1\}^n)^{m(n)} \setminus \overline{KS}$ the Unique Paths Corollary shows that the probability that there exists a key $\boldsymbol{K} \in KS$ such that $O_{\boldsymbol{K}}(\boldsymbol{x}_1, \boldsymbol{x}_2) = (\boldsymbol{y}_1, \boldsymbol{y}_2)$ is no more than $2^{n \cdot (m(n) - \ell(n)) - \ell(n)}$

Therefore, the probability of the case when $|\mathcal{K}| = 0$ is bounded by $\frac{2(i \cdot m(n))^2}{2^n} + \frac{2i}{2^n} + \frac{(\ell(n)+2)}{(2^n - i \cdot m(n))} + 2^{n \cdot (m(n) - \ell(n)) - \ell(n)} < \frac{1}{2^{n/2}}$ (for sufficiently large $n$), where the first two summands bound the probabilities of events **E** and **F** respectively.

**[Case $|\mathcal{K}| \geq 2$]:** We observe that in order for $|\mathcal{K}| \geq 2$ to occur, the event **E** of Claim 3 must occur at least $\ell(n)+2$ times: there must be at least one collision for each $y \in (\boldsymbol{y}_1, \boldsymbol{y}_2)$ in order for there to be two keys $\boldsymbol{K}_1, \boldsymbol{K}_2 \in \mathcal{K}$ such that $(\boldsymbol{y}_1, \boldsymbol{y}_2) = O_{\boldsymbol{K}_1}(\boldsymbol{x}_1, \boldsymbol{x}_2) = O_{\boldsymbol{K}_2}(\boldsymbol{x}_1, \boldsymbol{x}_2)$. Therefore, we can use the bound on the probability of **E** to bound the probability of incorrect output by $\frac{2(i \cdot m(n))^2}{2^n} < \frac{1}{2^{n/2}}$ (for sufficiently large $n$). □

*Claim 6. The probability that G gives incorrect output on line 10 is less than* $\frac{1}{2^{n/2}}$.

*Proof.* We begin by conditioning on $\overline{\mathbf{F}}$, so that responses from queries to $O_k$, for $k \in \{0,1\}^n$, are independent of the responses of queries from $f$. We consider two exclusive cases: first, when $\widehat{A}_{i+1}$ queries $f$ before it queries $R_2(\boldsymbol{x}_1, \boldsymbol{y}_1)$; and second, when $\widehat{A}_{i+1}$ queries $R_2(\boldsymbol{x}_1, \boldsymbol{y}_1)$ before it queries $f$.

**[Case that $f$ was queried before $R_2(\boldsymbol{x}_1, \boldsymbol{y}_1)$]:** The intuition behind this case is that the adversary needs to construct a key $\kappa = (k_1, .., k_{u-1}, f, k_{u+1}, .., k_{m(n)})$, and perform queries such that $(x_{\ell(n)+1}, y_{\ell(n)+1}), (x_{\ell(n)+2}, y_{\ell(n)+2}) \in Q_\kappa$. We will argue that it is very unlikely that the queries $x_{\ell(n)+1}$ or $x_{\ell(n)+2}$ were made to $O_k$ for any $k \in \{0,1\}^n$ or $f$ before the query $R_2(\boldsymbol{x}_1, \boldsymbol{y}_1)$. Assuming this to be true, a necessary condition to find a $\kappa$ satisfying our requirements is to make queries $O_k(\alpha) = \beta$, for $\alpha, k \in \{0,1\}^n$, for which there exists a $j, \gamma \in \{0,1\}^n$ such that there was a $(\beta, \gamma) \in Q_j$ at the time of the query to $R_2$. We show this is unlikely as well.

We begin by bounding the probability that there had been a query of the form $x_{\ell(n)+1}$ or $x_{\ell(n)+2}$ before the query to $R_2$, Assume the query $R_2(\boldsymbol{x}_1, \boldsymbol{y}_1)$ was the $j$th query ($j \leq i$), then the probability that there exists a $\beta, k \in \{0,1\}^n$ such that $(x_{\ell(n)+1}, \beta) \in Q_k, (x_{\ell(n)+2}, \beta) \in Q_k, (x_{\ell(n)+1}, \beta) \in Q_f$ or $(x_{\ell(n)+2}, \beta) \in Q_f$ is less than $\frac{2 \cdot j \cdot m(n)}{2^n - \ell(n) - 2}$. Next, we condition on that event not happening, and show that there is a small probability that any of the $(j+1)$st through $i$th queries are of the form $O_k(a) = b$, where there exists a $c, v \in \{0,1\}^n$ such that $(b, c) \in Q_v$ or $(b, c) \in Q_f$ is small. This probability can easily be bounded by $\sum_{s=j}^{i+1} \frac{i+1}{2^n - s \cdot m(n)}$ Therefore, the probability of the first case is less than $\frac{2 \cdot j \cdot m(n)}{2^n - \ell(n) - 2} + \sum_{s=j}^{i+1} \frac{(i+1)}{2^n - s \cdot m(n)} \leq \frac{2 \cdot i \cdot m(n)}{2^n - \ell(n) - 2} + \frac{(i+1)^2}{2^n - (i+1) \cdot m(n)} \leq 2^{-2n/3}$.

**[Case $R_2(\boldsymbol{x}_1, \boldsymbol{y}_1)$ was queried before $f$]:** In the second case when the adversary queries $f$ it has already queried $R_2$, and therefore it needs to find a key $\kappa = (\kappa_1, ..\kappa_{u-1}, f, \kappa_{u+1}, ..., \kappa_{m(n)})$ such that for each $t \leq \ell(n)$, $(x_t, y_t) \in Q_\kappa$. A necessary condition is for there to exist an $a \in \{0,1\}^n$ and $y_s \in \boldsymbol{y}_1$ such that $(a, y_s) \in Q_{f, \kappa_{j+1}, \ldots, \kappa_{m(n)}}$. We show the probability of this occurring is small. We begin by showing it is unlikely that after the query to $f$ there will exists an $a, b, c, k \in \{0,1\}^n$ where both $(a, b) \in Q_f$ and $(b, c) \in Q_k$. Likewise, it is unlikely that there will exist an $a \in \{0,1\}^n$ and $y \in \boldsymbol{y}_1$ where $(a, y) \in Q_f$. If neither of these cases hold then, in order to satisfy our necessary condition, a query to $O$ must be made after the query to $f$ in which there exists a $b, k \in \{0,1\}^n$ where $O_k(b) \in \boldsymbol{y}_1$. We show that the probability of this is also low, proving the lemma.

More formally, assume the query to $f$ is the $j$th query. There can be at most $i$ (parallel) queries made to $f$. The probability that the queries to $f$ collide with with any previously made queries is less than $\frac{i \cdot j}{2^n - i}$. The probability that the queries to $f$ will output a $y \in \boldsymbol{y}$ is bound by $\frac{i \cdot \ell(n)}{2^n - i}$. Finally, the probability that any queries to $O$ after the query to $f$ will result in $y \in \boldsymbol{y}$ is less than $\frac{m(n) \cdot i}{2^n - (i+1) \cdot m(n)}$. Therefore, by the union bound the probability of the second case can easily be bound by $\frac{3 \cdot i \cdot m(n)}{2^n - (i+1) \cdot m(n)}$.

Therefore, for all sufficiently large $n$ the probability that the entire claim holds is bound by $\frac{3 \cdot i \cdot m(n)}{2^n - (i+1) \cdot m(n)} + 1/2^{2n/3} + 2(i \cdot m(n))^2/2^n \leq 2^{-n/2}$, where the last summand accounts for our conditioning on $\overline{\mathbf{F}}$. □

*Claim 7. The probability that $G$ gives incorrect output on line 12 is less than $\frac{1}{2^{n/2}}$.*

*Proof.* The only reason we may have an incorrect output on line 12 is because the output to an actual query to $R_3(1^n, \boldsymbol{x}_1, \boldsymbol{x}_2, \boldsymbol{y}_1, \boldsymbol{y}_2)$ is $(x^*, y^*)$ for $\kappa \in_{\mathcal{U}} \left\{\kappa \in \{0,1\}^{n \cdot m(n)} | O_\kappa(\boldsymbol{x}_1, \boldsymbol{x}_2) = (\boldsymbol{y}_1, \boldsymbol{y}_2)\right\}$, $x^* \in_{\mathcal{U}} \{0,1\}^n$ and $y^* = O_\kappa(x^*)$; whereas, $G$ always outputs $O_{\boldsymbol{K}}(x^*)$ for $\boldsymbol{K} \in \mathcal{K}$ and $x^* \in_{\mathcal{U}} \{0,1\}^n$. Thus, even if there exists a $\boldsymbol{K}' \in (\{0,1\}^n)^{m(n)}$, where $\boldsymbol{K} \neq \boldsymbol{K}'$ and $O_{\boldsymbol{K}'}(\boldsymbol{x}_1, \boldsymbol{x}_2) = (\boldsymbol{y}_1, \boldsymbol{y}_2)$, there is no possibility for $(x^*, O_{\boldsymbol{K}'}(x^*))$ to be output by $G$. We show that it is highly unlikely that $\left|\left\{\kappa \in \{0,1\}^{n \cdot m(n)} | O_\kappa(\boldsymbol{x}_1, \boldsymbol{x}_2) = (\boldsymbol{y}_1, \boldsymbol{y}_2)\right\}\right| > 1$, and thus there is rarely an error in output of $G$ on line 12.

The result follows by conditioning on there being no collisions and then applying the Unique Paths Corollary. In particular, assuming $\overline{\mathbf{E}}$ holds then our sets $Q$ satisfy the requirements for the Unique Paths Corollary where $KS = (\{0,1\}^n)^{m(n)} \setminus \mathcal{K}$. Therefore, by the Unique Paths Corollary we can bound the probability by $2^{n \cdot (m(n) - \ell(n)) - \ell(n)}$, and we bound the probability of $\mathbf{E}$ by $2(i \cdot m(n))^2/2^n$. Therefore by the union bound, the probability of error is less than $2^{-n/2}$ for sufficiently large $n$. □

To finish proving Lemma 4 we simply take the union bound on the probability of errors in Claims 4,5,6 and 7, and this is less than $5/2^{n/2}$ proving the lemma. □

## 4.3 Proof of Lemma 5

For the convenience of the reader we restate Lemma 5.

**Lemma 9.** *For all sufficiently large $n$:*

$$\Pr[O \overset{O}{\leftarrow} \Pi; R \overset{R}{\leftarrow} \Psi(O); f \in_{\mathcal{U}} \Pi^n :: \widehat{A}_{i+1}^{O,R,f}(1^n) \neq \widehat{A}_i^{O,R,f}(1^n)] \leq \frac{5}{2^{n/2}}$$

*Proof.* We note that this proof is basically the same as the proof of Lemma 4 in the previous section. The only portion of the proof of Lemma 4 that relied on the fact that $f \in O$ as opposed to $f \in \Pi$ was Claim 2, which defines the event $\mathbf{F}$ and bound the probability of it occurring; and those claims that conditioned on the event $\overline{\mathbf{F}}$ and then later had to add in a small probability for error in the case that $\mathbf{F}$ held.

We remind the reader that definition of the event $\mathbf{F}$ is that $A_{i+1}$ makes any of its first $i$ queries to $O_k$, where $f = O_k$. Clearly, in the experiment for Lemma 5 the probability of the event $\mathbf{F}$ is 0, as $f \in \Pi$ and not $O$. Therefore, the probability of error in this lemma will be smaller than that of Lemma 4. □

## 5 Other Constructions, Concluding Remarks & Open Questions

The authors note that the basic design of this oracle and the proof techniques of this paper can be naturally lifted to at least one other natural construction: the XOR of functions. The important observation is that the construction needs to have some natural combinatorial property that corresponds to the Unique Paths Lemma, and with XOR such a property exists, although the notion needs a bit of massaging. The authors leave the proof of this claim to a later version of this paper.

The previous observation leads to the question of whether or not there is a simple combinatorial characterization of those constructions that require a non-relativizing proof technique to show they achieve adaptive security. It also leads to a natural quantitative question: what is the lower-bound on the number of calls to a non-adaptively secure function generator in an adaptively secure black-box construction? Recently, there has been some success in getting quantitative lower bounds in such black-box settings [5,6,13], and so it is conceivable one could be found in this setting as well.

As mentioned in the introduction, there is currently a known upper-bound of $\Omega(n/\log n)$ calls to a non-adaptive generator in order to achieving black-box adaptive security. Further, the same upper-bound is achieved by two independent constructions. It would be interesting to know whether or not the current constructions are effectively the best possible. A natural question along these lines is whether or not there are any constructions that would give a smaller upper-bound.

## 6 Acknowledgments

The author would like to thank Charles Rackoff, Omer Reingold, Vladimir Kolesnikov and the anonymous referees for comments and suggestions that substantially improved the presentation of this paper.

## References

1. W. Aiello, M. Bellare, G. Di Crescenzo, and R. Vekatesan. Security amplification by composition: The case of doubly-iterated, ideal ciphers. In *Advances in Cryptology - Crypto 98*, pages 390–407, 1998.
2. Theodore Baker, John Gill, and Robert Solovay. Relativizations of the $\mathcal{P} =? \mathcal{NP}$ question. *SIAM Journal on Computing*, 4(4):431–442, 1975.

3. B. Barak, O. Goldreich, S. Goldwasser, and Y. Lindell. Resettably-sound zero-knowledge and its applications. In *42nd IEEE Symposium on Foundations of Computer Science*, pages 116–125. IEEE Computer Society Press, 2001.
4. Manuel Blum, Paul Feldman, and Silvio Micali. Non-interactive zero-knowledge and its applications. In *Proceedings of the 20th Annual Symposium on Theory of Computing*, pages 103–112. ACM Press, 1988.
5. R. Gennaro and L. Trevisan. Lower bounds on the efficiency of generic cryptographic constructions. In *41st Annual Symposium on Foundations of Computer Science*, pages 305–313. IEEE Computer Society Press, 2000.
6. Rosario Gennaro, Yael Gertner, and Jonathan Katz. Lower bounds on the efficiency of encryption and digital signature schemes. In *Proceedings of the thirty-fifth ACM symposium on Theory of computing*, pages 417–425. ACM Press, 2003.
7. Y. Gertner, S. Kannan, T. Malkin, O. Reingold, and M. Viswanathan. The relationship between public key encryption and oblivious transfer. In IEEE, editor, *41st Annual Symposium on Foundations of Computer Science*, pages 325–335. IEEE Computer Society Press, 2000.
8. Y. Gertner, T. Malkin, and O. Reingold. On the impossibility of basing trapdoor functions on trapdoor predicates. In IEEE, editor, *42nd IEEE Symposium on Foundations of Computer Science*, pages 126–135. IEEE Computer Society Press, 2001.
9. O. Goldreich, S. Goldwasser, and S. Micali. How to construct random functions. *Journal of the ACM*, 33(4):792–807, 1986.
10. Oded Goldreich, Silvio Micali, and Avi Wigderson. Proofs that yield nothing but their validity or all languages in NP have zero-knowledge proof systems. *Journal of the ACM*, 38(3):690–728, 1991.
11. Shafi Goldwasser, Silvio Micali, and Charles Rackoff. The knowledge complexity of interactive proof systems. In *Proceedings of the 17th Annual Symposium on Theory of Computing*, pages 291–304. ACM Press, 1985.
12. R. Impagliazzo and S. Rudich. Limits on the provable consequences of one-way permutations. In *Proceedings of the 21st Annual ACM Symposium on Theory of Computing*, pages 44–61. ACM Press, 1989.
13. Jeong Han Kim, D. R. Simon, and P. Tetali. Limits on the efficiency of one-way permutation-based hash functions. In *40th Annual Symposium on Foundations of Computer Science*, pages 535–542. IEEE Computer Society Press, 1999.
14. M. Luby and C. Rackoff. Pseudo-random permutation generators and cryptographic composition. In *Proceedings of the 18th Annual Symposium on Theory of Computing*, pages 353–363. ACM, 1986.
15. M. Luby and C. Rackoff. How to construct pseudorandom permutations from pseudorandom functions. *SIAM Journal on Computing*, 17:373–386, 1988.
16. Ueli Maurer and Krzysztof Pietrzak. Composition of random systems: When two weak make one strong. In *The First Theory of Cryptography Conference*, 2004.
17. Moni Naor and Omer Reingold. Synthesizers and their application to the parallel construction of psuedo-random functions. In *36th Annual Symposium on Foundations of Computer Science*, pages 170–181. IEEE, 1995.
18. A. Sahai. Non-malleable non-interactive zero knowledge and adaptive chosen-ciphertext security. In *40th Annual Symposium on Foundations of Computer Science*, pages 543–553. IEEE Computer Society Press, 1999.
19. D. R. Simon. Finding collisions on a one-way street: Can secure hash functions be based on general assumptions? In *Advances in Cryptology – EUROCRYPT 98*, pages 334–345, 1998.

# Chosen-Ciphertext Security from Identity-Based Encryption

Ran Canetti[1], Shai Halevi[1], and Jonathan Katz[2*]

[1] IBM T. J. Watson Research Center, Hawthorne, NY.
{canetti,shaih}@watson.ibm.com

[2] Dept. of Computer Science, University of Maryland, College Park, MD.
jkatz@cs.umd.edu

**Abstract.** We propose a simple and efficient construction of a CCA-secure public-key encryption scheme from any CPA-secure identity-based encryption (IBE) scheme. Our construction requires the underlying IBE scheme to satisfy only a relatively "weak" notion of security which is known to be achievable without random oracles; thus, our results provide a new approach for constructing CCA-secure encryption schemes in the standard model. Our approach is quite different from existing ones; in particular, it avoids non-interactive proofs of "well-formedness" which were shown to underlie most previous constructions. Furthermore, applying our conversion to some recently-proposed IBE schemes results in CCA-secure schemes whose efficiency makes them quite practical.

Our technique extends to give a simple and reasonably efficient method for securing any binary tree encryption (BTE) scheme against adaptive chosen-ciphertext attacks. This, in turn, yields more efficient CCA-secure hierarchical identity-based and forward-secure encryption schemes in the standard model.

*Keywords:* Chosen-ciphertext security, Forward-secure encryption, Identity-based encryption, Public-key encryption.

## 1 Introduction

Security against adaptive chosen-ciphertext attacks (i.e., "CCA security") is a strong and very useful notion of security for public-key encryption schemes [RS91, DDN00, BDPR98] . This notion is known to suffice for many applications of encryption in the presence of *active* attackers, including secure communication, auctions, voting schemes, and many others. Indeed, CCA security is commonly accepted as the security notion of choice for encryption schemes that are to be "plugged in" to a protocol running in an arbitrary setting; see, e.g., [S98].

However, there are only a handful of known public-key encryption schemes that can be proven CCA-secure in the standard model (i.e., without the use of heuristics such as random oracles). In fact, only two main techniques have been

* Work supported by NSF Trusted Computing Grant #ANI-0310751.

C. Cachin and J. Camenisch (Eds.): EUROCRYPT 2004, LNCS 3027, pp. 207–222, 2004.

proposed for constructing such cryptosystems. The first follows the paradigm of Naor and Yung [NY90] (later extended by Sahai [S99] and simplified by Lindell [L03]), and the related scheme of Dolev, Dwork, and Naor [DDN00]. This technique uses as building blocks any CPA-secure public-key encryption scheme (i.e., any scheme secure against chosen-plaintext attacks [GM84]) along with any non-interactive zero-knowledge (NIZK) proof system [BFM88, FLS90]; in turn, each of these primitives may be constructed using any family of trapdoor permutations. The encryption schemes resulting from this approach, however, are highly inefficient precisely because they employ NIZK proofs which in turn use a generic Karp reduction from an instance of the encryption scheme to an instance of some NP-complete problem. Furthermore, there are currently no known efficient NIZK proof systems even under specific assumptions and for particular cryptosystems of interest. Thus, given current techniques, this methodology for constructing CCA-secure cryptosystems serves as a "proof of feasibility" but does not lead to practical constructions.

The second technique is due to Cramer and Shoup [CS98, CS02], and is based on algebraic constructs with particular homomorphic properties (namely, those which admit "smooth hash proof systems"; see [CS02]). Algebraic constructs of the appropriate type are known to exist based on some specific assumptions, namely the hardness of the decisional Diffie-Hellman problem [CS98] or the hardness of deciding quadratic residuosity or $N^{\text{th}}$ residuosity in certain groups [CS02]. More efficient schemes following the same basic technique have been given recently [GL03, CS03], and the technique leads to a number of possible instantiations which are efficient enough to be used in practice.

Interestingly, as observed by Elkind and Sahai [ES02], both of these techniques for constructing CCA-secure encryption schemes can be viewed as special cases of a *single* paradigm. In this, more general paradigm (informally) one starts with a CPA-secure cryptosystem in which certain "ill-formed" ciphertexts are indistinguishable from "well-formed" ones. A CCA-secure cryptosystem is then obtained by having the sender include a "proof of well-formedness" for the transmitted ciphertext. Both NIZK proofs and smooth hash proof systems were shown to meet the requirements for these proofs of well-formedness, and thus all the schemes mentioned above (with the possible exception of [DDN00]) may be viewed as instantiations of a single paradigm.

## 1.1 Our Contributions

We propose a new approach for constructing CCA-secure public-key encryption schemes. Instead of using "proofs of well-formedness" as in all previous schemes, we instead give a direct construction using identity-based encryption (IBE) schemes satisfying a "weak" notion of security. A number of IBE schemes meeting this weak notion of security in the standard model were recently proposed (see below); thus, our approach yields new constructions of CCA-secure encryption in the standard model. The resulting schemes are simple and reasonably efficient, and are quite different from the ones described above. In particular, they do not seem to fit within the characterization of Elkind and Sahai. We remark that our techniques may also be used to construct a non-adaptive (or "lunchtime") CCA1-secure encryption scheme [NY90, DDN00, BDPR98] based on any weak IBE scheme; interestingly, our conversion in this case adds (essentially) *no overhead* to the original IBE scheme.

Before sketching our construction, we first recall the notion of IBE. The concept of identity-based encryption was introduced by Shamir [S84], and provably-secure IBE schemes (in the random oracle model) were recently demonstrated by Boneh and Franklin [BF01] and Cocks [C01]. An IBE scheme is a public-key encryption scheme in which, informally, any string (i.e., identity) can serve as a public key. In more detail, a trusted private-key generator (PKG) initializes the system by running a key-generation algorithm to generate "master" public and secret keys. The public key is published, while the PKG stores the secret key. Given any string $id \in \{0,1\}^*$ (which can be viewed as a receiver's identity), the PKG can derive a "personal secret key" $SK_{id}$. Any sender can encrypt a message for this receiver using only the master public key and the string $id$. The resulting ciphertext can be decrypted using the derived secret key $SK_{id}$, but the message remains hidden from an adversary who does not know $SK_{id}$ even if that adversary is given $SK_{id'}$ for various identities $id' \neq id$.

In the definition of security for IBE given by Boneh and Franklin [BF01], the adversary is allowed to choose the "target identity" ($id$ in the above discussion) in an adaptive manner, possibly based on the master public key and any keys $SK_{id'}$ the adversary has obtained thus far. Boneh and Franklin construct a scheme meeting this definition of security based on the bilinear Diffie-Hellman (BDH) assumption in the random oracle model. A weaker notion of security for IBE, proposed by Canetti, Halevi, and Katz [CHK03], requires the adversary to specify the target identity *before* the public-key is published; we will refer to this notion of security as "weak" IBE. Canetti, et al. [CHK03] show that a weak IBE scheme can be constructed based on the BDH assumption *in the standard model.* Concurrent with the present work, more efficient constructions of weak IBE schemes in the standard model (including one based on the BDH assumption) were given by Boneh and Boyen [BB04]. Both of the above-mentioned constructions of weak IBE based on the BDH assumption build on earlier work of Gentry and Silverberg [GS02].

Our construction of CCA-secure encryption requires only an IBE scheme satisfying the weaker notion of security referred to above. The conversion of any such IBE scheme to a CCA-secure public-key encryption scheme proceeds as follows: The public key of the new scheme is simply the master public key of the IBE scheme, and the secret key is the corresponding master secret key. To encrypt a message, the sender first generates a key-pair $(vk, sk)$ for a one-time strong signature scheme, and then encrypts the message with respect to the "identity" $vk$. (A "strong" signature scheme has the property that it is infeasible to create new valid signature even for previously-signed messages.) The resulting ciphertext $C$ is then signed using $sk$ to obtain a signature $\sigma$. The final ciphertext consists of the verification key $vk$, the IBE ciphertext $C$, and the signature $\sigma$. To decrypt a ciphertext $\langle vk, C, \sigma\rangle$, the receiver first verifies the signature on $C$

with respect to $vk$, and outputs $\perp$ if the verification fails. Otherwise, the receiver derives the secret key $SK_{vk}$ corresponding to the "identity" $vk$, and uses $SK_{vk}$ to decrypt the ciphertext $C$ as per the underlying IBE scheme.

Security of the above scheme against adaptive chosen-ciphertext attacks can be informally understood as follows. Say a ciphertext $\langle vk, C, \sigma \rangle$ is *valid* if $\sigma$ is a valid signature on $C$ with respect to $vk$. Now consider a "challenge ciphertext" $c^* = \langle vk^*, C^*, \sigma^* \rangle$ given to the adversary. Any valid ciphertext $c = \langle vk, C, \sigma \rangle$ submitted by the adversary to a decryption oracle (implying $c \neq c^*$), must have $vk \neq vk^*$ by the (strong) security of the one-time signature scheme. The crux of the security proof then involves showing that (weak) security of the IBE scheme implies that decrypting $c$ does not give the adversary any further advantage in decrypting the challenge ciphertext. Intuitively, this is because the adversary would be unable to decrypt the underlying ciphertext $C^*$ *even if it had the secret key* $SK_{vk}$ *corresponding to* $vk$ (since $vk \neq vk^*$, and $C^*$ was encrypted for "identity" $vk^*$ using an IBE scheme).

A simple modification of the above gives a (non-adaptive) CCA1-secure scheme with virtually no overhead compared to the original IBE scheme. Namely, replace the verification key $vk$ by a randomly-chosen string $r \in \{0,1\}^k$ (and forego any signature); the resulting ciphertext is simply $\langle r, C \rangle$, where $C$ is encrypted with respect to the "identity" $r$. Since an adversary cannot guess in advance which $r$ a sender will use, an argument similar to the above shows that this scheme is secure against non-adaptive chosen-ciphertext attacks.

Straightforward implementation of the above ideas using the "weak IBE" construction from [CHK03] is still rather inefficient; in particular, decryption requires computation of (roughly) one bilinear mapping per bit of the verification key. (Using standard hashing techniques, however, one can obtain a signature scheme in which the length of the verification key is exactly the security parameter.) One can somewhat optimize this construction by working with trees of high degree instead of binary trees as in [CHK03]. Specifically, using a tree of degree $d$ results in a scheme requiring $n/\log_2 d$ mapping computations for an $n$-bit verification key; in this case we pay for these savings by having to increase the key size by a factor of $d$. (We speculate that using $d = 16$ results in a "borderline practical" scheme.) Alternatively, using one of the weak IBE schemes proposed by [BB04] results in a considerably more efficient scheme, including one which is nearly as efficient as the Cramer-Shoup cryptosystem [CS98].

**Further extensions and applications.** Canetti, Halevi, and Katz [CHK03] propose the notion of binary tree encryption (BTE), show how to construct a secure BTE scheme in the standard model, and furthermore show how to construct both hierarchical IBE (HIBE) schemes [HL02, GS02] and forward-secure encryption (FSE) schemes starting from any BTE scheme, again in the standard model. To obtain security against chosen-ciphertext attacks in each of these cases, they suggest using the technique of Naor and Yung [NY90] as adapted by Sahai and Lindell [S99, L03]. This involves the use of NIZK proofs, as noted above, which makes the resulting CCA-secure schemes highly inefficient.

Here, we extend our technique to obtain a simple conversion from any CPA-secure BTE scheme to a CCA-secure BTE scheme. The resulting BTE scheme is considerably more efficient than a scheme derived using the previously-suggested approach (based on NIZK); furthermore, the efficiency gain carries over immediately to yield improved constructions of CCA-secure HIBE and FSE schemes as well. Our techniques may also be used *directly* to convert any CPA-secure HIBE scheme to a CCA-secure HIBE scheme, with possibly improved efficiency.

**Implications for "black-box" separations.** Our construction of a CCA-secure encryption scheme from any weak IBE scheme is *black box* in the sense that it only uses the underlying IBE scheme by invoking its prescribed interface (and not, for example, by using the circuit which implements the scheme). A recent result of Aiello, et al. [AGMM04] rules out certain classes of black-box constructions of CCA-secure encryption schemes from CPA-secure ones. Combined with their result, the current work rules out the same classes of black-box constructions of IBE from CPA-secure encryption.

Although a result of this sort should not be viewed as a strict impossibility result (after all, the known constructions of CCA-secure encryption schemes based on trapdoor permutations [DDN00, S99] rely on NIZK and are inherently *non*-black box), it does rule out certain techniques for constructing IBE schemes based on general assumptions.

**Related work.** In recent and independent work, MacKenzie, Reiter, and Yang [MRY04] introduce the notion of tag-based non-malleability (tnm), give efficient constructions of "tnm-cca-secure" cryptosystems in the random oracle model, and show how to construct a CCA-secure cryptosystem from any tnm-cca-secure scheme. Interestingly, their conversion from tnm-cca security to (full) CCA security uses a one-time signature scheme in essentially the same way that we do. Viewed in the context of their results, our results of Section 3 give an efficient construction of a tnm-cca-secure scheme from any weak IBE scheme, and imply an efficient and novel construction of a tnm-cca-secure scheme *in the standard model*. Our results of Section 4 have no counterpart in [MRY04].

## 2 Definitions

### 2.1 Public-Key Encryption

**Definition 1.** *A public-key encryption scheme* PKE *is a triple of* PPT *algorithms* $(\mathsf{Gen}, \mathcal{E}, \mathcal{D})$ *such that:*

- *The randomized* key generation algorithm Gen *takes as input a security parameter* $1^k$ *and outputs a public key* $PK$ *and a secret key* $SK$. *We write* $(PK, SK) \leftarrow \mathsf{Gen}(1^k)$.
- *The randomized* encryption algorithm $\mathcal{E}$ *takes as input a public key* $PK$ *and a message* $m \in \{0,1\}^*$, *and outputs a ciphertext* $C$. *We write* $C \leftarrow \mathcal{E}_{PK}(m)$.
- *The* decryption algorithm $\mathcal{D}$ *takes as input a ciphertext* $C$ *and a secret key* $SK$. *It returns a message* $m \in \{0,1\}^*$ *or the distinguished symbol* $\perp$. *We write* $m \leftarrow \mathcal{D}_{SK}(C)$.

*We require that for all* $(PK, SK)$ *output by* Gen, *all* $m \in \{0,1\}^*$, *and all* $C$ *output by* $\mathcal{E}_{PK}(m)$ *we have* $\mathcal{D}_{SK}(C) = m$.

We recall the standard definition of security against adaptive chosen-ciphertext attacks (cf. [BDPR98]).

**Definition 2.** *A public-key encryption scheme* PKE *is secure against adaptive chosen-ciphertext attacks (i.e., "CCA-secure") if the advantage of any* PPT *adversary* $A$ *in the following game is negligible in the security parameter* $k$*:*

1. $\mathsf{Gen}(1^k)$ *outputs* $(PK, SK)$. *Adversary* $A$ *is given* $1^k$ *and* $PK$.
2. *The adversary may make polynomially-many queries to a decryption oracle* $\mathcal{D}_{SK}(\cdot)$.
3. *At some point,* $A$ *outputs two messages* $m_0, m_1$ *with* $|m_0| = |m_1|$. *A bit* $b$ *is randomly chosen and the adversary is given a "challenge ciphertext"* $C^* \leftarrow \mathcal{E}_{PK}(m_b)$.
4. $A$ *may continue to query its decryption oracle* $\mathcal{D}_{SK}(\cdot)$ *except that it may not request the decryption of* $C^*$.
5. *Finally,* $A$ *outputs a guess* $b'$.

*We say that* $A$ succeeds *if* $b' = b$, *and denote the probability of this event by* $\Pr_{A,\mathsf{PKE}}[\mathsf{Succ}]$. *The adversary's* advantage *is defined as* $|\Pr_{A,\mathsf{PKE}}[\mathsf{Succ}] - 1/2|$.

## 2.2 Identity-Based Encryption

In an IBE scheme, an arbitrary identity (i.e., bit string) can serve as a public key once some master parameters have been established by a (trusted) private key generator (PKG). We review the definitions of Boneh and Franklin [BF01].

**Definition 3.** *An identity-based encryption scheme* IBE *is a 4-tuple of* PPT *algorithms* $(\mathsf{Setup}, \mathsf{Der}, \mathcal{E}, \mathcal{D})$ *such that:*

- *The randomized* setup algorithm Setup *takes as input a security parameter* $1^k$ *and a value* $\ell$ *for the identity length. It outputs some system-wide parameters* $PK$ *along with a master secret key* msk. *(We assume that* $k$ *and* $\ell$ *are implicit in* $PK$.*)*
- *The (possibly randomized)* key derivation algorithm Der *takes as input the master key* msk *and an identity* $ID \in \{0,1\}^\ell$. *It returns the corresponding decryption key* $SK_{ID}$. *We write* $SK_{ID} \leftarrow \mathsf{Der}_{\mathsf{msk}}(ID)$.
- *The randomized* encryption algorithm $\mathcal{E}$ *takes as input the system-wide public key* $PK$, *an identity* $ID \in \{0,1\}^\ell$, *and a message* $m \in \{0,1\}^*$*; it outputs a ciphertext* $C$. *We write* $C \leftarrow \mathcal{E}_{PK}(ID, m)$.
- *The* decryption algorithm $\mathcal{D}$ *takes as input an identity* $ID$, *its associated decryption key* $SK_{ID}$, *and a ciphertext* $C$. *It outputs a message* $m \in \{0,1\}^*$ *or the distinguished symbol* $\perp$. *We write* $m \leftarrow \mathcal{D}_{SK_{ID}}(ID, C)$.

*We require that for all* $(PK, \mathsf{msk})$ *output by* Setup, *all* $ID \in \{0,1\}^\ell$, *all* $SK_{ID}$ *output by* $\mathsf{Der}_{\mathsf{msk}}(ID)$, *all* $m \in \{0,1\}^*$, *and all* $C$ *output by* $\mathcal{E}_{PK}(ID, m)$ *we have* $\mathcal{D}_{SK_{ID}}(ID, C) = m$.

We now give a definition of security for IBE. As mentioned earlier, this definition is weaker than that given by Boneh and Franklin and conforms to the "selective-node" attack considered by Canetti, et al. [CHK03]. Under this definition, the identity for which the challenge ciphertext is encrypted is selected by the adversary *in advance* (i.e., "non-adaptively") before the public key is generated. An IBE scheme satisfying this definition suffices for our purposes. Furthermore, schemes satisfying this definition of security in the standard model are known [CHK03, BB04]. (For the case of the original definition of Boneh and Franklin, only constructions in the random oracle model are known.)

**Definition 4.** *An identity-based scheme* IBE *is secure against selective-identity, chosen-plaintext attacks if for all polynomially-bounded functions $\ell(\cdot)$ the advantage of any* PPT *adversary $A$ in the following game is negligible in the security parameter $k$:*

1. *$A(1^k, \ell(k))$ outputs a target identity $ID^* \in \{0,1\}^{\ell(k)}$.*
2. $\mathsf{Setup}(1^k, \ell(k))$ *outputs $(PK, \mathsf{msk})$. The adversary is given $PK$.*
3. *The adversary $A$ may make polynomially-many queries to an oracle* $\mathsf{Der}_{\mathsf{msk}}(\cdot)$, *except that it may not request the secret key corresponding to the target identity $ID^*$.*
4. *At some point, $A$ outputs two messages $m_0, m_1$ with $|m_0| = |m_1|$. A bit $b$ is randomly chosen and the adversary is given a "challenge ciphertext" $C^* \leftarrow \mathcal{E}_{PK}(ID^*, m_b)$.*
5. *$A$ may continue to query its oracle* $\mathsf{Der}_{\mathsf{msk}}(\cdot)$, *but still may not request the secret key corresponding to the identity $ID^*$.*
6. *Finally, $A$ outputs a guess $b'$.*

*We say that $A$* succeeds *if $b' = b$, and denote the probability of this event by* $\Pr_{A,\mathsf{IBE}}[\mathsf{Succ}]$. *The adversary's* advantage *is defined as* $|\Pr_{A,\mathsf{IBE}}[\mathsf{Succ}] - 1/2|$.

The above definition may be extended in the obvious way to encompass security against (adaptive) chosen-ciphertext attacks. In this case, in addition to the game as outlined above, the adversary now additionally has access to an oracle $\widehat{\mathcal{D}}(\cdot)$ such that $\widehat{\mathcal{D}}(C)$ returns $\mathcal{D}_{SK_{ID^*}}(C)$, where $SK_{ID^*}$ is the secret key associated with the target identity $ID^*$ (computed using $\mathsf{Der}_{\mathsf{msk}}(ID^*)$).[3] As usual, the adversary has access to this oracle throughout the entire game, but cannot submit the challenge ciphertext $C^*$ to $\widehat{\mathcal{D}}$.

### 2.3 Binary Tree Encryption

Binary tree encryption (BTE) was introduced by Canetti, Halevi, and Katz [CHK03], and may be viewed as a relaxed variant of hierarchical identity-based encryption (HIBE) [HL02, GS02] in the following sense: in a BTE scheme, each node has two children (labeled "0" and "1") while in a HIBE scheme, each node

[3] Note that decryption queries for identities $ID' \neq ID^*$ are superfluous, as $A$ may make the corresponding $\mathsf{Der}$ query itself and thereby obtain $SK_{ID'}$.

has arbitrarily-many children labeled with arbitrary strings. Although BTE is seemingly weaker than HIBE, it is known [CHK03] that a BTE scheme supporting a tree of depth polynomial in the security parameter may be used to construct a full-fledged HIBE scheme (and thus, in particular, an ID-based encryption scheme). We review the relevant definitions of Canetti, et al. [CHK03].

**Definition 5.** *A binary tree encryption scheme* BTE *is a 4-tuple of* PPT *algorithms* $(\mathsf{Setup}, \mathsf{Der}, \mathcal{E}, \mathcal{D})$ *such that:*

- *The randomized* setup algorithm Setup *takes as input a security parameter* $1^k$ *and a value* $\ell$ *representing the maximum tree depth. It outputs some system-wide parameters* $PK$ *along with a master (root) secret key* $SK_\varepsilon$*. (We assume that* $k$ *and* $\ell$ *are implicit in* $PK$ *and all secret keys.)*
- *The (possibly randomized)* key derivation algorithm Der *takes as input the name of a node* $w \in \{0,1\}^{<\ell}$ *and its associated secret key* $SK_w$*. It returns secret keys* $SK_{w0}, SK_{w1}$ *for the two children of* $w$*.*
- *The randomized* encryption algorithm $\mathcal{E}$ *takes as input* $PK$*, the name of a node* $w \in \{0,1\}^{\leq \ell}$*, and a message* $m$*, and returns a ciphertext* $C$*. We write* $C \leftarrow \mathcal{E}_{PK}(w, m)$*.*
- *The* decryption algorithm $\mathcal{D}$ *takes as input the name of a node* $w \in \{0,1\}^{\leq \ell}$*, its associated secret key* $SK_w$*, and a ciphertext* $C$*. It returns a message* $m$ *or the distinguished symbol* $\perp$*. We write* $m \leftarrow \mathcal{D}_{SK_w}(w, C)$*.*

*We require that for all* $(PK, SK_\varepsilon)$ *output by* Setup*, any* $w \in \{0,1\}^{\leq \ell}$ *and any correctly-generated secret key* $SK_w$ *for this node, any message* $m$*, and all* $C$ *output by* $\mathcal{E}_{PK}(w, m)$ *we have* $\mathcal{D}_{SK_w}(w, C) = m$*.*

The following definition of security for BTE, due to [CHK03], is weaker than the notion of security originally considered by Gentry and Silverberg [GS02]. As in the definition of security for ID-based encryption given in the previous section, the following definition refers to a "non-adaptive" selection of the node for which the challenge ciphertext is encrypted. Again, however, this definition suffices for our application, and a construction meeting this definition of security in the standard model is known [CHK03]. (In contrast, a construction meeting the stronger security definition of [GS02] is known only in the random oracle model and only for trees of constant depth).

**Definition 6.** *A binary tree encryption scheme* BTE *is secure against selective-node, chosen-plaintext attacks if for all polynomially-bounded functions* $\ell(\cdot)$ *the advantage of any* PPT *adversary* $A$ *in the following game is negligible in the security parameter* $k$*:*

1. $A(1^k, \ell(k))$ *outputs a node label* $w^* \in \{0,1\}^{\leq \ell(k)}$*.*
2. $\mathsf{Setup}(1^k, \ell(k))$ *outputs* $(PK, SK_\varepsilon)$*. In addition, algorithm* $\mathsf{Der}(\cdots)$ *is used to generate the secret keys of all the nodes on the path* $P$ *from the root to* $w^*$*, and also the secret keys for the two children of* $w^*$ *(if* $|w^*| < \ell$*). The adversary is given* $PK$ *and the secret keys* $\{SK_w\}$ *for all nodes* $w$ *of the following form:*

*– $w = w'\bar{b}$, where $w'b$ is a prefix of $w^*$ and $b \in \{0,1\}$ (i.e., $w$ is a sibling of some node in $P$);*

*– $w = w^*0$ or $w = w^*1$ (i.e., $w$ is a child of $w^*$; this assumes $|w^*| < \ell$).*

*Note that this allows the adversary to compute $SK_{w'}$ for any node $w' \in \{0,1\}^{\leq \ell(k)}$ that is* not *a prefix of $w^*$.*

3. *At some point, $A$ outputs two messages $m_0, m_1$ with $|m_0| = |m_1|$. A bit $b$ is randomly chosen and the adversary is given a "challenge ciphertext" $C^* \leftarrow \mathcal{E}_{PK}(w^*, m_b)$.*
4. *Finally, $A$ outputs a guess $b'$.*

*We say that $A$* succeeds *if $b' = b$, and denote the probability of this event by $\Pr_{A,\mathsf{BTE}}[\mathsf{Succ}]$. The adversary's* advantage *is defined as $|\Pr_{A,\mathsf{BTE}}[\mathsf{Succ}] - 1/2|$.*

A BTE scheme meeting the above definition of security will be termed "secure in the sense of SN-CPA". The above definition may also be extended in the natural way to encompass security against (adaptive) chosen-ciphertext attacks. (We refer to schemes meeting this definition of security as "secure in the sense of SN-CCA".) Such a definition can be found in [CHK03], and we describe it informally here: the above game is modified so that the adversary additionally has access to an oracle $\widehat{\mathcal{D}}$ such that $\widehat{\mathcal{D}}(w, C)$ first computes the secret key $SK_w$ for node $w$ (using $SK_\varepsilon$ and repeated calls to to Der); the oracle then outputs $m \leftarrow \mathcal{D}_{SK_w}(w, C)$. The adversary has access to this oracle throughout the entire game, but may not query $\widehat{\mathcal{D}}(w^*, C^*)$ after receiving the challenge ciphertext $C^*$ (we stress that the adversary is allowed to query $\widehat{\mathcal{D}}(w, C^*)$ for $w \neq w^*$, as well as $\widehat{\mathcal{D}}(w^*, C)$ for $C \neq C^*$).

## 3 Chosen-Ciphertext Security from ID-based Encryption

Given an ID-based encryption scheme $\Pi' = (\mathsf{Setup}, \mathsf{Der}, \mathcal{E}', \mathcal{D}')$ secure against selective-identity chosen-plaintext attacks, we construct a (standard) public-key encryption scheme $\Pi = (\mathsf{Gen}, \mathcal{E}, \mathcal{D})$ secure against chosen-ciphertext attacks. In the construction, we use a one-time signature scheme $\mathsf{Sig} = (\mathcal{G}, \mathsf{Sign}, \mathsf{Vrfy})$ in which the verification key output by $\mathcal{G}(1^k)$ has length $\ell_s(k)$. We require that this scheme be secure in the sense of *strong* unforgeability (i.e., an adversary is unable to forge even a new signature on a previously-signed message). We note that such a scheme may be based on any one-way function [L79, R90] so, in particular, such a scheme exists given the existence of $\Pi'$. The construction of $\Pi$ proceeds as follows:

- $\mathsf{Gen}(1^k)$ runs $\mathsf{Setup}(1^k, \ell_s(k))$ to obtain $(PK, \mathsf{msk})$. The public key is $PK$ and the secret key is $\mathsf{msk}$.
- To encrypt message $m$ using public key $PK$, the sender first runs $\mathcal{G}(1^k)$ to obtain verification key $vk$ and signing key $sk$ (with $|vk| = \ell_s(k)$). The sender then computes $C \leftarrow \mathcal{E}'_{PK}(vk, m)$ (i.e., the sender encrypts $m$ with respect to "identity" $vk$) and $\sigma \leftarrow \mathsf{Sign}_{sk}(C)$. The final ciphertext is $\langle vk, C, \sigma \rangle$.

- To decrypt ciphertext $\langle vk, C, \sigma \rangle$ using secret key $\mathsf{msk}$, the receiver first checks whether $\mathsf{Vrfy}_{vk}(C, \sigma) \stackrel{?}{=} 1$. If not, the receiver simply outputs $\perp$. Otherwise, the receiver computes $SK_{vk} \leftarrow \mathsf{Der}_{\mathsf{msk}}(vk)$ and outputs $m \leftarrow \mathcal{D}'_{SK_{vk}}(ID, C)$.

We first give some intuition as to why $\Pi$ is secure against chosen-ciphertext attacks. Let $\langle vk^*, C^*, \sigma^* \rangle$ be the challenge ciphertext (cf. Definition 2). It should be clear that, without any decryption oracle queries, the value of the bit $b$ remains hidden to the adversary; this is so because $C^*$ is output by $\Pi'$ which is CPA-secure, $vk^*$ is independent of the message, and $\sigma^*$ is merely the result of applying the signing algorithm to $C^*$.

We claim that decryption oracle queries cannot further help the adversary is guessing the value of $b$. On one hand, if the adversary submits ciphertext $\langle vk', C', \sigma' \rangle$ different from the challenge ciphertext but with $vk' = vk^*$ then the decryption oracle will reply with $\perp$ since the adversary is unable to forge new, valid signatures with respect to $vk$. On the other hand, if $vk' \neq vk^*$ then (informally) the decryption query will not help the adversary since the eventual decryption using $\mathcal{D}'$ (in the underlying scheme $\Pi'$) will be done with respect to a different "identity" $vk'$. Below, we formally prove that this cannot help an adversary.

**Theorem 1.** *If $\Pi'$ is an* IBE *scheme which is secure against selective-identity, chosen-plaintext attacks and* Sig *is a strongly unforgeable one-time signature scheme, then $\Pi$ is a* PKE *scheme which is secure against adaptive chosen-ciphertext attacks.*

*Proof.* Given any PPT adversary $\mathcal{A}$ attacking $\Pi$ in an adaptive chosen-ciphertext attack, we construct a PPT adversary $\mathcal{A}'$ attacking $\Pi'$ in a selective-identity, chosen-plaintext attack. Relating the success probabilities of these adversaries gives the desired result.

Before specifying $\mathcal{A}'$, we first define event $\mathsf{Forge}$ and bound the probability of its occurrence. Let $\langle vk^*, C^*, \sigma^* \rangle$ be the challenge ciphertext received by $\mathcal{A}$, and let $\mathsf{Forge}$ denote the event that $\mathcal{A}$ submits to its decryption oracle a ciphertext $\langle vk^*, C, \sigma \rangle$ with $(C, \sigma) \neq (C^*, \sigma^*)$ but for which $\mathsf{Vrfy}_{vk^*}(C, \sigma) = 1$. (We include in this event the case when $\mathcal{A}$ submits such a query to its decryption oracle *before* receiving the challenge ciphertext; in this case, we do not require $(C, \sigma) \neq (C^*, \sigma^*)$.) It is easy to see that we can use $\mathcal{A}$ to break the underlying one-time signature scheme $\mathsf{Sig}$ with probability exactly $\Pr_{\mathcal{A}}[\mathsf{Forge}]$; since $\mathsf{Sig}$ is a strongly unforgeable one-time signature scheme, it must be the case that $\Pr_{\mathcal{A}}[\mathsf{Forge}]$ is negligible (in the security parameter $k$).

We now define adversary $\mathcal{A}'$ as follows:

1. $\mathcal{A}'(1^k, \ell_s(k))$ runs $\mathcal{G}(1^k)$ to generate $(vk^*, sk^*)$. It then outputs the "target identity" $ID^* = vk^*$.
2. $\mathsf{Setup}(1^k, \ell_s(k))$ outputs $(PK, \mathsf{msk})$ and $\mathcal{A}'$ is given $PK$. Adversary $\mathcal{A}'$, in turn, runs $\mathcal{A}$ on input $1^k$ and $PK$.
3. When $\mathcal{A}$ makes decryption oracle query $\mathcal{D}(\langle vk, C, \sigma \rangle)$, adversary $\mathcal{A}'$ proceeds as follows:

(a) If $\mathsf{Vrfy}_{vk}(C, \sigma) \neq 1$, then $\mathcal{A}'$ simply returns $\perp$.
(b) If $\mathsf{Vrfy}_{vk}(C, \sigma) = 1$ and $vk = vk^*$ (i.e., event Forge occurs), then $\mathcal{A}'$ halts and outputs a random bit.
(c) If $\mathsf{Vrfy}_{vk}(C, \sigma) = 1$ and $vk \neq vk^*$, then $\mathcal{A}'$ makes the oracle query $\mathsf{Der}_{\mathsf{msk}}(vk)$ to obtain $SK_{vk}$. It then computes $m \leftarrow \mathcal{D}'_{SK_{vk}}(vk, C)$ and returns $m$.

4. At some point, $\mathcal{A}$ outputs two equal-length messages $m_0, m_1$. These messages are output by $\mathcal{A}'$. In return, $\mathcal{A}'$ is given a challenge ciphertext $C^*$; adversary $\mathcal{A}'$ then computes $\sigma^* \leftarrow \mathsf{Sign}_{vk^*}(C^*)$ and returns $\langle vk^*, C^*, \sigma^* \rangle$ to $\mathcal{A}$.
5. $\mathcal{A}$ may continue to make decryption oracle queries, and these are answered as before. (Recall, $\mathcal{A}$ may not query the decryption oracle on the challenge ciphertext itself.)
6. Finally, $\mathcal{A}$ outputs a guess $b'$; this same guess is output by $\mathcal{A}'$.

Note that $\mathcal{A}'$ represents a legal adversarial strategy for attacking $\Pi'$ in a selective-identity, chosen-plaintext attack; in particular, $\mathcal{A}'$ never requests the secret key corresponding to "target identity" $vk^*$. Furthermore, $\mathcal{A}'$ provides a perfect simulation for $\mathcal{A}$ (and thus $\mathcal{A}'$ succeeds whenever $\mathcal{A}$ succeeds) unless event Forge occurs. We therefore have:

$$\Pr_{\mathcal{A}', \Pi'}[\mathsf{Succ}] \geq \Pr_{\mathcal{A}, \Pi}[\mathsf{Succ}] - \tfrac{1}{2} \cdot \Pr_{\mathcal{A}}[\mathsf{Forge}].$$

Since $\Pr_{\mathcal{A}', \Pi'}[\mathsf{Succ}]$ is negligibly close to $1/2$ (because $\Pi'$ is assumed to be secure in against selective-identity, chosen-plaintext attacks), and since $\Pr_{\mathcal{A}}[\mathsf{Forge}]$ is negligible, it must be the case that $\Pr_{\mathcal{A}, \Pi}[\mathsf{Succ}]$ is negligibly close to $1/2$ as well.

## 4 Chosen-Ciphertext Security for BTE Schemes

The techniques of the previous section may also be used to construct a BTE scheme secure in the sense of SN-CCA from any BTE scheme secure in the sense of SN-CPA. Roughly, we view the subtree of each node as a (hierarchical) IBE scheme, and use the scheme from the previous section for that subtree. We first give a high-level overview for the simpler case of a BTE scheme which only allows encryption to nodes at a single depth $\ell$ (as opposed to a full-fledged BTE scheme which allows encryption to nodes at *all* depths $\leq \ell$). To encrypt a message for node $w$, the sender generates keys $(vk, sk)$ for a one-time signature scheme (as in the previous section) and encrypts the message $m$ for "node" $w|vk$ to obtain ciphertext $C$; the sender additionally signs $C$ using $sk$ resulting in signature $\sigma$. The complete ciphertext is $\langle vk, C, \sigma \rangle$. When node $w$, holding secret key $SK_w$, receives a ciphertext of this form, it first verifies that the signature is correct with respect to $vk$. If so, the receiver computes secret key $SK_{w|vk}$ on its own (using repeated applications of the Der algorithm) and then uses this key to recover $m$ from $C$. As for the scheme from the previous section, the intuition here is that encryption to "node" $w|vk$ is secure even if an adversary can obtain secret keys for multiple "nodes" $w'|vk'$ with $(w', vk') \neq (w, vk)$ (recall we are assuming here that all nodes $w$ are at the same depth, so $w'|vk'$ cannot be a prefix of $w|vk$).

Thus, even more so, encryption to "node" $w|vk$ remains secure if the adversary can obtain (only) *decryptions* of ciphertexts intended for "nodes" $w'|vk'$ of this sort. And of course, the adversary is unable to obtain any decryptions for "node" $w|vk$ itself unless it can forge a new signature with respect to $vk$.

The construction is a bit more involved for the case of general BTE schemes (i.e., when encryption is allowed to nodes at arbitrary depth rather than at a single depth). The issue that we must resolve is the encoding of node names; for example, we must ensure that $w|vk$ is not mapped to the same node as some other $w'$. A simple way of resolving this issue is to encode each node name $w = w_1 w_2 \ldots w_t$ as $1w_1 1w_2 \ldots 1w_t$, and then encode $w|vk$ as $1w_1 1w_2 \ldots 1w_t 0|vk$. We describe the full construction in detail below.

Let $\Pi' = (\mathsf{Setup}', \mathsf{Der}', \mathcal{E}', \mathcal{D}')$ be a BTE scheme and let $\mathsf{Sig} = (\mathcal{G}, \mathsf{Sign}, \mathsf{Vrfy})$ be a one-time signature scheme in which the verification key output by $\mathcal{G}(1^k)$ has length $\ell_s(k)$. As in the previous section, we require this scheme to be secure in the sense of *strong* unforgeability. Next, define a function $\mathsf{Encode}$ on strings $w$ such that:

$$\mathsf{Encode}(w) = \begin{cases} \varepsilon & \text{if } w = \varepsilon \\ 1w_1 1w_2 \cdots 1w_t & \text{if } w = w_1 \cdots w_t \text{ (with } w_i \in \{0,1\}) \end{cases}.$$

(Note that $|\mathsf{Encode}(w)| = 2|w|$.) The construction of binary tree encryption scheme $\Pi = (\mathsf{Setup}, \mathsf{Der}, \mathcal{E}, \mathcal{D})$ proceeds as follows:

- $\mathsf{Setup}(1^k, \ell)$ runs $\mathsf{Setup}'(1^k, 2\ell + \ell_s(k) + 1)$ to obtain $(PK, SK_\varepsilon)$. The system-wide public key is $PK$ and the root secret key is $SK_\varepsilon$.
- $\mathsf{Der}(w, SK_w)$ proceeds as follows. First, set $w' = \mathsf{Encode}(w)$. Next, compute $SK'_{w'1}$ using $\mathsf{Der}'_{SK_w}(w')$ followed by $(SK_{w'10}, SK_{w'11}) \leftarrow \mathsf{Der}_{SK'_{w'1}}(w'1)$. Set $SK_{w0} = SK'_{w'10}$ and $SK_{w1} = SK'_{w'11}$ and output $(SK_{w0}, SK_{w1})$. (Note that $w'10 = \mathsf{Encode}(w0)$ and analogously for $w'11$.)

  (Intuitively, any node $w$ in scheme $\Pi$ corresponds to a node $w' = \mathsf{Encode}(w)$ in $\Pi'$. Thus, secret key $SK_w$ for node $w$ (in $\Pi$) corresponds to secret key $SK'_{w'}$ for node $w'$ (in $\Pi'$). So, to derive the secret keys for the children of $w$ (i.e., $w0, w1$) in $\Pi$, we must derive the keys for the (right) *grandchildren* of node $w'$ in $\Pi'$.)
- To encrypt message $m$ for a node $w \in \{0,1\}^{\leq \ell}$ using public parameters $PK$, the sender first runs $\mathcal{G}(1^k)$ to obtain verification key $vk$ and signing key $sk$. Next, the sender sets $w' = \mathsf{Encode}(w)$. The sender then computes $C \leftarrow \mathcal{E}'_{PK}(w'|0|vk, m)$ (i.e., the sender encrypts $m$ with respect to "node" $w'|0|vk$ using $\Pi'$) and $\sigma \leftarrow \mathsf{Sign}_{sk}(C)$. The final ciphertext is $\langle vk, C, \sigma \rangle$.
- Node $w$, with secret key $SK_w$, decrypts a ciphertext $\langle vk, C, \sigma \rangle$ as follows. First, check whether $\mathsf{Vrfy}_{vk}(C, \sigma) \stackrel{?}{=} 1$. If not, simply output $\perp$. Otherwise, let $w' = \mathsf{Encode}(w)$. The receiver then computes the secret key $SK'_{w'|0|vk}$ using repeated applications of $\mathsf{Der}'$, and outputs $m \leftarrow \mathcal{D}'_{SK'_{w'|0|vk}}(w'|0|vk, C)$.

**Remark 1.** The above approach can be used to derive a CCA-secure HIBE scheme from a CPA-secure HIBE scheme in the following way: CPA-secure HIBE trivially implies CPA-secure BTE; the conversion above yields CCA-secure BTE; and the latter implies CCA-secure HIBE (see [CHK03]). However, it will in general be much more efficient to apply the above techniques *directly*: In this case, we would simply encode the ID-vector $\boldsymbol{w} = w_1|\cdots|w_t$ as $\boldsymbol{w}' = 1w_1|\cdots|1w_t$, and encode $\boldsymbol{w}|vk$ as an ID-vector $\boldsymbol{w}'|0vk$.

We now state the main result of this section:

**Theorem 2.** *If $\Pi'$ is a* BTE *scheme which is secure in the sense of* SN-CPA *and* Sig *is a strongly unforgeable one-time signature scheme, then $\Pi$ is a* BTE *scheme which is secure in the sense of* SN-CCA.

*Proof.* The proof is largely similar to that of Theorem 1. Given any PPT adversary $\mathcal{A}$ attacking $\Pi$ in a selective node, chosen-ciphertext attack, we construct a PPT adversary $\mathcal{A}'$ attacking $\Pi'$ in a selective node, chosen-plaintext attack. Relating the success probabilities of these adversaries gives the desired result.

We first define event Forge; because we are working in the context of BTE, the definition is slightly different from the definition used in the proof of Theorem 1. Specifically, let $w^*$ denote the node initially output by $\mathcal{A}$, and let $\langle vk^*, C^*, \sigma^* \rangle$ be the challenge ciphertext received by $\mathcal{A}$. Now, let Forge denote the event that $\mathcal{A}$ makes a decryption query $\widehat{\mathcal{D}}(w^*, \langle vk^*, C', \sigma' \rangle)$ with $(C', \sigma') \neq (C^*, \sigma^*)$ but for which $\mathsf{Vrfy}_{vk^*}(C', \sigma') = 1$. (We include in this event the case when $\mathcal{A}$ submits such a query to its decryption oracle *before* receiving the challenge ciphertext; in this case, we do not require $(C', \sigma') \neq (C^*, \sigma^*)$.) It is easy to see that we can use $\mathcal{A}$ to break the underlying one-time signature scheme Sig with probability exactly $\Pr_{\mathcal{A}}[\mathsf{Forge}]$; since Sig is a strongly unforgeable one-time signature scheme, it must be the case that $\Pr_{\mathcal{A}}[\mathsf{Forge}]$ is negligible (in the security parameter $k$).

We now define adversary $\mathcal{A}'$ as follows:

1. $\mathcal{A}'(1^k, \ell')$ sets $\ell = (\ell' - \ell_s(k) - 1)/2$ and runs $\mathcal{A}(1^k, \ell)$ who, in turn, outputs a node $w^* \in \{0,1\}^{\leq \ell}$. Adversary $\mathcal{A}'$ sets $w' = \mathsf{Encode}(w^*)$, and runs $\mathcal{G}(1^k)$ to generate $(vk^*, sk^*)$. Finally, $\mathcal{A}'$ outputs the node $w^{*\prime} = w'|0|vk^*$.
2. $\mathcal{A}'$ is given $PK$ as well as a set of secret keys $\{SK'_w\}$ for all nodes $w$ of the following form:
   - $w = v\bar{b}$, where $vb$ is a prefix of $w^{*\prime}$ and $b \in \{0,1\}$;
   - $w = w^{*\prime}0$ or $w = w^{*\prime}1$ (in case $|w^{*\prime}| < \ell'$).

   Using these, $\mathcal{A}'$ can compute and give to $\mathcal{A}$ all the relevant secret keys that $\mathcal{A}$ expects.
3. When $\mathcal{A}$ makes decryption query $\widehat{\mathcal{D}}(w, \langle vk, C, \sigma \rangle)$, adversary $\mathcal{A}'$ proceeds as follows:
   (a) If $\mathsf{Vrfy}_{vk}(C, \sigma) \neq 1$, then $\mathcal{A}'$ simply returns $\perp$.
   (b) If $w = w'$, $\mathsf{Vrfy}_{vk}(C, \sigma) = 1$, and $vk = vk^*$ (i.e., event Forge occurs), then $\mathcal{A}'$ halts and outputs a random bit.
   (c) Otherwise, set $\tilde{w} = \mathsf{Encode}(w)$. Note that $\mathcal{A}'$ is able to derive the secret key corresponding to the "node" $\tilde{w}|0|vk$ using the secret keys it obtained in step 2 (this follows since $\tilde{w}|0|vk$ cannot be a prefix of $w^{*\prime}$). So, $\mathcal{A}'$ simply computes the necessary key, performs the decryption of $C$, and returns the result to $\mathcal{A}$.

4. When $\mathcal{A}$ outputs its two messages $m_0, m_1$, these same messages are output by $\mathcal{A}'$. In return, $\mathcal{A}'$ receives a ciphertext $C^*$. Adversary $\mathcal{A}'$ computes $\sigma^* \leftarrow \mathsf{Sign}_{sk^*}(C^*)$ and returns ciphertext $\langle vk^*, C^*, \sigma^* \rangle$ to $\mathcal{A}$.
5. Any subsequent decryption queries of $\mathcal{A}$ are answered as before.
6. Finally, $\mathcal{A}$ outputs a guess $b'$; this same guess is output by $\mathcal{A}'$.

Note that $\mathcal{A}'$ represents a legal adversarial strategy for attacking $\Pi'$. Furthermore, $\mathcal{A}'$ provides a perfect simulation for $\mathcal{A}$ (and thus $\mathcal{A}'$ succeeds whenever $\mathcal{A}$ succeeds) unless event Forge occurs. An analysis as in the proof of Theorem 1 shows that $\Pr_{\mathcal{A},\Pi}[\mathsf{Succ}]$ must be negligibly close to $1/2$.

The above construction requires only a one-time signature scheme in addition to the underlying BTE scheme; the existence of the former (which may be constructed from any one-way function) is implied by the existence of any BTE scheme secure in the sense of SN-CPA. Putting these observations together shows:

**Theorem 3.** *If there exists a BTE scheme secure in the sense of* SN-CPA, *then there exists a BTE scheme secure in the sense of* SN-CCA.

Note that an analogous result for the case of (standard) public-key encryption is not known.

**Further applications.** In [CHK03] it is shown that any BTE scheme can be used to construct both a forward-secure public-key encryption scheme as well as a "full-fledged" HIBE scheme (and, as a special case, an IBE scheme). Furthermore, if the original BTE scheme is secure against chosen-ciphertext attacks, then so are the derived schemes. Canetti, et al. further suggest [CHK03] that a BTE scheme secure in the sense of SN-CCA can be derived using the Naor-Yung paradigm [NY90] along with 1-time, simulation-sound NIZK proofs [S99]. As mentioned in the Introduction, the use of NIZK proofs results in a completely impractical scheme, at least using currently-known techniques. Thus, the approach of this section provides a more efficient way of achieving CCA security for any BTE scheme (as well as CCA security for forward-secure encryption or HIBE) in the standard model. (See also Remark 1.)

When our techniques are applied to a BTE/IBE/HIBE scheme secure against selective-node/identity attacks, the resulting CCA-secure scheme is also only resilient to selective-node/identity attacks. However, when our techniques are applied to stronger schemes which are CPA-secure against an *adaptive* choice of node/identity, the resulting CCA-secure scheme maintains this level of security as well.

We remark that when the transformation outlined in this section is applied to the recent constructions of Boneh and Boyen [BB04], we obtain truly practical constructions of IBE and HIBE schemes secure against selective-identity, chosen-ciphertext attacks in the standard model.

## Acknowledgments

We thank Eu-Jin Goh for pointing out that our techniques imply a conversion from weak IBE to "lunchtime" CCA1 security with essentially no overhead.

## References

[AGMM04] B. Aiello, Y. Gertner, T. Malkin, and S. Myers. Personal communication.

[BDPR98] M. Bellare, A. Desai, D. Pointcheval, and P. Rogaway. Relations Among Notions of Security for Public-Key Encryption Schemes. *Adv. in Cryptology — Crypto 1998*, LNCS vol. 1462, Springer-Verlag, pp. 26–45, 1998.

[BFM88] M. Blum, P. Feldman, and S. Micali. Non-Interactive Zero-Knowledge and its Applications. *20th ACM Symposium on Theory of Computing (STOC)*, ACM, pp. 103–112, 1988.

[BB04] D. Boneh and X. Boyen. Efficient Selective-ID Secure Identity Based Encryption Without Random Oracles. *Adv. in Cryptology — Eurocrypt 2004*, to appear.

[BF01] D. Boneh and M. Franklin. Identity-Based Encryption from the Weil Pairing. *Adv. in Cryptology — Crypto 2001*, LNCS vol. 2139, Springer-Verlag, pp. 213–229, 2001. Full version to appear in *SIAM J. Computing* and available at `http://eprint.iacr.org/2001/090`.

[CHK03] R. Canetti, S. Halevi, and J. Katz. A Forward-Secure Public-Key Encryption Scheme. *Adv. in Cryptology — Eurocrypt 2003*, LNCS vol. 2656, Springer-Verlag, pp. 255–271, 2003. Full version available at `http://eprint.iacr.org/2003/083`.

[C01] C. Cocks. An Identity-Based Encryption Scheme Based on Quadratic Residues. *Cryptography and Coding*, LNCS vol. 2260, Springer-Verlag, pp. 360–363, 2001.

[CS98] R. Cramer and V. Shoup. A Practical Public Key Cryptosystem Provably Secure Against Chosen Ciphertext Attack. *Adv. in Cryptology — Crypto 1998*, LNCS vol. 1462, Springer-Verlag, pp. 13–25, 1998.

[CS02] R. Cramer and V. Shoup. Universal Hash Proofs and a Paradigm for Adaptive Chosen Ciphertext Secure Public-Key Encryption. *Adv. in Cryptology — Eurocrypt 2002*, LNCS vol. 2332, Springer-Verlag, pp. 45–64, 2002.

[CS03] J. Camenisch and V. Shoup. Practical Verifiable Encryption and Decryption of Discrete Logarithms. *Adv. in Cryptology — Crypto 2003*, LNCS vol. 2729, Springer-Verlag, pp. 126–144, 2003.

[DDN00] D. Dolev, C. Dwork, and M. Naor. Non-Malleable Cryptography. *SIAM J. Computing* 30(2): 391–437, 2000.

[ES02] E. Elkind and A. Sahai. A Unified Methodology For Constructing Public-Key Encryption Schemes Secure Against Adaptive Chosen-Ciphertext Attack. *First Theory of Cryptography Conference (TCC)* 2004, to appear. Available from `http://eprint.iacr.org/2002/042/`.

[FLS90] U. Feige, D. Lapidot, and A. Shamir. Multiple Non-Interactive Zero-Knowledge Proofs Under General Assumptions. *SIAM J. Computing* 29(1): 1–28, 1999.

[GL03] R. Gennaro and Y. Lindell. A Framework for Password-Based Authenticated Key Exchange. *Adv. in Cryptology — Eurocrypt 2003*, LNCS vol. 2656, Springer-Verlag, pp. 524–543, 2003.

[GS02] C. Gentry and A. Silverberg. Hierarchical Identity-Based Cryptography. *Adv. in Cryptology — Asiacrypt 2002*, LNCS vol. 2501, Springer-Verlag, pp. 548–566, 2002.

[GM84] S. Goldwasser and S. Micali. Probabilistic Encryption. *J. Computer System Sciences* 28(2): 270–299, 1984.

[HL02] J. Horwitz and B. Lynn. Toward Hierarchical Identity-Based Encryption. *Adv. in Cryptology — Eurocrypt 2002*, LNCS vol. 2332, Springer-Verlag, pp. 466–481, 2002.

[L79] L. Lamport. Constructing Digital Signatures from a One-Way Function. Technical Report CSL-98, SRI International, Palo Alto, 1979.

[L03] Y. Lindell. A Simpler Construction of CCA-Secure Public-Key Encryption Under General Assumptions. *Adv. in Cryptology — Eurocrypt 2003*, LNCS vol. 2656, Springer-Verlag, pp. 241–254, 2003.

[MRY04] P. MacKenzie, M. Reiter, and K. Yang. Alternatives to Non-Malleability: Definitions, Constructions, and Applications. *First Theory of Cryptography Conference (TCC)* 2004, to appear.

[NY90] M. Naor and M. Yung. Public-Key Cryptosystems Provably-Secure against Chosen-Ciphertext Attacks. *22nd ACM Symposium on Theory of Computing (STOC)*, ACM, pp. 427–437, 1990.

[RS91] C. Rackoff and D. Simon. Non-Interactive Zero-Knowledge Proof of Knowledge and Chosen Ciphertext Attack. *Adv. in Cryptology — Crypto 1991*, LNCS vol. 576, Springer-Verlag, pp. 433–444, 1992.

[R90] J. Rompel. One-Way Functions are Necessary and Sufficient for Secure Signatures. *22nd ACM Symposium on Theory of Computing (STOC)*, ACM, pp. 387–394, 1990.

[S99] A. Sahai. Non-Malleable Non-Interactive Zero Knowledge and Adaptive Chosen-Ciphertext Security. *40th IEEE Symposium on Foundations of Computer Science (FOCS)*, IEEE, pp. 543–553, 1999.

[S84] A. Shamir. Identity-Based Cryptosystems and Signature Schemes. *Adv. in Cryptology — Crypto 1984*, LNCS vol. 196, Springer-Verlag, pp. 47–53, 1985.

[S98] V. Shoup. Why Chosen Ciphertext Security Matters. IBM Research Report RZ 3076, November, 1998. Available at `http://www.shoup.net/papers`.

# Efficient Selective-ID Secure Identity-Based Encryption Without Random Oracles

Dan Boneh[1*] and Xavier Boyen[2]

[1] Computer Science Department, Stanford University, Stanford CA 94305-9045
dabo@cs.stanford.edu
[2] Voltage Security, Palo Alto, California
xb@boyen.org

**Abstract.** We construct two efficient Identity Based Encryption (IBE) systems that are selective identity secure *without the random oracle model*. Selective identity secure IBE is a slightly weaker security model than the standard security model for IBE. In this model the adversary must commit ahead of time to the identity that it intends to attack, whereas in the standard model the adversary is allowed to choose this identity adaptively. Our first secure IBE system extends to give a selective identity Hierarchical IBE secure without random oracles.

## 1 Introduction

Boneh and Franklin [BF01, BF03] recently defined a security model for Identity Based Encryption [Sha84] and gave a construction using bilinear maps. Cocks [Coc01] describes another construction using quadratic residues. Proving security for these systems requires the random oracle model [BR93]. A natural open question is to construct a secure IBE system without random oracles. No such system is currently known.

In the Boneh-Franklin security model the adversary can issue both adaptive chosen ciphertext queries and adaptive chosen identity queries (i.e., the adversary can request the private key for identities of its choice). Eventually, the adversary adaptively chooses the identity it wishes to attack and asks for a semantic security challenge for this identity. Canetti et al. [CHK03, CHK04] recently proposed a slightly weaker security model, called selective identity IBE. In this model the adversary must commit ahead of time (non-adaptively) to the identity it intends to attack. The adversary can still issue adaptive chosen ciphertext and adaptive chosen identity queries. Canetti et al. are able to construct a provably secure IBE in this weaker model without the random oracle model. However, their construction views identities as bit strings, causing their system to require a bilinear map computation for every bit in the identity.

We construct two efficient IBE systems that are provably selective identity secure without the random oracle model. In both systems, encryption requires no bilinear map computation and decryption requires at most two. Our first

* Supported by NSF and the Packard Foundation.

C. Cachin and J. Camenisch (Eds.): EUROCRYPT 2004, LNCS 3027, pp. 223–238, 2004.

construction is based on the Decision Bilinear Diffie-Hellman (Decision BDH) assumption. This construction extends to give an efficient selective identity secure Hierarchical IBE (HIBE) without random oracles. Hierarchical IBE was defined in [HL02] and the first construction in the random oracle model was given by Gentry and Silverberg [GS02]. Our efficient HIBE construction is similar to the Gentry-Silverberg system, but we are able to prove security without using random oracles. Our second IBE construction is even more efficient, but is based on a new assumption we call Decision Bilinear Diffie-Hellman Inversion (Decision BDHI). Roughly speaking, the assumption says that no efficient algorithm can distinguish $e(g,g)^{1/x}$ from random, given $g, g^x, g^{(x^2)}, \ldots, g^{(x^q)}$ for some $q$.

Canetti et al. [CHK04] recently showed that any selective identity, chosen plaintext IBE gives a chosen ciphertext secure (CCA2) public key system. Consequently, both our IBE systems give efficient CCA2-secure public key systems without random oracles. In particular, using our second system we obtain a CCA2-secure public key system that has comparable efficiency to the Cramer-Shoup system based on DDH.

## 2 Preliminaries

Before presenting our results we briefly review the definition of security for an IBE system. We also review the definition of groups equipped with a bilinear map.

### 2.1 Selective Identity Secure IBE and HIBE Systems

Recall that an Identity Based Encryption system (IBE) consists of four algorithms [Sha84, BF01]: *Setup*, *KeyGen*, *Encrypt*, *Decrypt*. The *Setup* algorithm generates system parameters, denoted by *params*, and a master key *master-key*. The *KeyGen* algorithm uses the master key to generate the private key corresponding to a given identity. The encryption algorithm encrypts messages for a given identity (using the system parameters) and the decryption algorithm decrypts ciphertexts using the private key. In a Hierarchical IBE [HL02, GS02] identities are vectors. A vector of dimension $\ell$ represents an identity at depth $\ell$. Algorithm *KeyGen* takes as input an identity $\mathsf{ID} = (\mathrm{I}_1, \ldots, \mathrm{I}_\ell)$ at depth $\ell$ and the private key $d_{\mathsf{ID}|\ell-1}$ of the parent identity $\mathsf{ID}_{|\ell-1} = (\mathrm{I}_1, \ldots, \mathrm{I}_{\ell-1})$ at depth $\ell - 1$. It outputs the private key $d_{\mathsf{ID}}$ for identity $\mathsf{ID}$. We refer to the *master-key* as the private key at depth 0 and note that an IBE system is an HIBE where all identities are at depth 1.

Boneh and Franklin [BF01, BF03] define chosen ciphertext security for IBE systems under a chosen identity attack. In their model the adversary is allowed to adaptively chose the public key it wishes to attack (the public key on which it will be challenged). Canetti, Halevi, and Katz [CHK03, CHK04] define a weaker notion of security in which the adversary commits ahead of time to the public key it will attack. We refer to this notion as selective identity, chosen ciphertext secure IBE (IND-sID-CCA). More precisely, selective identity IBE and HIBE security is defined using the following game:

**Init:** The adversary outputs an identity $\mathsf{ID}^*$ where it wishes to be challenged.

**Setup:** The challenger runs the *Setup* algorithm. It gives the adversary the resulting system parameters *params*. It keeps the *master-key* to itself.

**Phase 1:** The adversary issues queries $q_1, \ldots, q_m$ where query $q_i$ is one of:

- Private key query $\langle \mathsf{ID}_i \rangle$ where $\mathsf{ID}_i \neq \mathsf{ID}^*$ and $\mathsf{ID}_i$ is not a prefix of $\mathsf{ID}^*$. The challenger responds by running algorithm *KeyGen* to generate the private key $d_i$ corresponding to the public key $\langle \mathsf{ID}_i \rangle$. It sends $d_i$ to the adversary.
- Decryption query $\langle C_i \rangle$ for identity $\mathsf{ID}^*$ or any prefix of $\mathsf{ID}^*$. The challenger responds by running algorithm *KeyGen* to generate the private key $d$ corresponding to $\mathsf{ID}^*$ (or the relevant prefix thereof as requested). It then runs algorithm *Decrypt* to decrypt the ciphertext $C_i$ using the private key $d$. It sends the resulting plaintext to the adversary.

These queries may be asked adaptively, that is, each query $q_i$ may depend on the replies to $q_1, \ldots, q_{i-1}$.

**Challenge:** Once the adversary decides that Phase 1 is over it outputs two equal length plaintexts $M_0, M_1 \in \mathcal{M}$ on which it wishes to be challenged. The challenger picks a random bit $b \in \{0,1\}$ and sets the challenge ciphertext to $C = \mathit{Encrypt}(\mathit{params}, \mathsf{ID}^*, M_b)$. It sends $C$ as the challenge to the adversary.

**Phase 2:** The adversary issues additional queries $q_{m+1}, \ldots, q_n$ where $q_i$ is one of:

- Private key query $\langle \mathsf{ID}_i \rangle$ where $\mathsf{ID}_i \neq \mathsf{ID}^*$ and $\mathsf{ID}_i$ is not a prefix of $\mathsf{ID}^*$. The challenger responds as in Phase 1.
- Decryption query $\langle C_i \rangle \neq \langle C \rangle$ for $\mathsf{ID}^*$ or any prefix of $\mathsf{ID}^*$. The challenger responds as in Phase 1.

These queries may be asked adaptively as in Phase 1.

**Guess:** Finally, the adversary outputs a guess $b' \in \{0,1\}$. The adversary wins if $b = b'$.

We refer to such an adversary $\mathcal{A}$ as an IND-sID-CCA adversary. We define the advantage of the adversary $\mathcal{A}$ in attacking the scheme $\mathcal{E}$ as

$$\mathrm{Adv}_{\mathcal{E},\mathcal{A}} = \left| \Pr[b = b'] - \frac{1}{2} \right|$$

The probability is over the random bits used by the challenger and the adversary.

**Definition 1.** *We say that an IBE or HIBE system $\mathcal{E}$ is $(t, q_{ID}, q_C, \epsilon)$-selective identity, adaptive chosen ciphertext secure if for any t-time* IND-sID-CCA *adversary $\mathcal{A}$ that makes at most $q_{ID}$ chosen private key queries and at most $q_C$ chosen decryption queries we have that $Adv_{\mathcal{E},\mathcal{A}} < \epsilon$. As shorthand, we say that $\mathcal{E}$ is $(t, q_{ID}, q_C, \epsilon)$* IND-sID-CCA *secure.*

*Semantic Security.* As usual, we define selective identity, chosen plaintext security for an IBE system as in the preceding game, except that the adversary is not allowed to issue any decryption queries. The adversary may still issue adaptive private key queries.

**Definition 2.** *We say that an IBE or HIBE system $\mathcal{E}$ is $(t, q_{ID}, \epsilon)$-selective identity, chosen plaintext secure if $\mathcal{E}$ is $(t, q_{ID}, 0, \epsilon)$-selective identity, chosen ciphertext secure. As shorthand, we say that $\mathcal{E}$ is $(t, q_{ID}, \epsilon)$ IND-sID-CPA secure.*

### 2.2 Bilinear Groups

We briefly review the necessary facts about bilinear maps and bilinear map groups. We follow the notation in [BF01]:

1. $\mathbb{G}$ and $\mathbb{G}_1$ are two (multiplicative) cyclic groups of prime order $p$;
2. $g$ is a generator of $\mathbb{G}$.
3. $e$ is a bilinear map $e : \mathbb{G} \times \mathbb{G} \to \mathbb{G}_1$.

Let $\mathbb{G}$ and $\mathbb{G}_1$ be two groups as above. A bilinear map is a map $e : \mathbb{G} \times \mathbb{G} \to \mathbb{G}_1$ with the following properties:

1. Bilinear: for all $u, v \in \mathbb{G}$ and $a, b \in \mathbb{Z}$, we have $e(u^a, v^b) = e(u, v)^{ab}$.
2. Non-degenerate: $e(g, g) \neq 1$.

We say that $\mathbb{G}$ is a bilinear group if the group action in $\mathbb{G}$ can be computed efficiently and there exists a group $\mathbb{G}_1$ and an efficiently computable bilinear map $e : \mathbb{G} \times \mathbb{G} \to \mathbb{G}_1$ as above. Note that $e(,)$ is symmetric since $e(g^a, g^b) = e(g, g)^{ab} = e(g^b, g^a)$.

## 3 Complexity Assumptions

Let $\mathbb{G}$ be a bilinear group of prime order $p$ and $g$ be a generator of $\mathbb{G}$. We review the standard Bilinear Diffie-Hellman (BDH) assumption and define the Bilinear Diffie-Hellman Inversion (BDHI) assumption.

### 3.1 Bilinear Diffie-Hellman Assumption

The BDH problem [Jou00, BF01] in $\mathbb{G}$ is as follows: given a tuple $g, g^a, g^b, g^c \in \mathbb{G}$ as input, output $e(g, g)^{abc} \in \mathbb{G}_1$. An algorithm $\mathcal{A}$ has advantage $\epsilon$ in solving BDH in $\mathbb{G}$ if

$$\Pr\left[\mathcal{A}(g, g^a, g^b, g^c) = e(g, g)^{abc}\right] \geq \epsilon$$

where the probability is over the random choice of $a, b, c$ in $\mathbb{Z}_p^*$ and the random bits used by $\mathcal{A}$. Similarly, we say that an algorithm $\mathcal{B}$ that outputs $b \in \{0, 1\}$ has advantage $\epsilon$ in solving the *decision* BDH problem in $\mathbb{G}$ if

$$\left|\Pr\left[\mathcal{B}(g, g^a, g^b, g^c, e(g, g)^{abc}) = 0\right] - \Pr\left[\mathcal{B}(g, g^a, g^b, g^c, T) = 0\right]\right| \geq \epsilon$$

where the probability is over the random choice of $a, b, c$ in $\mathbb{Z}_p^*$, the random choice of $T \in \mathbb{G}_1^*$, and the random bits of $\mathcal{B}$.

**Definition 3.** *We say that the (Decision) $(t, \epsilon)$-BDH assumption holds in $\mathbb{G}$ if no $t$-time algorithm has advantage at least $\epsilon$ in solving the (Decision) BDH problem in $\mathbb{G}$.*

Occasionally we drop the $t$ and $\epsilon$ and refer to the BDH and Decision BDH assumptions in $\mathbb{G}$.

### 3.2 Bilinear Diffie-Hellman Inversion Assumption

The $q$-BDHI problem in the group $\mathbb{G}$ is defined as follows: given the $(q+1)$-tuple $(g, g^x, g^{(x^2)}, \ldots, g^{(x^q)}) \in (\mathbb{G}^*)^{q+1}$ as input, compute $e(g,g)^{1/x} \in \mathbb{G}_1^*$. An algorithm $\mathcal{A}$ has advantage $\epsilon$ in solving $q$-BDHI in $\mathbb{G}$ if

$$\Pr\left[\mathcal{A}(g, g^x, \ldots, g^{(x^q)}) = e(g,g)^{1/x}\right] \geq \epsilon$$

where the probability is over the random choice of $x$ in $\mathbb{Z}_p^*$ and the random bits of $\mathcal{A}$. Similarly, we say that an algorithm $\mathcal{B}$ that outputs $b \in \{0,1\}$ has advantage $\epsilon$ in solving the *decision* $q$-BDHI problem in $\mathbb{G}$ if

$$\left|\Pr\left[\mathcal{B}(g, g^x, \ldots, g^{(x^q)}, e(g,g)^{1/x}) = 0\right] - \Pr\left[\mathcal{B}(g, g^x, \ldots, g^{(x^q)}, T) = 0\right]\right| \geq \epsilon$$

where the probability is over the random choice of $x$ in $\mathbb{Z}_p^*$, the random choice of $T \in \mathbb{G}_1^*$, and the random bits of $\mathcal{B}$.

**Definition 4.** *We say that the (Decision) $(t, q, \epsilon)$-BDHI assumption holds in $\mathbb{G}$ if no $t$-time algorithm has advantage at least $\epsilon$ in solving the (Decision) $q$-BDHI problem in $\mathbb{G}$.*

Occasionally we drop the $t$ and $\epsilon$ and refer to the $q$-BDHI and Decision $q$-BDHI assumptions. It is easy to show that the 1-BDHI assumption is equivalent to the standard Bilinear Diffie-Hellman assumption (BDH). It is not known if the $q$-BDHI assumption, for $q > 1$, is equivalent to BDH.

## 4 Efficient Selective Identity HIBE Based on BDH Without Random Oracles

We construct an efficient HIBE system that is selective identity secure without random oracles based on the Decision BDH assumption. In particular, this implies an efficient selective identity, chosen ciphertext secure IBE based on Decision BDH without random oracles.

### 4.1 Construction

Let $\mathbb{G}$ be a bilinear group of prime order $p$ and $g$ be a generator of $\mathbb{G}$ (the security parameter determines the size of $\mathbb{G}$). Let $e : \mathbb{G} \times \mathbb{G} \to \mathbb{G}_1$ be the bilinear map. For now, we assume public keys (ID) of depth $\ell$ are vectors of elements in $\mathbb{Z}_p{}^{\ell}$. We write $\mathsf{ID} = (\mathrm{I}_1, \ldots, \mathrm{I}_\ell) \in \mathbb{Z}_p{}^\ell$. The $j$-th component corresponds to the identity at level $j$. We later extend the construction to public keys over $\{0,1\}^*$ by first hashing each component $\mathrm{I}_j$ using a collision resistant hash $H : \{0,1\}^* \to \mathbb{Z}_p$. We also assume messages to be encrypted are elements in $\mathbb{G}_1$. The HIBE system works as follows:

***Setup***$(\ell)$**:** To generate system parameters for an HIBE of maximum depth $\ell$, select a random $\alpha \in \mathbb{Z}_p^*$ and set $g_1 = g^\alpha$. Next, pick random elements $h_1, \ldots, h_\ell \in \mathbb{G}$ and a generator $g_2 \in \mathbb{G}^*$. The public parameters *params* and the secret *master-key* are given by

$$\mathit{params} = (g, g_1, g_2, h_1, \ldots, h_\ell)\ , \qquad \mathit{master\text{-}key} = g_2^\alpha$$

For $j = 1, \ldots, \ell$, we define $F_j : \mathbb{Z}_p \to \mathbb{G}$ to be the function: $F_j(x) = g_1^x h_j$.

***KeyGen***$(d_{\mathsf{ID}|j-1}, \mathsf{ID})$**:** To generate the private key $d_{\mathsf{ID}}$ for an identity $\mathsf{ID} = (\mathrm{I}_1, \ldots, \mathrm{I}_j) \in \mathbb{Z}_p{}^j$ of depth $j \le \ell$, pick random $r_1, \ldots, r_k \in \mathbb{Z}_p$ and output

$$d_{\mathsf{ID}} = \left( g_2^\alpha \cdot \prod_{k=1}^{j} F_k(\mathrm{I}_k)^{r_k},\ g^{r_1},\ \ldots,\ g^{r_j} \right)$$

Note that the private key for $\mathsf{ID}$ can be generated just given a private key for $\mathsf{ID}_{|j-1} = (\mathrm{I}_1, \ldots, \mathrm{I}_{j-1}) \in \mathbb{Z}_p{}^{j-1}$, as required. Indeed, let $d_{\mathsf{ID}|j-1} = (d_0, \ldots, d_{j-1})$ be the private key for $\mathsf{ID}_{|j-1}$. To generate $d_{\mathsf{ID}}$ pick a random $r_j \in \mathbb{Z}_p$ and output $d_{\mathsf{ID}} = (d_0 \cdot F_j(\mathrm{I}_j)^{r_j},\ d_1, \ldots,\ d_{j-1},\ g^{r_j})$.

***Encrypt(params,* ID, M)**: To encrypt a message $M \in \mathbb{G}_1$ under the public key $\mathsf{ID} = (\mathrm{I}_1, \ldots, \mathrm{I}_j) \in \mathbb{Z}_p{}^j$, pick a random $s \in \mathbb{Z}_p$ and output

$$C = \left( e(g_1, g_2)^s \cdot M,\ \ g^s,\ \ F_1(\mathrm{I}_1)^s,\ \ \ldots,\ \ F_j(\mathrm{I}_j)^s \right)$$

Note that $e(g_1, g_2)$ can be precomputed once and for all so that encryption does not require any pairing computations. Alternatively, $e(g_1, g_2)$ can be included in the system parameters.

***Decrypt***$(d_{\mathsf{ID}}, C)$**:** Let $\mathsf{ID} = (\mathrm{I}_1, \ldots, \mathrm{I}_j)$ be an identity. To decrypt a ciphertext $C = (A, B, C_1, \ldots, C_j)$ using the private key $d_{\mathsf{ID}} = (d_0, d_1, \ldots, d_j)$, output

$$A \cdot \frac{\prod_{k=1}^{j} e(C_j, d_j)}{e(B, d_0)} = M$$

Indeed, for a valid ciphertext, we have

$$\frac{\prod_{k=1}^{j} e(C_j, d_j)}{e(B, d_0)} = \frac{\prod_{k=1}^{j} e(F_k(\mathrm{I}_k), g)^{s r_k}}{e(g, g_2)^{s\alpha} \prod_{k=1}^{j} e(g, F_k(\mathrm{I}_k))^{s r_k}} = \frac{1}{e(g_1, g_2)^s}$$

## 4.2 Security

The HIBE system above is reminiscent of the Gentry-Silverberg HIBE which is only known to be secure in the random oracle model. Surprisingly, our choice of functions $F_1, \ldots, F_\ell$ enables us to prove security *without random oracles.* We prove security of our HIBE system under the standard Decision BDH assumption in $\mathbb{G}$.

**Theorem 1.** *Suppose the $(t, \epsilon)$-Decision BDH assumption holds in $\mathbb{G}$. Then the previously defined $\ell$-HIBE system is $(t', q_S, \epsilon)$-selective identity, chosen plaintext* (IND-sID-CPA) *secure for arbitrary $\ell$ and $q_S$ and any $t' < t - o(t)$.*

*Proof.* Suppose $\mathcal{A}$ has advantage $\epsilon$ in attacking the HIBE system. We build an algorithm $\mathcal{B}$ that solves the Decision BDH problem in $\mathbb{G}$. On input $(g, g^a, g^b, g^c, T)$ algorithm $\mathcal{B}$'s goal is to output 1 if $T = e(g,g)^{abc}$ and 0 otherwise. Let $g_1 = g^a$, $g_2 = g^b$, $g_3 = g^c$. Algorithm $\mathcal{B}$ works by interacting with $\mathcal{A}$ in a selective identity game as follows:

**Initialization.** The selective identity game begins with $\mathcal{A}$ first outputting an identity $\mathsf{ID}^* = (\mathrm{I}_1^*, \ldots, \mathrm{I}_k^*) \in \mathbb{Z}_p{}^k$ of depth $k \leq \ell$ that it intends to attack. If necessary, $\mathcal{B}$ appends random elements in $\mathbb{Z}_p$ to $\mathsf{ID}^*$ so that $\mathsf{ID}^*$ is a vector of length $\ell$.

**Setup.** To generate the system parameters, algorithm $\mathcal{B}$ picks $\alpha_1, \ldots, \alpha_\ell \in \mathbb{Z}_p$ at random and defines $h_j = g_1^{-\mathrm{I}_j^*} g^{\alpha_j} \in \mathbb{G}$ for $j = 1, \ldots, \ell$. It gives $\mathcal{A}$ the system parameters $params = (g, g_1, g_2, h_1, \ldots, h_\ell)$. Note that the corresponding master-key, which is unknown to $\mathcal{B}$, is $g_2^a = g^{ab} \in \mathbb{G}^*$. As before, for $j = 1, \ldots, \ell$ we define $F_j : \mathbb{Z}_p \to \mathbb{G}$ to be the function

$$F_j(x) = g_1^x h_j = g_1^{x - \mathrm{I}_j^*} g^{\alpha_j}$$

**Phase 1.** $\mathcal{A}$ issues up to $q_S$ private key queries. Consider a query for the private key corresponding to $\mathsf{ID} = (\mathrm{I}_1, \ldots, I_u) \in \mathbb{Z}_p{}^u$ where $u \leq \ell$. The only restriction is that $\mathsf{ID}$ is not a prefix of $\mathsf{ID}^*$. Let $j$ be the smallest index such that $\mathrm{I}_j \neq \mathrm{I}_j^*$. Necessarily $1 \leq j \leq u$. To respond to the query, algorithm $\mathcal{B}$ first derives a private key for the identity $(\mathrm{I}_1, \ldots, \mathrm{I}_j)$ from which it then constructs a private key for the requested identity $\mathsf{ID} = (\mathrm{I}_1, \ldots, \mathrm{I}_j, \ldots, \mathrm{I}_u)$. Algorithm $\mathcal{B}$ picks random elements $r_1, \ldots, r_j \in \mathbb{Z}_p$ and sets

$$d_0 = g_2^{\frac{-\alpha_j}{\mathrm{I}_j - \mathrm{I}_j^*}} \prod_{v=1}^{j} F_v(\mathrm{I}_v)^{r_v}, \quad d_1 = g^{r_1}, \quad \ldots, \quad d_{j-1} = g^{r_{j-1}}, \quad d_j = g_2^{\frac{-1}{\mathrm{I}_j - \mathrm{I}_j^*}} g^{r_j}$$

We claim that $(d_0, d_1, \ldots, d_j)$ is a valid random private key for $(\mathrm{I}_1, \ldots, \mathrm{I}_j)$. To see this, let $\tilde{r}_j = r_j - b/(\mathrm{I}_j - \mathrm{I}_j^*)$. Then we have that

$$g_2^{\frac{-\alpha_j}{(\mathrm{I}_j - \mathrm{I}_j^*)}} F_j(\mathrm{I}_j)^{r_j} = g_2^{\frac{-\alpha_j}{(\mathrm{I}_j - \mathrm{I}_j^*)}} (g_1^{\mathrm{I}_j - \mathrm{I}_j^*} g^{\alpha_j})^{r_j} = g_2^a (g_1^{\mathrm{I}_j - \mathrm{I}_j^*} g^{\alpha_j})^{r_j - \frac{b}{\mathrm{I}_j - \mathrm{I}_j^*}} = g_2^a F_j(\mathrm{I}_j)^{\tilde{r}_j}$$

It follows that the private key $(d_0, d_1, \ldots, d_j)$ defined above satisfies

$$d_0 = g_2^a \cdot (\prod_{v=1}^{j-1} F_v(\mathrm{I}_v)^{r_v}) \cdot F_j(\mathrm{I}_j)^{\tilde{r}_j}, \quad d_1 = g^{r_1}, \quad \ldots, \quad d_{j-1} = g^{r_{j-1}}, \quad d_j = g^{\tilde{r}_j}$$

where $r_1, \ldots, r_{j-1}, \tilde{r}_j$ are uniform in $\mathbb{Z}_p$. This matches the definition for a private key for $(\mathrm{I}_1, \ldots, \mathrm{I}_j)$. Hence, $(d_0, d_1, \ldots, d_j)$ is a valid private key for $(\mathrm{I}_1, \ldots, \mathrm{I}_j)$. Algorithm $\mathcal{B}$ derives a private key for the requested ID from the private key $(d_0, d_1, \ldots, d_j)$ and gives the result to $\mathcal{A}$.

**Challenge.** When $\mathcal{A}$ decides that Phase 1 is over, it outputs two messages $M_0, M_1 \in \mathbb{G}_1$. Algorithm $\mathcal{B}$ picks a random bit $b \in \{0, 1\}$ and responds with the ciphertext $C = (M_b \cdot T,\ g_3,\ g_3^{\alpha_1}, \ldots,\ g_3^{\alpha_k})$. Since $F_i(\mathrm{I}_i^*) = g^{\alpha_i}$ for all $i$, we have that

$$C = (M_b \cdot T,\ \ g^c,\ \ F_1(\mathrm{I}_1^*)^c,\ \ \ldots,\ \ F_k(\mathrm{I}_k^*)^c)$$

Hence, if $T = e(g, g)^{abc} = e(g_1, g_2)^c$ then $C$ is a valid encryption of $M_b$ under the public key $\mathsf{ID}^* = (\mathrm{I}_1^*, \ldots, \mathrm{I}_k^*)$. Otherwise, $C$ is independent of $b$ in the adversary's view.

**Phase 2.** $\mathcal{A}$ issues its complement of private key queries not issued in Phase 1. Algorithm $\mathcal{B}$ responds as before.

**Guess.** Finally, $\mathcal{A}$ outputs a guess $b' \in \{0, 1\}$. Algorithm $\mathcal{B}$ concludes its own game by outputting a guess as follows. If $b = b'$ then $\mathcal{B}$ outputs 1 meaning $T = e(g, g)^{abc}$. Otherwise, it outputs 0 meaning $T \neq e(g, g)^{abc}$.

When $T = e(g, g)^{abc}$ then $\mathcal{A}$ must satisfy $|\Pr[b = b'] - 1/2| > \epsilon$. When $T$ is uniform in $\mathbb{G}_1^*$ then $\Pr[b = b'] = 1/2$. Therefore, when $a, b, c$ are uniform in $\mathbb{Z}_p^*$ and $T$ is uniform in $\mathbb{G}_1^*$ we have that

$$\left|\Pr\left[\mathcal{B}(g, g^a, g^b, g^c, e(g, g)^{abc}) = 0\right] - \Pr\left[\mathcal{B}(g, g^a, g^b, g^c, T) = 0\right]\right| \geq \left|(\frac{1}{2} \pm \epsilon) - \frac{1}{2}\right| = \epsilon$$

as required. This completes the proof of Theorem 1. □

## 4.3 Chosen Ciphertext Security

A recent result of Canetti et al. [CHK04] gives an efficient way to build a selective identity, chosen ciphertext $\ell$-HIBE from a selective identity, chosen plaintext $(\ell + 1)$-HIBE. In combination with the above construction, we obtain a selective identity, chosen ciphertext $\ell$-HIBE for any $\ell$. In particular, we can easily construct an efficient selective identity, chosen ciphertext secure IBE without random oracles.

## 4.4 Arbitrary Identities

We can extend our HIBE above to handle identities $\mathsf{ID} = (\mathrm{I}_1, \ldots, \mathrm{I}_\ell)$ with $\mathrm{I}_j \in \{0, 1\}^*$ (as opposed to $I_j \in \mathbb{Z}_p$) by first hashing each $I_j$ using a collision resistant hash function $H : \{0, 1\}^* \to \mathbb{Z}_p$ prior to key generation and encryption. A standard argument shows that if the scheme above is selective identity, chosen ciphertext secure then so is the scheme with the additional hash function. We note that there is no need for a full domain hash into $\mathbb{Z}_p$; for example, a collision resistant hash function $H : \{0, 1\}^* \to \{1, \ldots, 2^b\}$ where $2^b < p$ is sufficient for the security proof.

# 5 More Efficient Selective Identity IBE Based on BDHI Without Random Oracles

We construct an efficient IBE system that is selective identity, chosen plaintext secure without random oracles based on the Decision $q$-BDHI assumption (see Section 3.2). The resulting IBE system is more efficient that the IBE construction in the previous section.

## 5.1 Basic Construction

Let $\mathbb{G}$ be a bilinear group of prime order $p$ and $g$ be a generator of $\mathbb{G}$. For now, we assume that the public keys (ID) are elements in $\mathbb{Z}_p^*$. We show later that arbitrary identities in $\{0,1\}^*$ can be used by first hashing ID using a collision resistant hash $H : \{0,1\}^* \to \mathbb{Z}_p^*$. We also assume that the messages to be encrypted are elements in $\mathbb{G}_1$. The IBE system works as follows:

***Setup*:** To generate IBE parameters, select random elements $x, y \in \mathbb{Z}_p^*$ and define $X = g^x$ and $Y = g^y$. The public parameters *params* and the secret *master-key* are given by

$$params = (g, g^x, g^y)\,, \qquad master\text{-}key = (x, y)$$

***KeyGen*(*master-key*,ID):** To create a private key for the public key $\mathsf{ID} \in \mathbb{Z}_p^*$:

1. pick a random $r \in \mathbb{Z}_p$ and compute $K = g^{1/(\mathsf{ID}+x+ry)} \in \mathbb{G}$,
2. output the private key $d_{\mathsf{ID}} = (r, K)$.

In the unlikely event that $x + ry + \mathsf{ID} = 0 \pmod p$, try again with a new random value for $r$.

***Encrypt*(*params*, ID, M):** To encrypt a message $M \in \mathbb{G}_1$ under public key $\mathsf{ID} \in \mathbb{Z}_p^*$, pick a random $s \in \mathbb{Z}_p^*$ and output the ciphertext

$$C = \left(g^{s \cdot \mathsf{ID}} X^s,\ \ Y^s,\ \ e(g,g)^s \cdot M\right)$$

Note that $e(g,g)$ can be precomputed once and for all so that encryption does not require any pairing computations.

***Decrypt*($d_{\mathsf{ID}}, C$):** To decrypt a ciphertext $C = (A, B, C)$ using the private key $d_{\mathsf{ID}} = (r, K)$, output $C/e(AB^r, K)$. Indeed, for a valid ciphertext we have

$$\frac{C}{e(AB^r, K)} = \frac{C}{e(g^{s(\mathsf{ID}+x+ry)}, g^{1/(\mathsf{ID}+x+ry)})} = \frac{C}{e(g,g)^s} = M$$

*Performance.* In terms of efficiency, we note that the ciphertext size and encryption time are similar to the IBE system of the previous section. However, decryption requires only one pairing computation, as opposed to two in the previous section.

### 5.2 Proving Security

We prove security of the scheme under the Decision $q$-BDIII assumption from Section 3.2.

**Theorem 2.** *Suppose the $(t, q, \epsilon)$-Decision BDHI assumption holds in $\mathbb{G}$. Then the previously defined IBE system is $(t', q_S, \epsilon)$-selective identity, chosen plaintext* (IND-sID-CPA) *secure for any $q_S < q$ and $t' < t - o(t)$.*

*Proof.* Suppose $\mathcal{A}$ has advantage $\epsilon$ in attacking the IBE system. We build an algorithm $\mathcal{B}$ that uses $\mathcal{A}$ to solve the decision $q$-BDHI problem in $\mathbb{G}$. On input $(g, g^\alpha, g^{\alpha^2}, \ldots, g^{\alpha^q}, T) \in (\mathbb{G}^*)^{q+1} \times \mathbb{G}_1^*$ for some unknown $\alpha \in \mathbb{Z}_p^*$, the goal of $\mathcal{B}$ is to output 1 if $T = e(g, g)^{1/\alpha}$ and 0 otherwise. It does so by interacting with $\mathcal{A}$ in a selective identity game as follows:

**Preparation.** Algorithm $\mathcal{B}$ builds a generator $h \in \mathbb{G}^*$ for which it knows $q-1$ pairs of the form $(w_i, h^{1/(\alpha+w_i)})$ for random $w_1, \ldots, w_{q-1} \in \mathbb{Z}_p^*$. This is done as follows:

1. Pick random $w_1, \ldots, w_{q-1} \in \mathbb{Z}_p^*$ and let $f(z)$ be the polynomial $f(z) = \prod_{i=1}^{q-1}(z + w_i)$. Expand the terms of $f$ to get $f(z) = \sum_{i=0}^{q-1} c_i x^i$. The constant term $c_0$ is non-zero.
2. Compute $h = \prod_{i=0}^{q-1}(g^{(\alpha^i)})^{c_i} = g^{f(\alpha)}$ and $u = \prod_{i=1}^{q}(g^{(\alpha^i)})^{c_{i-1}} = g^{\alpha f(\alpha)}$. Note that $u = h^\alpha$.
3. Check that $h \in \mathbb{G}^*$. Indeed if we had $h = 1$ in $\mathbb{G}$ this would mean that $w_j = -\alpha$ for some easily identifiable $w_j$, at which point $\mathcal{B}$ would be able to solve the challenge directly. We thus assume that all $w_j \neq -\alpha$.
4. Observe that for any $i = 1, \ldots, q-1$, it is easy for $\mathcal{B}$ to construct the pair $(w_i, h^{1/(\alpha+w_i)})$. To see this, write $f_i(z) = f(z)/(z + w_i) = \sum_{i=0}^{q-2} d_i z^i$. Then $h^{1/(\alpha+w_i)} = g^{f_i(\alpha)} = \prod_{i=0}^{q-2}(g^{(\alpha^i)})^{d_i}$.
5. Next, $\mathcal{B}$ computes

$$T_h = T^{(c_0^2)} \cdot T_0 \qquad \text{where} \qquad T_0 = \prod_{i=0}^{q-1}\prod_{j=0}^{q-2} e\left(g^{(\alpha^i)}, g^{(\alpha^j)}\right)^{c_i c_{j+1}}$$

Observe that if $T = e(g, g)^{1/\alpha}$ then $T_h = e\left(g^{f(\alpha)/\alpha}, g^{f(\alpha)}\right) = e(h, h)^{1/\alpha}$. On the contrary, if $T$ is uniform in $\mathbb{G}_1^*$, then $T_h$ is uniform in $\mathbb{G}_1 \setminus \{T_0\}$.

We will be using the values $h, u, T_h$ and the pairs $(w_i, h^{1/(\alpha+w_i)})$ for $i = 1, \ldots, q-1$ throughout the simulation.

**Initialization.** The selective identity game begins with $\mathcal{A}$ first outputting an identity $\mathsf{ID}^* \in \mathbb{Z}_p^*$ that it intends to attack.

**Setup.** To generate the system parameters *params* $= (g, X, Y)$, algorithm $\mathcal{B}$ does the following:

1. Pick random $a, b \in \mathbb{Z}_p^*$ under the constraint that $ab = \mathsf{ID}^*$.
2. Compute $X = u^{-a}h^{-ab} = h^{-a(\alpha+b)}$ and $Y = u = h^\alpha$.
3. Publish *params* $= (h, X, Y)$ as the public parameters. Note that $X, Y$ are independent of $\mathsf{ID}^*$ in the adversary's view.

4. We implicitly define $x = -a(\alpha + b)$ and $y = \alpha$ so that $X = h^x$ and $Y = h^y$. Algorithm $\mathcal{B}$ does not know the value of $x$ or $y$, but does know the value of $x + ay = -ab = -\mathsf{ID}^*$.

**Phase 1.** $\mathcal{A}$ issues up to $q_S < q$ private key queries. Consider the $i$-th query for the private key corresponding to public key $\mathsf{ID}_i \neq \mathsf{ID}^*$. We need to respond with a private key $(r, h^{1/(\mathsf{ID}_i + x + ry)})$ for a uniformly distributed $r \in \mathbb{Z}_p$. Algorithm $\mathcal{B}$ responds to the query as follows:

1. Let $\left(w_i, h^{1/(\alpha+w_i)}\right)$ be the $i$-th pair constructed during the preparation step. Define $h_i = h^{1/(\alpha+w_i)}$.
2. $\mathcal{B}$ first constructs an $r \in \mathbb{Z}_p$ satisfying $(r - a)(\alpha + w_i) = \mathsf{ID}_i + x + ry$. Plugging in the values of $x$ and $y$ the equation becomes

$$(r - a)(\alpha + w_i) = \mathsf{ID}_i - a(\alpha + b) + r\alpha$$

We see that $\alpha$ cancels from the equation and we get $r = a + \frac{\mathsf{ID}_i - ab}{w_i} \in \mathbb{Z}_p$.

3. Now, $(r,\ h_i^{1/(r-a)})$ is a valid private key for $\mathsf{ID}_i$ for two reasons. First,

$$h_i^{1/(r-a)} = (h^{1/(\alpha+w_i)})^{1/(r-a)} = h^{1/(r-a)(\alpha+w_i)} = h^{1/(\mathsf{ID}_i+x+ry)}$$

as required. Second, $r$ is uniformly distributed among all elements in $\mathbb{Z}_p$ for which $\mathsf{ID}_i + x + ry \neq 0$ and $r \neq a$. This is true since $w_i$ is uniform in $\mathbb{Z}_p \setminus \{0, -\alpha\}$ and is currently independent of $\mathcal{A}$'s view. Algorithm $\mathcal{B}$ gives $\mathcal{A}$ the private key $(r,\ h_i^{1/(r-a)})$.

We point out that this procedure will fail to produce the private key for $\mathsf{ID}^*$ since in that case we get $r - a = 0$. Hence, $\mathcal{B}$ can generate private keys for *all* public keys except for $\mathsf{ID}^*$.

**Challenge.** $\mathcal{A}$ outputs two messages $M_0, M_1 \in \mathbb{G}_1$. Algorithm $\mathcal{B}$ picks a random bit $b \in \{0,1\}$ and a random $\ell \in \mathbb{Z}_p^*$. It responds with the ciphertext $CT = (h^{-a\ell},\ h^{\ell},\ T_h^{\ell} \cdot M_b)$. Define $s = \ell/\alpha$. On the one hand, if $T_h = e(h,h)^{1/\alpha}$ we have

$$\begin{aligned} h^{-a\ell} &= h^{-a\alpha(\ell/\alpha)} = h^{(x+ab)(\ell/\alpha)} = h^{(x+\mathsf{ID}^*)(\ell/\alpha)} = h^{s\mathsf{ID}^*} \cdot X^s \\ h^{\ell} &= Y^{\ell/\alpha} = Y^s \\ T_h^{\ell} &= e(h,h)^{\ell/\alpha} = e(h,h)^s \end{aligned}$$

It follows that $CT$ is a valid encryption of $M_b$ under $\mathsf{ID}^*$, with the uniformly distributed randomization value $s = \ell/\alpha \in \mathbb{Z}_p^*$. On the other hand, when $T_h$ is uniform in $\mathbb{G}_1 \setminus \{T_0\}$, then, in the adversary's view, $CT$ is independent of the bit $b$.

**Phase 2.** $\mathcal{A}$ issues more private key queries, for a total of at most $q_S < q$. Algorithm $\mathcal{B}$ responds as before.

**Guess.** Finally, $\mathcal{A}$ outputs a guess $b' \in \{0,1\}$. If $b = b'$ then $\mathcal{B}$ outputs 1 meaning $T = e(g,g)^{1/\alpha}$. Otherwise, it outputs 0 meaning $T \neq e(g,g)^{1/\alpha}$.

We showed that when the input $T$ satisfies $T = e(g,g)^{1/\alpha}$ then $T_h = e(h,h)^{1/\alpha}$ in which case $\mathcal{A}$ must satisfy $|\Pr[b = b'] - 1/2| > \epsilon$. On the other hand, when $T$ is uniform and independent in $\mathbb{G}_1^*$ then $T_h$ is uniform and independent in $\mathbb{G}_1 \setminus \{T_0\}$ in which case $\Pr[b = b'] = 1/2$. Therefore, when $x$ is uniform in $\mathbb{Z}_p^*$ and $P$ is uniform in $\mathbb{G}_1^*$ we have that

$$\left|\Pr\left[\mathcal{B}(g, g^x, \ldots, g^{(x^q)}, e(g,g)^{1/x}) = 0\right] - \Pr\left[\mathcal{B}(g, g^x, \ldots, g^{(x^q)}, P) = 0\right]\right| \geq |(\frac{1}{2} \pm \epsilon) - \frac{1}{2}| = \epsilon$$

as required. This completes the proof of Theorem 2. □

### 5.3 Chosen-Ciphertext Security and Arbitrary Identities

Canetti et al. [CHK03, Section 2.2] describe a general method for converting a selective identity, chosen plaintext secure IBE into a selective identity, chosen ciphertext secure IBE. The method is based on [NY90, Sah99, Lin03]. Since it is generic, it applies to our system as well. In particular, the method can be used to render the IBE system above secure against chosen ciphertext attacks. The result is an IND-sID-CCA secure IBE without random oracles. However, the resulting system is inefficient since it relies on generic non-interactive zero-knowledge (NIZK) constructions.

As before, a standard argument shows that we can extend the IBE above to handle arbitrary identities $\mathsf{ID} \in \{0,1\}^*$ by first hashing ID using a collision resistant hash function $H : \{0,1\}^* \to \mathbb{Z}_p^*$ prior to key generation and encryption. If the underlying scheme is selective identity, chosen plaintext (resp. ciphertext) secure, then so is the scheme with the additional hash function.

### 5.4 An Efficient CCA2-secure Public-Key System

A recent result of Canetti et al. [CHK04] gives a general method for constructing a CCA2 public key system from any selective identity, chosen plaintext IBE. Essentially the same result was used in Section 4 to transform our first HIBE construction into a chosen ciphertext secure HIBE of lesser depth.

When used on the construction of this section, we obtain a new efficient CCA2 public key system. We briefly summarize its characteristics:

1. Encryption time: Dominated by three exponentiations in $\mathbb{G}$.
2. Decryption time: Dominated by one pairing computation.
3. Ciphertext size: Composed of three elements of $\mathbb{G}$ plus a public key and signature of a one-time signature scheme.

In terms of performance, this is comparable to, though not quite as efficient as, the Cramer-Shoup [CS98] CCA2-secure public key system which is proven secure in the standard model.

The ciphertext size can be further reduced by using the short signature scheme recently proposed by Boneh and Boyen [BB04] instead of the one-time signatures suggested by Canetti et al. [CHK04]. The Boneh-Boyen signature scheme is existentially unforgeable in the strong sense (sUF-CMA) without random oracle, and thus satisfies the requirements of the CCA2 construction. Here, strong existential unforgeability means that it is infeasible for an adversary to forge a new signature even on messages for which one or more valid signatures are already known.

## 6 DHI and Generalized Diffie-Hellman

In Section 3.2 we defined the $q$-BDHI problem in a bilinear group. A closely related problem is the $q$-Diffie-Hellman Inversion ($q$-DHI) problem: given a tuple $(g, g^x, g^{(x^2)}, \ldots, g^{(x^q)}) \in \mathbb{G}^{q+1}$ as input, output $g^{1/x} \in \mathbb{G}$. Here, $\mathbb{G}$ need not be a bilinear group. Loosely speaking, the $q$-DHI assumption states that the $q$-DHI problem is intractable in $\mathbb{G}$. This assumption was previously used in [MSK02] where it was called weak Diffie-Hellman.

Many cryptographic constructions rely on the Generalized Diffie-Hellman assumption (GenDH) for security [MSW96, NR97, BBR99, Lys02, BS03]. In this section we show that the $q$-DHI assumption implies the $(q+1)$-Generalized Diffie-Hellman assumption. Thus, constructions that rely on Generalized Diffie-Hellman could instead rely on $q$-DHI which appears to be a more natural complexity assumption, and is easier to state since the problem description does not require an oracle.

We first review the GenDH assumption. The assumption says that given $g^{a_1}, \ldots, g^{a_q}$ in $\mathbb{G}$ and given all the subset products $g^{\prod_{i \in S} a_i} \in \mathbb{G}$ for any strict subset $S \subset \{1, \ldots, q\}$, it is hard to compute $g^{a_1 \cdots a_q} \in \mathbb{G}$. Since the number of subset products is exponential in $q$, access to all these subset products is provided through an oracle. For a vector $\boldsymbol{a} = (a_1, \ldots, a_q) \in \mathbb{Z}_p{}^q$, define $\mathcal{O}_{g,\boldsymbol{a}}$ to be an oracle that for any strict subset $S \subset \{1, \ldots, q\}$ responds with

$$\mathcal{O}_{g,\boldsymbol{a}}(S) = g^{\prod_{i \in S} a_i} \in \mathbb{G}.$$

Define the advantage of algorithm $\mathcal{A}$ in solving the generalized Diffie-Hellman problem to be the probability that $\mathcal{A}$ is able to compute $g^{a_1 \cdots a_q}$ given access to the oracle $\mathcal{O}_{g,\boldsymbol{a}}(S)$. In other words,

$$\mathrm{Adv}_{\mathcal{A},q} = \Pr[\mathcal{A}^{\mathcal{O}_{g,\boldsymbol{a}}} = g^{a_1 \cdots a_q} \; : \; \boldsymbol{a} = (a_1, \ldots, a_q) \leftarrow \mathbb{Z}_p{}^q]$$

Note that the oracle only answers queries for strict subsets of $\{1, \ldots, q\}$.

**Definition 5.** *We say that $\mathbb{G}$ satisfies the $(t, q, \epsilon)$-Generalized Diffie-Hellman assumption if for all $t$-time algorithms $\mathcal{A}$ we have $Adv_{\mathcal{A},q} < \epsilon$.*

**Theorem 3.** *Suppose the $(t, q-1, \epsilon)$-DHI assumption holds in $\mathbb{G}$. Then the $(t, q, \epsilon)$-GenDH assumption also holds in $\mathbb{G}$.*

*Proof.* Suppose $\mathcal{A}$ is an algorithm that has advantage $\epsilon$ in solving the $q$-GenDH problem. We construct an algorithm $\mathcal{B}$ that solves $(q-1)$-DHI with the same advantage $\epsilon$. Algorithm $\mathcal{B}$ is given $g, g^x, g^{(x^2)}, \ldots, g^{(x^{q-1})} \in \mathbb{G}$ and its goal is to compute $g^{1/x} \in \mathbb{G}$. Let $h = g^{(x^{q-1})}$ and $y = x^{-1} \in \mathbb{Z}_p$. Then the input to $\mathcal{B}$ can be re-written as $h, h^y, h^{(y^2)}, \ldots, h^{(y^{q-1})} \in \mathbb{G}$ and $\mathcal{B}$'s goal is to output $h^{(y^q)} = g^{1/x}$.

Algorithm $\mathcal{B}$ first picks $q$ random values $c_1, \ldots, c_q \in \mathbb{Z}_p$. It then runs algorithm $\mathcal{A}$ and simulates the oracle $\mathcal{O}_{h,\boldsymbol{a}}$ for $\mathcal{A}$. The vector $\boldsymbol{a}$ that $\mathcal{B}$ will use is $\boldsymbol{a} = (y + c_1, \ldots, y + c_q)$. Note that $\mathcal{B}$ does not know $\boldsymbol{a}$ explicitly since $\mathcal{B}$ does not have $y$. When $\mathcal{A}$ issues a query for $\mathcal{O}_{h,\boldsymbol{a}}(S)$ for some strict subset $S \subset \{1, \ldots, q\}$ algorithm $\mathcal{B}$ responds as follows:

1. Define the polynomial $f(z) = \prod_{i \in S}(z + c_i)$ and expand the terms to obtain $f(z) = \sum_{i=0}^{|S|} b_i z^i$.
2. Compute $t = \prod_{i=0}^{|S|}(h^{(y^i)})^{b_i} = h^{f(y)}$. Since $|S| < q$ all the values $h^{(y^i)}$ in the product are known to $\mathcal{B}$.
3. By construction we know that $t = h^{\prod_{i \in S}(y + c_i)}$. Algorithm $\mathcal{B}$ responds by setting $\mathcal{O}_{h,\boldsymbol{a}}(S) = t$.

The responses to all of the adversary's oracle queries are consistent with the hidden vector $\boldsymbol{a} = (y + c_1, \ldots, y + c_q)$. Therefore, eventually, $\mathcal{A}$ will output $T = h^{\prod_{i=1}^{q}(y+c_i)}$. Define the polynomial $f(z) = \prod_{i=1}^{q}(z + c_i)$ and expand the terms to get $f(z) = z^q + \sum_{i=0}^{q-1} b_i z^i$. To conclude, $\mathcal{B}$ outputs

$$T \Big/ \prod_{i=0}^{q-1} (h^{(y^i)})^{b_i} = h^{(y^q)}$$

which is the required value. □

The same property as in Theorem 3 also holds for the decision version of the DHI and GenDH problems. The $q$-DHI assumption is easier to state than the $q$-GenDH assumption since there is no need for an oracle. When appropriate, constructions that depend on GenDH for security could instead use the DHI assumption.

## 7 Conclusions

We constructed two IBE systems that are secure against selective identity attacks in the standard model, i.e., without using random oracles. The first construction is based on the now classic BDH assumption. It extends readily to give a selective identity HIBE without random oracles, that can efficiently be made chosen ciphertext secure using a technique of [CHK04]. The second construction is based on the Bilinear Diffie-Hellman Inversion assumption. The same technique of [CHK04] converts both our constructions into efficient CCA2-secure public key systems without random oracles that are almost as efficient as the Cramer-Shoup public key system.

Currently, the problem of constructing a fully secure IBE (against adaptive identity attacks) without resorting to random oracles is still open. We hope to see this question resolved soon.

## References


[BB04] Dan Boneh and Xavier Boyen. Short signatures without random oracles. In Christian Cachin and Jan Camenisch, editors, *Proceedings of Eurocrypt 2004*, LNCS. Springer-Verlag, 2004.

[BBR99] Eli Biham, Dan Boneh, and Omer Reingold. Breaking generalized Diffie-Hellman modulo a composite is no easier than factoring. *IPL*, 70:83–87, 1999.

[BF01] Dan Boneh and Matt Franklin. Identity-based encryption from the Weil pairing. In Joe Kilian, editor, *Proceedings of Crypto 2001*, volume 2139 of *LNCS*, pages 213–29. Springer-Verlag, 2001.

[BF03] Dan Boneh and Matt Franklin. Identity-based encryption from the Weil pairing. *SIAM J. of Computing*, 32(3):586–615, 2003.

[BR93] Mihir Bellare and Phil Rogaway. Random oracle are practical: A paradigm for designing efficient protocols. In *Proceedings of the First ACM Conference on Computer and Communications Security*, pages 62–73, 1993.

[BS03] Dan Boneh and Alice Silverberg. Applications of multilinear forms to cryptography. *Contemporary Mathematics*, 324:71–90, 2003.

[CHK03] Ran Canetti, Shai Halevi, and Jonathan Katz. A forward-secure public-key encryption scheme. In *Proceedings of Eurocrypt 2003*, volume 2656 of *LNCS*. Springer-Verlag, 2003.

[CHK04] Ran Canetti, Shai Halevi, and Jonathan Katz. Chosen-ciphertext security from identity-based encryption. In *Proceedings of Eurocrypt 2004*, LNCS, 2004. http://eprint.iacr.org/2003/182/.

[Coc01] Clifford Cocks. An identity based encryption scheme based on quadratic residues. In *Proceedings of the 8th IMA International Conference on Cryptography and Coding*, pages 26–8, 2001.

[CS98] Ronald Cramer and Victor Shoup. A practical public key cryptosystem provably secure against adaptive chosen ciphertext attacks. In Hugo Krawczyk, editor, *Proceedings of Crypto 1998*, volume 1462 of *LNCS*, pages 13–25. Springer-Verlag, 1998.

[GS02] Craig Gentry and Alice Silverberg. Hierarchical ID-based cryptography. In *Proceedings of Asiacrypt 2002*, 2002.

[HL02] J. Horwitz and B. Lynn. Towards hierarchical identity-based encryption. In *Proceedings of Eurocrypt 2002*, pages 466–481, 2002.

[Jou00] Antoine Joux. A one round protocol for tripartite Diffie-Hellman. In Wieb Bosma, editor, *Proceedings of ANTS IV*, volume 1838 of *LNCS*, pages 385–94. Springer-Verlag, 2000.

[Lin03] Yehuda Lindell. A simpler construction of CCA2-secure public-key encryption under general assumptions. In *Proceeings of Eurocrypt '03*, volume 2656 of *LNCS*, pages 241–254, 2003.

[Lys02] Anna Lysyanskaya. Unique signatures and verifiable random functions from the DH-DDH separation. In *Proceedings of Crypto 2002*, LNCS. Springer-Verlag, 2002.

[MSK02] Shigeo Mitsunari, Ryuichi Sakai, and Masao Kasahara. A new traitor tracing. *IEICE Trans. Fundamentals*, E85-A(2):481–484, 2002.

[MSW96] G. Tsudik M. Steiner and M. Waidner. Diffie-Hellman key distribution extended to groups. In *Proceedings 1996 ACM Conference on Computer and Communications Security*, 1996.

[NR97] Moni Naor and Omer Reingold. Number-theoretic constructions of efficient pseudo-random functions. In *Proceeings 38th IEEE Symp. on Foundations of Computer Science*, pages 458–467, 1997.

[NY90] Moni Naor and Moti Yung. Public key cryptosystems provable secure against chosen ciphertext attacks. In *STOC '90*, pages 427–437. ACM, 1990.

[Sah99] Amit Sahai. Non-malleable non-interactive zero knowledge and adaptive chosen-ciphertext security. In *Proceeings 40 IEEE Symp. on Foundations of Computer Science*, 1999.

[Sha84] Adi Shamir. Identity-based cryptosystems and signature schemes. In *Proceedings of Crypto '84*, volume 196 of *LNCS*, pages 47–53. Springer-Verlag, 1984.

# Construction of Secure Random Curves of Genus 2 over Prime Fields

Pierrick Gaudry[1] and Éric Schost[2]

[1] Laboratoire LIX, École polytechnique, France
gaudry@lix.polytechnique.fr
[2] Laboratoire STIX, École polytechnique, France
Eric.Schost@polytechnique.fr

**Abstract.** For counting points of Jacobians of genus 2 curves defined over large prime fields, the best known method is a variant of Schoof's algorithm. We present several improvements on the algorithms described by Gaudry and Harley in 2000. In particular we rebuild the symmetry that had been broken by the use of Cantor's division polynomials and design a faster division by 2 and a division by 3. Combined with the algorithm by Matsuo, Chao and Tsujii, our implementation can count the points on a Jacobian of size 164 bits within about one week on a PC.

## 1 Introduction

Genus 2 hyperelliptic curves provide an interesting alternative to elliptic curves for the design of discrete-log based cryptosystems. Indeed, for a similar security, the key or signature lengths are the same as for elliptic curves and furthermore the size of the base field in which the computations take place is twice smaller. During the last years, efforts in improving the group law algorithms made these cryptosystems quite competitive [19,25].

To ensure the security of the system, it is required to have a group of large prime order. Until recently, for the Jacobian of a genus 2 curve, only specific constructions provided curves with known Jacobian order, namely the complex multiplication (CM) method [34] and the Koblitz curves. These curves have a very special structure; although nobody knows if they are weaker than general curves, it is pertinent to consider random curves as well. This raises the problem of point-counting: given a random curve, find the group order of its Jacobian.

With today's state of the art, the complexity of the point counting task in genus 2 highly depends on the size of the characteristic of the base field: in short, the smaller the characteristic, the easier the task of point counting ("easy" means fast and does not mean that the theoretical tools are simple).

In the case of genus 2 curves in small characteristic $p$, the point counting problem was recently solved using $p$-adic methods [31,23,20]. The particular case where $p = 2$ is in fact treated almost as quickly as in genus 1. Unfortunately, these dramatic improvements do not apply when $p$ becomes too large (say, a few thousands [10]).

C. Cachin and J. Camenisch (Eds.): EUROCRYPT 2004, LNCS 3027, pp. 239–256, 2004.

For large $p$, the best known algorithms are variants of Schoof's algorithm, theoretical descriptions of which can be found in [26,18,1,16]. In 2000, Gaudry and Harley [11] designed and implemented the first practical genus 2 Schoof algorithm, making use of Cantor's division polynomials [8]. To reach reasonable sizes, however, it was necessary to combine the Schoof approach with a Pollard lambda method. Their record was a random genus 2 curve over a prime field of size about $10^{19}$, thus too small to be used in a cryptosystem. For "medium characteristic", they also proposed to use the Cartier-Manin operator to get additional information that can be combined with others. Therefore, for medium characteristic $p$ (say $10^9$, see [5]), point counting is easier than for very large $p$.

We mentioned that in the non-small characteristic case, once the group order has been computed modulo some large integer, the computation is finished using a Pollard lambda method. Matsuo, Chao and Tsujii [21] proposed a Baby-step/Giant-step algorithm that speeds up this last phase. With this device and using the Cartier-Manin trick, they performed a point counting computation of cryptographical size for a medium characteristic field.

In this paper, we improve on the methods of [11], so that, combined with the algorithm of [21], we can reach cryptographical size over prime fields. Our improvements are concerned with the construction and the manipulation of torsion elements in the Schoof-like algorithm of [11]. The impact of these improvements is asymptotically by a constant factor, but they yield significant speed-up in practice for the size of interest in cryptography. We now summarize them:

Our first contribution is the reintroduction of symmetries that were lost in [11]. Indeed, the use of Cantor's division polynomials to construct torsion elements is very efficient, but the resulting divisor is given as a sum of points instead of in Mumford representation. Therefore a factor of 2 in the degrees of the polynomials that are manipulated is lost. In Sections 3.2 and 3.3, we give algorithms to save this factor of 2 in the degrees.

In [11], it is proposed to build $2^k$-torsion elements using a halving algorithm based on Gröbner basis computations. Our second contribution is a faster division by 2, using a better representation of the system; in the same spirit we show that a division by 3 can also be done: this is described in Section 4. Another practical improvement is the ubiquitous use of an explicit action on the roots coming from the group law to speed-up the factorizations that occur at different stages. We explain it in details in the case of the division by 2 in Section 3.4.

To illustrate and to test the performance of our improvements, we implemented them in Magma or NTL and mixed them with the algorithm of [21] and an early abort strategy. Our main outcome is the first construction of secure random curves of genus 2 over a prime field, as we obtained Jacobians of prime order of size about $2^{164}$.

## 2 Generalities

In this work, $p$ denotes a fixed odd prime, $\mathbb{F}_p$ is the finite field with $p$ elements, and $C$ is a genus 2 curve defined by the equation $y^2 = f(x)$, where $f$ is a squarefree monic polynomial in $\mathbb{F}_p[X]$ of degree 5. The main object we consider is the Jacobian $\mathbf{J}(\mathcal{C})$ of $\mathcal{C}$. We handle elements of $\mathbf{J}(\mathcal{C})$ through their Mumford representation: each element of $\mathbf{J}(\mathcal{C})$ can be uniquely represented by a pair of polynomials $\langle u(x), v(x)\rangle$, where $u$ is monic of degree at most 2, $v$ is of degree less than the degree of $u$, and $u$ divides $v^2 - f$. The degree of the $u$-polynomial in Mumford's representation is called the *weight* of a divisor. If $K$ is an extension field of $\mathbb{F}_p$, we may distinguish the curves defined on $K$ and $\mathbb{F}_p$, by denoting them $\mathcal{C}/K$ and $\mathcal{C}/\mathbb{F}_p$; the Jacobians are correspondingly denoted by $\mathbf{J}(\mathcal{C}/K)$ and $\mathbf{J}(\mathcal{C}/\mathbb{F}_p)$. For precise definitions and algorithms for the group law, we refer to [22] and [7,19].

Let $\overline{\mathbb{F}_p}$ be an algebraic closure of $\mathbb{F}_p$ and let us consider the Frobenius endomorphism on $\mathbf{J}(\mathcal{C}/\overline{\mathbb{F}_p})$ denoted by $\pi$. By Weil's theorem (see [24]), the characteristic polynomial $\chi(T)$ of $\pi$ has the form $\chi(T) = T^4 - s_1T^3 + s_2T^2 - ps_1T + p^2$, where $s_1$ and $s_2$ are integers such that $|s_1| \leq 4\sqrt{p}$ and $|s_2| \leq 6p$. Furthermore $\#\mathbf{J}(\mathcal{C}) = \chi(1) = p^2 + 1 - s_1(p+1) + s_2$.

In point-counting algorithms based on Schoof's idea [27], the torsion elements of $\mathbf{J}(\mathcal{C})$ play an important role. If $N$ is a positive integer, the subgroup of $N$-torsion elements of $\mathbf{J}(\mathcal{C}/\overline{\mathbb{F}_p})$ is a finite group denoted by $\mathbf{J}(\mathcal{C})[N]$; it is isomorphic to $(\mathbb{Z}/N\mathbb{Z})^4$ and has the structure of a free $\mathbb{Z}/N\mathbb{Z}$-module of dimension 4 (see [24]). Furthermore, the characteristic polynomial of the restriction of $\pi$ to $\mathbf{J}(\mathcal{C})[N]$ is $\chi(T) \bmod N$. Applying this to different small primes or prime powers leads to the genus 2 Schoof algorithm that is sketched in Algorithm 1.

**Algorithm 1** Sketch of a genus 2 Schoof algorithm

1. For sufficiently many small primes or prime powers $\ell$:
   (a) Let $L = \{(s_1, s_2);\ \ s_1, s_2 \in [0, \ell - 1]\}$.
   (b) While $\#L > 1$ do
   - Construct a new $\ell$-torsion divisor $D$;
   - Eliminate those elements $(s_1, s_2)$ in $L$ such that

   $$\pi^4(D) - s_1\pi^3(D) + s_2\pi^2(D) - (ps_1 \bmod \ell)\pi(D) + (p^2 \bmod \ell)D \neq 0$$

   (c) Deduce $\chi(T) \bmod \ell$ from the remaining pair in $L$.
2. Deduce $\chi(T)$ from the pairs $(\ell, \chi(T) \bmod \ell)$ by Chinese remaindering, or using the algorithm of [21].

Our contribution is to improve the first part of the algorithm, the construction of $\ell$-torsion divisors; the computations for small primes and prime powers are respectively described in Sections 3 and 4.

We will frequently make genericity assumptions on the curve $\mathcal{C}$ and its torsion divisors. We assume that $\mathcal{C}$ is chosen randomly among genus 2 curves defined over a large field $\mathbb{F}_p$, so we can expect that with high probability, such assumptions are satisfied. The cases when our assumptions fail should require special treatments, which are not developed here.

For the complexity estimates, we denote by $\mathsf{M}(d)$ the number of $\mathbb{F}_p$-operations required to multiply two polynomials of degree $d$ defined over $\mathbb{F}_p$. We make the classical assumptions on $\mathsf{M}$ (see for instance [32, Definition 8.26]). In the sequel, if no precise reference is given for an algorithm, then it can be found in [32], together with a complexity analysis in terms of $\mathsf{M}$.

## 3 Computation Modulo a Small Prime $\ell$

In the classical Schoof algorithm for elliptic curves, a formal $\ell$-torsion point is used: the computations are made with a point $P = (x, y)$, where $x$ cancels the $\ell$-th division polynomial $\psi_\ell$ and $y$ is linked to $x$ by the equation of the curve. In other words, we work in a rank 2 polynomial algebra quotiented by two relations: $\mathbb{F}_p[x, y]/(\psi_\ell(x), y^2 - (x^3 + ax + b))$.

In genus 2, we imitate this strategy. According to [18], it is enough to consider the $\ell$-torsion divisors of weight 2 (this is not surprising since a generic divisor has weight 2). Let thus $D$ be a weight 2 divisor given in Mumford representation, $D = \langle x^2 + u_1 x + u_0, v_1 x + v_0 \rangle$. Then there exists a radical ideal $I_\ell$ of $\mathbb{F}_p[U_1, U_0, V_1, V_0]$ such that

$$D \in \mathbf{J}(\mathcal{C})[\ell] \iff \varphi(u_1, u_0, v_1, v_0) = 0, \ \forall \varphi \in I_\ell.$$

By analogy with elliptic division polynomials, this ideal $I_\ell$ is called the *$\ell$-th division ideal.* There are $\ell^4 - 1$ non-zero $\ell$-torsion elements, so that $I_\ell$ has dimension 0 and degree at most $\ell^4 - 1$; generically, by the Manin-Mumford conjecture [15, p. 435], all non-zero torsion divisors have weight 2, so the degree of $I_\ell$ is exactly $\ell^4 - 1$.

From the computational point of view, a good choice for a generating set of $I_\ell$ is a Gröbner basis for a lexicographic order. Using the order $U_1 < U_0 < V_1 < V_0$, we can actually predict the shape of this Gröbner basis. Indeed, if $D$ is an $\ell$-torsion divisor, then its opposite $-D$ is also $\ell$-torsion, so it has the same $u$-coordinates, and opposite $v$-coordinates. Furthermore, we make the genericity assumption that all the pairs $\{D, -D\}$ of $\ell$-torsion divisors have different values for $u_1$. Then, the Gröbner basis for the ideal $I_\ell$ takes the form

$$I_\ell = \begin{cases} V_0 - V_1 S_0(U_1) \\ V_1^2 - S_1(U_1) \\ U_0 - R_0(U_1) \\ R_1(U_1), \end{cases}$$

where $R_1$ is a squarefree polynomial of degree $(\ell^4 - 1)/2$ and $R_0, S_1, S_0$ are polynomials of degree at most $(\ell^4 - 1)/2 - 1$. If such a Gröbner basis for $I_\ell$ is known, then it is not difficult to imitate Schoof's algorithm, by working in the quotient algebra $\mathbb{F}_p[U_1, U_0, V_1, V_0]/I_\ell$. Unfortunately, no easy computable recurrence formulae are known that relate Gröbner bases of $\ell$-division ideals for different values of $\ell$, just like for division polynomials of elliptic curves. Therefore we shall start with the approach of [11] using Cantor's division polynomials and show that we can derive efficiently a multiple of $R_1$.

### 3.1 Cantor's Division Polynomials

Let us fix a prime $\ell$. Cantor's division polynomials [8] are polynomials in $\mathbb{F}_p[X]$, denoted by $d_0$, $d_1$, $d_2$, $e_0$, $e_1$, $\Delta$, with the following property: for a divisor $P = \langle x - x_P, y_P \rangle$ of weight 1, the multiplication of $P$ by $\ell$ in $\mathbf{J}(\mathcal{C})$ is given by

$$[\ell]P = \left\langle x^2 + \frac{d_1(x_P)}{d_2(x_P)}x + \frac{d_0(x_P)}{d_2(x_P)},\ y_P\left(\frac{e_1(x_P)}{\Delta(x_P)}x + \frac{e_0(x_P)}{\Delta(x_P)}\right)\right\rangle.$$

These polynomials have respective degrees $2\ell^2-1$, $2\ell^2-2$, $2\ell^2-3$, $3\ell^2-2$, $3\ell^2-3$, $3\ell^2-2$ and are easily computed by means of recurrence formulae. Even if a naive method is used, the cost of their computation is by far negligible compared to the subsequent operations.

Now, let $D = \langle x^2+U_1x+U_0, V_1x+V_0\rangle$ be a generic divisor of weight 2, where $U_1$, $U_0$, $V_1$, $V_0$ are indeterminates, subject to the condition that $x^2+U_1x+U_0$ divides $(V_1x+V_0)^2 - f$. The divisor $D$ can be written as the sum of two weight 1 divisors $P_1 = \langle x - X_1, Y_1\rangle$ and $P_2 = \langle x - X_2, Y_2\rangle$, where $U_1 = -(X_1+X_2)$, $U_0 = X_1X_2$, and where $Y_1$ and $Y_2$ satisfy $V_1X_1+V_0 = Y_1$ and $V_1X_2+V_0 = Y_2$. Since $D = P_1 + P_2$, then $D$ is $\ell$-torsion if and only if $[\ell]P_1 = -[\ell]P_2$.

Rewriting this equation using Cantor's division polynomials, we get four equations that must be satisfied for $D$ to be $\ell$-torsion. Some of these equations are multiples of $X_1 - X_2$: this is an artifact due to the splitting of $D$ into divisors of weight 1 and if this is the case one should divide out this factor. Hence we obtain the following system:

$$\begin{cases} E_1(X_1,X_2) &= (d_1(X_1)d_2(X_2) - d_1(X_2)d_2(X_1))/(X_1-X_2) = 0, \\ E_2(X_1,X_2) &= (d_0(X_1)d_2(X_2) - d_0(X_2)d_2(X_1))/(X_1-X_2) = 0, \\ F_1(X_1,X_2,Y_1,Y_2) = & Y_1e_1(X_1)e_0(X_2) + Y_2e_1(X_2)e_0(X_1) = 0, \\ F_2(X_1,X_2,Y_1,Y_2) = & Y_1e_2(X_1)e_0(X_2) + Y_2e_2(X_2)e_0(X_1) = 0. \end{cases}$$

Consider now the finite-dimensional $\mathbb{F}_p$-algebra

$$B = \mathbb{F}_p[X_1,X_2,Y_1,Y_2]/(E_1,E_2,F_1,F_2,Y_1^2 - f(X_1), Y_2^2 - f(X_2)).$$

In a generic situation, the minimal polynomial of $-(X_1+X_2)$ in $B$ is then precisely the polynomial $R_1$ that appears in the Gröbner basis of $I_\ell$ (failures could occur, *e.g.*, if there exists an $\ell$-torsion divisor $D = P_1 + P_2$, such that $[\ell]P_1$ is not of weight 2). We will see below that the whole Gröbner basis of $I_\ell$ is not necessary to the point-counting application we have in mind. Thus, we can start by working with the first two equations $E_1, E_2$, which involve $X_1, X_2$ only.

These polynomials were already considered in [11]. The strategy used in that paper consisted in computing the resultant of $E_1, E_2$ with respect to $X_2$ for a start, from which it was possible to deduce the coordinates of $[\ell]$-torsion divisors. This approach did not take into account the symmetry in $(X_1, X_2)$; we now show how to work directly in Mumford's coordinates $U_1 = -(X_1+X_2), U_0 = X_1X_2$, so as to compute resultants of lower degrees.

### 3.2 Resymmetrisation

The polynomials $E_1(X_1, X_2)$ and $E_2(X_1, X_2)$ are symmetric polynomials. It is well known that they can be expressed in terms of the two elementary symmetric polynomials $X_1X_2$ and $X_1 + X_2$. The heart of Mumford's representation is the use of this expression, but this had been broken in order to apply Cantor's division polynomials. We call *resymmetrisation* the method that we present now to come back to a representation of bivariate polynomials in terms of the elementary symmetric polynomials. This is not as trivial as it seems, since the naive schoolbook method to symmetrize a polynomial yields a complexity jump in our case.

Let us consider the unique polynomials $\mathfrak{E}_1$ and $\mathfrak{E}_2$ in $\mathbb{F}_p[U_0, U_1]$ such that $\mathfrak{E}_1(X_1X_2, -X_1 - X_2) = E_1(X_1, X_2)$ and $\mathfrak{E}_2(X_1X_2, -X_1 - X_2) = E_2(X_1, X_2)$ and let $\overline{R_1} \in \mathbb{F}_p[U_1]$ be their resultant with respect to $U_0$; then $R_1$ divides $\overline{R_1}$.

We want to use the following evaluation/interpolation techniques to compute $\overline{R_1}$: evaluate the variable $U_1$ at sufficiently many scalars $u_1$, compute the resultants of $\mathfrak{E}_1(U_0, u_1)$ and $\mathfrak{E}_2(U_0, u_1)$, and interpolate the results. Unfortunately, computing with $\mathfrak{E}_1$ and $\mathfrak{E}_2$ themselves has prohibitive cost, as these polynomials have $O(\ell^4)$ monomials. However, their specific shape yields the following workaround.

Let $h$ be a polynomial in $\mathbb{F}_p[X]$ and $X_1$ and $X_2$ be two indeterminates. Then the *divided differences* of $h$ are the bivariate symmetric polynomials

$$A_0(h) = \big(h(X_1) - h(X_2)\big)/(X_1 - X_2) \quad \text{and} \quad A_1(h) = \big(X_1h(X_2) - X_2h(X_1)\big)/(X_1 - X_2).$$

We let $\mathfrak{A}_0(h)$ and $\mathfrak{A}_1(h)$ be the unique polynomials in $\mathbb{F}_p[U_0, U_1]$ such that $\mathfrak{A}_0(h)(X_1X_2, -X_1 - X_2) = A_0(h)$ and $\mathfrak{A}_1(h)(X_1X_2, -X_1 - X_2) = A_1(h)$. Then a direct computation shows that

$$\begin{aligned} \mathfrak{E}_1 &= \mathfrak{A}_0(d_1)\,\mathfrak{A}_1(d_2) - \mathfrak{A}_0(d_2)\,\mathfrak{A}_1(d_1), \\ \mathfrak{E}_2 &= \mathfrak{A}_0(d_0)\,\mathfrak{A}_1(d_2) - \mathfrak{A}_0(d_2)\,\mathfrak{A}_1(d_0) \quad \text{in } \mathbb{F}_p[U_0, U_1]. \end{aligned}$$

Given an arbitrary polynomial $h$ in $\mathbb{F}_p[X]$ and $u_1 \in \mathbb{F}_p$, we show in the last paragraphs how to compute the polynomials $\mathfrak{A}_0(h)$ and $\mathfrak{A}_1(h)$ evaluated at $U_1 = u_1$ efficiently. Taking this operation for granted, we deduce Algorithm 2 for computing the resultant $\overline{R_1}$ of $\mathfrak{E}_1$ and $\mathfrak{E}_2$.

**Algorithm 2** Computation of the resultant $\overline{R_1}$

1. For $\deg(\overline{R_1}) + 1$ different values of $u_1 \in \mathbb{F}_p$, do
   (a) Compute $\mathfrak{A}_0(d_0), \mathfrak{A}_1(d_0), \mathfrak{A}_0(d_1), \mathfrak{A}_1(d_1), \mathfrak{A}_0(d_2), \mathfrak{A}_1(d_2)$ evaluated at $U_1 = u_1$.
   (b) Deduce $\mathfrak{E}_1$ and $\mathfrak{E}_2$, evaluated at $U_1 = u_1$.
   (c) Compute $\overline{R_1}(u_1)$ as the resultant in $U_0$ of $\mathfrak{E}_1$ and $\mathfrak{E}_2$.
2. Interpolate $\overline{R_1}$ from the pairs $(u_1, \overline{R_1}(u_1))$.

The classical estimates for the degrees of resultants imply that the degree of $\overline{R_1}$ is $6\ell^4 - 17\ell^2 + 12$; thus to be able to perform the interpolation, it is necessary to take at least $6\ell^4 - 17\ell^2 + 13$ different values of $u_1$. In practice, it is recommended to take a few more values of $u_1$, in order to check the computation. Note that the resultant of $E_1, E_2$ has degree $8\ell^4 - 22\ell^2 + 15$.

We finish this subsection by detailing our solution to the problem raised above: given $u_1$ in $\mathbb{F}_p$ and $h$ in $\mathbb{F}_p[X]$, how to compute the polynomials $\mathfrak{A}_0(h)$ and $\mathfrak{A}_1(h)$ evaluated at $U_1 = u_1$ efficiently? It is immediate to check the following identity:

$$h(X) = \mathfrak{A}_1(h)(U_0, u_1)X + \mathfrak{A}_0(h)(U_0, u_1) \mod (X^2 + u_1X + U_0).$$

Thus, the problem amounts to reduce $h$ modulo $X^2 + u_1X + U_0$ in $\mathbb{F}_p[U_0][X]$. Our solution relies on the following primitive: If $h$ is a polynomial of degree $N$ in $\mathbb{F}_p[X]$ and $a$ is a scalar in $\mathbb{F}_p$, then the coefficients of $h(X + a)$ can be deduced from the coefficients of $h(X)$ for one polynomial multiplication in degree $N$, see [2]. We call this primitive `var-shift`.

The main idea is now to rewrite the relation $X^2 + u_1X + U_0 = 0$ in the form $(X + u_1/2)^2 = u_1^2/4 - U_0$. Let $Y = X + u_1/2$, and $k$ in $\mathbb{F}_p[X]$ such that $h(X) = k(Y)$. We group the coefficients of $k$ according to the parity of their indices, forming the polynomials $k_{\text{odd}}$ and $k_{\text{even}}$ such that $k(Y) = k_{\text{even}}(Y^2) + Y k_{\text{odd}}(Y^2)$. Taking $h$ modulo $X^2 + u_1X + U_0$, we have

$$h(X) \equiv k_{\text{even}}\left(\frac{u_1^2}{4} - U_0\right) + \left(X + \frac{u_1}{2}\right) k_{\text{odd}}\left(\frac{u_1^2}{4} - U_0\right).$$

Thus, computing $\mathfrak{A}_0(h)$ and $\mathfrak{A}_1(h)$ can be done by Algorithm 3 below.

**Algorithm 3** Reduction of $h(X)$ modulo $X^2 + u_1X + U_0$ in $\mathbb{F}_p[U_0][X]$

1. Compute $k$ from $h$ using `var-shift`.
2. Decompose $k$ in $k_{\text{odd}}$ and $k_{\text{even}}$.
3. Compute $k_{\text{even}}\left(u_1^2/4 - U_0\right)$ and $k_{\text{odd}}\left(u_1^2/4 - U_0\right)$ using `var-shift`.
4. Recombine their coefficients to get $h(X) \bmod X^2 + u_1X + U_0$.

### 3.3 Parasites Prediction and Removal

In [11] it is shown that a factor of the resultant of $E_1, E_2$ can be predicted and used to speed-up the computation. This prediction is still possible in the context of the resymmetrisation, and the factor of $\overline{R_1}$ corresponding to such roots can be computed efficiently. The roots of this factor of $\overline{R_1}$ are called parasites: they are not the $U_1$-coordinates of an $\ell$-torsion divisor, and actually appear as a by-product of our elimination scheme. Thus, they can be safely factored out.

If $x_1$ and $x_2$ in $\overline{\mathbb{F}_p}$ cancel $d_2$, then $E_1(x_1, x_2) = E_2(x_1, x_2) = 0$. The $U_1$ coordinates corresponding to these solutions can be written as $-(x_1 + x_2)$ where $x_1$ and $x_2$ are roots of $d_2$. Hence we obtain the following factor $\rho$ of $\overline{R_1}$:

$$\rho(U_1) = \prod_{d_2(x_1)=0} \prod_{d_2(x_2)=0} (U_1 + x_1 + x_2).$$

The factor $\rho$ is a *parasite*, as generically it does not lead to any $\ell$-torsion divisor. Then $\rho$ divides $\overline{R_1}$ but not $R_1$, so we lose nothing in eliminating it from $\overline{R_1}$. The polynomial $\rho$ is computed using an algorithm of [4] dedicated to such questions. Then Step 2. in Algorithm 2 is replaced by the interpolation of $\overline{R_1}/\rho$ from the pairs $(u_1, \overline{R_1}(u_1)/\rho(u_1))$.

The degree of $\rho$ is $4\ell^4 - 12\ell^2 + 9$, so the degree of $\overline{R_1}/\rho$ is $2\ell^4 - 5\ell^2 + 3$, reducing by a factor of about 3 the number of values of $u_1$ that have to be considered in Algorithm 2. As an output, we now have at our disposal the polynomial $\mathfrak{R}_1 = \overline{R_1}/\rho$, which is a multiple of $R_1$. For comparison, the resultant computed in [11] has degree $4\ell^4 - 10\ell^2 + 6$, which is twice the degree of $\mathfrak{R}_1$.

### 3.4 Factorization and Reconstruction of a Torsion Element

Once the resultant $\mathfrak{R}_1$ has been computed, the task is not finished: indeed, what we want is the representation of an $\ell$-torsion divisor, so that we can plug it into the equation of the Frobenius endomorphism. Here, there are two possible strategies:

1. Refine $\mathfrak{R}_1$ to get exactly $R_1$ and reconstruct from it the whole Gröbner basis of $I_\ell$ describing a generic $\ell$-torsion divisor.
2. Look for small degree factors of $\mathfrak{R}_1$, check if they are indeed factors of $R_1$ and deduce the corresponding $\ell$-torsion divisors.

By analogy with Schoof's algorithm for elliptic curves, one would think that the first choice is the most pertinent. However, refining $\mathfrak{R}_1$ into $R_1$ can be a costly task, and if there exist indeed small factors of $\mathfrak{R}_1$, then the second solution is faster. That is the reason why we chose the second solution in our experiments described below. However, especially for $\ell = 17$ or 19, we could feel the limit of this choice. Therefore, for larger computations, we should probably switch to the first solution.

We now describe the second strategy with more details.

Let $u_1$ be a root of $\mathfrak{R}_1$ in an extension $\mathbb{F}_q$ of $\mathbb{F}_p$. We evaluate the polynomials $E_1$ and $E_2$ at $(X_1, -u_1 - X_1)$ in $\mathbb{F}_q[X_1]$, and obtain two univariate polynomials in $\mathbb{F}_q[X_1]$. Their GCD is (generically) a polynomial of degree 2 which might, or not, be the $u$-polynomial of an $\ell$-torsion divisor. To settle the question, we take into account the last two equations $F_1$ and $F_2$, and check that our candidate $u$-polynomial is compatible with them. If not, we try again and select another root of $\mathfrak{R}_1$. Otherwise, we deduce the $v$-polynomial, and build an $\ell$-torsion divisor defined over $\mathbb{F}_q$. It is then plugged into all possible candidates for the characteristic polynomial $\chi(T) \bmod \ell$ to detect the right one.

We now concentrate on the problem of finding irreducible factors of $\mathfrak{R}_1$, using classical ingredients of polynomial factorization. It is interesting to find the factors of small degree first, as it reduces the subsequent computation. Thus, we start by detecting the linear factors, given by $\gcd(X^p - X, \mathfrak{R}_1(X))$. If this GCD is non-trivial, then the corresponding roots are separated and processed before maybe continuing the factorization. Then factors of degree $d$ are detected for increasing $d$ by computing $\gcd(X^{p^d} - X, \mathfrak{R}_1(X))$, and when we find a root, it is used to try to build an $\ell$-torsion divisor that perhaps determines $\chi(T) \bmod \ell$.

This can be improved using the fact that the factorization pattern of $R_1$ is partly predictable. Indeed, due to the Galois structure induced by the group law in $\mathbf{J}(\mathcal{C})$, some factorization patterns are forbidden. We can then proceed as follows: we first precompute the list of all possible patterns corresponding to $\ell$ and $p$, and we start looking for irreducible factors by increasing degree as before. At each step, the number of factors we find eliminates some patterns in the list. Then we look in the remaining patterns for the next smallest possible degree and try directly to catch factors of that degree. If there is a large gap between the current degree and the next one, the Baby-step/Giant-step strategy of [30] using modular compositions can yield a significant speed-up compared to the classical powering algorithm.

As another application of the factorization patterns, we mention the influence of the choice of $p$: if $p = 1 \bmod \ell$, then we can infer that the smallest irreducible factor of $R_1$ has degree at most $(\ell^2 + 1)/2$, compared to possibly $O(\ell^4)$ in the general case. We do not give details on the determination of the possible patterns for lack of space. The idea is similar to the one used in [12] for modular equations.

### 3.5 Complexity

We start by evaluating the cost in $\mathbb{F}_p$-operations of one iteration of Step 1 in Algorithm 2. Using Algorithm 3, the cost of computing $\mathfrak{A}_0(d_0)$, $\mathfrak{A}_1(d_0)$, $\mathfrak{A}_0(d_1)$, $\mathfrak{A}_1(d_1)$, $\mathfrak{A}_0(d_2)$, $\mathfrak{A}_1(d_2)$ is $O(\mathsf{M}(\ell^2))$, since the $d_i$ have degree $O(\ell^2)$. Deducing $\mathfrak{E}_1$ and $\mathfrak{E}_2$ involves 4 more multiplications of polynomials of degree $O(\ell^2)$ at a cost of $O(\mathsf{M}(\ell^2))$. The resultant of $\mathfrak{E}_1$ and $\mathfrak{E}_2$ can then be computed using the HGCD algorithm at a cost of $O(\mathsf{M}(\ell^2) \log \ell)$.

Hence the resultant computation is dominating this step; this would not have been the case without the `var-shift` strategy. The loop in Step 1 must be repeated for $O(\ell^4)$ different values of $u_1$, so the cost of Step 1 is $O(\ell^4 \mathsf{M}(\ell^2) \log \ell)$ operations in $\mathbb{F}_p$. Step 2 is a degree $O(\ell^4)$ polynomial interpolation, which can be done using $O(\mathsf{M}(\ell^4) \log \ell)$ operations in $\mathbb{F}_p$.

We now evaluate the influence of the parasite prediction on the complexity. The polynomial $\rho$ is computed using the algorithm of [4] at a cost of $O(\mathsf{M}(\ell^4))$ operations. Then its evaluations at the $O(\ell^4)$ different values of $u_1$ can be deduced using $O(\mathsf{M}(\ell^4) \log \ell)$ operations in $\mathbb{F}_p$. Therefore, the cost of precomputing the effect of the parasite factor is negligible compared to the cost of computing $\overline{R_1}$.

Knowing the values of $\rho$ on the different values of $u_1$ allows to interpolate a polynomial of degree 3 times less. This yields a speed-up by a factor at least 3 (and even more in practice, depending on the function $\mathsf{M}$). Also the input of the factorization step is 3 times smaller, thus gaining a constant factor in that phase.

The factorization phase is less easy to analyze, since its complexity varies quite a lot depending on the degrees of the smallest irreducible factors. Denote by $d$ the degree of the smallest factor of $\mathfrak{R}_1$ that allows to deduce $\chi(T) \bmod \ell$.

By the powering algorithm, computing $\gcd(X^{p^d} - X, \mathfrak{R}_1(X))$ can be done using $O((d \log p + \log \ell)\mathsf{M}(\ell^4))$ operations in $\mathbb{F}_p$ and isolating one of the factors of degree $d$ has similar expected complexity. From an irreducible factor of degree $d$, the reconstruction of an $\ell$-torsion divisor $D$ defined over $\mathbb{F}_{p^d}$ requires to manipulate polynomials of degree $O(\ell^2)$ over $\mathbb{F}_{p^d}$, so it costs $O(\mathsf{M}(d\ell^2)\log \ell)$ operations in $\mathbb{F}_p$.

Finally, the detection of the invalid choices for $(s_1, s_2) \bmod \ell$ requires 4 applications of the Frobenius endomorphism to $D$ and $O(\ell)$ group operations in $\mathbf{J}(\mathcal{C}/\mathbb{F}_{p^d})$, that is $O((\ell \log d + \log p)\mathsf{M}(d))$ operations in $\mathbb{F}_p$.

If $d$ is small enough (say $d = O(\ell)$), this factoring strategy is satisfactory since its complexity is not worse than computing $\mathfrak{R}_1$, if $\log p$ is not too large. However, if $d$ is $O(\ell^4)$, then the above complexity estimate of the factoring step is catastrophic. Using the known factorization patterns is useful in this context, even if the precise analysis is complicated. We expect that working with the whole ideal $I_\ell$ (thus avoiding the factorization) is more suited for a proper analysis; cleaning all details of that approach is out of the scope of this article.

## 4 Computation Modulo Small Prime Powers

Given a prime $\ell$, from the knowledge of an $\ell$-torsion divisor in $\mathbf{J}(\mathcal{C})$, one can deduce $\ell^2$-torsion divisors by performing a division by $\ell$ in the Jacobian; iterating this process yields divisors of $\ell^3$-torsion, $\ell^4$-torsion, ... This can be used within Schoof's algorithm, so as to obtain modular information on the polynomial $\chi(T)$ modulo $\ell$, $\ell^2$, and so on. As appears below, there are many computational difficulties to overcome before this can be efficiently applied in practice. We mostly dedicated our efforts on the case $\ell = 2$, improving the techniques of [11], and spend much of this section describing this case. We thereafter briefly describe the case $\ell = 3$.

In the case $\ell = 2$, this lifting strategy was already used in [11]. It starts from the data of a 2-torsion divisor; then the iterative step is as follows. Suppose that a divisor $D_k$ of $2^k$-torsion is given; we denote by $\mathbb{F}_q$ the extension of the base field $\mathbb{F}_p$ over which $D_k$ is defined. We make the assumption that $D_k$ has weight 2, and write $D_k = \langle x^2 + u_1 x + u_0, v_1 x + v_0 \rangle$.

There are exactly $2^4 = 16$ divisors $D$ such that $[2]D = D_k$. Let us make the genericity assumption that all these divisors have weight 2, and introduce 4 indeterminates $U_1, U_0, V_1, V_0$ to denote the coordinates of $D$. Using doubling formulas coming from Cantor's addition algorithm, we obtain a system $\mathcal{F}_k$ that relates $D$ and $D_k$:

$$\mathcal{F}_k \left| \begin{array}{ll} H_1(U_1, U_0, V_1, V_0) = u_1, & G_1(U_1, U_0, V_1, V_0) = 0, \\ H_2(U_1, U_0, V_1, V_0) = u_0, & G_2(U_1, U_0, V_1, V_0) = 0, \\ H_3(U_1, U_0, V_1, V_0) = v_1, & \\ H_4(U_1, U_0, V_1, V_0) = v_0, & \end{array} \right.$$

where $H_1, H_2, H_3, H_4$ are rational functions, and $G_1, G_2$ are polynomials which specify that $x^2 + U_1 x + U_0$ divides $(V_1 x + V_0)^2 - f$. Cleaning denominators, we are left with a polynomial system in $U_1, U_0, V_1, V_0$, with $u_1, u_, v_1, v_0$ as parameters. We make the further genericity assumption that the ideal generated by $\mathcal{F}_k$ admits a Gröbner basis of the form

$$\begin{cases} V_0 - L_0(U_1) \\ V_1 - L_1(U_1) \\ U_0 - M_0(U_1) \\ M_1(U_1), \end{cases}$$

where $M_1 \in \mathbb{F}_q[U_1]$ has degree 16 and $L_0, L_1, M_0$ have degree at most 15. Since $D_k$ is $2^k$-torsion, this provides a description of 16 divisors of $2^{k+1}$-torsion.

The next step is to factorize the polynomial $M_1$ in $\mathbb{F}_q[U_1]$. Any factor of $M_1$ can be used to try and determine the characteristic polynomial $\chi \bmod 2^{k+1}$, but some of them might give no information. Let $r$ be one irreducible factor of lowest degree that allows the determination of $\chi \bmod 2^{k+1}$, $n$ its degree, and $u_1$ a root of $r$ in $\mathbb{F}_{q^n}$. Then the divisor $D_{k+1} = \langle x^2 + u_1 x + M_0(u_1), L_1(u_1)x + L_0(u_1)\rangle$ is of $2^{k+1}$-torsion. It can be used for the next loop of the algorithm.

From the computational point of view, the main tasks to perform at the $k$th step are the following: First, solve a zero-dimensional polynomial system of the form $[2]D = D_k$, then factorize a polynomial of degree 16. The following subsections detail our contributions on these questions. It should be clear that these computations are done with polynomials defined over an extension $\mathbb{F}_q$ of the base field $\mathbb{F}_p$, whose possibly high degree is the main cause of concern.

## 4.1 Performing a Division by 2

All the systems $\mathcal{F}_k$ that we consider are obtained in the same manner; as $k$ grows, only their right-hand sides vary. The difficulty comes from the fact that the field of definition of $u_1, u_0, v_1, v_0$ is an extension of $\mathbb{F}_p$ of possibly high degree.

The solution, suggested in [11], is to solve the system $\mathcal{F}_k$ for generic values $\mathbf{u_1}, \mathbf{u_0}, \mathbf{v_1}, \mathbf{v_0}$. There are of course only two degrees of freedom, as $x^2 + \mathbf{u_1}x + \mathbf{u_0}$ must divide $(\mathbf{v_1}x + \mathbf{v_0})^2 - f$. Working over the base field $\mathbb{F}_p(\mathbf{u_1}, \mathbf{u_0})$, we are thus led to consider the system $\mathcal{F}_{\text{gen}}$ in the unknowns $\mathbf{v_1}, \mathbf{v_0}, U_1, U_0, V_1, V_0$

$$\mathcal{F}_{\text{gen}} \left| \begin{array}{ll} H_1(U_1, U_0, V_1, V_0) = \mathbf{u_1}, & G_1(U_1, U_0, V_1, V_0) = 0, \\ H_2(U_1, U_0, V_1, V_0) = \mathbf{u_0}, & G_2(U_1, U_0, V_1, V_0) = 0, \\ H_3(U_1, U_0, V_1, V_0) = \mathbf{v_1}, & G_1(\mathbf{u_1}, \mathbf{u_0}, \mathbf{v_1}, \mathbf{v_0}) = 0, \\ H_4(U_1, U_0, V_1, V_0) = \mathbf{v_0}, & G_2(\mathbf{u_1}, \mathbf{u_0}, \mathbf{v_1}, \mathbf{v_0}) = 0, \end{array} \right.$$

where the last two equations express that $x^2 + \mathbf{u_1}x + \mathbf{u_0}$ divides $(\mathbf{v_1}x + \mathbf{v_0})^2 - f$. Generically, the solutions of this system can be represented the following way:

$$\mathcal{T} \begin{cases} V_0 - L_0(U_1, \mathbf{v_1}), \\ V_1 - L_1(U_1, \mathbf{v_1}), \\ U_0 - M_0(U_1, \mathbf{v_1}), \\ M_1(U_1, \mathbf{v_1}), \\ \mathbf{v_0} - N_1(\mathbf{v_1}), \\ N_0(\mathbf{v_1}). \end{cases}$$

All these polynomials have coefficients in $\mathbb{F}_p(\mathbf{u_1}, \mathbf{u_0})$. The polynomial $N_0$ has degree 4, $N_1$ has degree less than 4, $M_1$ has degree 16 in $U_1$ and less than 4 in $\mathbf{v_1}$ and $L_0, L_1, M_0$ have degree less than 16 (resp. 4) in $U_1$ (resp. $\mathbf{v_1}$).

Systems like $\mathcal{F}_{\text{gen}}$ that involve free variables are difficult to handle. A direct application of a Gröbner basis algorithm over $\mathbb{F}_p(\mathbf{u_1}, \mathbf{u_0})$ fails by lack of memory, so we used the algorithm of [28], dedicated to such situations, to compute $\mathcal{T}$. Once $\mathcal{T}$ is known, it can be specialized on the coordinates of the divisor $D_k$, realizing its division by 2.

The solution presented in [11] followed the same approach, with a notable difference: instead of considering the representation $\mathcal{T}$, another representation was used, which involved polynomials of degree 64. Our approach reduces this degree to 16, and makes the subsequent computations much easier.

In terms of complexity, the polynomials defining the system $\mathcal{F}_{\text{gen}}$ have degree bounded independently from $p$; thus, computing $\mathcal{T}$ takes a bounded number of operations in $\mathbb{F}_p$. Next, at each division step, we must specialize $\mathbf{u_1}, \mathbf{u_0}, \mathbf{v_1}, \mathbf{v_0}$ on the coordinate of the divisor $D_k$ in $\mathcal{T}$. If $D_k$ is defined in a degree $d$ extension of $\mathbb{F}_p$, then this substitution requires $O(\mathsf{M}(d))$ operations in $\mathbb{F}_p$.

### 4.2 Factorization Using the Action of the 2-Torsion

After performing the division by 2, we are left with a description of the solution set $V_k$ of the system $\mathcal{F}_k$ by means of the following representation:

$$M_1(U_1) = 0, \quad U_0 = M_0(U_1), \quad V_1 = L_1(U_1), \quad V_0 = L_0(U_1)$$

Now, we have to factorize the polynomial $M_1 \in \mathbb{F}_q[U_1]$. It has degree 16, which is moderate; the main issue is the degree of $\mathbb{F}_q$ over its prime field: in the computations presented below, $\mathbb{F}_q$ had degree up to 1280 on its prime field. We now show how to simplify this factorization, using the natural action of the 2-torsion group $\mathbf{J}(\mathcal{C})[2]$ on $V_k$, in the spirit of [14].

Let us see $U_1$ as a coordinate function on the set of weight 2 divisors (the choice of $U_1$ is arbitrary, but makes the computation easier). To any subgroup $G$ of $\mathbf{J}(\mathcal{C})[2]$, we associate the averaging operator $\mathbf{S}_G : D \mapsto \sum_{g \in G} U_1(D+g)$, which is defined as soon as all divisors $D + g$ have weight 2. Now, $G$ acts on $V_k$, and each orbit has cardinality $|G|$. The function $\mathbf{S}_G$ takes constant values on each orbit, so it takes at most $[\mathbf{J}(\mathcal{C})[2] : G]$ distinct values on $V_k$. By an additional genericity assumption, we may suppose that $\mathbf{S}_G$ takes precisely $[\mathbf{J}(\mathcal{C})[2] : G]$ distinct values on $V_k$.

To realize this algebraically, let us introduce the "divisor" $\mathbf{D}_0 = \langle x^2 + U_1 x + M_0, L_1 x + L_0 \rangle$, defined over $\mathbb{F}_q[U_1]/M_1$. Given any 2-torsion divisor $g$, we can apply the addition formulas to $\mathbf{D}_0$ and $g$, performing all operations in $\mathbb{F}_q[U_1]/M_1$ (the addition formulas require divisions, but if one of them fails it gives a proper factor of $M_1$). We obtain a "divisor" $\mathbf{D}_g = \langle x^2 + U_1^{(g)} x + U_0^{(g)}, V_1^{(g)} x + V_0^{(g)} \rangle$, where $U_1^{(g)}, U_0^{(g)}, V_1^{(g)}, V_0^{(g)}$ are in $\mathbb{F}_q[U_1]/M_1$; by construction, if $D$ is any divisor in $V_k$, then the $U_1$-coordinate of $D + g$ is obtained by evaluating $U_1^{(g)}$ on the $U_1$-coordinate of $D$. Let thus $s_G = \sum_{g \in G} U_1^{(g)} \in \mathbb{F}_q[U_1]/M_1$. Then for any $D \in V_k$, the value $\mathbf{S}_G(D)$ is obtained by evaluating $s_G$ on the $U_1$-coordinate of $D$. From the above discussion on the function $\mathbf{S}_G$, we deduce that the minimal polynomial of $s_G$ in $\mathbb{F}_q[U_1]/M_1$ has degree $[\mathbf{J}(\mathcal{C})[2] : G]$.

As an abstract group, $\mathbf{J}(\mathcal{C})[2]$ is isomorphic to $(\mathbb{Z}/2\mathbb{Z})^4$. Let us consider subgroups

$$G_1 \simeq (\mathbb{Z}/2\mathbb{Z}) \subset G_2 \simeq (\mathbb{Z}/2\mathbb{Z})^2 \subset G_3 \simeq (\mathbb{Z}/2\mathbb{Z})^3 \subset \mathbf{J}(\mathcal{C})[2] \simeq (\mathbb{Z}/2\mathbb{Z})^4.$$

Using the above construction, we associate to these subgroups the elements $s_1$, $s_2$, $s_3$ of $\mathbb{F}_q[U_1]/M_1$. Introducing their minimal polynomials, we deduce that the extension $\mathbb{F}_q \to \mathbb{F}_q[U_1]/M_1$ is isomorphic to the quotient of $\mathbb{F}_p[U_1, S_1, S_2, S_3]$ by some polynomials

$$\begin{cases} T_U(U_1, S_1, S_2, S_3) \\ T_1(S_1, S_2, S_3) \\ T_2(S_2, S_3) \\ T_3(S_3), \end{cases}$$

where all polynomials have degree 2 in their main variables, resp. $S_3, S_2, S_1, U_1$. Using this decomposition, we avoid the factorization of $M_1$: We start by factorizing $T_3$ over $\mathbb{F}_q$, and adjoin one of its roots to $\mathbb{F}_q$; then we factor $T_2$ over this new field, and so on. Thus, only the computation of $T_3, T_2, T_1, T_U$ and four square root extractions are needed.

Suppose that $q = p^d$; then all polynomials $T_3, T_2, T_1, T_U$ can be computed in $O(\mathsf{M}(d))$ operations in $\mathbb{F}_p$. For square-root extraction, we used a factorization algorithm quite similar to those of [33] and [17]. Using such algorithms, the expected complexity of extracting a square root in $\mathbb{F}_{p^d}$ is $O(C(d)\log(d) + \mathsf{M}(d)\log(p))$ operations in $\mathbb{F}_p$, where $C(d)$ denotes the cost of modular composition in degree $d$, so that $C(d) \in O(d^2 + \sqrt{d}\mathsf{M}(d))$, see [6]. One should note that this whole process only saves a constant factor over the factorization of $M_1$ from scratch; however, it was quite significant in practice.

In the worst case, after $k$ lifting steps, the degree $d$ might be of order $O(16^k)$. In this case, taking into account all previous estimates, the expected complexity to obtain a $2^k$-torsion is expected to be in $O(kC(16^k) + \mathsf{M}(16^k)\log(p))$ base field operations. However, our experiments showed that with a surprising amount of uniformity, the degree of this extension was actually in $O(2^k)$, so the above complexity bound was by far overestimated.

### 4.3 Performing a Division by 3

Most of what was described above extends *mutatis mutandis* to arbitrary $\ell$. Nevertheless, the computations become much more difficult: even for $\ell = 3$, we did not solve the system describing the division of a generic divisor by 3. Thus, we used the plain strategy to divide torsion divisors by 3, by means of successive Gröbner bases computations, over extensions of $\mathbb{F}_p$ of increasing degrees. As the tables below reveal, the time required for solving these polynomial systems makes this approach much more delicate than for 2-torsion. As a consequence, we did not implement the equivalent of our improved factorization process, and used a plain factorization strategy.

## 5 Implementation and Experiments

We implemented a whole point-counting algorithm including all the above-mentioned improvements and the MCT algorithm [21], first within the Magma computer algebra system [3]. Then, the critical parts of the computation modulo small primes and the MCT algorithm were implemented in C++ using the NTL library [29]. The communication between different parts of the program is done using files for small communications or named pipes in the case of a heavy interaction. For instance, the analysis of the factorization pattern of the resultant $R_1$ is implemented in a Magma program that sends elementary factoring tasks (like a modular composition) to a running NTL program.

To test our program we ran it on several randomly chosen curves defined over $\mathbb{F}_p$ with $p = 5 \times 10^{24} + 8503491$, with the hope to find some cryptographically secure Jacobians. An early abort strategy was used to eliminate curves $\mathcal{C}$ for which either the Jacobian order or the Jacobian order of the twisted curve was discovered to be non-prime. In particular, $f$ must be irreducible to ensure the oddity of the group orders.

We have computed the characteristic polynomials of 32 randomly chosen curves, that yield 64 group orders, taking into account the twists. Due to the early abort strategy, these group orders are not divisible by any prime less than or equal to 19. Among them, 7 were found to be primes, meaning that the corresponding Jacobians are secure against all known attacks. One particular curve has the nice feature that both itself and its twist have a prime order Jacobian. The data for that curve can be found in the appendix.

Table 1 gives statistics for the runtimes of the different steps of the algorithm. They are given in seconds on a Pentium IV at 2.26 GHz having 1 GB of central memory. Due to the early abort strategy, the statistics for the factoring phase are made on less curves for larger $\ell$, *e.g.*, 39 curves for $\ell = 5$, versus 21 curves for $\ell = 19$. More curves were computed on different computers and were not taken into account for the statistics.

The modular composition used for factorization is done using Brent and Kung's algorithm [6]. For $\ell = 17$ and $\ell = 19$, the precomputation (Baby steps) is not balanced with the Giant steps due to memory constraints. This explains why the runtimes for those values look so bad compared to other values.

As for the torsion lifting, 2-torsion was much easier to handle than 3-torsion, as we computed divisors of order $1024 = 2^{10}$, versus $27 = 3^3$ only. The curves we used were selected so that they have 8-torsion defined over $\mathbb{F}_{p^{10}}$ and 3-torsion defined over $\mathbb{F}_{p^4}$. Then in almost all cases, the $2^i$-torsion divisors, $i \geq 3$, were defined in extensions of degrees $10, 20, 40, 80, \ldots$, and the $3^i$-torsion divisors in extensions of degrees $4, 12, 36, \ldots$

After the modular computations, we know $\chi(T) \bmod 44696171520 = 2^{10} \cdot 3^3 \cdot 5 \cdot 7 \cdot 11 \cdot 13 \cdot 17 \cdot 19$ ; for comparison's sake, note that in [11], the modular computation went to $3843840 = 2^8 \cdot 3 \cdot 5 \cdot 7 \cdot 11 \cdot 13$. To conclude; we run the MCT algorithm. Due to memory requirements, we used a Xeon at 2.66 GHz with 2 GB of memory; this computation takes about 3 hours and 1.7 GB per curve.

| Computations Modulo Small Primes | | | | | | | |
|---|---|---|---|---|---|---|---|
| $\ell$ | 5 | 7 | 11 | 13 | 17 | 19 | Theory |
| generic degree of $\mathfrak{R}_1$ | 1,128 | 4,560 | 28,680 | 56,280 | 165,600 | 258,840 | $2\ell^4 - 5\ell^2 + 3$ |
| generic degree of $\rho$ | 2,209 | 9,025 | 57,121 | 112,225 | 330,625 | 516,961 | $4\ell^4 - 12\ell^2 + 9$ |
| Time computing $\rho$ | 0.3 | 2 | 23 | 52 | 256 | 374 | $O(\mathsf{M}(\ell^4)\log \ell)$ |
| Time Step 1, Algo 1 | 6.5 | 63 | 1,504 | 5,072 | 34,869 | 69,162 | $O(\ell^4\mathsf{M}(\ell^2)\log \ell)$ |
| Time Step 2, Algo 1 | 0.3 | 2 | 17 | 39 | 182 | 275 | $O(\mathsf{M}(\ell^4)\log \ell)$ |
| Total time Algo 1 | 7.1 | 67 | 1,544 | 5,163 | 35,307 | 69,811 | |
| Time $X^p$ mod $\mathfrak{R}_1$ | 3.3 | 15 | 77 | 280 | 2,251 | 2,294 | $O(\mathsf{M}(\ell^4)\log p)$ |
| Time Prec. Mod Comp | 0.8 | 8 | 105 | 525 | 3,267 | 2,122 | $O(\ell^2\mathsf{M}(\ell^4))$ |
| Time Apply Mod Comp | 1.1 | 11 | 225 | 976 | 20,768 | 51,710 | $O(\ell^8 + \ell^2\mathsf{M}(\ell^4))$ |
| Factoring Time (Min) | 12.5 | 59 | 524 | 1,055 | 23,537 | 15,061 | |
| Factoring Time (Max) | 61 | 353 | 5,021 | 23,083 | 206,860 | 359,330 | |
| Factoring Time (Avg) | 42 | 193 | 2,700 | 9,415 | 117,785 | 145,734 | |

| 2- and 3- Torsion Lifting | | | |
|---|---|---|---|
| Torsion | Total Time (Min) | Total Time (Max) | Total Time (Avg) |
| 27 | 10,901 | 11,317 | 11,511 |
| 1024 | 71,421 | 103,433 | 90,071 |

| Lifting to 1024-torsion Details for a sample curve | | | | |
|---|---|---|---|---|
| Generic Resolution: 5,104 sec | | | | |
| Torsion | Degree | Specialization | Factor | Deducing $\chi$ |
| 8 | 10 | 1 | 12 | 1 |
| 16 | 20 | 1 | 37 | 3 |
| 32 | 40 | 3 | 178 | 16 |
| 64 | 80 | 15 | 543 | 50 |
| 128 | 160 | 41 | 1,423 | 146 |
| 256 | 320 | 115 | 4,627 | 459 |
| 512 | 640 | 390 | 16,776 | 1,602 |
| 1024 | 1280 | 1301 | 58,408 | 6,590 |

| Lifting to 27-torsion Details for a sample curve | | |
|---|---|---|
| Torsion | 9 | 27 |
| Degree | 12 | 36 |
| Gröbner | 745 | 3,811 |
| Factor | 914 | 5,917 |
| Deducing $\chi$ | 3 | 19 |

**Table 1.** Runtimes in seconds for the torsion computation on a 2.26 GHz Pentium IV.

Putting all this together, a complete point-counting for a random curve over $\mathbb{F}_p$ takes on average about 1 week.

For comparison, in the record-curve computation in [11] the Schoof-like part was used up to $\ell = 13$. Just the modulo 13 computation had taken 205 hours on a Pentium II at 450 MHz. A crude estimation gives a runtime of about 40 hours on the same computer as the one we used in this paper. This has to be compared with the 4 hour runtime that we obtained with our improvements and our new implementation.

**Are the Curves "Random"?** In our computer experiments, the "pure randomness" is biased in several places. Due to the cryptographical requirements, the group order must be prime, so "random curve" should be understood as random among the curves with prime order Jacobians, but that is standard. Also a bias is introduced by our early abort strategy on both the curve and its twist.

A more important bias is in the choice of $p$. We choose a prime which is congruent to 1 modulo all the small primes $\ell$ for which we do the Schoof computation. This was meant to speed-up the factorization of the resultant $\mathfrak{R}_1$, as mentioned in Section 3.4. This dependency of the runtime of the algorithm in the form of $p$ can be avoided by working in the formal algebra instead of factoring. In fact, in more recent versions of our software, we implemented this and the runtimes are slightly better for large $\ell$. Hence, this bias could be removed.

The last bias that we introduced is the particular shape of the 8- and 3-torsion that we imposed. The goal was mostly to have the same kind of behavior for all the curves with respect to the division by 2 and by 3. Indeed, the division algorithms rely on Gröbner basis computations and are very hard to implement and to debug. The technical difficulty of handling our computation on many computers, with interactions between Magma and NTL led us to add this simplification that made our code more reliable.

Our NTL implementation of the Schoof-like part has been made freely available [9]. The Magma implementation of the division algorithms is not stable enough to be exported in the present state.

## 6 Conclusion and Perspectives

In this paper, we have detailed algorithms used to compute the cardinalities of Jacobians defined over prime fields of order about $10^{24}$. Most of our attention was aimed at improving the techniques for torsion computation introduced in [11].

We expect more improvements to be possible. For instance, for torsion index about 17 or 19, the factorization strategy of Subsection 3.4 becomes lengthy, and comparative tests with other strategies are necessary, possibly using the modular equations of [12]. Also, our techniques for lifting the 3-torsion are still quite crude, as we would like it to be as efficient as that of 2-torsion. We have designed a birthday paradox version of the MCT algorithm, to be described elsewhere [13], that loses a constant factor in runtime but is highly parallelizable and requires almost no memory. In future work, we also plan to use it on top of our torsion computation algorithms.

**Acknowledgments.** Many people have been of assistance when designing, implementing and running our algorithms. We thank Gérard Guillerm and Bogdan Tomchuk for letting classroom computers at École polytechnique at our disposal, Grégoire Lecerf for his fast evaluation and interpolation code, Allan Steel for releasing is HGCD implementation in Magma by our request, Nicolas M. Thiéry for sharing his insight on symmetric polynomials and Emmanuel Thomé for his help on handling large scale computations. The computations were performed on classroom computers at École polytechnique, on the machines of the MEDICIS center for computer algebra `http://www.medicis.polytechnique.fr/`, and on machines paid by ACI Cryptologie.

## References


1. L. Adleman and M.-D. Huang. Counting points on curves and abelian varieties over finite fields. *J. Symbolic Comput.*, 32:171–189, 2001.
2. A. Aho, K. Steiglitz, and J. D. Ullman. Evaluating polynomials at fixed sets of points. *SIAM J. Comput.*, 4(4):533–539, 1975.
3. W. Bosma and J. Cannon. *Handbook of Magma functions*, 1997. `http://www.maths.usyd.edu.au:8000/u/magma/`.
4. A. Bostan, P. Flajolet, B. Salvy, and É. Schost. Fast computation with two algebraic numbers. Technical Report 4579, INRIA, 2002.
5. A. Bostan, P. Gaudry, and É. Schost. Linear recurrences with polynomial coefficients and computation of the Cartier-Manin operator on hyperelliptic curves. 2003. To appear in Proceedings Fq'7.
6. R. Brent and H. Kung. Fast algorithms for manipulating formal power series. *J. ACM*, 25:581–595, 1978.
7. D. G. Cantor. Computing in the Jacobian of an hyperelliptic curve. *Math. Comp.*, 48(177):95–101, 1987.
8. D. G. Cantor. On the analogue of the division polynomials for hyperelliptic curves. *J. Reine Angew. Math.*, 447:91–145, 1994.
9. P. Gaudry. *NTLJac2, Tools for genus 2 Jacobians in NTL.* `http://www.lix.polytechnique.fr/Labo/Pierrick.Gaudry/NTLJac2/`.
10. P. Gaudry and N. Gürel. Counting points in medium characteristic using Kedlaya's algorithm. To appear in *Experiment. Math.*
11. P. Gaudry and R. Harley. Counting points on hyperelliptic curves over finite fields. In W. Bosma, editor, *ANTS-IV*, volume 1838 of *Lecture Notes in Comput. Sci.*, pages 313–332. Springer–Verlag, 2000.
12. P. Gaudry and É. Schost. Modular equations for hyperelliptic curves. To appear in *Math. Comp.*
13. P. Gaudry and Éric Schost. A low-memory parallel version of Matsuo, Chao and Tsujii's algorithm. To appear in *ANTS VI.*
14. G. Hanrot and F. Morain. Solvability of radicals from an algorithmic point of view. In *ISSAC'01*, pages 175–182. ACM Press, 2001.
15. M. Hindry and J. Silverman. *Diophantine geometry. An introduction*, volume 201 of *Graduate Texts in Mathematics.* Springer–Verlag, 2000.
16. M.-D. Huang and D. Ierardi. Counting points on curves over finite fields. *J. Symbolic Comput.*, 25:1–21, 1998.
17. E. Kaltofen and V. Shoup. Fast polynomial factorization over high algebraic extensions of finite fields. In W. Kuchlin, editor, *ISSAC-97*, pages 184–188. ACM Press, 1997.
18. W. Kampkötter. *Explizite Gleichungen für Jacobische Varietäten hyperelliptischer Kurven.* PhD thesis, Universität Gesamthochschule Essen, August 1991.
19. T. Lange. Formulae for arithmetic on genus 2 hyperelliptic curves, 2003. Preprint.
20. R. Lercier and D. Lubicz. A quasi quadratic time algorithm for hyperelliptic curve point counting. Preprint.
21. K. Matsuo, J. Chao, and S. Tsujii. An improved baby step giant step algorithm for point counting of hyperelliptic curves over finite fields. In C. Fiecker and D. Kohel, editors, *ANTS-V*, volume 2369 of *Lecture Notes in Comput. Sci.*, pages 461–474. Springer-Verlag, 2002.
22. A. Menezes, Y.-H. Wu, and R. Zuccherato. An elementary introduction to hyperelliptic curves. In *Algebraic aspects of cryptography*, by N. Koblitz, pages 155–178, Springer-Verlag, 1997.

23. J.-F. Mestre. Utilisation de l'AGM pour le calcul de $E(\mathbb{F}_{2^n})$. Letter to Gaudry and Harley, December 2000.
24. J. S. Milne. Abelian varieties. In G. Cornell and J. H. Silverman, editors, *Arithmetic Geometry*, pages 103–150. Springer–Verlag, 1986.
25. J. Pelzl, T. Wollinger, J. Guajardo, and C. Paar. Hyperelliptic curve cryptosystems: Closing the performance gap to elliptic curves. Preprint, 2003.
26. J. Pila. Frobenius maps of abelian varieties and finding roots of unity in finite fields. *Math. Comp.*, 55(192):745–763, October 1990.
27. R. Schoof. Elliptic curves over finite fields and the computation of square roots mod $p$. *Math. Comp.*, 44:483–494, 1985.
28. É. Schost. Complexity results for triangular sets. *J. Symbolic Comput.*, 36:555–594, 2003.
29. V. Shoup. *NTL: A library for doing number theory.* `http://www.shoup.net/ntl/`.
30. V. Shoup. A new polynomial factorization algorithm and its implementation. *J. Symbolic Comput.*, 20:363–397, 1995.
31. F. Vercauteren. Computing Zeta functions of hyperelliptic curves over finite fields of characteristic 2. In M. Yung, editor, *Advances in Cryptology – CRYPTO 2002*, volume 2442 of *Lecture Notes in Comput. Sci.*, pages 369–384. Springer-Verlag, 2002.
32. J. von zur Gathen and J. Gerhard. *Modern computer algebra.* Cambridge University Press, 1999.
33. J. von zur Gathen and V. Shoup. Computing Frobenius maps and factoring polynomials. *Comput. Complexity*, 2:187–224, 1992.
34. A. Weng. Constructing hyperelliptic curves of genus 2 suitable for cryptography. *Math. Comp.*, 72:435–458, 2003.


## Appendix: A Cryptographically Secure Curve

Let $\mathcal{C}$ be defined by $y^2 = f(x)$ over $\mathbb{F}_p$ with $p = 5 \times 10^{24} + 8503491$, and

$$f(x) = x^5 + 2682810822839355644900736x^3 + 226591355295993102902116x^2 + 2547674715952929717899918x + 4797309959708489673059350.$$

Then its characteristic polynomial is $\chi(T) = T^4 - s_1 T^3 + s_2 T^2 - p s_1 T + p^2$, where

$$s_1 = 1173929286783 \quad \text{and} \quad s_2 = 4402219446392186881834853.$$

Thus the cardinality of its Jacobian is

$$N_{\mathbf{J}} = \chi(1) = 24999999999994130438600999402209463966197516075699,$$

which is a 164-bit prime number. Furthermore the quadratic twist of $\mathcal{C}$ has a Jacobian with group order

$$N_{\tilde{\mathbf{J}}} = \chi(-1) = 25000000000005869731468829402229428962794965968171,$$

which is also a prime number.

# Projective Coordinates Leak

David Naccache[1], Nigel P. Smart[2], and Jacques Stern[3]

[1] Gemplus Card International,
Applied Research & Security Centre,
34 rue Guynemer,
Issy-les-Moulineaux, F-92447, France
david.naccache@gemplus.com
[2] Department of Computer Science,
University of Bristol,
Merchant Venturers Building, Woodland Road,
Bristol, BS8 1UB, United Kingdom
nigel@cs.bris.ac.uk
[3] École Normale Supérieure,
Département d'Informatique,
45 rue d'Ulm,
F-75230, Paris 05, France
jacques.stern@ens.fr

**Abstract.** Denoting by $P = [k]G$ the elliptic-curve double-and-add multiplication of a public base point $G$ by a secret $k$, we show that allowing an adversary access to the projective representation of $P$, obtained using a particular double and add method, may result in information being revealed about $k$.

Such access might be granted to an adversary by a poor software implementation that does not erase the $Z$ coordinate of $P$ from the computer's memory or by a computationally-constrained secure token that sub-contracts the affine conversion of $P$ to the external world.

From a wider perspective, our result proves that the choice of representation of elliptic curve points *can reveal* information about their underlying discrete logarithms, hence casting potential doubt on the appropriateness of blindly modelling elliptic-curves as generic groups.

As a conclusion, our result underlines the necessity to sanitize $Z$ after the affine conversion or, alternatively, randomize $P$ before releasing it out.

## 1 Introduction

There are various systems of projective coordinates that are used in conjunction with elliptic curves: the usual (classical) system replaces the affine coordinates $(x, y)$ by any triple $(X, Y, Z) = (\lambda x, \lambda y, \lambda)$, where $\lambda \neq 0$ is an element of the base field.

C. Cachin and J. Camenisch (Eds.): EUROCRYPT 2004, LNCS 3027, pp. 257–267, 2004.

From such a $(X, Y, Z)$, the affine coordinates are computed back as

$$\left(x = \frac{X}{Z}, y = \frac{Y}{Z}\right) = \text{Affine}(X, Y, Z)$$

A variant of the above, often called Jacobian Projective coordinates, replaces the affine coordinates $(x, y)$ by any triple $(\lambda^2 x, \lambda^3 y, \lambda)$, where $\lambda$ is a non zero element of the base field. From $(X, Y, Z)$, the affine coordinates are computed as

$$\left(x = \frac{X}{Z^2}, y = \frac{Y}{Z^3}\right) = \text{Affine}(X, Y, Z)$$

These coordinates are widely used in practice, see for example [1] and [4].

This paper explores the following question:

*Denoting by $P = [k]G$ the elliptic-curve multiplication of a public base point $G$ by a secret $k$, does the projective representation of $P$ result in information being revealed about $k$?*

From a practical perspective access to $P$'s $Z$ coordinate might stem from a poor software implementation that does not erase the $Z$ coordinate of $P$ from the computer's memory or caused by a computationally-constrained secure token that sub-contracts the affine conversion of $P$ to the external world.

We show that information may leaks-out and analyse the leakage in two different settings: Diffie-Hellman key exchange and Schnorr signatures.

Moreover, our paper seems to indicate that *point representation matters*: The generic group model is often used to model elliptic curve protocols, see [2], [10], [11]. In this model one assumes that the representation of the group elements gives no benefit to an adversary. This approach allows cryptographic schemes built from elliptic curves to be supported by some form of provable security. However, it has some pitfalls. In [11], it was shown that using encodings which do not adequately distinguish an elliptic curve from its opposite, as done in ECDSA, open the way to potential flaws in the security proofs. In this paper we show that using projective coordinates to represent elliptic curve points rather than affine coordinates may leak some information to an attacker. Thus, we can conclude that modelling elliptic curves as generic groups is not appropriate in this case, so that the generic model methodology only applies under the assumption that affine points are made available to an external viewer/adversary of the protocol.

We note that our results imply that projective coordinates should be used with care when they could be made available to an adversary. Our results do not however imply that using projective coordinates for *internal* calculations has any security implications.

## 2 Elliptic Curve Addition Formulae

In the following, we will restrict our attention to elliptic curves over fields of large prime characteristic. We will also focus on projective coordinates of the second kind (the situation being quite similar *mutatis mutandis*, in the other cases).

In our prime field case, the reduced equation of the curve $\mathcal{C}$ is:

$$y^2 = x^3 + ax + b \mod p$$

Jacobian projective coordinates yield the equation:

$$Y^2 = X^3 + aXZ^4 + bZ^6 \mod p$$

Projective coordinates allow a smooth representation of the infinity point $\mathcal{O}$ on the curve: $(0, 1, 0)$ in the first system, $(1, 1, 0)$ in the other. They also provide division-free formulae for addition and doubling.

Standard (affine) addition of two distinct elliptic curve points, $(x_0, y_0)$ and $(x_1, y_1)$ yields $(x_2, y_2)$, with:

$$x_2 = \left(\frac{y_1 - y_0}{x_1 - x_0}\right)^2 - x_0 - x_1$$

Note that $x_1 - x_0$ equals:

$$\frac{X_1}{Z_1^2} - \frac{X_0}{Z_0^2} = \frac{W}{(Z_0Z_1)^2}$$

where $W$ is $X_1Z_0^2 - X_0Z_1^2$. From this it readily follows, that $(WZ_0Z_1)^2x_2$ is a polynomial in $X_0, Y_0, Z_0, X_1, Y_1, Z_1$, since the further factors coming from $Z_0$ and $Z_1$ cancel the denominators for $x_0$ and $x_1$.

The affine coordinate $y_2$ is given by:

$$y_2 = -y_0 + \left(\frac{y_1 - y_0}{x_1 - x_0}\right)(x_0 - x_2)$$

Expanding in projective coordinates yields a denominator equal to $W^3Z_0^3Z_1^3$. Thus, $(WZ_0Z_1)^3y_2$ is a polynomial in $X_0, Y_0, Z_0, X_1, Z_1$. Finally, we see that setting:

$$Z_2 = WZ_0Z_1$$

we can obtain division-free formulae. Such formulae are given in [4] and [1], and we simply reproduce them here:

$$\begin{array}{llll} U_0 \leftarrow X_0Z_1^2, & S_0 \leftarrow Y_0Z_1^3, & U_1 \leftarrow X_1Z_0^2, & S_1 \leftarrow Y_1Z_0^3, \\ W \leftarrow U_0 - U_1, & R \leftarrow S_0 - S_1, & T \leftarrow U_0 + U_1, & M \leftarrow S_0 + S_1, \\ Z_2 \leftarrow WZ_0Z_1, & X_2 \leftarrow R^2 - TW^2, & V \leftarrow TW^2 - 2X_2, & 2Y_2 \leftarrow VR - MW^3. \end{array}$$

There is a similar analysis for doubling; again, we simply provide the corresponding formulae:

$$\begin{array}{lll} M \leftarrow 3X_1^2 + aZ_1^4, & Z_2 \leftarrow 2Y_1Z_1, & S \leftarrow 4X_1Y_1^2, \\ X_2 \leftarrow M^2 - 2S, & T \leftarrow 8Y_1^4, & Y_2 \leftarrow M(S - X_2) - T. \end{array}$$

# 3 The Attack

Throughout this section we let $G$ be an element of prime order $r$ on an elliptic curve $C$ over a prime field, given by its regular coordinates $(x_G, y_G)$. Let $k$ be a secret scalar and define $P = [k]G$. Let $(X, Y, Z)$ be Jacobian projective coordinates for $P$, computed by the formulae introduced in Section 2, when the standard double-and-add algorithm is used.

## 3.1 Grabbing a Few Bits of $k$

Let $t$ be a small integer and guess the last $t$ bits of $k$. Once this is done, it is possible to compute a set of candidates for the coordinates of the sequence of intermediate values handled by the double-and-add algorithm while processing $k$'s $t$ trailing bits (appearing at the end of the algorithm). This is achieved by 'reversing' computations: reversing doubling is halving, *i.e.* by reversing the formulae for doubling ; reversing an addition amounts to subtracting $G$. Thus, we obtain a set of sequences,

$$\{s_1, s_2, \ldots, s_m\} \quad \text{where} \quad s_j = \{M_0^{(j)} \dashrightarrow M_1^{(j)} \dashrightarrow \cdots \dashrightarrow M_\ell^{(j)}\}$$

of intermediate points, with $M_\ell^{(j)} = P$. Let $M_i = (x_i, y_i)$ in affine coordinates. The corresponding projective coordinates which occur we denote by $(X_i, Y_i, Z_i)$. There are two cases:

- When the step $M_i \dashrightarrow M_{i+1}$ is an addition, we have

  $$Z_{i+1} = (X_i - x_G Z_i^2) Z_i \quad \text{which yields} \quad \frac{Z_{i+1}}{Z_i^3} = (x_i - x_G)$$

  Here, we need to compute a cubic root to get $Z_i$ from $Z_{i+1}$. This is impossible in some cases when $p \equiv 1 \bmod 3$, and when possible, it leads to one of three possible $Z_i$ values. When $p \equiv 2 \bmod 3$ taking the cubic root is always possible and leads to a unique value of $Z_i$. In either case once a set of possible values of $Z_i$ are determined from $Z_{i+1}$ we can obtain $X_i$ and $Y_i$.
- When the step $M_i \dashrightarrow M_{i+1}$ is a doubling, we have

  $$Z_{i+1} = 2Y_i Z_i \quad \text{which yields} \quad \frac{Z_{i+1}}{Z_i^4} = 2y_i$$

  Here, we need to compute a fourth root to get $Z_i$ from $Z_{i+1}$, which is impossible in some cases. Assume for example that $p \equiv 3 \bmod 4$. Then extracting a fourth root is possible for one half of the inputs and, when possible, yields two values. When $p \equiv 1 \bmod 4$ then this is possible in around one quarter of all cases and yields four values.

We can now take advantage of the above observation to learn a few bits of $k$.

More precisely, we observe that, with probability at least $1/2$, one can spot values of $k$ for which the least significant trailing bit is one. Suppose we consider such a $k$ and make the wrong guess that the last bit is zero. This means that the final operation $M_{\ell-1} \dashrightarrow M_\ell$ is a doubling. The error can be spotted when the value

$$\frac{Z_\ell}{2y_{\ell-1}}$$

is not a fourth power, which happens with probability at most 1/2. We can then iterate this to (potentially) obtain a few further bits of $k$. In the case of the least significant bit being zero a similar analysis can be performed.

### 3.2 Applicability to Different Coordinate Systems

Consider Jacobian projective coordinates:

$$(X, Y) \mapsto (\lambda^2 X, \lambda^3 Y, \lambda Z),$$

over a field $\mathbb{F}_q$ of characteristic $q > 3$. For a point $P = (x, y) \in \mathcal{C}(\mathbb{F}_q)$ let $S_P$ denote the set of all equivalent projective representations

$$S_P = \{(\lambda^2 x, \lambda^3 y, \lambda) : \lambda \in \mathbb{F}_q^*\}.$$

The standard addition formulae for computing $P + Q$, for a fixed value of $Q$ (by fixed we mean a fixed projective representation of $Q$, including an affine representation of $Q$) gives a map

$$\Psi_{P,P+Q} : S_P \longrightarrow S_{P+Q}.$$

The doubling formulae for Jacobian projective coordinates also gives us a map

$$\Psi_{P,[2]P} : S_P \longrightarrow S_{[2]P}.$$

The crucial observations from the previous subsection are summarized in the following Lemma

**Lemma 1.** *The following holds, for Jacobian projective coordinates in large prime characteristics:*

$$\begin{array}{llll} \textit{If } q \equiv 1 \mod 3 & \textit{then } \Psi_{P,P+Q} & \textit{is a } 3 \rightsquigarrow 1 \textit{ map.} \\ \textit{If } q \equiv 2 \mod 3 & \textit{then } \Psi_{P,P+Q} & \textit{is a } 1 \rightsquigarrow 1 \textit{ map.} \\ \textit{If } q \equiv 1 \mod 4 & \textit{then } \Psi_{P,[2]P} & \textit{is a } 4 \rightsquigarrow 1 \textit{ map.} \\ \textit{If } q \equiv 3 \mod 4 & \textit{then } \Psi_{P,[2]P} & \textit{is a } 2 \rightsquigarrow 1 \textit{ map.} \end{array}$$

Note: It is easy given an element in the image of either $\Psi_{P,P+Q}$ or $\Psi_{P,[2]P}$ to determine whether it has pre-images, and if so to compute all of them.

The attack is then simply to consider when a point could have arisen from an application of $\Psi_{P,P+Q}$ or $\Psi_{P,[2]P}$ and if so to compute all the pre-images and then recurse. The precise tests one applies at different points will depend on the precise exponentiation algorithm implemented by the attacked device, a subject we shall return to in a moment.

For the sake of completeness we present in the following *lemmata* similar results for other characteristics and other forms of projective representation. We concentrate on the most common and the most used coordinate systems and keep the same conventions and notation as above:

**Lemma 2.** *The following holds, for classical projective coordinates on elliptic curves over fields of large prime characteristic:*

$$\begin{array}{llll} \textit{If } q \equiv 1 \mod 4 & \textit{then } \Psi_{P,P+Q} & \textit{is a } 4 \leadsto 1 \textit{ map.} \\ \textit{If } q \equiv 3 \mod 4 & \textit{then } \Psi_{P,P+Q} & \textit{is a } 2 \leadsto 1 \textit{ map.} \\ \textit{If } q \equiv 1 \mod 3 & \textit{then } \Psi_{P,[2]P} & \textit{is a } 6 \leadsto 1 \textit{ map.} \\ \textit{If } q \equiv 2 \mod 3 & \textit{then } \Psi_{P,[2]P} & \textit{is a } 2 \leadsto 1 \textit{ map.} \end{array}$$

**Lemma 3.** *The following holds, for Jacobian projective coordinates on elliptic curves over fields of characteristic two:*

$$\begin{array}{llll} \textit{If } q \equiv 1 \mod 3 & \textit{then } \Psi_{P,P+Q} & \textit{is a } 3 \leadsto 1 \textit{ map.} \\ \textit{If } q \equiv 2 \mod 3 & \textit{then } \Psi_{P,P+Q} & \textit{is a } 1 \leadsto 1 \textit{ map.} \\ & \forall q \quad \Psi_{P,[2]P} & \textit{is a } 1 \leadsto 1 \textit{ map.} \end{array}$$

**Lemma 4.** *The following holds, for López-Dahab projective coordinates [6] on elliptic curves over fields of characteristic two:*

$$\begin{array}{llll} \textit{If } q \equiv 1 \mod 3 & \textit{then } \Psi_{P,[2]P} & \textit{is a } 3 \leadsto 1 \textit{ map.} \\ \textit{If } q \equiv 2 \mod 3 & \textit{then } \Psi_{P,[2]P} & \textit{is a } 1 \leadsto 1 \textit{ map.} \\ & \forall q \quad \Psi_{P,P+Q} & \textit{is a } 1 \leadsto 1 \textit{ map.} \end{array}$$

## 4 Application: Breaking Projective Schnorr Signatures

Assume now that one wishes to use the protocol described in Figure 1, mimicking Schnorr's basic construction [12]. The algorithm is a natural division-free version of Schnorr's original scheme, and might hence appear both safe and computationally attractive.

It should be stressed that while we are *not* aware of any suggestion to use this variant in practice it is still not evident, at a first glance, why this algorithm could be insecure.

We show how to attack this scheme using the observations from the previous subsection. This is based on recent work by Howgrave-Graham, Smart [3], Nguyen and Shparlinski [7].

From a sample of $N$ signatures, the attacker obtains around $\frac{N}{2^{2t}}$ signatures for which he knows that the $t$ low order bits of the hidden nonce $k$ are ones. Next, for each such $k$, he considers the relation:

$$d + xH(m, P_X, P_Y, P_Z) = k \mod r$$

| PARAMETERS AND KEYS | |
|---|---|
| | An elliptic-curve $\mathcal{C}$ |
| | $G \in_R \mathcal{C}$ of order $r$ |
| | A collision-resistant hash-function $H : \{0,1\}^* \longrightarrow \mathbb{Z}_r^\star$ |
| **Private** | $x \in_R \mathbb{Z}_r^\star$ |
| **Public** | $Q \leftarrow [x]G$ |
| SIGNATURE GENERATION | |
| | Pick $k \in_R \mathbb{Z}_r^\star$ |
| | Compute $(P_X, P_Y, P_Z) \leftarrow [k]G = \text{DoubleAdd}(k, G)$ |
| | $d \leftarrow k - x \times H(m, P_X, P_Y, P_Z) \bmod r$ |
| | If $d = 0$ or $H(m, P_X, P_Y, P_Z) = 0$ resume signature generation |
| | Output $\{P_X, P_Y, P_Z, d\}$ as the signature of $m$ |
| SIGNATURE VERIFICATION | |
| | $P \leftarrow [d]G + [H(m, P_X, P_Y, P_Z)]Q$ |
| | If $P \neq \text{Affine}((P_X, P_Y, P_Z))$ or $d \notin \mathbb{Z}_r^\star$ output invalid |
| | else output valid |

**Fig. 1.** Division-Free Projective Schnorr Signatures

Using the information he has, the attacker rewrites the above as:

$$d - (2^t - 1) + xH(m, P_X, P_Y, P_Z) = k - (2^t - 1) \bmod r$$

Dividing by $2^t$, he gets a final relation:

$$a + bx = u \bmod r$$

where $a$, $b$ are known but $x$ is unknown as well as $u$. Still the attacker knows that $u$ is small ($\leq \frac{r}{2^t}$). When the attacker has $n \approx \frac{N}{2^{2t}}$ such relations, he writes

$$\mathbf{a} + \mathbf{b}x = \mathbf{u} \bmod r$$

and considers the lattice $L = (\mathbf{b})^\perp$, consisting of all integer vectors orthogonal to $\mathbf{b}$ and applies lattice reduction. Let $\mathbf{\Lambda}$ be an element of $L$ with small Euclidean norm. We have:

$$\mathbf{\Lambda}(\mathbf{a}) = \mathbf{\Lambda}(\mathbf{u}) \bmod r$$

Now, the norm of the right-hand side is bounded by $||\mathbf{\Lambda}||||\mathbf{u}||$, which is $\leq ||\mathbf{\Lambda}||\frac{r}{2^t}\sqrt{n}$. The order of $||\mathbf{\Lambda}||$ is $r^{\frac{1}{n}}$ and, for $n$ large enough and $t$ not too small, this estimate provides a bound for the right-hand side $< r/2$. Thus, the modular equations are actual equations over the integers:

$$\mathbf{\Lambda}(\mathbf{a}) \bmod r = \mathbf{\Lambda}(\mathbf{u})$$

| PARAMETERS | |
|---|---|
| **Input** | $k \in \mathbb{Z}_r^\star, G \in \mathcal{C}$ |
| **Output** | $P \leftarrow [k]G$ |
| ALGORITHM DoubleAdd$(k, G)$ | |

```
P ← O
for j = ℓ − 1 downto 0:
    P ← [2]P
    if k_j = 1 then P ← P + G
return(P)
```

**Fig. 2.** Double-and-Add Exponentiation

The attacker can hope for at most $n-1$ such relations, since $L$ has dimension $n-1$. This defines $\mathbf{u}$ up to the addition of an element from a one-dimensional lattice. The correct value is presumably the element in this set closest to the origin. Once $\mathbf{u}$ has been found, the value of $x$ follows.

Lattice reduction experiments reported in [7] show that, with elliptic curves of standard dimensions, the attack will succeed as soon as $t$ reaches 5 digits. The deep analysis of Nguyen and Shparlinski, shows that the significant theoretical bound is related to $\sqrt{\log r}$.

## 5 Practical Experiments

The double-and-add exponentiation's case is the simplest to analyse: given the projective representation of the result $P$, we can try and 'unwind' the algorithm with respect to the fixed point $G$.

In other words, we can check whether there is a value $P'$ such that

$$\Psi_{P',P'+G}(P') = P$$

and if so compute all the pre-images $P'$. Then for all pre-images $P'$ we can check whether this was the result of a point doubling. We also need to check whether $P$ itself was the output of a point doubling. This results in a backtracking style algorithm which investigates all possible execution paths through the algorithm.

There are two factors at work here. For each testing of whether $\Psi_{P,P+G}$ (resp. $\Psi_{P,[2]P}$) was applied we have a representation-dependent probability of $\mathfrak{p}$ (from the above *lemmata*), this acts in the attacker's favour. However, each success for this test yields $1/\mathfrak{p}$ pre-images, which increases the attacker's workload. The result is that, while practical, the attack against the double-and-add algorithm is not as efficient as one might initially hope.

We ran one thousand experiments in each prime characteristic modulo 12. Table 1 presents the success of determining the parity of the secret exponent. One should interpret the entries in the table as follows: For example with $q \equiv 5 \pmod{12}$, we found that in 71 percent of all cases in which $k$ was even we where able to determine this using the above backtracking algorithm. This means that in these cases the execution path which started with assuming $P$ was the output of a point addition was eventually determined to be invalid.

**Table 1.** Probability of Determining the Secret's Parity Using Double-and-Add Exponentiation

| $q$ mod 12 | 1 | 5 | 7 | 11 |
|---|---|---|---|---|
| Pr[parity determined\|$k$ even] | 0.98 | 0.71 | 0.80 | 0.50 |
| Pr[parity determined\|$k$ odd] | 0.95 | 0.74 | 0.50 | 0.47 |
| Pr[parity determined] | 0.96 | 0.72 | 0.65 | 0.48 |

Only in the cases $q \equiv 1 \bmod 12$ and $q \equiv 7 \bmod 12$ did we have any success in determining the value of the secret exponent modulo 8 precisely (around 50 percent of the time when $q \equiv 1 \bmod 12$ and 8 percent of the time when $q \equiv 7 \bmod 12$).

We did a similar experiment using the signed sliding window method, with a window width of 5 (see also Algorithm IV.7 of [1]) assuming that the pre-computed table of multiples of the base point is known to the attacker. In this case we had a much lower probability of determining the parity, but could still determine the value of the exponent modulo 32 in a significant number of cases (Table 2).

**Table 2.** Probability of Determining the Secret's Parity Using Signed Sliding Window Exponentiation

| $q$ mod 12 | 1 | 5 | 7 | 11 |
|---|---|---|---|---|
| Pr[parity determined\|$k$ even] | 0.86 | 0.00 | 0.05 | 0.00 |
| Pr[parity determined\|$k$ odd] | 0.81 | 0.75 | 0.49 | 0.53 |
| Pr[parity determined] | 0.81 | 0.37 | 0.27 | 0.26 |
| Pr[$k$ mod 32 determined] | 0.42 | 0.01 | 0.01 | 0.00 |

Note that this means that if $q \equiv 1 \bmod 12$ then we will be successful in determining the full private key for the division free signature algorithm of Section 4 using lattice reduction.

PARAMETERS

**Input** $k \in \mathbb{Z}_r^\star, G \in \mathcal{C}$
**Output** $P \leftarrow [k]G$

PRECOMPUTATION

$G_1 \leftarrow G$
$G_2 \leftarrow [2]G$
for $j = 1$ to $2^{r-2} - 1$:
  $G_{2j+1} \leftarrow G_{2j-1} + G_2$
$P \leftarrow G_{k_\ell - 1}$

EXPONENT ENCODING

set $k = \sum_{i=0}^{\ell-1} k_i 2^{e_i}$
with $e_{i+1} - e_i \geq r$ and $k_i \in \{\pm 1, \pm 3, \ldots, \pm 2^{r-1} - 1\}$

ALGORITHM SlidingWindow$(k, G)$

for $j = \ell - 2$ downto 0:
  $P \leftarrow [2^{e_{j+1} - e_j}]P$
  if $\ell_j > 0$ then $P \leftarrow P + G_{k_j}$ else $P \leftarrow P + G_{-k_j}$
$P \leftarrow [2^{e_0}]P$
return$(P)$

**Fig. 3.** Signed Sliding Window Exponentiation

## 6 Thwarting the Attack

There is a simple trick that avoids the attacks described in the previous sections. It consists in randomly replacing the output $(X, Y, Z)$ of the computation by $(X, \epsilon Y, \epsilon Z)$, with $\epsilon = \pm 1$. This makes it impossible for an attacker to spot projective coordinates, which cannot be obtained by squaring. It should be underlined that this countermeasure (that we regard as a challenge for the research community) thwarts *our* specific attack but does not lend itself to a formal security proof. Note, such a defence only appears to need to be done at the end of the computation as our attack model assume the attacker does not obtain any intermediate points from the multiplication algorithm.

A more drastic method replaces $(X, Y, Z)$ by $(\lambda^2 x, \lambda^3 y, \lambda)$, where $\lambda$ is randomly chosen among the non zero elements of the base field (with ordinary projective coordinates, one uses $(\lambda x, \lambda y, \lambda)$). This method provides a randomly chosen set of projective coordinates for the result and, therefore, cannot leak additional information.

With this new protection, the division-free signature scheme of Section 4 can be shown to be secure in the random oracle model, against adaptive attackers trying to achieve existential forgery. We outline the proof. As usual (see [9]), one uses the attacker to solve the discrete logarithm problem (here, on $\mathcal{C}$). The public key of the scheme is set to $Q$, the curve element for which we want to compute the discrete logarithm in base $G$. Signature queries are answered by randomly creating $P = [d]G + [h]Q$, picking random projective coordinates for $P$, say $(X, Y, Z)$ and setting the hash value of $\{m, X, Y, Z\}$ as any element $= h \mod r$. Thus fed, the attacker should create a forged message signature pair, with significant probability. We let $m$ be the corresponding message and $\{X, Y, Z, d\}$ be the signature. With significant probability, $\{m, X, Y, Z\}$ is queried from the random oracle. Replaying the attack with a different answer modulo $r$ to this question, one gets, with significant probability, another forgery $\{m, X, Y, Z, d'\}$, with $h$ replaced by $h'$. From the relation

$$[d]G + [h]Q = [d']G + [h']Q$$

one finally derives the discrete logarithm of $Q$.

## References


1. I.F. Blake, G. Seroussi and N.P. Smart, *Elliptic Curves in Cryptography*, Cambridge University Press, 1999.
2. D. Brown, *Generic Groups, Collision Resistance, and ECDSA*, ePrint Report 2002/026, **http://eprint.iacr.org/**.
3. N.A. Howgrave-Graham and N.P. Smart, *Lattice attacks on digital signature schemes, Designs, Codes and Cryptography*, **23**, pp. 283–290, 2001.
4. IEEE 1363, IEEE standard specifications for public key cryptography, 2000.
5. A. Joux and J. Stern, *Lattice Reduction: a Toolbox for the Cryptanalyst*, In *Journal of Cryptology*, vol. 11, pp. 161–186, 1998.
6. J. López and R. Dahab, *Improved algorithms for elliptic curve arithmetic in* $GF(2^n)$, In *Selected Areas in Cryptography* - SAC*'98*, Springer-Verlag LNCS 1556, pp. 201–212, 1999.
7. P. Nguyen and I. Shparlinski, *The Insecurity of the Digital Signature Algorithm with Partially Known Nonces*, In *Journal of Cryptology*, vol. 15, pp. 151–176, 2002.
8. P. Nguyen and J. Stern, *The hardness of the subset sum problem and its cryptographic implications*, In *Advances in Cryptology* CRYPTO*'99*, Santa Barbara, Lectures Notes in Computer Science 1666, pp. 31–46, Springer-Verlag, 1999.
9. D. Pointcheval and J. Stern, *Security Arguments for Digital Signatures and Blind Signatures*, In *Journal of Cryptology*, vol. 13, pp. 361–396, 2000.
10. N. P. Smart, *The Exact Security of ECIES in the Generic Group Model* In B. Honary (Ed.), Cryptography and Coding 8-th IMA International Conference Cirencester, LNCS 2260, Springer Verlag, pp. 73–84, 2001.
11. J. Stern, D. Pointcheval, J. Malone-Lee and N. P. Smart, *Flaws in Applying Proof Methodologies to Signature Schemes*, In *Advances in Cryptology* CRYPTO*'02*, Santa Barbara, Lectures Notes in Computer Science 2442, pp. 93–110, Springer-Verlag, 2002.
12. C. P. Schnorr, *Efficient Signature Generation by Smart Cards*, In *Journal of Cryptology*, vol. 4, pp. 161–174, 1991.
13. U.S. Department of Commerce, National Institute of Standards and Technology. Digital Signature Standard. Federal Information Processing Standard Publication 186, 1994.

# Security Proofs for Identity-Based Identification and Signature Schemes

Mihir Bellare[1], Chanathip Namprempre[2], and Gregory Neven[3]

[1] Department of Computer Science & Engineering, University of California, San Diego, 9500 Gilman Drive, La Jolla, CA 92093, USA
mihir@cs.ucsd.edu
http://www-cse.ucsd.edu/users/mihir

[2] Electrical Engineering Department, Thammasat University, Klong Luang, Patumtani 12121, Thailand
cnamprem@engr.tu.ac.th
http://www.engr.tu.ac.th/~nchanath

[3] Department of Computer Science, Katholieke Universiteit Leuven, Celestijnenlaan 200A, 3001 Heverlee-Leuven, Belgium
Gregory.Neven@cs.kuleuven.ac.be
http://www.cs.kuleuven.ac.be/~gregory/

**Abstract.** This paper provides either security proofs or attacks for a large number of identity-based identification and signature schemes defined either explicitly or implicitly in existing literature. Underlying these are a framework that on the one hand helps explain how these schemes are derived, and on the other hand enables modular security analyses, thereby helping to understand, simplify and unify previous work.

## 1 Introduction

CURRENT STATE OF THE AREA. The late eighties and early nineties saw the proposal of many identity-based identification (IBI) and identity-based signature (IBS) schemes. These include the Fiat-Shamir IBI and IBS schemes [11], the Guillou-Quisquater IBI and IBS schemes [16], the IBS scheme in Shamir's paper [29] introducing identity-based cryptography, and others [21, 13, 6]. Now, new, pairing-based IBS schemes are being proposed [26, 17, 23, 8, 32].

Prompted by the renewed interest in identity-based cryptography that has followed identity-based encryption (IBE) [7], we decided to revisit the IBI and IBS areas. An examination of past work revealed the following.

Although there is a lot of work on proving security in the identification domain, it pertains to standard rather than identity-based schemes. (For example, security proofs have been provided for standard identification schemes related to the Fiat-Shamir and Guillou-Quisquater IBI schemes [10, 4], but not for the IBI schemes themselves.) In fact, a provable-security treatment of IBI schemes is entirely lacking: there are no security definitions, and none of the existing schemes is proven secure. Given the practical importance and usage of IBI schemes, this is an important (and somewhat surprising) gap.

C. Cachin and J. Camenisch (Eds.): EUROCRYPT 2004, LNCS 3027, pp. 268–286, 2004.

The situation for IBS is somewhat better. Cha and Cheon provide a definition of security for IBS schemes and prove their scheme secure [8]. Dodis, Katz, Xu, and Yung [9] define a class of standard signature (SS) schemes that they call trapdoor, and then present a random-oracle-using transform (let us call it tSS-2-IBS) that turns any secure trapdoor SS (tSS) scheme into a secure IBS scheme. Security proofs for several existing IBS schemes, including those of [11, 16], are obtained by observing that these are the result of applying tSS-2-IBS to underlying tSS schemes already proven secure in the literature [24, 20, 1]. However, as we will see, there are several IBS schemes not yet proven secure (one example is Shamir's IBS scheme [29]), either because they are not the result of applying tSS-2-IBS to a tSS scheme, or because, although they are, the tSS scheme in question has not yet been analyzed.

The goal of this paper is to fill the above-mentioned gaps in the IBI and IBS areas.

PRELIMINARIES. The first step, naturally, is definitions. We extend to the IBI setting the three notions of security for standard identification (SI) schemes, namely security against impersonation under passive attacks (imp-pa), active attacks (imp-aa) [10], and concurrent attacks (imp-ca) [4]. Our model allows the adversary to expose user (prover) keys, and to mount either passive, active, or concurrent attacks on the provers, winning if it succeeds in impersonating a prover of its choice. We remark that although existing security definitions for other identity-based primitives [7, 8, 9] give us some guidance as to what adversary capabilities to consider, there are some issues in the definition for IBI that need thought, mainly related to what capabilities the adversary gets in what stage of its two-stage attack. See Section 2.

The security notion for SS schemes is the standard unforgeability under chosen-message attack (uf-cma) [15]. An appropriate extension of it for IBS schemes exists [8, 9] and we refer to it also as uf-cma. These definitions are recalled in the full version of the paper [2].

CERTIFICATION-BASED IBI AND IBS. Before executing the main task of analyzing practical IBI and IBS schemes, we pause to consider the following natural design of an IBI scheme, based on any given SI scheme, via the certification paradigm. The authority picks a public and secret key pair $(pk, sk)$ for a SI scheme, and provides these to prover $I$ along with a certificate *cert* consisting of the authority's signature on $I, pk$. The prover can now flow $pk$, *cert* to the verifier and then identify itself via the SI scheme under $pk$. The verifier needs to know only $I$ and the public key of the authority in order to authenticate the prover.

In [2], we prove that the above yields a secure IBI scheme. An analogous result holds in the IBS case. We believe that this is worth noting because it highlights the fact that, unlike IBE [7], IBI and IBS are trivial to achieve (and in particular do not require random-oracles), and enables us to better understand what the practical schemes are trying to do, namely to beat the trivial certification-based schemes in performance.

MAIN CONTRIBUTIONS AND APPROACH. This paper delivers security proofs for a large number of practical IBI and IBS schemes, including not only the ones mentioned above, but many more that we surface as having been, with hindsight, implicit in the literature.

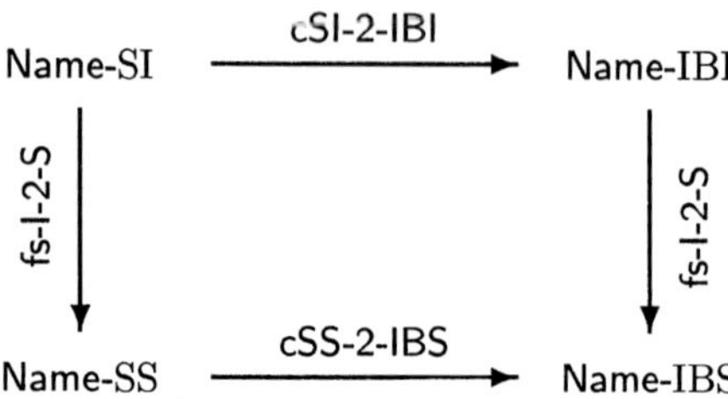


**Fig. 1.** Family of schemes associated to a cSI scheme Name-SI. If Name-SI is imp-atk secure then Name-IBI is also imp-atk secure, for all atk $\in \{\mathrm{pa}, \mathrm{aa}, \mathrm{ca}\}$. If Name-SI is imp-pa secure then Name-IBS is uf-cma secure. Implicit in drawing the diagram this way is that fs-I-2-S(cSI-2-IBI(Name-SI)) = cSS-2-IBS(fs-I-2-S(Name-SI)).

We do this in two steps. In the first step, we provide a framework that (in most cases) reduces proving security of IBI or IBS schemes to proving security of an underlying SI scheme. In a few cases, we found that the SI schemes in question were already analyzed in the literature, but in many cases they were not. The second step, where lies the main technical work of the paper, is to provide security proofs for those SI schemes not already proven secure, and then provide direct security proofs for the few exceptional IBI or IBS schemes that escape being captured by our framework.

The framework, we believe, is of value beyond its ability to reduce proving security of IBI and IBS schemes to proving security of SI schemes. It helps understand how schemes are being derived, and in the process surfaces the implicit schemes we mentioned above. Overall, the framework contributes to simplifying and unifying our picture of the area. We now explain the framework, which is based on a set of transforms, and then summarize the results for specific schemes.

THE TRANSFORMS. We introduce (cf. Definition 2) a class of SI schemes that we call convertible. The idea is that their key-generation process be underlain by a primitive called a trapdoor samplable relation that we introduce in Definition 1. We then present a random-oracle-using transform cSI-2-IBI that transforms a convertible SI (cSI) scheme into an IBI scheme (cf. Construction 1). Theorem 1 shows that cSI-2-IBI is security-preserving, meaning that if the starting cSI scheme is imp-atk secure then so is the resulting IBI scheme (in the random oracle model), for each atk $\in \{\mathrm{pa}, \mathrm{aa}, \mathrm{ca}\}$. This will be our main tool for proving security of IBI schemes.

It is useful to analogously define convertible standard signature (cSS) schemes and a transform cSS-2-IBS that turns a uf-cma secure cSS scheme into a uf-cma secure IBS scheme. These extend [9] in the sense that any tSS scheme is also a cSS scheme, and cSS-2-IBS coincides with tSS-2-IBS when the starting scheme is a tSS scheme, but the class of cSS schemes is larger than the class of tSS schemes.

Now let fs-I-2-S denote the (random-oracle using) Fiat-Shamir transform [11] which turns a SI scheme into a SS scheme. We know that if the former is imp-pa secure then the latter is uf-cma secure [1]. (Application of the transform and this last result requires that the starting SI scheme be a three-move public-coin protocol satisfying a certain technical condition, but all this will always be true for the applications we consider.)

Putting the above together yields Corollary 1, which says that, as long as a cSI scheme $X$ is imp-pa secure, the IBS scheme cSS-2-IBS(fs-I-2-S($X$)) is uf-cma secure. This will be our main tool for proving security of IBS schemes.

We note that fs-I-2-S also transforms a given IBI scheme into an IBS scheme. Furthermore, cSS-2-IBS(fs-I-2-S($X$)) = fs-I-2-S(cSI-2-IBI($X$)) for any cSI scheme $X$. In other words, the diagram of Figure 1 "commutes."

As an aside, we remark that the analogue of the result of [1] does *not* hold for fs-I-2-S as a transform of IBI schemes to IBS schemes: Proposition 1 shows that there exists an imp-pa secure IBI scheme $Y$ which under fs-I-2-S yields an insecure IBS scheme. This does not contradict the above since this $Y$ is not the result of cSI-2-IBI applied to a cSI scheme, but it makes things more difficult in a few exception cases (that we will see later) in which we need to consider an IBS scheme $Z$ = fs-I-2-S($Y$) where $Y$ is an IBI scheme that is not equal to cSI-2-IBI($X$) for any cSI scheme $X$. See the end of Section 3 for more information.

SCHEME FAMILIES. We seek to explain any IBI scheme $Y$ in the literature by surfacing a cSI scheme $X$ such that cSI-2-IBI($X$) = $Y$. We seek to explain any IBS scheme $Z$ in the literature by surfacing a cSI scheme $X$ such that cSS-2-IBS(fs-I-2-S($X$)) = $Z$. We are able to do this for the schemes in [11, 16, 29, 13, 17, 8, 32, 6] and for the RSA-based IBI scheme in [21], which, by Theorem 1 and Corollary 1, reduces the task of showing that $Y$, $Z$ are secure to showing that $X$ is secure in these cases.

We remark that the above gives rise to numerous schemes that are "new" in the sense that they were not provided explicitly in the literature. For example, Shamir [29] defined an IBS scheme but no IBI scheme. (He even says providing an IBI scheme is an open question.) Denoting Shamir's IBS scheme by Sh-IBS, we surface the cSI scheme Sh-SI such that cSS-2-IBS(fs-I-2-S(Sh-SI)) = fs-I-2-S(cSI-2-IBI(Sh-SI)) = Sh-IBS. As a consequence, we surface the IBI scheme Sh-IBI = cSI-2-IBI(Sh-SI) that is related in a natural way to Sh-IBS, namely by the fact that fs-I-2-S(Sh-IBI) = Sh-IBS. In an analogous way we surface IBI schemes Hs-IBI and ChCh-IBI underlying the IBS schemes of [17] and [8, 32], respectively.

Beside explaining existing IBI or IBS schemes, we are able to derive some new ones. We found papers in the literature [19, 22, 12] not defining IBI or IBS schemes, but defining SI schemes that we can show are convertible. Our transforms then yield new IBI and IBS schemes that we analyze.

We feel that this systematic surfacing of implicit schemes helps to homogenize, unify, and simplify the area. Figure 1 summarizes the perspective that emerges. We view schemes as occurring in families. Each family has a family name Name. At the core of the family is a cSI scheme Name-SI. The other schemes are related to it via Name-IBI = cSI-2-IBI(Name-SI), Name-SS = fs-I-2-S(Name-SI), and Name-IBS = cSS-2-IBS(Name-SS). If Name-SI is secure, so are all other schemes in the family.

RESULTS FOR SPECIFIC SCHEMES. In order to complete the task of obtaining security proofs for the existing and new IBI and IBS schemes we have discussed, it remains to analyze the cSI schemes underlying the families in question. This turns out to be a large task, for although in a few cases the cSI scheme is one already analyzed in the

| Name | Origin | Name-SI | | | Name-IBI | | | Name-SS | Name-IBS |
|---|---|---|---|---|---|---|---|---|---|
| | | imp-pa | imp-aa | imp-ca | imp-pa | imp-aa | imp-ca | uf-cma | uf-cma |
| FS | IBI,IBS [11, 10] | [11] | [10] | **I** | **I** | **I** | **I** | [24] | [9] |
| ItR | SI, SS [19, 22] | [28] | [28] | **U** | **I** | **I** | **U** | [24] | [9] |
| FF | SI,SS [12] | [12] | [12] | [12] | **I** | **I** | **I** | [12] | [9] |
| GQ | IBI, IBS [16] | [16] | [4] | [4] | **I** | **I** | **I** | [24] | [9] |
| Sh | IBS [29] | **P** | **A** | **A** | **I** | **A** | **A** | **I** | **I** |
| Sh* | SI | **P** | **P** | **P** | **I** | **I** | **I** | **I** | **I** |
| OkRSA | SI, IBI, SS [21] | [21] | [21] | **I** | **I** | **I** | **I** | [24] | [9] |
| Gir | SI, IBI [13, 25] | **A** | **A** | **A** | **A** | **A** | **A** | **A** | **A** |
| SOK | IBS [26] | **P** | **A** | **A** | **I** | **A** | **A** | **I** | **I** |
| Hs | IBS [17] | **P** | **P** | **P** | **I** | **I** | **I** | [17] | [9] |
| ChCh | IBS [8, 32] | **P** | **P** | **P** | **I** | **I** | **I** | [8] | [8] |
| Beth | IBI [6] | **P** | **U** | **U** | **I** | **U** | **U** | **I** | **I** |
| OkDL | IBI [21] | **I** | **I** | **I** | **P** | **P** | **P** | **I** | **I** |
| BNN | SI,IBI | **I** | **I** | **I** | **P** | **P** | **P** | **I** | **I** |

**Fig. 2.** Summary of security results. Column 1 is the family name of a family of schemes. Column 2 indicates which of the four member-schemes of the family existed in the literature. (The others we surface.) In the security columns, a known result is indicated via a reference to the paper establishing it. The marks **I**, **P**, and **A** all indicate new results obtained in this paper. An **I** indicates a proof of security obtained by implication. (If under Name-IBI it means we obtain it via Theorem 1, if under Name-IBS it means we obtain it either via Corollary 1 or via our modified fs-I-2-S transform, if elsewhere it means it follows easily from, or is an easy extension of, existing work.) A **P** indicates a new security proof, such as a from-scratch analysis of some SI or IBI scheme. An **A** indicates an attack that we have found. A **U** indicates that the security status is unknown. In all but the last two rows, the SI scheme is convertible. The first set of schemes are factoring based, the next RSA based, the next pairing based, and the last DL based. For each of the schemes above except for the last two, Name-IBS is obtained through the fs-I-2-S transform. OkDL-IBS and BNN-IBS are obtained through a modified version of the fs-I-2-S transform.

literature, we found (perhaps surprisingly) that in many cases it is not. Additionally, we need to directly analyze two IBI schemes not underlain by cSI schemes, namely the DL-based scheme in [21], and a somewhat more efficient Schnorr-based [27] variant that we introduce.

A summary of our results is in Figure 2. Section 4 and the full version of the paper [2] provide scheme descriptions and more precise result statements. Note all security proofs for SS, IBI, and IBS schemes are in the random-oracle (RO) model of [5]. Proofs are in [2]. Here, we highlight some of the important elements of these results.

CASES CAPTURED BY OUR FRAMEWORK. Section 4 begins by surfacing SI schemes underlying the first 12 (i.e. all but the last two) families of Figure 2 and shows that they are convertible, so that the picture of Figure 1 holds in all these cases and we need only consider security of the cSI schemes. The analysis of these schemes follows.

Easy cases are FS, ItR (the iterated-root, also called $2^t$-th root, family), FF, GQ, and OkRSA (an RSA-based family from [21]) where the SI schemes are already present and analyzed in the literature [10, 28, 12, 4, 21].

The Sh-SI scheme turns out to be a mirror-image of GQ-SI, and is interesting technically because we show that it is honest-verifier zero-knowledge (HVZK) even though it might not at first appear to be so. Based on this, we prove that it is imp-pa (cf. Theorem 3), but simple attacks show that imp-aa and imp-ca do not hold. A slight modification Sh*-SI of this scheme however is not only imp-pa but also proven imp-aa and imp-ca secure under the one-more-RSA assumption of [3] (cf. Theorem 4), so that its security is like that of GQ-SI [4].

An attack and a fix for Girault's IBI scheme [13] were proposed in [25], but we find attacks on the fixed scheme as well, breaking all schemes in the family.

We prove imp-pa security of the pairing-based SOK-SI, Hs-SI and ChCh-SI schemes under a computational DH assumption and imp-aa, imp-ca security under a one-more computational DH assumption (cf. Theorems 5 and 6). We remark that the SOK-IBS scheme defined via our transforms is not the one of [26], but is slightly different. This suggests the value of our framework, for it is unclear whether the IBS scheme of [26] can be proved uf-cma secure, whereas Corollary 1 implies that SOK-IBS is uf-cma secure.

Since the discrete-log function has no known trapdoor it is not an obvious starting point for IBI schemes, but some do exist. Beth's (unproven) IBI scheme [6] is based on ElGamal signatures. The proof of convertibility of the Beth-SI scheme we surface is interesting in that it exploits the existential forgeability of ElGamal signatures. Theorem 7 says that Beth-SI is imp-pa secure if the hashed-message ElGamal signature scheme is universally unforgeable under no-message attack in the random-oracle model.

EXCEPTIONS. The last two rows of Figure 2 represent cases where our framework does not apply and direct analyses are needed. The first such case is an unproven DL-based IBI scheme OkDL-IBI due to Okamoto [21], which introduces an interesting SS-based method for constructing IBI schemes and instantiates it with his own DL-based SS scheme. We were unable to surface any cSI scheme which under cSI-2-IBI maps to OkDL-IBI. (OkDL-IBI can be "dropped" in a natural way to a SI scheme OkDL-SI, but the latter does not appear to be convertible.) However, we show in [2] that OkDL-IBI is nevertheless imp-pa, imp-aa, and imp-ca secure assuming hardness of the DL problem. This direct proof is probably the most technical in the paper and uses the security of Okamoto's DL-based SS scheme under a weakened notion of non-malleability [31], which is established via an extension of the result of [1] combined with results from [21]. We also present a new IBI scheme BNN-IBI that is based on the paradigm underlying OkDL-IBI but uses Schnorr signatures [27] instead of Okamoto signatures. It is slightly more efficient than OkDL-IBI. Security results are analogous to those above. See [2] for descriptions of the schemes and our results.

Proposition 1 precludes proving security of the IBS schemes fs-I-2-S(OkDL-IBI) and fs-I-2-S(BNN-IBI) based merely on the security properties of the IBI schemes. However, we slightly modify the classical fs-I-2-S transform and obtain a transform that yields a secure uf-cma IBS scheme when applied to an imp-pa IBI scheme. We can then apply this transform to OkDL-IBI or BNN-IBI to obtain uf-cma IBS schemes.

RELATED WORK. Independent of our work, Kurosawa and Heng [18] recently presented a transform from a certain class of "zero-knowledge" SS schemes to IBI schemes. However, the IBI scheme resulting from their transform is only shown to be secure against impersonation under *passive* attacks.

## 2 Security Notions for Identification Schemes

NOTATION. We let $\mathbb{N} = \{1, 2, 3, \ldots\}$ denote the set of natural numbers. If $k \in \mathbb{N}$, then $1^k$ is the string of $k$ ones. The empty string is denoted $\varepsilon$. If $x, y$ are strings, then $|x|$ is the length of $x$ and $x \| y$ is the concatenation of $x$ and $y$. If $S$ is a set, then $|S|$ is its cardinality. If $A$ is a randomized algorithm, then $A(x_1, x_2, \ldots : \mathrm{O}_1, \mathrm{O}_2, \ldots)$ means that $A$ has inputs $x_1, x_2, \ldots$ and access to oracles $\mathrm{O}_1, \mathrm{O}_2, \ldots$, and $y \xleftarrow{\$} A(x_1, x_2, \ldots : \mathrm{O}_1, \mathrm{O}_2, \ldots)$ means that the output of $A$'s run is assigned to $y$. We denote the set of all possible outputs by $[A(x_1, x_2, \ldots : \mathrm{O}_1, \mathrm{O}_2, \ldots)]$, the running time of $A$ by $\mathbf{T}_A$, and the number of times $A$ queried the $\mathrm{O}_i$ oracle by $\mathbf{Q}_A^{\mathrm{O}_i}$. We define $\mathbf{Q}_A = \sum_i \mathbf{Q}_A^{\mathrm{O}_i}$.

An interactive algorithm (modelling a party such as prover or verifier in a protocol) is a stateful algorithm that on input an incoming message $M_{in}$ (this is $\varepsilon$ if the party is initiating the protocol) and state information $\mathrm{St}$ outputs an outgoing message $M_{out}$ and updated state $\mathrm{St}'$. For an interactive algorithm $A$ that has access to oracles $\mathrm{O}_1, \mathrm{O}_2, \ldots$, this is written as $(M_{out}, \mathrm{St}') \xleftarrow{\$} A(M_{in}, \mathrm{St} : \mathrm{O}_1, \mathrm{O}_2, \ldots)$. The initial state of $A$ contains its inputs and optionally a random tape $\rho$; if no random tape is explicitly given in the initial state, $A$ is assumed to toss its own coins.

STANDARD IDENTIFICATION SCHEMES. A *standard identification (SI) scheme* is a tuple $\mathcal{SI} = (\mathsf{Kg}, \mathsf{P}, \mathsf{V})$ where $\mathsf{Kg}$ is the randomized polynomial-time key generation algorithm, and $\mathsf{P}$ and $\mathsf{V}$ are polynomial-time interactive algorithms called the prover and verifier algorithms, respectively. In an initialization step, the prover runs $\mathsf{Kg}(1^k)$, where $k$ is a security parameter, to obtain a key pair $(pk, sk)$, and publishes the public key $pk$ while keeping the secret key $sk$ private. In the interactive identification protocol, the prover runs $\mathsf{P}$ with initial state $sk$, and the verifier runs $\mathsf{V}$ with initial state $pk$. The first and last messages of the protocol belong to the prover. The protocol ends when $\mathsf{V}$ enters either the $\mathsf{acc}$ or $\mathsf{rej}$ state. We require that for all $k \in \mathbb{N}$ and for all $(pk, sk) \in [\mathsf{Kg}(1^k)]$, the result of the interaction between $\mathsf{P}$ (initialized with $sk$) and $\mathsf{V}$ (initialized with $pk$) is $\mathsf{acc}$ with probability one.

SECURITY OF SI SCHEMES. An adversary $A$ is a pair of algorithms $(CV, CP)$ called the *cheating verifier* and the *cheating prover* [10]. We briefly recall the notions of imp-pa, imp-aa [10], and imp-ca [4]. The experiment first chooses keys $(pk, sk)$ via $\mathsf{Kg}(1^k)$ and then runs $CV$ on $pk$. For a passive attack (pa), $CV$ gets a conversation oracle, which, upon a query, returns a transcript of the conversation between $\mathsf{P}$ (with initial state $sk$) and $\mathsf{V}$ (with initial state $pk$), each time generated under fresh coins for both parties. For an active attack (aa) or concurrent attack (ca), $CV$ gets a prover oracle PROV. Upon a query $(M, s)$ where $M$ is a message and $s$ is a session number, the PROV oracle runs the prover algorithm using $M$ as an incoming message and returns the prover's outgoing message while maintaining the prover's state associated with the session $s$ across the invocations. (For each new session, PROV uses fresh random coins to start the prover, initializing it with $sk$.) The difference between active and concurrent attacks is that the former allows only a single prover to be active at a time. Eventually, $CV$ halts with some output that is given to $CP$, and $A$ wins if the interaction between $CP$ and $\mathsf{V}$ (initialized with $pk$) leads the latter to accept. For atk $\in \{\mathrm{pa}, \mathrm{aa}, \mathrm{ca}\}$, the *imp-atk advantage* of $A$ in attacking $\mathcal{SI}$ is written as $\mathbf{Adv}^{\mathrm{imp\text{-}atk}}_{\mathcal{SI},A}(k)$ and is defined to be the probability of $A$ winning in the above experiment. We say that $\mathcal{SI}$ is an *imp*-atk-*secure SI scheme* if $\mathbf{Adv}^{\mathrm{imp\text{-}atk}}_{\mathcal{SI},A}(\cdot)$ is negligible for every polynomial-time $A$.

IDENTITY-BASED IDENTIFICATION SCHEMES. An *identity-based identification (IBI) scheme* is a four-tuple $\mathcal{IBI} = (\mathsf{MKg}, \mathsf{UKg}, \overline{\mathsf{P}}, \overline{\mathsf{V}})$ of polynomial-time algorithms. The trusted, key-issuing authority runs the *master-key generation* algorithm MKg on input $1^k$, where $k$ is a security parameter, to obtain a master public and secret key pair $(mpk, msk)$. It can then run the *user-key generation* algorithm UKg on $msk$ and the identity $I \in \{0,1\}^*$ of a user to generate for this user a secret key $usk$ which is then assumed to be securely communicated to the user in question. In the interactive identification protocol, the prover with identity $I$ runs interactive algorithm $\overline{\mathsf{P}}$ with initial state $usk$, and the verifier runs $\overline{\mathsf{V}}$ with initial state $mpk, I$. The first and last messages of the protocol belong to the prover. The protocol ends when $\overline{\mathsf{V}}$ enters either the acc or rej state. In the random oracle model, $\mathsf{UKg}, \overline{\mathsf{P}}, \overline{\mathsf{V}}$ additionally have oracle access to a function H whose range may depend on $mpk$. We require that for all $k \in \mathbb{N}$, $I \in \{0,1\}^*$, $(mpk, msk) \in [\mathsf{MKg}(1^k)]$, functions H with appropriate domain and range, and $usk \in [\mathsf{UKg}(msk, I : \mathrm{H})]$, the interaction between $\mathsf{P}$ (initialized with $usk$) and $\mathsf{V}$ (initialized with $mpk, I$) is acc with probability one.

SECURITY OF IBI SCHEMES. The security definition for IBI schemes is similar to that of SI schemes. We highlight only the differences here. An adversary $\overline{A}$ is a pair of a cheating verifier $\overline{CV}$ and a cheating prover $\overline{CP}$. It is given a conversation oracle for passive attacks or a prover oracle for active and concurrent attacks as before except that here it can ask for transcripts or for interactions with respect to identities of its choice. For all three types of attacks, it is additionally given access to an initialization oracle and a corrupt oracle with which it can initialize and corrupt an identity, respectively. The former causes the new identity to receive a newly generated user secret key while the latter exposes the identity's user secret key to $\overline{A}$ then marks the identity as corrupted. As before, $\overline{CV}$ is run first. At its completion, it returns an uncorrupted identity $J$ to be impersonated (along with other state information). Then, $\overline{CP}$ attempts the impersonation for $J$. Throughout, $\overline{A}$ is not allowed to submit queries involving corrupted identities (other than the original corrupting queries). Additionally, $\overline{CP}$ is not allowed to submit queries involving $J$. For atk $\in \{\mathrm{pa}, \mathrm{aa}, \mathrm{ca}\}$, the *imp-atk advantage* of $\overline{A}$ in attacking $\mathcal{IBI}$ is written as $\mathbf{Adv}^{\mathrm{imp\text{-}atk}}_{\mathcal{IBI},\overline{A}}(k)$ and is defined to be the probability of $\overline{A}$ winning in the above experiment. We say that $\mathcal{IBI}$ is an *imp*-atk-*secure IBI scheme* if $\mathbf{Adv}^{\mathrm{imp\text{-}atk}}_{\mathcal{IBI},\overline{A}}(\cdot)$ is negligible for every polynomial-time $\overline{A}$. Details are in [2].

## 3 Convertible Schemes and Our Transforms

In analogy with the definition of trapdoor signature schemes [9], we define the concept of *convertible identification schemes* and show how to transform these into IBI schemes. We use a slightly more general concept than the trapdoor one-way permutations used by [9] that we will call *trapdoor samplable relations*. A relation $\mathbf{R}$ is a set of ordered pairs $(x, y) \in \mathrm{Dom}(\mathbf{R}) \times \mathrm{Ran}(\mathbf{R})$. We write the set of images of $x \in \mathrm{Dom}(\mathbf{R})$ as $\mathbf{R}(x) = \{y \mid (x, y) \in \mathbf{R}\}$ and the set of inverses of $y \in \mathrm{Ran}(\mathbf{R})$ as $\mathbf{R}^{-1}(y) = \{x \mid (x, y) \in \mathbf{R}\}$.

**Definition 1.** *A family of* trapdoor samplable relations $F$ *is a triplet of polynomial-time algorithms* $(\mathsf{TDG}, \mathsf{Sample}, \mathsf{Inv})$ *such that the following properties hold: (1)* Efficient generation: *On input* $1^k$, *where* $k \in \mathbb{N}$ *is the security parameter,* $\mathsf{TDG}$ *outputs the description* $\langle\mathbf{R}\rangle$ *of a relation* $\mathbf{R}$ *in the family together with its trapdoor information* $t$*; (2)* Samplability: *The output of the algorithm* $\mathsf{Sample}$ *on an input* $\langle\mathbf{R}\rangle$ *is uniformly distributed over* $\mathbf{R}$*; (3)* Inversion: *On input a relation description* $\langle\mathbf{R}\rangle$*, the corresponding trapdoor* $t$*, and an element* $y \in \mathrm{Ran}(\mathbf{R})$*, the randomized algorithm* $\mathsf{Inv}$ *outputs a random element of* $\mathbf{R}^{-1}(y)$*; (4)* Regularity: *Every relation* $\mathbf{R}$ *in the family is regular, meaning that the number of inverses* $|\mathbf{R}^{-1}(y)|$ *is the same for all* $y \in \mathrm{Ran}(\mathbf{R})$. ∎

Note that this definition does not ask that any computational problem relating to the family be hard. (For example, there is no "one-wayness" requirement.) We do not need any such assumption.

**Definition 2.** *A SI scheme* $\mathcal{SI} = (\mathsf{Kg}, \mathsf{P}, \mathsf{V})$ *is said to be* convertible *if there exists a family of trapdoor samplable relations* $F = (\mathsf{TDG}, \mathsf{Sample}, \mathsf{Inv})$ *such that for all* $k \in \mathbb{N}$ *the output of the following is distributed identically to the output of* $\mathsf{Kg}(1^k)$*:*

$$(\langle\mathbf{R}\rangle, t) \xleftarrow{\$} \mathsf{TDG}(1^k)\,;\ (x, y) \xleftarrow{\$} \mathsf{Sample}(\langle\mathbf{R}\rangle)\,;$$
$$pk \leftarrow (\langle\mathbf{R}\rangle, y)\,;\ sk \leftarrow (\langle\mathbf{R}\rangle, x)\,;\ \textit{Return}\ (pk, sk)$$ ∎

The following describes the cSI-2-IBI transform of a convertible SI (cSI) scheme into an IBI scheme. The idea is that to each identity $I$ we can associate a value that is derivable from the master public key and $I$. This value plays the role of a public key for the underlying cSI scheme. This "pseudo-public-key" is $(\langle\mathbf{R}\rangle, \mathrm{H}(I))$, where H is a random oracle.

**Construction 1.** Let $\mathcal{SI} = (\mathsf{Kg}, \mathsf{P}, \mathsf{V})$ be a cSI scheme, and let $F = (\mathsf{TDG}, \mathsf{Sample}, \mathsf{Inv})$ be the family of trapdoor samplable relations that underlies it as per Definition 2. The cSI-2-IBI transform associates to $\mathcal{SI}$ the random-oracle model IBI scheme $\mathcal{IBI} = (\mathsf{MKg}, \mathsf{UKg}, \overline{\mathsf{P}}, \overline{\mathsf{V}})$ whose components we now describe. The master and user key generation algorithms are defined as

| Algorithm $\mathsf{MKg}(1^k)$ | Algorithm $\mathsf{UKg}(msk, I : \mathrm{H})$ |
|---|---|
| $(\langle\mathbf{R}\rangle, t) \xleftarrow{\$} \mathsf{TDG}(1^k)$ | Parse $msk$ as $(\langle\mathbf{R}\rangle, t)$ |
| $mpk \leftarrow \langle\mathbf{R}\rangle$ ; $msk \leftarrow (\langle\mathbf{R}\rangle, t)$ | $x \xleftarrow{\$} \mathsf{Inv}(\langle\mathbf{R}\rangle, t, \mathrm{H}(I))$ ; $usk \leftarrow (\langle\mathbf{R}\rangle, x)$ |
| Return $(mpk, msk)$ | Return $usk$ |

where $\mathrm{H} : \{0,1\}^* \rightarrow \mathrm{Ran}(\mathbf{R})$ is a random oracle. The prover algorithm $\overline{\mathsf{P}}$ is identical to $\mathsf{P}$. The verifier algorithm $\overline{\mathsf{V}}(\cdot, \cdot : \mathrm{H})$ parses its initial state as $(\langle \mathbf{R} \rangle, I)$ and runs $\mathsf{V}$ on initial state $(\langle \mathbf{R} \rangle, \mathrm{H}(I))$. ∎

The following theorem, proved in [2], says that cSI-2-IBI is security-preserving.

**Theorem 1.** *Let $\mathcal{SI}$ be a cSI scheme and let $\mathcal{IBI}$ = cSI-2-IBI($\mathcal{SI}$) be the associated IBI scheme as per Construction 1. For any atk $\in \{pa, aa, ca\}$, if $\mathcal{SI}$ is imp-atk secure then $\mathcal{IBI}$ is imp-atk secure.*

Convertibility of a standard signature (SS) scheme $\mathcal{SS} = (\mathsf{Kg}, \mathsf{Sign}, \mathsf{Vf})$ is defined by analogy to Definition 2. (The condition is only on the key-generation algorithm.) The cSS-2-IBS transform is defined analogously to the cSI-2-IBI transform: given a convertible SS (cSS) scheme $\mathcal{SS} = (\mathsf{Kg}, \mathsf{Sign}, \mathsf{Vf})$, the transform yields an IBS scheme $\mathcal{IBS} = (\mathsf{MKg}, \mathsf{UKg}, \overline{\mathsf{Sign}}, \overline{\mathsf{Vf}})$ where the master and the user key generators are exactly as in Construction 1, and $\overline{\mathsf{Sign}}(usk, \cdot)$ and $\overline{\mathsf{Vf}}(mpk, I, \cdot, \cdot : \mathrm{H})$ are identical to $\mathsf{Sign}(usk, \cdot)$ and $\mathsf{Vf}((mpk, \mathrm{H}(I)), \cdot, \cdot)$, respectively. The proof of the following analogue of Theorem 1 is similar to the proof of Theorem 1 and is thus omitted.

**Theorem 2.** *Let $\mathcal{SS}$ be a cSS scheme and let $\mathcal{IBS}$ = cSS-2-IBS($\mathcal{SS}$) be the associated IBS scheme as defined above. If $\mathcal{SS}$ is uf-cma secure then $\mathcal{IBS}$ is also uf-cma secure.*

One can check that any trapdoor SS (tSS) scheme as defined in [9] is a cSS scheme, and their tSS-2-IBS transform coincides with cSS-2-IBS in case the starting cSS scheme is trapdoor. Thus, Theorem 2 represents a (slight) extension of their result. However, the extension is important, for we will see cases of cSS schemes that are not trapdoor and where the extension is needed.

We know that, if $\mathcal{SI}$ is an imp-pa secure SI scheme, then fs-I-2-S($\mathcal{SI}$) is a uf-cma secure SS scheme [1]. It is also easy to see that the fs-I-2-S transform of a cSI scheme is a cSS scheme. Combining this with Theorem 2 yields the following, which will be our main tool to prove security of IBS schemes.

**Corollary 1.** *Let $\mathcal{SI}$ be a cSI scheme, and let $\mathcal{IBS}$ = cSS-2-IBS(fs-I-2-S($\mathcal{SI}$)). If $\mathcal{SI}$ is imp-pa secure then $\mathcal{IBS}$ is uf-cma secure.*

Above, it is assumed that $\mathcal{SI}$ is a three-move, public coin protocol (so that one can apply fs-I-2-S to it) and also that the commitment (first move of the prover) is drawn from a space of super-polynomial size (so that the result of [1] applies). An SI or IBI scheme having these properties is called *canonical.*

One can also apply the fs-I-2-S transform to a canonical IBI scheme to obtain an IBS scheme, and one can check that cSS-2-IBS(fs-I-2-S($\mathcal{SI}$)) = fs-I-2-S(cSI-2-IBI($\mathcal{SI}$)) for any canonical cSI scheme $\mathcal{SI}$. It follows that fs-I-2-S yields a uf-cma secure IBS scheme if it is applied to a *converted* IBI scheme, meaning one that is obtained as the result of applying cSI-2-IBI to some (canonical) cSI scheme. However, one can also apply fs-I-2-S to a canonical IBI scheme that is not converted and get an IBS scheme, and there will be instances later where we would like to do this. Unfortunately, the IBS scheme so obtained need not be secure, in the sense that the analogue of the result of [1] does not hold, as stated below and proved in [2].

**Proposition 1.** *Assume there exists an imp-pa secure canonical IBI scheme. Then, there exists an imp-pa secure canonical IBI scheme $\mathcal{IBI}$ such that* fs-I-2-S($\mathcal{IBI}$) *is not uf-cma secure.*

We now provide a remedy for the above. We consider a modified version of the fs-I-2-S transform that hashes the identity of the signer (prover) along with the commitment and message, rather than merely hashing the commitment and message as in fs-I-2-S. We can show (by an extension of the proof of [1] that we omit) that, if this transform is applied to a canonical imp-pa secure IBI scheme, then the outcome is a uf-cma secure IBS scheme. We apply this in [2] to obtain uf-cma secure IBS schemes from the two unconverted IBI schemes we consider, namely OkDL-IBI and BNN-IBI.

## 4 Applying the Framework

We now apply the above transform-based framework to prove security of existing and new IBI and IBS schemes. To do this, we consider numerous SI schemes. (Some are known. Some are new.) We show that they are convertible, and then analyze their security. The implications for corresponding IBI and IBS schemes, obtained via the transforms discussed above, follow from Theorem 1 and Corollary 1. Figure 3 presents the key generation algorithms of the SI schemes we consider, and Figure 4 presents the corresponding identification protocols.

GENERATORS. The key generation algorithms shown in Figure 3 make use of parameter generation algorithms: $\mathcal{K}_{\text{fact}}$ for factoring-based schemes, $\mathcal{K}_{\text{rsa}}$ for RSA-based schemes, $\mathcal{K}_{\text{dlog}}$ for DL-based schemes and $\mathcal{K}_{\text{pair}}$ for pairing based schemes. These are randomized polynomial-time algorithms that on input $1^k$ produce the following outputs: $\mathcal{K}_{\text{fact}}$ generates tuples $(N, p, q)$ such that $p, q$ are primes and $N = pq$; $\mathcal{K}_{\text{rsa}}$ outputs $(N, e, d)$ such that $N$ is the product of two primes and $ed \equiv 1 \bmod \varphi(N)$; $\mathcal{K}_{\text{dlog}}$ outputs the description of a multiplicative group $\mathbb{G}$, its prime order $q$ and a generator $g$; $\mathcal{K}_{\text{pair}}$ generates the description of an additive group $\mathbb{G}_1$ and a multiplicative $\mathbb{G}_2$ of the same prime order $q$, a generator $P$ of $\mathbb{G}_1$ and a non-degenerate, polynomial-time computable bilinear map $\hat{e}\colon\ \mathbb{G}_1 \times \mathbb{G}_1 \to \mathbb{G}_2$. We say that $\mathcal{K}_{\text{rsa}}$ is a prime-exponent generator if $e$ is always a prime. Security results will make various assumptions about the computational problems underlying these generators.

HASH FUNCTION RANGES. In applying cSI-2-IBI to FS-SI, we assume the hash function in Construction 1 has range the set of quadratic residues modulo $N$ where $N$ is the modulus in the public key. This is a convenient abstraction in the random-oracle model, but note that implementing such a hash function is difficult since the range is not decidable in polynomial-time. However, this is a standard problem in this domain and various standard changes to the scheme take care of it. The same problem arises for several other schemes below as well, and also arises in [9]. We will not mention it again, but instead assume our random-oracle hash functions have whatever ranges we need. Those usually being obvious from the scheme are not discussed explicitly.

FS AND ItR. Since FS-SI is the special case of ItR-SI in which $m = 1$, it suffices to show that the latter is convertible. This is easily seen by considering the relation $\mathbf{R} =$

| FS | ItR | GQ, Sh, Sh* |
|---|---|---|
| $(N,p,q) \stackrel{\$}{\leftarrow} \mathcal{K}_{\mathrm{fact}}(1^k)$<br>For $i = 1 \ldots t$ do<br>$x_i \stackrel{\$}{\leftarrow} \mathbb{Z}_N^*$<br>$X_i \leftarrow {x_i}^{-2} \bmod N$<br>$pk \leftarrow (N,(X_1,\ldots,X_t))$<br>$sk \leftarrow (N,(x_1,\ldots,x_t))$ | $(N,p,q) \stackrel{\$}{\leftarrow} \mathcal{K}_{\mathrm{fact}}(1^k)$<br>For $i = 1 \ldots t$ do<br>$x_i \stackrel{\$}{\leftarrow} \mathbb{Z}_N^*$<br>$X_i \leftarrow {x_i}^{-2^m} \bmod N$<br>$pk \leftarrow (N,(X_1,\ldots,X_t))$<br>$sk \leftarrow (N,(x_1,\ldots,x_t))$ | $(N,e,d) \stackrel{\$}{\leftarrow} \mathcal{K}_{\mathrm{rsa}}(1^k)$<br>$x \stackrel{\$}{\leftarrow} \mathbb{Z}_N^*$<br>$X \leftarrow x^e \bmod N$<br>$pk \leftarrow ((N,e),X)$<br>$sk \leftarrow ((N,e),x)$ |
| **FF** | **Gir** | **OkRSA** |
| $(N,p,q) \stackrel{\$}{\leftarrow} \mathcal{K}_{\mathrm{fact}}(1^k)$<br>Choose $\tau \geq \eta(p,q) - 1$<br>$g \stackrel{\$}{\leftarrow} \mathrm{HQR}_N$<br>$x_1 \stackrel{\$}{\leftarrow} \mathbb{Z}_{2^m}$ ; $x_2 \stackrel{\$}{\leftarrow} \mathbb{Z}_N^*$<br>$X \leftarrow g^{x_1} x_2^{2^{m+\tau}} \bmod N$<br>$pk \leftarrow ((N,\tau,g),X)$<br>$sk \leftarrow ((N,\tau,g),(x_1,x_2))$ | $(N,e,d,f) \stackrel{\$}{\leftarrow} \mathcal{K}_{\mathrm{rsa}}(1^k)$<br>Choose $g \in \mathbb{Z}_N^*$ of order $f$<br>$h \leftarrow g^e \bmod N$ ; $s \stackrel{\$}{\leftarrow} \mathbb{Z}_f$<br>$X \stackrel{\$}{\leftarrow} \mathbb{Z}_N^*$<br>$S \leftarrow g^{-s} \bmod N$<br>$P \leftarrow X^{-d} S \bmod N$<br>$pk \leftarrow ((N,e,h,f),X)$<br>$sk \leftarrow ((N,e,h,f),(P,s))$ | $(N,e,d) \stackrel{\$}{\leftarrow} \mathcal{K}_{\mathrm{rsa}}(1^k)$<br>$g \stackrel{\$}{\leftarrow} \mathbb{Z}_N^*$<br>$x_1 \stackrel{\$}{\leftarrow} \mathbb{Z}_e$ ; $x_2 \stackrel{\$}{\leftarrow} \mathbb{Z}_N^*$<br>$X \leftarrow g^{-x_1} x_2^{-e} \bmod N$<br>$pk \leftarrow ((N,e,g),X)$<br>$sk \leftarrow ((N,e,g),(x_1,x_2))$ |

| SOK, Hs, ChCh | Beth |
|---|---|
| $(\mathbb{G}_1,\mathbb{G}_2,q,P,\hat{e}) \leftarrow \mathcal{K}_{\mathrm{pair}}(1^k)$<br>$s,u \stackrel{\$}{\leftarrow} \mathbb{Z}_q$ ; $S \leftarrow sP$<br>$U \leftarrow uP$ ; $V \leftarrow suP$<br>$pk \leftarrow ((\mathbb{G}_1,\mathbb{G}_2,q,P,\hat{e},S),U)$<br>$sk \leftarrow ((\mathbb{G}_1,\mathbb{G}_2,q,P,\hat{e},S),V)$ | $(\mathbb{G},q,g) \stackrel{\$}{\leftarrow} \mathcal{K}_{\mathrm{dlog}}(1^k)$<br>$r \stackrel{\$}{\leftarrow} \mathbb{Z}_q$ ; $R \leftarrow g^r$ ; $x,h \stackrel{\$}{\leftarrow} \mathbb{Z}_q$ ; $X \leftarrow g^x$<br>$s \leftarrow (h - Rx)r^{-1} \bmod q$<br>$pk \leftarrow ((\mathbb{G},q,g,X),h)$<br>$sk \leftarrow ((\mathbb{G},q,g,X),(R,s))$ |

**Fig. 3. Key generation algorithms of the 12 cSI schemes that we consider.** Each takes input $1^k$ and returns $(pk, sk)$. The integers $m, t \geq 1$ where used are scheme parameters. See the text for notation used above.

$\{((x_1,\ldots,x_t),(X_1,\ldots,X_t)) \mid X_i \equiv x_i^{-2^m} \bmod N \text{ for } i = 1,\ldots,t\}$ with description $\langle \mathbf{R} \rangle = N$ and trapdoor $(p,q)$. Pair sampling involves selecting random elements from $\mathbb{Z}_N^*$, raising them to the $2^m$-th power, and inverting them modulo $N$.

We note that FS-IBI = cSI-2-IBI(FS-SI) is exactly the IBI scheme in [11] and FS-IBS = cSS-2-IBS(fs-I-2-S(FS-SI)) is exactly the IBS scheme in [11]. We know that FS-SI is imp-pa and imp-aa secure assuming factoring is hard [10], and this easily extends to imp-ca. Theorem 1 implies that FS-IBI inherits these security attributes. (Corollary 1 implies uf-cma security of FS-IBS assuming factoring is hard, but this was known [9].)

We know that ItR-SI is imp-pa and imp-aa secure assuming factoring is hard [30, 28]. Theorem 1 implies that ItR-IBI = cSI-2-IBI(ItR-SI) is imp-pa and imp-aa secure assuming factoring is hard. (Corollary 1 implies that ItR-IBS = cSS-2-IBS(fs-I-2-S( ItR-SI)) is uf-cma assuming factoring is hard, but this was known [9].) Whether ItR-SI is imp-ca secure, and hence whether ItR-IBI is imp-ca secure, remains open.

| Scheme | Cmt / Ch | Rsp / Accept condition |
|---|---|---|
| FS | $y \xleftarrow{\$} \mathbb{Z}_N^*; Y \leftarrow y^2 \bmod N$ | $z \leftarrow y \prod_i x_i^{c_i} \bmod N$ |
| | $c = (c_1, \ldots, c_t) \xleftarrow{\$} \mathbb{Z}_2^t$ | Accept iff $Y \equiv z^2 \prod_i X_i^{c_i} \bmod N$ |
| ItR | $y \xleftarrow{\$} \mathbb{Z}_N^*; Y \leftarrow y^{2^m} \bmod N$ | $z \leftarrow y \prod_i x_i^{c_i} \bmod N$ |
| | $c = (c_1, \ldots, c_t) \xleftarrow{\$} \mathbb{Z}_{2^m}^t$ | Accept iff $Y \equiv z^{2^m} \prod_i X_i^{c_i} \bmod N$ |
| FF | $y_1 \xleftarrow{\$} \mathbb{Z}_{2^{m+\tau}}$<br>$y_2 \xleftarrow{\$} \mathbb{Z}_N^*$<br>$Y \leftarrow g^{y_1} y_2^{2^{m+\tau}} \bmod N$ | $z_1 \leftarrow y_1 + cx_1 \bmod 2^{m+\tau}$<br>$\alpha \leftarrow \lfloor (y_1 + cx_1)/2^{m+\tau} \rfloor$<br>$z_2 \leftarrow g^{\alpha} y_2 x_2^c \bmod N$ ; $z \leftarrow z_1, z_2$ |
| | $c \xleftarrow{\$} \mathbb{Z}_{2^m}$ | Accept iff $g^{z_1} z_2^{2^{m+\tau}} \equiv YX^c \bmod N$ |
| Sh | $y \xleftarrow{\$} \mathbb{Z}_N^*; Y \leftarrow y^e \bmod N$ | $z \leftarrow xy^c \bmod N$ |
| | $c \xleftarrow{\$} \{0, \ldots, 2^{l(k)} - 1\}$ | Accept iff $z^e \equiv XY^c \bmod N$ |
| Sh* | $y \xleftarrow{\$} \mathbb{Z}_N^*; Y \leftarrow y^e \bmod N$ | $z \leftarrow xy^c \bmod N$ |
| | $c \xleftarrow{\$} \{1, \ldots, 2^{l(k)}\}$ | Accept iff $z^e \equiv XY^c \bmod N$ |
| GQ | $y \xleftarrow{\$} \mathbb{Z}_N^*; Y \leftarrow y^e \bmod N$ | $z \leftarrow x^c y \bmod N$ |
| | $c \xleftarrow{\$} \{0,1\}^{l(k)}$ | Accept iff $z^e \equiv X^c Y \bmod N$ |
| OkRSA | $y_1 \xleftarrow{\$} \mathbb{Z}_e$<br>$y_2 \xleftarrow{\$} \mathbb{Z}_N^*$<br>$Y \leftarrow g^{y_1} y_2^e \bmod N$ | $z_1 \leftarrow y_1 + cx_1 \bmod e$<br>$\alpha \leftarrow \lfloor (y_1 + cx_1)/e \rfloor$<br>$z_2 \leftarrow g^{\alpha} y_2 x_2^c \bmod N$ |
| | $c \xleftarrow{\$} \{0,1\}^{l(k)}$ | Accept iff $Y \equiv g^{z_1} z_2^e X^c \bmod N$ |
| Gir | $y \xleftarrow{\$} \mathbb{Z}_f; Y \leftarrow h^y \bmod N$<br>Cmt $\leftarrow (P, Y)$ | $z \leftarrow y + sc \bmod f$ |
| | $c \xleftarrow{\$} \{0,1\}^{l(k)}$ | Accept iff $h^z (P^e X)^c \equiv Y \bmod N$ |
| SOK | $y \xleftarrow{\$} \mathbb{Z}_q$ ; $Y \leftarrow yP$ | $z \leftarrow yc + V$ |
| | $c \xleftarrow{\$} \mathbb{G}_1$ | Accept iff $\hat{e}(z, P) = \hat{e}(U, S)\hat{e}(c, Y)$ |
| Hs | $y \xleftarrow{\$} \mathbb{Z}_q; Y \leftarrow \hat{e}(P, P)^y$ | $z \leftarrow yP + cV$ |
| | $c \xleftarrow{\$} \mathbb{Z}_q$ | Accept iff $\hat{e}(z, P) = Y \cdot \hat{e}(U, S)^c$ |
| ChCh | $y \xleftarrow{\$} \mathbb{Z}_q$ ; $Y \leftarrow yU$ | $z \leftarrow (y + c)V$ |
| | $c \xleftarrow{\$} \mathbb{Z}_q$ | Accept iff $\hat{e}(z, P) = \hat{e}(Y + cU, S)$ |
| Beth | $y \xleftarrow{\$} \mathbb{Z}_q$ ; $Y \leftarrow R^{-y}$<br>Cmt $\leftarrow (R, Y)$ | $z \leftarrow y + cs \bmod q$ |
| | $c \xleftarrow{\$} \{0,1\}^{l(k)}$ | Accept iff $g^c h \equiv R^z Y X^{cR}$ |

**Fig. 4. Identification protocols of the 12 cSI schemes that we consider.** We show the first commitment message Cmt sent by the prover, the challenge Ch sent by the verifier, the response Rsp returned by the prover, and the condition under which the verifier accepts. All schemes use Cmt $= Y$, Ch $= c$ and Rsp $= z$ unless explicitly defined otherwise. The prover is initialized with *sk* and the verifier with *pk*. The integers $m, t \geq 1$, and the challenge length $l$: $\mathbb{N} \rightarrow \mathbb{N}$, where used, are scheme parameters. In Sh-SI, Sh*-SI, GQ-SI, and Gir-SI, it is assumed that $2^{l(k)} < e$ for all $e$ output by $\mathcal{K}_{\text{rsa}}(1^k)$. All security results assume $l$ is super-logarithmic. $\mathcal{K}_{\text{rsa}}$ is a prime-exponent generator in Sh-SI, Sh*-SI, and GQ-SI.

FF. The FF-SI scheme was introduced by [12] as a fix to an attack they found on a scheme in [21]. In the key-generation algorithm of Figure 3, $\eta(p)$ denotes the largest integer such that $2^{\eta(p)}$ divides $p-1$ and $\eta(p,q) = \max(\eta(p), \eta(q))$. FF-SI is shown in [12] to be imp-pa, imp-aa, and imp-ca secure assuming factoring is hard. The authors defined no IBI or IBS schemes. We can show that FF-SI is convertible, and we thus obtain FF-IBI = cSI-2-IBI(FF-SI) and FF-IBS = cSS-2-IBS(fs-I-2-S(FF-SI)), and these are secure if factoring moduli generated by $\mathcal{K}_{\text{fact}}$ is hard.

Let $\text{HQR}_N = \{x^{2^{\eta(p,q)}} \bmod N \mid x \in \mathbb{Z}_N^*\}$ denote the set of higher quadratic residues modulo $N$, which is also the subset of elements of $\mathbb{Z}_N^*$ of odd order. To show convertibility of FF-SI we consider the relation $\mathbf{R} \subseteq (\mathbb{Z}_{2^m} \times \mathbb{Z}_N^*) \times \text{HQR}_N$ described by $(N, g, \tau)$ and containing tuples $((x_1, x_2), X)$ such that $g^{x_1} x_2^{2^{\tau+m}} \equiv X \bmod N$. The trapdoor is the factorization of $N$. Regularity holds since squaring is a permutation over $\text{HQR}_N$ and since each higher quadratic residue has exactly $2^{\eta(p)+\eta(q)}$ different $2^{\tau+m}$-th roots modulo $N$. Pair sampling involves choosing $x_1, x_2$ at random and computing $X = g^{x_1} x_2^{2^{\tau+m}}$.

GQ. The GQ-SI scheme defined via Figures 3 and 4 is the standard one considered in the literature. Convertibility is easily seen by considering the relation $\mathbf{R} = \{(x, X) \mid x^e \equiv X \bmod N\}$, relation description $\langle \mathbf{R} \rangle = (N, e)$, and trapdoor $d$. Pair sampling involves choosing $x \stackrel{\$}{\leftarrow} Z_N^*$ and computing $X \leftarrow x^e \bmod N$. We note that GQ-IBI = cSI-2-IBI(GQ-SI) is exactly the IBI scheme in [16], and GQ-IBS = cSS-2-IBS(fs-I-2-S(GQ-SI)) is exactly the IBS scheme in [16]. We know that GQ-SI is imp-pa secure assuming RSA is one-way, and imp-aa and imp-ca secure assuming hardness of the one-more-RSA problem [4]. Theorem 1 says that these results extend to GQ-IBI. (Also Corollary 1 says that GQ-IBS is uf-cma assuming RSA is one-way, but this was known [9].)

Sh AND Sh*. Shamir [29] defined an IBS scheme, but no SI or IBI schemes. He gave no security proof for his IBS scheme, and none has been provided until now.

We surface the SI scheme Sh-SI defined via Figures 3 and 4. One can check that Sh-IBS = cSS-2-IBS(fs-I-2-S(Sh-SI)) is exactly the IBS scheme in [29]. Sh-SI is interesting both historically and technically. It turns out to be a "mirror-image" of GQ-SI that closely resembles the latter. Convertibility of Sh-SI follows from the convertibility of GQ-SI since the two schemes have the same key-generation algorithm. Coming to consider security, the first question to ask is whether Sh-SI is honest-verifier zero-knowledge (HVZK). While this was obvious for GQ-SI (and in fact, if true for an SI scheme, is usually obvious), it is in fact not apparent at first glance for Sh-SI, and one might suspect that the scheme is not HVZK. However, using a trick involving gcds, we show that Sh-SI is statistical (not perfect) HVZK. We also show, in [2], that it is a proof of knowledge and thereby obtain the following:

**Theorem 3.** *The* Sh-SI *is imp-pa secure assuming one-wayness of the underlying RSA key generator* $\mathcal{K}_{\text{rsa}}$.

Corollary 1 now implies that Sh-IBS is uf-cma secure under the same assumptions.

However, Sh-SI scheme is trivially insecure under active attacks, since the cheating verifier can learn the secret key by sending a zero challenge. But this minor weakness is easily fixed by "removing" the zero challenge. We define via Figures 3 and 4 a modified scheme we denote Sh*-SI. This scheme turns out to have security attributes analogous to those of GQ-SI in that we can show the following:

**Theorem 4.** *The* Sh*-SI *scheme is imp-pa secure assuming one-wayness of the underlying RSA key generator $\mathcal{K}_{\rm rsa}$, and imp-aa and imp-ca secure assuming the one-more-RSA problem relative to $\mathcal{K}_{\rm rsa}$ is hard.*

The proof of this theorem is in [2]. We obtain the usual consequences for Sh*-IBI = cSI-2-IBI(Sh*-SI) and Sh*-IBS = cSS-2-IBS(fs-I-2-S(Sh*-SI)).

OkRSA. Okamoto [21] presented an RSA-based SI scheme and a related RSA-based IBI scheme. He proved the former imp-pa and imp-aa secure assuming factoring is hard, and the proofs extend to establish imp-ca as well. However, he did not prove the IBI scheme secure, a gap we fill.

The OkRSA-SI scheme defined via Figures 3 and 4 is the above-mentioned SI scheme. Notice that OkRSA-IBI = cSI-2-IBI(OkRSA-SI) is exactly the RSA-based IBI scheme in [21]. To show security of OkRSA-IBI and OkRSA-IBS = cSS-2-IBS (fs-I-2-S(OkRSA-SI)), it suffices to show that OkRSA-SI is convertible. For this, the relation has description $\langle \mathbf{R} \rangle = (N, e, g)$, and contains tuples $((x_1, x_2), X) \in (\mathbb{Z}_e \times \mathbb{Z}_N^*) \times \mathbb{Z}_N^*$ such that $X \equiv g^{x_1} x_2^e \bmod N$. The trapdoor is $d$ such that $ed \equiv 1 \bmod \varphi(N)$. Pair sampling involves choosing $x_1, x_2$ at random and computing $X \equiv g^{x_1} x_2^e$.

Gir. In [13], Girault proposed an SI scheme that we have defined via Figures 3 and 4 and named Gir-SI. He also proposed a related IBI scheme. (These schemes are inspired by the Schnorr identification scheme [27] but use a modulus $N = pq$ where $p, q$ are of the special form $p = 2fp' + 1$ and $q = 2fq' + 1$ such that $f, p', q', p, q$ are all primes.) This IBI scheme did not use hash functions, which lead to an attack and later a fix [25]. The fixed IBI scheme turns out to be exactly Gir-IBI = cSI-2-IBI(Gir-SI).

Gir-SI is convertible with relation $\mathbf{R} = \{((P, s), X) \mid P^e \equiv X^{-1} h^{-s} \bmod N\}$ described by $(N, e, h, f)$. The trapdoor is $d \equiv e^{-1} \bmod \varphi(N)$. Pair sampling involves choosing $P$ and $s$ at random and computing $X$ as $P^{-e} h^{-s} \bmod N$. However, this does not help here because we found that all schemes in the family are insecure. In particular, Gir-SI is not even imp-pa secure, and neither is the fixed IBI scheme Gir-IBI. The signature scheme Gir-IBS = cSS-2-IBS(fs-I-2-S(Gir-IBI)) is not uf-cma secure either.

We attack only the Gir-IBS scheme, since the insecurity of the SI, IBI, and SS schemes then follows. In the Gir-IBS scheme, a signature of a user $I$ on a message $M$ under the master public key $mpk = (N, e, h, f)$ is a tuple $(P, Y, z)$ such that $Y \equiv h^z (P^e \cdot \mathrm{H}_1(I))^{\mathrm{H}_2(P \| Y \| M)} \bmod N$. Given a valid signature $(P_1, Y_1, z_1)$ for message $M_1$ and identity $I$, an adversary can forge $I$'s signature for any message $M_2$ as follows. If first computes $d_2 \leftarrow e^{-1} \bmod f$, $g \leftarrow h^{d_2} \bmod N$, and $S \leftarrow (P^e \cdot \mathrm{H}_1(I))^{d_2} \bmod N$. Then, it chooses $s_2$ from $\mathbb{Z}_f$ and computes $P_2 \leftarrow P_1 S^{-1} g^{-s_2} \bmod N$. To obtain the forgery, it chooses $y_2$ from $\mathbb{Z}_q$, lets $Y_2 \leftarrow h^{y_2} \bmod N$, computes $z_2 \leftarrow y_2 + s_2 \mathrm{H}_2(P_2 \| Y_2 \| M_2) \bmod f$. The forgery is $(P_2, Y_2, z_2)$.

It is natural to consider counteracting the above attack by removing $f$ from the public key. While this might work for the SI scheme, it does not for the IBI (or IBS) scheme. The reason is that, since $f$ still has to be included in each user's secret key, an adversary can easily extract it by corrupting one identity.

We stress that the scheme broken here is *not* the (perhaps better-known) SI scheme by Girault based on discrete logarithms [14].

PAIRING-BASED SCHEMES. Many recent papers propose pairing-based IBS schemes [26, 8, 32, 23, 17] (the schemes independently published by [8] and [32] are actually equivalent). Barring [8], none of these papers prove their scheme secure. (Some proofs in weak models were however provided in [17, 32].) However, the scheme of [17] was proven secure in [9].

None of these papers define SI or IBI schemes. We surface SOK-SI (from [26]), ChCh-SI (from [8, 32]) and Hs-SI (from [17]), as defined by Figures 3 and 4. The ChCh-IBS = cSS-2-IBS(fs-I-2-S(ChCh-SI)) and Hs-IBS = cSS-2-IBS(fs-I-2-S(Hs-SI)) schemes are exactly the original IBS schemes, while SOK-IBS = cSS-2-IBS(fs-I-2-S (SOK-SI)) is slightly different from the scheme of [26].

We now show that all these pairing-based SI schemes are convertible. Since they all have the same key-generation algorithm, a common argument applies. The relation is $\{(V, U) \in \mathbb{G}_1 \times \mathbb{G}_1 \mid \hat{e}(V, P) = \hat{e}(U, S)\}$, described by $\langle \mathbf{R} \rangle = (\mathbb{G}_1, \mathbb{G}_2, q, P, \hat{e}, S)$. The trapdoor is $s$ such that $S = sP$. Pair sampling involves choosing $r \stackrel{\$}{\leftarrow} \mathbb{Z}_q$ and computing the pair $(rP, rS)$. The following is proved in [2].

**Theorem 5.** SOK-SI *and* ChCh-SI *are imp-pa secure assuming that the computational Diffie-Hellman problem in the group* $\mathbb{G}_1$ *associated to* $\mathcal{K}_{\text{pair}}$ *is hard.*

Corollary 1 implies that ChCh-IBS, SOK-IBS and Hs-IBS are uf-cma secure IBS schemes, but of these only the result about SOK-IBS is new. However, we prove the following in [2]:

**Theorem 6.** ChCh-SI *and* Hs-SI *are imp-aa and imp-ca secure assuming that the one-more computational Diffie-Hellman problem in the group* $\mathbb{G}_1$ *associated to* $\mathcal{K}_{\text{pair}}$ *is hard.*

Theorem 1 implies that the ChCh-IBI and Hs-IBI schemes are imp-aa and imp-ca secure assuming that the one-more computational Diffie-Hellman problem in the group $\mathbb{G}_1$ associated to $\mathcal{K}_{\text{pair}}$ is hard. Thus, we obtain new, pairing-based IBI schemes with proofs of security.

SOK-SI and SOK-IBI are insecure under active or concurrent attacks: upon receiving a commitment $Y$, an adversary can choose $c' \stackrel{\$}{\leftarrow} \mathbb{Z}_q$, submit $c \leftarrow c'P$ as the challenge, and compute the prover's secret key from the response $z$ as $V \leftarrow z - cY$.

Beth. The Beth-SI scheme defined via Figures 3 and 4 was surfaced from [6]. Beth-IBI = cSI-2-IBI(Beth-SI) is a more efficient version of the IBI scheme actually presented in [6]. In these schemes, the prover proves knowledge of an ElGamal signature of his identity. Beth [6] gives no security proofs, but here we obtain one for Beth-IBI.

The Beth-SI scheme is convertible with the relation $\{((R, s), h) \in (\mathbb{G} \times \mathbb{Z}_q) \times \mathbb{Z}_q \mid X^R R^s \equiv g^h\}$ described by $\langle \mathbf{R} \rangle = (\mathbb{G}, q, g, X)$. The trapdoor is $x$ such that $g^x \equiv X$. Pair sampling involves choosing $a, b$ at random from $\mathbb{Z}_q$ and letting $R \leftarrow X^a g^b$, $s \leftarrow a^{-1}R \bmod q$ and $h \leftarrow bs \bmod q$. In [2],we prove the following:

**Theorem 7.** Beth-SI *is imp-pa secure assuming that the hashed-message ElGamal signature scheme associated to* $\mathcal{K}_{\mathrm{dlog}}$ *is universally unforgeable under no-message attacks in the random oracle model.*

While the hashed-message ElGamal signature scheme has never been formally proven secure, we note that *universal* forgery under *no-message* attacks is a very weak security notion for signature schemes and that a close variant of hashed-message ElGamal was proven uf-cma secure under the discrete log assumption in [24]. Now, Theorem 1 implies that Beth-IBI inherits the above security attributes, and Corollary 1 implies that Beth-IBS = cSS-2-IBS(fs-I-2-S(Beth-SI)) is uf-cma secure under the same assumptions. The imp-aa and imp-ca security of Beth-SI remains open.

## Acknowledgments

We thank Marc Fischlin for pointing out that the Sh-SI scheme is zero-knowledge. The first author is supported in part by NSF grants CCR-0098123, ANR-0129617, CCR-0208842, and an IBM Faculty Partnership Development Award. The second author is supported in part by the above-mentioned grants of the first author. The third author is supported by a Research Assistantship and travel grant from the Fund for Scientific Research – Flanders (Belgium).

## References

[1] M. Abdalla, J.H. An, M. Bellare, and C. Namprempre. From identification to signatures via the Fiat-Shamir transform: Minimizing assumptions for security and forward-security. In L. Knudsen, editor, *EUROCRYPT 2002*, volume 2332 of *LNCS*, pages 418–433. Springer-Verlag, April 2002.

[2] M. Bellare, C. Namprempre, and G. Neven. Security proofs for identity-based identification and signature schemes. http://www.cse.ucsd.edu/users/mihir/crypto-research-papers.html, February 2004.

[3] M. Bellare, C. Namprempre, D. Pointcheval, and M. Semanko. The one-more-RSA-inversion problems and the security of Chaum's blind signature scheme. *J. Cryptology*, 16(3):185–215, June 2003.

[4] M. Bellare and A. Palacio. GQ and Schnorr identification schemes: Proofs of security against impersonation under active and concurrent attack. In M. Yung, editor, *CRYPTO 2002*, volume 2442 of *LNCS*, pages 162–177. Springer-Verlag, August 2002.

[5] M. Bellare and P. Rogaway. Random oracles are practical: A paradigm for designing efficient protocols. In ACM, editor, *Proc. of the 1st CCS*, pages 62–73. ACM Press, November 1993.

[6] T. Beth. Efficient zero-knowledged identification scheme for smart cards. In C. Gunther, editor, *EUROCRYPT 1988*, volume 330 of *LNCS*, pages 77–86. Springer-Verlag, May 1988.

[7] D. Boneh and M. Franklin. Identity-based encryption from the Weil Pairing. In J. Kilian, editor, *CRYPTO 2001*, volume 2139 of *LNCS*, pages 213–229. Springer-Verlag, August 2001.

[8] J.C. Cha and J.H. Cheon. An identity-based signature from gap diffie-hellman groups. In Y. Desmedt, editor, *PKC 2003*, volume 2567 of *LNCS*, pages 18–30. Springer-Verlag, January 2003.

[9] Y. Dodis, J. Katz, S. Xu, and M. Yung. Strong key-insulated signature schemes. In Y. Desmedt, editor, *PKC 2003*, volume 2567 of *LNCS*, pages 130–144. Springer-Verlag, January 2003.

[10] U. Feige, A. Fiat, and A. Shamir. Zero knowledge proofs of identity. *J. Cryptology*, 1(2):77–94, 1988.

[11] A. Fiat and A. Shamir. How to prove yourself: Practical solutions to identification and signature problems. In A. Odlyzko, editor, *CRYPTO 1986*, volume 263 of *LNCS*, pages 186–194. Springer-Verlag, August 1986.

[12] M. Fischlin and R. Fischlin. The representation problem based on factoring. In B. Preneel, editor, *CT-RSA 2002*, volume 2271 of *LNCS*, pages 96–113. Springer-Verlag, February 2002.

[13] M. Girault. An identity-based identification scheme based on discrete logarithms modulo a composite number. In I. Damgård, editor, *EUROCRYPT 1990*, volume 473 of *LNCS*, pages 481–486. Springer-Verlag, May 1990.

[14] M. Girault. Self-certified public keys. In D. Davies, editor, *EUROCRYPT 1991*, volume 547 of *LNCS*, pages 490–497. Springer-Verlag, April 1991.

[15] S. Goldwasser, S. Micali, and R. Rivest. A digital signature scheme secure against adaptive chosen-message attacks. *SIAM J. Computing*, 17(2):281–308, April 1988.

[16] L. Guillou and J. J. Quisquater. A "paradoxical" identity-based signature scheme resulting from zero-knowledge. In S. Goldwasser, editor, *CRYPTO 1988*, volume 403 of *LNCS*, pages 216–231. Springer-Verlag, August 1989.

[17] F. Hess. Efficient identity based signature schemes based on pairings. In K. Nyberg and H. Heys, editors, *Selected Areas in Cryptography, SAC 2002*, pages 310–324. Springer-Verlag, February 2003.

[18] K. Kurosawa and S.-H. Heng. From digital signature to ID-based identification/signature. In *PKC 2004*. Springer-Verlag, 2004.

[19] K. Ohta and T. Okamoto. A modification of the Fiat-Shamir scheme. In S. Goldwasser, editor, *CRYPTO 1988*, volume 403 of *LNCS*, pages 232–243. Springer-Verlag, August 1990.

[20] K. Ohta and T. Okamoto. On concrete security treatment of signatures derived from identification. In H. Krawczyk, editor, *CRYPTO 1998*, volume 1462 of *LNCS*, pages 354–370. Springer-Verlag, August 1998.

[21] T. Okamoto. Provably secure and practical identification schemes and corresponding signature schemes. In E. Brickell, editor, *CRYPTO 1992*, volume 740 of *LNCS*, pages 31–53. Springer-Verlag, August 1992.

[22] H. Ong and C. Schnorr. Fast signature generation with a Fiat Shamir–like scheme. In I. Damgård, editor, *EUROCRYPT 1990*, volume 473 of *LNCS*, pages 432–440. Springer-Verlag, May 1990.

[23] K.G. Paterson. ID-based signatures from pairings on elliptic curves. Technical Report 2002/004, IACR ePrint Archive, January 2002.

[24] D. Pointcheval and J. Stern. Security arguments for digital signatures and blind signatures. *J. Cryptology*, 13(3):361–396, 2000.

[25] S. Saeednia and R. Safavi-Naini. On the security of girault's identification scheme. In H. Imai and Y. Zheng, editors, *PKC 1998*, volume 1431 of *LNCS*, pages 149–153. Springer-Verlag, February 1998.

[26] R. Sakai, K. Ohgishi, and M. Kasahara. Cryptosystems based on pairing. In *SCIS 2000*, Okinawa, Japan, January 2000.

[27] C. Schnorr. Efficient identification and signatures for smartcards. In G. Brassard, editor, *CRYPTO 1989*, volume 435 of *LNCS*, pages 239–252. Springer-Verlag, August 1990.

[28] C. Schnorr. Security of $2^t$-root identification and signatures. In N. Koblitz, editor, *CRYPTO 1996*, volume 1109 of *LNCS*, pages 143–156. Springer-Verlag, August 1996.

[29] A. Shamir. Identity-based cryptosystems and signature schemes. In G.R. Blakely and D. Chaum, editors, *CRYPTO 1984*, volume 196 of *LNCS*, pages 47–53. Springer-Verlag, 1984.

[30] V. Shoup. On the security of a practical identification scheme. *J. Cryptology*, 12(4):247–260, 1999.

[31] J. Stern, D. Pointcheval, J. Malone-Lee, and N.P. Smart. Flaws in applying proof methodologies to signature schemes. In M. Yung, editor, *CRYPTO 2002*, volume 2442 of *LNCS*, pages 93–110. Springer-Verlag, August 2002.

[32] X. Yi. An identity-based signature scheme from the weil pairing. *IEEE Communications Letters*, 7(2):76–78, 2003.

# Concurrent Signatures

Liqun Chen[1], Caroline Kudla[2*], and Kenneth G. Paterson[2**]

[1] Hewlett-Packard Laboratories, Bristol, UK
liqun.chen@hp.com
[2] Information Security Group
Royal Holloway, University of London, UK
{c.j.kudla, kenny.paterson}@rhul.ac.uk

**Abstract.** We introduce the concept of *concurrent signatures*. These allow two entities to produce two signatures in such a way that, from the point of view of any third party, both signatures are ambiguous with respect to the identity of the signing party until an extra piece of information (the keystone) is released by one of the parties. Upon release of the keystone, both signatures become binding to their true signers concurrently.

Concurrent signatures fall just short of providing a full solution to the problem of fair exchange of signatures, but we discuss some applications in which concurrent signatures suffice. Concurrent signatures are highly efficient and require neither a trusted arbitrator nor a high degree of interaction between parties. We provide a model of security for concurrent signatures, and a concrete scheme which we prove secure in the random oracle model under the discrete logarithm assumption.

**Keywords:** Concurrent signatures, fair exchange, Schnorr signatures, ring signatures.

## 1 Introduction

The problem of fair exchange of signatures is a fundamental and well-studied problem in cryptography, with potential application in a wide range of scenarios in which the parties involved are mutually distrustful. Ideally, we would like the exchange of signatures to be done in a *fair* way, so that by engaging in a protocol, either each party obtains the other's signature, or neither party does. It should not be possible for one party to terminate the protocol at some stage leaving the other party committed when they themselves are not.

The literature contains essentially two different approaches to solving the problem of fair exchange of signatures.

Early work on solving the problem was based on the idea of timed release or timed fair exchange of signatures [BN00, EGL85, G83]. Here, the two parties sign their respective messages and exchange their signatures "little-by-little" using

* This author is funded by Hewlett-Packard Laboratories.
** This author supported by the Nuffield Foundation NUF-NAL02.

C. Cachin and J. Camenisch (Eds.): EUROCRYPT 2004, LNCS 3027, pp. 287–305, 2004.

a protocol. Typically, such protocols are highly interactive with many message flows. Moreover, one party, say $B$, may often be at an advantage in that he sometimes has (at least) one more bit of $A$'s signature than she has of $B$'s. This may not be a significant issue if the computing power of the two parties are roughly equivalent. But if $B$ has superior computing resources, this may put him at a significant advantage since he may terminate the protocol early and use his resources to compute the remainder of $A$'s signature, while it may be infeasible for $A$ to do the same. Even if the fairness of such protocols could be guaranteed, they may still be too interactive for many applications. See [GP03] for further details and references for such protocols.

An alternative approach to solving the problem of fair exchange of signatures involves the use of a (semi-trusted) third party or arbitrator $T$ who can be called upon to handle disputes between signers. The idea is that $A$ registers her public key with $T$ in a one-time registration, and thereafter may perform many fair exchanges with other entities. To take part in a fair exchange with $B$, $A$ creates a partial signature which she sends to $B$. Entity $B$ can be convinced that the partial signature is valid (perhaps via a protocol interaction with $A$) and that $T$ can extract a full, binding signature from the partial signature. However, the partial signature on its own is not binding for $A$. $B$ then fulfils his commitment by sending $A$ his signature, and if valid, $A$ releases the full version of her signature to $B$. The protocol is fair since if $B$ does not sign, then $A$'s partial signature is worthless to $B$, and if $B$ does sign but $A$ refuses to release her full signature then $B$ can obtain it from $T$. The third party is only required in case of dispute; for this reason, protocols of this type are commonly referred to as optimistic fair exchange protocols. See [ASW98, ASW00, BGLS03, BW00, CS03, DR03, GJM99, PCS03] for further details of such schemes.

The main problem with such an approach is the requirement for a dispute-resolving third party with functions beyond those required of a normal Certification Authority. In general, appropriate third parties may not be available.

It is our thesis that the *full* power of fair exchange is not necessary in many application scenarios. This paper introduces a somewhat weaker concept, which we name *concurrent signatures*. The cost of concurrent signatures is that they do not provide the full security guarantees of a fair exchange protocol. Their benefit is that they have none of the disadvantages of previous solutions: they do not require a special trusted third party[3], and they do not rely on a computational balance between the parties. Moreover, our concrete realization is computationally and bandwidth efficient. Informally, concurrent signatures appear to be as close to fair exchange as it's possible to get whilst staying truly practical and not relying on special third parties.

### 1.1 Our Contributions

We introduce the notion of *concurrent signatures* and *concurrent signature protocols*. In a concurrent signature protocol, two parties $A$ and $B$ interact without

[3] Our concurrent signatures will still require a conventional CA for the distribution of public keys, but not a trusted third party with any other special functions.

the help of a third party to sign (possibly identical) messages $M_A$ and $M_B$ in such a way that both $A$ and $B$ become publicly committed to their respective messages at the same moment in time (i.e. concurrently). This moment is determined by one of the parties through the release of an extra piece of information $k$ which we call a *keystone*. Before the keystone's release, neither party is publicly committed through their signatures, while after this point, both are. In fact, from a third party's point of view, before the keystone is released, both parties could have produced both signatures, so the signatures are completely ambiguous.

Note that the party who controls the keystone $k$ has a degree of extra power: it controls the timing of the keystone release and indeed whether the keystone is released at all. Upon receipt of $B$'s signature $\sigma_B$, $A$ might privately show $\sigma_B$ and $k$ to a third party $C$ and gain some advantage from doing so. This is the main feature that distinguishes concurrent signatures from fair exchange schemes. In a fair exchange scheme, each signer $A$ should either have recourse to a third party to release the other party $B$'s signature or be assured that the $B$ cannot compute $A$'s signature significantly more easily than $A$ can compute $B$'s. With concurrent signatures, only when $A$ releases the keystone do both signatures become simultaneously binding, and there is no guarantee that $A$ will do so. However, in the real world, there are often existing mechanisms that can naturally be used to guarantee that $B$ will receive the keystone should his signature be used. These existing mechanisms can provide a more natural dispute resolution process than reliance on a special trusted party. We argue that concurrent signatures are suited to any fair exchange application where:

- There is no sense in $A$ withholding the keystone because she needs it to obtain a service from $B$. For example, suppose $B$ sells computers. $A$ signs a payment instruction to pay $B$ the price of a computer, and $B$ signs that he authorizes her to pick one up from the depot ($B$'s signature may be thought of as a receipt). Then $A$ can withhold the keystone, but as soon as she tries to pick up her computer, $B$ will ask for a copy of his signature authorizing her to collect one. In this way $B$ can obtain the keystone which validates $A$'s payment signature. In this example, the application itself forces the delivery of the keystone to $B$.
- There is no possibility of $A$ keeping $B$'s signature private in the long term. For example, consider the routine "four corner" credit card payment model. Here $C$ may be $A$'s acquiring bank, and $B$'s signature may represent a payment to $A$ that $A$ must channel via $C$ to obtain payment. Bank $C$ would then communicate with $B$'s issuing bank $D$ to obtain payment against $B$'s signature and $D$ could ensure that $B$'s signature, complete with keystone, reaches $B$ (perhaps via a credit card statement). As soon as $B$ has the keystone, $A$ becomes bound to her signature. In this application, the back-end banking system provides a mechanism by which keystones would reach $B$ if $A$ were to withhold them.
- There is a single third party $C$ who verifies both $A$ and $B$'s signature. Now, if $A$ tries to present $B$'s signature along with $k$ to $C$ whilst withholding $k$ from

$B$, $B$ will be able to present $A$'s signature to $C$ and have it verified. As an application, consider the (perhaps somewhat artificial) scenario where $A$ and $B$ are two politicians from different parties who want to form a coalition to jointly release a piece of information to the press $C$ in such a way that neither of them is identified as being the sole signatory to the release. Concurrent signatures seem just right for this task. Here the keystone is not necessarily returned to $B$, but it does reach the third party to whom $B$ wishes to show $A$'s signature.

We also consider an example where concurrent signatures provide a novel solution to an old problem: that of fair tendering of contracts (our signatures can also be used in a similar way in auction applications). Suppose that $A$ has a bridge-building contract that she wishes to put out to tender, and suppose companies $B$ and $C$ wish to put in proposals to win the contract and build the bridge. This process is sometimes open to abuse by $A$ since she can privately show $B$'s signed proposal to $C$ to enable $C$ to better the proposal. Using concurrent signatures, $B$ would sign his proposal to build the bridge for an amount X, but keep the keystone private. If $A$ wishes to accept the proposal, she returns a payment instruction to pay $B$ amount X. She knows that if $B$ attempts to collect the payment, then $A$ will obtain the keystone through the banking system. But $A$ may also wish to examine $C$'s proposal before deciding which to accept. However there is no advantage for $A$ to show $B$'s signature to $C$ since at this point $B$'s signature is ambiguous and so $C$ will not be convinced of anything at all by seeing it. We see that the tendering process is immune to abuse by $A$. We note that this example makes use of the ambiguity of our signatures prior to the keystone release, and although the solution can be realized by using standard fair exchange protocols, such protocols do not appear to previously have been suggested for this purpose.

Our schemes are not abuse-free in the sense of [BW00, GJM99], since the party $A$ who holds the keystone can always determine whether to complete or abort the exchange of signatures, and can demonstrate this by showing an outside party $C$ the signature from $B$ with the keystone before revealing the keystone to $B$. However the above example shows that abuse can be addressed by our schemes in certain applications.

### 1.2 Technical Approach

We briefly explain how a concurrent signature protocol can be built using the *ambiguity* property enjoyed by ring signatures [RST01, AOS02] and designated verifier signatures [JSI96]. This introduces the key technical idea of our paper.

A two-party ring signature has the property that it could have been produced by either of the two parties. A similar property is shared by designated verifier signatures. We will refer to any signature scheme with this property as an *ambiguous* signature scheme and we will formalize the notion of ambiguity for signatures in the sequel. Since either of two parties could have produced such an ambiguous signature, both parties can deny having produced it. However, we note that if $A$ creates an ambiguous signature which only either $A$ or $B$ could have created, and sends this to $B$, then $B$ is convinced of the authorship of the signature (since he knows that he did not create it himself). However $B$ cannot prove this to a third party. The same situation applies when the roles of $A$ and $B$ are reversed.

Suppose now that the ambiguous signature scheme has the property that, when $A$ computes an ambiguous signature, she must choose some random bits $h_B$ to combine with $B$'s public key, but that the signing process is otherwise deterministic. Likewise, suppose the same is true for $B$ with random bits $h_A$ (when the roles of $A$ and $B$ are interchanged). Suppose $A$ creates an ambiguous signature $\sigma_A$ on $M_A$ using bits $h_B$ that are derived by applying a hash function to a string $k$ that is secret to $A$; $h_B$ is then a commitment to $k$. $B$ can verify that $A$ created the signature $\sigma_A$ but not demonstrate this to a third party. Now $B$ can create an ambiguous signature $\sigma_B$ on $M_B$ using as its input $h_A$ the same $h_B$ that $A$ used. Again, $A$ can verify that $B$ is the signer. As long as $k$ remains secret, neither party can demonstrate authorship to a third party.

But now if $A$ publishes the *keystone* $k$, then any third party can be convinced of the authorship of both signatures. The reason for this is that the only way that $B$ could produce $\sigma_B$ is by following his signing algorithm, choosing randomness $h_A$ and deterministically producing $\sigma_B$. The existence of a pre-image $k$ of $B$'s randomness $h_A$ determines $B$ as being the only party who could have conducted the signature generation process to produce $\sigma_B$. The same holds true for $A$ and $\sigma_A$. Thus the pairs $\langle k, \sigma_A \rangle$ and $\langle k, \sigma_B \rangle$ amount to a simultaneously binding pair of signatures on $A$ and $B$'s messages. We call these pairs *concurrent* signatures.

We point out that Rivest *et al.* in their pioneering work on ring signatures [RST01] considered the situation in which an anonymous signer $A$ wants to have the option of later proving his authorship of a ring signature. Their solution was to choose the bits $h_B$ pseudo-randomly and later to reveal the seed used to generate $h_B$. In this work, we use the same trick for a new purpose: to ensure that either both or neither of the parties can be identified as signers of messages.

We note that any suitably ambiguous signature scheme can be used to produce a concurrent signature protocol. We choose to base our concrete scheme on the non-separable ring signature scheme of [AOS02]. This scheme is, in turn, an adaptation of the Schnorr signature scheme. A second concrete scheme can be built from the short ring signature scheme of [BGLS03] using our ideas. An earlier version of our scheme used the designated verifier signatures of [JSI96] instead, however it achieved slightly weaker ambiguity properties than our concrete scheme.

We give generic definitions of concurrent signatures and concurrent signature protocols, a suitably powerful multi-party adversarial model for this setting, and give a formal definition of what it means for such schemes and protocols to be secure. Security is defined via the notions of unforgeability, ambiguity and fairness.

Because our concrete scheme is ultimately based on the Schnorr signature scheme [S91], we are able to directly relate its security to the hardness of the discrete logarithm problem in an appropriate group. In doing this, we make use of the forking lemma methodology of [PS96, PS00]; for this reason, our security proof will be in the random oracle model.

# 2 Formal Definitions

## 2.1 Concurrent Signature Algorithms

We now give a more formal definition of a concurrent signature scheme. Our protocols are naturally multi-party ones, so our model assumes a system with a number of different participants that is polynomial in the security parameter $l$.

**Definition 1.** *A concurrent signature scheme is a digital signature scheme comprised of the following algorithms:*

**SETUP:** *A probabilistic algorithm that on input a security parameter $l$, outputs descriptions of: the set of participants $\mathcal{U}$, the message space $\mathcal{M}$, the signature space $\mathcal{S}$, the keystone space $\mathcal{K}$, the keystone fix space $\mathcal{F}$, and a function $KGEN : \mathcal{K} \rightarrow \mathcal{F}$. The algorithm also outputs the public keys $\{X_i\}$ of all the participants, each participant retaining their private key $x_i$, and any additional system parameters $\pi$.*

**ASIGN:** *A probabilistic algorithm that on inputs $\langle X_i, X_j, x_i, h_2, M\rangle$, where $h_2 \in \mathcal{F}$, $X_i$ and $X_j \neq X_i$ are public keys, $x_i$ is the private key corresponding to $X_i$, and $M \in \mathcal{M}$, outputs an* ambiguous signature *$\sigma = \langle s, h_1, h_2\rangle$ on $M$, where $s \in \mathcal{S}$, $h_1, h_2 \in \mathcal{F}$.*

**AVERIFY:** *An algorithm which takes as input $S = \langle \sigma, X_i, X_j, M\rangle$, where $\sigma = \langle s, h_1, h_2\rangle$, $s \in \mathcal{S}$, $h_1, h_2 \in \mathcal{F}$, $X_i$ and $X_j$ are public keys, and $M \in \mathcal{M}$, outputs accept or reject. We also require that if $\sigma' = \langle s, h_2, h_1\rangle$, then $AVERIFY(\sigma', X_j, X_i, M) = AVERIFY(\sigma, X_i, X_j, M)$. We call this the* symmetry *property of AVERIFY.*

**VERIFY:** *An algorithm which takes as input $\langle k, S\rangle$ where $k \in \mathcal{K}$ is a keystone and $S$ is of the form $S = \langle \sigma, X_i, X_j, M\rangle$, where $\sigma = \langle s, h_1, h_2\rangle$ with $s \in \mathcal{S}$, $h_1, h_2 \in \mathcal{F}$, $X_i$ and $X_j$ are public keys, and $M \in \mathcal{M}$. The algorithm checks if KGEN(k)$= h_2$. If not, it terminates with output reject. Otherwise it runs AVERIFY(S) (in which case the output of VERIFY is just that of AVERIFY).*

We call a signature $\sigma$ an *ambiguous signature* and any pair $\langle k, \sigma\rangle$, where $k$ is a valid keystone for $\sigma$, a *concurrent signature*. The obvious correctness properties for ambiguous and concurrent signatures are formalized in Section 3.

## 2.2 Concurrent Signature Protocol

We will describe a concurrent signature protocol between two parties $A$ and $B$ (or Alice and Bob). Since one party needs to create the keystone and send the first ambiguous signature, we call this party the *initial signer*. A party who responds to this initial signature by creating another ambiguous signature with the same keystone fix we call a *matching signer*. Without loss of generality, we assume $A$ to be the initial signer, and $B$ the matching signer. From here on, we will use subscripts $A$ and $B$ to indicate initial signer $A$ and matching signer $B$. The signature protocol works as follows:

$A$ and $B$ run SETUP to determine the public parameters of the scheme. We assume that $A$'s public and private keys are $X_A$ and $x_A$, and $B$'s public and private keys are $X_B$ and $x_B$.

*1:* $A$ picks a random keystone $k \in \mathcal{K}$, and computes $f$= KGEN($k$). $A$ takes her own public key $X_A$ and $B$'s public key $X_B$ and picks a message $M_A \in \mathcal{M}$ to sign. $A$ then computes her ambiguous signature to be

$$\sigma_A = \langle s_A, h_A, f \rangle = \text{ASIGN}(X_A, X_B, x_A, f, M_A),$$

and sends this to $B$.

*2:* Upon receiving $A$'s ambiguous signature $\sigma_A$, $B$ verifies the signature by checking that AVERIFY($\langle s_A, h_A, f \rangle, X_A, X_B, M_A$)= accept. If not $B$ aborts, otherwise $B$ picks a message $M_B \in \mathcal{M}$ to sign and computes his ambiguous signature

$$\sigma_B = \langle s_B, h_B, f \rangle = \text{ASIGN}(X_B, X_A, x_B, f, M_B)$$

and sends this back to $A$. Note that $B$ uses the same value $f$ in his signature as $A$ did to produce $\sigma_A$.

*3:* Upon receiving $B$'s signature $\sigma_B$, $A$ verifies that AVERIFY($\langle s_B, h_B, f \rangle, X_B, X_A, M_B$) = accept, where $f$ is the same keystone fix as $A$ used in Step 1. If not, $A$ aborts, otherwise $A$ sends keystone $k$ to $B$.

Note that inputs $\langle k, S_A \rangle$ and $\langle k, S_B \rangle$ will now both be accepted by VERIFY, where $S_A = \langle \langle s_A, h_A, f \rangle, X_A, X_B, M_A \rangle$ and $S_B = \langle \langle s_B, h_B, f \rangle, X_B, X_A, M_B \rangle$.

## 3 Formal Security Model

We present a formal security model for concurrent signatures in this section.

### 3.1 Correctness

We give a formal definition of correctness for a concurrent signature scheme.

**Definition 2.** *We say that a concurrent signature scheme is* correct *if the following conditions hold.*

*If $\sigma = \langle s, h_1, f \rangle$ = ASIGN($X_i, X_j, x_i, f, M$), and $S = \langle \sigma, X_i, X_j, M \rangle$, then AVERIFY($S$)= accept. Moreover, if KGEN($k$) = $f$ for some $k \in \mathcal{K}$, then VERIFY($k, S$)= accept.*

### 3.2 Unforgeability

We give a formal definition of existential unforgeability of a concurrent signature scheme under a chosen message attack in the multi-party setting. To do this, we extend the definition of existential unforgeability against a chosen message attack of [GMR88] to the multi-party setting. Our extension is similar to that of [B03] and is strong enough to capture an adversary who can simulate and observe concurrent signature protocol runs between any pair of participants. It is defined using the following game between an adversary $E$ and a challenger $C$.

**Setup:** $C$ runs SETUP for a given security parameter $l$ to obtain descriptions of $\mathcal{U}$, $\mathcal{M}$, $\mathcal{S}$, $\mathcal{K}$, $\mathcal{F}$, and KGEN : $\mathcal{K} \rightarrow \mathcal{F}$. SETUP also outputs the public and private keys $\{X_i\}$ and $\{x_i\}$ and any additional public parameters $\pi$. $E$ is given all the public parameters and the public keys $\{X_i\}$ of all participants. $C$ retains the private keys $\{x_i\}$.
$E$ can make the following types of query to the challenger $C$:

**KGen Queries:** $E$ can request that $C$ select a keystone $k \in \mathcal{K}$ and return the keystone fix $f = \text{KGEN}(k)$. If $E$ wishes to choose his own keystone, then he can compute his own keystone fix using KGEN directly.

**KReveal Queries:** $E$ can request that $C$ reveal the keystone $k$ that it used to produce a keystone fix $f \in \mathcal{F}$ in a previous KGEN query. If $f$ was not a previous KGEN output then $C$ outputs invalid, otherwise $C$ outputs $k$ where $f = \text{KGEN}(k)$.

**ASign Queries:** $E$ can request an ambiguous signature for any input of the form $\langle X_i, X_j, h_2, M\rangle$ where $h_2 \in \mathcal{F}$, $X_i$ and $X_j \neq X_i$ are public keys and $M \in \mathcal{M}$. $C$ responds with an ambiguous signature $\sigma = \langle s, h_1, h_2\rangle = \text{ASIGN}(X_i, X_j, x_i, h_2, M)$. Note that using ASign queries in conjunction with KGen queries, $E$ can obtain concurrent signatures $\langle k, \sigma\rangle$ for messages and pairs of users of his choice.

**AVerify and Verify Queries:** Answers to these queries are not provided by $C$ since $E$ can compute them for himself using the AVERIFY and VERIFY algorithms.

**Private Key Extract Queries:** $E$ can request the private key corresponding to the public key of any participant $X_i$. In response $C$ outputs $x_i$.

**Output:** Finally $E$ outputs a tuple $\sigma = \langle s, h_1, f\rangle$ where $s \in \mathcal{S}$, $h_1, f \in \mathcal{F}$, along with public keys $X_c$ and $X_d$, and a message $M \in \mathcal{M}$. The adversary wins the game if $\text{AVERIFY}(\langle s, h_1, f\rangle, X_c, X_d, M) =$ accept, and if either of the following two cases hold:

1. No ASign query with input either of the tuples $\langle X_c, X_d, f, M\rangle$ or $\langle X_d, X_c, h_1, M\rangle$ was made by $E$, and no Private Key Extract query was made by $E$ on either $X_c$ or $X_d$.
2. No ASign query with input $\langle X_c, X_i, f, M\rangle$ was made by $E$ for any $X_i \neq X_c, X_i \in \mathcal{U}$, no Private Key Extract query with input $X_c$ was made by $E$, and either $f$ was a previous output from a KGen query or $E$ produces a keystone $k$ such that $f = \text{KGEN}(k)$.

**Definition 3.** *We say that a concurrent signature scheme is* existentially unforgeable under a chosen message attack in the multi-party model *if the probability of success of any polynomially bounded adversary in the above game is negligible (as a function of the security parameter $l$).*

Case 1 of the output conditions in the above game models forgery of an ambiguous signature in the situation where the adversary does not have knowledge of either of the respective private keys. This condition is required for our protocol so that the matching signer $B$ is convinced that $A$'s ambiguous signature can only originate from $A$. Case 2 models forgery in the situation where the adversary knows one of the private keys and so applies to the situation in our protocol where one of the two parties attempts to cheat the other. More specifically, it covers attacks where an initial signer forges a concurrent signature by a matching signer, and where a matching signer has access to an initial signer's ambiguous signature and keystone fix (but not the actual keystone) and forges a concurrent signature of the initial signer.

A further point to note is that in case 2, we insist that no ASign query of the form $\langle X_c, X_i, f, M\rangle$ is made, for any $X_i \neq X_c, X_i \in \mathcal{U}$. This is because, given a valid ambiguous signature $\sigma = \langle s, h_1, f\rangle$ for public keys $X_c$ and $X_i$, and the private keys of both $X_i$ and $X_d$, it may be possible to create a valid ambiguous signature $\sigma' = \langle s', h_1, f\rangle$ with public keys $X_c$ and $X_d$ on a message $M$. This is certainly the case for our concrete scheme, but should not be considered as a useful forgery because an attacker does not succeed in changing who is actually bound by the signature: in this case $X_c$.

### 3.3 Ambiguity

Ambiguity for a concurrent signature is defined by the following game between an adversary $E$ and a challenger $C$.

**Setup:** This is as before in the game of Section 3.2.

**Phase 1:** $E$ makes a sequence of KGen, KReveal, ASign and Private Key Extract queries. These are answered by $C$ as in the unforgeability game of Section 3.2.

**Challenge:** Then $E$ selects a challenge tuple $\langle X_i, X_j, M\rangle$ where $X_i$ and $X_j$ are public keys, and $M \in \mathcal{M}$ is the message to be signed. In response, $C$ randomly selects $k \in \mathcal{K}$ and computes $f = \text{KGEN}(k)$, then randomly selects a bit $b \in \{0,1\}$. $C$ outputs $\sigma_1 = \langle s_1, h_1, f\rangle = \text{ASIGN}(X_i, X_j, x_i, f, M)$ if $b = 0$; otherwise $C$ computes $\sigma_2' = \langle s_2, h_2, f\rangle = \text{ASIGN}(X_j, X_i, x_j, f, M)$ and outputs $\sigma_2 = \langle s_2, f, h_2\rangle$.

**Phase 2:** $E$ may make another sequence of queries as in Phase 1; these are handled by $C$ as before.

**Output:** Finally $E$ outputs a guess bit $b' \in \{0,1\}$. $E$ wins if $b' = b$ and $E$ has not made a KReveal query on any of the values $f, h_1$ or $h_2$.

**Definition 4.** *We say that a concurrent signature scheme is* ambiguous *if no polynomially bounded adversary has advantage that is non-negligibly greater than 1/2 of winning in the above game.*

We note that ambiguity in our concrete concurrent signature scheme will come directly from the ambiguity property of an underlying ring signature scheme. However the definition for ambiguity (or anonymity) in two-party ring signatures [RST01, BSS02, ZK02] states that an unbounded adversary should have probability exactly 1/2 of guessing $b$ correctly. Our definition must be slightly weaker because in our ambiguous signatures, one of our $h$ values is generated by KGEN and is therefore at best pseudorandom. However, since we model KGEN by a random oracle when proving ambiguity for our concrete scheme, we achieve perfect ambiguity as in the stronger definition for ring signatures.

### 3.4 Fairness

We require the concurrent signature scheme and protocol to be fair for both an initial signer $A$, and a matching signer $B$. This concept is defined via the following game between an adversary $E$ and a challenger $C$:

**Setup:** This is as before in the game of Section 3.2.

**KGen, KReveal, ASign and Private Key Extract Queries:** These queries are answered by $C$ as in the unforgeability game of Section 3.2.

**Output:** Finally $E$ chooses the challenge public keys $X_c$ and $X_d$, outputs a keystone $k \in \mathcal{K}$, and $S = \langle \sigma, X_c, X_d, M \rangle$ where $\sigma = \langle s, h_1, f \rangle$, $s \in \mathcal{S}$, $h_1, f \in \mathcal{F}$, and $M \in \mathcal{M}$, and where AVERIFY($S$) = accept. The adversary wins the game if either of the following cases hold:

1. If $f$ was a previous output from a KGen query, no KReveal query on input $f$ was made, and if $\langle k, S \rangle$ is accepted by VERIFY.
2. If $E$ also produces $S' = \langle \sigma', X_d, X_c, M' \rangle$, with $\sigma' = \langle s', h'_1, f \rangle$, $s' \in \mathcal{S}$, $h'_1, f \in \mathcal{F}$, message $M' \in \mathcal{M}$, where AVERIFY($S'$) = accept, and $\langle k, S \rangle$ is accepted by VERIFY, but $\langle k, S' \rangle$ is not accepted by VERIFY.

**Definition 5.** *We say that a concurrent signature scheme is* fair *if a polynomially bounded adversary's probability of success in the above game is negligible.*

Our definition of fairness formalizes our intuitive understanding of fairness for $A$ in the protocol of Section 2.2 (in case 1 of the output conditions), since it guarantees that only the entity who generates a keystone can use it to create a binding signature (by revealing it). It also captures fairness for $B$ (in case 2 of the output conditions), since it guarantees that any valid ambiguous signatures produced using the same keystone fix will all become binding. Thus $B$ cannot be left in a position where a keystone binds his signature to him while $A$'s initial signature is not also bound to $A$. However note that our definition does not guarantee that $B$ will ever receive the necessary keystone.

### 3.5 Security

**Definition 6.** *We say that a correct concurrent signature scheme is* secure *if it is existentially unforgeable under a chosen message attack in the multi-party setting, ambiguous, and fair.*

## 4 A Concrete Concurrent Signature Scheme

We present a concrete concurrent signature scheme in which the underlying ambiguous signatures and the resulting concurrent signatures are obtained by modifying signatures in the basic scheme of Schnorr [S91]. The scheme's algorithms (SETUP, ASIGN, AVERIFY, VERIFY) are as follows:

**SETUP:** On input a security parameter $l$, two large primes $p$ and $q$ are selected such that $q|p-1$. These are published along with an element $g$ of $(\mathbb{Z}/p\mathbb{Z})^*$ of order $q$, where $q$ is exponential in $l$. The spaces $\mathcal{S}, \mathcal{F}, \mathcal{M}, \mathcal{K}$ are defined as follows: $\mathcal{S} \equiv \mathcal{F}=\mathbb{Z}_q$ and $\mathcal{M} \equiv \mathcal{K}=\{0,1\}^*$. Two cryptographic hash functions $H_1, H_2 : \{0,1\}^* \rightarrow \mathbb{Z}_q$ are also selected and we define KGEN to be $H_1$. Private keys $x_i, 1 \leq i \leq n$ are chosen uniformly at random from $\mathbb{Z}_q$, where $n$ is polynomial in $l$. The public keys are computed as $X_i = g^{x_i} \bmod p$ and are made public.

**ASIGN:** This algorithm takes as input $\langle X_i, X_j, x_i, h_2, M\rangle$, where $X_i, X_j \neq X_i$ are public keys, $x_i \in \mathbb{Z}_q$ is the private key corresponding to $X_i$, $h_2 \in \mathcal{F}$ and $M \in \mathcal{M}$ is a message. The algorithm picks a random value $t \in \mathbb{Z}_q$ and then computes the values:

$$\begin{aligned} h &= H_2(g^t {X_j}^{h_2} \bmod p||M), \\ h_1 &= h - h_2 \bmod q, \\ s &= t - h_1 x_i \bmod q. \end{aligned}$$

Here "||" denotes concatenation. The algorithm outputs $\sigma = \langle s, h_1, h_2\rangle$.

**AVERIFY:** This algorithm takes as input $\langle \sigma, X_i, X_j, M\rangle$ where $\sigma = \langle s, h_1, h_2\rangle$, $s \in \mathcal{S}$, $h_1, h_2 \in \mathcal{F}$, $X_i$ and $X_j$ are public keys, and $M \in \mathcal{M}$ is a message. The algorithm checks that the equation

$$h_1 + h_2 = H_2(g^s X_i^{h_1} X_j^{h_2} \bmod p \; ||M) \quad \bmod q$$

holds, and if so, outputs accept. Otherwise, it outputs reject.

**VERIFY:** This algorithm is defined in terms of KGEN and AVERIFY, as in Section 2.1.

The ASIGN algorithm is a direct modification of the ring signature algorithm of [AOS02], and guarantees our property of ambiguity before the keystone is revealed. We require that $X_j \neq X_i$ since otherwise the signature would be a standard Schnorr signature [S91] and would not be ambiguous. It is also easily checked that the scheme satisfies the definition of correctness and that AVERIFY has the required symmetry property.

A concrete concurrent signature protocol can be derived directly from the algorithms defined above and the generic protocol in Section 2.2.

## 5 Security of the Concrete Concurrent Signature Scheme

We now state some security results for the concrete scheme of Section 4. The proofs of Lemmas 1 and 3 are proved in Appendix A. The proof of Lemma 2 is routine, and the details are left to the reader. Our proofs of security are in the random oracle model [BR93].

**Lemma 1.** *The concurrent signature scheme of Section 4 is existentially unforgeable under a chosen message attack in the random oracle model, assuming the hardness of the discrete logarithm problem.*

**Lemma 2.** *The concurrent signature scheme of Section 4 is ambiguous in the random oracle model.*

**Lemma 3.** *The concurrent signature scheme of Section 4 is fair in the random oracle model.*

**Theorem 1.** *The concurrent signature scheme of Section 4 is secure in the random oracle model, assuming the hardness of the discrete logarithm problem.*

*Proof.* The proof follows directly from Lemmas 1, 2 and 3. □

## 6 Extensions and Open Problems

### 6.1 The Scheme Can Use a Variety of Keys

Our concurrent signature scheme can be based on any ring signature scheme, as long as it is compatible with the keystone fix idea. Thus it is feasible to build concrete concurrent signature schemes using a variety of key types, and therefore the security of such schemes may be based on a variety of underlying hard problems. Furthermore, the key pairs in a single concurrent signature scheme do not have to be of the same type. The techniques to be used for achieving concurrent signatures from a variety of keys are the same as the key separability techniques for ring signatures as described in [AOS02].

### 6.2 The Multi-party Case

It would be interesting to see if concurrent signatures could be extended to the multi-party case, that is, where many entities can fairly exchange signatures concurrently. The existing two party scheme can trivially be extended to include multiple matching signers. However we do not as yet have a model for fairness for such a scheme. It would also be interesting to investigate methods whereby the revelation of keystones did not depend entirely on the initial signer, but on the other signing parties as well.

# 7 Conclusion

We introduced the notion of concurrent signatures, presented a concurrent signature scheme and related its security to the hardness of the discrete logarithm problem in an appropriate security model. We have also discussed some applications for concurrent signatures, and the advantages they have over previous work. In particular, we have compared concurrent signatures to techniques for fair exchange of signatures, and presented some applications in which the full security of fair exchange may not be necessary and the more pragmatic solution of concurrent signatures suffice.

## Acknowledgements

We would like to thank Alex Dent for useful comments on the paper. We also thank Keith Harrison, who suggested the name "keystone", and Brian Monahan for some useful discussions.

## References

[AOS02] M. Abe, M. Ohkubo, and K. Suzuki, 1-out-of-n signatures from a variety of keys, In *Advances in Cryptology - ASIACRYPT 2002*, LNCS vol. 2501, pp. 415-432. Springer-Verlag, 2002.

[ASW98] N. Asokan, V. Shoup, and M. Waidner, Optimistic fair exchange of signatures, In *Advances in Cryptology - EUROCRYPT 1998*, LNCS vol. 1403, pp. 591-606. Springer-Verlag, 1998.

[ASW00] N. Asokan, V. Shoup, and M. Waidner, Optimistic fair exchange of signatures, In *IEEE Journal on Selected Areas in Communication* vol. 18(4), pp. 593-610, 2000.

[B03] X. Boyen, Multipurpose identity-based signcryption. A Swiss army knife for identity-based cryptography. In *Advances in Cryptology - CRYPTO 2003*, LNCS vol. 2729, pp. 383-399, Springer-Verlag, 2003.

[BGLS03] D. Boneh, C. Gentry, B. Lynn and H. Shacham, Aggregrate and verifiably encrypted signatures from bilinear maps. In *Advances in Cryptology - EUROCRYPT 2003*, LNCS vol. 2656, pp. 416-432, Springer-Verlag, 2003.

[BN00] D. Boneh, and M. Naor, Timed commitments (extended abstract). In *Advances in Cryptology - CRYPTO 2000*, LNCS vol. 1880, pp. 236-254. Springer-Verlag, 2000.

[BR93] M. Bellare, and P. Rogaway, Random oracles are practical: a paradigm for designing efficient protocols. In *Proc. of the 1st CCCS*, pp. 62-73. ACM press, 1993.

[BSS02] E. Bresson, J. Stern and M. Szydlo, Threshold ring signatures for adhoc groups. In *Advances in Cryptology - CRYPTO 2002*, LNCS vol. 2442, pp. 465-480, Springer-Verlag, 2002.

[BW00] B. Baum-Waidner and M. Waidner, Round-optimal and abuse free optimistic multi-party contract signing. In Proc. of *Automata, Languages and Programming*, pp. 524-535, 2000.

[CS03] J. Camenisch and V. Shoup, Practical verifiable encryption and decryption of discrete logarithms. In *Advances in Cryptology - CRYPTO 2003*, LNCS vol. 2729, pp. 126-144, Springer-Verlag, 2002.

[DR03] Y. Dodis, and L. Reyzin, Breaking and repairing optimistic fair exchange from PODC 2003, In *ACM Workshop on Digital Rights Management (DRM)*, October 2003.

[EGL85] S. Even, O. Goldreich, and A. Lempel, A randomized protocol for signing contracts. In *Commun. ACM*, vol. 28(6), pp. 637-647, June 1985.

[G83] O. Goldreich, A simple protocol for signing contracts. In *Advances in Cryptology - CRYPTO 1983*, pp. 133-136, Springer-Verlag, 1983.

[GJM99] J. Garay, M. Jakobsson and P. MacKenzie, Abuse-free optimistic contract signing, In *Advances in Cryptology - CRYPTO 1999*, LNCS vol. 1666, pp. 449-466, Springer-Verlag, 1999.

[GMR88] S. Goldwasser, S. Micali and R. Rivest, A digital signature scheme secure against adaptive chosen message attacks. In *SIAM Journal of Computing*, vol. 17(2), pp. 281-308, April 1988.

[GP03] J. Garay, and C. Pomerance, Timed fair exchange of standard signatures, In *Proc. Financial Cryptography 2003*, LNCS vol. 2742, pp. 190-207, Springer-Verlag, 2003.

[JSI96] M. Jakobsson, K. Sako, and R. Impagliazzo, Designated verifier proofs and their applications, In *Advances in Cryptology - EUROCRYPT 1996*, LNCS Vol.1070, pp. 143-154, Springer-Verlag, 1996.

[PCS03] J. Park, E. Chong, and H. Siegel, Constructing fair exchange protocols for e-commerce via distributed computation of RSA signatures. In *22nd Annual ACM Symp. on Principles of Distributed Computing* pp. 172-181, July 2003.

[PS96] D. Pointcheval and J. Stern, Security proofs for signature schemes, In *Advances in Cryptology - EUROCRYPT 1996*, LNCS vol. 1070, pp. 387-398, Springer-Verlag, 1996.

[PS00] D. Pointcheval and J. Stern, Security arguments for digital signatures and blind signatures. In *Journal of Cryptology*, vol. 13, pp. 361-396, 2000.

[RST01] R. Rivest, A. Shamir and Y. Tauman, How to leak a secret. In *Advances in Cryptology - ASIACRYPT 2001*, LNCS 2248, pp. 552-565, Springer-Verlag, 2001.

[S91] C.P. Schnorr, Efficient signature generation by smart cards, In *Journal of Cryptology*, vol. 4, no. 3, pp. 161-174, 1991.

[ZK02] F. Zhang and K. Kim, ID-based blind signature and ring signature from pairings. In *Advances in Cryptology - ASIACRYPT 2002*, LNCS vol. 2501, pp. 533-547, Springer-Verlag, 2002.

## Appendix A

*Proof of Lemma 1.* The proof is similar to the proof of unforgeability of the Schnorr signature scheme [S91] by Pointcheval and Stern [PS96], and makes use of the forking lemma [PS96,PS00].

*The Forking Lemma [PS96, PS00]:* The forking lemma applies in particular to signature schemes which on input a message $M$ produce signatures of the form $(r_1, h, r_2)$ where $r_1$ takes its value randomly from a large set, $h$ is the hash of $M$ and $r_1$, and $r_2$ depends only on $r_1, M$ and $h$.

The forking lemma in [PS00] states that if $E$ is a polynomial time Turing machine with input only public data, which produces, in time $\tau$ and with probability $\eta \geq 10(\mu_s+1)(\mu_s+\mu)/2^l$ (where $l$ is a security parameter) a valid signature $(m, r_1, h, r_2)$, where $\mu$ is the number of hash queries, and $\mu_s$ is the number of signature queries, and if triples $r_1, m, r_2$ are simulatable with indistinguishable probability distribution without knowledge of the secret key, then there exists an algorithm $A$, which controls $E$ and replaces $E$'s interaction with the signer by the simulation, and which produces two valid signatures $(m, r_1, h, r_2)$ and $(m, r_1, h', r_2')$ such that $h \neq h'$ in expected time at most $\tau' = 120686\mu_s\tau/\eta$.

Firstly, we note that our concurrent signature scheme in Section 4 on input a message $M$, public keys $X_i$ and $X_j$ and a value $h_2$, produces signatures of the required form $\langle r_1, h, r_2\rangle$, where $r_1 = g^t X_j^{h_2} \bmod p$ which takes its values randomly from $\mathbb{Z}_q$, $h = h_1 + h_2$ is the hash of $M$ and $r_1$, and $r_2 = s$ depends on $r_1, M$ and $h$. Although the actual output of the signature is the tuple $\langle s, h_1, h_2\rangle$, the values $r_1, h$ and $r_2$ can easily be derived from the output. We also note that if by the forking methodology, we have two valid signatures $(r_1, h, r_2)$ and $(r_1, h', r_2')$ on the same message $M$ with $h \neq h'$, then provided that the value $h_2$ is computed before the relevant $H_2$ query, then this would be equivalent to two concurrent signatures $\langle s, h_1, h_2\rangle$ and $\langle s', h_1', h_2\rangle$ with $h_1 \neq h_1'$.

We suppose that $H_1$ and $H_2$ are random oracles, and suppose there exists an algorithm $E$ who is able to forge concurrent signatures. So we assume that $E$ is an attacker that makes at most $\mu_i$ queries to the random oracles $H_i, i = \{1, 2\}$, at most $\mu_s$ queries to the signing oracle, and wins the unforgeability game of Section 3.2 in time at most $\tau$ with probability at least $\eta = 10(\mu_s+1)(\mu_s+\mu_2)/q$, where $q$ is exponential in security parameter $l$.

We show how to construct an algorithm $B$ that uses $E$ to solve the discrete logarithm problem. $B$ will simulate the random oracles and the challenger $C$ in a game with $E$. $B$'s goal is to solve the discrete logarithm problem on input $\langle g, X, p, q\rangle$, that is to find $x \in \mathbb{Z}_q$ such that $g^x = X \bmod p$, where $g$ is of prime order $q$ modulo prime $p$.

*Simulation:* $B$ gives the parameters $\langle g, p, q\rangle$ to $E$. $B$ generates a set of participants $U$, where $|U| = \rho(l)$ and $\rho$ is a polynomial function of the security parameter $l$. Each participant has a public key $X_i$ and private key $x_i$. $B$ guesses that $E$ will choose $X_\alpha$ in the position of $X_c$ in its output. $B$ sets $X_\alpha = X$, and for each $i \neq \alpha$, $x_i$ is chosen randomly from $\mathbb{Z}_q$, and $B$ sets $X_i = g^{x_i} \bmod p$. $E$ is given all the public keys $X_i$. $B$ now simulates the challenger by simulating all the oracles which $E$ can query as follows:

***$H_1$*-Queries:** $E$ can query the random oracle $H_1$ at any time. $B$ simulates the random oracle by keeping a list of tuples $\langle M_i, r_i\rangle$ which is called the $H_1$-List. When the oracle is queried with an input $M \in \{0,1\}^*$, $B$ responds as follows:

1. If the query $M$ is already on the $H_1$-List in the tuple $\langle M, r_i\rangle$, then $B$ outputs $r_i$.
2. Otherwise $B$ selects a random $r \in \mathbb{Z}_q$, outputs $r$ and adds $\langle M, r\rangle$ to the $H_1$-List.

**$H_2$-Queries:** $E$ can query the random oracle $H_2$ at any time. $B$ simulates the $H_2$ oracle in the same way as the $H_1$ oracle by keeping an $H_2$-List of tuples.

**KGen Queries:** $E$ can request that the challenger select a keystone $k \in \mathcal{K}$ and return a keystone fix $f = H_1(k)$. $B$ maintains a K-List of tuples $\langle k, f \rangle$, and answers queries by choosing a random keystone $k \in \mathcal{K}$ and computing $f = H_1(k)$. $B$ outputs $f$ and adds the tuple $\langle k, f \rangle$ to the K-List. Note that K-List is a sublist of $H_1$-List, but is required to answer KReveal queries.

**KReveal Queries:** $E$ can request the keystone of any keystone fix $f \in \mathcal{F}$ produced by a previous KGen Query. If there exists a tuple $\langle k, f \rangle$ on the K-List, then $B$ returns $k$, otherwise it outputs invalid.

**ASign Queries:** $B$ simulates the signature oracle by accepting signature queries of the form $\langle X_i, X_j, h_2, M \rangle$ where $h_2 \in \mathcal{F}$, $X_i$ and $X_j \neq X_i$ are public keys, and $M \in \mathcal{M}$ is the message to be signed. If $X_i \neq X_\alpha$ then $B$ computes the signature as normal and outputs $\sigma = \langle s, h_1, h_2 \rangle = \text{ASIGN}(X_i, X_j, x_i, h_2, M)$. If $X_i = X_\alpha$ then $B$ answers the query as follows:

1. $B$ picks a random $h_1$ and $s$ in $\mathbb{Z}_q$, computes $T = g^s X_i^{h_1} X_j^{h_2} \bmod p$, and forms the string "$T||M$".
2. If $h = h_1 + h_2$ is equal to some previous output for the $H_2$ oracle, or if "$T||M$" was some previous input, then return to step 1.
3. Otherwise add the tuple $\langle T||M, h \rangle$ to the $H_2$-List.
4. $B$ outputs $\sigma = \langle s, h_1, h_2 \rangle$ as the signature for message $M$ with public keys $X_i$ and $X_j$.

**Private Key Extract Queries:** $E$ can request the private key for any public key $X_i$. If $X_i = X_\alpha$, then $B$ terminates the simulation with $E$ having failed to guess the correct challenge public key. Otherwise $B$ returns the appropriate private key $x_i$.

**Output:** Finally, with non-negligible probability, $E$ outputs a signature $\sigma = \langle s, h_1, f \rangle$ where $s \in \mathcal{S}$, $h_1, f \in \mathcal{F}$, along with public keys $X_c$ and $X_d$, and a message $M \in \mathcal{M}$, where AVERIFY($\langle s, h_1, f \rangle, X_c, X_d, M$)= accept, and one of the following two cases holds:

1. No ASign query with input either of the tuples $\langle X_c, X_d, f, M \rangle$ or $\langle X_d, X_c, h_1, M \rangle$ was made by $E$, and no Private Key Extract query was made by $E$ on either $X_c$ or $X_d$.
2. No ASign query with input $\langle X_c, X_i, f, M \rangle$ was made by $E$ for any $X_i \neq X_c, X_i \in \mathcal{U}$, no Private Key Extract query with input $X_c$ was made by $E$, and either $f$ was a previous output from a KGen query or $E$ produces a keystone $k$ such that $f = \text{KGEN}(k)$.

It is easy to show that case 1 of the output conditions can occur only with negligible probability $\delta$. This follows immediately from the unforgeability of the underlying ring signature [AOS02], assuming the hardness of the discrete logarithm problem. An outline of the ring signature unforgeability proof is given in [AOS02], hence we omit the details here. Since the adversary wins the game with non-negligible probability, we assume that case 2 must have occurred.

If $X_c \neq X_\alpha$ then $B$ aborts, having failed to guess the correct challenge public key. Henceforth, we assume that $X_c = X_\alpha = X$ (this occurring with probability $1/\rho(l)$ where $\rho$ is a polynomial function). Given that $B$ does not abort for any reason, it can be seen that, because of the way $B$ handles oracle queries, the simulation seen by $E$ is indistinguishable from a real interaction with a challenger.

Because in case 2 algorithm AVERIFY with $E$'s signature as input returns accept, we have the equation $h = h_1 + f = H_2(g^s X_c^{h_1} X_d^f \bmod p \,||M)$. We now analyze two further cases.

*Case 1.* We recall that we can rewrite the signature above in the form $(r_1, h, r_2)$. If $h = h_1 + f$ has never appeared in any previous signature query before, then by the forking lemma, $B$ can repeat its simulation so that $E$ produces another such signature $(r_1, h', r_2')$, with $h \neq h'$.

Note that $E$ has in fact produced two signatures $\sigma = \langle s, h_1, f\rangle$ and $\sigma' = \langle s', h_1', f'\rangle$, with $h = h_1 + f \neq h_1' + f' = h'$. If $h_1 = h_1'$, then $B$ aborts. However, if $h_1 = h_1'$, then the $h_1$ values must have been computed before the relevant $H_2$ queries (which produced $h$ and $h'$), or $h_1$ and $h_1'$ are independent of $h$ and $h'$ respectively. Also, if $h_1 = h_1'$, then $f \neq f'$, so these values must have been computed after the relevant $H_2$ queries, and satisfy the equations $f = h - h_1$ and $f' = h' - h_1'$. But we know that $f$ is also an output of $H_1$, either from a direct $H_1$ query, or via a KGen query, and the probability that an output from $H_1$ query matches (some function of) an output from some $H_2$ query is at most $\mu_2\mu_1/q$. This is negligible, so we assume that $f = f'$, and therefore that $h_1 \neq h_1'$.

Now, since $h$ and $h'$ resulted from different oracle queries on the same input, we know that $g^s X^{h_1} X_d^f = g^{s'} X^{h_1'} X_d^f \bmod p$. So taking the exponents from both sides we get $s + xh_1 = s' + xh_1' \bmod q$. Since $h_1 \neq h_1'$, $B$ can now solve for $x$, the discrete logarithm of $X$, using the equation $x = \frac{s-s'}{h_1'-h_1} \bmod q$.

So in Case 1, the probability that $B$ does not have to abort at some point in the simulation is at least

$$\gamma = (1-\delta)\cdot\frac{1}{\rho(l)}\cdot(1-\frac{\mu_1\mu_2}{q}),$$

which is non-negligible in security parameter $l$. So $B$ solves the discrete logarithm of $X$ by the forking lemma, in expected time at most $\tau'/\gamma = 120686\mu_s\tau/\eta\gamma$. This contradicts the hardness of the discrete logarithm problem.

*Case 2.* However suppose that $h = h'$, where $h' = h_1' + f'$ was the output in some previous signature query $\langle X_{c'}, X_{d'}, f', M'\rangle$. Say the previous signature was $\sigma' = \langle s', h_1', f'\rangle$ with public keys $X_{c'}$ and $X_{d'}$ on message $M'$. Now $h = H_2(g^s X^{h_1} X_d^f \bmod p||M)$, $h' = H_2(g^{s'} X_{c'}^{h_1'} X_{d'}^{f'} \bmod p||M')$ and $h = h'$. If the inputs to $H_2$ are not equal, then $B$ aborts. This occurs with probability $\mu_2\mu_s/q$. Otherwise we have that the inputs to the random oracle are equal, so $M = M'$ and $g^s X^{h_1} X_d^f = g^{s'} X_{c'}^{h_1'} X_{d'}^{f'} \bmod p$.

If $X_{c'}, X_{d'} \neq X$, or $X_{c'} = X$ or $X_{d'} = X$ but their exponents are different (e.g. if $X_{c'} = X$ but $h_1' \neq h_1$), then it is easy to see that $B$ can extract $x$ directly from the equation $g^s X^{h_1} X_d^f = g^{s'} X_{c'}^{h_1'} X_{d'}^{f'} \bmod p$.

However suppose that either $X_{c'} = X$ or $X_{d'} = X$, and their exponents are equal. If $X_{c'} = X$ and $h'_1 = h_1$, then since $h_1 + f = h'_1 + f'$, we have that $f' = f$. But this is impossible since it contradicts the assumption that no tuple of the form $\langle X_c, X_i, f, M\rangle$ was queried on the signing oracle before.

If $X_{d'} = X$ and $h_1 = f'$, then $f = h'_1$, where $h'_1 = h' - f'$ was generated in a previous signature query, and is determined by the outputs of the random oracles $H_1$ and $H_2$. But we know that $f$ is also a direct output of $H_1$, perhaps via a KGen query. However the probability that an output from $H_1$ matches an $h'_1$ from some signature query is $\mu_1\mu_s/q$. This probability is negligible and if this case occurs, then $B$ aborts.

So for Case 2, the probability that $B$ is not forced to abort at some point is at least

$$\gamma = (1-\delta) \cdot \frac{1}{\rho(l)} \cdot (1 - \frac{\mu_2\mu_s}{q}) \cdot (1 - \frac{\mu_1\mu_s}{q}),$$

which is non-negligible in security parameter $l$. If $B$ is not forced to abort, then $B$ can solve the discrete logarithm of $X$ directly from $E$'s output. Our analysis therefore shows that in Case 2, $B$ can extract the discrete logarithm of $X$ within expected time at most $\tau/\gamma$. This again contradicts the hardness of the discrete logarithm problem. □

*Proof of Lemma 3.* We suppose that $H_1$ and $H_2$ are random oracles as before, and suppose that there exists an algorithm $E$ that with non-negligible probability wins the game in Section 3.4. In this game, the challenger runs the SETUP algorithm to initialize all the public parameters as usual, choosing all the private keys $x_i$ randomly from $\mathbb{Z}_q$, generating the public keys as $X_i = g^{x_i} \bmod p$, and giving these public keys to $E$. Also as part of this game, $C$ responds to $H_1$, $H_2$, KGen and KReveal queries as usual, and responds to ASign and Private Key Extract queries using its knowledge of the private keys.

In the final stage of the game, $E$ chooses challenge public keys $X_c$, $X_d$ and with non-negligible probability $\eta$ outputs keystone $k$ and $S = \langle\sigma, X_c, X_d, M\rangle$ with $\sigma = \langle s, h_1, f\rangle$ for which one of the following cases holds:

1. $f$ was a previous output from a KGen query, $f$ was not queried on the KReveal oracle, and $\langle k, S\rangle$ is accepted by VERIFY.
2. $E$ also produces $S' = \langle\sigma', X_d, X_c, M'\rangle$, where $\sigma' = \langle s', h'_1, f\rangle$ is an ambiguous signature on $M'$ with public keys $X_d$, $X_c$, both $S$ and $S'$ are accepted by AVERIFY, $\langle k, S\rangle$ is accepted by VERIFY, but $\langle k, S'\rangle$ is not accepted by VERIFY.

We now analyse $E$'s output.

*Case 1.* Suppose case 1 of the output conditions occurs. Then $E$ has found a keystone $k$ and an output of a KGen query $f$ such that $f = H_1(k)$, but without making a KReveal query on input $f$. However, since $H_1$ is a random oracle, $E$'s probability of producing such a $k$ is at most $\mu_1\mu_2/q$, where $\mu_1$ is the number of $H_1$ queries made by $E$ and $\mu_2$ is the number of KGen queries made by $E$. Since both $\mu_1$ and $\mu_2$ are polynomially bounded in the security parameter $l$ and $q$ is exponential in $l$, this probability is negligible. This contradicts our assumption that $E$ wins the game with non-negligible probability.

*Case 2.* Suppose case 2 of the output conditions occurs. Since $S$ is accepted by AVERIFY and $\langle k, S \rangle$ is accepted by VERIFY, we must have KGEN($k$)=$f$. But then, since $S$ and $S'$ share the value $f$, we must also have that $\langle k, S' \rangle$ is accepted by VERIFY. This is a contradiction. □

# The Hierarchy of Key Evolving Signatures and a Characterization of Proxy Signatures

Tal Malkin[1], Satoshi Obana[2], and Moti Yung[1]

[1] Columbia University
{tal,moti}@cs.columbia.edu
[2] NEC and Columbia University
obana@bx.jp.nec.com

**Abstract.** For the last two decades the notion and implementations of *proxy signatures* have been used to allow transfer of digital signing power within some context (in order to enable flexibility of signers within organizations and among entities). On the other hand, various notions of the key-evolving signature paradigms (forward-secure, key-insulated, and intrusion-resilient signatures) have been suggested in the last few years for protecting the security of signature schemes, localizing the damage of secret key exposure.

In this work we relate the various notions via direct and concrete security reductions that are tight. We start by developing the first formal model for fully hierarchical proxy signatures, which, as we point out, also addresses vulnerabilities of previous schemes when self-delegation is used. Next, we prove that proxy signatures are, in fact, equivalent to key-insulated signatures. We then use this fact and other results to establish a tight hierarchy among the key-evolving notions, showing that intrusion-resilient signatures and key-insulated signatures are equivalent, and imply forward-secure signatures. We also introduce other relations among extended notions.

Besides the importance of understanding the relationships among the various notions that were originally designed with different goals or with different system configuration in mind, our findings imply new designs of schemes. For example, many proxy signatures have been presented without formal model and proofs, whereas using our results we can employ the work on key-insulated schemes to suggest new provably secure designs of proxy signatures schemes.

## 1 Introduction

Characterizing relationships among cryptographic notions is an important task that increases our understanding of the notions and can contribute to concrete designs. In this work we look at two paradigms, proxy signatures and key-evolving signatures, that were suggested at different times for totally different purposes. After developing the first formal model for fully hierarchical proxy signatures and addressing a vulnerability in previous proxy schemes, we prove that proxy signatures are equivalent in a very strong sense to key-insulated

C. Cachin and J. Camenisch (Eds.): EUROCRYPT 2004, LNCS 3027, pp. 306–322, 2004.

signatures (one of the key-evolving notions). We also relate the various notions within the key-evolving paradigm, that were originally suggested for different system architecture settings and adversarial assumptions, establishing a tight hierarchy among them (tight in the sense of no security loss in the reductions). In the rest of the introduction we elaborate on these primitives, our results, and their significance.

*Proxy Signatures and Our Contributions in Modeling them.* The paradigm of proxy signature is a method for an entity to delegate signing capabilities to other participants so that they can sign on behalf of the entity within a given context (the context and limitations on proxy signing capabilities are captured by a certain warrant issued by the delegator which is associated with the delegation act). For example, Alice the executive might want to empower Bob the secretary to sign on her behalf for a given week when Alice is out of town. Such proxy capability transfer may be defined recursively to allow high flexibility in assigning limited entitlements. The notion is motivated by real life flexibility of "power of attorney" and other mechanisms of proxy.

The notion has been suggested and implemented in numerous works for about 20 years now: one of the early works to be published was presented in [6], whereas for a cryptographic treatment see [14]. Most of the past work is informal and without complete proofs. The first (and to the best of our knowledge, only) work to formally define the model of proxy signatures, is the recent work of Boldyreva, Palacio, and Warinschi [3]. Their definition is of proxy signature, with only one level of delegation, and without using the warrants as part of the model (though warrants are used in the common scheme of delegation by certificate, a notion that was analyzed by [3]).

We provide the first definition of fully hierarchical proxy signatures with warrants, supporting chains of several levels of delegation. Furthermore, the fully hierarchical approach illuminates an important aspect of proxy signatures, regarding self-delegations, which was previously overlooked. Specifically, we identify a vulnerability in previous solutions (both in existing proxy signature implementations such as the delegation by certificate, and in the formal model which rendered them secure). This weakness, which results in enabling a delegatee to possibly take "rogue actions" on behalf of a delegator, does not exist in our model, and we point out how the delegation by certification implementation (and other schemes with the same problem) can be modified in a simple way so as to avoid such attacks, and satisfy our strong notion of security.

*Key Evolving Signatures.* The paradigm of key evolving signatures started with Anderson's suggestion in [1], towards mitigating the damage caused by key exposure, one of the biggest threats to security of actual cryptographic schemes. Indeed, if the secret key in a standard signature scheme is exposed, this allows for forgery, invalidation of past and future signatures, and thus repudiation through leaking of the secret key. To limit the damage, the key evolving paradigm splits the time for which the signature is valid (say, 5 years) into well defined short periods (say months, days, or a period per signature, as required by the application). The secret key can then evolve with the periods (see details below), while maintaining the same public key. This idea gave rise to three well-defined notions of protection against key exposure, compartmentalizing the damage. The three notions have different configurations and different adversarial settings, achieving different properties:

1. Forward-Secure Signature Schemes (FS) [1,2]: Here the system is comprised of a single agent holding the private signing key, and at each period the key is evolved (via a one-way transformation) so that the exposure does not affect past periods. This notion has the advantage that even if *all* the key material is completely exposed, past signatures are still valid, and cannot be forged or repudiated. On the other hand, such a complete exposure necessarily compromises the validity of all future signatures, and the public key cannot be used any more.
2. Key-Insulated Signature Scheme (KI) [5]: Here the system is made out of two entities: the signer and a helper (base). At the start of the period the signer is updated by the helper to produce the next period's key. The helper is involved only in the updates. In fact, the helper can give the signer access to any period at any time (random access capability). The exposure of up to $t$ of the $N$ periods, chosen adaptively by the adversary, still keeps any period that was not exposed secure. The limitation of necessarily exposing all future keys, as in forward security does not apply anymore; this limitation is removed by the introduction of the helper (base) which is never exposed. The optimal $t$ achieved by some of the schemes is $N-1$ where the remaining period is still secure. Note that here the keys at the helper and the signer are not forward-secure. This model was first considered in [4]. We remark that the notion of strong KI which protects the signer from the helper is irrelevant here (and there is a simple transformation from KI to strong KI).
3. Intrusion-Resilient Signature Scheme (IR) [9]: Here the scheme is also made out of a signer and a helper (base). Now the exposures of both the helper and the signer are allowed. If the exposure is alternating (i.e., at each period at most one of the signer or the helper is exposed) then the scheme remains secure for all unexposed signing periods. If the exposure is of both the helper and the signer simultaneously, then the system becomes forward-secure from that period on: the past is protected (excluding the periods where the signer was directly exposed) but the future is now necessarily insecure. Note that unlike KI, this notion allows exposure of the helper, and that both the helper's key and the signer's key are forward-secure.

*Our Reductions: A Characterization of Proxy Signatures, and The Hierarchy of Key Evolving Signatures.* Our goal is to explore the relations among the key evolving signature notions and proxy signatures, towards gaining a better understanding of the primitives, and obtaining practical constructions. From a complexity-theoretic point of view, one can establish equivalences using the fact that these notions have implementations based on a generic signature scheme (typically less efficient than implementations based on specific number theoretic assumptions). For example, see the generic constructions of [2,12,5,7] for key evolving signatures, and the delegation by certificate scheme for proxy signatures that was suggested with different variations in numerous works (see Section 2.1). Thus, the notions are equivalent to the existence of one-way functions in terms of computational assumptions [15,16]. However, our goal is to establish *direct* reductions, both from a practical point of view (namely, given an implementation of one primitive, construct the other primitive using the first almost "as-is", with a straight-forward and efficient transformation), and from a theoretical point of view: analyzing the efficiency and the *concrete* security guarantees. In particular, we consider direct reductions between paradigms so that there is a concrete security evaluation of one scheme based on the concrete security of the related scheme to which it is reduced, while minimizing the loss of the concrete security value, and minimizing overhead in efficiency. Under this notion of direct reduction we found that:

- Proxy signatures are equivalent to KI signatures. In particular, we show that proxy signatures imply KI signatures via a tight reduction achieving the same concrete security, and that KI signatures imply proxy signatures via a tight security reduction. Our characterization of proxy signatures immediately provides a suite of provably secure proxy signature schemes, based on the previous (and future) schemes for KI signatures. For example, all the schemes of [5] can be used, including the efficient ones based on trapdoor-signature schemes, and their instantiations (based on RSA, identity-based signatures from the Gap Diffie-Hellman group, etc.). This is a significant contribution, since only few provably secure proxy schemes were known before (e.g., [3] for the non-hierarchical case).
- We show a direct and tight hierarchy for key evolving signature schemes. Specifically, we show that IR implies KI implies FS, and KI implies IR without loss in concrete security. The implication KI $\rightarrow$ FS was left as an open problem in [5], and our proof of it utilizes our result about the equivalence of KI and proxy signatures.[1] Note that while proving IR$\rightarrow$ FS is trivial, relating them to KI is not. For example, the naive approach of unifying the signer and helper of the KI model into the single signer entity of the FS model, does not work. This is because the keys of the signer and helper together are *not* forward-secure, by definition. In fact, the opposite is true since the helper keys with the signing key for any period should be able to provide the signing key for all other periods through the random-access property.

The relationships we establish are summarized in Figure 1 on the left side. In addition, on the right side is a diagram summarizing our technical results which are employed in the derivation of these relationships, showing the structure of our proofs (and may be helpful to obtain the best constructions from an engineering

[1] Once we established this result through the connection to proxy signatures, we also succeeded in showing that KI $\rightarrow$ IR, which together with the trivial IR $\rightarrow$ FS gave an alternative proof that KI $\rightarrow$ FS directly within key evolving signatures.

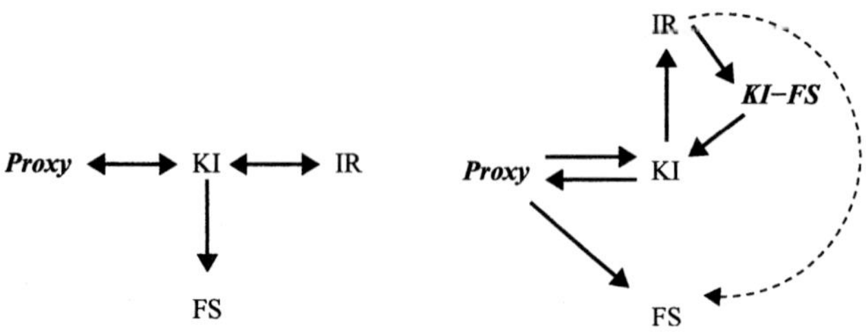


**Fig. 1.** The left diagram is a summary of our main results, and the right diagram is a summary of our technical reductions used to established them.

point of view). In particular, we introduce an intermediate notion between IR and KI, denoted KI-FS, which has helped us to conceptualize the IR $\rightarrow$ KI relation (and it may be of independent interest for certain applications). The dashed line refers to the trivial implication of FS from IR, which together with our result that KI implies IR gives an alternative proof that KI implies FS. We believe that directly relating proxy signing (which is a trust management and flexibility mechanism) to that of key evolving signatures (which are mechanisms to protect against key exposure attacks) is somewhat surprising and conceptually interesting. This was also a crucial step in answering the question about the relation between KI and FS, a recognized open question about seemingly closer notions.

*Organization* We provide the definitions for proxy signature schemes in Section 2.1, together with motivations and discussions of the model. This includes the differences and generalizations of our model compared with the previous single-level model, the weakness of previous schemes, how it is addressed by our model, and how to modify previous schemes to achieve security. In Section 2.2 we briefly review definitions for the key-evolving notions of IR, KI, and FS. In Section 3 we present the characterization of proxy signatures as equivalent to KI. Finally, in Section 4 we present the hierarchy of key evolving signatures, by showing that IR implies KI (which is a consolidation of our proofs that IR implies KI-FS and that KI-FS implies KI, given in the full version), KI implies IR, and by showing that Proxy implies FS (and therefore KI implies FS).

## 2 Definitions of Proxy Signatures and Key Evolving Signatures

### 2.1 Proxy Signature

**Model** Proxy signature scheme $\Pi_{\mathsf{PS}} = (\mathsf{Gen}_{\mathsf{PS}}, \mathsf{Sign}_{\mathsf{PS}}, \mathsf{Vrfy}_{\mathsf{PS}}, (\mathsf{Dlg}_{\mathsf{DPS}}, \mathsf{Dlg}_{\mathsf{PPS}}), \mathsf{PSig}_{\mathsf{PS}}, \mathsf{PVrf}_{\mathsf{PS}}, \mathsf{ID}_{\mathsf{PS}})$ consists of the following eight algorithms.

$\mathsf{Gen}_{\mathsf{PS}}$, the key generation algorithm, which takes security parameters $k \in \mathbb{N}$ as input, output an signing key $SK$ and a public key $PK$.

$\mathsf{Sign}_{\mathsf{PS}}$, the signing algorithm, which takes a signing key $SK$ and a message $M$ as input, outputs a signature $sig$ on $M$.

$\mathsf{Vrfy}_{\mathsf{PS}}$, the verification algorithm, which takes the public key $PK$, a message $M$, and a candidate signature $sig$ as input, outputs a bit $b$, where $b = 1$ iff the signature is accepted.

$(\mathsf{Dlg}_{\mathsf{DPS}}, \mathsf{Dlg}_{\mathsf{PPS}})$, (interactive) proxy-designation algorithms (where $\mathsf{Dlg}_{\mathsf{DPS}}$ and $\mathsf{Dlg}_{\mathsf{PPS}}$ are owned by the designator $i_{L-1}$ and the proxy signer $i_L$, respectively.)

$\mathsf{Dlg}_{\mathsf{DPS}}$ takes public keys of a designator $PK_{i_{L-1}}$ and a proxy signer $PK_{i_L}$, the signing key of which the designator delegates its signing right (i.e., the signing key is either a signing key $SK_{i_{L-1}}$ or a proxy signing key $SKP_{i_0 \dashrightarrow i_{L-1}}$ depending on whether $i_{L-1}$ is original signer or proxy signer), a warrant up to previous delegation $W_{L-1}$ and a warrant $\omega_L$ set in current delegation as inputs. $\mathsf{Dlg}_{\mathsf{DPS}}$ has no local output. Note that the warrant usually contains the information on "valid period", "limitation", etc. We say that a message *violates* a warrant if the message is not compliant with the contents of the warrant.

$\mathsf{Dlg}_{\mathsf{PPS}}$ takes public keys of a designator $PK_{i_{L-1}}$ and a proxy signer $PK_{i_L}$, the secret key of the proxy signer $SK_{i_L}$ as inputs and outputs a proxy signing key $SKP_{i_0 \dashrightarrow i_L}$ and a warrant $W_L$. Note that no secret key is given when the type of the designation is "self delegation" in which the designator designates its signing right to itself with limited capability[2].

$\mathsf{PSig}_{\mathsf{PS}}$, the proxy signing algorithm, which takes a proxy signing key $SKP_{i_0 \dashrightarrow i_L}$, a message $M$ and a warrant $W$ as input, outputs a proxy signature $psig$.

$\mathsf{PVrf}_{\mathsf{PS}}$, the proxy verification algorithm, which takes a public key $PK_{i_0}$ of the original designator, a message $M$, a warrant $W$, and a proxy signature $psig$ as input, outputs a bit $b$, where $b = 1$ iff the proxy signature is accepted.

$\mathsf{ID}_{\mathsf{PS}}$, the proxy identification algorithm, which takes a warrant $W$ and a proxy signature $psig$ as input, outputs a list of identity (i.e., public key) $PK^*$ in the delegation chain.

CORRECTNESS: We require that all message $M$ and any delegation chain $j_0 \to j_2 \to \cdots \to j_L$, $\mathsf{PVrf}_{\mathsf{PS}}(PK_{i_0}, M, W_L, \mathsf{PSig}_{\mathsf{PS}}(SKP_{i_0 \dashrightarrow i_L}, M, W_L)) = 1$ and $\mathsf{ID}_{\mathsf{PS}}(W, \mathsf{PSig}_{\mathsf{PS}}(SKP_{i_0 \dashrightarrow i_L}, M, W_L)) = (PK_{i_0}, \ldots, PK_{i_L})$ if the proxy signing key $SKP_{i_0 \dashrightarrow i_L}$ and the warrant $W_L$ is the output of consecutive executions of $(SKP_{i_0 \dashrightarrow i_l}, W_l) \leftarrow \begin{bmatrix} \mathsf{Dlg}_{\mathsf{DPS}}(PK_{i_{l-1}}, PK_{i_l}, SK_{i_{l-1}}, W_{l-1}, \omega_l), \\ \mathsf{Dlg}_{\mathsf{DPS}}(PK_{i_{l-1}}, PK_{i_l}, SK_{i_l}) \end{bmatrix}$; and the message $M$ does not violate the warrant $W_L$.

**Definition of Security** Let $F$ be a probabilistic polynomial-time oracle Turing machine with the following oracles:

[2] This is significant since if the proxy signer (or device) has the original signing key in self delegation it is impossible for the designator to limit the signing capability of the proxy signer.

- $O_{\mathsf{sig}}$, the signing oracle, which
  1. on input $(\text{"s"}, M, j)$, outputs $\mathsf{Sign}_{\mathsf{PS}}(SK_j, M)$.
  2. on input $(\text{"p"}, M, (j_1, \ldots, j_L), W)$, outputs $\mathsf{PSig}_{\mathsf{PS}}(SKP_{j_1 \dashrightarrow j_L}, M, W)$.
- $O_{\mathsf{sec}}$, the key exposure oracle, which on input
  1. $(\text{"s"}, j)$, outputs $(SK_j, PK_j)$.
  2. $(\text{"sd"}, j, L, (\omega_1, \ldots, \omega_{L-1}))$, outputs the pair of self proxy signing key and the warrant $(SKP_{j \dashrightarrow j}, W)$ where the length of the delegation chain is $L$.
- $O_{\mathsf{Dlg}}$, the designation oracle, which
  1. on input $(\text{"d"}, (j_1, \ldots, j_L, W, \omega))$, interacts with $\mathsf{Dlg}_{\mathsf{PPS}}(PK_{j_{L-1}}, PK_{j_L}, SK_{j_L})$ on behalf of $\mathsf{Dlg}_{\mathsf{DPS}}(PK_{j_{L-1}}, PK_{j_L}, SKP_{j_1 \dashrightarrow j_{L-1}}, W, \omega)$.
  2. on input $(\text{"p"}, (j_1, \ldots, j_L))$, interacts with $\mathsf{Dlg}_{\mathsf{DPS}}(PK_{j_{L-1}}, PK_{j_L}, SKP_{j_1 \dashrightarrow j_{L-1}}, W, \omega)$ on behalf of $\mathsf{Dlg}_{\mathsf{PPS}}(PK_{j_{L-1}}, PK_{j_L}, SK_{j_L})$.

Let $Q = (Q_{\mathsf{sec}}, Q_{\mathsf{Dlg}})$ where $Q_{\mathsf{sec}}$ and $Q_{\mathsf{Dlg}}$ be the set of $F$'s valid query to the key exposure oracle and designation oracle, respectively. We say that the scheme is

- $(j, Q)$-signable if and only if $(\text{"s"}, j) \in Q_{\mathsf{sec}}$.
- $((j_1, \ldots, j_L), W, Q)$-proxy-signable if and only if either of the following holds
  1. $(\text{"s"}, j) \in Q_{\mathsf{sec}}$ (for all $j$ such that $1 \le j \le L$)
  2. there exists $L' (\le L)$ such that
     - $(\text{"d"}, (j_1, \ldots, j_{L'}, W', \omega')) \in Q_{\mathsf{Dlg}}$
     - $W', (W', \omega')$ do not contradict $W$
     - $j_l = j_{l-1}$ or $(\text{"s"}, j_l) \in Q_{\mathsf{sec}}$ (for $L' \le l \le L$)
  3. there exists $L' (\le L)$ such that
     - $j_1 = \ldots, j_{L'}$ and $(\text{"sd"}, L', (\omega_1, \ldots, \omega_{L'-1})) \in Q_{\mathsf{sec}}$
     - $\omega_i$ do not contradict $W$
     - $j_l = j_{l-1}$ or $(\text{"s"}, j_l) \in Q_{\mathsf{sec}}$ (for $L' < l \le L$)

Let $\mathsf{Succ}_F^{\Pi_{\mathsf{PS}}}(k)$ be defined as follows,

$$\mathsf{Succ}_F^{\Pi_{\mathsf{FS}}}(k) =$$
$$\Pr\left[\begin{array}{l|c} (\sigma = (M, s, PK) \wedge \mathsf{Vrfy}_{\mathsf{PS}}(PK, M, s) = 1) \quad \vee & (SK_j, PK_j) \leftarrow \mathsf{Gen}_{\mathsf{PS}}(1^k) \\ (\sigma = (M, W, ps, PK) \wedge \mathsf{PVrf}_{\mathsf{PS}}(PK, M, ps) = 1) & \sigma \leftarrow F^{O_{\mathsf{sig}}, O_{\mathsf{sec}}, O_{\mathsf{Dlg}}}(1^k) \end{array}\right]$$

where

- $M$ is never queried to $O_{\mathsf{sig}}$ and the scheme is not $(j, Q)$-signable if $\sigma = (M, s, PK_j)$.
- $((i_1, \ldots, i_L), M, W)$ is never queried to $O_{\mathsf{sig}}$ and the scheme is not $((i_1, \ldots, i_L), W, Q)$-proxy-signable if $(\text{"s"}, i_L) \notin Q_{\mathsf{sec}}$ where $\sigma = (M, W, ps, PK)$ and $\mathsf{ID}_{\mathsf{PS}}(W, ps) = (PK_{i_1}, \ldots, PK_{i_L})$.

We say $\Pi_{\mathsf{PS}}$ is $(\tau, \epsilon, q)$-secure proxy signature if $\mathsf{Succ}_F^{\Pi_{\mathsf{PS}}}(k) < \epsilon$ for any probabilistic polynomial time Turing machine $F$ with running time at most $\tau$ and the number of the queries to $O_{\mathsf{sig}}$ is upper bounded by $q$.

**Discussion: Delegation by Certificate, and the Self-delegation Attack** Delegation by certificate is a well-known simple notion. It achieves delegation by the fact that the delegator computes a warrant $W = \mathsf{Sign}(SK_d, (PK_p, \mathtt{limitation}))$ with its secret key where $SK_d$ is the secret key of the delegator and $PK_p$ is the public key of the proxy signer. The proxy signer can computes a proxy signature $ps$ for the message $M$ simply by $ps = \mathsf{Sign}(SK_p, (W, M))$.

Delegation by certificate works well in many setting, however, we must be aware that a naive implementation leads to an attack, even on the delegation by certificate scheme. Specifically, we must take care of implementing *self-delegation* securely. For example, the scheme in [3] is not secure under our security definition, and it can be easily broken simply by querying ("sd", $2, \Lambda$) from the $Q_{\mathsf{sig}}$ oracle (we will use $\Lambda$ to denote null data.) Since the scheme of [3] is constructed in such a way that the proxy signing key is exactly the same as the original signing key of the proxy signer even in the case of self-delegation, an adversary can forge (non-proxy) signature for any message simply by querying the self-delegation signing key. We must carefully consider the meaning of the self-delegation, which is usually used for delegating *limited* signing capability.

The model proposed in [3] also possesses the problem of self-delegation. Namely, the oracles defined by [3] only allows giving transcript of $\mathsf{Dlg}_{\mathsf{DPS}}$ and $\mathsf{Dlg}_{\mathsf{PPS}}$. Therefore, there is no way for the adversary to get the self-delegation key. This is not the case in real life since self-delegation is needed when the signing key is stored in insecure environment (e.g. laptop PC get delegation from a host). Therefore, the scheme must be secure even if the self-delegation key is exposed. In contrast, our model allows the adversary to gain self-delegation keys to reflect this real life setting. Our implementation of proxy signature based on KI also takes care of this problem. Namely, in our implementation, new key pair is always generated in self-delegation, which prevents the attack above.

In defining the model of proxy signatures the most crucial point is how to treat the semantics of the warrant since the warrant usually contains application specific information. Therefore, in the model level, it is desirable not to define the detailed semantics. In our model no semantics is defined for the warrant, it is only defined as input and output of the algorithm and a messages can be in agreement or in violation with the warrant. Further, not having access to a warrant prevents the usage of the delegated key, which is part of our model.

We also note that, in the general case, the chain of warrants may have arbitrary information in it and one needs to read it to understand whether a message is in agreement with the warrant. In this cases the length of verification of a proxy signature must be linear in the size of the delegation chain. (Of course, if warrants are of special semantics, e.g if they are not present at all, then this may be improved, e.g using aggregate signatures as suggested by [3].)

## 2.2 Definitions of Key-Evolving Signatures

In this section we briefly review the definition of key-evolving signatures. These definitions are the same ones as introduced in the original papers, except that we unify them following the notations of [9]. The complete definition of each notion is given in the full version of this paper [13].

**Model** The three key-evolving signature notions, Forward-Secure signatures (FS), Key-Insulated signatures (KI), and Intrusion-Resilient signatures (IR) consist of subsets of the following seven algorithms, as indicated below.

Gen, the key generation algorithm, which takes security parameters $k \in \mathbb{N}$ and the total number of periods $N$ as input, outputs an initial signing key $SK_0$ and public key $PK$. Gen also outputs initial key for the helper (base) $SK^*$ in KI and IR.

Upd*, the update algorithm of the base, which takes a base key $SK^*$ as inputs, key update message $SKU$. Key update message is used to update the signing key of the signer $SK$. In KI, indices $i, j$ is also taken as input where $i$ denotes the current time period of $SK$ and $j$ denotes the time period of $SK$ after update. In IR, base key is also updated by Upd*.

Upd, the signer key update algorithm, which takes a signer key $SK_i$ of the previous time period $i$ and a key update message $SKU$, outputs the signer key $SK_j$ of the time period $j$. In IR, $j$ is always $i+1$ whereas, in KI, $i$ and $j$ can be chosen arbitrary within the condition $0 \leq i, j \leq N$.

Refr*, the base-key refresh algorithm, which takes a base key $SK^*$ of the current time period, outputs new base key of the current time period and a key refresh message $SKR$. Only IR has Refr*.

Refr, the signer-key refresh algorithm, which takes a signer key $SK$ of the current time period and a key refresh message $SKR$, outputs new signer key of the current time period. Only IR has Refr.

Sign, the signing algorithm, which takes a signer key $SK_j$, an index of a time period $j$ and a message $M$ as input, outputs a signature $\langle j, sig \rangle$ on $M$ for time period $j$.

Vrfy, the verification algorithm, which takes the public key $PK$, a message $M$, a pair $\langle j, s \rangle$, outputs a bit $b$, where $b = 1$ iff the signature is accepted.

FS consists of four algorithms $\Pi_{\mathsf{FS}} = (\mathsf{Gen}_{\mathsf{FS}}, \mathsf{Upd}_{\mathsf{FS}}, \mathsf{Sign}_{\mathsf{FS}}, \mathsf{Vrfy}_{\mathsf{FS}})$, KI consists of five algorithms $\Pi_{\mathsf{KI}} = (\mathsf{Gen}_{\mathsf{KI}}, \mathsf{Upd}^*_{\mathsf{KI}}, \mathsf{Upd}_{\mathsf{KI}}, \mathsf{Sign}_{\mathsf{KI}}, \mathsf{Vrfy}_{\mathsf{KI}})$ and IR consists of seven algorithms $\Pi_{\mathsf{IR}} = (\mathsf{Gen}_{\mathsf{IR}}, \mathsf{Upd}^*_{\mathsf{IR}}, \mathsf{Upd}_{\mathsf{IR}}, \mathsf{Refr}^*_{\mathsf{IR}}, \mathsf{Refr}_{\mathsf{IR}}, \mathsf{Sign}_{\mathsf{IR}}, \mathsf{Vrfy}_{\mathsf{IR}})$, with appropriate (and natural) correctness requirements.

**Definition of Security** To define the definition of security, we consider a probabilistic polynomial-time oracle Turing machine $F$ with the following oracles:

- $O_{\mathsf{sig}}$, the signing oracle, which take the message $M$ the time period $i$ outputs the signature for $M$ of the designated time period.
- $O_{\mathsf{sec}}$, the key exposure oracle, which on input the name of the target key (e.g. signing key, base key, etc.) and the time period, outputs the key of the designated period.

Let $Q$ be the set of valid key exposure query of $F$. Then the successful probability of the adversary $\mathsf{Succ}_F^{\Pi_{\mathsf{FS}}}(k)$ can be defined as follows,

$$\mathsf{Succ}_F^{\Pi}(k) = \Pr\left[\mathsf{Vrfy}_{\mathsf{FS}}(PK, M, \langle i, s\rangle) = 1 \,\middle|\, \begin{array}{c}(PK, SK_0) \leftarrow \mathsf{Gen}(1^k),\\ (M, \langle i, s\rangle) \leftarrow F^{O_{\mathsf{sec}}, O_{\mathsf{sig}}}(PK)\end{array}\right]$$

where $(M, i)$ is never queried to $O_{\mathsf{sig}}$ and

- (In FS), $F$ never gets the signing key before the time period $i$.
- (In KI), $F$ never gets the signing key of the time period $i$.
- (In IR), $F$ never gets the signer key of the time period $i$ and $F$ never gets the signer key and the base key simultaneously in time time period before $i$.

We refer the reader to our full version [13] (and to the original papers [2,5,9]) for more complete definitions.

## 3 Characterization of Proxy Signatures

In this section we give the characterization of proxy signature. Namely, we prove that proxy signatures are equivalent to key-insulated signatures by constructing a key-insulated signature based on any proxy signature with concrete security reduction and vice versa.

### 3.1 Proxy $\rightarrow$ $(N-1, N)$ KI

We construct $(N-1, N)$ key-insulated signature as follows. The signing key of time period $j$ corresponds to proxy signing key with delegation chain of length $j+1$. The important point is that the proxy signer is changed every time when the period changes, which prevents the attacker who gets the signing key of period $j$ from forging the signature of the other periods.

The complete construction of $\Pi_{\mathsf{KI}} = (\mathsf{Gen}_{\mathsf{KI}}, \mathsf{Upd}^*_{\mathsf{KI}}, \mathsf{Upd}_{\mathsf{KI}}, \mathsf{Sign}_{\mathsf{KI}}, \mathsf{Vrfy}_{\mathsf{KI}})$ from proxy signature $\Pi_{\mathsf{PS}} = (\mathsf{Gen}_{\mathsf{PS}}, \mathsf{Sign}_{\mathsf{PS}}, \mathsf{Vrfy}_{\mathsf{PS}}, (\mathsf{Dlg}_{\mathsf{DPS}}, \mathsf{Dlg}_{\mathsf{PPS}}), \mathsf{PSig}_{\mathsf{PS}}, \mathsf{PVrf}_{\mathsf{PS}}, \mathsf{ID}_{\mathsf{PS}})$ is as follows.

---

$\mathsf{Gen}_{\mathsf{KI}}(1^k, N)$

$(SK_*^{(\mathsf{PS})}, PK_*^{(\mathsf{PS})}) \leftarrow \mathsf{Gen}_{\mathsf{PS}}(1^k)$; $\quad (SK_0^{(\mathsf{PS})}, PK_0^{(\mathsf{PS})}) \leftarrow \mathsf{Gen}_{\mathsf{PS}}(1^k)$;

$$(SKP_{*\rightarrow 0}^{(\mathsf{PS})}, W_0) \leftarrow \begin{bmatrix}\mathsf{Dlg}_{\mathsf{DPS}}(PK_*^{(\mathsf{PS})}, PK_0^{(\mathsf{PS})}, SK_*^{(\mathsf{PS})}, \Lambda, \Lambda),\\ \mathsf{Dlg}_{\mathsf{PPS}}(PK_*^{(\mathsf{PS})}, PK_0^{(\mathsf{PS})}, \Lambda)\end{bmatrix};$$

$SK^{*(\mathsf{KI})} \leftarrow (PK_*^{(\mathsf{PS})}, SK_*^{(\mathsf{PS})})$; $\quad SK_0^{(\mathsf{KI})} \leftarrow (SKP_{*\rightarrow 0}^{(\mathsf{PS})}, W_0)$;
$PK^{(\mathsf{KI})} \leftarrow PK_*^{(\mathsf{PS})}$;

**output** $(SK^{*(\mathsf{KI})}, SK_0^{(\mathsf{KI})}, PK^{(\mathsf{KI})})$;

---

$\underline{\mathsf{Upd}^*_{\mathsf{KI}}(SK^{*(\mathsf{KI})}, i, j)}$

$(PK_*^{(\mathsf{PS})}, SKP) \leftarrow SK^{*(\mathsf{KI})}$;
$W \leftarrow \Lambda$;
`for` $n = 0$ `to` $j$ `do`

$$(SKP, W) \leftarrow \begin{bmatrix} \mathsf{Dlg}_{\mathsf{DPS}}(PK_*^{(\mathsf{PS})}, PK_*^{(\mathsf{PS})}, SKP, W, \Lambda), \\ \mathsf{Dlg}_{\mathsf{PPS}}(PK_*^{(\mathsf{PS})}, PK_*^{(\mathsf{PS})}, \Lambda) \end{bmatrix};$$

$(SK_j^{(\mathsf{PS})}, PK_j^{(\mathsf{PS})}) \leftarrow \mathsf{Gen}_{\mathsf{PS}}(1^k)$;

$$(SKP, W) \leftarrow \begin{bmatrix} \mathsf{Dlg}_{\mathsf{DPS}}(PK_*^{(\mathsf{PS})}, PK_j^{(\mathsf{PS})}, SKP, W, \Lambda), \\ \mathsf{Dlg}_{\mathsf{PPS}}(PK_*^{(\mathsf{PS})}, PK_j^{(\mathsf{PS})}, SK_j^{(\mathsf{PS})}) \end{bmatrix};$$

$SK'^{(\mathsf{KI})}_{i,j} \leftarrow (SKP, W)$;

**output** $SK'^{(\mathsf{KI})}_{i,j}$;

---

$\underline{\mathsf{Upd}_{\mathsf{KI}}(SK_i^{(\mathsf{KI})}, SK'^{(\mathsf{KI})}_{i,j})}$

**output** $SK'^{(\mathsf{KI})}_{i,j}$;

---

$\underline{\mathsf{Sign}_{\mathsf{KI}}(SK_j^{(\mathsf{KI})}, j, M)}$

$(SKP^{(\mathsf{PS})}_{*\dashrightarrow j}, W) \leftarrow SK_j^{(\mathsf{KI})}$;
$ps \leftarrow \mathsf{PSig}_{\mathsf{PS}}(SKP^{(\mathsf{PS})}_{*\dashrightarrow j}, M, W)$;
**output** $\langle j, (W, ps)\rangle$;

$\underline{\mathsf{Vrfy}_{\mathsf{KI}}(PK^{(\mathsf{KI})}, M, \langle j, s\rangle)}$

$PK_* \leftarrow PK^{(\mathsf{KI})}$;
$(W, ps) \leftarrow s$; $PK^* \leftarrow \mathsf{ID}_{\mathsf{PS}}^{(\mathsf{PS})}(W, ps)$;
`if` $(PK^* \neq (\underbrace{PK_*, \ldots, PK_*}_{j+1}, \cdot))$ `then`
  **output** $0$;
`else`
  **output** $\mathsf{PVrf}_{\mathsf{PS}}(PK_*, M, W, ps)$;

---

The following theorem holds for the above construction.

**Theorem 1.** *Suppose there exists* $(\tau_{\mathsf{KI}}, \epsilon_{\mathsf{KI}}, q_{\mathsf{KI}}^{\mathsf{sig}}, q_{\mathsf{KI}}^{\mathsf{sec}})$*-Adversary* $F_{\mathsf{KI}}$ *against* KI *as constructed above with probability* $\epsilon_{\mathsf{KI}}$*, with running time* $\tau_{\mathsf{KI}}$*,* $q_{\mathsf{KI}}^{\mathsf{sig}}$ *queries to the signing oracle,* $q_{\mathsf{KI}}^{\mathsf{sec}}$ *queries to the key exposure oracle then there exists* $(\tau_{\mathsf{PS}}, \epsilon_{\mathsf{PS}}, q_{\mathsf{PS}}^{\mathsf{sig}}, q_{\mathsf{PS}}^{\mathsf{sec}}, q_{\mathsf{PS}}^{\mathsf{Dlg}})$*-Adversary* $F_{\mathsf{PS}}$ *against* PS *with* $\tau_{\mathsf{PS}} = \tau_{\mathsf{KI}}$*,* $\epsilon_{\mathsf{PS}} = \epsilon_{\mathsf{PS}}$*,* $q_{\mathsf{PS}}^{\mathsf{sig}} = q_{\mathsf{KI}}^{\mathsf{sig}}$*,* $q_{\mathsf{PS}}^{\mathsf{sec}} = q_{\mathsf{KI}}^{\mathsf{sec}}$*,* $q_{\mathsf{PS}}^{\mathsf{Dlg}} = q_{\mathsf{KI}}^{\mathsf{sec}}$*.*

*Proof.* We construct the signing oracle $O_{\mathsf{sig}}^{\mathsf{KI}}$ and the key exposure oracle $O_{\mathsf{sec}}^{(\mathsf{KI})}$ from $O_{\mathsf{sig}}^{(\mathsf{PS})}$, $O_{\mathsf{sec}}^{(\mathsf{PS})}$ and $O_{\mathsf{Dlg}}^{(\mathsf{PS})}$ as follows.

---

$\underline{O_{\mathsf{sig}}^{(\mathsf{KI})}(M, j)}$

**output** $O_{\mathsf{sig}}^{(\mathsf{PS})}(\text{“p”}, (\underbrace{*, *, \ldots, *}_{j+1}, j).M, W_j)$;

$\underline{O^{(\mathsf{KI})}_{\mathsf{sec}}(\mathtt{query})}$

```
if (query = ("s", j)) then
    (SK_j, PK_j) ← O_sec^(PS)("s", j);
    (SKP_{*-->j}, W_{j+1}) ← [ O_Dlg^(PS)("d", (*, *, ..., *)_{j+1}, W_j, Λ),
                               Dlg_PPS(PK_*, PK_j, SK_j) ];
    output (SKP_{*-->j}, W_{j+1});
else
    output ⊥;
```

Then $F_{\mathsf{PS}}^{O^{(\mathsf{PS})}_{\mathsf{sig}},O^{(\mathsf{PS})}_{\mathsf{sec}},O^{(\mathsf{PS})}_{\mathsf{Dlg}}}(PK^{(\mathsf{PS})}_*) = (M, W, \sigma, PK^{(\mathsf{PS})}_*)$ where $(M, \langle j, (W, \sigma)\rangle) = F_{\mathsf{KI}}^{O^{(\mathsf{KI})}_{\mathsf{sig}},O^{(\mathsf{KI})}_{\mathsf{sec}}}(PK^{(\mathsf{PS})}_*)$ is the adversary as desired. Since if $F_{\mathsf{KI}}$ can forge a valid signature $\langle j, \sigma\rangle$ for the message $M$ then it is easy to see from the construction that $\sigma = (W, ps)$ is also a valid pair of a warrant and a proxy signature for the message $M$. Further, the scheme $\Pi_{\mathsf{PS}}$ is not $((\underbrace{*, \ldots, *}_{j+1}, j), W, Q^{(\mathsf{PS})})$-proxy-signable where $Q_{\mathsf{PS}} = (Q^{(\mathsf{PS})}_{\mathsf{sig}}, Q^{(\mathsf{PS})}_{\mathsf{sec}} Q^{(\mathsf{PS})}_{\mathsf{Dlg}})$ is a set of valid query to the oracles of $\Pi_{\mathsf{PS}}$ and $(\text{“p”}, (\underbrace{*, \ldots, *}_{j+1}, j), M, W)$ is never queried to $O^{(\mathsf{PS})}_{\mathsf{sig}}$.

Further, the scheme is $(N-1, N)$ KI since if the adversary who gets the signing key of periods $j_1, \ldots, j_{N-1}$ can compute the signature of the period $j_N \notin \{j_1, \ldots, j_{N-1}\}$ then the adversary can compute the proxy signature which is not proxy signable. □

EFFICIENCY: The running time of each algorithm $\mathsf{Gen}_{\mathsf{KI}}$, $\mathsf{Upd}_{\mathsf{KI}}$, $\mathsf{Sign}_{\mathsf{KI}}$ and $\mathsf{Vrfy}_{\mathsf{KI}}$ becomes as follows, where $\tau^{(\mathsf{SIG})}_{\mathsf{Alg}}$ denotes the running time of the algorithm Alg for the signature scheme SIG.

$$\tau^{(\mathsf{KI})}_{\mathsf{Gen}} = 2\cdot\tau^{(\mathsf{PS})}_{\mathsf{Gen}} + \tau^{(\mathsf{PS})}_{\mathsf{Dlg_P}} + \tau^{(\mathsf{PS})}_{\mathsf{Dlg_D}}, \qquad \tau^{(\mathsf{KI})}_{\mathsf{Upd*}} = (N+1)\cdot\left(\tau^{(\mathsf{PS})}_{\mathsf{Dlg_D}} + \tau^{(\mathsf{PS})}_{\mathsf{Dlg_D}}\right) + \tau^{(\mathsf{PS})}_{\mathsf{Gen}},$$
$$\tau^{(\mathsf{KI})}_{\mathsf{Upd}} = \mathcal{O}(1), \qquad \tau^{(\mathsf{KI})}_{\mathsf{Sign}} = \tau^{(\mathsf{PS})}_{\mathsf{PSig}}, \qquad \tau^{(\mathsf{KI})}_{\mathsf{Vrfy}} = \tau^{(\mathsf{PS})}_{\mathsf{PVrf}} + \tau^{(\mathsf{PS})}_{\mathsf{ID}}$$

### 3.2 KI → Proxy

PS with $n$ designators can be constructed constructed from $(c\cdot n-1, c\cdot n)$ KI as follows (where $c$ is the total number of self delegation allowed for each delegator.) In key generation phase, $c$ signer keys $SK_{j\cdot c}, SK_{j\cdot c+1}, \ldots, SK_{(j-1)\cdot c-1}$ is assigned to designator $j$. the signer key $SK_{j\cdot c}$ is used for (ordinary) signing, proxy signing and delegation. The other key is used for self proxy signing and self delegation.

Delegation is simply based on so-called “certificate chain”. That is, to delegate the signing right of user $i$ to user $j$, the user $i$ simply compute the *warrant* containing information of the public key of user $i$, the limitation of the delegation and the signature of user $j$. In our construction the warrant $W$ is of the form $W = (W', \omega, \mathsf{Sign}_{\mathsf{KI}}(SK, (W', \omega)))$ where $W'$ is the warrant of previous delegation and $\omega = (l_1, l_2, \mathtt{usage})$ describes the limitation of the current delegation, namely, $l_1$ and $l_2$ denote the range of possible secret keys used for self proxy signing (therefore, $l_1, l_2$ only make sense in the self delegation.) This type of warrant prevents the user $i$ with warrants $W_1, \ldots, W_n$ from computing a valid proxy signature of any warrant other than $W_1, \ldots, W_n$.

Note that different signer key of KI is assigned for each self delegation. This prevents the attacker who gets a signer key which can be used with some self delegation from computing a valid proxy signature for the other self delegation. The concrete security reduction can be shown by the following theorem.

**Theorem 2.** *It is possible to construct* PS *(with $n$ designators and the total number of self delegation allowed for each delegator is less than a constant $c$) from* $(c \cdot n - 1, c \cdot n)$ KI *in such a way that if there exists* $(\tau_{\mathsf{PS}}, \epsilon_{\mathsf{PS}}, q^{\mathsf{sig}}_{\mathsf{PS}}, q^{\mathsf{sec}}_{\mathsf{PS}}, q^{\mathsf{Dlg}}_{\mathsf{PS}})$*-Adversary* $F_{\mathsf{PS}}$ *against* PS *then there exists* $(\tau_{\mathsf{KI}}, \epsilon_{\mathsf{KI}}, q^{\mathsf{sig}}_{\mathsf{KI}}, q^{\mathsf{sec}}_{\mathsf{KI}})$*-Adversary* $F_{\mathsf{KI}}$ *against* KI *with* $\tau_{\mathsf{KI}} = \tau_{\mathsf{PS}}$, $\epsilon_{\mathsf{KI}} = \epsilon_{\mathsf{PS}}$, $q^{\mathsf{sig}}_{\mathsf{KI}} = q^{\mathsf{sig}}_{\mathsf{PS}} + q^{\mathsf{Dlg}}_{\mathsf{PS}}$ *and* $q^{\mathsf{sec}}_{\mathsf{KI}} \leq q^{\mathsf{sec}}_{\mathsf{PS}} + c \cdot q^{\mathsf{Dlg}}_{\mathsf{PS}}$

The proof is given in the full version [13].

EFFICIENCY: The running time of $\mathsf{Gen}_{\mathsf{PS}}, \mathsf{Sign}_{\mathsf{PS}}, \mathsf{Vrfy}_{\mathsf{PS}}, \mathsf{Dlg}_{\mathsf{DPS}}, \mathsf{Dlg}_{\mathsf{PPS}}, \mathsf{PSig}_{\mathsf{PS}}$ and $\mathsf{PVrf}_{\mathsf{PS}}$ in the construction of the above theorem, become as follows where $L$ denotes the length of the delegation chain.

$$\tau^{(\mathsf{PS})}_{\mathsf{Gen}} = \tau^{(\mathsf{KI})}_{\mathsf{Gen}} + c\left(\tau^{(\mathsf{KI})}_{\mathsf{Upd}^*} + \tau^{(\mathsf{KI})}_{\mathsf{Upd}}\right), \qquad \tau^{(\mathsf{PS})}_{\mathsf{Sign}} = \tau^{(\mathsf{KI})}_{\mathsf{Sign}}, \qquad \tau^{(\mathsf{PS})}_{\mathsf{Vrfy}} = \tau^{(\mathsf{KI})}_{\mathsf{Vrfy}},$$
$$\tau^{(\mathsf{PS})}_{\mathsf{Dlg_D}} = \tau^{(\mathsf{KI})}_{\mathsf{Sign}}, \qquad \tau^{(\mathsf{PS})}_{\mathsf{Dlg_P}} = \mathcal{O}(1),$$
$$\tau^{(\mathsf{PS})}_{\mathsf{PSig}} = \tau^{(\mathsf{KI})}_{\mathsf{Sign}}, \qquad \tau^{(\mathsf{PS})}_{\mathsf{PVrf}} = L \cdot \tau^{(\mathsf{KI})}_{\mathsf{Vrfy}}, \qquad \tau^{(\mathsf{PS})}_{\mathsf{ID}} = \mathcal{O}(L)$$

## 4 The Hierarchy of Key Evolving Signatures

In this section we show the hierarchy among the key evolving signatures. Namely, we show that intrusion-resilient signatures imply $(N-1, N)$ key-insulated signatures and vice versa, and that proxy signatures (and thus $(N-1, N)$ key-insulated signatures) imply forward-secure signatures. The results are summarized below, each followed by a brief overview of the proof. In some cases the complete formal constructions and proofs are omitted from this extended abstract, and can be found in the full version of our paper [13].

**Theorem 3** (IR $\rightarrow$ KI)**.** *It is possible to construct* KI *from* IR *in such a way that if there exists* $(\tau_{\mathsf{KI}}, \epsilon_{\mathsf{KI}}, q^{\mathsf{sig}}_{\mathsf{KI}}, q^{\mathsf{sec}}_{\mathsf{KI}})$*-Adversary* $F_{\mathsf{KI}}$ *which breaks* KI *then there exists* $(\tau_{\mathsf{IR}}, \epsilon_{\mathsf{IR}}, q^{\mathsf{sig}}_{\mathsf{IR}}, q^{\mathsf{sec}}_{\mathsf{IR}})$*-Adversary* $F_{\mathsf{IR}}$ *which breaks* IR *with* $\tau_{\mathsf{IR}} = \tau_{\mathsf{KI}}$, $\epsilon_{\mathsf{IR}} = \epsilon_{\mathsf{KI}}$, $q^{\mathsf{sig}}_{\mathsf{IR}} = q^{\mathsf{sig}}_{\mathsf{KI}}$ *and* $q^{\mathsf{sec}}_{\mathsf{IR}} = q^{\mathsf{sec}}_{\mathsf{KI}}$.

The reduction is based on the following idea: all the initial data of IR is stored in the base of KI and the signer of the KI only stores signer key of the current period. Then the random access to the key is possible by simply computing the signer key of any period from the initial state. The formal details are given below.

*Proof.* We construct $(N-1, N)$ key-insulated signature $\Pi_{\mathsf{KI}} = (\mathsf{Gen}_{\mathsf{KI}}, \mathsf{Upd}^*_{\mathsf{KI}}, \mathsf{Upd}_{\mathsf{KI}}, \mathsf{Sign}_{\mathsf{KI}}, \mathsf{Vrfy}_{\mathsf{KI}})$ from intrusion-resilient signature $\Pi_{\mathsf{IR}} = (\mathsf{Gen}_{\mathsf{IR}}, \mathsf{Upd}^*_{\mathsf{IR}}, \mathsf{Upd}_{\mathsf{IR}}, \mathsf{Refr}^*_{\mathsf{IR}}, \mathsf{Refr}_{\mathsf{IR}}, \mathsf{Sign}_{\mathsf{IR}}, \mathsf{Vrfy}_{\mathsf{IR}})$ as follows.

---

$\mathsf{Gen}_{\mathsf{KI}}(1^k, N)$

$(SKB^{(\mathsf{IR})}_{0.0}, SKS^{(\mathsf{IR})}_{0.0}, PK^{(\mathsf{IR})}) \leftarrow \mathsf{Gen}_{\mathsf{IR}}(1^k, N)$;
$SK^{*(\mathsf{KI})} \leftarrow (SKS^{(\mathsf{IR})}_{0.0}, SKB^{(\mathsf{IR})}_{0.0})$; $SK^{(\mathsf{KI})}_0 \leftarrow SKS^{(\mathsf{IR})}_{0.0}$; $PK^{(\mathsf{KI})} \leftarrow PK^{(\mathsf{IR})}$;

**output** $(SK^{*(\mathsf{KI})}, SK^{(\mathsf{KI})}_0, PK^{(\mathsf{KI})})$;

---

$\mathsf{Upd}^*_{\mathsf{KI}}(SK^{*(\mathsf{KI})}, i, j)$

$(SKB, SKS) \leftarrow SK^{*(\mathsf{KI})}$;
`for` $n=0$ `to` $j-1$ `do`
  $(SKB, SKU) \leftarrow \mathsf{Upd}^*_{\mathsf{IR}}(SKB)$;
  $SKS \leftarrow \mathsf{Upd}_{\mathsf{IR}}(SKS, SKU)$;
  $(SKB, SKR) \leftarrow \mathsf{Refr}^*_{\mathsf{IR}}(SKB)$;
  $SKS \leftarrow \mathsf{Refr}_{\mathsf{IR}}(SKS, SKR)$;
$SK'^{(\mathsf{KI})}_{i,j} \leftarrow SKS$;

**output** $SK'^{(\mathsf{KI})}_{i,j}$;

$\mathsf{Upd}_{\mathsf{KI}}(SK^{(\mathsf{KI})}_i, SK'^{(\mathsf{KI})}_{i,j})$

$SK^{(\mathsf{KI})}_j \leftarrow SK'^{(\mathsf{KI})}_{i,j}$;

**output** $SK^{(\mathsf{KI})}_j$;

---

$\mathsf{Sign}_{\mathsf{KI}}(SK^{(\mathsf{KI})}_j, j, M)$

**output** $\mathsf{Sign}_{\mathsf{IR}}(SK^{(\mathsf{KI})}_j, j, M)$;

$\mathsf{Vrfy}_{\mathsf{KI}}(PK^{(\mathsf{KI})}, M, \langle j, s\rangle)$

**output** $\mathsf{Vrfy}_{\mathsf{IR}}(PK^{(\mathsf{KI})}, M, \langle j, s\rangle)$;

---

We also construct the signing oracle $O^{(\mathsf{KI})}_{\mathsf{sig}}$ and the key exposure oracle $O^{(\mathsf{KI})}_{\mathsf{sec}}$ of KI from $O^{(\mathsf{IR})}_{\mathsf{sig}}$ and $O^{(\mathsf{IR})}_{\mathsf{sec}}$ as follows.

---

$O^{(\mathsf{KI})}_{\mathsf{sig}}(M, j)$

**output** $O^{(\mathsf{IR})}_{\mathsf{sig}}(M, j.1)$;

$O^{(\mathsf{KI})}_{\mathsf{sec}}(\mathtt{query})$

`if (query =` ("s", $j$)`) then`
  **output** $O^{(\mathsf{IR})}_{\mathsf{sec}}$("s", $j.1$);
`else`
  **output** $\perp$;

---

Then $F_{\mathsf{IR}}^{O^{(\mathsf{IR})}_{\mathsf{sig}}, O^{(\mathsf{IR})}_{\mathsf{sec}}}(PK^{(\mathsf{IR})}) = F_{\mathsf{KI}}^{O^{(\mathsf{KI})}_{\mathsf{sig}}, O^{(\mathsf{KI})}_{\mathsf{sec}}}(PK^{(\mathsf{IR})})$ is the adversary as desired. This is because KI and two oracles for KI are constructed in such a way that $SK^{(\mathsf{KI})}_j = SK^{(\mathsf{IR})}_{j.1}$ holds and the signing algorithm and the verification algorithm are exactly the same as those of IR. Therefor, if $F_{\mathsf{KI}}$ can produce a valid signature $(M, \langle j, sig\rangle)$

such that the scheme is not $(j, Q^{\mathsf{KI}})$-compromised and $(M, j)$ is never queried to $O^{\mathsf{KI}}_{\mathsf{sig}}$ then $\langle j, sig\rangle$ is also valid in IR and the scheme is not $(j, Q^{\mathsf{IR}})$-compromised and $(M, j.1)$ is never queried to $O^{\mathsf{IR}}_{\mathsf{sig}}$. Further, the resulting KI is $(N-1, N)$ KI since the key exposure of $N-1$ point in KI is corresponding to the key exposure of $N-1$ signer secret key of IR and no base key of IR is compromised. Therefore the security of the remaining signing key can be guaranteed by the IR property. □

We note that this construction is in fact a consolidation of earlier proofs we got regarding intermediate constructions, namely showing IR implies KI-FS and KI-FS implies KI. This intermediate notion of KI-FS is defined, and the corresponding reductions are proved, in the full version of our paper [13].

EFFICIENCY: The running time of $\mathsf{Gen}_{\mathsf{KI}}, \mathsf{Upd}^*_{\mathsf{KI}}, \mathsf{Upd}_{\mathsf{KI}}, \mathsf{Sign}_{\mathsf{KI}}$ and $\mathsf{Vrfy}_{\mathsf{KI}}$ in the above construction become as follows.

$$\tau^{(\mathsf{KI})}_{\mathsf{Gen}} = \tau^{(\mathsf{IR})}_{\mathsf{Gen}}, \qquad \tau^{(\mathsf{KI})}_{\mathsf{Upd}^*} = N \cdot \left(\tau^{(\mathsf{IR})}_{\mathsf{Upd}^*} + \tau^{(\mathsf{IR})}_{\mathsf{Upd}} + \tau^{(\mathsf{IR})}_{\mathsf{Refr}^*} + \tau^{(\mathsf{IR})}_{\mathsf{Refr}}\right),$$
$$\tau^{(\mathsf{KI})}_{\mathsf{Upd}} = \mathcal{O}(1), \qquad \tau^{(\mathsf{KI})}_{\mathsf{Sign}} = \tau^{(\mathsf{IR})}_{\mathsf{Sign}}, \qquad \tau^{(\mathsf{KI})}_{\mathsf{Vrfy}} = \tau^{(\mathsf{IR})}_{\mathsf{Vrfy}}.$$

**Theorem 4** (KI $\to$ IR). *It is possible to construct* IR *from* $(N-1, N)$ KI *in such a way that if there exists* $(\tau_{\mathsf{IR}}, \epsilon_{\mathsf{IR}}, q^{\mathsf{sig}}_{\mathsf{IR}}, q^{\mathsf{sec}}_{\mathsf{IR}})$*-Adversary* $F_{\mathsf{IR}}$ *which breaks* IR *then there exists* $(\tau_{\mathsf{IR}}, \epsilon_{\mathsf{KI}}, q^{\mathsf{sig}}_{\mathsf{KI}}, q^{\mathsf{sec}}_{\mathsf{KI}})$*-Adversary* $F_{\mathsf{IR}}$ *which breaks* KI *with* $\tau_{\mathsf{KI}} = \tau_{\mathsf{IR}}$, $\epsilon_{\mathsf{KI}} = \epsilon_{\mathsf{IR}}$, $q^{\mathsf{sig}}_{\mathsf{KI}} = q^{\mathsf{sig}}_{\mathsf{IR}}$ *and* $q^{\mathsf{sec}}_{\mathsf{KI}} = q^{\mathsf{sec}}_{\mathsf{IR}}$.

The reduction is constructed as follows. In key generation phase the key generation algorithm of KI outputs the secret keys $SK_0, \ldots, SK_N$ of *all* the time periods. Then $(SK_0, SK_1 \oplus R_1, SK_2 \oplus R_2, \ldots, SK_N \oplus R_N)$ is given to the signer as the signing key $SKS$ and $(R_1, R_2, \ldots, R_N)$ is given to the base as its base key $SKB$ where $R_1, R_2, \ldots, R_N$ are random data. $SKS$ and $SKB$ for time period $j$ are of the form $(SK_j, SK_{j+1} \oplus R_{j+1}, SK_{j+2} \oplus R_{j+2}, SK_N \oplus R_N)$ and $(R_{j+1}, R_{j+2}, \ldots, R_N)$, respectively and the signature for the message $M$ in the time period $j$ is simply computed by $\mathsf{Sign}_{\mathsf{KI}}(SK_j, M)$. Further, random date $R_i$s are updated by the refresh algorithms. By this simple construction we can construct IR since

- The adversary knows only the secret key of the time period $j$ if the adversary can successfully attack the signer in the time period $j$. Further, the knowledge of the signing key of the time period $j$ does not help to forge the signature for the other time period.
- The adversary knows no information about the signing key of any period even if the adversary successfully attack the base.
- The adversary knows no information about the past key even if the adversary successfully attack the signer and the base in the same time period.

**Theorem 5** (PS $\to$ FS). *It is possible to construct* FS *from* PS *in such a way that if there exists* $(\tau_{\mathsf{FS}}, \epsilon_{\mathsf{FS}}, q^{\mathsf{sig}}_{\mathsf{FS}}, q^{\mathsf{sec}}_{\mathsf{FS}})$*-Adversary* $F_{\mathsf{FS}}$ *against* FS *then there exists*

$(\tau_{\mathsf{PS}}, \epsilon_{\mathsf{PS}}, q_{\mathsf{PS}}^{\mathsf{sig}}, q_{\mathsf{PS}}^{\mathsf{sec}}, q_{\mathsf{PS}}^{\mathsf{Dlg}})$*-Adversary* $F_{\mathsf{PS}}$ *against* PS *with* $\tau_{\mathsf{PS}} = \tau_{\mathsf{FS}}$, $\epsilon_{\mathsf{PS}} = \epsilon_{\mathsf{PS}}$, $q_{\mathsf{PS}}^{\mathsf{sig}} = q_{\mathsf{FS}}^{\mathsf{sig}}$, $q_{\mathsf{PS}}^{\mathsf{sec}} = q_{\mathsf{FS}}^{\mathsf{sec}}(=1)$, $q_{\mathsf{PS}}^{\mathsf{Dlg}} = q_{\mathsf{FS}}^{\mathsf{sec}}(=1)$.

The reduction is constructed in such a way that the signing key of the time period $j$ corresponds to the self-delegation key of delegation level $j+1$. Though this is a simple construction, forward-security can be achieved since an attacker is not able to get the signing key of lower delegation level even if the attacker gets the self delegation key of some delegation level.

Efficiency: The running time of $\mathsf{Gen}_{\mathsf{FS}}, \mathsf{Upd}_{\mathsf{FS}}, \mathsf{Sign}_{\mathsf{FS}}$ and $\mathsf{Vrfy}_{\mathsf{FS}}$ in the above construction become as follows.

$$\tau_{\mathsf{Gen}}^{(\mathsf{FS})} = \tau_{\mathsf{Gen}}^{(\mathsf{PS})} + \tau_{\mathsf{Dlg_D}}^{(\mathsf{PS})} + \tau_{\mathsf{Dlg_P}}^{(\mathsf{PS})}, \qquad \tau_{\mathsf{Upd}}^{(\mathsf{FS})} = \tau_{\mathsf{Dlg_D}}^{(\mathsf{PS})} + \tau_{\mathsf{Dlg_P}}^{(\mathsf{PS})},$$
$$\tau_{\mathsf{Sign}}^{(\mathsf{FS})} = \tau_{\mathsf{PSig}}^{(\mathsf{PS})}, \qquad \tau_{\mathsf{Vrfy}}^{(\mathsf{FS})} = \tau_{\mathsf{PVrf}}^{(\mathsf{PS})} + \tau_{\mathsf{ID}}^{(\mathsf{PS})}$$

The following corollary is immediate from Theorem 2 and Theorem 5.

**Corollary 1 (KI $\rightarrow$ FS).** *It is possible to construct* FS *from* KI *in such a way that if there exists* $(\tau_{\mathsf{FS}}, \epsilon_{\mathsf{FS}}, q_{\mathsf{FS}}^{\mathsf{sig}}, q_{\mathsf{FS}}^{\mathsf{sec}})$*-Adversary* $F_{\mathsf{FS}}$ *against* FS *then there exists* $(\tau_{\mathsf{KI}}, \epsilon_{\mathsf{KI}}, q_{\mathsf{KI}}^{\mathsf{KI}}, q_{\mathsf{KI}}^{\mathsf{KI}})$*-Adversary* $F_{\mathsf{KI}}$ *against* KI *with* $\tau_{\mathsf{KI}} = \tau_{\mathsf{FS}}$, $\epsilon_{\mathsf{KI}} = \epsilon_{\mathsf{FS}}$, $q_{\mathsf{KI}}^{\mathsf{sig}} = q_{\mathsf{FS}}^{\mathsf{sig}} + q_{\mathsf{FS}}^{\mathsf{sec}}$ *and* $q_{\mathsf{KI}}^{\mathsf{sec}} = N \cdot q_{\mathsf{FS}}^{\mathsf{sec}}$.

## References


1. R. Anderson, *Two remarks on public key cryptology*, available at `http://www.cl.cam.ac.uk/users/rja14/`, 2001.
2. M. Bellare and S. Miner, *A forward-secure digital signature scheme*, Proc. Crypto'92, Lecture Notes in Computer Science, vol. 1666, pp. 15–19, 1999.
3. A. Boldyreva, A. Palacio and B. Warinschi, *Secure Proxy Signature Scheme for Delegation of Signing Rights*, IACR ePrint Archive, available at `http://eprint.iacr.org/2003/096/`, 2003.
4. Y. Dodis, J. Katz, S. Xu and M. Yung, *Key-Insulated Public Key Cryptosystems*, Proc. Eurocrypt 2002, Lecture Notes in Computer, vol. 2332, pp. 65–82, 2002.
5. Y. Dodis, J. Katz, S. Xu and M. Yung, *Strong Key-Insulated Signature Schemes*, Proc. PKC2003, Lecture Notes in Computer Science, vol. 2567, pp. 130–144, 2003.
6. M. Gassr, A. Goldstein, C. Kaufman and B. Lampson, *The Digital Distributed Security Architecture*, Proc. National Computer Security Conference, 1989.
7. G. Itkis, *Intrusion-Resilient Signatures: Generic Constructions, or Defeating Strong Adversary with Minimal Assumptions*, Proc. SCN2002, Lecture Notes in Computer Science, vol. 2576, pp. 102–118, 2002.
8. G. Itkis and L. Reyzin, *Forward-secure signatures with optimal signing and verifying*, Proc. Crypto2001, Lecture Notes in Computer Science, vol. 2139, pp. 332-354, 2001.
9. G. Itkis and L. Reyzin, *SiBIR: Signer-Base Intrusion-Resilient Signatures*, Proc. Crypto2002, Lecture Notes in Computer Science, vol. 2442, pp. 499–514, 2002.
10. A. Kozlov and L. Reyzin, *Forward-Secure Signatures with Fast Key Update*, Proc. SNC2002, Lecture Notes in Computer Science, vol. 2576, pp. 241–256, 2002.

11. H. Krawczyk, *Simple forward-secure signatures from any signature scheme.*, Proc. the 7th ACM Conference on Computer and Communications Security, pp. 108–115, 2000.
12. T. Malkin, D. Micciancio and S. Miner, *Efficient generic forward-secure signatures with an unbounded number of time periods*, Proc. Eurocrypt2002, Lecture Notes in Computer Science, vol. 2332, pp. 400–417, 2002.
13. T. Malkin, S. Obana and M. Yung, *The Hierarchy of Key Evolving Signatures and a Characterization of Proxy Signatures*, full version of this paper available at the IACR ePrint Archive, `http://eprint.iacr.org/2004/052`, 2004.
14. M. Mambo, K. Usuda and E. Okamoto, *Proxy signatures for delegating signing operation*, Proc. the 3rd ACM Conference on Computer and Communications Security, pp. 48–57, 1996.
15. M. Naor and M. Yung, *Universal One-Way Hash Functions and their Cryptographic Applications*, In Proceedings of the ACM Symposium on Theory of Computing, 1989, pp. 33–43..
16. J. Rompel, *One-way functions are necessary and sufficient for secure signatures.* In Proceedings of the ACM Symposium on Theory of Computing, 1990, pp. 387–394.

# Public-Key Steganography

Luis von Ahn and Nicholas J. Hopper

Computer Science Dept, Carnegie Mellon University, Pittsburgh PA 15213 USA

**Abstract.** Informally, a public-key steganography protocol allows two parties, who have never met or exchanged a secret, to send hidden messages over a public channel so that an adversary cannot even detect that these hidden messages are being sent. Unlike previous settings in which provable security has been applied to steganography, public-key steganography is information-theoretically *impossible*. In this work we introduce computational security conditions for public-key steganography similar to those introduced by Hopper, Langford and von Ahn [7] for the private-key setting. We also give the first protocols for public-key steganography and steganographic key exchange that are provably secure under standard cryptographic assumptions. Additionally, in the random oracle model, we present a protocol that is secure against adversaries that have access to a decoding oracle (a steganographic analogue of Rackoff and Simon's attacker-specific adaptive chosen-ciphertext adversaries from CRYPTO 91 [10]).

## 1 Introduction

Steganography refers to the problem of sending messages hidden in "innocent-looking" communications over a public channel so that an adversary eavesdropping on the channel cannot even detect the presence of the hidden messages. Simmons [11] gave the most popular formulation of the problem: two prisoners, Alice and Bob, wish to plan an escape from jail. However, the prison warden, Ward, can monitor any communication between Alice and Bob, and if he detects any hint of "unusual" communications, he throws them both in solitary confinement. Alice and Bob must then transmit their secret plans so that nothing in their communication seems "unusual" to Ward.

There have been many proposed solutions to this problem, ranging from rudimentary schemes using invisible ink to a protocol which is provably secure assuming that one-way functions exist [7]. However, the majority of these protocols have focused on the case where Alice and Bob share a secret or private key. If Alice and Bob were incarcerated before the need for steganography arose, these protocols would not help them. In contrast, public-key steganography allows parties to communicate steganographically with no prior exchange of secrets. As with public-key encryption, the sender of a message still needs to know the recipient's public key or otherwise participate in a key exchange protocol. While it is true that if there is no global PKI, the use of public keys might raise suspicion, in many cases it is the sender of a message who is interested in concealing his communication and there is no need for him to publish any keys.

C. Cachin and J. Camenisch (Eds.): EUROCRYPT 2004, LNCS 3027, pp. 323–341, 2004.

In this paper we consider the notion of public-key steganography against adversaries that do not attempt to disrupt the communication between Alice and Bob (i.e., the goal of the adversary is only to detect whether steganography is being used and not to disrupt the communication between the participants). We show that secure public-key steganography exists if any of several standard cryptographic assumptions hold (each of these assumptions implies semantically secure public-key cryptography). We also show that secure steganographic key exchange is possible under the Integer Decisional Diffie-Hellman (DDH) assumption. Furthermore, we introduce a protocol that is secure in the random oracle model against adversaries that have access to a decoding oracle (a steganographic analogue of attacker-specific adaptive chosen-ciphertext adversaries [10]).

**Related Work.** There has been very little work work on provably secure steganography (either in the private or the public key settings). A critical first step in this field was the introduction of an information-theoretic model for steganography by Cachin [4], and several papers have since given similar models [8,9,14]. Unfortunately, these works are limited in the same way that information theoretic cryptography is limited. In particular, in any of these frameworks, secure steganography between two parties with no shared secret is impossible. Hopper, Langford, and von Ahn [7] have given a theoretical framework for steganography based on computational security. Our model will be substantially similar to theirs, but their work addresses only the shared-key setting, which is already possible information-theoretically. Although one of their protocols can be extended to the public-key setting, they do not consider formal security requirements for public-key steganography, nor do they consider the notions of steganographic-key exchange or adversaries that have access to both encoding and decoding oracles.

Anderson and Petitcolas [1], and Craver [5], have both previously described ideas for public-key steganography with only heuristic arguments for security. Since our work has been distributed, others have presented ideas for improving the efficiency of our basic scheme [12] and proposing a modification which makes the scheme secure against a more powerful active adversary [2].

To the best of our knowledge, we are the first to provide a formal framework for public-key steganography and to *prove* that public-key steganography is possible (given that standard cryptographic assumptions hold). We are also the first to consider adversaries that have access to decoding oracles (in a manner analogous to attacker-specific adaptive chosen-ciphertext adversaries [10]); we show that security against such adversaries can be achieved in the random oracle model. We stress, however, that our protocols are not robust against adversaries wishing to render the steganographic communication channel useless. Throughout the paper, the goal of the adversary is detection, not disruption.

## 2 Definitions

**Preliminaries.** A function $\mu : \mathbb{N} \to [0,1]$ is said to be *negligible* if for every $c > 0$, for all sufficiently large $n$, $\mu(n) < 1/n^c$. We denote the length (in bits) of a string or integer $s$ by $|s|$. The concatenation of string $s_1$ and string $s_2$ will be denoted by $s_1||s_2$. We also assume the existence of efficient, unambiguous *pairing* and *un-pairing* operations, so $(s_1, s_2)$ is not the same as $s_1||s_2$. We let $U_k$ denote the uniform distribution on $k$ bit strings. If $X$ is a finite set, we let $U(X)$ denote the uniform distribution on $X$.

If $\mathcal{C}$ is a distribution with finite support $X$, we define the *minimum entropy* of $\mathcal{C}$, $H_\infty(\mathcal{C})$, as $H_\infty(\mathcal{C}) = \min_{x \in X}\{\log_2(1/\Pr_{\mathcal{C}}[x])\}$. We say that a function $f : X \to \{0,1\}$ is *$\epsilon$-biased* if $|\Pr_{x \leftarrow \mathcal{C}}[f(x) = 0] - 1/2| < \epsilon$. We say $f$ is *unbiased* if $f$ is $\epsilon$-biased for $\epsilon$ a negligible function of the appropriate security parameter. We say $f$ is *perfectly unbiased* if $\Pr_{x \leftarrow \mathcal{C}}[f(x) = 0] = 1/2$.

**Integer Decisional Diffie-Hellman.** Let $P$ and $Q$ be primes such that $Q$ divides $P - 1$, let $\mathbb{Z}_P^*$ be the multiplicative group of integers modulo $P$, and let $g \in \mathbb{Z}_P^*$ have order $Q$. Let $\mathbf{A}$ be an adversary that takes as input three elements of $\mathbb{Z}_P^*$ and outputs a single bit. Define the *DDH advantage of* $\mathbf{A}$ *over* $(g, P, Q)$ as:

$$\mathbf{Adv}^{\mathsf{ddh}}_{g,P,Q}(\mathbf{A}) = \left| \Pr_{a,b,r}[\mathbf{A}_r(g^a, g^b, g^{ab}) = 1] - \Pr_{a,b,c,r}[\mathbf{A}_r(g^a, g^b, g^c) = 1] \right| ,$$

where $\mathbf{A}_r$ denotes the adversary $\mathbf{A}$ running with random tape $r$, $a, b, c$ are chosen uniformly at random from $\mathbb{Z}_Q$ and all the multiplications are over $\mathbb{Z}_P^*$. Define *the DDH insecurity of* $(g, P, Q)$ as $\mathbf{InSec}^{\mathsf{ddh}}_{g,P,Q}(t) = \max_{\mathbf{A} \in \mathcal{A}(t)} \left\{ \mathbf{Adv}^{\mathsf{ddh}}_{g,P,Q}(\mathbf{A}) \right\}$, where $\mathcal{A}(t)$ denotes the set of adversaries $\mathbf{A}$ that run for at most $t$ time steps.

**Trapdoor One-Way Permutations.** A trapdoor one-way permutation family $\Pi$ is a sequence of sets $\{\Pi_k\}_k$, where each $\Pi_k$ is a set of bijective functions $\pi : \{0,1\}^k \to \{0,1\}^k$, along with a triple of algorithms $(G, E, I)$. $G(1^k)$ samples an element $\pi \in \Pi_k$ along with a *trapdoor* $\tau$; $E(\pi, x)$ evaluates $\pi(x)$ for $x \in \{0,1\}^k$; and $I(\tau, y)$ evaluates $\pi^{-1}(y)$. For a PPT $\mathbf{A}$ running in time $t(k)$, denote the advantage of $\mathbf{A}$ against $\Pi$ by

$$\mathbf{Adv}^{\mathsf{ow}}_{\Pi}(\mathbf{A}, k) = \Pr_{(\pi,\tau) \leftarrow G(1^k), x \leftarrow U_k}[\mathbf{A}(\pi(x)) = x] .$$

Define the insecurity of $\Pi$ by $\mathbf{InSec}^{\mathsf{ow}}_{\Pi}(t, k) = \max_{\mathbf{A} \in \mathcal{A}(t)} \{\mathbf{Adv}^{\mathsf{ow}}_{\Pi}(\mathbf{A}, k)\}$, where $\mathcal{A}(t)$ denotes the set of all adversaries running in time $t(k)$. We say that $\Pi$ is a trapdoor one-way permutation family if for every probabilistic polynomial-time (PPT) $\mathbf{A}$, $\mathbf{Adv}^{\mathsf{ow}}_{\Pi}(\mathbf{A}, k)$ is negligible in $k$.

## 3 Channels

We seek to define steganography in terms of indistinguishability from a "usual" or innocent-looking distribution on communications. In order to do so, we must characterize this innocent-looking distribution. We follow [7] in using the notion of a channel, which models a prior distribution on the entire sequence of communication from one party to another:

**Definition.** Let $D$ be an efficiently recognizable, prefix-free set of strings, or *documents*. A *channel* is a distribution on sequences $s \in D^*$.

Any particular sequence in the support of a channel describes one possible outcome of all communications from Alice to Bob. The process of drawing from the channel, which results in a *sequence* of documents, is equivalent to a process that repeatedly draws a single "next" document from a distribution consistent with the history of already drawn documents. Therefore, we can think of communication as a series of these partial draws from the channel distribution, conditioned on what has been drawn so far. Notice that this notion of a channel is more general than the typical setting in which every symbol is drawn independently according to some fixed distribution: our channel explicitly models the dependence between symbols common in typical real-world communications.

Let $\mathcal{C}$ be a channel. We let $\mathcal{C}_h$ denote the marginal channel distribution on a single document from $D$ conditioned on the history $h$ of already drawn documents; we let $\mathcal{C}_h^l$ denote the marginal distribution on sequences of $l$ documents conditioned on $h$. When we write "sample $x \leftarrow \mathcal{C}_h$" we mean that a single document should be returned according to the distribution conditioned on $h$. We use $\mathcal{C}_{A \rightarrow B,h}$ to denote the distribution on the communication from party $A$ to party $B$.

We will require that a channel satisfy a minimum entropy constraint for all histories. Specifically, we require that there exist constants $L > 0$, $b > 0$, $\alpha > 0$ such that for all $h \in D^L$, either $\Pr_{\mathcal{C}}[h] = 0$ or $H_\infty(\mathcal{C}_h^b) \geq \alpha$. If a channel does not satisfy this property, then it is possible for Alice to drive the information content of her communications to 0, so this is a reasonable requirement. We say that a channel satisfying this condition is *$L$-informative*, and if a channel is $L$-informative for all $L > 0$, we say it is *always informative*. Note that this definition implies an additive-like property of minimum entropy for marginal distributions, specifically, $H_\infty(\mathcal{C}_h^{lb}) \geq l\alpha$ . For ease of exposition, we will assume channels are always informative in the remainder of this paper; however, our theorems easily extend to situations in which a channel is $L$-informative.

In our setting, each ordered pair of parties $(P, Q)$ will have their own channel distribution $\mathcal{C}_{P \rightarrow Q}$. In these cases, we assume that among the legitimate parties, only party $A$ has oracle access to marginal channel distributions $\mathcal{C}_{A \rightarrow B,h}$ for every other party $B$ and history $h$. On the other hand, we will allow the adversary oracle access to marginal channel distributions $\mathcal{C}_{P \rightarrow Q,h}$ for every pair $P, Q$ and every history $h$. This allows the adversary to learn as much as possible about any channel distribution but does not require any legitimate participant to know the distribution on communications from any other participant. We will assume that each party knows the history of communications it has sent and received from every other participant. We will also assume that cryptographic primitives remain secure with respect to oracles which draw from the marginal channel distributions $\mathcal{C}_{A \rightarrow B,h}$.

# 4 Pseudorandom Public-Key Encryption

We will require public-key encryption schemes that are secure in a slightly non-standard model, which we will denote by IND\$-CPA in contrast to the more standard IND-CPA. The main difference is that security against IND\$-CPA requires the output of the encryption algorithm to be indistinguishable from uniformly chosen random bits. Let $\mathcal{E} = (G, E, D)$ be a probabilistic public-key encryption scheme, where $E : \mathcal{PK} \times \mathcal{R} \times \mathcal{P} \to \mathcal{C}$. Consider a game in which an adversary $\mathbf{A}$ is given a public key drawn from $G(1^k)$ and chooses a message $m_{\mathbf{A}}$. Then $\mathbf{A}$ is given either $E_{PK}(m_{\mathbf{A}})$ or a uniformly chosen string of the same length. Let $\mathcal{A}(t, l)$ be the set of adversaries $\mathbf{A}$ which produce a message of length at most $l(k)$ bits and run for at most $t(k)$ time steps. Define the IND\$-CPA advantage of $\mathbf{A}$ against $\mathcal{E}$ as

$$\mathbf{Adv}_{\mathcal{E}}^{\mathsf{cpa}}(\mathbf{A}, k) = \left| \Pr_{PK}[\mathbf{A}(PK, E_{PK}(m_{\mathbf{A}})) = 1] - \Pr_{PK}[\mathbf{A}(PK, U_{|E_{PK}(m_{\mathbf{A}})|}) = 1] \right|$$

Define the insecurity of $\mathcal{E}$ as $\mathbf{InSec}_{\mathcal{E}}^{\mathsf{cpa}}(t, l, k) = \max_{\mathbf{A} \in \mathcal{A}(t,l)} \{\mathbf{Adv}_{\mathcal{E}}^{\mathsf{cpa}}(\mathbf{A}, k)\}$. $\mathcal{E}$ is $(t, l, k, \epsilon)$-*indistinguishable from random bits under chosen plaintext attack* if $\mathbf{InSec}_{\mathcal{E}}^{\mathsf{cpa}}(t, l, k) \leq \epsilon(k)$. $\mathcal{E}$ is called *indistinguishable from random bits under chosen plaintext attack* (IND\$-CPA) if for every probabilistic polnyomial-time (PPT) $\mathbf{A}$, $\mathbf{Adv}_{\mathcal{E}}^{\mathsf{cpa}}(\mathbf{A}, k)$ is negligible in $k$. For completeness, we show how to construct IND\$-CPA public-key encryption schemes from the RSA and Decisional Diffie-Hellman assumptions. We omit detailed proofs of security for the constructions below, as they are standard modifications to existing schemes. In the full version of this paper, we show that much more general assumptions suffice for IND\$-CPA security.

## 4.1 RSA-based Construction

The RSA function $E_{N,e}(x) = x^e \bmod N$ is believed to be a trapdoor one-way permutation family. The following construction uses Young and Yung's Probabilistic Bias Removal Method (PBRM) [13] to remove the bias incurred by selecting an element from $\mathbb{Z}_N^*$ rather than $U_k$.

**Construction 1.** (RSA-based Pseudorandom Encryption Scheme)

| **Procedure** Encrypt: | **Procedure** Decrypt: |
|---|---|
| **Input:** plaintext $m$; public key $N, e$ | **Input:** $x', c$; $(N, d)$ |
| let $k = \|N\|$, $l = \|m\|$ | let $l = \|c\|$, $k = \|N\|$ |
| repeat: | if $(x' > N)$ set $x_l = x'$ |
|   Sample $x_0 \leftarrow \mathbb{Z}_N^*$ | else set $x_l = 2^k - x'$ |
|   for $i = 1 \ldots l$ do | for $i = l \ldots 1$ do |
|     set $b_i = x_{i-1} \bmod 2$ |   set $x_{i-1} = x_i^d \bmod N$ |
|     set $x_i = x_{i-1}^e \bmod N$ |   set $b_i = x_{i-1} \bmod 2$ |
|   sample $c \leftarrow U_1$ | **Output:** $c \oplus b$ |
| until $(x_l \leq 2^k - N)$ OR $c = 1$ | |
| if $(x_1 \leq 2^k - N)$ and $c = 0$ set $x' = x$ | |
| if $(x_1 \leq 2^k - N)$ and $c = 1$ set $x' = 2^k - x$ | |
| **Output:** $x', b \oplus m$ | |

The IND\$-CPA security of the scheme follows from the correctness of PBRM and the fact that the least-significant bit is a hardcore bit for RSA. Notice that the expected number of repeats in the encryption routine is at most 2.

### 4.2 DDH-based Construction

Let $E_{(\cdot)}(\cdot)$, $D_{(\cdot)}(\cdot)$ denote the encryption and decryption functions of a *private-key* encryption scheme satisfying IND\$-CPA, keyed by $\kappa$-bit keys, and let $\kappa \leq k/3$ (private-key IND\$-CPA encryption schemes have appeared in the literature; see, for instance, [7]). Let $\mathcal{H}_k$ be a family of pairwise-independent hash functions $H : \{0,1\}^k \to \{0,1\}^\kappa$. We let $P$ be a $k$-bit prime (so $2^{k-1} < P < 2^k$), and let $P = rQ + 1$ where $(r, Q) = 1$ and $Q$ is also a prime. Let $g$ generate $\mathbb{Z}_P^*$ and $\hat{g} = g^r \bmod P$ generate the unique subgroup of order $Q$. The security of the following scheme follows from the Decisional Diffie-Hellman assumption, the leftover-hash lemma, and the security of $(E, D)$:

**Construction 2.** (ElGamal-based random-bits encryption)

| **Procedure** `Encrypt`: | **Procedure** `Decrypt`: |
|---|---|
| **Input:** $m \in \{0,1\}^*$; $(g, \hat{g}^a, P)$ | **Input:** $(H, s, c)$; private key $(a, P, Q)$ |
| Sample $H \leftarrow \mathcal{H}_k$ | let $r = (P-1)/Q$ |
| repeat: | set $K = H(s^{ra} \bmod P)$ |
|   Sample $b \leftarrow \mathbb{Z}_{P-1}$ | **Output:** $D_K(c)$ |
| until $(g^b \bmod P) \leq 2^{k-1}$ | |
| set $K = H((\hat{g}^a)^b \bmod P)$ | |
| **Output:** $H, g^b, E_K(m)$ | |

The security proof considers two hybrid encryption schemes: $H_1$ replaces the value $(\hat{g}^a)^b$ by a random element of the subgroup of order $Q$, $\hat{g}^c$, and $H_2$ replaces $K$ by a random draw from $\{0,1\}^\kappa$. Clearly distinguishing $H_2$ from random bits requires distinguishing some $E_K(m)$ from random bits. The Leftover Hash Lemma gives that the statistical distance between $H_2$ and $H_1$ is at most $2^{-\kappa}$. Finally, any distinguisher $\mathbf{A}$ for $H_1$ from the output of `Encrypt` with advantage $\epsilon$ can be used to solve the DDH problem with advantage at least $\epsilon/2$, by transforming $\hat{g}^b$ to $B = (\hat{g}^b)^{\hat{r}} g^{\beta Q}$, where $\hat{r}$ is the least integer such that $r\hat{r} = 1 \bmod Q$ and $\beta \leftarrow \mathbb{Z}_r$, outputting 0 if $B > 2^{k-1}$, and otherwise drawing $H \leftarrow \mathcal{H}_k$ and running $\mathbf{A}$ on $\hat{g}^a, H||B||E_{H(\hat{g}^c)}(m)$.

## 5 Public-Key Steganography

**Definition 1.** (Stegosystem) A public-key stegosystem is a triple of probabilistic algorithms $S = (SG, SE, SD)$. $SG(1^k)$ generates a key pair $(PK, SK) \in \mathcal{PK} \times \mathcal{SK}$. $SE$ takes a (public) key $PK \in \mathcal{PK}$, a string $m \in \{0,1\}^*$ (the *hiddentext*), and a message history $h$. $SE$ also has access to a channel oracle for some channel $\mathcal{C}$, which can sample from $\mathcal{C}_h$ for any $h$. $SE(PK, m, h)$ returns a sequence of documents $s_1, s_2, \ldots, s_l$ (the *stegotext*) from the support of $\mathcal{C}_h^l$. $SD$ takes a (secret) key $SK \in \mathcal{SK}$, a sequence of documents $s_1, s_2, \ldots, s_l$, and a message history $h$, and returns a hiddentext $m$. Additionally, for every polynomial $p$ there must exist a negligible $\mu$ such that

$$\forall m \in \{0,1\}^{p(k)} : \Pr_{(PK,SK) \leftarrow SG(1^k)}[SD(SK, SE(PK, m, h), h) = m] \geq 1 - \mu(k)$$

where the randomization is also over any coin tosses of $SE$, $SD$, $SG$ and the oracle to $\mathcal{C}_h$.

**Remarks.** We emphasize the terminology: the secret message that Alice wants to send to Bob is called the *hiddentext*; documents from the channel are called *covertexts*, and documents that are output by $SE$ are called *stegotexts*. We also stress that *$SE$ need not know the exact probabilities of documents in $\mathcal{C}_h$*. This is important to mention, as it is unreasonable to assume that the probabilities in $\mathcal{C}_h$ are known, whereas anybody communicating can be thought of as an oracle for the channel distribution $\mathcal{C}_h$.

### 5.1 Steganographic Secrecy against Chosen Hiddentext Attack

In a chosen hiddentext attack, $W$ is allowed to supply hiddentexts to Alice, in hopes of detecting whether or not she is using steganography in her communications with Bob. (Note that a chosen hiddentext attack is the steganographic equivalent of a chosen plaintext attack against an encryption scheme.)

We will model a warden attacking a stegosystem as an efficient oracle machine which plays the following game. First, a pair $(PK, SK)$ is drawn from $SG(1^k)$ and $W$ is allowed to access $PK$ and $\mathcal{C}_h$ for any history $h$, to produce a message $m_W$ and a history $h_W$. Then $W$ is given access to a sequence $s \in D^*$, which is drawn either from $SE(PK, m_W, h_W)$ or $\mathcal{C}_h^{|s|}$, and continues to access $\mathcal{C}_h$ for any history $h$ (This oracle allows $W$ to learn the *covertext* distribution on all communications). Finally $W$ outputs a bit which is a "guess" about whether $s$ is a stegotext or a covertext.

We define the *CHA advantage of $W$ against stegosystem $S$ over channel $\mathcal{C}$* by

$$\mathbf{Adv}^{\mathsf{cha}}_{S,\mathcal{C}}(W,k) = \left|\Pr[W^{\mathcal{C}}(PK,s) = 1] - \Pr[W^{\mathcal{C}}(PK, \mathcal{C}_{h_W}^{|s|}) = 1]\right| ,$$

where the probabilities are taken over $(PK, SK) \leftarrow SG(1^k)$, $(m_W, h_W) \leftarrow W^{\mathcal{C}}(PK)$, $s \leftarrow SE(PK, m_W, h_W)$, the random bits of $W$ and the responses of the oracle $\mathcal{C}$. Define *the insecurity of $S$ over channel $\mathcal{C}$* by $\mathbf{InSec}^{\mathsf{cha}}_{S,\mathcal{C}}(t,l,k) = \max_{W \in \mathcal{W}(t,l)} \left\{\mathbf{Adv}^{\mathsf{cha}}_{S,\mathcal{C}}(W,k)\right\}$, where $\mathcal{W}(t,l)$ denotes the set of all adversaries which choose a message of length at most $l(k)$ bits, and run in time at most $t(k)$. For ease of notation, we will denote $W^{\mathcal{C}}$ by $W$ for the remainder of the paper.

Thus, $S$ is secure against chosen-hiddentext attacks if the warden, even given Bob's public key, cannot tell whether Alice's message to Bob encodes any hiddentext at all, even one of his choice.

### 5.2 Chosen-Stegotext Attacks

In a Chosen-Stegotext Attack, Ward is monitoring the communications from Alice to Bob, trying to detect any hint that the pair are using steganography. In particular, he not only has some partial knowledge of the messages Alice might be encoding, he can pretend to be Alice, sending messages (presumably encoding hiddentexts) to Bob and observing Bob's reaction to see if Bob is trying to decode them. Ward can even make small changes to Alice's messages to see how Bob reacts to them. If Bob is decoding Alice's messages, he will output hiddentexts, whereas otherwise he will have no special reaction, which we signify by the output $\perp$. In order to protect against this attack, there must be something Alice knows that Ward doesn't. So we extend the definition of a stegosystem to include a fourth algorithm $SSG(1^k)$ which outputs a pair $(KV, KS)$. If Alice wants to send messages that are protected in this way, she chooses $(KV, KS) \leftarrow SSG(1^k)$ and publishes $KV$ in the same way that Bob would publish $PK$. We modify the encoding algorithm to take $KS$ as an additional argument, and the decoding algorithm to require $KV$. (So in a system designed to resist this attack, Bob must know Alice's public value $KV$ to decode her messages.)

We formally model a chosen-stegotext attacker as an oracle PPT $W$ which plays the following game. First, draw $(PK, SK) \leftarrow SG(1^k)$, and $(KV, KS) \leftarrow SSG(1^k)$. Then $W$ is given $(PK, KV)$ and accesses an oracle, which is either $ST$ (for StegoText) or $CT$ (for CoverText), and which respond to queries using these programs: (where $\phi = \{\}$ initially):

| | |
|---|---|
| $ST(b \in \{\texttt{enc}, \texttt{dec}\}, m, h)$<br>if(b = $\texttt{enc}$) then:<br>  Sample $s \leftarrow SE(PK, KS, m, h)$<br>  Set $\phi = \phi \cup \{(s,h)\}$; return $s$<br>else if $(m,h) \in \phi$ return “”<br>else return $SD(SK, KV, m, h)$ | $CT(b \in \{\texttt{enc}, \texttt{dec}\}, m, h)$<br>if(b = $\texttt{enc}$) then:<br>  Sample $s \leftarrow \mathcal{C}_h^{\lvert SE(PK,KS,m,h)\rvert}$<br>  Set $\phi = \phi \cup \{(s,h)\}$; return $s$<br>else if $(m,h) \in \phi$ return “”<br>else return $\perp$ |

Suppose $W$ runs in time $t$, and makes $q_e$ queries of the form $(\texttt{enc}, *, *)$, totalling $l$ bits, and $q_d$ queries of the form $(\texttt{dec}, *, *)$. We define $W$'s CSA advantage against $S$ for channel $\mathcal{C}$ by

$$\mathbf{Adv}^{\mathsf{csa}}_{S,\mathcal{C}}(W,k) = \left|\Pr[W^{ST}(PK,KV)=1] - \Pr[W^{CT}(PK,KV)=1]\right| ,$$

and the insecurity by $\mathbf{InSec}^{\mathsf{csa}}_{S,\mathcal{C}}(t, q_e, q_d, l, k) = \max_{w \in \mathcal{W}(t,q_e,q_d,l)} \left\{\mathbf{Adv}^{\mathsf{csa}}_{S,\mathcal{C}}(W,k)\right\}$.

Thus, in a chosen-stegotext attack, Ward may ask Alice to encode any (message, history) pair of his choosing, as often as he likes, and may query Bob on any pair $(s, h)$ where $s$ was not a result of an encoding query for history $h$. Notice that if Ward can ask Bob to decode any message output by Alice for the same history it was encoded with, he can detect the use of steganography between Alice and Bob; this is why we do not allow Ward to query Bob on such stegotexts. This restriction is roughly analogous to the standard restriction that an adaptive chosen-ciphertext attacker may not query his decryption oracle on the challenge ciphertext. The use of a secret by the encryptor makes this security condition roughly analogous to the notion of "attacker-specific" adaptive chosen-ciphertext attack from [10], in which an attacker who tries to access the decryption of a message he didn't send is given the response $\perp$. This "attacker-specific" notion of security is stronger than CCA-1 security but weaker than the now-standard CCA-2 security [10]. Note, however, that chosen-stegotext security protects *both* the sender and the receiver.

### 5.3 Relationships among Notions

We formalize the notions of steganographic secrecy as follows.

**Definition 2.** A stegosystem $S$ is called steganographically secret for channel $\mathcal{C}$ against attack model $\mathsf{atk}$ (SS-ATK) if for every PPT $\mathbf{A}$, $\mathbf{Adv}^{\mathsf{atk}}_{S,\mathcal{C}}(\mathbf{A}, k)$ is negligible in $k$.

A natural question is: what are the relationships between these security notions and the standard notions from public-key cryptography? In this section we give the key relationships between these notions.

**SS-CHA is strictly stronger than IND-CPA.** By a standard argument based on the triangle inequality, if $A$ can distinguish $SE(m_0)$ from $SE(m_1)$ with advantage $\epsilon$, he must be able to distinguish one of these from $\mathcal{C}_h$ with advantage at least $\epsilon/2$. Thus every SS-CHA secure stegosystem must also be IND-CPA secure. On the other hand, let $S$ be any IND-CPA secure cryptosystem. Then $S'$ which prepends a known, fixed sequence of documents $m \in D^k$ to the output of $S$ is still IND-CPA secure but has an SS-CHA distinguisher with advantage $1 - o(1)$ for any $L$-informative channel.

**SS-CSA is strictly stronger than SS-CHA.** Suppose that we take a SS-CSA-secure stegosystem $S = (SG, SSG, SE, SD)$ and define $SE'(PK, m, h)$ to draw a random $(KV, KS) \leftarrow SSG(1^k)$ and return $SE(PK, KS, m, h)$. Then any CHA warden against $SE'$ is also a single-query CSA warden against $S$. (However, whether there is a corresponding modification $SD'$ so that $S'$ is sound may be dependent on the construction; such modification is possible for our construction.) On the other hand, SS-CSA is strictly stronger than SS-CHA: if $(SG, SE, SD)$ is SS-CHA secure, then so is $S' = (SG, SE', SD')$ where $SE'(m, h)$ draws $s \leftarrow SE(m, h)$ and $s' \leftarrow \mathcal{C}_{(h,s)}$, and returns $(s, s')$, while $SD'((s, s'), h)$ returns $SD(s, h)$. But $S'$ is trivially vulnerable to a chosen-stegotext attack with advantage 1: query $(\mathtt{enc}, m, h)$ to get $(s, s')$, draw $s'' \leftarrow \mathcal{C}_{(h,s)}$ and query $(\mathtt{dec}, (s, s''), h)$. If the result is not $\perp$, return 1, otherwise return 0.

## 6 Constructions

Most of our protocols build on the following construction, a generalization of Construction 2 in [7] and similar to a protocol given by Cachin [4]. Let $f : D \rightarrow \{0, 1\}$ be a public function (recall that $\mathcal{C}$ is a distribution on sequences of elements of $D$). If $f$ is is perfectly unbiased on $\mathcal{C}_h$ for all $h$, then the following encoding procedure, on uniformly distributed $l$-bit input $c$, produces output distributed exactly according to $\mathcal{C}_h^l$:

**Construction 3.** (Basic encoding/decoding routines)

**Procedure** `Basic_Encode`:
**Input:** $c_1, \ldots, c_l \in \{0,1\}^l$, $h \in D^*$, $k$
for $i = 1 \ldots l$ do
  Let $j = 0$
  repeat:
    sample $s_i \leftarrow \mathcal{C}_h$, increment $j$
  until $f(s_i) = c_i$ OR $(j > k)$
  set $h = h||s_i$
**Output:** $s_1, s_2, \ldots, s_l$

**Procedure** `Basic_Decode`:
**Input:** Stegotext $s_1, s_2, \ldots, s_l$
for $i = 1 \ldots l$ do
  set $c_i = f(s_i)$
set $c = c_1||c_2||\cdots||c_l$.
**Output:** $c$

Note that for infinitely many $\mathcal{C}_h$ there is no perfectly unbiased function $f$. In appendix B, we prove Proposition 1, which together with Proposition 2, justifies our use of unbiased functions. The proof for Proposition 2 is straightforward and is omitted from the paper.

**Proposition 1.** *Any channel $\mathcal{C}$ which is always informative can be compiled into a channel $\mathcal{C}^{(k)}$ which admits an efficiently computable function $f$ such that for any polynomial-length sequence $h_1, \ldots, h_n$ all satisfying* $\Pr_{\mathcal{C}}[h_i] \neq 0$*, the bias* $\left|\Pr[f(\mathcal{C}_{h_i}^{(k)}) = 1] - \frac{1}{2}\right|$ *is negligible in $k$ for all $1 \leq i \leq n$.*

**Proposition 2.** *If $f$ is $\epsilon$-biased on $\mathcal{C}_h$ for all $h$, then for any $k$ and $s_1, s_2, \ldots, s_l$:*

$$\left| \Pr_{c \leftarrow U_l}[\texttt{Basic\_Encode}(c, h, k) = s_1, s_2, \ldots, s_l] - \Pr_{s \leftarrow \mathcal{C}_h^l}[s = s_1, s_2, \ldots, s_l] \right| \leq \epsilon l \; .$$

Using the previous two propositions, we will show that public-key steganography is possible in any channel that is always informative. We note that procedure `Basic_Encode` has a small probability of failure: it is possible to have $f(s_i) \neq c_i$. This probability of failure, however, is negligible in $k$.

### 6.1 Chosen Hiddentext Security

Let $E_{PK}(\cdot)$ and $D_{SK}(\cdot)$ denote the encryption and decryption algorithms for a public-key cryptosystem $\mathcal{E}$ which is indistinguishable from random bits under chosen plaintext attack (IND\$-CPA). Let $\ell$ be the expansion function of $\mathcal{E}$, i.e., $|E_{PK}(m)| = \ell(|m|)$. The following procedures allow encoding and decoding of messages in a manner which is steganographically secret under chosen hiddentext attack for the channel distribution $\mathcal{C}$:

**Construction 4.** (Chosen Hiddentext Security)

**Procedure** `CHA_Encode`:
**Input:** key $PK$, $m \in \{0,1\}^*$, $h \in D^*$
Let $c = E_{PK}(m)$
**Output:** `Basic_Encode`$(c, h, k)$

**Procedure** `CHA_Decode`:
**Input:** secret key $SK$, $s \in D^l$
Let $c =$ `Basic_Decode`$(s)$
**Output:** $D_{SK}(c)$

**Lemma 1.** *Assume $f$ is $\epsilon$-biased on $\mathcal{C}_h$ for all $h$. For any warden $W \in \mathcal{W}(t,l)$, there is an* IND\$-CPA *adversary* **A** *with* $\mathbf{Adv}^{\mathsf{cha}}_{\mathtt{CHA},\mathcal{C}}(W,k) \le \mathbf{Adv}^{\mathsf{cpa}}_{\mathcal{E}}(\mathbf{A},k) + \ell(l)\epsilon$. *The running time of* **A** *is at most $t + O(kl)$ ($k$ is the security parameter of $(PK, SK)$ and is used as a bound for* `Basic_Encode`*), and* **A** *submits a plaintext of length at most $l$.*

*Proof.* We are given a program $W$ for distinguishing between the output of Construction 4 and the ordinary traffic of the channel, and an oracle for sampling blocks from that channel. We construct a program **A** which plays the IND\$-CPA game: distinguishing $E_{PK}(m_W)$ from $U_{\ell(l)}$. $\mathbf{A}(PK)$ simply runs $W(PK)$, to get $(m_W, h_W)$ and returns $m_W$ for $m_A$. Then $\mathbf{A}(PK, c)$ uses the oracle $\mathcal{C}_h$ to compute $s = \mathtt{Basic\_Encode}(c, h_W, k)$, and returns the output of $W(PK, s)$. Consider the cases for **A**'s input. If $c \leftarrow E_{PK}(m_W)$, then $s \leftarrow \mathtt{CHA\_Encode}(PK, m_W, h_W)$; so we have $\Pr[\mathbf{A}(PK, E_{PK}(m_A)) = 1] = \Pr[W(PK, SE(PK, m_W, h_W)) = 1]$. If $c \leftarrow U_\ell$, then $s$ is distributed identically to $\mathtt{Basic\_Encode}(U_\ell, h_W, k)$; thus $|\Pr[\mathbf{A}(PK, U_\ell) = 1] - \Pr[W(PK, \mathcal{C}^\ell_{h_W}) = 1]| \le \ell\epsilon$ by Proposition 2.

Combining the cases, we have $\mathbf{Adv}^{\mathsf{cha}}_{\mathtt{CHA},\mathcal{C}}(W,k) \le |\Pr[\mathbf{A}(PK, E_{PK}(m_A)) = 1] - \Pr[\mathbf{A}(PK, U_\ell) = 1]| + \ell\epsilon = \mathbf{Adv}^{\mathsf{cpa}}_{\mathcal{E}}(\mathbf{A},k) + \ell(l)\epsilon$, as claimed.

**Theorem 1.** *If $f$ is $\epsilon$-biased on $\mathcal{C}_h$ for all $h$, then*

$$\mathbf{InSec}^{\mathsf{cha}}_{\mathtt{CHA},\mathcal{C}}(t,l,k) \le \mathbf{InSec}^{\mathsf{cpa}}_{\mathcal{E}}(t + O(kl), l, k) + \ell(l)\epsilon \ .$$

### 6.2 Chosen-Stegotext Security

We will construct a stegosystem which is SS-CSA-secure in the Random Oracle model, for any channel which is *efficiently sampleable*: that is, for which there exists a PPT C such that for all $h$, $\mathtt{C}(h; U_k)$ and $\mathcal{C}_h$ are indistinguishable. We assume that $\pi_A, \pi_B$ are elements of trapdoor one-way permutation family $\Pi_k$, where Alice knows $\pi_A^{-1}$ and Bob knows $\pi_B^{-1}$. In addition, we assume all parties have access to random oracles $F : \{0,1\}^* \to \{0,1\}^k$, $G : \{0,1\}^* \to \{0,1\}^k$, $H_1 : \{0,1\}^k \to \{0,1\}^*$, and $H_2 : \{0,1\}^* \to \{0,1\}^k$. The following construction slightly modifies techniques from [3], using the random oracles $H_1$ and $H_2$ with $\pi_B$ to construct a pseudorandom non-malleable encryption scheme and the oracle $F$ in conjunction with $\pi_A$ to construct a strongly unforgeable signature scheme.

**Construction 5.** (Chosen Stegotext Security)

**Procedure** `UEncode`:
**Input:** $c \in \{0,1\}^l$, $r$, $h$
for $i = 1 \ldots l$ do
  Let $j = 0$
  repeat:
    set $r_j = G(h, r, c, j)$
    set $s_i = \mathtt{C}(h; r_j)$
    increment $j$
  until $f(s_i) = c_i \vee (j > k)$
  set $h = (h, s_i)$
**Output:** $s_1, s_2, \ldots, s_l$

**Procedure** `CSA_Encode`:
**Input:** $m$, $h$, $\pi_A^{-1}$, $\pi_B$
Choose $r \leftarrow U_k$
Let $\sigma = \pi_A^{-1}(F(r, m, h))$
Let $e = H_1(r) \oplus (m, \sigma)$
Let $\tau = H_2(r, m, h)$
Let $y = \pi_B(r)$
Let $c = y||e||\tau$
**Output:** $\mathtt{UEncode}^G(c, r, h)$

**Procedure** `CSA_Decode`:
**Input:** $s$, $h$, $\pi_A$, $\pi_B^{-1}$
Let $c = \mathtt{Basic\_Decode}(s)$
Parse $c$ as $y||e||\tau$.
Set $r = \pi_B^{-1}(y)$.
Let $(m, \sigma) = e \oplus H_1(r)$
If $s \ne \mathtt{UEncode}^G(c, r, h) \vee$
  $\tau \ne H_2(r, m, h) \vee$
  $\pi_A(\sigma) \ne F(r, m, h)$
  return $\perp$
**Output:** $m$

**Theorem 2.** *If $f$ is $\epsilon$-biased for $\mathcal{C}$, then*

$$\mathbf{InSec}^{\mathsf{csa}}_{\mathsf{CSA},\mathcal{C}}(t,q,l,k) \le (2q_e + q_F)\mathbf{InSec}^{\mathsf{ow}}_{\pi}(t',k) + (l + 3q_e k)\epsilon + (q_e^2 + 2q_d)/2^k \ ,$$

*where $t' \le t + (q_G + q_F + q_H)(q_e + q_d)T_\pi + k(l + 3q_e k)T_{\mathsf{C}}$, $T_\pi$ is the time to evaluate members of $\pi$, and $T_{\mathsf{C}}$ is the running time of* $\mathsf{C}$.

Intuitively, this stegosystem is secure because the encryption scheme employed is non-malleable, the signature scheme is strongly unforgeable, and each triple of hiddentext, history, and random-bits has a unique valid stegotext, which contains a signature on $(m, h, r)$. Thus any adversary making a valid decoding query which was not the result of an encoding query can be used to forge a signature for Alice — that is, invert the one-way permutation $\pi_A$. The full proof is omitted for space considerations; see Appendix A for details.

## 7 Steganographic Key Exchange

Consider the original motivating scenario: Alice and Bob are prisoners, in an environment controlled by Ward, who wishes to prevent them from exchanging messages he can't read. Then the best strategy for Ward, once he has read the preceding sections, is to ban Alice and Bob from publishing public keys. In this case, a natural alternative to public-key steganography is *steganographic key exchange*: Alice and Bob exchange a sequence of messages, indistinguishable from normal communication traffic, and at the end of this sequence they are able to compute a shared key. So long as this key is indistinguishable from a random key to the warden, Alice and Bob can proceed to use their shared key in a secret-key stegosystem. In this section, we will formalize this notion.

**Definition 3.** (Steganographic Key Exchange Protocol) A *steganographic key exchange protocol,* or SKEP, is a quadruple of efficient probabilistic algorithms $S_{KE} = (SE_A, SE_B, SD_A, SD_B)$. $SE_A$ and $SE_B$ take as input a security parameter $1^k$ and a string of random bits, and output a sequence of documents of length $l(k)$; $SD_A$ and $SD_B$ take as input a security parameter, a string of random bits, and a sequence of documents of length $l(k)$, and output an element of the key space $\mathcal{K}$. Additionally, these algorithms satisfy the property that there exists a negligible function $\mu(k)$ satisfying:

$$\Pr_{r_A,r_B}[SD_A(1^k, r_A, SE_B(1^k, r_B)) = SD_B(1^k, r_B, SE_A(1^k, r_A))] \ge 1 - \mu(k) \ .$$

We call the output of $SD_A(1^k, r_A, SE_B(1^k, r_B))$ the *result* of the protocol, we denote this result by $S_{KE}(r_A, r_B)$, and we denote by $\mathcal{T}_{r_A,r_B}$ (for transcript) the pair $(SE_A(1^k, r_A), SE_B(1^k, r_B))$.

Alice and Bob perform a key exchange using $S_{KE}$ by sampling private randomness $r_A, r_B$, asynchronously sending $SE_A(1^k, r_A)$ and $SE_B(1^k, r_B)$ to each other, and using the result of the protocol as a key. Notice that in this definition a SKEP must be an asynchronous single-round scheme, ruling out multi-round key exchange protocols. This is for ease of exposition only.

Let $W$ be a warden running in time $t$. We define $W$'s *SKE advantage against* $S_{KE}$ on channels $\mathcal{C} = (\mathcal{C}_{A\to B}, \mathcal{C}_{B\to A})$ with security parameter $k$ by:

$$\mathbf{Adv}^{\mathsf{ske}}_{S_{KE},\mathcal{C}}(W,k) = |\Pr[W(\mathcal{T}_{r_A,r_B}, S_{KE}(r_A,r_B)) = 1] - \Pr[W((\sigma_A,\sigma_B),K) = 1]|$$

where $\sigma_A \leftarrow \mathcal{C}^{l(k)}_{A\to B,h_A}, \sigma_B \leftarrow \mathcal{C}^{l(k)}_{B\to A,h_B}$, and $K \leftarrow \mathcal{K}$. We remark that, as in our other definitions, $W$ also has access to channel oracles $\mathcal{C}_{A\to B,h}$ and $\mathcal{C}_{B\to A,h}$. Let $\mathcal{W}(t)$ denote the set of all wardens running in time $t$. The *SKE insecurity of $S_{KE}$ on $\mathcal{C}$ with security parameter $k$* is given by $\mathbf{InSec}^{\mathsf{ske}}_{S_{KE},\mathcal{C}}(t,k) = \max_{W\in\mathcal{W}(t)}\left\{\mathbf{Adv}^{\mathsf{ske}}_{S_{KE},\mathcal{C}}(W,k)\right\}$.

**Definition 4.** (Secure Steganographic Key Exchange) A SKEP $S_{KE}$ is said to be *$(t,\epsilon)$-secure for channels $\mathcal{C}_{A\to B}$ and $\mathcal{C}_{B\to A}$* if $\mathbf{InSec}^{\mathsf{ske}}_{S_{KE}}(t,k) \leq \epsilon(k)$. $S_{KE}$ is said to be secure if for all polynomials $p$, $S_{KE}$ is $(p(k),\epsilon(k))$-secure for some negligible function $\epsilon$.

**Construction.** The idea behind behind the construction for steganographic key exchange is simple: let $g$ generate $\mathbb{Z}^*_P$, let $Q$ be a large prime with $P = rQ + 1$ and $r$ coprime to $Q$, and let $\hat{g} = g^r$ generate the subgroup of order $Q$. Alice picks random values $a \in \mathbb{Z}_{P-1}$ uniformly at random until she finds one such that $g^a \mod P$ has its most significant bit (MSB) set to 0 (so that $g^a \mod P$ is uniformly distributed in the set of bit strings of length $|P|-1$). She then uses `Basic_Encode` to send all the bits of $g^a \mod P$ except for the MSB (which is zero anyway). Bob does the same and sends all the bits of $g^b \mod P$ except the most significant one (which is zero anyway) using `Basic_Encode`. Bob and Alice then perform `Basic_Decode` and agree on the key value $\hat{g}^{ab}$:

**Construction 6.** (Steganographic Key Exchange)

| | |
|---|---|
| **Procedure** `SKE_Encode`$_A$:<br>**Input:** $(P,Q,h,g)$<br>repeat:<br>  sample $a \leftarrow U(\mathbb{Z}_{P-1})$<br>until $g^a \mod P < 2^{k-1}$<br>Let $c_a = g^a \mod 2^{k-1}$<br>**Output:** `Basic_Encode`$(c_a,h,k)$ | **Procedure** `SKE_Decode`$_A$:<br>**Input:** $s \in D^l$, exponent $a$<br>Let $c_b$ = `Basic_Decode`$(s)$<br>**Output:** $c_b^{ra} \mod P = \hat{g}^{ab}$ |

(`SKE_Encode`$_B$ and `SKE_Decode`$_B$ are analogous)

**Lemma 2.** *Let $f$ be $\epsilon$-biased on $\mathcal{C}_{A\to B,h_A}$ and $\mathcal{C}_{B\to A,h_B}$ for all $h_A,h_B$. Then for any warden $W \in \mathcal{W}(t)$, we can construct a DDH adversary $\mathbf{A}$ where $\mathbf{Adv}^{\mathsf{ddh}}_{\hat{g},P,Q}(\mathbf{A}) \geq \frac{1}{4}\mathbf{Adv}^{\mathsf{ske}}_{\mathsf{SKE}}(W,k) - \epsilon|P|$. The running time of $\mathbf{A}$ is at most $t + O(k|P|)$.*

*Proof.* (Sketch) Define $\hat{r}$ to be the least element such that $r\hat{r} = 1 \mod Q$. The algorithm **A** works as follows. Given elements $(\hat{g}^a, \hat{g}^b, \hat{g}^c)$ of the subgroup of order $Q$, we uniformly choose elements $k_a, k_b \leftarrow \mathbb{Z}_r$, and set $c_a = (\hat{g}^a)^{\hat{r}} g^{k_a Q}$, and $c_b = (\hat{g}^b)^{\hat{r}} g^{k_b Q}$. If $MSB(c_a) = MSB(c_b) = 0$, we then return $W(\texttt{Basic\_Encode}(c_a, h_A, k), \texttt{Basic\_Encode}(c_b, h_B, k), \hat{g}^c)$, otherwise we return 0. Notice that the key computed by `SKE_Decode` would be $c_a^{rb} = \left((\hat{g}^a)^{\hat{r}} g^{k_a Q}\right)^{rb} = (\hat{g}^{ab})^{r\hat{r}} g^{rQk_a b} = \hat{g}^{ab}$.

The decrease in $W$'s advantage comes from the fact that **A** excludes some elements of $\mathbb{Z}_P^*$ by sampling to get the MSB $= 0$, but we never exclude more than 1/2 of the cases for either $c_a$ or $c_b$. The $\epsilon|P|$ difference follows from Proposition 2 and the fact that $c_a$,$c_b$ are uniformly distributed on $U_{|P|-1}$.

**Theorem 3.** *If $f$ is $\epsilon$-biased on $\mathcal{C}_{A\to B,h_A}$ and $\mathcal{C}_{B\to A,h_B}$ for all $h_A, h_B$, then*

$$\mathbf{InSec}^{\mathsf{ske}}_{\mathsf{SKE},\mathcal{C}}(t,k) \leq 4\epsilon|P| + 4\mathbf{InSec}^{\mathsf{ddh}}_{\hat{g},P,Q}(t + O(k|P|)) \ .$$

## 8 Discussion and Open Problems

**Need for a PKI.** A potential stumbling block for public-key steganography is the need for a system which allows Alice and Bob to publish public keys for encryption and signatures without raising suspicion. The most likely source of a resolution to this issue is the existence of a global public-key infrastructure which publishes such public keys for every party in any case. In many cases (those modeled by the chosen hiddentext attack), however, it may be Alice who is trying to avoid suspicion while it is Bob who publishes the public key. For example Alice may be a government employee who wishes to leak a story and Bob a newspaper reporter, who may publish his public key daily.

In case Alice and Bob are both trying to avoid suspicion, it may be necessary to perform SKE instead. Even in this case, there is a need for a one-bit "secret channel" which alerts Bob to the fact that Alice is attempting key exchange. However, as long as Bob and Alice assume key exchange is occurring, it is easy to check at completion that it has indeed occurred by using `Basic_Encode` to exchange the messages $F_K(A, h_A), F_K(B, h_B)$ for $F$ a pseudorandom function.

**Stegosystems with Backdoors.** Suppose we wish to design steganography software which will be used as a black box by many users. Then as long as there is some entropy in the stegosystem of choice, we can use public-key steganography to implement a backdoor into the stegosystem which is provably undetectable via input/output behavior, by using the encoding routine as an oracle for Construction 4, with a fixed hiddentext ($1^k$, for instance). This will make it possible, with enough intercepted messages, to detect the use of the steganography software. If a total break is desired and the software implements private-key steganography, we can replace $1^k$ by the user's private key.

**Relationship to PKC: Complexity-Theoretic Implications.** In contrast to the private-key results of [7], we are not aware of a general result showing that the existence of any semantically secure public-key cryptosystem implies the existence of secure public-key steganography. However, our results allow construction of provably secure public-key steganography based on the security of any popular public-key cryptosystem.

**Acknowledgements.** The authors wish to thank Manuel Blum for his constant support and encouragement. We thank Christian Cachin for his helpful comments on our submitted draft, as well as Adam Bender, Andrew Bortz, Tri Van Le, Ke Yang, and several anonymous CRYPTO and Eurocrypt referees. This material is based upon work partially supported by the National Science Foundation under Grants CCR-0122581 and CCR-0058982 (The Aladdin Center). This work was also partially supported by the Army Research Office (ARO) and the Center for Computer and Communications Security (C3S) at Carnegie Mellon University. The first author is partially supported by a Microsoft Research Fellowship and the second author has been partially supported by a NSF Graduate Research Fellowship and a Siebel Scholarship.

## References

1. R. J. Anderson and F. A. P. Petitcolas. On The Limits of Steganography. *IEEE Journal of Selected Areas in Communications*, 16(4), pages 474-481, 1998.
2. M. Backes and C. Cachin. Public-Key Steganography with Active Attacks. Cryptology ePrint Archive, Report 2003/231, November 6, 2003. Available electronically: `http://eprint.iacr.org/2003/231`.
3. M. Bellare and P. Rogaway. Random Oracles are Practical. *Computer and Communications Security: Proceedings of ACM CCS 93*, pages 62–73, 1993.
4. C. Cachin. An Information-Theoretic Model for Steganography. *Proceedings of Second International Information Hiding Workshop*, Springer LNCS 1525, pages 306-318, 1998.
5. S. Craver. On Public-key Steganography in the Presence of an Active Warden. *Proceedings of Second International Information Hiding Workshop*, Springer LNCS 1525, pages 355-368, 1998.
6. J. Hastad, R. Impagliazzo, L. Levin, and M. Luby. A Pseudorandom generator from any one-way function. *SIAM Journal of Computing*, 28(4), pages 1364-1396, 1999.
7. N. J. Hopper, J. Langford, and L. Von Ahn. Provably Secure Steganography. *Advances in Cryptology: CRYPTO 2002*, Springer LNCS 2442, pages 77-92, 2002.
8. T. Mittelholzer. An Information-Theoretic Approach to Steganography and Watermarking. *Proceedings of the Third International Information Hiding Workshop*, Springer LNCS 1768, pages 1-16, 2000.
9. J. A. O'Sullivan, P. Moulin, and J. M. Ettinger. Information-theoretic analysis of Steganography. *Proceedings ISIT 98*, 1998.
10. C. Rackoff and D. Simon. Non-interactive Zero-Knowledge Proof of Knowledge and Chosen Ciphertext Attack. *Advances in Cryptology: CRYPTO 91*, Springer LNCS 576, pages 433-444, 1992.

11. G. J. Simmons. The Prisoner's Problem and the Subliminal Channel. *Advances in Cryptology: CRYPTO 83*, pages 51-67, 1983.
12. T. Van Le. Efficient Proven Secure Public Key Steganography. Cryptology ePrint Archive, Report 2003/156, September 3, 2003. Available electronically: `http://eprint.iacr.org/2003/156`.
13. A. Young and M. Yung. Kleptography: Using Cryptography against Cryptography. *Advances in Cryptology: Eurocrypt 97*, Springer LNCS 1233, pages 62-74, 1997.
14. J. Zollner, H. Federrath, H. Klimant, A. Pftizmann, R. Piotraschke, A. Westfield, G. Wicke, G. Wolf. Modeling the security of steganographic systems. *Proceedings of the Second International Information Hiding Workshop*, Springer LNCS 1525, pages 344-354, 1998.

# A Proof of Chosen-Stegotext Security

We define the following sequence of hybrid oracle distributions:

1. $\mathsf{P0}(b,m,h) = CT_{\mathsf{csa}}$, the covertext oracle.
2. $\mathsf{P1}(b,m,h)$ responds to `dec` queries as in $\mathsf{P0}$, and responds to `enc` queries using `CSA_Encode` but with calls to $\texttt{UEncode}^G$ replaced by calls to `Basic_Encode`.
3. $\mathsf{P2}(b,m,h)$ responds to `dec` queries as in $\mathsf{P1}$, and responds to `enc` queries using `CSA_Encode`.
4. $\mathsf{P3}(b,m,h) = ST_{\mathsf{csa}}$, the stegotext oracle.

We are given a CSA attacker $W \in \mathcal{W}(t, q_e, q_d, q_F, q_H, q_{H_1}, q_{H_2}, l)$ and wish to bound his advantage. Notice that $\mathbf{Adv}^{\mathsf{csa}}_{\mathsf{CSA},\mathcal{C}}(W,k) \leq |\Pr[W^{P0}=1] - \Pr[W^{P1}=1]| + |\Pr[W^{P1}=1] - \Pr[W^{P2}=1]| + |\Pr[W^{P2}=1] - \Pr[W^{P3}=1]|$ (for ease of notation, we omit the arguments $\pi_A, \pi_B$ to $W$). Hence, we can bound the advantage of $W$ by the sum of its advantages in distinguishing the successive hybrids. For hybrids $\mathsf{P},\mathsf{Q}$ we let $\mathbf{Adv}^{\mathsf{P},\mathsf{Q}}(W,k) = |\Pr[W^{\mathsf{P}}=1] - \Pr[W^{\mathsf{Q}}=1]|$.

**Lemma 3.** $\mathbf{Adv}^{\mathsf{P0},\mathsf{P1}}(W,k) \leq q_e \mathbf{InSec}^{\mathsf{ow}}_{\Pi}(t',k) + 2^{-k}(q_e^2/2 - q_e/2) + (l + 3q_e k)\epsilon$

*Proof.* Assume WLOG that $\Pr[W^{P1}=1] > \Pr[W^{P0}=1]$. Let $E_r$ denote the event that, when $W$ queries $\mathsf{P1}$, the random value $r$ never repeats, and let $E_q$ denote the event that $W$ never makes random oracle queries of the form $H_1(r)$ or $H_2(r,*,*)$ for an $r$ used by `CSA_Encode`, and let $E \equiv E_r \wedge E_q$.

$$\begin{aligned}
\Pr[W^{P1}=1] - \Pr[W^{P0}=1] &= \Pr[W^{P1}=1|E](1-\Pr[\overline{E}]) + \Pr[W^{P1}=1|\overline{E}]\Pr[\overline{E}] \\
&\quad - \Pr[W^{P0}=1] \\
&= \Pr[\overline{E}]\left(\Pr[W^{P1}=1|\overline{E}] - \Pr[W^{P1}=1|E]\right) \\
&\quad + \left(\Pr[W^{P1}=1|E] - \Pr[W^{P0}=1]\right) \\
&\leq \Pr[\overline{E}] \ + (l+3q_e k)\epsilon \\
&\leq \Pr[\overline{E_r}] + \Pr[\overline{E_q}] + (l+3q_e k)\epsilon \\
&\leq 2^{-k}\frac{q_e(q_e-1)}{2} + \Pr[\overline{E_q}] + (l+3q_e k)\epsilon
\end{aligned}$$

because if $r$ never repeats and $W$ never queries $H_1(r)$ or $H_2(r,*,*)$ for some $r$ used by CSA_Encode, then $W$ cannot distinguish between the ciphertexts passed to Basic_Encode and random bit strings.

It remains to bound $\Pr[\overline{E_q}]$. Given $W \in \mathcal{W}(t, q_e, q_d, q_F, q_G, q_{H_1}, q_{H_2}, l)$ we construct a one-way permutation adversary $\mathbf{A}$ against $\pi_B$ which is given a value $\pi_B(x)$ and uses $W$ in an attempt to find $x$, so that $\mathbf{A}$ succeeds with probability at least $(1/q_e)\Pr[\overline{E_q}]$. $\mathbf{A}$ picks $(\pi_A, \pi_A^{-1})$ from $\Pi_k$ and $i$ uniformly from $\{1,\ldots,q_e\}$, and then runs $W$ answering all its oracle queries as follows:

- enc queries are answered as follows: on query $j \neq i$, respond using CSA_Encode but with calls to $\text{UEncode}^G$ replaced by calls to Basic_Encode. On the $i$-th query respond with $s = \text{Basic\_Encode}(\pi_B(x)||e_1||\tau_1, h)$ where $e_1 = h_1 \oplus (m, \sigma_1)$ and $h_1, \sigma_1, \tau_1$ are chosen uniformly at random from the set of all strings of the appropriate length ($|e_1| = |m| + k$ and $|\tau_1| = k$), and set $\phi = \phi \cup \{(s,h)\}$.
- dec queries are answered using $CT_{\mathsf{csa}}$.
- Queries to $G, F, H_1$ and $H_2$ are answered in the standard manner: if the query has been made before, answer with the same answer, and if the query has not been made before, answer with a uniformly chosen string of the appropriate length. If a query contains a value $r$ for which $\pi_B(r) = \pi_B(x)$, halt the simulation and output $r$.

It should be clear that $\Pr[\mathbf{A}(\pi_B(x)) = x] \geq \frac{1}{q_e}(\Pr[\overline{E_q}])$.

**Lemma 4.** $\mathbf{Adv}^{\mathsf{P1,P2}}(W,k) \leq q_e \mathbf{InSec}^{\mathsf{ow}}_{\Pi}(t',k) + 2^{-k}(q_e^2/2 - q_e/2)$

*Proof.* Assume WLOG that $\Pr[W^{P2} = 1] > \Pr[W^{P1} = 1]$. Denote by $E_r$ the event that, when answering queries for $W$, the random value $r$ of CSA_Encode never repeats, and by $E_q$ the event that $W$ never queries $G(*, r, \pi_B(r)||*, *)$ for some $r$ used by CSA_Encode, and let $E \equiv E_r \wedge E_q$. Then:

$$
\begin{aligned}
\Pr[W^{P2}=1] - \Pr[W^{P1}=1] = & \left(\Pr[W^{P2}=1|E]\Pr[E] + \Pr[W^{P2}=1|\overline{E}]\Pr[\overline{E}]\right) \\
& - \Pr[W^{P1}=1|E]\Pr[E] - \Pr[W^{P1}=1|\overline{E}]\Pr[\overline{E}] \\
= & \Pr[\overline{E}]\left(\Pr[W^{P2}=1|\overline{E}] - \Pr[W^{P1}=1|\overline{E}]\right) \\
\leq & \Pr[\overline{E}] \\
\leq & 2^{-k}\frac{q_e(q_e-1)}{2} + \Pr[\overline{E_q}]
\end{aligned}
$$

Given $W \in \mathcal{W}(t, q_e, q_d, q_F, q_G, q_{H_1}, q_{H_2}, l)$ we construct a one-way permutation adversary $\mathbf{A}$ against $\pi_B$ which is given a value $\pi_B(x)$ and uses $W$ in an attempt to find $x$. $\mathbf{A}$ picks $(\pi_A, \pi_A^{-1})$ from $\Pi_k$ and $i$ uniformly from $\{1,\ldots,q_E\}$, and then runs $W$ answering all its oracle queries as follows:

- enc queries are answered as follows: on query $j \neq i$, respond using CSA_Encode. On the $i$-th query respond with $s = \text{UEncode}^G(\pi_B(x)||e_1||\tau_1, r_1, h)$ where $e_1 = h_1 \oplus (m, \sigma_1)$ and $h_1, \sigma_1, \tau_1, r_1$ are chosen uniformly at random from the set of all strings of the appropriate length ($|e_1| = |m| + k$ and $|\tau_1| = k$), and set $\phi = \phi \cup \{(s,h)\}$.

- dec queries are answered using $CT'_{\mathsf{csa}}$.
- Queries to $G, F, H_1$ and $H_2$ are answered in the standard manner: if the query has been made before, answer with the same answer, and if the query has not been made before, answer with a uniformly chosen string of the appropriate length. If a query contains a value $r$ for which $\pi_B(r) = \pi_B(x)$, halt the simulation and output $r$.

It should be clear that $\Pr[\mathbf{A}(\pi_B(x)) = x] \geq \frac{1}{q_e}(\Pr[\overline{E_q}])$.

**Lemma 5.** $\mathbf{Adv}^{\mathsf{P2,P3}}(W, k) \leq q_F \mathbf{InSec}^{\mathsf{ow}}_{\Pi}(t', k) + q_d/2^{k-1} + q_e/2^k$

*Proof.* Given $W \in \mathcal{W}(t, q_e, q_d, q_F, q_G, q_{H_1}, q_{H_2}, l)$ we construct a one-way permutation adversary $\mathbf{A}$ against $\pi_A$ which is given a value $\pi_A(x)$ and uses $W$ in an attempt to find $x$. $\mathbf{A}$ chooses $(\pi_B, \pi_B^{-1})$ from $\Pi_k$ and $i$ uniformly from $\{1, \ldots, q_F\}$, and then runs $W$ answering all its oracle queries as follows:

- enc queries are answered using CSA_Encode except that $\sigma$ is chosen at random and $F(r, m, h)$ is set to be $\pi_A(\sigma)$. If $F(r, m, h)$ was already set, fail the simulation.
- dec queries are answered using CSA_Decode, with the additional constraint that we reject any stegotext for which there hasn't been an oracle query of the form $H_2(r, m, h)$ or $F(r, m, h)$.
- Queries to $G, F, H_1$ and $H_2$ are answered in the standard manner (if the query has been made before, answer with the same answer, and if the query has not been made before, answer with a uniformly chosen string of the appropriate length) except that the $i$-th query to $F$ is answered using $\pi_A(x)$.

$\mathbf{A}$ then searches all the queries that $W$ made to the decryption oracle for a value $\sigma$ such that $\pi_A(\sigma) = \pi_A(x)$. This completes the description of $\mathbf{A}$.
Notice that the simulation has a small chance of failure: at most $q_e/2^k$. For the rest of the proof, we assume that the simulation doesn't fail. Let $E$ be the event that $W$ makes a decryption query that is rejected in the simulation, but would not have been rejected by the standard CSA_Decode. It is easy to see that $\Pr[E] \leq q_d/2^{k-1}$. Since the only way to differentiate P3 from P2 is by making a decryption query that P3 accepts but P2 rejects, and, conditioned on $\overline{E}$, this can only happen by inverting $\pi_A$ on some $F(r, m, h)$, we have that: $\mathbf{Adv}^{\mathsf{P2,P3}}(W, k) \leq q_F \mathbf{InSec}^{\mathsf{ow}}_{\Pi}(t', k) + q_d/2^{k-1} + q_e/2^k$.

## B Negligibly Biased Functions for Any Channel

Let $l(k) = \omega(\log k)$. Then the channel $\mathcal{C}^{(k)}$ is simply a distribution on sequences of documents which are elements of $D^{l(k)}$ and the marginal distributions $\mathcal{C}_h^{(k)}$ are simply $\mathcal{C}_h^{l(k)}$. The minimum entropy requirement from Section 3 then gives us that for any $h$ which has non-zero probability, $H_\infty(\mathcal{C}_h^{(k)}) = \omega(\log k)$.

Let $h_1, h_2, \ldots, h_m$ be any sequence of histories which all have non-zero probability under $\mathcal{C}^{(k)}$ and let $f : \{0,1\}^{m(k)} \times D \times \{0,1\}$ be a universal hash function.

Let $Y, Z \leftarrow U_{m(k)}, B \leftarrow U_m$, and $D_i \leftarrow C_{h_i}^{(k)}$. Let $L(k) = \min_i H_\infty(D_i)$, and note that $L(k) = \omega(\log k)$. Then the "Leftover Hash Lemma" (see, e.g., [6]) implies that

$$\Delta(\langle Y, f_Y(D_1), ..., f_Y(D_m)\rangle, \langle Y, B\rangle) \leq m2^{-L(k)/2+1} ,$$

where $\Delta(X, Y) = \frac{1}{2}\sum_x |\Pr[X = x] - \Pr[Y = x]|$ is the statistical distance, from which it is immediate that if we choose $Y \leftarrow U_{m(k)}$ once and publicly, then for all $1 \leq i \leq m$, $f_Y$ will have negligible bias for $C_{h_i}$ except with negligible probability.

# Immunizing Encryption Schemes from Decryption Errors

Cynthia Dwork[1], Moni Naor[2⋆], and Omer Reingold[2⋆⋆]

[1] Microsoft Research, SVC
1065 L'Avenida
Mountain View, CA 94043
dwork@microsoft.com

[2] Weizmann Institute of Science
Rehovot 76100, Israel
{moni.naor,omer.reingold}@weizmann.ac.il

**Abstract.** We provide methods for transforming an encryption scheme susceptible to decryption errors into one that is immune to these errors. Immunity to decryption errors is vital when constructing non-malleable and chosen ciphertext secure encryption schemes via current techniques; in addition, it may help defend against certain cryptanalytic techniques, such as the attack of Proos [33] on the NTRU scheme.

When decryption errors are very infrequent, our transformation is extremely simple and efficient, almost free. To deal with significant error probabilities, we apply amplification techniques translated from a related information theoretic setting. These techniques allow us to correct even very weak encryption schemes where in addition to decryption errors, an adversary has substantial probability of breaking the scheme by decrypting random messages (without knowledge of the secret key). In other words, under these weak encryption schemes, the only guaranteed difference between the legitimate recipient and the adversary is in the frequency of decryption errors. All the above transformations work in a standard cryptographic model; specifically, they do not rely on a random oracle.

We also consider the random oracle model, where we give a simple transformation from a one-way encryption scheme which is error-prone into one that is immune to errors.

We conclude that error-prone cryptosystems can be used in order to create more secure cryptosystems.

## 1 Introduction

In their seminal paper on semantic security Goldwasser and Micali defined a public key encryption scheme as one where the decryption is perfect, i.e., given

⋆ Incumbent of the Judith Kleeman Professorial Chair. Research supported in part by a grant from the Israel Science Foundation. Part of this work was done while visiting Microsoft Research, SVC.

⋆⋆ Most of this research was performed while at AT&T Labs - Research and while visiting the Institute for Advanced Study, Princeton, NJ.

C. Cachin and J. Camenisch (Eds.): EUROCRYPT 2004, LNCS 3027, pp. 342–360, 2004.

a properly formed ciphertext the answer is always the *unique* corresponding plaintext [20]. More formally, let the encryption algorithm be $E$ and the corresponding decryption algorithm be $D$. If $E$ maps a message $m$ with random coins $r$ to a ciphertext $c = E(m, r)$, then it is always the case that $D(E(m, r)) = m$. However, some cryptosystems do not satisfy this condition, two notable examples being the Ajtai-Dwork cryptosystem [1] and NTRU [21]. (In fact, sometimes a cryptosystem is deliberately designed to have ambiguous decryption; see more in Section 6.)

One might think that an encryption scheme with small probability of decryption error is merely an aesthetic nuisance, since the event of a decryption error can be compared to the event of an adversary guessing the secret key, which should be rare. However, serious difficulties arise in trying to construct cryptosystems secure under more stringent notions of security, such as non-malleability and chosen-ciphertext immunity, based on systems with ambiguous decryption. In fact, all known "bootstrapping" methods for constructing strong cryptosystems fail when the underlying one is susceptible to errors[3]. Furthermore, Proos was able to exploit decryption errors in his attack on the NTRU scheme [33]. Our goal in this work is to discuss general methods for eliminating errors and constructing secure cryptosystems based on less than perfect underlying schemes.

### 1.1 Random Oracles and the Real World

The literature contains constructions for cryptographic primitives in two well studied models: the random oracle world as described below, and the real world, where the assumption of a random oracle may not be justified. In general it is more difficult and involved to provide and prove correct constructions in the real world model.

If one makes the simplifying assumption that a specific function behaves as an idealized random function (random oracle), then it is possible to obtain simple and efficient constructions of public-key encryption schemes that are secure against chosen ciphertext attacks in the post-processing mode ("cca-post", also known as CCA2); these include OAEP and its variants [5,3,32,15,6], Fujisaki-Okamoto [14] and REACT [31][4]. However, it is not known if any one of these methods (or some other method) can be used to convert every public-key cryptosystem – including systems with decryption errors – that is semantically secure (or that satisfies even some weaker property such as one-wayness on the messages) against chosen plaintext attacks into one that is secure against stronger attacks, such as cca-post attacks (see below for more information on attacks). Among the problems in applying these approaches are that in the underlying "input" cryptosystem (1) there can exist ciphertexts which are valid encryptions

[3] One reason for the failure of those methods is that when the adversary chooses the input to the decryption algorithm, this input can have a distribution completely different from that of correctly encrypted messages and so the error probability may be large instead of small

[4] The meaning of such results is the subject of much debate (see e.g., [8,13,2]).

of two different plaintext messages; and (2) the decryption mechanism may sometimes fail to return "invalid" on an invalid ciphertext. As mentioned above, these problems were exploited by Proos [33] to attack various paddings of NTRU [30].

In the real world we have no idealized function, and we must do with what nature gives us. An important idea used either explicitly or at least implicitly in the construction of chosen ciphertext secure cryptosystem in the real world is to add some redundancy to the encryption and provide a proof of consistency of the ciphertext. The most general form of the proof of consistency is via a *non-interactive zero-knowledge* proof system (NIZKs) [11,27,29,34], but there are also more specific methods [9,10]. Here too a cryptosystem with possible decryption errors may cause problems in the construction. Take for instance the method that is based on a pair of keys together with a NIZK of consistency (this is the one suggested by Naor and Yung [29] and also a subsystem of the Dolev, Dwork, and Naor scheme [11]). A central idea in the proof of security is that knowing any of several private keys is sufficient for the decryption, and which one (of the several) is known is indistinguishable to the adversary. However, if there is no unique decryption, then seeing which plaintext is returned may leak which key is known, and the proof of security collapses.

### Our Results

We suggest methods for dealing with errors in both worlds described above:

*In the land of random oracles:* We provide a generic and efficient method for converting any public-key cryptosystem where decryption errors may occur, but where an adversary cannot retrieve the plaintext of a randomly chosen message (sometimes known as one-way cryptosystem), into one that is secure against chosen ciphertext attack in the post-processing mode. This is done in Section 5.

*The real world:* We show two transformations from cryptosystem with errors to ones without. When decryption errors are very infrequent, our transformation is extremely simple and efficient, almost free. The case of significant error probabilities is technically more involved. Our transformation for this case corrects even very weak encryption schemes where in addition to decryption errors, an adversary has substantial probability of breaking the scheme by decrypting random messages (without knowledge of the secret key). In other words, under these weak encryption schemes, the only guaranteed difference between the legitimate recipient (holder of the secret key) and the adversary is in the frequency of decryption errors: the legitimate recipient experiences fewer errors than does the adversary.

To demonstrate the subtleties of this task, consider the case where the legitimate recipient decrypts correctly with probability $9/10$ (and let us assume for simplicity that otherwise he gets an error message), but the adversary decrypts correctly with probability $1/10$. A natural approach is to use error correcting codes, setting the parameters in such a way that the legitimate recipient will have enough information to decode, whereas the adversary will get no information. This approach indeed works in the information theoretic counterpart of a channel where the receiver gets the piece of information with certain probability and the eavesdropper with another. But it is not clear how to carry it through in the computational setting. Therefore, the solutions given in this paper use a different approach: we apply amplification techniques translated from the related information theoretic setting of [35]. We note that here, too, the computational setting introduces additional complications.

The conclusion we reach is that once provided with noninteractive zero knowledge proof systems, one can convert essentially *any* public-key cryptosystem with decryption errors into one that is secure against chosen ciphertext attack in the postprocessing mode.

**Related Work:** In addition to the work mentioned above we should point out two specific papers that converted an error-prone scheme into an error free one. Goldreich, Goldwasser and Halevi [18] showed how to eliminate decryption errors in the Ajtai-Dwork [1] cryptosystem. Our methods, especially those of Section 3, can be seen as a general way of achieving that goal. In the papers of Howgrave-Graham et al. [23,24] the problem of constructing an CCA-post-secure NTRU-based method in the random oracles world is considered.

## 2 Preliminaries

### Notation and Conventions

We will abbreviate "probabilistic polynomial time Turing Machine" with PPTM. We use the notation $poly(\cdot)$ to refer to some polynomially bounded function and $neg(\cdot)$ to refer to some function that is smaller than $1/p(\cdot)$ for any polynomial $p(\cdot)$ (for all sufficiently large inputs). For any integer $n$, we let $U_n$ denote the uniform distribution over $\{0,1\}^n$. We let the operation $\oplus$ on two bit-strings denote their bit-wise XOR.

### 2.1 Public-Key Encryption – Correctness

A public-key encryption scheme consists of three probabilistic polynomial time algorithms $(G, E, D)$, for key generation, encryption and decryption respectively. For simplicity we fix $n$ to be both the security parameter and input length, and assume that the message space is $\{0,1\}^n$. Algorithm $G$, for the key generation is given $1^n$ as input (as well as internal random coins), and outputs the public key and secret key pair $(pk, sk)$. We have that $|pk| = |sk| = poly(n)$. $E$ and $D$ are, respectively, the encryption and decryption algorithms. $E$ takes as input a public key $pk$, an $n$-bit plaintext message $m$, and uses internal random coins. We refer to the output $c \in E_{pk}(m)$ as the ciphertext. When we want to refer to $E$'s additional $poly(n)$-long random input $r$ explicitly, we will use the notation $E_{pk}(m; r)$. Finally, $D$ takes as input a secret key $sk$ and a ciphertext. The output of $D$ is either a message $m'$ (which may fail to equal the original message $m$) or $\perp$ to indicate invalid (we are deliberately not attaching semantics to a response of "invalid"). The standard definition of public-key encryption schemes requires perfect correctness. Namely, that if the input $c$ to $D_{sk}$ is well constructed using $E_{sk}$, then the output $D_{sk}(c)$ is supposed to retrieve the original plaintext. We make this explicit in the next definition.

**Definition 1.** *A public-key encryption scheme $(G, E, D)$ is* perfectly correct *if the following holds:*

- *For every message $m$ of length $n$, for every pair $(pk, sk)$ generated by $G$ on input $1^n$, and all possible coin tosses of $E$ and $D$, it should hold that $D_{sk}(E_{pk}(m)) = m$.*

Although we allowed $D$ to output $\perp$ we made no assumption on the probability of $\perp$ being the output in case the ciphertext is indeed *invalid* (where invalid means that there do not exist $m$ and $r$ such that $c = E_{pk}(m; r)$).

We now want to relax the notion of public key-encryption so as to allow decryption errors. We define an encryption scheme to be $\alpha$-correct, if the probability of decryption error is at most $1 - \alpha$.

**Definition 2.** *For any function $\alpha : \mathbf{N} \mapsto [0, 1]$, a public-key $(G, E, D)$ encryption scheme is* $\alpha$-correct *if $Pr[D_{sk}(E_{pk}(m)) \neq m] \leq 1 - \alpha(n)$, where the probability is taken over the random coins of $G$ used to generate $(pk, sk)$ on input $1^n$, over the choice of $m \in U_n$, and over the random coins of $E$ and $D$.*

In the above definition the error probability is taken over the random choice of the message (uniformly at random), the randomness of the encryption and decryption *and the choice of the key.* In particular, some keys may be completely useless as they don't allow decryption at all. We now consider the case that the bound on the decryption error holds for all keys or for all but a negligible fraction of the keys. These definitions are relevant here for two reasons: (1) Our transformations will be a bit more efficient if we only try to immunize against this kind of errors. (In the sense that the key of the revised scheme will only include a single key of the original scheme.) (2) Our transformations will produce schemes that are "almost-all-keys perfectly correct" rather than perfectly correct encryptions. This means that decryption errors can only occur with a negligible probability *over the choice of the key.* Note that such errors are usually much less harmful, and in particular such schemes can be made non-malleable using "standard" techniques (unlike the case where errors may occur for a substantial fraction of the keys).

**Definition 3.** *Let $(G, E, D)$ be any public-key encryption scheme and $\alpha : \mathbf{N} \mapsto [0, 1]$ an arbitrary function.*

- *$(G, E, D)$ is* all-keys $\alpha$-correct *if for every pair $(pk, sk)$ generated by $G$ on input $1^n$, $Pr[D_{sk}(E_{pk}(m)) \neq m] \leq 1 - \alpha(n)$, where the probability is taken over the choice of $m \in U_n$, and over the random coins of $E$ and $D$.*

- $(G, E, D)$ *is* almost-all-keys $\alpha$-correct *if with probability* $(1-neg(n))$ *over the random coins of* $G$ *used to generate* $(pk, sk)$ *on input* $1^n$, $Pr[D_{sk}(E_{pk}(m)) \neq m] \leq 1-\alpha(n)$, *where the probability is taken over the choice of* $m \in U_n$, *and over the random coins of* $E$ *and* $D$.
- $(G, E, D)$ *is* almost-all-keys perfectly correct *if with probability* $(1-neg(n))$ *over the random coins of* $G$ *used to generate* $(pk, sk)$ *on input* $1^n$, $Pr[D_{sk}(E_{pk}(m)) \neq m] = 0$, *where the probability is taken over the choice of* $m \in U_n$, *and over the random coins of* $E$ *and* $D$.

## 2.2 Public-Key Encryption – Security

Semantic security [20] has established itself as essentially the minimal desired notion of security for encryption schemes. Intuitively, a public-key encryption scheme is semantically secure if anything that a polynomial-time adversary can compute about the plaintext $m$ given the ciphertext $c = E_{pk}(m)$, it can also compute without access to $c$. Semantic security was shown in [20] to be equivalent to the indistinguishability of ciphertexts, which intuitively means that ciphertexts which correspond to different plaintexts are indistinguishable. Three basic modes of attack for which semantic security was considered are: chosen plaintext attack (which for public-key encryption essentially amounts to giving the adversary the public-key $pk$ and allowing the adversary to decide the challenge distribution), and chosen ciphertext attack in the preprocessing and the postprocessing modes (in both the adversary also gets access to a decryption oracle; in the preprocessing mode this access ends when the ciphertext challenge is published). Semantic security under these attacks is denoted IND-CPA, IND-CCA-Post and IND-CCA-Pre respectively. An even stronger notion of security than semantic security is that of non-malleability [11]. Intuitively, here the adversary should not even gain a (non-negligible) advantage in creating an encryption of a message that relates to $m$. Non malleability with respect to the above attacks is denoted NM-CPA, NM-CCA-Post and NM-CCA-Pre respectively. For the formal definitions of the above notions we rely on [11].

Both semantic security and non-malleability were originally defined for perfectly correct encryption schemes. Nevertheless they are just as meaningful for schemes with decryption errors. Section 3 gives a very simple way of eliminating decryption errors (as long as they are very rare) while preserving each one of the above six notions of security. Section 4 shows how to immunize much weaker encryption schemes. Here decryption errors will be more likely (may even happen with probability $1 - poly$). In addition, we will make much weaker security assumptions: we will only bound the success probability of the adversary in "inverting $E$" and completely retrieving the plaintext message $m$. (Therefore, the only advantage the legitimate recipient has over the adversary is in the probability of decryption.) This notion of weak security is captured by the following definition.

**Definition 4.** *For any function* $\beta : \mathbf{N} \mapsto [0,1]$, *a public-key encryption scheme is* $\beta$-one-way *($\beta$-OW) if for every PPTM* $A$, $Pr[A((E_{pk}(m)) = m] \leq \beta(n) +$

*$neg(n)$, where the probability is taken over the random coins of $G$ used to generate $(pk, sk)$ on input $1^n$, over the choice of $m \in U_n$, and over the random coins of $E$ and $A$.*

We note that unlike semantic security and non-malleability, this notion of security allows the encryption scheme $E$ to be deterministic.

*Pseudorandom Generators* One of the transformations of this paper uses pseudorandom generators as a main tool. A pseudorandom generator is a function $prg : \{0,1\}^* \mapsto \{0,1\}^*$ such that on $n$-bit input $x$, the output $prg(x)$ is $\ell(n) > n$ bits long and such that $prg(U_n)$ is *computationally* indistinguishable from $U_{\ell(n)}$. See [17,16] for a formal definition.

## 3 The Case of Infrequent Errors

This section describes a very efficient way for eliminating decryption errors when errors are very rare. If errors are too frequent to apply this technique directly, then one can first apply the amplification methods described in Section 4.

Let $E$ be an encryption scheme where for every message $m$, the probability over the randomness $r$ of $E$ that $D_{sk}(E_{pk}(m; r)) \neq m$ is tiny. To correct this scheme we use the "reverse randomization" trick from the construction of Zaps [12] and commitment protocols [28] (which can be traced back to Lautemann's proof that BPP is in the polynomial time hierarchy [26]). The idea is very simple: by assumption, only a tiny fraction of "bad" random strings $r$ lead to ciphertexts with decryption errors. Thus, we will arrange that the ciphertexts are constructed using only a rather small fraction of the possible values for $r$; the particular set of values will depend on the choice of public key. Very minimal independence in the selection of this subset will already assure that we are avoiding the bad strings with very high probability. In addition, the subset will be constructed to be pseudorandom, which will guarantee that the semantic security of the original scheme is preserved. Finally, the construction will ensure that the error probability is *only on the choice of encryption key* – if the encryption key is good, no ciphertext created with this encryption key will suffer a decryption error. The only significant computational cost incurred by this transformation is a single invocation of a pseudorandom generator (and in fact, this may already be performed to save on random bits, in which case the transformation is essentially for free).

For simplicity we state the next construction (and the corresponding theorem) under the assumption that the decryption algorithm $D$ is deterministic. In the case of chosen-plaintext attack (which is probably the most interesting setting of the theorem), this can be obtained simply by fixing the randomness of $D$ as part of the key. The case of chosen-ciphertext attacks is a bit more delicate but still the construction can be easily extended to randomized $D$.

**Construction 31** *Let $(G, E, D)$ be any public-key encryption scheme. Let $\ell(n)$ be the (polynomially bounded) number of bits used by $E$ to encrypt $n$-bit messages. Without loss of generality assume that $\ell(n) > n$ (as $E$ can always ignore* *part of its random input). Let prg be a pseudorandom generator that expands $n$ bits to $\ell(n)$ bits.*

*Define the public-key encryption scheme $(G', E', D')$ as follows: on input $1^n$, the generation algorithm $G'$ outputs $((pk, \bar{r}), sk)$ where $(pk, sk)$ is obtained by invoking $G$ on the same input and $\bar{r} \in U_{\ell(n)}$. On an $n$-bit input $m$, the encryption function $E'$ uses an $n$-bit random string $s$ and outputs $E_{pk}(m; prg(s) \oplus \bar{r})$. The decryption function $D'$ is identical to $D$.*

**Theorem 1.** *Let $(G, E, D)$ be any $(1 - 2^{-4n})$ correct public-key encryption scheme with $D$ being deterministic. Define $(G', E', D')$ as in Construction 31. Then $(G', E', D')$ is an almost-all-key perfectly correct public-key encryption scheme. Furthermore, if $(G, E, D)$ is NN-AAA secure with NN-AAA $\in$ {IND-CPA, IND-CCA-Post, IND-CCA-Pre, NM-CPA, NM-CCA-Post, NM-CCA-Pre} then so is $(G', E', D')$.*

*Proof.* For any fixed value of $\bar{r}$, the distribution $prg(U_n) \oplus \bar{r}$ is pseudorandom. Therefore, it easily follows that $(G', E', D')$ is NN-AAA secure (otherwise we could construct a distinguisher that breaks the pseudorandom generator).

It remains to prove the correctness of $(G', E', D')$, i.e. that with high probability over the choice of keys the scheme is perfectly correct. First, with probability at least $(1-2^{-n})$ over the choice of $(pk, sk)$, the value $\Pr_{m,r}[D_{sk}(E_{pk}(m; r)) \neq m]$ is at most $2^{-3n}$. Assume that $(pk, sk)$ satisfies this property. Since $\bar{r}$ is uniformly distributed we also have that $\Pr_{m,s,\bar{r}}[D_{sk}(E_{pk}(m; prg(s) \oplus \bar{r})) \neq m] \leq 2^{-3n}$. As $m$ and $s$ are only $n$-bit long, we get by a union bound that the probability over $\bar{r}$ that for *some* $m$ and $s$ a decryption error $D_{sk}(E_{pk}(m; prg(s) \oplus \bar{r})) \neq m$ will occur is at most $2^{-n}$. We can therefore conclude that for all but at most a $2^{-n+1}$ fraction of $(G', E', D')$ keys $((pk, \bar{r}), sk)$ the scheme is perfectly correct.

*Remark 1.* The existence of the pseudorandom generator needed for Construction 31, follows from the security of $(G, E, D)$ (under any one of the notions considered by the theorem). This is because the security of $(G, E, D)$ implies the existence of one-way functions [25] which in turn imply the existence of pseudorandom generators [22].

Consider the construction of [11] for NM-CCA-post secure public key cryptosystems. This requires (i) a perfectly correct public-key cryptosystem which is semantically secure against chosen plaintext attacks (ii) A non-interactive zero-knowledge (NIZK) proof system for NP (that is for some specific language in NP) (iii) other primitives that can be based on one-way functions. Furthermore, if we replace in that construction the perfectly correct cryptosystem with one that is almost-all-keys-perfectly-correct, then all that happens is that the resulting construction is also of a similar nature. Therefore we can conclude

**Corollary 2** *If $(1 - 2^{-4n})$-correct public-key encryption schemes semantically secure against chosen plaintext attacks exist and NIZK proof system for NP exist, then almost-all-key perfectly correct public-key encryption schemes which are NM-CCA-post secure public key cryptosystems exist.*

# 4 Immunizing Very Weak Encryption Schemes

We now consider much weaker encryption schemes than in Section 3. Here the encryption may only be $\alpha$-correct and $\beta$-OW where $\alpha$ and $\beta$ may be as small as $1/poly$. Naturally, $\alpha$ has to be larger than $\beta$ as otherwise the legitimate recipient of a message will have no advantage over the adversary (and such a scheme is useless and trivial to construct). The transformation given here works under the assumption that $\beta < \alpha^4/c$ for some fixed constant $c$. An interesting open problem is to give a transformation that works for even smaller gaps. Nevertheless, as we discuss below, having the transformation work for a gap $\beta - \alpha$ that is larger than an arbitrary constant, may involve improving the corresponding transformation in the related information-theoretic setting of [35].

## 4.1 Polarization in the Statistical Setting

Sahai and Vadhan [35], give an efficient transformation of a pair of distributions $(X_0, X_1)$ (encoded by the circuits that sample them) into a new pair of distributions $(Y_0, Y_1)$. The transformation "polarizes" the statistical distance between $X_0$ and $X_1$. If this distance is below some threshold $\beta'$ then the statistical distance between $Y_0$ and $Y_1$ is exponentially small. If on the other hand the distance between $X_0$ and $X_1$ is larger than another threshold $\alpha'$ then the statistical distance between $Y_0$ and $Y_1$ is exponentially close to 1. The condition for which this transformation works is that $\beta' < \alpha'^2$.

What is the relation between this problem and ours? Consider an $\alpha$-correct and $\beta$-OW encryption scheme, for one-bit messages. Let $X_0$ be the distribution of encryptions of 0 and $X_1$ the distribution of encryptions of 1. Intuitively we have that the legitimate recipient can distinguish these distributions with advantage $\alpha - (1 - \alpha) = 2\alpha - 1$ (recall that $\alpha > 1/2$), while the adversary cannot distinguish the distributions with advantage better than $2\beta - 1 < 2\alpha - 1$. Our transformation produces a new encryption scheme; let $Y_0$ and $Y_1$ be the corresponding distributions. We now have that the ability of the adversary to distinguish between $Y_0$ and $Y_1$ shrinks (to negligible), whereas the legitimate recipient distinguishes with probability that is exponentially close to 1. In fact, this intuitive similarity can be formalized to show that any transformation in the computational setting that is "sufficiently black box" implies a transformation in the statistical setting. This in particular implies that for our transformations to work for any constant gap $\alpha - \beta$, we may need to improve the transformation of [35] (or to use non black-box techniques).

What about the other direction? It seems much harder in general to translate transformations from the statistical setting to the computational one. Nevertheless, the transformations given in this section are heavily influenced by [35]. However, the computational versions of the amplification tools used in [35] are significantly weaker, which imposes additional complications and implies somewhat weaker bounds than those of [35].

## 4.2 Tools and Basic Transformations

To improve an $\alpha$-correct and $\beta$-OW encryption scheme $(G, E, D)$, we will use three basic transformations:

**Parallel Repetition** The encryption $E^k$ of a $k$-tuple of messages $m_1, \ldots, m_k$ will be defined as $E^k(m_1, \ldots, m_k) = E(m_1), \ldots, E(m_k)$. A negative effect of this transformation is that the probability of correct decryption of the entire $k$-tuple is reduced to $\alpha^k$. The gain of the transformation is that the probability of the adversary to break the one-wayness of $E^k$ will also decrease below $\beta$ (usually in an exponential rate as well). To bound this probability we apply a result of Bellare et al. [4] on the amplification of games in parallel execution. To conclude, this transformation makes decryption harder both for the legitimate recipient and for the adversary. As the adversary has a weaker starting point (success probability $\beta \ll \alpha$), it will be hurt more by the transformation.

**Hard Core Bit** Here we will transform an encryption scheme for strings to one that encrypts single bits. This will employ a hard core predicate in a rather standard fashion. The gain from this transformation is in turning the one-wayness of an encryption scheme into indistinguishability (which is easier to work with and is also our final goal).

**Direct Product** The encryption $E^{\otimes k}$ of a message $m$ will be the concatenation of $k$ independent encryptions of $m$ under $E$. This transformation has the reverse affect to $E^k$: Decryption becomes easier both for the legitimate recipient and for the adversary. As the legitimate recipient has a better starting point (success probability $\alpha \gg \beta$), it will gain more by the transformation.

In the formal definition of $E^k$ and $E^{\otimes k}$, we use *independently generated* keys for each one of the invocations of $E$ by these schemes. This is necessary as a large fraction of the keys of $E$ may be completely useless (i.e., do not allow decryption at all or completely reveal the message). So in order to amplify the security and correctness, we should use more than a single key. This can be avoided if we assume that $(G, E, D)$ is $\alpha$-correct and $\beta$-OW even after we fix the key of $E$ (for all but negligible fraction of the keys). In such a case, the transformations of this paper will become much more efficient (in terms of key size). We now turn to the formal definition of the basic transformations.

### Parallel Repetition

**Definition 5.** *Let $(G, E, D)$ be any public-key encryption scheme, and let $k : \mathbf{N} \mapsto \mathbf{N}$ be any polynomially bounded function. Define $(G^k, E^k, D^k)$ as follows: On input $1^n$, the key-generating algorithm $G^k$ invokes $G$, with input $1^n$, $k = k(n)$ times using independent random coins for each invocation. The output of $G^k$ is $(\bar{pk}, \bar{sk})$ where $\bar{pk} = pk_1, \ldots pk_k$, $\bar{sk} = sk_1, \ldots sk_k$, and $(pk_i, sk_i)$ is the output of $G$ in its $i^{th}$ invocation. On input $\bar{m} = m_1, \ldots m_k$ the output $E^k_{\bar{pk}}(\bar{m})$ is defined by $E^k_{\bar{pk}}(\bar{m}) = E_{pk_1}(m_1), \ldots E_{pk_k}(m_k)$, where the $k$ encryptions are performed* *with independent random coins. Finally, on input* $\bar{c} = c_1, \ldots c_k$*, the decryption algorithm* $D^k_{\bar{sk}}$ *tries to decrypt each* $c_i$ *by applying* $D_{sk_i}(c_i)$*. It outputs* $\perp$ *if one of these invocations of* $D$ *returned* $\perp$ *and otherwise* $D^k_{\bar{sk}}$ *outputs the sequence* $D_{sk_1}(c_1) \ldots D_{sk_k}(c_k)$*.*

**Lemma 1.** *Let* $(G, E, D)$ *be any public-key encryption scheme, and let* $k : \mathbf{N} \mapsto \mathbf{N}$ *be any polynomially bounded function. If* $(G, E, D)$ *is* $\alpha$*-correct and* $\beta$*-OW with* $\beta < 1 - 1/poly$*, then* $(G^k, E^k, D^k)$ *is* $\alpha^k$*-correct and* $\beta'$*-OW for any* $\beta' > 1/poly$ *that satisfies* $\beta' > 32/(1-\beta) \cdot e^{-k(1-\beta)^2/256}$*.*

*Proof.* The correctness of $(G^k, E^k, D^k)$ follows immediately from the definition. The security is much more delicate. Fortunately, it can be obtained as a simple corollary of a theorem of Bellare, Impagliazzo, and Naor regarding error probability in parallel execution of protocols of up to three rounds ([4] Theorem 4.1). Thus, we need to translate the breaking of $(G^k, E^k, D^k)$ into winning the parallel execution of a game that is composed of at most three messages. Specifically, consider the following game between $P$ and (an honest) $V$, where $V$ invokes $G$ to select $(pk, sk)$, it selects a uniform message $m$ and sends $pk$ and $E_{pk}(m)$ to $P$. In return, $P$ sends a message $m'$ and wins if $m = m'$. From the one-wayness of $(G, E, D)$ we get that the best *efficient* strategy of $P$ can win with probability at most $\beta + neg$. Note that the probability of winning the $k$-times parallel repetition of this game is the same as breaking the one-wayness of $(G^k, E^k, D^k)$. The lemma now follows from Theorem 4.1 of [4].

### Hard Core Bit

For concreteness we will use the Goldreich-Levin (inner product) bit [19]. This could be replaced with hard-core bits implied by other error-correcting codes that have strong list-decoding properties.

**Definition 6.** *Let* $(G, E, D)$ *be any public-key encryption scheme, where the encryption function operates on plaintexts of length* $\ell \geq 1$*, and let* $k : \mathbf{N} \mapsto \mathbf{N}$ *be any polynomially bounded function. Define* $(G^{\odot}, E^{\odot}, D^{\odot})$ *as follows:* $G^{\odot}$ *is simply identical to* $G$*. On a one-bit message* $\sigma$*, the encryption function* $E^{\odot}_{pk}$ *samples two* $\ell$*-bit strings* $m$ *and* $r$ *uniformly at random and outputs* $E_{pk}(m), r, \langle m, r\rangle \oplus \sigma$*, where* $\langle m, r\rangle$ *is the inner product of* $m$ *and* $r$ *(mod 2). On input* $c, r, \sigma'$ *the decryption function* $D^{\odot}_{pk}$ *evaluates* $m' = D_{pk}(c)$*. If* $m' \neq \perp$*, then* $D^{\odot}_{pk}$ *outputs* $\langle m', r\rangle \oplus \sigma'$*, otherwise* $D^{\odot}_{pk}$ *outputs a random bit.*

**Lemma 2.** *Let* $(G, E, D)$ *be any public-key encryption scheme. If* $(G, E, D)$ *is* $\alpha$*-correct and* $\beta$*-OW, then* $(G^{\odot}, E^{\odot}, D^{\odot})$ *is* $(1/2+\alpha/2)$*-correct and* $1/2+O(\sqrt{\beta})$*-OW. In particular, if* $\beta$ *is negligible then* $(G^{\odot}, E^{\odot}, D^{\odot})$ *is IND-CPA secure.*

*Proof.* For correctness, note that if $m' = D_{pk}(c) = m$ (as in Definition 6), then $D^{\odot}_{pk}$ decrypts correctly with probability one. Otherwise $D^{\odot}_{pk}$ decrypts correctly with probability half (since the probability over $r$ that for any $m' \neq m$ we have that $\langle m', r\rangle = \langle m, r\rangle$ is half). We can therefore conclude that the probability of correct decryption is at least $\alpha \cdot 1 + (1-\alpha) \cdot 1/2 = 1/2 + \alpha/2$.

For security, let us first assume that $\beta$ is negligible. In this case $(G^{\odot}, E^{\odot}, D^{\odot})$ is $(1/2)$-OW and equivalently is IND-CPA secure. Assume for the sake of contradiction that there exists an efficient adversary that decrypts $D^{\odot}_{pk}$ with probability $1/2 + 1/poly$ without access to $sk$. In this case, there is an efficient adversary that given $E_{pk}(m)$ and $r$ guesses $\langle m, r\rangle$ with probability $1/2 + 1/poly$. Now we obtain from [19] that there exists an efficient adversary that given $E_{pk}(m)$ outputs $m$ with probability $1/poly$. This contradicts the assumption that $(G, E, D)$ is *neg*-OW.

Finally, let us consider the case where $\beta$ is non-negligible. Assume for the sake of contradiction that there exists an efficient adversary that decrypts $D^{\odot}_{pk}$ with probability $1/2 + \epsilon$, where $\epsilon = c \cdot \sqrt{\beta}$ for some large constant $c$ (note that $\epsilon > 1/poly$). This again implies the existence of an efficient adversary that given $E_{pk}(m)$ and $r$ guesses $\langle m, r\rangle$ with the same probability. Using a tight enough version of the reconstruction algorithm for the Goldreich-Levin hard-core bit, we can conclude that there exists an efficient adversary that given $E_{pk}(m)$ computes a list of $O(1/\epsilon^2)$ candidates that include $m$ with probability $1/2$. This means that this adversary can also guess $m$ with probability $\Omega(\epsilon^2)$ which can be made say $2\beta$ by setting the constant $c$ to be large enough. This contradicts the $\beta$-one-wayness of $(G, E, D)$ and completes the proof of the lemma.

## Direct Product

**Definition 7.** *Let $(G, E, D)$ be any public-key encryption scheme, and let $k : \mathbf{N} \mapsto \mathbf{N}$ be any polynomially bounded function. Define $(G^{\otimes k}, E^{\otimes k}, D^{\otimes k})$ as follows: On input $1^n$, the key-generating algorithm $G^{\otimes k}$ invokes $G$, with input $1^n$, $k = k(n)$ times using independent random coins for each invocation. The output of $G^{\otimes k}$ is $(\bar{pk}, \bar{sk})$ where $\bar{pk} = pk_1, \ldots pk_k$, $\bar{sk} = sk_1, \ldots sk_k$, and $(pk_i, sk_i)$ is the output of $G$ in its $i^{th}$ invocation. On input $m$ the output $E^{\otimes k}_{\bar{pk}}(m)$ is defined by $E^{\otimes k}_{\bar{pk}}(m) = E_{pk_1}(m), \ldots E_{pk_k}(m)$, where the $k$ encryptions are performed with independent random coins. Finally, on input $\bar{c} = c_1, \ldots c_k$, the decryption algorithm $D^{\otimes k}_{\bar{sk}}$ tries to decrypt each $c_i$ by applying $D_{sk_i}(c_i)$. It outputs the value that is obtained the largest number of times (ties are resolved arbitrarily).*

We will use the direct product transformation only for encryptions of single bits. In this case, it is convenient to express correctness and security in terms of the advantage over half.

**Lemma 3.** *Let $(G, E, D)$ be any public-key encryption scheme over the message space $\{0,1\}$, and let $k : \mathbf{N} \mapsto \mathbf{N}$ be any polynomially bounded function. If $(G, E, D)$ is $(1/2 + \alpha)$-correct and $(1/2 + \beta)$-OW, then $(G^{\otimes k}, E^{\otimes k}, D^{\otimes k})$ is $(1/2 + k\beta)$-OW and for every $\epsilon > 0$, it is $(1-\epsilon)$-correct as long as $k > c \cdot 1/\alpha^2 \cdot \log 1/\epsilon$ for some fixed constant $c$.*

*Proof.* The one-wayness of $(G^{\otimes k}, E^{\otimes k}, D^{\otimes k})$ is obtained by a standard hybrid argument. Correctness is also simple to show using Chernoff bound. We note that we assume here that decryption errors occur with roughly the same probability for encryptions of zero and encryptions of one. For example, it is sufficient to assume that both $Pr[D_{sk}(E_{pk}(0)) = 0] > 1/2 + \alpha/2$ and $Pr[D_{sk}(E_{pk}(1)) = 1] > 1/2 + \alpha/2$. This is with no loss of generality as biases of $D$ (towards outputting zero or towards one) can always be corrected.

### 4.3 Combining the Basic Transformations

The three basic transformations defined above can be combined in various ways to improve $\alpha$-correct and $\beta$-OW encryption schemes. The most efficient combination depends on the particular values of $\alpha$ and $\beta$. We will not attempt to optimize the efficiency of our transformations but rather to demonstrate their effectiveness. For that we consider two settings of the parameters: (1) $\beta$ is an arbitrary constant smaller than one and $\alpha$ is also a constant smaller than one (that depends on $\beta$). (2) $\alpha$ is as small as $1/poly$ and $\beta$ is non-negligible ($\beta = \Omega(\alpha^4)$).

#### Constant Decryption Errors

**Theorem 3.** *For any constant $\beta < 1$ there exists a constant $\alpha < 1$ such that if there exists an $\alpha$-correct and $\beta$-OW public-key encryption scheme then there exists an almost-all-keys perfectly-correct IND-CPA secure public-key encryption scheme.*

*Proof.* Set $\alpha$ to be a constant such that $e^{-(1-\beta)^2/256} < \alpha^8$ and let $(G_0, E_0, D_0)$ be an $\alpha$-correct and $\beta$-OW public-key encryption scheme. Define the following systems:

- $(G_1, E_1, D_1) = (G_0^{k_1}, E_0^{k_1}, D_0^{k_1})$ where $k_1 = log_\alpha(1/n)$. Lemma 1 implies that $(G_1, E_1, D_1)$ is $(1/n)$-correct and $O(1/n^8)$-OW.
- $(G_2, E_2, D_2) = (G_1^{\odot}, E_1^{\odot}, D_1^{\odot})$. Lemma 2 implies that $(G_2, E_2, D_2)$ is $(1/2 + n/2)$-correct and $(1/2 + O(1/n^4))$-OW.
- $(G_3, E_3, D_3) = (G_2^{\otimes k_2}, E_2^{\otimes k_2}, D_2^{\otimes k_2})$ where $k_2 = O(n^3)$, for which Lemma 3 implies that $(G_3, E_3, D_3)$ is $(1 - 2^{-5n})$-correct and $(1/2 + O(1/n))$-OW.
- $(G_4, E_4, D_4) = (G_3^n, E_3^n, D_0^n)$. Lemma 1 implies that $(G_1, E_1, D_1)$ is $(1 - 2^{-5n})^n$-correct, which means that it is also $(1 - n \cdot 2^{-5n})$-correct. In addition it is $(1/p)$-OW *for any polynomial p*. Thus it is also *neg*-OW.
- $(G_5, E_5, D_5) = (G_4^{\odot}, E_4^{\odot}, D_4^{\odot})$. Lemma 2 implies that $(G_5, E_5, D_5)$ is $(1 - (n/2) \cdot 2^{-5n})$-correct and IND-CPA secure.

Theorem 3 now follows as a corollary of Theorem 1.

#### Very Frequent Decryption Errors

**Theorem 4.** *There exists some positive constant c such that for any functions $\alpha > 1/poly$ and $\beta < \alpha^4/c$ the following holds: If there exists an $\alpha$-correct and $\beta$-OW public-key encryption scheme then there exists an almost-all-keys perfectly-correct IND-CPA secure public-key encryption scheme.*

*Proof.* Let $(G_0, E_0, D_0)$ be an $\alpha$-correct and $\beta$-OW public-key encryption scheme. The conditions of the theorem imply that it is also $(\alpha^4/c)$-OW.

Define $(G_1, E_1, D_1) = (G_0^{\odot}, E_0^{\odot}, D_0^{\odot})$. Lemma 2 implies that $(G_1, E_1, D_1)$ is $(1/2 + \alpha/2)$-correct and $(1/2 + O(\alpha^2/\sqrt{c}))$-OW.

Define $(G_2, E_2, D_2) = (G_1^{\otimes k}, E_1^{\otimes k}, D_1^{\otimes k})$. For any constant $\epsilon > 0$ we can let $k = O(1/\alpha^2)$ (with the constant hidden in the big $O$ notation depending on $\epsilon$), such that Lemma 3 will imply that $(G_2, E_2, D_2)$ is $(1 - \epsilon)$-correct and $(1/2 + O(1/\sqrt{c}))$-OW. Setting $c$ to be a large enough constant implies that $(G_2, E_2, D_2)$ is $(3/4)$-OW. In other words, for any constant $\epsilon > 0$, if $c$ is a large enough constant, there exists a $(1-\epsilon)$-correct and $(3/4)$-OW encryption scheme. Theorem 4 now follows as a corollary of Theorem 3.

### 4.4 Conclusion – Obtaining Non-malleability

As discussed in the introduction, one of the main motivations in dealing with decryption errors is obtaining non-malleability and chosen ciphertext security. As with Corollary 2 we now get from Theorem 4 the following corollary.

**Corollary 5** *There exists some positive constant $c$ such that for any functions $\alpha > 1/poly$ and $\beta < \alpha^4/c$ the following holds: If there exists an $\alpha$-correct and $\beta$-OW public-key encryption scheme and NIZK proof system for NP exist, then there exists an almost-all-keys perfectly-correct NM-CCA-post secure public-key encryption scheme.*

## 5 Dealing with Errors Using Random Oracles

In this section we provide an integrated construction for transforming error-prone public-key encryption schemes with some negligible probability of error that are not necessarily secure against chosen ciphertext attacks into schemes that enjoy non-malleability against a chosen ciphertext attack of the post-processing kind. The advantage over the construction of Section 3 is that it works for *any* negligible probability of error (no need to first decrease the error probability to $2^{-\Omega(n)}$ where $n$ is the message length).

Let $(G, E, D)$ be a public-key encryption scheme that for public key $pk$ maps a message $m \in \{0,1\}^n$ and random coins string $r \in \{0,1\}^\ell$ into a ciphertext $c = E_{pk}(m, r)$ (since we may start with a scheme that is not necessarily semantically secure, we consider also the case of deterministic encryption, so $\ell$ may be 0). We assume without loss of generality that the decryption algorithm $D$ is deterministic[5]. The properties that we assume $E$ satisfies are:

$\alpha$ **correctness and few bad pairs** For a random message $m$ and random $r$ we have $Pr[D_{sk}(E_{pk}(m,r)) \neq m] \leq 1 - \alpha(n)$, where $1 - \alpha(n)$ is negligible. The probability is over the choice of $m, r$. We call a pair $(m, r)$ where $D_{sk}(E_{pk}(m,r)) \neq m$ a *bad* pair. The set of bad pairs is sparse in $\{0,1\}^{n+\ell}$

[5] This may justified, for instance by applying a pseudo-random function to the message in order to obtain the random bits and adding the seed of the function to the secret key.

**One-wayness** For any polynomial time adversary $\mathcal{A}$ and for $c = E_{pk}(m, r)$ for random $m$ and $r$ we have $\Pr_{m,r}[\mathcal{A}(c, pk) = m]$ is negligible. In other words, $E$ is 0-OW.

In addition to the public-key cryptosystem $E$ satisfying the above conditions, we require (i) a shared-key encryption scheme $F_S$ which is NM-CCA-post secure. The keys $S$ are of length $k$ bits. Note that such schemes are easy to construct from pseudo-random functions (see [11]); and (ii) Four functions $H_1 : \{0,1\}^{n/2} \mapsto \{0,1\}^{n/2}$, $H_2 : \{0,1\}^{n/2} \mapsto \{0,1\}^{n/2}$, $H_3 : \{0,1\}^{n/2} \mapsto \{0,1\}^{\ell}$ and $H_4 : \{0,1\}^{n/2} \mapsto \{0,1\}^k$ which will be modelled as ideal random functions. We assume that $n$ is sufficiently large so that $2^{n/2}$ is infeasible.

**Construction 51** *Let $(G, E, D)$ be a public-key encryption scheme, $H_1, H_2, H_3, H_4$ be idealized random functions as above and $F_S$ be shared-key encryption scheme as above.*

**Generation** *$G'$ operates the same as G and generates a public key pk and secret key sk.*

**Encryption** $E'$: *Choose $t \in_R \{0,1\}^{n/2}$. Compute $z = H_1(t)$ and $w = H_2(z) \oplus t$ and $r = H_3(z \circ w)$. The encrypted message is composed of two parts $(c_1, c_2)$:*
- *The generated $c_1 = E_{pk}(z \circ w, r)$*
- *The plaintext m itself is encrypted with the shared-key encryption scheme $F_s$ with key $s = H_4(t)$, i.e. $c_2 = F_s(m)$.*

**Decryption** $D'$: *Given ciphertext $(c_1, c_2)$:*
1. *Apply $D$ to $c_1$ and obtain candidates for $z$ and $w$. Set $t = H_2(z) \oplus w$ and $r = H_3(z \circ w)$.*
2. *Check that $H_1(t) = z$ and that for $r = H_3(z \circ w)$ we have that $c_1 = E(z \circ w, r)$.*
3. *Check, using $s = H_4(t)$, that $c_2$ is a valid ciphertext under $F_s$.*
4. *If any of the tests fails, output invalid ($\perp$). Otherwise, output the decryption of $c_2$ using $s$.*

Note that once $t \in \{0,1\}^{n/2}$ has been chosen, there is unique ciphertext $(c_1, c_2)$ generated from $t$ and encrypting $m$, which we denote $E'_{pk}(m, t)$. Furthermore, for any ciphertext, once the corresponding $t \in \{0,1\}^{n/2}$ is known, it is easy to decrypt the ciphertext *without access to sk*. This is the key for obtaining security against chosen ciphertext attacks (since it is possible to follow the adversary calls to $H_1$).

Why does this process immunize against decryption errors? The point is *not* that the decryption errors have disappeared, but that it is hard to find them. We can partition all strings (of length equal to $|E_{pk}(z \circ w, r)|$) into those that are in the range of $E$ (i.e., such that there exist $m$ and $r$ such that the string is equal to $E_{pk}(m, r)$) and those that are not. Consider a candidate ciphertext $(c_1, c_2)$ that is given to the decryption procedure $D'$. If the prefix of the ciphertext (i.e. $c_1$) is not in the range of $E$, then it is going to be rejected by $D'$ (at Step 2). So the security rests on the hardness of finding among the *bad* pairs $(z \circ w, r)$ one where $r = H_3(z \circ w)$ and $H_1(H_2(z) \oplus w) = z$. This is difficult for any *fixed* (but sparse) set of bad pairs and a random set of functions $H_1, H_2$, and $H_3$ even for an all powerful adversary who is simply restricted in the number of calls to $H_1, H_2$, and $H_3$. In particular, as we will explain, if there are $q_1$ calls to $H_1$ and $q_2$ calls to $H_2$ then the probability that the adversary finds a bad pair that passes the test is bounded by $q_1(1-\alpha) + q_1 q_2/2^{n/2}$. The first term comes from the "natural" method for constructing a pair that satisfies the constraints: Choose an arbitrary $y$. Apply $H_1$ to $y$ and call the result $z$, so that $z = H_1(y)$. Define $w = H_2(z) \oplus y$. Then $r = H_3(z \circ w)$, and we have the pair $(z \circ w, r)$ satisfying the necessary constraints. Note that the pair is completely determined by $y$, once the random oracles are fixed, and the pair is random, because the oracles are random. So for any method of choosing $y$ the probability of hitting a bad pair is $(1-\alpha)$. This gives us the first term. For the second term, suppose during its history the adversary invokes $H_2$ a total of $q_2$ times, say, on inputs $x_1, x_2, \ldots, x_{q_2}$. Let $y$ be arbitrary. Define $w_i = y \oplus H_2(x_i)$, for $i = 1, \ldots, q_2$. We now check to see if $H_1(y) \in \{x_1, \ldots, x_{q_2}\}$. Suppose indeed that $H_1(y) = x_i$ (an event that occurs with probability at most $q_2/2^{n/2}$). Let $z = x_i$. Then we have that $z = H_1(y) = H_1(w_i \oplus H_2(x_i)) = H_1(w_i \oplus H_2(z))$. We let $r = H_3(z \oplus w_i)$ and again we have a pair satisfying the constraints. The total number of pairs we can hope to generate this way is $q_1 q_2/2^{n/2}$.

Why does this process protect against chosen ciphertext attacks? This is very much for the same reason that the Fujisaki-Okamoto [14] scheme is secure. Note that hardness of finding a bad pair is true also for someone knowing the private key $sk$ of $E$, that is *even the creator of the cryptosystem cannot find a bad pair.* Therefore, even under a chosen ciphertext attack w.h.p. a bad pair will not be found. So w.h.p. on all queries given during the attack there is only one response. Furthermore, this response can be given by someone who is aware of the attacker's calls to $H_1$ (by going over all candidates for $t$). The addition of the function $H_4$ and the shared key scheme $F_S$ transforms the system from a one-way scheme into one that is non-malleably secure against chosen ciphertext attacks. From these sketched arguments we get:

**Theorem 6.** *If $(G, E, D)$ is $(1-neg)$-correct and neg-one-way then $(G', E', D')$ is $(1-neg)$-correct and NM-CCA-post secure.*

## 6 Conclusions and Open Problems

We have shown how to eliminate decryption errors in encryption schemes (and even handle non-negligible success probability of the adversary). It is interesting to note that sometimes such ambiguity is actually desirable. This is the case with *deniable encryption* [7], where the goal is, in order to protect the privacy of the conversation, to allow a sender to claim that the plaintext corresponding to a given ciphertext is different than the one actually sent.

As discussed in Section 4, an interesting open problem is to give a transformation that deals with $\alpha$-correct and $\beta$-OW encryption schemes when the gap between $\alpha$ and $\beta$ is very small. For example, we may hope to have $\beta - \alpha$ be an arbitrary constant or even $1/poly$. Nevertheless, as discussed there, having such a strong transformation may involve improving the corresponding transformation in the related information-theoretic setting of [35].

## Acknowledgments

We thank Eran Tromer for initially pointing us to Proos's work, Shafi Goldwasser for raising our interest in the problem and Russell Impagliazzo, Adam Smith and Salil Vadhan for conversations concerning amplification. We thank the anonymous referees for helpful comments.

## References

1. M. Ajtai and C. Dwork, *A public-key cryptosystem with worst-case/average-case equivalence*, Proceedings 29th Annual ACM Symposium on the Theory of Computing, El Paso, TX, 1997, pp. 284–293.
2. M. Bellare, A. Boldyreva and A. Palacio, *A Separation between the Random-Oracle Model and the Standard Model for a Hybrid Encryption Problem*, Cryptology ePrint Archive.
3. M. Bellare, A. Desai, D. Pointcheval and P. Rogaway. *Relations among notions of security for public-key encryption schemes,* Advances in Cryptology – CRYPTO'98, LNCS 1462, Springer, pp. 26–45.
4. M. Bellare, R. Impagliazzo and M. Naor, *Does parallel repetition lower the error in computationally sound protocols?*, in Proceedings 38th Annual IEEE Symposium on Foundations of Computer Science, Miami Beach, FL, 1997, pp. 374–383.
5. M. Bellare and P. Rogaway, P. Optimal Asymmetric Encryption.In Advances in Cryptology - EUROCRYPT '94 (1995), vol. 950 of LNCS, Springer-Verlag, pp. 92111.
6. D. Boneh, *Simplified OAEP for the RSA and Rabin Functions*, Advances in Cryptology - CRYPTO 2001, LNCS2139, Springer 2001, pp. 275–291.
7. R. Canetti, C. Dwork, M. Naor and R. Ostrovsky, *Deniable Encryption,* Advances in Cryptology - CRYPTO'97, LNCS 1294, Springer, 1997, pp. 90–104.
8. R. Canetti, O. Goldreich, and S. Halevi, *The random oracle methodology,* in Proceedings 30th Annual ACM Symposium on the Theory of Computing, Dallas, TX, 1998, pp. 209–218.
9. R. Cramer and V. Shoup, *A practical public key cryptosystem provable secure against adaptive chosen ciphertext attack*, in Advances in Cryptology—Crypto '98, Lecture Notes in Comput. Sci. 1462, Springer-Verlag, New York, 1998, pp. 13–25.
10. R. Cramer and V. Shoup, Universal Hash Proofs and a Paradigm for Adaptive Chosen Ciphertext Secure Public-Key Encryption, Advances in Cryptology – EUROCRYPT 2002, LNCS 2332, pp. 45–64, Springer Verlag, 2002.
11. D. Dolev, C. Dwork and M. Naor, *Non-malleable Cryptography*, Siam J. on Computing, vol 30, 2000, pp. 391–437.
12. C. Dwork, M. Naor, *Zaps and Their Applications*, Proc. 41st IEEE Symposium on Foundations of Computer Science, pp. 283–293. Full version: ECCC, Report TR02-001, `www.eccc.uni-trier.de/eccc/`.
13. C. Dwork, M. Naor, O. Reingold, L. J. Stockmeyer, Magic Functions, Proc. IEEE FOCS 1999, pp. 523–534.

14. E. Fujisaki and T. Okamoto, How to Enhance the Security of Public-Key Encryption at Minimum Cost. In PKC '99 (1999), vol. 1560 of LNCS, Springer-Verlag, pp. 5368.
15. E. Fujisaki, T. Okamoto, D. Pointcheval, J. Stern, *RSA-OAEP Is Secure under the RSA Assumption* Advances in Cryptology – CRYPTO 2001, Springer, 2001, pp. 260–274.
16. O. Goldreich, **Foundations of Cryptography**, Cambridge, 2001.
17. O. Goldreich, **Modern cryptography, probabilistic proofs and pseudorandomness**. *Algorithms and Combinatorics*, vol. 17, Springer-Verlag, 1998.
18. O. Goldreich, S. Goldwasser, and S. Halevi, *Eliminating decryption errors in the Ajtai-Dwork cryptosystem*, Advances in Cryptology – CRYPTO'97, Springer, Lecture Notes in Computer Science, 1294, 1997, pp. 105–111.
19. O. Goldreich and L. Levin, A hard-core predicate for all one-way functions, *Proc. 21st Ann. ACM Symp. on Theory of Computing*, 1989, pp. 25-32.
20. S. Goldwasser and S. Micali, Probabilistic Encryption, *Journal of Computer and System Sciences*, vol. 28, 1984, pp. 270-299.
21. J. Hoffstein, J. Pipher, J. H. Silverman, NTRU: A Ring-Based Public Key Cryptosystem, Algorithmic Number Theory (ANTS III), Portland, OR, June 1998, J.P. Buhler (ed.), Lecture Notes in Computer Science 1423, Springer-Verlag, Berlin, 1998, pp. 267–288.
22. J. Hastad, R. Impagliazzo, L. A. Levin and M. Luby, Construction of a pseudorandom generator from any one-way function, *SIAM Journal on Computing*, vol 28(4), 1999, pp. 1364-1396.
23. N. Howgrave-Graham, P. Nguyen, D. Pointcheval, J. Proos, J. H. Silverman, A. Singer, W. Whyte *The Impact of Decryption Failures on the Security of NTRU Encryption*, Proc. Crypto 2003
24. N. Howgrave-Graham, J. H. Silverman, A. Singer and W. Whyte *NAEP: Provable Security in the Presence of Decryption Failures*, Available: `http://www.ntru.com/cryptolab/pdf/NAEP.pdf`
25. R. Impagliazzo and M. Luby, *One-way functions are essential to computational based cryptography*, in Proceedings 30th IEEE Symposium on the Foundation of Computer Science, Research Triangle Park, NC, 1989, pp. 230–235.
26. C. Lautemann, *BPP and the Polynomial-time Hierarchy*, Information Processing Letters vol. 17(4), 1983, pp. 215–217.
27. Y. Lindell, *A Simpler Construction of CCA2-Secure Public-Key Encryption Under General Assumptions*, Advances in Cryptology—Proceedings Eurocrypt 2003, LNCS 2656, 2003, pp. 241–254.
28. M. Naor, *Bit Commitment Using Pseudorandomness*, J. of Cryptology vol. 4(2), 1991, pp. 151–158.
29. M. Naor and M. Yung, *Public-key cryptosystems provably secure against chosen ciphertext attacks* in Proceedings 22nd Annual ACM Symposium on the Theory of Computing, Baltimore, MD, 1990, pp. 427–437.
30. P. Q. Nguyen, D. Pointcheval: *Analysis and Improvements of NTRU Encryption Paddings*, Advances in Cryptology—Proceedings Crypto'2002, Lecture Notes in Computer Science 2442 Springer 2002, pp. 210–225.
31. T. Okamoto and D. Pointcheval, *REACT: Rapid Enhanced-security Asymmetric Cryptosystem Transform*, In Proc. of CT-RSA'01 (2001), vol. 2020 of LNCS, Springer-Verlag, pp. 159175.
32. V. Shoup, *OAEP Reconsidered*, Journal of Cryptology 15(4): 223-249 (2002).
33. J. Proos, *Imperfect Decryption and an Attack on the NTRU Ecnryption Scheme*, IACR Cryptlogy Archive, Report 02/2003.

34. A. Sahai, *Non-Malleable Non-Interactive Zero Knowledge and Achieving Chosen-Ciphertext Security*, Proc. 40th IEEE Symposium on Foundations of Computer Science, 1999, pp. 543–553.
35. A. Sahai and S. Vadhan, *A Complete Promise Problem for Statistical Zero-Knowledge*, *Proceedings of the 38th Annual Symposium on the Foundations of Computer Science*, 1997, pp.~448–457. Full version: Electronic Colloquium on Computational Complexity TR00-084.

# Secure Hashed Diffie-Hellman over Non-DDH Groups

Rosario Gennaro[1], Hugo Krawczyk[2], and Tal Rabin[1]

[1] IBM T.J.Watson Research Center, PO Box 704, Yorktown Heights, NY 10598, USA
{rosario,talr}@watson.ibm.com

[2] Department of Electrical Engineering, Technion, Haifa 32000, Israel, and
IBM T.J. Watson Research Center, New York, USA
hugo@ee.technion.ac.il

**Abstract.** The Diffie-Hellman (DH) transform is a basic cryptographic primitive used in innumerable cryptographic applications, most prominently in discrete-log based encryption schemes and in the Diffie-Hellman key exchange. In many of these applications it has been recognized that the direct use of the DH output, even over groups that satisfy the strong Decisional Diffie-Hellman (DDH) assumption, may be insecure. This is the case when the application invoking the DH transform requires a value that is pseudo-randomly distributed over a set of strings of some length rather than over the DH group in use. A well-known and general solution is to hash (using a universal hash family) the DH output; we refer to this practice as the "hashed DH transform".

The question that we investigate in this paper is to what extent the DDH assumption is required when applying the hashed DH transform. We show that one can obtain a secure hashed DH transform over a non-DDH group $G$ (i.e., a group in which the DDH assumption does not hold); indeed, we prove that for the hashed DH transform to be secure it suffices that $G$ contain a sufficiently large DDH *subgroup*. As an application of this result, we show that the hashed DH transform is secure over $Z_p^*$ for random prime $p$, provided that the DDH assumption holds over the large prime-order subgroups of $Z_p^*$. In particular, we obtain the same security working directly over $Z_p^*$ as working over prime-order subgroups, without requiring any knowledge of the prime factorization of $p-1$ and without even having to find a generator of $Z_p^*$.

Further contributions of the paper to the study of the DDH assumption include: the introduction of a DDH relaxation, via computational entropy, which we call the "$t$-DDH assumption" and which plays a central role in obtaining the above results; a characterization of DDH groups in terms of their DDH subgroups; and the analysis of of the DDH (and $t$-DDH) assumptions when using short exponents.

## 1 Introduction

**The Diffie-Hellman Transform and DDH Assumption.** The *Diffie-Hellman transform* is one of the best-known and fundamental cryptographic primitives. Its discovery by Whitfield Diffie and Martin Hellman [DH76] revolutionized

C. Cachin and J. Camenisch (Eds.): EUROCRYPT 2004, LNCS 3027, pp. 361–381, 2004.

the science of cryptography and marked the birth of Modern Cryptography. Even today, almost 30 years later, the DH transform remains the basis of some of the most widely used cryptographic techniques. In particular, it underlies the Diffie-Hellman key exchange and the ElGamal encryption scheme [ElG85], and is used over a large variety of mathematical groups. In its basic form the Diffie-Hellman (or DH for short) transform maps a pair of elements $g^a, g^b$ drawn from a cyclic group $G$ generated by the element $g$ into the group element[3] $g^{ab}$. The usefulness of this transform was originally envisioned under the conjecture, known as the *Computational Diffie-Hellman (CDH)* assumption, that states the infeasibility of computing the value $g^{ab}$ given only the exponentials $g^a$ and $g^b$. Namely, the value $g^{ab}$ should be computable only by those knowing one of the exponents $a$ or $b$. Note that the CDH assumption implies the difficulty of computing discrete logarithms over the group $G$ (the converse, however, is unknown for most practical groups).

Over time it was realized that the CDH assumption is insufficient to guarantee the security of most DH applications (in particular those mentioned above). For this reason a much stronger assumption was introduced: the *Decisional Diffie-Hellman (DDH)* assumption postulates that given the values $g^a$ and $g^b$ not only it is computationally hard to derive the value $g^{ab}$ but even the seemingly much easier task of distinguishing $g^{ab}$ from random group elements is infeasible [Bra93] (see [Bon98] for a survey on the DDH assumption). On the basis of this assumption one can consider the DH transform as a good generator of pseudorandomness as required in key-exchange, encryption and other cryptographic applications. Hereafter we refer to groups in which the DDH assumption holds as *DDH groups*. The need to rely on the DDH disqualifies many natural groups where the assumption does not hold. For example, any group whose order is divisible by small factors, such as the classic groups $Z_p^*$ of residues modulo a large prime $p$; in this case the group's order, $p-1$, is always divisible by 2 and thus the DDH assumption does not hold. Moreover, for randomly generated primes $p$, $p-1$ has (with very high probability) additional small factors. Due to the *perceived* need to work over DDH groups it is often recommended in the cryptographic literature that one work over subgroups of large prime order where no attacks are known on the DDH assumption.

**The Need for Hashing the Diffie-Hellman Result.** Interestingly, the DDH assumption, while apparently necessary, turns out to be insufficient for guaranteeing the security of some of the most basic applications of the DH transform. Consider for example the ElGamal encryption scheme: Given a public key $y = g^a$ (for secret $a$), a message $m \in G$ is encrypted by the pair $(g^b, my^b)$ where the value $b$ is chosen randomly anew for each encryption. In this case, the DDH assumption guarantees the semantic security ([GM84]) of the scheme (against chosen-plaintext attacks) *provided that the plaintexts $m$ are elements of the group $G$*. However, if the message space is different, e.g. the set of strings of some length

[3] Here we use the exponential notation that originates with multiplicative groups but our treatment applies equally to additive groups such as Elliptic Curves.

smaller than $\log(|G|)$), then the above encryption scheme becomes problematic. First of all, you need to encode messages $m$ as group elements in $G$ and that could be cumbersome. If $G$ is a subgroup of prime order of $Z_p^*$, a naive (and common) approach would be to trivially encode $m$ as an integer and perform the multiplication $my^b$ modulo $p$. But now the scheme is *insecure even if the group $G$ does satisfy the DDH assumption.* A good illustration of the potential weaknesses of this straightforward (or "textbook") application of ElGamal is presented in [BJN00]. It is shown that if the space of plaintexts consists of random strings of length shorter than $|G|$ (e.g., when using public key encryption to encrypt symmetric keys) the above scheme turns out to be insecure even under a ciphertext-only attack and, as said, even if the group $G$ is DDH. For example, if the plaintexts to be encrypted are keys of length 64, an attacker that sees a ciphertext has a significant probability of finding the plaintext with a work factor in the order of $2^{32}$ operations and comparable memory; for encrypted keys of length 128 the complexity of finding the key is reduced to $2^{64}$.

A general and practical approach to solving these serious security weaknesses is to avoid using the DH value itself to "mask" $m$ via multiplication, but rather to *hash* the DH value $g^{ab}$ to obtain a pseudorandom key $K$ of suitable length which can then be used to encrypt the message $m$ under a particular encryption function (in particular, $K$ can be used as a one-time pad). In this case the hash function is used to *extract the (pseudo) randomness* present in the DH value. Suitable hash functions with provable extraction properties are known, for example *universal hash functions* [CW79,HILL99]. The above considerations are common to many other applications of the DH transform, including encryption schemes secure against chosen-ciphertext attacks [CS98] and, most prominently, the Diffie-Hellman key-exchange protocol (in the latter case one should not use the DH output as a cryptographic key but rather derive the agreed shared keys via a hashing of the DH result); see Section 3.2 for a discussion on how these applications choose a random hash function out of a given family. For additional examples and justification of the need for hashing the DH output see [Bon98,NR97,CS98,ABR01]. In the sequel we refer to the combination of the DH transform with a (universal) hash function as the *hashed DH transform*.

### 1.1 Our Results

**The Security of the Hashed DH Transform over non-DDH Groups.** In light of the need to hash the DH value, some natural questions arise: when applying the hashed DH transform, is it still necessary to work over groups where the DDH assumption holds, or can this requirement be relaxed? Can one obtain a secure (hashed) DH transform over a non-DDH group, and specifically, is doing hashed DH over $Z_p^*$ secure? In this paper we provide answers to these questions. Our main result can be informally stated as follows: *For any cyclic group $G$, applying the hashed DH transform over $G$ has the same security as applying the hashed DH transform directly over the maximal DDH subgroup of $G$.* In particular, one can obtain secure applications of the hashed DH transform over non-DDH groups; the only requirement is that $G$ contain a (sufficiently large) DDH subgroup (see below for the exact meaning of "sufficiently large" and other parameter size considerations). A significant point is that we are only concerned with the *existence* of such a subgroup; there is no need to know the exact size or structural properties of, nor to be able to construct, this specific (maximal) DDH subgroup.

A particularly interesting consequence of the above result is that assuming that DDH holds on large subgroups of $Z_p^*$ (we will see later that it is sufficient to assume that DDH holds on large prime-order subgroups of $Z_p^*$), one can build secure (hashed) DH applications working directly over $Z_p^*$, where $p$ is an unconstrained random prime. Only the length of the prime is specified, while other common requirements such as the knowledge of the partial or full factorization of $p-1$, insisting that $p-1$ has a prime factor of a particular size, or disqualifying primes for which $(p-1)/2$ has a smooth part, are all avoided here. Moreover, we show that there is no need to find a generator of $Z_p^*$; instead we prove that a randomly chosen element from $Z_p^*$ will span a (probably non-DDH) subgroup with a large enough DDH subgroup. In particular, the DH security is preserved even if the order of the chosen element has small factors or if it misses some prime divisors of $p-1$. Note that avoiding the need to find a generator for $Z_p^*$ allows us to work with primes $p$ with unknown factorization of $p-1$ (which is otherwise required to find a $Z_p^*$ generator).

**The $t$-DDH Assumption.** In order to prove our main result (i.e., that the hashed DH transform is secure over any group $G$, not necessarily a DDH group, that contains a large enough DDH subgroup), we introduce a relaxation of the DDH assumption which we call the *$t$-DDH assumption*. Informally, a group $G$ satisfies the $t$-DDH assumption (where $0 \leq t \leq |G|$) if given the pair $(g^a, g^b)$ (where $g$ is a generator of $G$) the value $g^{ab}$ contains *$t$ bits of computational entropy*. The notion of computational entropy, introduced in [HILL99], captures the amount of computational hardness present in a probability distribution. In other words, we relax the "full hardness" requirement at the core of the DDH assumption, and assume partial hardness only. Moreover, we do not care about the exact subsets of bits or group elements where this hardness is contained, but only assume their existence. On this basis, and using the entropy-smoothing theorem from [HILL99] (also known as the leftover hash lemma), we obtain a way to efficiently transform (via universal hashing) DH values over groups in which the $t$-DDH assumption holds into shorter outputs that are computationally indistinguishable from the uniform distribution. The maximal length of (pseudorandom) strings that one can obtain as output from the hashed DH transform depends on the maximum value of $t$ for which the $t$-DDH holds in $G$. In particular, in order to be $2^{-k}$-computationally close to uniform one can output up to $t-2k$ pseudorandom bits (e.g., to produce 128-bit keys with a security parameter of $k = 80$ the group $G$ should be 288-DDH, while for $k = 128$, $G$ is to be 384-DDH).

After defining the $t$-DDH assumption and showing its usefulness in extracting random bits from $t$-DDH groups, we show that *if $G$ contains a DDH subgroup of order $m$ then $G$ is $|m|$-DDH*. This forms the basis for our main result as stated above. Indeed, it suffices that $G$ has a suitably large-order DDH subgroup to ensure that hashing the DH output results in pseudorandom outputs of the required length. Again, it is important to stress that we do not need to know the specific DDH subgroup or its order, only (assume) its existence.

**A Direct Product Characterization of the DDH Assumption.** A further contribution of our work is in providing a characterization of the DDH assumption in a given group in terms of its DDH subgroups. Specifically, we show that *a group is DDH if and only if it is the direct product of (disjoint) prime power DDH groups.* In other words, a group $G$ is DDH if and only if all its prime power subgroups are DDH. Moreover, for any cyclic group $G$, the maximal DDH group in $G$ is obtained as the product of all prime power DDH subgroups in $G$. Beyond its independent interest, this result plays a central role in our proof that the hashed DH transform over $Z_p^*$ is secure as long as the DDH assumption holds in the subgroups of $Z_p^*$ of large prime order. In particular, this allows us to expand significantly the groups in which one can work securely with the hashed DH transform without having to strengthen the usual assumption that DDH holds in large prime order subgroups.

**Some Practical Considerations.** Beyond the theoretical interest in understanding the role of the DDH assumption and proving the usefulness of relaxed assumptions, our results point out some practical issues that are worth discussing. In this respect, one significant contribution is the justification of the use of non-DDH groups in applications of DH that hash their output. It needs to be noted that in spite of an extensive crypto literature regarding the use of prime order subgroups for performing DH, many real-world instantiations of this primitive work over non-DDH groups (e.g. $Z_p^*$). Examples include the widespread SSH and IPsec standards. Interestingly, the latter has standardized a set of groups for use with the IKE Diffie-Hellman key-exchange protocols [RFC2409], none of which constitute a DDH group. However, since the IKE protocol takes care of hashing the output of the DH transform before generating the cryptographic keys (see [Kra03]), then our results serve to justify the security of this mechanism[4].

In addition, and as pointed out before, our results also show that under the sole assumption that the DDH holds in groups of large prime order one can work directly over $Z_p^*$ for a random prime $p$, without having to know the factorization of $p-1$ and without having to find a generator of $Z_p^*$. Moreover, the ability to work over non-prime order groups has the benefit of eliminating the attacks on the DH transform described in [LL97], without having to search for primes of a special form (and without necessitating special parameter checks when certifying public keys [LL97]).

[4] In IKE, the family of hash functions used for extracting a pseudorandom key from the DH value are implemented using common pseudorandom function families keyed with random, but known, keys. The randomness extraction properties of the latter families are studied in [GHKR04].

**Short-Exponent Diffie-Hellman.** One important practical consideration is the length of exponents used when applying the DH transform. Full exponents when working over $Z_p^*$ are, typically, of size 1024 or more. Even if one works over a prime-order subgroup, one still needs to use relatively large orders (e.g. 288-bit long primes), with their correspondingly large exponents, to ensure a hashed output (say of 128 bits) that is indistinguishable from uniform. (This requirement for large computational entropy is often overlooked; indeed, the usual practice of using 160-bit prime-order groups, which originates with Schnorr's signatures, is inappropriate for hashed DH-type applications.)

Motivated by the significant cost of exponentiation using long exponents, we investigate whether one can use short exponents (e.g. as in [RFC2409]) and still preserve the security of the hashed DH transform. An obviously necessary requirement for the short exponent practice to be secure is the assumption that the discrete log problem is hard when exponents are restricted to a short length (say of $s$ bits). We show that this requirement (called the $s$-DLSE assumption) is sufficient for the secure use of short exponents in the setting of the DH transform; more precisely, we prove (based on [Gen00]) that if the $s$-DLSE assumption holds in a group $G$, then the hashed DH transform in $G$ is as secure with full exponents as with $s$-bit exponents. As a consequence, one can analyze the security of the hashed DH transform in the group $G$ with full exponents and later replace the full exponents with much shorter ones without sacrificing security. In this case the important parameter is $s$; we note that the appropriate value of $s$ depends on the underlying group. See [vOW96] for an extensive study of the plausible value of $s$ for different groups.

**Paper's Organization.** In Section 2 we recall the DDH Assumption and prove the DDH Characterization Theorem. In Section 3 we introduce the $t$-DDH Assumption and its application to the hashed DH transform, and prove the central Max-Subgroup Theorem. In Section 4 we investigate the security of the hashed DH transform when using short exponents. We conclude in Section 5 by describing the applicability of our results to the hashed DH transform over non-DDH groups.

**Notation.** The formal treatment in this paper often involves sequences of probability distributions $\{\mathcal{D}_n\}_{n\in\mathbb{N}}$ to which we refer as *probability ensembles* (or simply as "ensembles"). We adopt the convention that by the "probability distribution $\mathcal{D}_n$" we mean the specific element (distribution) $\mathcal{D}_n$ in the above sequence, while the term "probability ensemble $\mathcal{D}_n$" is short for "probability ensemble $\{\mathcal{D}_n\}_{n\in\mathbb{N}}$". We also assume that each distribution $\mathcal{D}_n$ is taken over a set $A_n \subset \{0,1\}^{n'}$ where $n'$ is polynomial in $n$ (i.e., each ensemble has a fixed polynomial in $n$ that determines the value $n'$). The notation $x \in_{\mathcal{D}_n} A_n$ is to be read as $x$ chosen in $A_n$ according to the distribution $\mathcal{D}_n$, and $x \in_R S$ means choosing $x$ with uniform distribution over the set $S$. Finally if $m$ is an integer, we denote with $|m|$ its binary length.

## 2 A Direct-Product DDH Characterization

We consider a (infinite) family of cyclic groups $\mathcal{G} = \{G_n\}_n$. Denote with $g_n$ and $m_n$ a generator and the order of $G_n$, respectively, where $|m_n|$ is bounded by a polynomial in $n$.

Consider the following problem: Given a pair $g_n^a, g_n^b$ compute the value $g_n^{ab}$. If this problem is intractable over a family $\mathcal{G}$ then we say that the Computational Diffie-Hellman (CDH) assumption holds (over $\mathcal{G}$).

A much stronger, but also more useful, assumption is the following. Consider the family of sets $G_n^3 = G_n \times G_n \times G_n$ and the following two probability ensembles over it:

$$\mathcal{R}_n = \{(g_n^a,\, g_n^b,\, g_n^c) \;\text{ for }\; a, b, c \in_R [0..m_n]\}$$

and

$$\mathcal{DH}_n = \{(g_n^a,\, g_n^b,\, g_n^{ab}) \;\text{ for }\; a, b \in_R [0..m_n]\}$$

**Definition 1.** *We say that the* Decisional Diffie-Hellman (DDH) Assumption *holds over $\mathcal{G}$ if the ensembles $\mathcal{R}_n$ and $\mathcal{DH}_n$ are computationally indistinguishable (with respect to non-uniform distinguishers)*[5]*. If $\mathcal{G}$ satisfies the DDH assumption, we call $\mathcal{G}$ a* DDH group (family).

Informally what the above assumption requires is that no polynomial time judge can *decide* if the third element of the triple $(g_n^a, g_n^b, g_n^c)$ is the result of the Diffie-Hellman transform applied to $g_n^a, g_n^b$ or a randomly chosen group element. Clearly this is a much weaker requirement from the attacker than computing the value $g_n^{ab}$ from $g_n^a, g_n^b$. And therefore, as a general hardness assumption, DDH is (much) stronger than the CDH.

The group family $\mathcal{G}$ over which the two distributions $\mathcal{R}_n$ and $\mathcal{DH}_n$ are defined is very important and indeed it makes a difference for the validity of the assumption.

**Example 1:** *A group where the DDH assumption does not hold.* Consider the following group family; for each $n$ take an $n$-bit prime $p_n$ and the group $G_n = Z_{p_n}^*$. Since testing for quadratic residuosity over $Z_{p_n}^*$ is easy, by computing $(\frac{\cdot}{p_n})$ (the Legendre symbol), then we immediately get a distinguisher against DDH in this group: by mapping the Legendre symbol of 1 (i.e. quadratic residues) to 0, and the Legendre symbol of -1 to 1, we can simply check that $(\frac{g_n^a}{p_n})(\frac{g_n^b}{p_n}) = (\frac{g_n^c}{p_n})$, and output "$\mathcal{DH}_n$" if it holds and "$\mathcal{R}_n$" otherwise. Clearly, if the triple is a legal DH triple then the distinguisher outputs $\mathcal{DH}_n$ with probability 1, while in the other case the probability is only 1/2.

**Example 2:** *A group where the DDH is conjectured to hold.* For each integer $n$ consider an $n$-bit prime $q_n$ and $poly(n)$-bit prime $p_n$ such that $q_n$ divides $p_n - 1$.

[5] The notion of computational indistinguishability is recalled in Appendix A; see also the remark below regarding our non-uniform formalism.

The group $G_n$ is the subgroup of prime order $q_n$ in $Z^*_{p_n}$. In this case no efficient distinguisher against the DDH is known.

**An important remark about our formalism.** We assume a notion of computational indistinguishability under non-uniform distinguishers. In particular, such a distinguisher may be given an "auxiliary input" for each group $G_n$ in the family $\mathcal{G}$. This approach allows us to keep the simplicity of arguments in the asymptotic polynomial-time model while capturing the fact that we are interested in the security of individual groups for which the attacker may have some side information. A particularly important example of such "side information" is the possible knowledge by the attacker of the group order and its factorization. Our results do assume that such factorization may be given to the attacker (as part of the non-uniform auxiliary input). In particular, this assumption plays an important role in the proof of the following theorem, which does not necessarily hold when the factorization of $ord(G)$ is unknown (as it may be the case when working over $Z^*_N$ where $N = pq$ is a modulus of unknown factorization).
Due to our focus on the security of specific groups we will often omit the subscript $n$ in the notation of groups, generators, etc.

The next theorem provides a full characterization of DDH groups in terms of their prime order subgroups (as remarked above, the proof of this theorem assumes that the distinguisher is given the factorization of $ord(G)$).

**Theorem 1 (Direct Product Characterization Theorem.).** *A cyclic group $G$ is DDH if and only if all its prime power subgroups are DDH.*

The proof follows from Lemmas 1 and 2.

**Lemma 1.** *If the DDH assumptions holds in a group $G$ then it holds in all the subgroups of $G$.*

*Proof.* Let $G$ be a DDH (cyclic) group of order order $m = m_1m_2$, and let $G_1$ be a subgroup of $G$ of order $m_1$. Let $g$ be a generator of $G$ and $g_1 = g^{m_2}$ be a generator of $G_1$. Assume by contradiction that the DDH does not hold in $G_1$, i.e. there is a distinguisher $D_1$ that upon receiving a triple $(A_1 = g_1^{a_1}, B_1 = g_1^{b_1}, C_1 = g_1^{c_1}) \in G_1^3$, can distinguish whether it came from the distribution $\mathcal{R}_{G_1}$ or $\mathcal{DH}_{G_1}$ with non-negligible advantage $\epsilon$. We build a distinguisher $D$ for $G$ which distinguishes between the distributions $\mathcal{DH}_G$ and $\mathcal{R}_G$ with the same probability $\epsilon$.

Upon receiving a triple $(A = g^a, B = g^b, C = g^c)$, where $a, b \in_R Z_{m_1m_2}$ and $c$ is either the product of $ab$ or picked uniformly at random in $Z_{m_1m_2}$, the distinguisher $D$ :

1. Computes $(A_1, B_1, C_1)$ by setting $A_1 = A^{m_2}, B_1 = B^{m_2}$, and $C_1 = C^{m_2}$.
2. Passes the triple $(A_1, B_1, C_1)$ to $D_1$
3. Outputs the same output bit as $D_1$.

Note that by construction the values $A_1, B_1, C_1$ equal $g_1^{a_1}, g_1^{b_1}, g_1^{c_1}$, respectively, where $a_1 = a \bmod m_1, b_1 = b \bmod m_1, c_1 = c \bmod m_1$. Since $a, b \in_R Z_{m_1m_2}$

then $a_1, b_1 \in_R Z_{m_1}$. Also, if $c = ab \bmod m_1 m_2$ then $c_1 = a_1 b_1 \bmod m_1$, while if $c \in_R Z_{m_1 m_2}$ then $c_1 \in_R Z_{m_1}$. In other words, whenever the triple $(A, B, C)$ is distributed according to $\mathcal{DH}_G$ then the triple $(A_1, B_1, C_1)$ is distributed according to $\mathcal{DH}_{G_1}$, while if $(A, B, C)$ is distributed according to $\mathcal{R}_G$ then the triple $(A_1, B_1, C_1)$ is distributed according to $\mathcal{R}_{G_1}$. Therefore, $D$ distinguishes between the distributions $\mathcal{DH}_G$ and $\mathcal{R}_G$ with the same probability $\epsilon$ that $D_1$ distinguishes between $\mathcal{DH}_{G_1}$ and $\mathcal{R}_{G_1}$. □

**Lemma 2.** *Let $G$ be a cyclic group of order $m = m_1 m_2$, where $(m_1, m_2) = 1$, and let $G_1$ and $G_2$ be the subgroups of $G$ of orders $m_1, m_2$ resp. If DDH holds in $G_1$ and $G_2$ then DDH holds in $G$.*

*Proof.* Let $g, g_1, g_2$ be generators of $G, G_1$, and $G_2$, respectively; in particular, $g_1 = g^{m_2}$ and $g_2 = g^{m_1}$. Given a triple $t_1 = (A_1 = g_1^{a_1}, B_1 = g_1^{b_1}, C_1 = g_1^{c_1}) \in G_1^3$ and a triple $t_2 = (A_2 = g_2^{a_2}, B_2 = g_2^{b_2}, C_2 = g_2^{c_2}) \in G_2^3$ we define the following transformation $T$ which "lifts" this pair of triples into a triple in $G^3$. ($T$ is the standard isomorphism between the group $G$ and its product group representation as determined by the Chinese Reminder Theorem.) On input $t_1, t_2$, $T(t_1, t_2)$ outputs a triple $(A = g^a, B = g^b, C = g^c) \in G^3$ defined as follows:

1. Let $r_1, r_2$ be such that $r_1 m_1 + r_2 m_2 = 1$ (i.e., $r_1 = m_1^{-1} \bmod m_2$ and $r_2 = m_2^{-1} \bmod m_1$)
2. Set $A = A_1^{r_2} A_2^{r_1} = g^{a_1 m_2 r_2 + a_2 m_1 r_1} \in G$, i.e. $a = a_1 m_2 r_2 + a_2 m_1 r_1 \bmod m$
3. Set $B = B_1^{r_2} B_2^{r_1} = g^{b_1 m_2 r_2 + b_2 m_1 r_1} \in G$, i.e. $b = b_1 m_2 r_2 + b_2 m_1 r_1 \bmod m$
4. Set $C = C_1^{m_2 r_2^2} C_2^{m_1 r_1^2} = g^{c_1 m_2^2 r_2^2 + c_2 m_1^2 r_1^2} \in G$, i.e. $c = c_1 m_2^2 r_2^2 + c_2 m_1^2 r_1^2 \bmod m$

Note the following facts about the triple $(A, B, C)$ which result from the above transformation:

**Fact 1** If $a_1, b_1 \in_R Z_{m_1}$, and $a_2, b_2 \in_R Z_{m_2}$, then $a, b \in_R Z_m$.
**Fact 2** $c - ab \equiv c_1 - a_1 b_1 \bmod m_1$ and $c - ab \equiv c_2 - a_2 b_2 \bmod m_2$
**Fact 3** Following Facts 1 and 2, if the triple $t_1$ is chosen according to distribution $\mathcal{DH}_{G_1}$ and $t_2$ according to distribution $\mathcal{DH}_{G_2}$, then the triple $(A, B, C)$ is distributed according to the distribution $\mathcal{DH}_G$. Similarly, if $t_1, t_2$ are distributed according to $\mathcal{R}_{G_1}$ and $\mathcal{R}_{G_2}$, respectively, then $(A, B, C)$ is distributed according to $\mathcal{R}_G$.

For probability distributions $\mathcal{P}_1, \mathcal{P}_2$ we denote by $T(\mathcal{P}_1, \mathcal{P}_2)$ the probability distribution induced by the random variable $T(x_1, x_2)$ where $x_1, x_2$ are random variables distributed according to $\mathcal{P}_1, \mathcal{P}_2$, respectively, and $T$ is the above defined transform. Using this notation and Fact 3 we get: $\mathcal{DH}_G = T(\mathcal{DH}_{G_1}, \mathcal{DH}_{G_2})$ and $\mathcal{R}_G = T(\mathcal{R}_{G_1}, \mathcal{R}_{G_2})$.

Let us now consider the "hybrid" probability distribution $T(\mathcal{R}_{G_1}, \mathcal{DH}_{G_2})$. Note that this distribution is computationally indistinguishable from $T(\mathcal{DH}_{G_1}, \mathcal{DH}_{G_2})$. Indeed, since the distribution $\mathcal{DH}_{G_2}$ is efficiently samplable and the transformation $T$ is efficiently computable, then one can transform any efficient distinguisher between the above two distributions into an efficient distinguisher between $\mathcal{R}_{G_1}$ and $\mathcal{DH}_{G_1}$, in contradiction to the Lemma's premise that the distributions $\mathcal{R}_{G_1}$ and $\mathcal{DH}_{G_1}$ are indistinguishable. Similarly, we have that the hybrid distribution $T(\mathcal{R}_{G_1}, \mathcal{DH}_{G_2})$ is indistinguishable from $T(\mathcal{R}_{G_1}, \mathcal{R}_{G_2})$. Summarizing, we have that:

$$\mathcal{DH}_G = T(\mathcal{DH}_{G_1}, \mathcal{DH}_{G_2}) \overset{c}{\approx} T(\mathcal{R}_{G_1}, \mathcal{DH}_{G_2}) \overset{c}{\approx} T(\mathcal{R}_{G_1}, \mathcal{R}_{G_2}) = \mathcal{R}_G$$

where $\overset{c}{\approx}$ denotes computational indistinguishability. Therefore by a standard hybrid argument (or the triangle inequality for computational indistinguishability) we get that, provided that the DDH assumption holds in $G_1$ and $G_2$, then $\mathcal{DH}_G$ and $\mathcal{R}_G$ are computationally indistinguishable, i.e. $G$ is DDH. □

## 3 The $t$-DDH Assumption and the Hashed DH Transform

In this section we introduce an intractability assumption that is, in general, weaker than the DDH assumption, yet it suffices for ensuring DH outputs from which a large number of pseudorandom bits can be extracted. We start by recalling the notions of computational entropy and entropy smoothing. We use the notations introduced at the end of Section 1.

### 3.1 Computational Entropy and Entropy Smoothing

**Definition 2.** *Let $\mathcal{X}_n$ be a probability ensemble over $A_n$. The min-entropy of $\mathcal{X}_n$ is the value*

$$\textsf{min-ent}(\mathcal{X}_n) = \textsf{min}_{x \in A_n : Prob_{\mathcal{X}_n}[x] \neq 0}(-\log(Prob_{\mathcal{X}_n}[x]))$$

Note that if $\mathcal{X}_n$ has min-entropy $t(n)$ then for all $x \in A_n$, $Prob_{\mathcal{X}_n}[x] \leq 2^{-t(n)}$.

The notion of min-entropy provides a measurement of the amount of randomness present in a probability distribution. Indeed, the Entropy Smoothing Theorem (see below) shows that if $\mathcal{X}_n$ has min-entropy $t(n)$ it is possible to construct from $\mathcal{X}_n$ an (almost) uniform distribution over (almost) $t(n)$ bits, by simply hashing elements chosen according to $\mathcal{X}_n$. The basic hashing tool to do this uses the following notion of universal hashing.

**Definition 3.** *Let $\mathcal{H}_n$ be a family of functions, where each $H \in \mathcal{H}_n$ is defined as $H : A_n \to \{0,1\}^{m(n)}$. We say that $\mathcal{H}_n$ is a* family of (pairwise-independent) universal hash functions *if, for all $x, x' \in A_n$, $x \neq x'$, and for all $a, a' \in \{0,1\}^{m(n)}$ we have*

$$Prob_{H \in \mathcal{H}_n}[H(x) = a \text{ and } H(x') = a'] = 2^{-2m(n)}.$$

*That is, a randomly chosen $H$ will map any pair of distinct elements independently and uniformly.*

Our techniques use as a central tool the following Entropy Smoothing Theorem from [HILL99] (see also [Lub96]). The definition of statistical distance used below is recalled in Appendix A.

**Theorem 2 (Entropy Smoothing Theorem** [HILL99]**.).** *Let $t$ be a positive integer and let $\mathcal{X}$ be a random variable defined on $\{0,1\}^n$ such that* $\mathsf{min\text{-}ent}(\mathcal{X}) > t$. *Let $k > 0$ be an integer parameter. Let $\mathcal{H}$ be a family of universal hash functions such that $h \in \mathcal{H}$, $h : \{0,1\}^n \to \{0,1\}^{t-2k}$. Let $\mathcal{U}$ be the uniform distribution over $\{0,1\}^{t-2k}$. Then, the distributions $[< h(\mathcal{X}), h >]_{h \in_R \mathcal{H}}$ and $[< \mathcal{U}, h >]_{h \in_R \mathcal{H}}$ have statistical distance at most $2^{-(k+1)}$.*

Thus, the Entropy Smoothing Theorem guarantees that if $\mathcal{X}_n$ is a probability ensemble over $A_n$ with min-entropy of at least $t(n)$, and $\mathcal{H}_n$ a family of universal hash functions from $A_n$ to $\{0,1\}^{t(n)-2k(n)}$, then the random variable $H(x)$, where $H \in_R \mathcal{H}_n$ and $x$ is chosen according to the distribution $\mathcal{X}_n$, is "almost" uniformly distributed over $\{0,1\}^{t(n)-2k(n)}$ even when the hash function $H$ is given. Here, "almost" means a statistical distance of at most $2^{-k(n)}$. Therefore, if one sets $k(n) = \omega(\log n)$, then the statistical distance of $H(x)$ from uniform becomes negligible.

The following notion represents a computational analogue of the notion of min-entropy; it is due to [HILL99].

**Definition 4.** *A probability ensemble $\mathcal{Y}_n$ has* computational entropy *$t(n)$ if there exists a probability ensemble $\mathcal{X}_n$ such that*

- $\mathsf{min\text{-}ent}(\mathcal{X}_n) \geq t(n)$
- *$\mathcal{X}_n$ and $\mathcal{Y}_n$ are computationally indistinguishable*

Using a standard hybrid argument it is easy to show that the Entropy Smoothing Theorem, as discussed above, can be generalized to probability ensembles $\mathcal{X}_n$ that have computational entropy $t(n)$. In this case, applying a (randomly chosen) universal hash function with $k(n) = \omega(\log n)$ to $\mathcal{X}_n$ results in a *pseudorandom* ensemble, namely, an ensemble which is computationally indistinguishable from the uniform distribution.

### 3.2 *t*-DDH: A Relaxed DDH Assumption

We proceed to define the $t$-DDH assumption. The intuition behind this assumption is that if the Computational Diffie-Hellman Assumption holds in a group $G$ generated by a generator $g$, then the DH value $g^{ab}$ must have some degree of unpredictability (or "partial hardness") even when $g^a$ and $g^b$ are given. Specifically, we say that the $t$-DDH Assumption holds in the group $G$ if the Diffie-Hellman output $g^{ab}$ has $t$ bits of computational entropy (here $0 \leq t \leq \log(G)$). Formally:

**Definition 5.** *We say that the $t(n)$-DDH Assumption holds over a group family $\mathcal{G} = \{G_n\}_n$ if for all $n$ there exists a family of probability distributions $\mathcal{X}_n(g_n^a, g_n^b)$ over $G_n$ (one distribution for each pair $g_n^a, g_n^b$) such that*

– $\text{min-ent}(\mathcal{X}_n(g_n^a, g_n^b)) \geq t(n)$
– *The probability ensemble $\mathcal{DH}_n$ (see Section 2) is computationally indistinguishable from the ensemble*

$$\mathcal{R}_n^* = \{(g_n^a, g_n^b, C) \text{ for } a, b \in_R \text{ord}(G_n) \text{ and } C \in_{\mathcal{X}_n(g_n^a, g_n^b)} G_n\}$$

It is important to note that the distributions $\mathcal{X}_n(g^a, g^b)$ in the above definition may be different for each pair of values $g^a, g^b$. Requiring instead a single distribution $\mathcal{X}$ for all pairs $g^a, g^b$ (as may seem more natural at first glance) results in a significantly stronger, and consequently less useful, assumption.

Consider Example 1 from Section 2: over $Z_p^*$ one can break the DDH by detecting if the quadratic residuosity character of $C$ is consistent with the one induced by $g^a, g^b$. Yet, $Z_p^*$ can satisfy the $t$-DDH assumption even for high values of $t$. For example, if for all $a, b$ for which one of $a, b$ is even we define $\mathcal{X}_n(g^a, g^b)$ to be the set of quadratic residues in $Z_p^*$, and for all other pairs $g^a, g^b$ we define $\mathcal{X}_n(g^a, g^b)$ to be the set of quadratic non-residues in $Z_p^*$, then the trivial break of DDH in the above example does not hold against these distributions. More generally, if we consider a prime $p$ of the form $2^u q + 1$ where $q$ is a prime then we can get that (given current knowledge) the $t$-DDH assumption holds for $Z_p^*$ for $t = |p| - u$, while clearly the DDH assumptions does not hold over this group.

Note that the DDH assumption can also be stated in terms of computational entropy. Indeed the DDH assumption over a group $G$ is equivalent to the $t$-DDH assumption over $G$ for $t = \log(\text{ord}(G))$.

**Sampling $\mathcal{X}_n(g^a, g^b)$.** The $t$-DDH Assumption as stated above makes no requirement of efficient samplability for $\mathcal{X}_n(g^a, g^b)$. It is possible to strengthen the assumption by requiring that $\mathcal{X}_n(g^a, g^b)$ be efficiently samplable. We say that the samplable [resp. semi-samplable] $t$-DDH Assumption holds over $\mathcal{G}$, if the $t$-DDH Assumption holds over $\mathcal{G}$ and the underlying distributions $\mathcal{X}_n(g^a, g^b)$ are polynomial-time samplable [resp. polynomial-time samplable when either exponent $a$ or $b$ is known].

As a direct consequence of the Entropy Smoothing Theorem and the definition of $t$-DDH we have:

**Lemma 3.** *Let $\mathcal{G} = \{G_n\}_n$ be a group family in which the $t(n)$-DDH Assumption holds, and let $\{\mathcal{H}_n\}_n$ be a family of universal hash functions such that for all $h \in \mathcal{H}_n$, $h : G_n \to \{0,1\}^{t'(n)}$ where $t'(n) = t(n) - \omega(\log n)$. Then the induced distribution of $h(g_n^{ab})$, for $a, b \in_R [1..\text{ord}(G_n)]$ and $h \in_R \mathcal{H}_n$, is computationally indistinguishable from the uniform distribution over $\{0,1\}^{t'(n)}$ even when $h$, $g_n^a$ and $g_n^b$ are given to the distinguisher.*

Notice that the above lemma requires the hash function $h$ to be chosen at random for each application. This is the case in several practical protocols (such as the case of IKE [RFC2409], mentioned in the Introduction, in which a key to the hash function is chosen by the communicating parties anew with each run of the protocol). However, it is also possible to fix a randomly chosen hash function and apply it repeatedly to different DH values. An example of such an application would be its use in the context of the Cramer-Shoup CCA-secure cryptosystem [CS98] (also discussed in the Introduction) in which the specific hash function $h$ would be chosen at random from the family $\mathcal{H}$ by the owner of the decryption key, and published as part of the public key parameters. In this case, the security of the repeated use of the same hash function $h$ can be proved via a standard simulation argument.

Finally we point out that for groups of prime order, the $t$-DDH Assumption is equivalent to the full DDH. The proof of this fact can be obtained by a standard random self-reducibility argument.

**Lemma 4.** *Let $G$ be a group of prime order $q$. If the $t$-DDH Assumption holds in $G$ for $t > 0$ then the DDH Assumption holds in $G$ as well.*

This yields an interesting 0-1 law for prime order groups, in which either the DDH Assumption holds, and thus the DH output has $\log(q)$ bits of computational entropy, or we cannot claim that the DH output has *any* bits of computational entropy.

### 3.3 The Max-Subgroup Theorem

We now proceed to prove our main theorem concerning the $t$-DDH assumption. The significance of the theorem below is that we can claim that a cyclic group is $t$-DDH if $t$ is the order of the maximal subgroup of $G$ where the DDH holds.

**Theorem 3.** *Let $G$ be a cyclic group of order $m = m_1 m_2$ where $(m_1, m_2) = 1$, and $G_1$ be a sub-group of order $m_1$ in $G$. If the DDH Assumption holds over $G_1$ then the $|m_1|$-DDH Assumption holds in $G$.*

*Proof.* An initial intuition behind the correctness of the theorem is that the hardness hidden in $G_1$ could be "sampled" when applying a hash function to the DH values over $G$. This however is incorrect: the size of $G_1$ may be negligible in relation to $|G|$ and as such the probability to sample a triple $(g^a, g^b, g^{ab})$ from $G_1$ is negligible too. The actual argument, presented next, uses the observation that the "hardness" present in $G_1$ can be extended to its cosets in $G$.

Let $g$ be a generator of $G$ and $g_1 = g^{m_2}$ be a generator of order $m_1$ of $G_1$. Given $g^a, g^b \in G$, we define the distribution $\mathcal{X}(g^a, g^b)$ to be the uniform distribution over $\{C = g^c \in G$ such that $c \in Z_m$ and $c \equiv ab \bmod m_2\}$ Thus, it is easy to see that $\mathcal{X}(g^a, g^b)$ has $|m_1|$ bits of min-entropy (since the above set has $m_1$ elements). Let $\mathcal{R}^*$ denote the probability distribution $\{(g^a, g^b, C) : a, b \in_R Z_m$ and $C \in_{\mathcal{X}(g^a, g^b)} G\}$.

We assume by contradiction that the $|m_1|$-DDH assumption does not hold in $G$, and thus we have a distinguisher $D$ between the distributions $\mathcal{DH}_G$ and $\mathcal{R}^*$. Using $D$ we build a distinguisher $D_1$ that distinguishes between the distributions $\mathcal{DH}_{G_1}$ and $\mathcal{R}_{G_1}$.

Given a triple $(A_1, B_1, C_1)$ where $A_1 = g_1^{a_1}, B_1 = g_1^{b_1}$, and $C_1$ either equals $g_1^{a_1 b_1}$ or $g_1^{c_1}$ for $c_1 \in_R Z_{m_1}$, the distinguisher $D_1$ does the following:

1. Chooses $i, j \in_R Z_m$
2. Sets $A = A_1 g^i, B = B_1 g^j$ and $C = C_1^{m_2} A_1^j B_1^i g^{ij}$ computed in $G$
3. Hands $D$ the triple $(A, B, C)$
4. Outputs the same output bit as $D$.

Let's examine the distribution of the triple $(A, B, C)$. Consider first $A$. This value is set to $A = A_1 g^i = g_1^{a_1} g^i = g^{m_2 a_1 + i}$ thus $a = m_2 a_1 + i$. Since $i \in_R Z_m$ then also $a \in_R Z_m$. Similarly for $B = g^b$ we get $b \in_R Z_m$. In the case of $C$ we have $C = C_1^{m_2} A_1^j B_1^i g^{ij} = g^{c_1 m_2^2 + m_2 a_1 j + m_2 b_1 i + ij}$, thus $c = c_1 m_2^2 + m_2 a_1 j + m_2 b_1 i + ij$. In addition, we have that $ab = (m_2 a_1 + i)(m_2 b_1 + j) = m_2^2 a_1 b_1 + m_2 a_1 j + m_2 b_1 i + ij$. Thus

$$c - ab = m_2^2 c_1 + m_2 a_1 j + m_2 b_1 i + ij - (m_2^2 a_1 b_1 + m_2 a_1 j + m_2 b_1 i + ij) = m_2^2 c_1 - m_2^2 a_1 b_1$$

which implies $c = m_2^2(c_1 - a_1 b_1) + ab \bmod m$. Therefore, if $c_1 = a_1 b_1$ then $c = ab$, while if $c_1 \in_R Z_{m_1}$ then $c_1 - a_1 b_1 \in_R Z_{m_1}$, and consequently $C$ is distributed according to the distribution $\mathcal{X}(g^a, g^b)$. In other words, the triple $(A, B, C)$ is distributed according to $\mathcal{DH}_G$ if $(A_1, B_1, C_1)$ came from $\mathcal{DH}_{G_1}$, and it is distributed according to $\mathcal{R}^*$ if $(A_1, B_1, C_1)$ came from $\mathcal{R}_{G_1}$. Therefore, $D_1$ distinguishes between $\mathcal{DH}_{G_1}$ and $\mathcal{R}_{G_1}$ with the same probability that $D$ distinguishes between $\mathcal{DH}_G$ and $\mathcal{R}^*$. Since we assumed the latter probability to be non-negligible we reached a contradiction with the premise that $G_1$ is a DDH group. □

**Remark on samplability.** The distributions $\mathcal{X}(g^a, g^b)$ defined in the above proof are efficiently samplable given $m_1, m_2$ and at least one of $a, b$. Indeed given, say, $a, B = g^b$ we can sample $\mathcal{X}(g^a, g^b)$ by choosing $k \in_R Z_{m_1}$ and setting $C = g^{k m_2} B^a$. In other words, provided that $m_1, m_2$ are given, Theorem 3 (and its corollary below) can be strengthened to claim that the *semi-samplable* $|m_1|$-DDH Assumption holds in $G$. We will use this stronger version of the theorem in Section 5.

From the above theorem and the Characterization Theorem we get:

**Corollary 1.** *For any cyclic group $G$, $G$ is $|m|$-DDH where $m$ is the order of the* maximal *DDH subgroup of $G$.*

## 4 DDH and $t$-DDH with Short Exponents

In this section we investigate the use of the DDH and $t$-DDH assumptions in conjunction with the so called "short-exponent discrete-log" assumption.

**The Short-Exponent Discrete-Log Assumption.** A common practice for increasing the efficiency of exponentiation in cryptographic applications based on the hardness of computing discrete logarithms, and in particular those using the Diffie-Hellman transform, is to replace full-length exponents (i.e. of length logarithmic in the group order) with (significantly) shorter exponents. The security of this practice cannot be justified by the usual assumption that computing discrete logarithms (with full-length exponents) is hard, but rather requires a specific assumption first analyzed in [vOW96] and formalized (as follows) in [PS98].

**Assumption 4 ($s$-DLSE [PS98])** *Let $\mathcal{G} = \{G_n\}_n$ be a family of cyclic groups where each $G_n$ has a generator $g_n$ and $ord(G_n) = m(n) > 2^n$. We say that the $s$-DLSE Assumption holds in $G$ if for every probabilistic polynomial time Turing machine $I$, for every polynomial $P(\cdot)$ and for all sufficiently large $n$ we have that $Prob_{x \in_R [1..2^s]}(I(g_n, m(n), s, g_n^x) = x) \leq 1/P(n)$.*

Current knowledge points to the plausibility of the above assumption even for exponents $s$ significantly shorter than $\log(ord(g))$. The exact values of $s$ for which the assumption seems to hold depends on the group generated by the element $g$. An obvious lower bound on $s$, if one wants to achieve security against $2^n$-complexity attacks, is $s \geq 2n$ which is necessary to thwart the usual square-root attacks such as Shanks and Pollard methods. However, as it was pointed out in [vOW96], there are cases where $s$ needs to be chosen larger than $2n$. Specifically, they show how to use a Pohlig-Hellman decomposition to obtain some of the bits of the exponent. The power of the attack depends on the (relatively) small prime factors of the group order. For example, when working over $Z_p^*$ with a random prime $p$, the [vOW96] results indicate the use of $s \approx 4n$ (e.g., with a security parameter of 80 one should use $s = 320$ which is much shorter than the 1024 or 2048 bits of $p$, yet twice as much as the bare minimum of $s = 160$). If one wants to use $s = 2n$ (i.e. assume the $2n$-DLSE), it is necessary to work in special groups such as those of prime order or $Z_p^*$ with $p$ a safe prime (i.e. $p = 2q + 1$, and $q$ prime).

**From Hardness to Indistinguishability.** Gennaro [Gen00] proves that if the $s$-DLSE assumption holds in $G = Z_p^*$ with $p$ a safe prime then the distribution over $G$ generated by $g^x$ for $x \in_R [1..2^s]$ is computationally indistinguishable from the uniform distribution over $G$. Here we use a generalization of this result that we summarize in the following proposition (see the full version of this paper [GKR04] for a proof of this Proposition).

**Proposition 1.** *Let $G$ be a cyclic group of order $m$ generated by $g$, such that $m$ is odd or $m/2$ is odd. If the $s$-DLSE Assumption holds in $G$, then the following two distributions $S_G = \{g^x : x \in_R [1..2^s]\}$ and $\mathcal{U}_G = \{g^x : x \in_R Z_m\}$ are computationally indistinguishable.*

Next we show that if in a group $G$, both the $s$-DLSE and the $t$-DDH Assumptions hold, then performing the Diffie-Hellman transform with short exponents $a$ and $b$, yields a DH output with $t$ bits of computational entropy. In other words, the security of the hashed DH transform over such groups when using $s$-bit long exponents is essentially equivalent to that of using full exponents.

**Theorem 5.** *Lct $G$ be a cyclic group of order $m$ generated by $g$, such that $m$ is odd, or $m/2$ is odd. Let $s, t$ be such that the $s$-DLSE and the semi-samplable $t$-DDH Assumptions hold in $G$. Denote with $\mathcal{X}(g^a, g^b)$ the family of distributions induced by the $t$-DDH assumption over $G$ (see Def. 5). Then the following two distributions*

$$\mathcal{SDH} = \{(g^a, g^b, g^{ab}) \ \ for \ \ a, b \in_R [1..2^s]\}$$

*and*

$$\mathcal{SR}^* = \{(g^a, g^b, C) \ \ for \ \ a, b \in_R [1..2^s] \ \ and \ C \in_{\mathcal{X}(g^a, g^b)} G\}$$

*are computationally indistinguishable.*

*Proof.* Recall that if the $t$-DDH Assumption holds over the group $G$ of order $m$, then there exists a family of probability distributions $\mathcal{X}(g^a, g^b)$ with min-entropy $t$ (one distribution for each pair $g^a, g^b$) over $G$ such that the distributions

$$\mathcal{DH} = \{(g^a, g^b, g^{ab}) \text{ for } a, b \in_R Z_m\}$$

and

$$\mathcal{R}^* = \{(g^a, g^b, C) \text{ for } a, b \in_R Z_m \text{ and } C \in_{\mathcal{X}(g^a, g^b)} G\}$$

are computationally indistinguishable.

The following standard hybrid argument yields the proof of the Theorem. Consider the intermediate distributions

$$\mathcal{D}_0 = \{(g^a, g^b, g^{ab}) \ \ \text{for} \ \ a, b \in_R [1..2^s]\}$$

$$\mathcal{D}_1 = \{(g^\alpha, g^b, g^{\alpha b}) \ \ \text{for} \ \ \alpha \in_R Z_m, b \in_R [1..2^s]\}$$

$$\mathcal{D}_2 = \{(g^\alpha, g^\beta, g^{\alpha\beta}) \ \ \text{for} \ \ \alpha, \beta \in_R Z_m\}$$

$$\mathcal{D}_3 = \{(g^\alpha, g^\beta, C) \ \ \text{for} \ \ \alpha, \beta, \in_R Z_m \text{ and } C \in_{\mathcal{X}(g^\alpha, g^\beta)} G\}$$

$$\mathcal{D}_4 = \{(g^\alpha, g^b, C) \ \ b \in_R [1..2^s], \alpha \in_R Z_m \text{ and } C \in_{\mathcal{X}(g^\alpha, g^b)} G\}$$

$$\mathcal{D}_5 = \{(g^a, g^b, C) \ : \ a, b \in_R [1..2^s] \text{ and } C \in_{\mathcal{X}(g^a, g^b)} G\}$$

Clearly $\mathcal{D}_0 = \mathcal{SDH}$ while $\mathcal{D}_5 = \mathcal{SR}^*$. If there is an efficient distinguisher between these distributions then, by a standard hybrid argument, there is an efficient distinguisher between $\mathcal{D}_i$ and $\mathcal{D}_{i+1}$ for some $i \in \{0, 1, 2, 3, 4\}$. But under the $t$-DDH Assumption we know that $\mathcal{D}_2$ is computationally indistinguishable from $\mathcal{D}_3$. Also, under the $s$-DLSE Assumption we know that $\mathcal{D}_i$ is computationally indistinguishable from $\mathcal{D}_{i+1}$ for $i = 0, 1, 3, 4$ by reduction to Proposition 1 (in the case $i = 3, 4$ one needs $\mathcal{X}(g^a, g^b)$ to be semi-samplable). □

Note that, as a particular case, when $t = \log(m)$ the theorem states that if $G$ is a DDH group in which the $s$-DLSE assumption holds, then performing the DH transform over $G$ with exponents of size $s$ yields values that are indistinguishable from random elements in $G$.

# 5 Hashed DH over $Z_p^*$ and Its Subgroups

Here we discuss the security of the hashed DH transform over groups and subgroups of $Z_p^*$ for prime $p$. Throughout this section we assume that the DDH assumption holds over the large prime-order subgroups of $Z_p^*$. Under this assumption we immediately get that it is secure to use the hashed DH transform over a subgroup $G_q$ of $Z_p^*$ of order $q$, provided that $q$ is a sufficiently large prime that divides $p-1$. By sufficiently large we mean that the DDH assumption (plausibly) holds in $G_q$ (for a given security parameter $k$), and that the computational entropy of $q$ is sufficient for the application. Specifically, if the application requires a pseudorandom output of $\ell$ bits then $q$ needs to satisfy $|q| \geq \ell + 2k$. Similarly, we get that it is secure to work in any subgroup of $Z_p^*$ whose order $m$ is the product of large primes (each of which divides $p-1$); also here it is required that $|m| \geq \ell + 2k$, although note that each of the prime factors of $m$ may be smaller than that bound (one usually assumes the DDH to hold on groups of prime order $q$ with $|q| \geq 2k$).

Moreover, one of the most significant contributions of our work is in showing the security of the hashed DH transform also over groups (or subgroups) whose order is divisible by small prime factors (and therefore not satisfying the DDH assumption). In particular, this is necessarily the case for the group $Z_p^*$ with prime $p$ (the order $m = p-1$ of this group is always divisible by small prime factors, e.g., 2). Our results show that the hashed DH is secure over $Z_p^*$ provided that $p-1$ has enough prime divisors whose product is larger than the entropy bound $2^{\ell+2k}$, and for which the subgroups of corresponding prime order are DDH. (In particular, the fact that $p-1$ has additional smaller prime factors does not invalidate the security of the hashed DDH in $Z_p^*$.)

A particularly interesting group is $Z_p^*$ for $p = 2q+1$ and $q$ prime. In this case, working directly with the hashed DH over $Z_p^*$ is secure since we are assuming that its subgroup of order $q$ is DDH, and therefore the whole $Z_p^*$ group is $|\frac{p-1}{2}|$-DDH. Working over $Z_p^*$ in this case has several important advantages: (i) one can produce a large (actually, largest) number of pseudorandom bits (specifically, $|p| - 1 - 2k$ bits); (ii) $p$ can be chosen such that 2 is a generator of $Z_p^*$ (which speeds up exponentiation); (iii) the $2k$-DLSE Assumption (see Section 4) is conjectured to hold in these groups [vOW96] and therefore one can use minimal-length exponents (i.e., of length $2k$) in these groups, obtaining yet another significant exponentiation speedup without sacrificing the security of the (hashed) DH transform; and (iv) these groups are free from the potentially serious attacks described in [LL97] (that affect subgroups of prime order $q$ where $(p-1)/q$ has a relatively large smooth factor). Note that items (i) and (iii) follow essentially from our results. The only drawback working over such a group is the cost of generating $p$'s of the above form; this, however is insignificant in typical applications (e.g., IKE [RFC2409]) in which this generation is very rare, and usually done at the set-up of the system and used for a large period of time.

Note that in all of the above examples it is assumed that one knows the full or partial factorization of $p-1$; in particular, the knowledge of this factorization is essential for selecting a generator of the group. It is a theoretically and practically important question to establish whether the knowledge of the factorization of $p-1$ is essential for working securely over $Z_p^*$ or over one of its subgroups. In the rest of this section we show that this knowledge is not essential. Specifically, it follows from our results that if one chooses a random prime $p$ (of a pre-specified size such that the Discreet Logarithm Problem is hard in $Z_p^*$) and a random element $e$ in $Z_p^*$, then performing the hashed DH transform over the group generated by $e$ is secure.[6]

Let $p$ be a random prime such that $p-1 = p_1 p_2 ... p_n$ and $p_1 \leq p_2 \leq ... \leq p_n$ are all (not necessarily different and possibly unknown) primes. Let $e$ be an element randomly chosen from $Z_p^*$, and let $G_e$ denote the subgroup of $Z_p^*$ generated by $e$. We first claim that with overwhelming probability the large prime factors of $p-1$ divide the order of $G_e$.

**Lemma 5.** *Let $Z_p^*$ and $p-1 = p_1..p_n$ be as described above. Then for all $1 \leq i \leq n$: $Pr_{e \in_R Z_p^*}[p_i \not| \ ord(e)] \leq 1/p_i$.*

*Proof.* Let $g$ be a generator of $Z_p^*$. There are at most $(p-1)/p_i$ elements whose order is not divisible by $p_i$, and they are the elements of the form $g^{jp_i}$ for $1 \leq j \leq (p-1)/p_i$. When $p_i^2 | p-1$ this is a strict upper bound, otherwise this is an exact bound. Thus, the probability to choose $e$ such that $p_i \not| \ ord(e)$ is at most $\frac{(p-1)/p_i}{p-1} = \frac{1}{p_i}$. □

**Corollary 2.** *For a given bound $B$, let $p-1 = \Pi_{i=1}^n p_i$ where $p_j, p_{j+1}, ..., p_n > B$. Then*

$$Pr_{e \in_R Z_p^*}[\Pi_{i=j}^n \ p_i \mid ord(e)] \geq 1 - \sum_{i=j}^{n} \frac{1}{p_i} \geq 1 - \frac{n-j}{B} \geq 1 - \frac{\log p}{B}.$$

Thus, for large values of $B$, the order of a random element $e$ is divisible, with overwhelming probability, by all the prime factors of $p-1$ which are larger than $B$. Or, equivalently, $G_e$ has as subgroups all the prime-order subgroups of $Z_p^*$ whose order is larger than $B$.

Now, if we set our security parameter to $k$, define $B = 2^{2k}$, and assume that the DDH holds in subgroups of prime order larger than $B$, then we have that, with overwhelming probability, $G_e$ contains all the prime order DDH subgroups of $Z_p^*$. In other words, if we denote by $P$ the product of all prime factors of $p-1$ larger than $B$, we have that $G_e$ contains, by virtue of our DDH Characterization Theorem (Theorem 1) a DDH subgroup of size $P$, and then by the Max-Subgroup Theorem (Theorem 3) we get that $G_e$ is $|P|$-DDH.

All it is left to argue is that $|P|$ is large enough. For this we use the following Lemma from [vOW96] that provides an upper bound on the expected size of

[6] We stress that while the legitimate users of such a scheme do not need to know the factorization of $p-1$, the scheme remains secure even if this factorization is known to the attacker.

the product of all prime divisors of $p-1$ that are smaller than $B$ (and thus, it provides a lower bound on the expected size of $|P|$).

**Lemma 6 ([vOW96]).** *For a random prime p (as above) and a fixed bound B, the expected length of* $\Pi_i p_i$ *where* $p_i < B$ *is* $\log B + 1$.

In other words, the lemma states that the expected size of $|P|$ is $|p| - |B| = |p| - 2k$.

If, for the sake of illustration, we set $|p| = 1024$ and $k = 80$ we get that we expect $G_e$ to be 864-DDH. However, note that this expected size may vary for specific $p$'s. Yet, note that even if $p$ happens to have a $B$-smooth part that is 4 times larger than expected (!) we are still left with a 384-DDH subgroup $G_e$ with enough computational entropy for most DH applications (such as deriving a 128-bit pseudorandom key). If one considers 2048-bits and $k = 160$ then the expected amount of entropy is 2048-320=1728 bits which, again, leaves plenty room to compensate for "unlucky choices" of $p$.

Notice that in order to use short exponents in this case (i.e. random prime $p$ and random generator $e$), one must make sure that the order $m$ of the group generated by $e$ is either odd, or $m/2$ is odd (so that we can invoke Theorem 5). This can be easily achieved by choosing first a random element $e$ in $Z_p^*$ and then using as the group generator the element $e^{2^f} \bmod p$ where $f$ is the maximal integer such that $2^f|(p-1)$ (the value $f$ is, of course, trivial to obtain without requiring of any significant factorization of $p-1$).

**Remark (semi-samplability).** In the above discussion we have justified the usage of short exponents on the basis of Theorem 5. Note, however, that this theorem assumes the semi-samplability of the distributions $\mathcal{X}(g^a, g^b)$. Therefore, we need to verify that this semi-samplability property holds for the above applications. This is indeed the case since these applications use the distributions defined in the proof of Theorem 3, which are semi-samplable when the factorization of the group order is known (see the remark following the proof of Theorem 3). Therefore, we obtain that, even though the honest parties can perform the hashed DH transform securely with short exponents, and *without* requiring the knowledge of the factorization of $p-1$, the DH transform remains secure *even if* such factorization is available to the attacker.

## Acknowledgment

We thank the anonymous referees for their useful comments.

## References

[ABR01] M. Abdalla, M. Bellare, and P. Rogaway. DHIES: An Encryption Scheme Based on the Diffie-hellman Problem. In *CT-RSA '01*, pages 143–158, 2001. LNCS No. 2020.

[BJN00] D. Boneh, A. Joux, and P. Nguyen. Why Textbook ElGamal and RSA Encryption are Insecure . In *AsiaCrypt '00*, pages 30–44, 2000. LNCS No. 1976.

[Bon98] D. Boneh. The Decision Diffie-Hellman Problem. In *Third Algorithmic Number Theory Symposium*, pages 48–63, 1998. LNCS No. 1423.

[Bra93] S. Brands. An Efficient Off-Line Electronic Cash System Based on the Representation Problem. TR CS-R9323, CWI, Holland, 1993.

[CW79] L. Carter and M. N. Wegman. Universal Classes of Hash Functions. *JCSS*, 18(2):143–154, April 1979.

[CS98] R. Cramer and V. Shoup. A Practical Public Key Cryptosystem Provable Secure Against Adaptive Chosen Ciphertext Attack. In *Crypto '98*, pages 13–25, 1998. LNCS No. 1462.

[DH76] W. Diffie and M. E. Hellman. New Directions in Cryptography. *IEEE Transactions on Information Theory*, 22(6):644–654, 1976.

[ElG85] T. ElGamal. A Public Key Cryptosystem and a Signature Scheme Based on Discrete Logarithms. *IEEE Trans. Info. Theory*, IT 31:469–472, 1985.

[Gen00] R. Gennaro. An Improved Pseudo Random Generator Based on Discrete Log. In *Crypto '00*, pages 469–481, 2000. LNCS No. 1880.

[GHKR04] R. Gennaro, J. Hastad, H. Krawczyk and T. Rabin. Randomness Extraction and Key Derivation Using the CBC, Cascade and HMAC modes. Manuscript, 2004.

[GKR04] R. Gennaro, H. Krawczyk and T. Rabin. Secure Hashed Diffie-Hellman over Non-DDH Groups. Full version available at `http://eprint.iacr.org/2004/`

[GM84] S. Goldwasser and S. Micali. Probabilistic Encryption. *JCSS*, 28(2):270–299, April 1984.

[HILL99] J. Hastad, R. Impagliazzo, L. Levin, and M. Luby. Construction of a Pseudo-random Generator from any One-way Function. *SIAM. J. Computing*, 28(4):1364–1396, 1999.

[Kra03] H. Krawczyk. SIGMA: The 'SiGn-and-MAc' Approach to Authenticated Diffie-Hellman and Its Use in the IKE Protocols. In *Crypto '03*, pages 400–425, 2003. LNCS No. 2729. `http://www.ee.technion.ac.il/~hugo/sigma.html`

[LL97] C.H. Lim and P.J. Lee. A Key Recovery Attack on Discrete Log-Based Schemes Using a Prime Order Subgroup. In *Crypto '97*, pages 249–263, 1997. LNCS No. 1294.

[Lub96] M. Luby. *Pseudorandomness and Cryptographic Applications*. Princeton Computer Science Note, Princeton University Press, January 1996.

[NR97] M. Naor and O. Reingold. Number-theoretic Constructions of Efficient Pseudorandom Functions. In *Proc. 38th FOCS*, pages 458–467. IEEE, 1997.

[PS98] S. Patel and G. Sundaram. An Efficient Discrete Log Pseudo Random Generator. In *Crypto '98*, pages 304–317, 1998. LNCS No. 1462.

[RFC2409] RFC2409. The Internet Key Exchange (IKE). Authors: D. Harkins and D. Carrel, Nov 1998.

[Sta96] M. Stadler. Publicly Verifiable Secret sharing. In *Eurocrypt '96*, pages 190–199, 1996. LNCS No. 1070.

[vOW96] P.C. van Oorschot and M. Wiener. On Diffie-Helman Key Agreement with Short Exponents. In *Eurocrypt '96*, pages 332–343, 1996. LNCS No. 1070.

# A Indistinguishability of Probability Distributions

**Definition 6.** *Let $\mathcal{X}_n, \mathcal{Y}_n$ be two probability distributions over a support set $A_n$. We say that $\mathcal{X}_n$ and $\mathcal{Y}_n$ have* statistical distance *bounded by $\Delta(n)$ if $\sum_{x \in A_n} |Prob_{\mathcal{X}_n}[x] - Prob_{\mathcal{Y}_n}[x]| \leq \Delta(n)$. We say that the ensembles $\mathcal{X}_n$ and $\mathcal{Y}_n$ are* statistically indistinguishable *if for every polynomial $P(\cdot)$ and for all sufficiently large $n$ we have that $\Delta(n) \leq \frac{1}{P(n)}$.*

**Definition 7.** *Let $\mathcal{X}_n, \mathcal{Y}_n$ be two probability ensembles. Given a family of circuits $D = \{D_n\}_n$ (called the distinguisher) consider the following quantities*

$$\delta_{D,\mathcal{X}_n} = Prob_{x \in \mathcal{X}_n}[D_n(x) = 1] \quad \text{and} \quad \delta_{D,\mathcal{Y}_n} = Prob_{y \in \mathcal{Y}_n}[D_n(y) = 1]$$

*We say that the probability ensembles $\mathcal{X}_n$ and $\mathcal{Y}_n$ are* computationally indistinguishable *(by non-uniform distinguishers) if for every polynomial-size distinguisher family $D$, for every polynomial $P(\cdot)$, and for all sufficiently large $n$ we have that $|\delta_{D,\mathcal{X}_n} - \delta_{D,\mathcal{Y}_n}| \leq \frac{1}{P(n)}$*

# On Simulation-Sound Trapdoor Commitments

Philip MacKenzie[1] and Ke Yang[2]

[1] Bell Laboratories, Lucent Technologies
Murray Hill, NJ 07040,
philmac@lucent.com

[2] Computer Science Department, Carnegie Mellon University,
Pittsburgh, PA 15213,
yangke@cs.cmu.edu

**Abstract.** We study the recently introduced notion of a *simulation-sound trapdoor commitment (SSTC)* scheme. In this paper, we present a new, simpler definition for an SSTC scheme that admits more efficient constructions and can be used in a larger set of applications. Specifically, we show how to construct SSTC schemes from any one-way functions, and how to construct very efficient SSTC schemes based on specific number-theoretic assumptions. We also show how to construct simulation-sound, non-malleable, and universally-composable zero-knowledge protocols using SSTC schemes, yielding, for instance, the most efficient universally-composable zero-knowledge protocols known. Finally, we explore the relation between SSTC schemes and non-malleable commitment schemes by presenting a sequence of implication and separation results, which in particular imply that SSTC schemes are non-malleable.

## 1 Introduction

The notion of a *commitment* is one of the most important and useful notions in cryptography. Intuitively, a commitment is the digital equivalent of a "locked combination safe." A party Alice would commit to a value by placing it into the safe, closing the safe, and spinning the lock, so that the value may later be revealed by Alice divulging the combination of the safe. Obviously, the value cannot be viewed by any other party prior to this opening (this is known as the "secrecy" or "hiding" property), and cannot be altered (this is known as the "binding" property). Commitments have been useful in a wide range of applications, from zero-knowledge protocols (e.g., [4,12,26]) to electronic commerce (e.g., remote electronic bidding), and have been studied extensively (e.g., [3,32,33]). In many cases, however, one needs commitment schemes with *additional properties* besides hiding and binding, such as those described below.

A *trapdoor commitment (TC) scheme* is a commitment scheme with an additional "equivocability" property. Roughly speaking, for such a commitment scheme there is some *trapdoor* information whose knowledge would allow one to open a commitment in more than one way (and thus "equivocate"). Naturally, without the trapdoor, equivocation would remain computationally infeasible [4,20,2].

C. Cachin and J. Camenisch (Eds.): EUROCRYPT 2004, LNCS 3027, pp. 382–400, 2004.

A *non-malleable commitment (NMC) scheme* is a commitment scheme with the property that (informally) not only is the value $v$ placed inside a commitment secret, but seeing this commitment does not give another party any advantage in generating a new commitment that, once $v$ is revealed, can then be opened to a value related to $v$ [18,16,23,17,14].[3]

Finally, a *universally composable commitment (UCC) scheme* is a commitment scheme with a very strong property that intuitively means that the security of a commitment is guaranteed even when commitment protocols are concurrently composed with arbitrary protocols [5,6,15]. To achieve universal composability, a commitment scheme seems to require equivocability, non-malleability, and furthermore, *extractability*. Roughly speaking, an extractable commitment scheme has a modified secrecy definition, which states that there is a *secret key* whose knowledge would allow one to extract the value placed in a commitment. Naturally, without this knowledge, the value would remain hidden. We note that the notion of a UCC scheme appears to be strictly stronger than the other notions of commitment schemes. In particular, Damgård and Groth [14] show that a UCC scheme implies secure key exchange, while both TC schemes and NMC schemes can be constructed from one-way functions.

### 1.1 Simulation-Sound Trapdoor Commitments

In this paper, we focus our attention on another extension of commitment schemes, namely simulation-sound trapdoor commitment (SSTC) schemes. An SSTC scheme is a TC scheme with a strengthened binding property, called simulation-sound binding. Roughly speaking, in an SSTC scheme, an adversary cannot equivocate on a commitment with a certain tag, even after seeing the equivocation of an unbounded number of commitments with different tags (i.e., the adversary may request an equivocation oracle to generate an unbounded number of commitments with different tags, and then to open them to arbitrary values). Here, a tag for a commitment is simply a binary string associated with the commitment. We will discuss tags in more detail below.

The term "simulation soundness" was first used to describe a property of zero-knowledge proofs by Sahai [37], and intuitively meant that even though an adversary could see simulated proofs of incorrect statements, it could not itself produce a new proof of any incorrect statement. Garay *et al.* [24] first applied this term to trapdoor commitments. They gave a slightly stronger, although more complicated, simulation-sound binding property and an efficient construction based on DSA signatures [29]. Their definition was specifically tailored to the goal of developing a universally-composable zero-knowledge (UCZK) proof that was secure in the presence of adversaries that could adaptively corrupt parties.[4]

---

[3] The original definition of [18] states (informally) that another party does not even have any advantage in creating a new commitment to a value related to $v$, regardless of the ability to open the new commitment. However, we will use the definition based on opening.

[4] They use the term *identifier* in place of the term *tag*, and intuitively, in their definition [24], a commitment made by the adversary using identifier *id* is binding, even if

## 1.2 Summary of Results

*Simpler Definition* We provide a simpler definition of SSTC schemes than the one by Garay *et. al.* [24]. Though the binding property in our definition is weaker, it is still sufficient in many applications (e.g., to construct UCZK protocols that are secure in the presence of adversaries that can adaptively corrupt parties).

We also discuss various design issues in the definition, and most notably, the choice between definitions based on the tag of the commitment and on the body of the commitment. Informally, a tag-based definition requires that an adversary cannot equivocate a commitment com with a certain tag so long as it does not see the equivocation of any commitment with the same tag. On the other hand, a body-based definition requires that the adversary cannot equivocate a commitment com so long as the commitment com itself has not been equivocated. (Note that we use the term "body" to refer to the bit-string that is the commitment.)

In our paper, we choose to focus on tag-based schemes since they admit simpler constructions and seem to be the most appropriate for our applications. For example, in constructing secure zero-knowledge protocols in the UC framework, where the communication is normally assumed to be authenticated, it is natural to use a tag-based scheme, setting the tag to be the pair of the identities of the prover and the verifier.

*Efficient Constructions* We present various constructions of SSTC schemes. The first construction is a generic one based on the (minimal) assumption that one-way functions exist. Our construction is similar to that of a UCC commitment scheme in Canetti *et. al.* [7]. However, because SSTC schemes do not require the extractability property, we are able to simplify the construction, and have it rely on a weaker assumption. The second construction is based on the DSA assumption, and is very efficient, involving only a small constant number of modular exponentiations. It is similar to the construction from Garay *et. al.* [24], but is about twice as efficient. The third assumption is based on Cramer-Shoup signatures [11], and relies on the strong RSA assumption [1]. It is also very efficient, again requiring only a small constant number of modular exponentiations.

We remark here that our most efficient SSTC schemes are more efficient than all known UCC schemes. For instance, the UCC constructions of [6,7] are for bit commitments, and thus have an expansion factor of at least the security parameter. The UCC construction of [15] has constant expansion factor, but requires a CRS of length proportional to the number of parties times the security parameter. Recently and independent from this work, Damgård and Groth [14] presented a UCC scheme with a constant expansion factor with a CRS whose length is independent of the number of parties. However, their scheme is still quite complicated, since it requires interaction, and uses two different types of

the adversary has seen any commitment using identifier $id$ opened (using an oracle that knows a trapdoor) once to any arbitrary value, and moreover, any commitment using identifier $id' \neq id$ opened (again using the oracle) an unbounded number of times to any arbitrary values.

commitments, one a non-malleable commitment scheme, and the other a special "mixed commitment scheme."

*Applications* We show constructions of unbounded simulation-sound, unbounded non-malleable, and universally composable zero-knowledge (ZK) protocols using SSTC schemes in the common reference string (CRS) model. In particular, we show how to (1) convert a $\Sigma$-protocol [10] (which is a special three-round, honest-verifier protocol where the verifier only sends random bits) into an unbounded simulation-sound ZK protocol; and (2) convert an $\Omega$-protocol [24] (which is a $\Sigma$-protocol with a straight-line extractor) into an unbounded non-malleable ZK protocol, and further into a universally-composable ZK protocol. The constructions are conceptually very simple. In fact, they all share the same structure, and all use a technique from Damgård [13] and Jarecki and Lysyanskaya [28]. The same technique was also used in Garay *et. al.* [24] in constructing a universally-composable ZK protocol that is secure against adaptive corruptions.

Our constructions are very efficient, and in particular our construction of a universally-composable ZK protocol is more efficient than previous constructions, at least when starting with a $\Sigma$-protocol. Compared to UCZK protocols based on universally-composable commitment schemes [6,7,14,15], our efficiency gain comes mainly from the fact that we avoid the Cook-Levin theorem [8,30],[5] but also from the fact that some of our SSTC schemes are more efficient than any UCC schemes, as discussed above. Compared to the UCZK protocol in Garay *et. al.* [24], our savings are twofold: the simpler SSTC construction (with a weaker definition) cuts the overhead of the SSTC commitments by half, and the direct use of the identities as tags eliminates the need for one-time signatures on the protocol transcripts.

In recent and independent work, Gennaro [25] presented an SSZK protocol[6] that is similar to our construction in Section 4. It uses a new type of commitment scheme called *multi-trapdoor commitments*, and an efficient implementation of this scheme based on the strong RSA assumption and a special hash property. A multi-trapdoor commitment scheme is similar to an SSTC scheme, except that it requires the existence of a different trapdoor (i.e., secret key) corresponding to each tag, and its security property corresponding to simulation-sound binding requires tags to be pre-chosen by the adversary.[7]

*Relation to Non-malleable Commitments* We discuss the relation between SSTC schemes and NMC schemes [18,16,17,14].[8] At first glance, binding and non-

[5] In previous constructions, they build a UCZK protocol $\Pi^L$ for an NP-complete language $L$ (e.g. Hamiltonian Cycle or Satisfiability), and then the UCZK protocols for any NP language is reduced to $\Pi^L$ via the Cook-Levin theorem, which is not very efficient.

[6] It is also concurrent non-malleable ZK, if rewinding is allowed in witness extraction

[7] We have recently defined a *static SSTC* scheme as a commitment scheme with only the second requirement, and note that it is also sufficient in our SSZK and NMZK constructions.

[8] Technically, when we refer to an NMC scheme, we will always mean an $\epsilon$-non-malleable commitment scheme, following the notation proposed in [17].

malleability (or analogously, equivocation and malleability) seem like very different notions: while the former concerns the adversary's ability to open a commitment to multiple values, the latter concerns the adversary's ability to produce and open a commitment to a single value related to a previously committed value. However, they are actually closely related, and we shall show that simulation-sound binding implies non-malleability (when both are appropriately defined). In fact, a similar observation was used implicitly in [16,17,14] to construct NMC schemes. In particular, these NMC schemes are all based on trapdoor commitment schemes that satisfy a weak notion of simulation-sound binding. (Note that these results all use body-based definitions instead of tag-based definitions.) However, the *exact* relationship between the notions of simulation-sound binding and non-malleability was not known, e.g., if simulation-sound binding is strictly stronger than non-malleability, or if they are equivalent.

We study the exact relationship between these two notions in this paper. To do this, we need to resolve some technical issues. First, just as SSTC schemes can be tag-based or body-based, NMC schemes can also be tag-based or body-based, where a tag-based NMC scheme is informally defined as one in which seeing a commitment (to some value $v$) with a certain tag does not give an adversary any advantage in generating a new commitment with a different tag that can later be opened to a value related to $v$. Since we focus on tag-based SSTC schemes, we will focus on their relation to tag-based NMC schemes.[9] (Analogous results could be obtained for the relationship between body-based SSTC schemes and body-based NMC schemes.) Second, an SSTC scheme is a TC scheme, so to make a useful comparison, we consider non-malleable trapdoor commitment (NMTC) schemes. Third, since an adversary for an SSTC scheme is allowed to query an equivocation oracle, we will also consider NMTC schemes in which an adversary is allowed to query an equivocation oracle.

Finally, we refine our definitions of SSTC schemes and NMTC schemes by specifying the number of equivocation oracle queries an adversary is allowed to make. An equivocation oracle, on a commit query, produces a commitment $\widetilde{\mathsf{com}}$ and on an decommit query, opens $\widetilde{\mathsf{com}}$ to an arbitrary value. We say a TC scheme is $\mathsf{SSTC}(\ell)$, if it remains secure if the adversary is allowed to make at most $\ell$ commit queries to the oracle (with no restriction on the number of decommit queries). We define $\mathsf{NMTC}(\ell)$ schemes similarly. We use $\mathsf{SSTC}(\infty)$ and $\mathsf{NMTC}(\infty)$ to denote the schemes where the adversary can make an unlimited number of commit queries. With the refined definitions (except for those related to the definition in [14], discussed below), we shall then prove that, for any constant $\ell$, $\mathsf{SSTC}(\ell+1)$ is strictly stronger than $\mathsf{NMTC}(\ell)$ and $\mathsf{NMTC}(\ell)$ is strictly stronger than $\mathsf{SSTC}(\ell)$. (In particular, note that even an $\mathsf{SSTC}(1)$ scheme is strictly stronger than an NMC scheme, since an $\mathsf{NMTC}(0)$ scheme is at least as strong as an NMC scheme.) Furthermore, $\mathsf{SSTC}(\infty)$ is equivalent to $\mathsf{NMTC}(\infty)$.

---

[9] Tag-based NMC schemes are also related to UCC schemes. In particular, it can be shown that a UCC scheme is also a tag-based NM commitment scheme in which the tag is the identity of the committing party.

See Figure 1. This makes it clear that the two notions, simulation-sound binding and non-malleability, are very closely related.

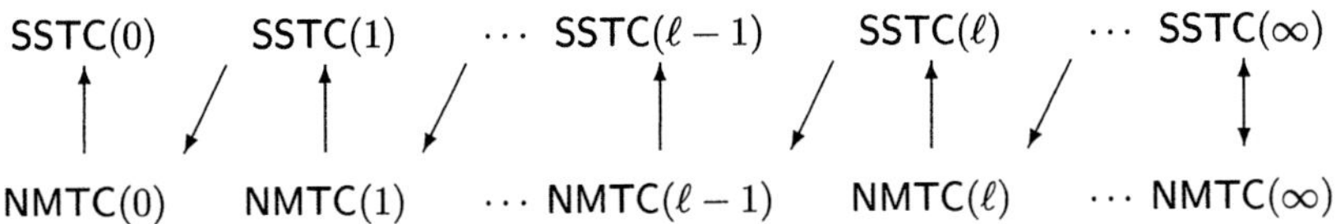


**Fig. 1.** The relation between SSTC and NMTC schemes, with one-sided arrows denoting strict implication and two-sided arrows denoting equivalence

The definition of non-malleable commitments in Damgård and Groth [14] (which they call *reusable* non-malleable commitments) does not quite fit into the equivalence and separation results above. Their definition states that seeing one *or more* commitments does not give another party any advantage in generating one *or more* commitments that can later be opened to values related to the values in the original commitments. However it can be shown that $\mathsf{SSTC}(\infty)$ implies a reusable NMC scheme. As mentioned above, one can characterize their construction of a reusable NMC scheme as constructing a trapdoor commitment schemes that satisfies a slightly weaker notion of simulation-sound binding, and showing that this implies a reusable NMC scheme.

Due to space limitations, some proofs to our theorems are omitted, and can be found in the full version [31].

## 2 Preliminaries and Definitions

We will use signature schemes that are existentially unforgeable against adaptive chosen-message attacks [27]. However, some of these may only be used for a single signature, and for these, more efficient *one-time* signature scheme constructions may be used [19].

A commitment scheme is a two-phase protocol[10] between a sender and a receiver, both probabilistic polynomial-time Turing machines, that runs as follows. In the commitment phase, the sender commits to a value $v$ by computing a pair $(\mathsf{com}, \mathsf{dec})$ and sending $\mathsf{com}$ to the receiver, and in the decommitment phase, the sender reveals $(v, \mathsf{dec})$ to the receiver, who checks whether the pair is valid.

Informally, a commitment scheme satisfies the hiding property, meaning that for any $v_1 \neq v_2$ of the same length, a commitment to $v_1$ is indistinguishable from a commitment to $v_2$, and the binding property, meaning that once the receiver receives $\mathsf{com}$, the sender cannot open $\mathsf{com}$ to two different values, except with negligible probability.

[10] We define a standard non-interactive commitment scheme. We do not consider relaxations to interactive commitment schemes.

We will always assume that commitments are labeled with a tag. While this is not a factor in the security of basic commitment schemes, it will be useful in defining certain enhanced commitment schemes, as will be obvious below. We also assume that there is a commitment generator function that generates a set of parameters for the commitment scheme. In other papers this is often referred to as a *trusted third party* or as the *common reference string* generation,[11] and it is especially important when we define trapdoor commitment schemes below. (We include it in the basic definition to more conveniently define trapdoor commitment schemes.)

Formally, we define a commitment scheme as follows.

**Definition 1. [Commitment Scheme]** $\mathsf{CS} = (\mathsf{Cgen}, \mathsf{Ccom}, \mathsf{Cver})$ *is a commitment scheme if* Cgen, Ccom, *and* Cver *are probabilistic polynomial-time algorithms such that*

- **Completeness** *For all v and tag,*

$$\Pr[pk \leftarrow \mathsf{Cgen}(1^k); (\mathsf{com}, \mathsf{dec}) \leftarrow \mathsf{Ccom}(pk, v, tag) : \\ \mathsf{Cver}(pk, \mathsf{com}, v, tag, \mathsf{dec}) = 1] = 1.$$

- **Binding** *There is a negligible function $\alpha(k)$ such that for all non-uniform probabilistic polynomial-time adversaries $\mathcal{A}$,*

$$\Pr[pk \leftarrow \mathsf{Cgen}(1^k); (\mathsf{com}, tag, v_1, v_2, \mathsf{dec}_1, \mathsf{dec}_2) \leftarrow \mathcal{A}(pk) : \\ (\mathsf{Cver}(pk, \mathsf{com}, v_1, tag, \mathsf{dec}_1) = \mathsf{Cver}(pk, \mathsf{com}, v_2, tag, \mathsf{dec}_2) = 1) \\ \wedge (v_1 \neq v_2)] \leq_{\mathrm{ev}} \alpha(k).$$

- **Hiding** *For all pk generated with non-zero probability by* $\mathsf{Cgen}(1^k)$*, for all $v_1, v_2$ of equal length, and for all tag, the following probability distributions are computationally indistinguishable:*

$$\{(\mathsf{com}_1, \mathsf{dec}_1) \leftarrow \mathsf{Ccom}(pk, v_1, tag) : \mathsf{com}_1\} \text{ and} \\ \{(\mathsf{com}_2, \mathsf{dec}_2) \leftarrow \mathsf{Ccom}(pk, v_2, tag) : \mathsf{com}_2\}.$$

Next, we define trapdoor commitment schemes. (We borrow some notation from Reyzin [35].) Informally a trapdoor commitment scheme has the property that there exists a trapdoor that would allow one to generate a "fake" commitment along with information that would later allow to decommit to any subsequently given value $v$, and that this commitment/decommitment pair is indistinguishable from an actual commitment to $v$ and a subsequent decommitment

[11] We do not use the term "common reference string" in our definition, since these parameters may be generated in a number of ways, and in particular, they may be generated by the receiver. In protocols where this value actually comes from a common reference string, we will make this clear.

**Definition 2.** [**Trapdoor Commitment Scheme**]
$\mathsf{TC} = (\mathsf{TCgen}, \mathsf{TCcom}, \mathsf{TCver}, \mathsf{TCfakeCom}, \mathsf{TCfakeDecom})$ *is a* trapdoor commitment scheme *if* $\mathsf{TCgen}(1^k)$ *outputs a public/secret key pair* $(pk, sk)$*,* $\mathsf{TCgen}_{pk}$ *is a function that restricts the output of* $\mathsf{TCgen}$ *to the public key,* $(\mathsf{TCgen}_{pk}, \mathsf{TCcom}, \mathsf{TCver})$ *is a commitment scheme and* $\mathsf{TCfakeCom}$ *and* $\mathsf{TCfakeDecom}$ *are probabilistic polynomial-time algorithms such that*

– **Trapdoor Property** *For all identifiers tag and values v, the following probability distributions are computationally indistinguishable:*

$$\{(pk, sk) \leftarrow \mathsf{TCgen}(1^k); (\widetilde{\mathsf{com}}, \xi) \leftarrow \mathsf{TCfakeCom}(pk, sk, tag); \widetilde{\mathsf{dec}} \leftarrow \mathsf{TCfakeDecom}(\xi, v) : (pk, tag, v, \widetilde{\mathsf{com}}, \widetilde{\mathsf{dec}})\}$$

*and*

$$\{(pk, sk) \leftarrow \mathsf{TCgen}(1^k); (\mathsf{com}, \mathsf{dec}) \leftarrow \mathsf{TCcom}(pk, v, tag) : (pk, tag, v, \mathsf{com}, \mathsf{dec})\}.$$

## 3 Simulation-Sound Trapdoor Commitments

In [24], simulation-sound trapdoor commitment (SSTC) schemes were introduced, in order to construct a universally-composable zero-knowledge (UCZK) protocol secure against adaptive corruptions. Intuitively, they defined an SSTC scheme as a trapdoor commitment scheme with a *simulation-sound binding* property that guarantees that a commitment made by the adversary using *tag* is binding, even if the adversary has seen any commitment using *tag* opened (using a simulator that knows a trapdoor) once to any arbitrary value, and moreover, any commitment using $tag' \neq tag$ opened (again using the simulator) an unbounded number of times to any arbitrary values.

Here we introduce a new definition for an SSTC scheme where the simulation-sound binding property only guarantees that a commitment made by the adversary using *tag* is binding, if the adversary has *never* seen the simulator open a commitment using *tag* (i.e., not even once, as is allowed in the previous definition).[12] Obviously this is a weaker property. However, we will show that it also suffices for the desired application in [24], namely, for constructing UCZK protocols secure against adaptive adversaries.

**Definition 3.** [$\mathsf{SSTC}$ **Scheme**]
$\mathsf{TC} = (\mathsf{TCgen}, \mathsf{TCcom}, \mathsf{TCver}, \mathsf{TCfakeCom}, \mathsf{TCfakeDecom})$ *is an* $\mathsf{SSTC}$ scheme *if* $\mathsf{TC}$ *is a trapdoor commitment scheme such that*

– **Simulation-Sound Binding** *There is a negligible function $\alpha(k)$ such that for all non-uniform probabilistic polynomial-time adversaries $\mathcal{A}$,*

[12] Note that in addition to the simulation-sound binding property being modified, our definition of the underlying trapdoor commitment scheme is slightly different than the one given in [24].

$$\Pr[(pk, sk) \leftarrow \mathsf{TCgen}(1^k); (\mathsf{com}, tag, v_1, v_2, \mathsf{dec}_1, \mathsf{dec}_2) \leftarrow \mathcal{A}^{\mathcal{O}_{pk,sk}}(pk) : \\ (\mathsf{TCver}(pk, \mathsf{com}, v_1, tag, \mathsf{dec}_1) = \mathsf{TCver}(pk, \mathsf{com}, v_2, tag, \mathsf{dec}_2) = 1) \\ \wedge (v_1 \neq v_2) \wedge tag \notin Q] \\ \leq_{\text{ev}} \alpha(k),$$

*where $\mathcal{O}_{pk,sk}$ operates as follows, with $Q$ initially set to $\emptyset$:*

- *On input* (commit, $tag$)*:*
  *compute* $(\widetilde{\mathsf{com}}, \xi) \leftarrow \mathsf{TCfakeCom}(pk, sk, tag)$, *store* $(\widetilde{\mathsf{com}}, tag, \xi)$, *and add* $tag$ *to* $Q$. *Return* $\widetilde{\mathsf{com}}$.
- *On input* (decommit, $\widetilde{\mathsf{com}}, v$)*:*
  *if for some tag and some* $\xi$, *a tuple* $(\widetilde{\mathsf{com}}, tag, \xi)$ *is stored, compute* $\widetilde{\mathsf{dec}} \leftarrow \mathsf{TCfakeDecom}(\xi, v)$. *Return* $\widetilde{\mathsf{dec}}$.

For the remainder of the paper, SSTC will refer to this new definition, and SSTC(GMY) will refer to the old definition of [24].

Now we construct SSTC schemes based on specific cryptographic assumptions, and sketch the proofs showing that they achieve simulation-sound binding.

*SSTC scheme based on any one-way function* Here we present an efficient SSTC scheme TC based on a signature scheme, which in turn may be based on any one-way function [36]. TC is the aHC scheme from Canetti *et al.* [7] with the following changes:

1. The underlying commitment scheme based on one-way permutations is replaced by the commitment scheme of Naor [32] based on pseudorandom generators (which can be built from any one-way function).
2. An extra parameter $tag$ is included, and the one-way function $f$ and corresponding NP language $\{y | \exists\, x \text{ s.t. } y = f(x)\}$ used in the underlying non-interactive Feige-Shamir trapdoor commitment [21] is replaced by the signature verification relation $\{((\mathsf{sig\_vk}, tag), \sigma) | 1 = \mathsf{sig\_verify}(\mathsf{sig\_vk}, tag, \sigma)\}$.

We omit the detailed description and proof of the the simulation-soundness of the scheme in this extended abstract.

*SSTC scheme based on DSA* Here we present an efficient SSTC scheme TC based on DSA. It is a simplified version of the DSA-based SSTC(GMY) scheme from [24]. $\mathsf{TCgen}(1^k)$ generates a DSA public/private key pair $(pk, sk)$, where $pk = (g, p, q, y)$ and $sk = (g, p, q, x)$. For a message $m \in \mathbb{Z}_q$, $\mathsf{TCcom}((g, p, q, y), m, tag)$ first computes $\alpha \stackrel{R}{\leftarrow} \mathbb{Z}_q$, $g' \leftarrow g^{\alpha} \bmod p$, and $h \leftarrow g^{H(tag)} y^{g'} \bmod p$. (Note that if $s$ is the discrete log of $h$ over $g'$, then $(g' \bmod q, s)$ is a DSA signature for $tag$.) Then it generates a Pedersen commitment [34] to $m$ over bases $(g', h)$, i.e., it generates $\beta \stackrel{R}{\leftarrow} \mathbb{Z}_q$ and computes the commitment/decommitment pair $((g', c), \beta)$, where $c \leftarrow (g')^{\beta} h^m$. $\mathsf{TCver}((g, p, q, y), (g', c), m, tag, \beta)$ verifies that $c \equiv (g')^{\beta} h^m$, where $h \equiv g^{H(tag)} y^{g'} \bmod p$. $\mathsf{TCfakeCom}((g, p, q, y), (g, p, q, x), tag')$ computes a DSA signature $(g'', s)$ on $tag'$ using the secret key (g,p,q,x), computes the values $g' \leftarrow (g^{H(tag')} y^{g''})^{s^{-1}} \bmod p$ and $h \leftarrow (g')^s \bmod p$, generates $\beta' \stackrel{R}{\leftarrow} \mathbb{Z}_q$, and sets $c \leftarrow h^{\beta'} \bmod p$. It outputs commitment $(g', c)$ and auxiliary information $(q, \beta', s)$. Then $\mathsf{TCfakeDecom}((q, \beta', s), m)$ outputs $(m, (\beta' - m)s \bmod q)$, which is a decommitment to $m$.

To show the simulation-sound binding property, we show that if an adversary can break this property, we can break DSA as follows. (We assume that DSA is existentially unforgeable against an adaptive chosen-message attack.) Take a DSA key $vk_0$ and its corresponding DSA signature oracle (from the definition of existential unforgeability against an adaptive chosen-message attack). It is easy to see that the equivocation oracle, and in particular the commit queries to that oracle, may be implemented using the DSA signature oracle on the requested *tag*'s.

Now say the adversary gives a double opening with *tag*, for which no commitment was requested, and thus no call to the DSA signature oracle was made. In particular, say it gives openings $(m, \beta)$ and $(m', \beta')$ of $(g', c)$. Then $(g' \bmod q, (\beta' - \beta)/(m - m') \bmod q)$ is a signature on *tag*, breaking DSA.

*SSTC scheme based on Cramer-Shoup signatures* Here we present an efficient SSTC scheme $\mathsf{TC}$ based on Cramer-Shoup signatures [11] and as secure as strong RSA. (We note that the more efficient version of the Cramer-Shoup signature scheme in Fischlin [22] could be used here as well to obtain an even more efficient SSTC scheme.) $\mathsf{TCgen}(1^k)$ generates a public/private key pair $(pk, sk)$ for Cramer-Shoup signatures, where $pk = (N, h, x, e', H)$ and $sk = (p, q)$. For a message $m \in \{0,1\}^k$, $\mathsf{TCcom}((N, h, x, e', H), m, tag)$ first computes $(y', x', e)$ as in the Cramer-Shoup signature protocol for *tag*, and sets $x'' \leftarrow xh^{H(x')} \bmod N$. (Note that if $y$ is $e$th root of $x''$ modulo $N$, then $\langle e, y, y' \rangle$ is a Cramer-Shoup signature for *tag*.) Then it uses the unconditionally-hiding commitment scheme from [9] based on $e$-one-way homomorphisms (specifically, based on the RSA encryption function with public key $(e, N)$, i.e., $f(a) : a^e \bmod N$) over base $x''$ to commit to $m$. That is, it chooses $\beta \stackrel{R}{\leftarrow} \mathbb{Z}_N^*$ and computes the commitment/decommitment pair $((y', e, c), \beta)$, where $c \leftarrow (x'')^m \beta^e \bmod N$. $\mathsf{TCver}((N, h, x, e', H), (y', e, c), m, tag, \beta)$ verifies that $e$ is an odd $k+1$-bit integer different from $e'$, $c \equiv (x'')^m \beta^e \bmod N$, and $x'' \equiv xh^{H(x')} \bmod N$, where $x'$ is computed from $y'$ and $e$ as in the Cramer-Shoup signature protocol.

$\mathsf{TCfakeCom}((N, h, x, e', H), (p, q), tag')$ first computes a signature $\langle e, y, y' \rangle$ on $tag'$ using the secret key. Then it computes $x' \leftarrow (y')^{e'} h^{-H(tag')} \bmod N$ and $x'' \leftarrow xh^{H(x')} \bmod N$, generates $\beta' \stackrel{R}{\leftarrow} \mathbb{Z}_N^*$, and sets $c \leftarrow (\beta')^e \bmod N$. It outputs commitment $(y', e, c)$ and auxiliary information $(N, \beta', y)$. Finally, the function $\mathsf{TCfakeDecom}((N, \beta', y), m)$ outputs $(m, \beta' y^{-m} \bmod N)$, which is a decommitment to $m$.

To show the simulation-sound binding property, we show that if an adversary can break this property, we can break the Strong RSA assumption. The proof basically follows the proof of security (i.e., existential unforgeability against adaptive chosen-message attack) of the Cramer-Shoup signature scheme from [11], which for brevity we will call the *CSSig proof.* As in the CSSig proof, we divide adversaries into Types I, II, and III. For each type, we respond to commit queries to the equivocation oracle using signatures as computed in the responses to the corresponding signature queries in the CSSig proof. Finally, instead of the adversary producing a forged signature, the adversary gives a double opening of a commitment with some *tag* for which no commit query was made (and thus for which no corresponding signature query was necessary). In particular, say the adversary gives openings $(m, \beta)$ and $(m', \beta')$ of $(y', e, c)$ with $m > m'$. Then $(x'')^{m-m'} \equiv (\beta'\beta^{-1})^e \bmod N$. In the case of Type I and Type II adversaries, i.e., when $e$ is produced in response to a commit query, $e$ is prime and $e > m - m'$. Therefore the value $y$ such that $y^e \equiv x \bmod N$ may be computed (e.g., using the Extended Euclidean Algorithm) and $\langle e, y, y' \rangle$ is a signature on *tag*. Then as in the corresponding cases in the CSSig proof, this can be shown to break the standard RSA assumption. In the case of a Type III adversary, $e$ is not necessarily prime, so we may not necessarily obtain a signature on *tag*. However, the CSSig proof simply uses the fact that $x'' \equiv (\beta'\beta^{-1})^e \bmod N$ to show that Strong RSA can be broken, and the equation $(x'')^{m-m'} \equiv (\beta'\beta^{-1})^e \bmod N$ that we obtain can be used in a similar way to show that Strong RSA can be broken. We omit the details.

# 4 Application to ZK Proofs

We show how an SSTC scheme can be used to construct unbounded simulation-sound ZK protocols, unbounded non-malleable ZK protocols, and universally composable ZK protocols. Our constructions are conceptually simpler than those given by Garay *et al.* [24].

All our results will be in the *common reference string* (CRS) model, which assumes that there is a string uniformly generated from some distribution and is available to all parties at the start of a protocol. Note that this is a generalization of the *public random string* model, where a uniform distribution over fixed-length bit strings is assumed.

## 4.1 Unbounded Simulation Sound and Non-malleable ZK

Intuitively, a ZK protocol is unbounded simulation sound if an adversary cannot convince the verifier of a false statement with non-negligible probability, even after interacting with an arbitrary number of (simulated) provers. We refer the readers to [24] for a formal definition.

Our construction starts with a class of three-round, public-coin, honest-verifier zero-knowledge protocols, also known as $\Sigma$-protocols [10].

Consider a binary relation $R(x, w)$ that is computable in polynomial time. A $\Sigma$-protocol $\Pi$ for the relation $R$ proves membership of $x$ in the language $L_R = \{x | \exists w, \text{ s.t. } R(x, w) = 1\}$. For a given $x$, let $(a, c, z)$ denote the conversation between the prover and the verifier. To compute the first and the final messages, the prover invokes efficient algorithms $a_\Pi(x, w, r)$ and $z_\Pi(x, w, r, c)$, respectively, where $w$ is the witness, $r$ is the random bits, and $c$ is the challenge from the verifier (as the second message). Using an efficient predicate $\phi(x, a, c, z)$, the verifier decides whether the conversation is accepting with respect to $x$. The relation $R$, and the algorithms $a(\cdot)$, $z(\cdot)$ and $\phi(\cdot)$, are public.

We assume the protocol $\Pi$ has a simulator $\mathcal{S}_\Pi$ that, taking the challenge as input, generates an accepting conversation. More precisely, $(a, c, z) \leftarrow \mathcal{S}_\Pi(c)$, where that the distribution of $(a, c, z)$ is computationally indistinguishable from the real conversation.

The protocol $\mathsf{USS}^R_{[pk]}(x)$ is shown in Figure 2, and uses an SSTC scheme $\mathsf{TC}$. Say $\Pi$ is a $\Sigma$-protocol for relation $R$. The prover generates a pair $(\mathsf{sig\_vk}, \mathsf{sig\_sk})$ for a strong one-time signature scheme and sends $\mathsf{sig\_vk}$ to the verifier. Then the prover generates the first message $a$ of $\Pi$ and sends its commitment $\mathsf{com}$ to the verifier, using the signature verification key $\mathsf{sig\_vk}$ as the commitment *tag*. After receiving the challenge $c$, the prover generates and sends the third message $z$ of $\Pi$, opens the commitment $\mathsf{com}$, signs the entire transcript using the signing key $\mathsf{sig\_sk}$, and sends the signature on the transcript to the verifier. (To be specific, the *transcript* consists of all values sent or received by the prover in the protocol, except the final signature.)

| prover | | verifier |
|---|---|---|
| $(\mathsf{sig\_vk}, \mathsf{sig\_sk}) \leftarrow \mathsf{sig\_gen}_1(1^k)$ | | |
| $a \leftarrow a_\Pi(x, w, r)$ | | |
| $(\mathsf{com}, \mathsf{dec}) \leftarrow \mathsf{TCcom}(pk, a, \mathsf{sig\_vk})$ | $\xrightarrow{\mathsf{sig\_vk},\, \mathsf{com}}$ | |
| | $\xleftarrow{c}$ | |
| $z \leftarrow z_\Pi(x, w, r, c)$ | | |
| $s \leftarrow \mathsf{sig\_sign}_1(\mathsf{sig\_sk}, \text{transcript})$ | $\xrightarrow{a,\, \mathsf{dec},\, z,\, s}$ | $\mathsf{TCver}(pk, \mathsf{com}, a, \mathsf{sig\_vk}, \mathsf{dec})$ |
| | | $\phi_\Pi(x, a, c, z)$ |
| | | $\mathsf{sig\_verify}_1(\mathsf{sig\_vk}, \text{transcript}, s)$ |

**Fig. 2.** $\mathsf{USS}^R_{[pk]}(x)$: An unbounded simulation-sound ZK protocol for relationship $R$ with common input $x$ and common reference string $pk$, where $pk$ is drawn from the distribution $\mathsf{TCgen}(1^k)$. The prover also knows the witness $w$ such that $R(x, w) = 1$.

**Theorem 1.** *The protocol* $\mathsf{USS}^R_{[pk]}(x)$ *is a USSZK argument.*

Intuitively, a ZK protocol is unbounded non-malleable if an efficient witness extractor successfully extracts a witness from any adversary that causes the verifier to accept, even when the adversary is also allowed to interact with any number of (simulated) provers. Again, we refer the readers to [24] for a formal definition.

Our construction of the NMZK protocol is very similar to that of the USSZK protocol presented above, where the only difference is that the $\Sigma$-protocol is replaced by an $\Omega$-protocol. Recall that an $\Omega$-protocol [24] is like a $\Sigma$-protocol with the additional property that it admits a polynomial-time, straight-line extractor (an $\Omega$-protocol works in the CRS model).

The protocol $\mathsf{NM}^R_{[pk,\sigma]}(x)$ is very similar to the protocol in Figure 2, but note that here we assume that $\Pi$ is an $\Omega$-protocol with $\sigma$ being the CRS.

**Theorem 2.** *The protocol* $\mathsf{NM}^R_{[pk,\sigma]}(x)$ *is an NMZK argument of knowledge for the relation* $R$.

### 4.2 Universally Composable ZK

The universal composability paradigm was proposed by Canetti [5] for defining the security and composition of protocols. To define security one first specifies an *ideal functionality* using a trusted party that describes the desired behavior of the protocol. Then one proves that a particular protocol operating in a real-life model securely realizes this ideal functionality, as defined below. Here we briefly summarize the framework.

A (real-life) protocol $\pi$ is defined as a set of $n$ interactive Turing Machines $P_1, \ldots, P_n$, designating the $n$ parties in the protocol. It operates in the presence of an environment $\mathcal{Z}$ and an adversary $\mathcal{A}$, both of which are also modeled as interactive Turing Machines. The environment $\mathcal{Z}$ provides inputs and receives outputs from honest parties, and may communicate with $\mathcal{A}$. $\mathcal{A}$ controls (and may view) all communication between the parties. (Note that this models asynchronous communication on open point-to-point channels.) We will assume that messages are authenticated, and thus $\mathcal{A}$ may not insert or modify messages between honest parties.[13] $\mathcal{A}$ also may corrupt parties, in which case it obtains the internal state of the party. (In the non-erasing model, the internal state would encompass the complete internal history of the party.)

The ideal process with respect to a functionality $\mathcal{F}$, is defined for $n$ parties $P_1, \ldots, P_n$, an environment $\mathcal{Z}$, and an (ideal-process) adversary $\mathcal{S}$. However, $P_1, \ldots, P_n$ are now dummy parties that simply forward (over secure channels) inputs received from $\mathcal{Z}$ to $\mathcal{F}$, and forward (again over secure channels) outputs received from $\mathcal{F}$ to $\mathcal{Z}$. Thus the ideal process is a trivially secure protocol with the input-output behavior of $\mathcal{F}$.

*The zero-knowledge functionality.* The (multi-session) ZK functionality as defined by Canetti [5] is given in Figure 3. In the functionality, parameterized by a relation $R$, the prover sends to the functionality the input $x$ together with a witness $w$. If $R(x, w)$ holds, then the functionality forwards $x$ to the verifier. As pointed out in [5], this is actually a proof of knowledge in that the verifier is assured that the prover actually knows $w$.

Garay *et al.* [24] proved that any "augmentable" NMZK protocol can be easily converted to a UCZK protocol in the $\mathcal{F}^{\mathcal{D}}_{\mathrm{CRS}}$-hybrid model, assuming static corruptions. Intuitively, an NMZK protocol is augmentable if the first message sent by the prover contains the common input $x$ and a special field $\mathsf{aux}$ in which the prover can fill with an arbitrary string without compromising security. (In

[13] This feature could be added to an unauthenticated model using a message authentication functionality as described in [5].

$\hat{\mathcal{F}}_{\mathrm{ZK}}^{R}$ proceeds as follows, running parties $P_1, \ldots, P_n$, and an adversary $\mathcal{S}$:
- Upon receiving (zk-prover, $sid, ssid, P_i, P_j, x, w$) from $P_i$: If $R(x, w)$ then send (ZK-PROOF, $sid, ssid, P_i, P_j, x$) to $P_j$ and $\mathcal{S}$. Otherwise, ignore.

**Fig. 3.** The (multi-session) zero-knowledge functionality (for relation $R$)

| $P_i$ (prover) | | $P_j$ (verifier) |
|---|---|---|
| $a \leftarrow a_\Pi(x, w, r, \sigma)$ | | |
| $tag \leftarrow \langle P_i, P_j \rangle$ | | |
| $(\mathsf{com}, \mathsf{dec}) \leftarrow \mathsf{TCcom}(pk, a, tag)$ | $\xrightarrow{x, \mathsf{com}}$ | |
| | $\xleftarrow{c}$ | |
| $z \leftarrow z_\Pi(x, w, r, c, \sigma)$ | $\xrightarrow{a, \mathsf{dec}, z}$ | $tag \leftarrow \langle P_i, P_j, \rangle$ |
| | | $\mathsf{TCver}(pk, \mathsf{com}, a, tag, \mathsf{dec})$ |
| | | $\phi_\Pi(x, a, c, z)$ |

**Fig. 4.** $\mathsf{MYZK}^{R}_{[pk,\sigma]}(x)$: A UCZK protocol for relationship $R$ with common reference string $(pk, \sigma)$ where $pk$ is drawn from the distribution $\mathsf{TCgen}(1^k)$ and $\sigma$ is drawn from the distribution of the CRS for protocol $\Pi$.

the conversion to UCZK in [24], the auxiliary string contains the *sid*, the *ssid*, and the identities of the prover and verifier.)

It can be readily verified that the protocol $\mathsf{NM}^{R}_{[pk,\sigma]}(x)$ can be easily made augmentable by adding $x$ and aux in the first message. We denote the slightly modified protocol where the aux field is set to $(sid, ssid, P_i, P_j)$ by $\mathsf{ANM}^{R}_{[pk,\sigma]}(x)$. Then it follows that $\mathsf{ANM}^{R}_{[pk,\sigma]}(x)$ is a UCZK protocol for relation $R$, assuming static corruptions.

However, one can simplify this protocol by removing the one-time signature scheme, only including the identities of the prover and verifier in the auxiliary string, and using this auxiliary string as the tag of the commitment scheme. This simplified scheme, $\mathsf{MYZK}^{R}_{[pk,\sigma]}(x)$, is shown in Figure 4. (Note that since we are assuming authenticated communication in the UC framework, the identities $P_i$ and $P_j$ will be known to both parties, and thus do not need to be explicitly sent in our protocol.) Furthermore, this protocol can be easily modified into one that remains secure against adaptive corruption in the erasing model. In fact, all that is needed is to have the prover erase the randomness used in the $\Omega$-protocol before sending the final message.

**Theorem 3.** *The protocol* $\mathsf{MYZK}^{R}_{[pk,\sigma]}(x)$ *is a UCZK protocol for relation* $R$*, assuming static corruptions. By erasing the randomness* $(r)$ *used in the* $\Omega$*-protocol before the final message, it is a UCZK protocol for relation* $R$*, assuming adaptive corruption (in the erasing model).*

## 5 Comparison to Non-malleable Commitments

We explore the exact relation between SSTC schemes and NMC schemes.

Our definition for non-malleable (NM) commitments is based on the definition in [17], which, technically speaking, defines the notion of $\epsilon$-non-malleability, instead of strict non-malleability. For the clarity of presentation, we shall use the term "non-malleability" to mean $\epsilon$-non-malleability, and will note any places where our results have application to strict non-malleability.

Informally, similar to the definition in [17], we say a commitment scheme is non-malleable if when an adversary sees a commitment $\mathsf{com}_1$, generates its own commitment $\mathsf{com}_2$, and sees $\mathsf{com}_1$ opened, it cannot then open $\mathsf{com}_2$ to a value related to $\mathsf{com}_1$ with any greater probability than a simulator that never saw $\mathsf{com}_1$ in the first place.[14] Note that this is also called *non-malleability with respect to opening* [16] and differs from the original definition of [18] that was discussed in the introduction, and which is also called *non-malleability with respect to commitment.* Our definition differs from the definition in [17] as follows.

- We only define NM *trapdoor* commitment (NMTC) schemes, since that is what will be of most interest in comparisons to SSTC schemes. Non-trapdoor versions of these definitions are straightforward.
- We use tag-based definitions instead of body-based definitions. Again this is what will be of most interest in comparisons to SSTC schemes. Body-based definitions are straightforward. In fact, most of our results relating SSTC schemes and NMTC schemes also hold when these schemes are defined using body-based definitions. We will discuss this later.

Due to space limitations, we omit the formal definition of an NMTC scheme. It may be obtained in a straightforward manner from the formal definition in [17] and the changes described above.

As mentioned in the introduction, the recent work of Damgård and Groth [14] generalizes and strengthens the definition of non-malleable commitments to be reusable, i.e., to have the property that seeing one *or more* commitments does not give another party any advantage in generating one *or more* commitments that can later be opened to values related to the values in the original commitments. Their definition also stipulates that the distribution of committed messages is dependent on the public key. However, we will continue to use the simpler definition, since it exemplifies the relation between SSTC schemes and NMTC schemes. Later we will discuss how to obtain similar relations to reusable NMTC schemes.

Note that we can generalize the definition of NMTC to NMTC($\ell$) schemes, which are NMTC schemes in which the adversary is allowed to query an oracle $\mathcal{O}_{pk,sk}$ as defined in the SSTC definition, but with at most $\ell$ commit queries allowed, and with the restriction that the commitment produced by the adversary has a tag that is not used in any of the commit queries. Note that an NMTC

[14] Slightly more formally, we say that it is $\epsilon$-non-malleable if for all $\epsilon$ it cannot do this with probability non-negligibly greater than $\epsilon$.

scheme is an $\mathsf{NMTC}(0)$ scheme. We use $\ell = \infty$ to denote an oracle which accepts an unbounded number of $\mathsf{commit}$ queries.

We similarly generalize the definition of SSTC schemes and consider $\mathsf{SSTC}(\ell)$ schemes. Then an $\mathsf{SSTC}(0)$ scheme is just a TC scheme, and an $\mathsf{SSTC}(\infty)$ scheme is what we have called an SSTC scheme.

As mentioned above, we have defined NMTC schemes as tag-based, as opposed to body-based, as usually seen in literature [18,16,23,17,14]. However, this is not a significant distinction since there exists fairly generic reductions from one to the other. Our next theorem shows such a reduction from body-based NMTC schemes to tag-based ones.

Here, we assume the commitment scheme allows commitments to strings of arbitrary length. A similar theorem could be shown for commitment schemes which allow only fixed length commitments, say of length equal to the security parameter.

**Theorem 4.** *Let* $\mathsf{TC}$ *be a body-based* $\mathsf{NMTC}$ *scheme. Let* $\mathsf{TC}'$ *be* $\mathsf{TC}$*, but with the tag added to the message being committed. That is,* $\mathsf{TCgen}'(1^k)$ *returns the result of* $\mathsf{TCgen}(1^k)$*,* $\mathsf{TCcom}'(pk, v, tag)$ *returns the result of* $\mathsf{TCcom}(pk, \langle v, tag\rangle, tag)$*, and* $\mathsf{TCver}'(pk, \mathsf{com}, v, tag, \mathsf{dec})$ *returns the result of* $\mathsf{TCver}(pk, \mathsf{com}, \langle v, tag\rangle, tag, \mathsf{dec})$*. Then* $\mathsf{TC}'$ *is a tag-based* $\mathsf{NMTC}$ *scheme.*

Considering the problem of converting tag-based $\mathsf{SSTC}$ or $\mathsf{NMTC}$ schemes to body-based $\mathsf{SSTC}$ or $\mathsf{NMTC}$ schemes, it seems that a simple construction like the one in Theorem 4 does not suffice. Instead, one could construct a body-based scheme by generating a verification/signing key pair for a strong one-time signature scheme, using the verification key as the tag in the tag-based commitment, signing the tag-based commitment using the signing key, and giving the pair (the tag-based commitment and the associated signature) as the full commitment. As this is a fairly standard technique, used in, e.g. [24], we omit the analysis here.

### 5.1 Relations between SSTC and NMTC

First we show that for all $\ell \geq 0$, an $\mathsf{SSTC}(\ell+1)$ scheme is also an $\mathsf{NMTC}(\ell)$ scheme, and an $\mathsf{NMTC}(\ell)$ scheme is also an $\mathsf{SSTC}(\ell+1)$ scheme.

**Theorem 5.** *Let* $\mathsf{TC}$ *be an* $\mathsf{SSTC}(\ell+1)$ *scheme. Then* $\mathsf{TC}$ *is an* $\mathsf{NMTC}(\ell)$ *scheme.*

**Theorem 6.** *Let* $\mathsf{TC}$ *be an* $\mathsf{NMTC}(\ell)$ *scheme. Then* $\mathsf{TC}$ *is an* $\mathsf{SSTC}(\ell)$ *scheme.*

To relate our results to reusable non-malleable commitment schemes as defined in [14], we need to consider adversaries that input a vector of commitments (and later decommitments), and output a vector of commitments (and later decommitments). To be specific, let $(t, u)$-$\mathsf{NMTC}(\ell)$ denote a reusable NMTC commitment scheme with an input vector of size $t$ and an output vector of size $u$.

Then using a proof similar to above, but with some additional ideas from [14], we can prove the following theorem.[15]

**Theorem 7.** *Let* TC *be an* SSTC$(\ell + t)$ *scheme. Then* TC *is a* $(t, u)$-NMTC$(\ell)$ *scheme.*

Finally, we show the following separation results.

**Theorem 8.** *Assuming the hardness of the discrete logarithm problem, there exists an* SSTC$(\ell)$ *scheme that is not* NMTC$(\ell)$*, for every* $\ell \geq 0$.

**Theorem 9.** *If there exists an* NMTC$(\ell)$ *scheme, then there exists an* NMTC$(\ell)$ *scheme that is not* SSTC$(\ell + 1)$.

## References

1. N. Barić and B. Pfitzmann. Collision-free accumulators and fail-stop signature schemes without trees. In *Advances in Cryptology – EUROCRYPT '97* (LNCS 1233), 480–494, 1997.
2. D. Beaver. Adaptive zero-knowledge and computational equivocation. In *28th ACM Symp. on Theory of Computing*, 629–638, 1996.
3. M. Blum. Coin flipping by telephone. In *IEEE Spring COMPCOM*, pp. 133–137, 1982.
4. G. Brassard, D. Chaum, and C. Crépeau. Minimum Disclosure Proofs of Knowledge. *JCSS*, 37(2):156–189, 1988.
5. R. Canetti. Universally composable security: A new paradigm for cryptographic protocols. In *42nd IEEE Symp. on Foundations of Computer Sci.*, 136–145, 2001.
6. R. Canetti and M. Fischlin. Universally composable commitments. In *Advances in Cryptology – CRYPTO 2001* (LNCS 2139), 19–40, 2001.
7. R. Canetti, Y. Lindell, R. Ostrovsky and A. Sahai. Universally composable two-party computation. In 34th ACM Symp. on Theory of Computing, 494–503, 2002. Full version in *ePrint archive*, Report 2002/140. `http://eprint.iacr.org/`, 2002.
8. S. A. Cook. The complexity of theorem-proving procedures. In *3rd IEEE Symp. on Foundations of Computer Sci.*, 151–158, 1971.
9. R. Cramer and I. Damgård. Zero-Knowledge Proofs for Finite Field Arithmetic, or: Can Zero-Knowledge Be for Free? In *Advances in Cryptology – CRYPTO '98* (LNCS 1462), pages 424–441, 1998.
10. R. Cramer, I. Damgård, and B. Schoenmakers. Proofs of partial knowledge and simplified design of witness hiding protocols. In *Advances in Cryptology – CRYPTO '94* (LNCS 839), pages 174–187, 1994.
11. R. Cramer and V. Shoup. Signature scheme based on the strong RSA assumption. In *ACM Trans. on Information and System Security* 3(3):161-185, 2000.
12. I. Damgård. On the existence of bit commitment schemes and zero-knowledge proofs. In *Advances in Cryptology – CRYPTO '89* (LNCS 435), 17–29, 1989.

[15] As in [14], we change the definition of a valid relation (over vectors of messages) to one in which all messages including $\perp$ are allowed, but where the probability of the relation being true cannot be increased by changing a message in the second (adversarially-chosen) vector to $\perp$.

13. I. Damgård. Efficient Concurrent Zero-Knowledge in the Auxiliary String Model. In *Advances in Cryptology – EUROCRYPT 2000* (LNCS 1807), 418–430, 2000.
14. I. Damgård and J. Groth. Non-interactive and reusable non-malleable commitment schemes. In *35th ACM Symp. on Theory of Computing*, 426–437, 2003.
15. I. Damgård and J. Nielsen. Perfect hiding and perfect binding universally composable commitment schemes with constant expansion factor. In *Advances in Cryptology – CRYPTO 2002* (LNCS 2442), 581–596, 2002. Full version in *ePrint Archive*, report 2001/091. `http://eprint.iacr.org/`, 2001.
16. G. Di Crescenzo, Y. Ishai, and R. Ostrovsky. Non-interactive and non-malleable commitment. In *30th ACM Symp. on Theory of Computing*, 141–150, 1998.
17. G. Di Crescenzo, J. Katz, R. Ostrovsky, and A. Smith. Efficient and Non-Interactive Non-Malleable Commitment. In *Advances in Cryptology – EUROCRYPT 2001* (LNCS 2045), 40–59, 2001.
18. D. Dolev, C. Dwork and M. Naor. Non-malleable cryptography. *SIAM J. on Comput.*, 30(2):391–437, 2000. Also in *23rd ACM Symp. on Theory of Computing*, 542–552, 1991.
19. S. Even, O. Goldreich, and S. Micali. On-line/Off-line digital signatures. *J. Cryptology* 9(1):35-67 (1996).
20. U. Feige and A. Shamir. Witness Indistinguishable and Witness Hiding Protocols. In *22nd ACM Symp. on Theory of Computing*, 416–426, 1990.
21. U. Feige and A. Shamir. Zero-Knowledge Proofs of Knowledge in Two Rounds. In *Advances in Cryptology – CRYPTO '89* (LNCS 435), 526–544, 1989.
22. M. Fischlin. The Cramer-Shoup strong-RSA signature scheme revisited. In *Public Key Cryptography – PKC 2003* (LNCS 2567), 116–129, 2003.
23. M. Fischlin and R. Fischlin. Efficient non-malleable commitment schemes. In *Advances in Cryptology – CRYPTO 2000* (LNCS 1880), 413–431, 2000.
24. J. A. Garay, P. MacKenzie, and K. Yang. Strengthening Zero-Knowledge Protocols using Signatures. In *Advances in Cryptology – EUROCRYPT 2003* (LNCS 2656), 177–194, 2003.
25. R. Gennaro. Improved Proofs of Knowledge Secure under Concurrent Man-in-the-middle Attacks and their Applications. In *ePrint Archive*, report 2003/214. `http://eprint.iacr.org/`, 2003.
26. O. Goldreich, S. Micali and A. Wigderson. Proofs that yield nothing but their validity or All languages in NP have zero-knowledge proof systems. *J. ACM*, 38(3):691–729, 1991.
27. S. Goldwasser, S. Micali and R. Rivest. A digital signature scheme secure against adaptive chosen-message attacks. *SIAM J. Comput.*, 17:281–308, 1988.
28. S. Jarecki and A. Lysyanskaya. Adaptively Secure Threshold Cryptography: Introducing Concurrency, Removing Erasures. In *Advances in Cryptology – EUROCRYPT 2000* (LNCS 1807), 221–242, 2000.
29. D. W. Kravitz. Digital signature algorithm. U.S. Patent 5,231,668, 27 July 1993.
30. L. A. Levin. Universal sorting problems. *Problemy Peredaci Informacii*, 9:115–116, 1973. In Russian. Engl. trans.: *Problems of Information Transmission* 9:265–266.
31. P. MacKenzie and K. Yang. On simulation-sound trapdoor commitments (full version). Available on the Cryptology ePrint Archive: `http://eprint.iacr.org/2003/252`.
32. M. Naor. Bit commitment Using Pseudo-Randomness. *J. Cryptology* 4(2):151–158 (1991).
33. M. Naor, R. Ostrovsky, R. Venkatesan, and M. Yung. Perfect zero-knowledge arguments for NP can be based on general complexity assumptions. In *Advances in Cryptology – CRYPTO '92* (LNCS 740), 196–214, 1992.

34. T. P. Pedersen. Non-Interactive and Information-Theoretic Secure Verifiable Secret Sharing. In *Advances in Cryptology – CRYPTO '91* (LNCS 576), 129–140, 1991.
35. L. Reyzin. Zero-knowledge with public keys. Ph.D. Thesis, MIT, 2001.
36. J. Rompel. One-way functions are necessary and sufficient for secure signatures. In *22nd ACM Symp. on Theory of Computing*, 387–394, 1990.
37. A. Sahai. Non-malleable non-interactive zero knowledge and adaptive chosen-ciphertext security. In *40th IEEE Symp. on Foundations of Computer Sci.*, 543–553, 1999.

# Hash Function Balance and Its Impact on Birthday Attacks

Mihir Bellare and Tadayoshi Kohno

Dept. of Computer Science & Engineering, University of California, San Diego
9500 Gilman Drive, La Jolla, CA 92093, USA
{mihir,tkohno}@cs.ucsd.edu
http://www-cse.ucsd.edu/users/{mihir,tkohno}

**Abstract.** Textbooks tell us that a birthday attack on a hash function $h$ with range size $r$ requires $r^{1/2}$ trials (hash computations) to find a collision. But this is quite misleading, being true only if $h$ is regular, meaning all points in the range have the same number of pre-images under $h$; if $h$ is not regular, *fewer* trials may be required. But how much fewer? This paper addresses this question by introducing a measure of the "amount of regularity" of a hash function that we call its balance, and then providing estimates of the success-rate of the birthday attack, and the expected number of trials to find a collision, as a function of the balance of the hash function being attacked. In particular, we will see that the number of trials can be significantly less than $r^{1/2}$ for hash functions of low balance. This leads us to examine popular design principles, such as the MD (Merkle-Damgård) transform, from the point of view of balance preservation, and to mount experiments to determine the balance of popular hash functions.

## 1 Introduction

BIRTHDAY ATTACKS. Let $h\colon D \to R$ be a hash function. In a birthday attack, we pick points $x_1, \ldots, x_q$ from $D$ and compute $y_i = h(x_i)$ for $i = 1, \ldots, q$. The attack is successful if there exists a collision, i.e. a pair $i, j$ such that $x_i \neq x_j$ but $y_i = y_j$. We call $q$ the number of *trials*.

There are several variants of this attack which differ in the way the points $x_1, \ldots, x_q$ are chosen (cf. [4,8,9,10]). The one we consider is that they are chosen independently at random from $D$.[1]

Textbooks (eg. Stinson [8, Section 7.3]) say that (due to the birthday phenomenon which gives the attack its name) a collision is expected within $r^{1/2}$ trials, where $r$ denotes the size of the range of $h$. In particular, they say that

[1] One might ask how to mount the attack (meaning how to pick random domain points) when the domain is a very large set as in the case of a hash function like SHA-1 whose domain is the set of all strings of length at most $2^{64}$. We would simply let $h$ be the restriction of SHA-1 to inputs of some reasonable length, like 161 bits or 320 bits. A collision for $h$ is a collision for SHA-1, so it suffices to attack the restricted function.

C. Cachin and J. Camenisch (Eds.): EUROCRYPT 2004, LNCS 3027, pp. 401–418, 2004.

collisions in a hash function with output length $m$ bits can be found in about $2^{m/2}$ trials. This estimate is the basis for the choice of hash function length $m$, which is typically made just large enough to make $2^{m/2}$ trials infeasible.

However Stinson's analysis [8, Section 7.3], as well as all others that we have seen, are misleading, for they assume the hash function is *regular*, meaning all points in the range have the same number of pre-images under $h$.[2] It turns out that if $h$ is not regular, it takes *fewer* than $r^{1/2}$ trials to find a collision, meaning the birthday attack would succeed sooner than expected.

This could be dangerous, for we do not know that popular hash functions are regular. In fact they are usually designed to have "random" behavior and thus would not be regular. Yet, one might say, they are probably "almost" regular. But what exactly does this mean, and how does the "amount of regularity" affect the number of trials to success in the birthday attack? Having answers to such questions will enable us to better assess the *true* impact of birthday attacks.

THIS PAPER. To help answer questions such as those posed above, this paper begins by introducing a measure of the "amount of regularity" that we call the *balance* of a hash function. This is a real number between 0 and 1, with balance 1 indicating that the hash function is regular and balance 0 that it is a constant function, meaning as irregular as can be. We then provide quantitative estimates of the success-rate, and number of trials to success, of the birthday attack, as a function of the balance of the hash function being attacked.

This yields a tool that has a variety of uses, and lends insight into various aspects of hash function design and parameter choices. For example, by analytically or experimentally estimating the balance of a particular hash function, we can tell how quickly the birthday attack on this hash function will succeed. Let us now look at all this in more detail.

THE BALANCE MEASURE. View the range $R$ of hash function $h: D \to R$ as consisting of $r \geq 2$ points $R_1, \ldots, R_r$. For $i = 1, \ldots, r$ we let $h^{-1}(R_i)$ be the pre-image of $R_i$ under $h$, meaning the set of all $x \in D$ such that $h(x) = R_i$, and let $d_i = |h^{-1}(R_i)|$ be the size of the pre-image of $R_i$ under $h$. We let $d = |D|$ be the size of the domain. We define the balance of $h$ as

$$\mu(h) = \log_r \left[ \frac{d^2}{d_1^2 + \cdots + d_r^2} \right] ,$$

where $\log_r(\cdot)$ denotes the logarithm in base $r$. Proposition 1 says that for any hash function $h$, the balance of $h$ is a real number in the range from 0 to 1. Furthermore, the maximum balance of 1 is achieved when $h$ is regular (meaning $d_i = d/r$ for all $i$) and the minimum balance of 0 is achieved when $h$ is a constant function (meaning $d_i = d$ for some $i$ and $d_j = 0$ for all $j \neq i$). Thus regular functions are well-balanced and constant functions are poorly balanced, but there are lots of possibilities in between these extremes.

[2] They regard $x_i$ as a ball thrown into bin $h(x_i)$ and then apply the standard birthday analysis. But the latter assumes each ball is equally likely to land in each bin. If $R_1, \ldots, R_r$ denote the range points then the probability that a ball lands in bin $R_j$ is $|h^{-1}(R_j)|/d$ where $d = |D|$. These values are all the same only if $h$ is regular.

RESULTS. We are interested in the probability $C$ of finding a collision in $q$ trials of the birthday attack, and also in the *threshold* $Q$, defined as the number of trials required for the expected number of collisions to be one. (Alternatively, the expected number of trials to find a collision.) Corollary 1 and Theorem 2, respectively, say that, up to constant factors,[3]

$$C = \binom{q}{2} \cdot \frac{1}{r^{\mu(h)}} \qquad \text{and} \qquad Q = r^{\mu(h)/2} . \tag{1}$$

These results indicate that the performance of the birthday attack can be characterized, quite simply and accurately, via the balance of the hash function $h$ being attacked.

REMARKS. Note that when $\mu(h) = 1$ (meaning, $h$ is regular) then Equation (1) says that, up to constant factors, $Q = r^{1/2}$, which agrees with the above-discussed standard estimate for this case. At the other extreme, when $\mu(h) = 0$, meaning $h$ is a constant function, the attack finds collisions in $O(1)$ trials so $Q = 1$. The value of the general results of Equation (1) is that they show the full spectrum in between the extremes of regular and constant functions. As the balance of the hash function drops, the threshold $Q$ of the attack decreases, meaning collisions are found faster. For example a birthday attack on a hash function of balance $\mu(h) = 1/2$ will find a collision in about $Q = r^{1/4}$ trials, which is significantly less than $r^{1/2}$. Thus, we now have a way to quantitatively assess how irregularity in $h$ impacts the success-rate of the birthday attack.

We clarify that the attacker does not need to know the balance of the hash function in order to mount the attack. (The attack itself remains the birthday attack outlined above.)

BOUNDS RATHER THAN APPROXIMATE EQUALITIES. Corollary 1 provides both upper and lower bounds on $C$ that are tight in the sense of being within a constant factor (specifically, a factor of four) of each other. (And Theorem 1 does even better, but the expressions are a little more complex.) Similarly, Theorem 2 provides upper and lower bounds on $Q$ that are within a constant factor of each other.

We claim bounds are important. The estimates of how long the birthday attack takes to succeed, and the ensuing choices of output-lengths of hash functions, have been based so far on textbook approximate equality calculations of the threshold that are usually upper bounds but not lower bounds on the exact value. Yet, from a design perspective, the relevant parameter is actually a lower bound on the threshold since otherwise the attack might be doing better than we estimate.

The quality (ie. tightness) of the bounds is also important. Deriving a good lower bound on $C$ required significantly more analytical work than merely producing a rough estimate of approximate equality. With regard to $Q$ we remark that our upper bound, although within a constant factor of the lower bound, is not as tight as would like, and it is an interesting question to improve it.

[3] This assumes $d \geq 2r$ and, in the case of $C$, that $q \leq O(r^{\mu(h)/2})$.

IMPACT ON OUTPUT LENGTHS. Suppose we wish to design a hash function $h$ for which the birthday attack threshold is $2^{80}$ trials. A consequence of our results above is that we must have $r^{\mu(h)/2} = 2^{80}$, meaning must choose the output-length of the hash function to be $160/\mu(h)$ bits. Thus to minimize output-length we must maximize balance, meaning we would usually want to design hash functions that are almost regular (balance close to one).

The general principle that hash functions should be as close to regular as possible is, we believe, well-known as a heuristic. Our results, however, provide a way of quantifying the loss in security as a function of deviations from regularity.

RANDOM HASH FUNCTIONS. Designers of hash functions often have as target to make the hash function have "random" behavior. Proposition 2 together with Equation (1) enable us to estimate the impact of this design principle on birthday attacks. As an example, they imply that if $h$ is a random hash function with $d = 2r$ then the expected probability of a collision in $q$ trials is about $3/2$ times what it would be for a regular function, while the expected threshold is about $\sqrt{2/3}$ times what it would be for a regular function. In particular, random functions are *worse* than regular functions from the point of view of protection against birthday attacks, though the difference between random and regular functions decrease as the ratio $d/r$ increases.

Thus, if one wants the best possible protection against both birthday and cryptanalytic attacks, one should design a function that is not entirely random but random subject to being regular. This is true both of the hash function itself, and of the hash function restricted to domains from which the adversary may draw points in its attack (eg. a restriction of SHA-1 to all 161-bit strings). This, however, may be more difficult than designing a hash function that has entirely random behavior, so that the latter remains the design goal, and in this case it is useful to have tools like ours that enable designers to estimate the impact of deviations from regularity on the birthday attack and fine tune output lengths if necessary.

DOES THE MD TRANSFORM PRESERVE BALANCE? Given the above results we would like to be building hash functions that have high balance. We look at some elements of current design to see how well they reflect this requirement.

Hash functions like MD5 [7], SHA-1 [6] and RIPEMD-160 [3] are designed by applying the Merkle-Damgård (MD) [5,2] transform to an underlying compression function. Designers could certainly try to ensure that the compression function is regular or has high balance, but this turns out not to be enough to ensure high balance of the hash function because Proposition 3 shows that the MD transform does not preserve regularity or maintain balance. (We give an example of a compression function that has balance one, yet the hash function resulting from the MD transform applied to this compression function has balance zero.)

Proposition 4 is more positive, showing that regularity not only of the compression function but also of certain associated functions does suffice to guarantee regularity of the hash function. But Proposition 5 notes that if the compression and associated functions have even minor deviations from regularity, meaning balance that is high but not equal to one, then the MD transform can amplify the imbalance and result in a hash function with very low balance.

Given that a random compression function has balance close to but not equal to one, and we expect practical compression functions to be similar, our final conclusion is that we cannot recommend, as a general design principle, attempting to ensure high balance of a hash function by only establishing some properties of the compression function and hoping the MD transform does the rest.

We stress that none of this implies any weaknesses in specific existing hash functions such as those mentioned above. But it does indicate a weakness in the MD transform based design principle from the point of view of ensuring high balance, and means that if we want to ensure or verify high balance of a hash function we might be forced to analyze it directly rather than being able to concentrate on the possibly simpler task of analyzing the compression function. We turn next to some preliminary experimental work in this vein with SHA-1.

EXPERIMENTING WITH SHA-1. The hash function SHA-1 was designed with the goal that the birthday attack threshold is about $2^{80}$ trials. As per the above, this goal would only be met if the balance of the hash function was close to one. More precisely, letting $\mathrm{SHA}_n$: $\{0,1\}^n \to \{0,1\}^{160}$ denote the restriction of SHA-1 to inputs of length $n < 2^{64}$, we would like to know whether $\mathrm{SHA}_n$ has balance close to one for practical values of $n$, since otherwise a birthday attack on $\mathrm{SHA}_n$ will find a collision for SHA-1 in less than $2^{80}$ trials.

The balance of $\mathrm{SHA}_n$ is however hard to compute, and even to estimate experimentally, when $n$ is large. Section 6 however reports on some experiments that compute $\mu(\mathrm{SHA}_{32;t_1\ldots t_2})$ for small values of $t_2-t_1$, where $\mathrm{SHA}_{n;t_1\ldots t_2}$: $\{0,1\}^n \to \{0,1\}^{t_2-t_1+1}$ is the function which returns the $t_1$-th through $t_2$-th output bits of $\mathrm{SHA}_n$. The computed values for $\mu(\mathrm{SHA}_{32;t_1\ldots t_2})$ are extremely close to what one would expect from a random function with the same domain and range. Toward estimating the balance of $\mathrm{SHA}_n$ for larger values of $n$, Section 6 reports on some experiments on $\mathrm{SHA}_{n;t_1\ldots t_2}$ for larger $n$. Broadly speaking, the experiments indicate that these functions have high balance. This can be taken as some indication that $\mathrm{SHA}_n$ also has high balance, meaning SHA-1 is well-designed from the balance point of view.

REMARKS. We clarify that while high balance is a necessary requirement for a collision-resistant hash function, it is certainly not sufficient. It is easy to give examples of high-balance hash functions for which it easy to find collisions. High balance is just one of many design criteria that designers should consider.

We also clarify that this paper does not uncover any weaknesses, or demonstrate improved performance of birthday attacks, on any specific, existing hash functions such as those mentioned above. However it provides analytical tools that contribute toward the goal of better understanding the security of existing hash functions or building new ones, and suggests a need to put more effort into estimating the balance of existing hash functions to see whether weaknesses exist.

## 2 Notation and Terminology

If $n$ is a non-negative integer then we let $[n] = \{1, \ldots, n\}$. If $S$ is a set then $|S|$ denotes its size. We denote by $h\colon D \to R$ a function mapping domain $D$ to range $R$, and throughout the paper we assume that $R$ has size at least two. We usually denote $|D|$ by $d$ and $|R|$ by $r$. A *collision* for $h$ is a pair $x_1, x_2$ of points in $D$ such that $x_1 \neq x_2$ but $h(x_1) = h(x_2)$. For any $y \in R$ we let

$$h^{-1}(y) = \{ x \in D : h(x) = y \} .$$

We say that $h$ is *regular* if $|h^{-1}(y)| = d/r$ for every $y \in R$, where $d = |D|$ and $r = |R|$.

## 3 The Balance Measure and Its Properties

We introduce a measure that we call the *balance*, and establish some of its basic properties.

**Definition 1.** *Let $h\colon D \to R$ be a function whose domain $D$ and range $R = \{R_1, \ldots, R_r\}$ have sizes $d, r \geq 2$, respectively. For $i \in [r]$ let $d_i = |h^{-1}(R_i)|$ denote the size of the pre-image of $R_i$ under $h$. The balance of $h$, denoted $\mu(h)$, is defined as*

$$\mu(h) = \log_r \left[ \frac{d^2}{d_1^2 + \cdots + d_r^2} \right] , \tag{2}$$

*where $\log_r(\cdot)$ denotes the logarithm in base $r$.* ▮

It is easy to see that a regular function has balance 1 and a constant function has balance 0. The following says that these are the two extremes: In general, the balance is a real number that could fall somewhere in the range between 0 and 1. The proof is based on standard facts and provided in the full version of this paper [1] for completeness.

**Proposition 1.** *Let $h$ be a function. Then*

$$0 \leq \mu(h) \leq 1 . \tag{3}$$

*Furthermore, $\mu(h) = 0$ iff $h$ is a constant function, and $\mu(h) = 1$ iff $h$ is a regular function.* ▮

The following lemma, which we prove in [1], will be useful later.

**Lemma 1.** *Let $h\colon D \to R$ be a function. Let $d = |D|$ and $r = |R|$ and assume $d \geq r \geq 2$. Then*

$$r^{-\mu(h)} - \frac{1}{d} \geq \left(1 - \frac{r}{d}\right) \cdot r^{-\mu(h)} , \tag{4}$$

*where $\mu(h)$ is the balance of $h$ as per Definition 1.* ▮

```
For i = 1,...,q do      // q is the number of trials
    Pick x_i at random from the domain of h
    y_i ← h(x_i)      // Hash x_i to get y_i
    If there exists j < i such that y_i = y_j but x_i ≠ x_j then
        return x_i, x_j      // collision found
    EndIf
EndFor
Return ⊥      // No collision found
```

**Fig. 1.** Birthday attack on a hash function $h\colon\ D \rightarrow R$. The attack is successful in finding a collision if it does not return $\perp$. We call $q$ the number of *trials*.

## 4 Balance-Based Analysis of the Birthday Attack

The attack is presented in Figure 1. (Note that it picks the points $x_1, \ldots, x_q$ independently at random, rather than picking them at random subject to being distinct as in some variants of the attack [8]. The difference in performance is negligible as long as the domain is larger than the range.)

We are interested in two quantities: the probability $C$ of finding a collision in a given number $q$ of trials, and the *threshold* $Q$, defined as the expected number of trials to get a collision. Both will be estimated in terms of the balance of the hash function being attacked. Note that although $Q$ is a simpler metric it is less informative than $C$ since the latter shows how the success-rate of the attack grows with the number of trials. We begin with Theorem 1 below, which gives both upper and lower bounds on $C$ that are within constant factors of each other. The proof of Theorem 1 is in Section 4.1 below.

**Theorem 1.** *Let $h\colon\ D \rightarrow R$ be a hash function. Let $d = |D|$ and $r = |R|$ and assume $d > r \geq 2$. Let $C$ denote the probability of finding a collision for $h$ in $q \geq 2$ trials of the birthday attack of Figure 1. Let $\mu(h)$ be the balance of $h$ as per Definition 1. Then*

$$C \leq \binom{q}{2} \cdot \left[\frac{1}{r^{\mu(h)}} - \frac{1}{d}\right] . \tag{5}$$

*Additionally, if $\alpha$ is any real number, we have*

$$\left(1 - \frac{\alpha^2}{4} - \alpha\right) \cdot \binom{q}{2} \cdot \left[\frac{1}{r^{\mu(h)}} - \frac{1}{d}\right] \leq C \tag{6}$$

*under the assumption that*

$$q \leq \alpha \cdot \left(1 - \frac{r}{d}\right) \cdot r^{\mu(h)/2} . \quad \blacksquare \tag{7}$$

The above may be a bit hard to interpret. The following, which simply picks a particular value for the parameter $\alpha$ and applies the above, may be easier to understand. It provides upper and lower bounds on $C$ that are within a factor of four of each other assuming $q = O(r^{\mu(h)/2})$. The proof of Corollary 1 is in [1].

**Corollary 1.** *Let $h: D \to R$ be a hash function. Let $d = |D|$ and $r = |R|$ and assume $d \geq 2r \geq 4$. Let $C$ denote the probability of finding a collision for $h$ in $q \geq 2$ trials of the birthday attack of Figure 1. Let $\mu(h)$ be the balance of $h$ as per Definition 1. Then*

$$C \leq \binom{q}{2} \cdot \frac{1}{r^{\mu(h)}} . \tag{8}$$

*Additionally,*

$$\frac{1}{4} \cdot \binom{q}{2} \cdot \frac{1}{r^{\mu(h)}} \leq C \tag{9}$$

*under the assumption that $q \leq (1/5) \cdot r^{\mu(h)/2}$.* ▌

As we mentioned before, we believe it is important to have close upper and lower bounds rather than approximate equalities when it comes to computing the success rate of attacks since we are making very specific choices of parameters, such as hash function output lengths, based on these estimates, and if our estimates of the success rates are not specific too we might choose parameters incorrectly.

*Remark 1.* The lower bound in Equation (9) is only valid when $2 \leq q \leq (1/5) \cdot r^{\mu(h)/2}$. The upper bound on $q$ here is not particularly restrictive since we know that as $q$ approaches $r^{\mu(h)/2}$, the probability $C$ gets close to 1. However, note that we are implicitly assuming $2 \leq (1/5) \cdot r^{\mu(h)/2}$, meaning we are assuming a lower bound on $\mu(h)$. However the result only excludes functions of tiny balance. ▌

Next, we show that the threshold is $\Theta(r^{\mu(h)/2})$. Again, we provide explicit upper and lower bounds that are within a constant factor of each other. The proof of Theorem 2 is in Section 4.2.

**Theorem 2.** *Let $h: D \to R$ be a hash function. Let $d = |D|$ and $r = |R|$ and assume $d \geq 2r \geq 4$. Let $Q$ denote the threshold, meaning the expected number of trials, in the birthday attack of Figure 1, to get a collision. Let $\mu(h)$ be the balance of $h$ as per Definition 1 and assume $((\sqrt{7}-2)/3) \cdot r^{\mu(h)/2} \geq 2$. Then*

$$(1/2) \cdot r^{\mu(h)/2} \leq Q \leq 72 \cdot r^{\mu(h)/2} \quad ▌ \tag{10}$$

Designers of hash functions often have as target to make the hash function have "random" behavior. We now state a result which will enable us to gage how well random functions fare against the birthday attack. (Consequences are discussed after the statement). Proposition 2 below says that if $h$ is chosen at random then the expectation of $r^{-\mu(h)}$ is more than $1/r$ (what it would be for a regular function) by a factor equal to about $1 + r/d$. The proof of Proposition 2 is in the full version of this paper [1].

**Proposition 2.** *Let $D, R$ be sets of sizes $d, r$ respectively, where $d \geq r \geq 2$. If we choose a function $h: D \to R$ at random then*

$$\mathbf{E}\left[r^{-\mu(h)}\right] = \frac{1}{r} \cdot \left(1 + \frac{r-1}{d}\right) . \quad ▌$$

As an example, suppose $d = 2r$. Then the above implies that if $h$ is chosen at random then

$$\mathbf{E}\left[r^{-\mu(h)}\right] \approx \frac{3}{2} \cdot \frac{1}{r} .$$

As per Theorem 1 and Theorem 2 this means that if $h$ is chosen at random then the probability of finding a collision in $q$ trials is expected to rise to about $3/2$ times what it would be for a regular function, while the threshold is expected to fall to about $\sqrt{2/3}$ times what it would be for a regular function. Although the difference in the efficacy of birthday attacks against regular and random functions becomes less as $d/r$ increases, the above example with $d = 2r$ suggests that although hash functions are often designed to be "random", in terms of resistance to birthday attacks a more desirable goal is to have randomness subject to regularity. This also applies to all restrictions of the hash function to domains from which an adversary may draw during a birthday attack (eg. SHA-1 restricted to 161-bit inputs).

### 4.1 Proof of Theorem 1

We let $[q]_2$ denote the set of all two-element subsets of $[q]$. Recall that the attack picks $x_1, \ldots, x_q$ at random from the domain $D$ of the hash function. We associated to any two-element set $I = \{i, j\} \in [q]_2$ the random variable $X_I$ which takes value 1 if $x_i, x_j$ form a collision (meaning $x_i \neq x_j$ and $h(x_i) = h(x_j)$), and 0 otherwise. We let

$$X \;=\; \textstyle\sum_{I \in [q]_2} X_I \, .$$

The random variable $X$ is the number of collisions. (We clarify that in this manner of counting the number of collisions, if $n$ distinct points have the same hash value, they contribute $n(n-1)/2$ toward the value of $X$.) For any $I \in [q]_2$ we have

$$\mathbf{E}\left[X_I\right] \;=\; \Pr\left[\, X_I = 1 \,\right] \;=\; \sum_{i=1}^{r} \frac{d_i(d_i - 1)}{d^2} \;=\; \sum_{i=1}^{r} \frac{d_i^2}{d^2} - \sum_{i=1}^{r} \frac{d_i}{d^2} \;=\; r^{-\mu(h)} - \frac{1}{d} \, . \tag{11}$$

By linearity of expectation we have

$$\mathbf{E}\left[X\right] \;=\; \sum_{I \in [q]_2} \mathbf{E}\left[X_I\right] \;=\; \binom{q}{2} \cdot \left[r^{-\mu(h)} - \frac{1}{d}\right] . \tag{12}$$

Let

$$p \;=\; r^{-\mu(h)} - \frac{1}{d} \, .$$

The upper bound of Theorem 1 is a simple application of Markov's inequality and Equation (12):

$$\Pr\left[\, C \,\right] \;=\; \Pr\left[\, X \geq 1 \,\right] \;\leq\; \frac{\mathbf{E}\left[X\right]}{1} \;=\; \binom{q}{2} \cdot p \, .$$

We proceed to the lower bound. Let $[q]_{2,2}$ denote the set of all two-elements subsets of $[q]_2$. Via the inclusion-exclusion principle we have

$$\begin{aligned}\Pr[C] &= \Pr\left[\bigvee_{I\in[q]_2} X_I = 1\right] \\ &\geq \sum_{I\in[q]_2} \Pr[X_I = 1] - \sum_{\{I,J\}\in[q]_{2,2}} \Pr[X_I = 1 \wedge X_J = 1] \,. \qquad (13)\end{aligned}$$

Equation (12) tells us that the first sum above is

$$\sum_{I\in[q]_2} \Pr[X_I = 1] = \sum_{I\in[q]_2} \mathbf{E}[X_I] = \mathbf{E}[X] = \binom{q}{2}\cdot p \,. \qquad (14)$$

We now claim that

$$\sum_{\{I,J\}\in[q]_{2,2}} \Pr[X_I = 1 \wedge X_J = 1] \leq \left(\frac{\alpha^2}{4} + \alpha\right)\cdot\binom{q}{2}\cdot p \,. \qquad (15)$$

This completes the proof because from Equations (13), (14) and (15) we obtain Equation (6) as follows:

$$\begin{aligned}\Pr[C] &\geq \binom{q}{2}\cdot p - \sum_{\{I,J\}\in[q]_{2,2}} \Pr[X_I = 1 \wedge X_J = 1] \\ &\geq \binom{q}{2}\cdot p - \left(\frac{\alpha^2}{4} + \alpha\right)\cdot\binom{q}{2}\cdot p \\ &= \left(1 - \frac{\alpha^2}{4} - \alpha\right)\cdot\binom{q}{2}\cdot p \,.\end{aligned}$$

It remains to prove Equation (15).

Let $E$ be the set of all $\{I,J\} \in [q]_{2,2}$ such that $I \cap J = \emptyset$, and let $N$ be the set of all $\{I,J\} \in [q]_{2,2}$ such that $I \cap J \neq \emptyset$. Then

$$\begin{aligned}&\sum_{\{I,J\}\in[q]_{2,2}} \Pr[X_I = 1 \wedge X_J = 1] \\ &= \underbrace{\sum_{\{I,J\}\in E} \Pr[X_I = 1 \wedge X_J = 1]}_{S_E} + \underbrace{\sum_{\{I,J\}\in N} \Pr[X_I = 1 \wedge X_J = 1]}_{S_N} \,. \qquad (16)\end{aligned}$$

We now claim that

$$S_E \leq \binom{q}{2}\cdot\frac{1}{4}\cdot\alpha^2\cdot p \qquad (17)$$

$$S_N \leq \binom{q}{2}\cdot\alpha\cdot p \,, \qquad (18)$$

Equation (15) follows from Equations (16), (17) and (18). We now prove Equations (17) and (18).

To upper bound $S_E$, we note that if $\{I, J\} \in E$ then the random variables $X_I$ and $X_J$ are independent. Using Equation (11) we get

$$S_E = \sum_{\{I,J\} \in E} \Pr[X_I = 1 \wedge X_J = 1]$$
$$= \sum_{\{I,J\} \in E} \Pr[X_I = 1] \cdot \Pr[X_J = 1] = |E| \cdot p^2 .$$

Computing the size of the set $E$ and simplifying, we get

$$S_E = \frac{1}{2}\binom{q}{2}\binom{q-2}{2} \cdot p^2 = \binom{q}{2} \cdot p \cdot \frac{q^2 - 5q + 6}{4} \cdot p .$$

We now upper bound this as follows:

$$S_E < \binom{q}{2} \cdot p \cdot q^2 \cdot \frac{p}{4} \leq \binom{q}{2} \cdot p \cdot \alpha^2 \cdot r^{\mu(h)} \cdot \frac{p}{4} \leq \frac{1}{4} \cdot \alpha^2 \cdot \binom{q}{2} \cdot p .$$

Above the first inequality is true because Theorem 1 assumes $q \geq 2$. The second inequality is true because of the assumption made in Equation (7). The third inequality is true because $r^{\mu(h)} \cdot p < 1$. We have now obtained Equation (17).

The remaining task is to upper bound $S_N$. The difficulty here is that for $\{I, J\} \in N$ the random variables $X_I$ and $X_J$ are not independent. We let $d_i = |h^{-1}(R_i)|$ for $i \in [r]$ where $R = \{R_1, \ldots, R_r\}$ is the range of the hash function. If $\{I, J\} \in N$ then the two-elements sets $I$ and $J$ intersect in exactly one point. (They cannot be equal since $I, J$ are assumed distinct.) Accordingly we have

$$S_N = \sum_{\{I,J\} \in N} \Pr[X_I = 1 \wedge X_J = 1]$$
$$= |N| \cdot \sum_{i=1}^{r} \frac{d_i(d_i - 1)^2}{d^3}$$
$$< \frac{|N|}{d^3} \cdot \sum_{i=1}^{r} d_i^3 . \tag{19}$$

We now compute the size of the set $N$:

$$|N| = \frac{1}{2}\binom{q}{2}\binom{q}{2} - \frac{1}{2}\binom{q}{2} - \frac{1}{2}\binom{q}{2}\binom{q-2}{2}$$
$$= \binom{q}{2} \cdot (q-2) .$$

Putting this together with Equation (19) we have

$$S_N < \binom{q}{2} \cdot q \cdot \left[\frac{1}{d^3} \cdot \sum_{i=1}^{r} d_i^3\right] . \tag{20}$$

To upper bound the sum of Equation (20), we view $d_1, \ldots, d_r$ as variables and consider the problem of maximizing $d_1^3 + \cdots + d_r^3$ subject to the constraint

$\sum_{i=1}^{r} d_i^2 = d^2 \cdot r^{-\mu(h)}$. The maximum occurs when $d_1 = d \cdot r^{-\mu(h)/2}$ and $d_i = 0$ for $i = 2, \ldots, r$, meaning that

$$\sum_{i=1}^{r} d_i^3 \le d^3 r^{-3\mu(h)/2} .$$

Returning to Equation (20) with this information we get

$$S_N < \binom{q}{2} \cdot q \cdot \left[ \frac{1}{d^3} \cdot \sum_{i=1}^{r} d_i^3 \right] \le \binom{q}{2} \cdot q \cdot \frac{1}{d^3} \cdot d^3 r^{-3\mu(h)/2} = \binom{q}{2} \cdot q \cdot r^{-3\mu(h)/2} .$$

We now use the assumption made in Equation (7), and finally use Lemma 1, to get

$$\begin{aligned} S_N &< \binom{q}{2} \cdot \alpha \cdot \left(1 - \frac{r}{d}\right) \cdot r^{\mu(h)/2} \cdot r^{-3\mu(h)/2} \\ &\le \binom{q}{2} \cdot \alpha \cdot \left(1 - \frac{r}{d}\right) r^{-\mu(h)} \le \binom{q}{2} \cdot \alpha \cdot p . \end{aligned}$$

This proves Equation (18) and thus concludes the proof of Theorem 1.

### 4.2 Proof of Theorem 2

We begin by proving the lower bound. Let the random variable $Y$ denote the number of trials to collision. Let $C(q)$ denote the probability of finding a collision for $h$ in $q \ge 2$ trials of the birthday attack in Figure 1, and let $D(q)$ denote the probability of finding the first collision on the $q$-th trial. Let $Q = r^{\mu(h)/2}$. From the definition of $Y$:

$$\mathbf{E}[Y] = \sum_{x=1}^{\infty} x \cdot D(x) \ge Q \cdot \sum_{x=Q}^{\infty} D(x) = Q \cdot (1 - C(Q-1)) .$$

We claim that

$$C(Q-1) < \frac{1}{2} . \qquad (21)$$

It follows that

$$\mathbf{E}[Y] \ge Q \cdot (1/2) \ge (1/2) \cdot r^{\mu(h)/2} ,$$

as desired. We now justify Equation (21). From Equation (8) of Corollary 1 we know that

$$C(Q-1) \le \binom{Q-1}{2} \cdot \frac{1}{r^{\mu(h)}} = \frac{1}{2} \cdot \left((Q-1)^2 - (Q-1)\right) \cdot \frac{1}{r^{\mu(h)}} .$$

Since $Q = r^{\mu(h)/2} \ge 2$ by assumption,

$$(Q-1)^2 - (Q-1) = Q^2 - 3 \cdot Q + 2 < Q^2 = r^{\mu(h)}$$

and

$$C(Q-1) < \frac{1}{2} \cdot r^{\mu(h)} \cdot \frac{1}{r^{\mu(h)}} = \frac{1}{2}$$

as desired

For the upper bound, we must be careful since there is an upper restrictions on $q$ in Equation (9) and Equation (6). Fix $\alpha = (2\sqrt{7} - 4)/3$ and $q = (\alpha/2) \cdot r^{\mu(h)/2}$. First note that

$$q = \frac{\alpha}{2} \cdot r^{\mu(h)/2} \le \alpha \cdot \left(1 - \frac{r}{d}\right) \cdot r^{\mu(h)/2}$$

since we assume that $d \ge 2r$ and therefore that $1 - r/d \ge 1/2$. This means that we can use Theorem 1 with $\alpha$ and $q$ defined as above. Combining Theorem 1 with Lemma 1 and the assumptions that $d \ge 2r$ and $q = (\alpha/2) \cdot r^{\mu(h)/2} \ge 2$, we have

$$\begin{aligned} C(q) &\ge \left(1 - \frac{\alpha^2}{4} - \alpha\right) \cdot \binom{q}{2} \cdot \frac{1}{2} \cdot \frac{1}{r^{\mu(h)}} \\ &\ge \left(1 - \frac{\alpha^2}{4} - \alpha\right) \cdot q^2 \cdot \frac{1}{8} \cdot \frac{1}{r^{\mu(h)}} . \end{aligned}$$

Replacing $q$ with $(\alpha/2) \cdot r^{\mu(h)/2}$ we get

$$\begin{aligned} C(q) &\ge \left(1 - \frac{\alpha^2}{4} - \alpha\right) \cdot \left(\frac{\alpha}{2} \cdot r^{\mu(h)/2}\right)^2 \cdot \frac{1}{8} \cdot \frac{1}{r^{\mu(h)}} \\ &= \frac{1}{32} \cdot \left(\alpha^2 - \frac{\alpha^4}{4} - \alpha^3\right) . \end{aligned} \tag{22}$$

Now consider the following experiment that repeatedly runs the birthday attack, using $q = (\alpha/2) \cdot r^{\mu(h)/2}$ trials, until a collision is found:

```
For j = 1, 2, ... do
    For i = 1, ..., q do
        Pick x_{q(j-1)+i} at random from the domain of h
        y_{q(j-1)+i} ← h(x_{q(j-1)+i})
        If there exists k such that q(j − 1) < k < q(j − 1) + i
        and y_{q(j-1)+i} = y_k but x_{q(j-1)+i} ≠ x_k then
            return x_{q(j-1)+i}, x_k      // collision found in this block of q trials
        EndIf
    EndFor
EndFor
```

Let the random variable $A$ denote the number of trials to success in the above experiment. We claim that

$$\mathbf{E}[Y] \le \mathbf{E}[A] \tag{23}$$

and

$$\mathbf{E}[A] \le \frac{q}{C(q)} , \tag{24}$$

and combining with Equation (22), it follows that

$$\mathbf{E}[Y] \le \frac{q}{C(q)} \le \frac{(\alpha/2) \cdot r^{\mu(h)/2}}{(1/32) \cdot (\alpha^2 - (\alpha^4/4) - \alpha^3)} < 72 \cdot r^{\mu(h)/2} ,$$

giving the upper bound in the theorem statement.

To prove Equation (23) it is sufficient to note that, for any random tape $T$,

$$Y(T) \leq A(T)$$

since any collision in the above experiment is immediately a collision for the birthday attack in Figure 1.

To prove Equation (24), consider each inner loop of the above experiment an independent Bernoulli trial, and let $Z$ denote the expected number of Bernoulli trials (inner loop executions) to collision. Since each inner loop has a success probability $C(q)$, standard results tell us that

$$\mathbf{E}[Z] \leq \frac{1}{C(q)} . \tag{25}$$

Let $F(i)$ denote the probability that the first collision in the above experiment occurs on the $i$-th trial. Let $G(j)$ denote the probability that the first collision is found in the $j$-th execution of the inner loop in the above experiment. Then

$$\begin{aligned}
\mathbf{E}[A] &= \sum_{i=1}^{\infty} i \cdot F(i) \\
&= \sum_{j=1}^{\infty}\sum_{i=1}^{q} (q \cdot (j-1) + i) \cdot F(q \cdot (j-1) + i) \\
&\leq q \cdot \sum_{j=1}^{\infty} \left( j \cdot \sum_{i=1}^{q} F(q \cdot (j-1) + i) \right)
\end{aligned}$$

Since, by the definition of $G(j)$, for any $j \geq 1$

$$\sum_{i=1}^{q} F(q \cdot (j-1) + i) = G(j) ,$$

it follows that

$$\mathbf{E}[A] \leq q \cdot \sum_{j=1}^{\infty} j \cdot G(j) = q \cdot \mathbf{E}[Z] . \tag{26}$$

Combining Equation (25) with Equation (26) yields Equation (24), completing the proof.

## 5 Does the MD Transform Preserve Balance?

We consider the following popular paradigm for the construction of hash functions. First build a *compression function* $H$: $\{0,1\}^{b+c} \to \{0,1\}^c$, where $b \geq 1$ is called the *block-length* and $c \geq 1$ is called the *chaining-length.* Then transform $H$ into a hash function $\overline{H}$: $D_b \to \{0,1\}^c$, where

$$D_b = \{ M \in \{0,1\}^* : |M| = nb \text{ for some } 1 \leq n < 2^b \} ,$$

```
Function H̄(M)
    Break M into b-bit blocks M_1 || ··· || M_n
    M_{n+1} ← ⟨n⟩_b ; C_0 ← 0^c
    For i = 1, ..., n+1 do C_i ← H(M_i || C_{i-1}) EndFor
    Return C_{n+1}
```

**Fig. 2.** Hash function $\overline{H}$: $D_b \to \{0,1\}^c$ obtained via the MD transform applied to compression function $H$: $\{0,1\}^{b+c} \to \{0,1\}^c$.

---

via the Merkle-Damgård (MD) [5,2] transform depicted in Figure 2. (In this description and below, we let $\langle i \rangle_b$ denote the representation of integer $i$ as a string of length *exactly* $b$ bits for $i = 0, \ldots, 2^b - 1$.) In particular, modulo details, this is the paradigm used in the design of popular hash functions including MD5 [7], SHA-1 [6] and RIPEMD-160 [3].

For the considerations in this section, we will focus on the restriction of $\overline{H}$ to strings of some particular length. For any integer $1 \leq n < 2^b$ (the number of blocks) we let $\overline{H}_n$: $D_{b,n} \to \{0,1\}^c$ denote the restriction of $\overline{H}$ to the domain $D_{b,n}$, defined as the set of all strings in $D_b$ that have length exactly $nb$ bits.

Our results lead us to desire that $\overline{H}_n$ has high balance for all practical values of $n$. Designers could certainly try to ensure that the compression function is regular or has high balance, but to be assured that $\overline{H}_n$ has high balance it would need to be the case that the MD transform is "balance preserving." Unfortunately, the following shows that this is not true. It presents an example of a compression function $H$ which has high balance (in fact is regular, with balance one) but $\overline{H}_n$ has low balance (in fact, balance zero) even for $n = 2$.

**Proposition 3.** *Let $b, c$ be positive integers. There exists a compression function $H$: $\{0,1\}^{b+c} \to \{0,1\}^c$ such that $H$ is regular ($\mu(H) = 1$) but $\overline{H}_2$ is a constant function ($\mu(\overline{H}_2) = 0$).* ▮

*Proof (Proposition 3).* Let $H$: $\{0,1\}^{b+c} \to \{0,1\}^c$ map $B \| C$ to $C$ for all $b$-bit strings $B$ and $c$-bit strings $C$. Clearly $\mu(H) = 1$ since each point in $\{0,1\}^c$ has exactly $2^b$ pre-images under $H$. Because the initial vector (IV) in the MD transform is the constant $C_0 = 0^c$, and by the definition of $H$, the function $\overline{H}_2$ maps all inputs to $0^c$. □

This example might be viewed as contrived particularly because the compression function $H$ above is not collision-resistant (although it is very resistant to birthday attacks), but in fact it still serves to illustrate an important point. The popularity of the MD paradigm arises from the fact that it *provably* preserves collision-resistance [5,2]. However, the above shows that it does not provably preserve balance. Even though Proposition 3 does not say that the transform will *always* be poor at preserving balance, it says that we cannot count on the transform to preserve balance in general. This means that simply ensuring high balance of the compression function is not a suitable general design principle.

(We also remark that there exist adversaries capable of finding collisions for any unkeyed compression function, including the compression functions in MD5, SHA-1, and RIPEMD-160, using exactly two trials. We just do not know what these adversaries are.)

Is there any other design principle whereby some properties of the compression function suffice to ensure high balance of the hash function? Toward finding one we note that the behavior exhibited by the function $\overline{H}_2$ in the proof of Proposition 3 arose because the initial vector (IV) of the MD transform was $C_0 = 0^c$, and although $H$ was regular, the restriction of $H$ to inputs having the last $c$ bits 0 was not regular, and in fact was constant. Accordingly we consider requiring regularity conditions not just on the compression function but on certain related functions as well. If $H\colon \{0,1\}^{b+c} \to \{0,1\}^c$ then define $H_0\colon \{0,1\}^b \to \{0,1\}^c$ via $M \mapsto H(M\|0^c)$ for all $M \in \{0,1\}^b$, and for $n \geq 1$ define $H_n\colon \{0,1\}^c \to \{0,1\}^c$ via $M \mapsto H(\langle n \rangle_b\|M)$ for all $M \in \{0,1\}^c$. The following shows that if $H, H_0, H_n$ are all regular, meaning have balance one, then $\overline{H}_n$ is also regular.

**Proposition 4.** *Let $b, c, n$ be positive integers. Let $H\colon \{0,1\}^{b+c} \to \{0,1\}^c$ and let $H_0, H_n$ be as above. Assume $H$, $H_0$, and $H_n$ are all regular. Then $\overline{H}_n$ is regular.* ▌

*Proof (Proposition 4).* The computation of $\overline{H}_n$ can be written as

Function $\overline{H}_n(M)$
  Break $M$ into $b$-bit blocks $M_1\|\cdots\|M_n$ ; $C_1 \leftarrow H_0(M_1)$
  For $i = 2, \ldots, n$ do $C_i \leftarrow H(M_i\|C_{i-1})$ EndFor
  $C_{n+1} \leftarrow H_n(C_n)$ ; Return $C_{n+1}$

It is not hard to check that the assumed regularity of $H_0, H$ and $H_n$ imply the regularity of $\overline{H}_n$. □

Unfortunately Proposition 4 is not "robust." Although $\overline{H}_n$ has balance one if $H, H_0, H_n$ have balance one, it turns out that if $H, H_0, H_n$ have balance that is high but not quite one, we are *not* assured that $\overline{H}_n$ has high balance. Proposition 5 shows that even a slight deviation from the maximum balance of one in $H, H_0, H_n$ can be amplified, and result in $\overline{H}_n$ having very low balance. The proof of the following is in the full version of this paper [1].

**Proposition 5.** *Let $b, c$ be integers, $b \geq c \geq 2$, and let $n \geq c$. Then there exists a compression function $H\colon \{0,1\}^{b+c} \to \{0,1\}^c$ such that $\mu(H) \geq 1 - 1/c$, $\mu(H_0) = 1$, and $\mu(H_n) \geq 1 - 2/c$, but $\mu(\overline{H}_n) \leq 1/c$, where the functions $H_0, H_n$ are defined as above.* ▌

As indicated by Proposition 2, a random compression function will have expected balance that is high but not quite 1. We expect that practical compression functions are in the same boat. Furthermore it seems harder to build compression functions that have balance exactly one than close to one. So the lack of robustness of Proposition 4, as exhibited by Proposition 5, means that Proposition 4 is of limited use.

The consequence of the results in this section is that we are unable to recommend any design principle that, to ensure high balance, focuses solely on establishing properties of the compression function. It seems one is forced to look directly at the hash function. We endeavor next to do this for SHA-1.

## 6 Experiments on SHA-1

Let $\mathrm{SHA}_n$: $\{0,1\}^n \to \{0,1\}^{160}$ denote the restriction of SHA-1 to inputs of length $n < 2^{64}$. Because SHA-1's range is $\{0,1\}^{160}$, it is commonly believed that the expected number of trials necessary to find a collision for $\mathrm{SHA}_n$ is approximately $2^{80}$. As Theorem 2 shows, however, this is only true if the balance of $\mathrm{SHA}_n$ is one or close to one for all practical values of $n$. If the balance is not close to one, then we expect to be able to find collisions using less work. It therefore seems desirable to calculate (or approximate) the balance of $\mathrm{SHA}_n$ for reasonable values of $n$ (eg. $n = 256$). A direct computation of $\mu(\mathrm{SHA}_n)$ based on Definition 1 is however infeasible given the size of the domain and range of $\mathrm{SHA}_n$. Accordingly we focus on a more achievable goal. We look at properties of $\mathrm{SHA}_n$ that one can reasonably test and whose absence might indicate that $\mathrm{SHA}_n$ does not have high balance. Our experiments are not meant to be exhaustive, but rather representative of the types of feasible experiments one can perform with SHA-1.

Let $\mathrm{SHA}_{n;t_1\ldots t_2}$: $\{0,1\}^n \to \{0,1\}^{t_2-t_1+1}$ denote the function that returns the $t_1$-th through $t_2$-th output bits of $\mathrm{SHA}_n$. We ask what exactly is the balance of $\mathrm{SHA}_{32;t_1\ldots t_2}$ when $t_2 - t_1 + 1 \in \{8, 16, 24\}$. And we ask whether the functions $\mathrm{SHA}_{m;t_1\ldots t_2}$, $m \in \{160, 256, 1024, 2048\}$, appear regular when $t_2 - t_1 + 1 \in \{8, 16, 24\}$. (Note that $\mathrm{SHA}_{256}$ is SHA-1 restricted to the domain $\{0,1\}^{256}$, not NIST's SHA-256 hash algorithm.)

BALANCE OF $\mathrm{SHA}_{32;t_1\ldots t_2}$. We calculate the balance of $\mathrm{SHA}_{32;t_1\ldots t_2}$ for all pairs $t_1, t_2$ such that $t_2 - t_1 + 1 \in \{8, 16, 24\}$ and $t_1$ begins on a byte boundary (ie. we look at all 1-, 2-, and 3-byte portions of the SHA-1 output). The calculated values of $\mu(\mathrm{SHA}_{32;t_1\ldots t_2})$ appear in the full version of this paper [1]. Characteristic values are $\mu(\mathrm{SHA}_{32;1\ldots 8}) = 0.99999998893$, $\mu(\mathrm{SHA}_{32;1\ldots 16}) = 0.999998623$ and $\mu(\mathrm{SHA}_{32;1\ldots 24}) = 0.99976567$, indicating that, for the specified values of $t_1, t_2$, the balance of $\mathrm{SHA}_{32;t_1\ldots t_2}$ is high.

These results do not imply that the functions $\mathrm{SHA}_{n;t_1\ldots t_2}$ or $\mathrm{SHA}_n$, $n > 32$ and $t_1, t_2$ as before, are regular. But it is encouraging that $\mu(\mathrm{SHA}_{32;t_1\ldots t_2})$ are high, and in fact very close to what one would expect from a random function (cf. Proposition 2), since a small value for $\mu(\mathrm{SHA}_{32;t_1\ldots t_2})$ for any of the specified $t_1, t_2$ pairs might indicate some unusual property of the SHA-1 hash function.

EXPERIMENTS ON $\mathrm{SHA}_{160}$, $\mathrm{SHA}_{256}$, $\mathrm{SHA}_{1024}$, AND $\mathrm{SHA}_{2048}$. Let $n \in \{160, 256, 1024, 2048\}$. Although we cannot calculate the balance of $\mathrm{SHA}_n$, we can compare the behavior of $\mathrm{SHA}_{n;t_1\ldots t_2}$, $t_2 - t_1 + 1 \in \{8, 16, 24\}$, on random inputs to what one would expect from a regular or random function. There are several possible approaches to take. Knowing that the balance of $\mathrm{SHA}_{n;t_1\ldots t_2}$ directly affects the expected number of trials to collision, the approach we take is to compute the average, over 10000 runs, of the number of trials to collision in a birthday attack against $\mathrm{SHA}_{n;t_1\ldots t_2}$.

If the average number of trials to collision against $\mathrm{SHA}_{n;t_1\ldots t_2}$ on random bits is approximately the same as what one would expect from a regular function, it would support the view that $\mathrm{SHA}_n$ has high balance. However, a significant difference between the results for $\mathrm{SHA}_{n;t_1\ldots t_2}$ on random inputs and what one would expect from a regular function might indicate some unusual behavior with SHA-1, and this unusual behavior would deserve further investigation. Our experimental results are consistent with $\mathrm{SHA}_n$ having high balance. However, we again point out that these tests were only designed to uncover gross anomalies and are not exhaustive. Details are in [1].

## Acknowledgments

Mihir Bellare is supported in part by NSF grants CCR-0098123, ANR-0129617 and CCR-0208842, and by an IBM Faculty Partnership Development Award. Tadayoshi Kohno is supported by a National Defense Science and Engineering Graduate Fellowship.

## References

1. M. Bellare and T. Kohno. Hash function balance and its impact on birthday attacks. IACR ePrint archive, `http://eprint.iacr.org/2003/065/`. Full version of this paper
2. I. Damgård. A design principle for hash functions. *Advances in Cryptology – CRYPTO '89*, Lecture Notes in Computer Science Vol. 435, G. Brassard ed., Springer-Verlag, 1989.
3. H. Dobbertin, A. Bosselaers and B. Preneel. RIPEMD-160, a strengthened version of RIPEMD. *Fast Software Encryption '96*, Lecture Notes in Computer Science Vol. 1039, D. Gollmann ed., Springer-Verlag, 1996.
4. A. Menezes, P. van Oorschot and S. Vanstone. Handbook of applied cryptography. CRC Press, 1997.
5. R. Merkle. One way hash functions and DES. *Advances in Cryptology – CRYPTO '89*, Lecture Notes in Computer Science Vol. 435, G. Brassard ed., Springer-Verlag, 1989.
6. National Institute of Standards. FIPS 180-2, Secure hash standard. August 1, 2000.
7. R. Rivest. The MD5 message-digest algorithm. IETF RFC 1321, April 1992.
8. D. Stinson. Cryptography theory and practice, 1st Edition. CRC Press, 1995.
9. P. van Oorschot and M. Wiener. Parallel collision search with cryptanalytic applications, *Journal of Cryptology* 12(1), Jan 1999, 1–28.
10. G. Yuval. How to swindle Rabin. *Cryptologia* (3), 1979, 187–190.

# Multi-party Computation with Hybrid Security

Matthias Fitzi[1], Thomas Holenstein[2], and Jürg Wullschleger[3]

[1] Department of Computer Science
University of California, Davis
fitzi@cs.ucdavis.edu

[2] Department of Computer Science
ETH Zurich, Switzerland
holenst@inf.ethz.ch

[3] Départernent d'Informatique et Recherche Opŕationnelle
Université de Montréal, Canada
wullschj@iro.umontreal.ca

**Abstract.** It is well-known that $n$ players connected only by pairwise secure channels can achieve multi-party computation secure against an active adversary if and only if

- $t < n/2$ of the players are corrupted with respect to computational security, or
- $t < n/3$ of the players are corrupted with respect to unconditional security.

In this paper we examine to what extent it is possible to achieve conditional (such as computational) security based on a given intractability assumption with respect to some number $T$ of corrupted players while simultaneously achieving unconditional security with respect to a smaller threshold $t \leq T$. In such a model, given that the intractability assumption cannot be broken by the adversary, the protocol is secure against $T$ corrupted players. But even if it is able to break it, the adversary is still required to corrupt more than $t$ players in order to make the protocol fail.

For an even more general model involving three different thresholds $t_p$, $t_\sigma$, and $T$, we give tight bounds for the achievability of multi-party computation. As one particular implication of this general result, we show that multi-party computation computationally secure against $T < n/2$ actively corrupted players (which is optimal) can additionally guarantee unconditional security against $t \leq n/4$ actively corrupted players "for free."

**Keywords:** Broadcast, computational security, multi-party computation, unconditional security.

## 1 Introduction

Secure distributed cooperation among mutually distrusting players can be achieved by means of general multi-party computation (MPC). Typically, the goal of such a cooperation consists of jointly computing a function on the players' inputs in a way that guarantees correctness of the computation result while keeping

C. Cachin and J. Camenisch (Eds.): EUROCRYPT 2004, LNCS 3027, pp. 419–438, 2004.

thc players' inputs private — even if some of the players are corrupted by an adversary.

Different models for MPC have been proposed in the literature with respect to communication, corruption flavor, and adversarial power. In this paper, we restrict our view to the following parameters.

COMMUNICATION: We exclusively consider *synchronous networks* meaning that, informally speaking, the players are synchronized to common communication rounds with the guarantee that a sent message will be delivered still during the same communication round.

CORRUPTION: We assume a *central threshold adversary* with respect to a given, fixed threshold $t$, meaning that it can select up to arbitrary $t$ out of the $n$ players and corrupt them (meaning to take control over them). Such a player is then said to be *corrupted* whereas a non-corrupted player is called *correct.* An *active adversary* (*active corruption*) corrupts players by making them deviate from the protocol in an arbitrarily malicious way.

SECURITY: A protocol achieves *unconditional security* if even a computationally unbounded adversary cannot make the protocol fail — except for some negligible error probability. A protocol achieves *computational security* if an adversary restricted to probabilistic polynomial time computations cannot make the protocol fail except for some negligible error probability. In this paper, we consider both kinds of security.

## 1.1 Previous Work

The MPC problem was first stated by Yao [Yao82]. Goldreich, Micali, and Wigderson [GMW87] gave the first complete solution to the problem with respect to computational security. For the model with a passive adversary (passive model, for short), and given pairwise communication channels, they gave an efficient protocol that tolerates any number of corrupted players, $t < n$. For the model with an active adversary (active model), and given both pairwise and broadcast channels, they gave an efficient protocol that tolerates any faulty minority, $t < n/2$, which is optimal in the sense that no protocol exists for $t \geq n/2$. Note that when not demanding security to the full extent, computationally secure MPC is also achievable in presence of an active adversary that corrupts $t \geq n/2$ players [GMW87, GHY87, BG89, GL90, FGH+02, GL02, FHHW03]. However, in this case, robustness cannot be guaranteed, i.e., it can not be guaranteed that every player receives a result [Cle86].

With respect to unconditional security, Ben-Or, Goldwasser, and Wigderson [BGW88], and independently, Chaum, Crépeau, and Damgård [CCD88] gave efficient protocols for the passive model that tolerate $t < n/2$ and protocols for the active model that tolerate $t < n/3$ — assuming only pairwise communication channels in both cases. Both bounds are tight. Beaver [Bea89], Rabin, and Ben-Or [RB89] considered the active model when given both pairwise and broadcast channels among the players. They gave efficient protocols that achieve unconditional security for $t < n/2$ which is optimal. A more efficient protocol for this model was given by Cramer et al. [CDD+99].

With lack of better knowledge, protocols with computational security must be based on unproven intractability assumptions, i.e., they must build up on cryptographic primitives such as trapdoor permutations that are not known to exist. Furthermore, even if such primitives existed, the particular choice of a candidate implementation of such a primitive might be a bad one.

In order to prevent complete failure in these cases, Chaum [Cha89] considered a "hybrid" security model for MPC that achieves computational security for some large threshold $T$ but, at the same time, unconditional security for some smaller threshold $t \leq T$ — meaning that, in order to make the protocol fail, the adversary must either corrupt more than $T$ players, or corrupt more than $t$ players but additionally be able to break the underlying computational hardness assumption. In the passive model, given pairwise communication channels, Chaum's protocol achieves computational security with respect to $T < n$ and unconditional security with respect to $t < n/2$. Thus, this protocol simultaneously achieves the optimal bounds for computational and unconditional security.

In the active model, given pairwise and broadcast channels, his protocol achieves computational security with respect to $T < n/2$ and additionally provides unconditional privacy for all players' inputs as long as up to $t < n/3$ players are corrupted. Note that the later results in [Bea89, RB89, CDD$^+$99] strictly imply this result: unconditional security for $t < n/2$ when assuming broadcast.[4] In [WP89], the same "hybrid" model was considered with respect to the simulation of broadcast when given only pairwise communication channels.

### 1.2 Multi-party Computation beyond t < n/3 without Broadcast

The active model for MPC tolerating at least $n/3$ corrupted players typically assumes broadcast channels [GMW87, Bea89, RB89]. This is a very strong assumption and might not always be appropriate. Rather, broadcast has to be simulated by the players using the bilateral channels. But, without further assumptions, this simulation is only possible if $t < n/3$ [LSP82, DFF$^+$82].

The only known way to allow for the simulation of broadcast beyond $t < n/3$ is to use digital signatures [LSP82]. However, it is important to note that digital signatures by themselves are not enough. It must be additionally guaranteed that all correct players verify each player's signatures in the same way, i.e., that all players hold the same list of public keys. Otherwise, the transfer of a signature would not be conclusive. We call such a setup a *consistent public-key infrastructure (PKI)*. Such a PKI allows to efficiently simulate broadcast among the players secure against any number of corrupted players, $t < n$ [DS82]. Not only can a PKI be based on a computationally secure digital signature scheme but also on unconditionally secure pseudo-signatures [PW96] and thus allowing for the simulation of *unconditionally* secure broadcast. Thus the results

[4] Note that Chaum's protocol still completely relies on cryptography since the protocol's correctness is only protected by cryptographic means, i.e., by breaking the cryptographic assumption the adversary can make the protocol fail by only corrupting one single player.

in [GMW87, Bea89, RB89] are equally achievable without broadcast channels but with an appropriate PKI to be set up among the players — computationally secure for [GMW87] and unconditionally secure for the other cases.

We believe that assuming a PKI is more realistic than assuming broadcast channels among the players and thus follow this model.

It should be noted, though, that the use of unconditional pseudo-signatures is not very practical. The cost of broadcasting a single bit based on (computational) digital signatures is $t+1$ communication rounds and $O(n^3 s)$ bits to be sent by all players during the protocol overall — where $s$ is the size of a signature [DS83]. The cost of broadcasting a bit using unconditional pseudo-signatures is $\Omega(n^3)$ rounds and $\Omega(n^{17})$ bits to be sent overall [PW96] — which is still polynomial but nevertheless quite impractical.[5]

### 1.3 Contributions

Typical ways of setting up a PKI are to run a setup protocol among the players or to involve trust management over the Internet such as, e.g., the one in PGP. Evidently, both methods can fail to achieve a consistent PKI, namely, when to many players are corrupted, or, respectively, when the trust management is built on wrong assumptions. Thus, analogously to relying on computational security, relying on the consistency of a previously set-up PKI also imposes a potential security threat since the adversary might have been able to make the PKI inconsistent.

This raises the natural question of whether MPC relying on the consistency of a PKI and/or the security of a particular signature scheme can additionally guarantee unconditional security for the case where only a small number of the players are corrupted. Thus, in this paper, we extend the considerations in [Cha89] (regarding the active model) to the case where not only the adversary might be able to break the underlying hardness assumption but where also the PKI might be inconsistent.

In particular, we consider the following model for *hybrid MPC* involving three thresholds $t_p$, $t_\sigma$, and $T$, where $t_p, t_\sigma \leq T$ with the following properties (see also Figure 1).

- If at most $f \leq \min(t_p, t_\sigma)$ players are corrupted then we demand *unconditional security*.
- If $f > t_p$ then we assume that the PKI is consistent, i.e., for $t_p < f \leq T$ the computation is only as secure as the PKI.
- If $f > t_\sigma$ then we assume that the adversary cannot forge signatures (except for some non-negligible probability), i.e., for $t_\sigma < f \leq T$ the computation is only as secure as the underlying signature scheme.

[5] Note that there is also a $(t+1)$-round variant with an overall bit complexity of approximately $\Theta(n^6)$. However, this variant is only a one-time signature scheme which basically means that the PKI only allows for a very limited number of signatures to be issued.

| $f$ players corrupted | Security |
|---|---|
| $f \leq \min(t_p, t_\sigma)$ | unconditional |
| $f \leq t_\sigma \wedge t_p < f \leq T$ | as secure as PKI, independent of signature scheme |
| $f \leq t_p \wedge t_\sigma < f \leq T$ | as secure as signature scheme, independent of PKI |
| $t_p, t_\sigma < f \leq T$ | as secure as PKI and signature scheme together |

**Fig. 1.** Threshold conditions for hybrid MPC.

Or, in other words, if $f \leq t_p$ then the protocol must be secure even if the PKI is inconsistent, and, if $f \leq t_\sigma$ then the protocol must be secure even if the adversary is able to forge signatures. Thus, in order to make such a hybrid protocol fail with non-negligible probability, the adversary would have to corrupt more than $f = \min(t_p, t_\sigma)$ players and; having made for a bad PKI if $f > t_p$, or be able to forge signatures if $f > t_\sigma$.

*Result.* We show that hybrid MPC is achievable if and only if

$$(2T + t_p < n) \quad \wedge \quad (T + 2t_\sigma < n) \tag{1}$$

implying that, without loss of generality, we can always assume that $t_p \leq t_\sigma \leq T$.[6] See Figure 2 for a graphical representation of the tight bound. Achievability for all cases will be demonstrated by efficient protocols that neither rely on any particular signature scheme nor on any particular way of setting up a PKI.

As an interesting special case, the optimal result of [GMW87] (assuming a consistent PKI instead of broadcast — allowing to drop parameter $t_p$ since the consistency of the PKI is granted) computationally secure against $T < n/2$ corrupted players additionally allows to guarantee unconditional security against $t_\sigma \leq n/4$ corrupted players "for free." On the other hand, when requiring optimality with respect to unconditional security, $t_\sigma = \lfloor (n-1)/3 \rfloor$, then practically no higher computational bound $T$ can be simultaneously tolerated on top.

Finally, when basing the PKI on an unconditional pseudo-signature scheme (which is not our focus), forgery becomes impossible by definition and the tight bound collapses to $2T + t_p < n$.

*Constructions.* Our final MPC protocol is obtained by simulating broadcast in the unconditional MPC protocol of [CDD+99]. Thus the main technical contribution in this paper is to simulate broadcast (aka Byzantine agreement) in the given models with respect to the required security aspects. The (efficient) protocol in [CDD+99] is unconditionally secure against $t < n/2$ corrupted players. So, obviously, the final MPC protocol wherein broadcast is simulated is as secure as the given broadcast protocol.

[6] That is, additional forgery gives the adversary no additional power when the PKI is inconsistent.

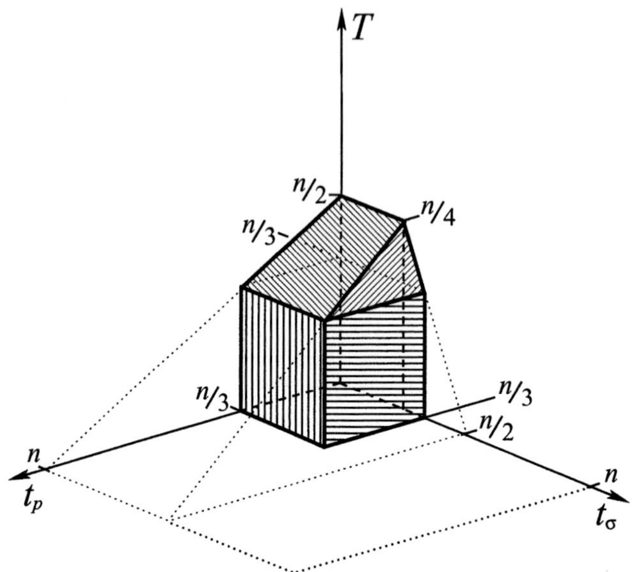


**Fig. 2.** Graphical representation of Bound (1).

## 2 Definitions and Notations

### 2.1 Multi-party Computation

In MPC a set of players want to distributedly evaluate some agreed function(s) on their inputs in a way preserving privacy of their inputs and correctness of the computed result. More precisely, in an MPC among a player set $P$ with respect to a collection of functions $(f_1, \ldots, f_n)$, every player $p_i \in P$ holds a secret input (vector) $x_i$ and secretly receives an output (vector) $y_i = f_i(x_1, \ldots, x_n)$.

From a qualitative point of view, the security of MPC is often broken down to the conditions "privacy", "correctness", "robustness", and "fairness", and ideally, a protocol should satisfy all these properties.

PRIVACY. A protocol achieves *privacy* if the adversary cannot learn more about the correct players' inputs than given by the inputs and outputs of the corrupted players.

CORRECTNESS. A protocol achieves *correctness* if the correct players' outputs are indeed computed as defined by the functions $f_i$.

ROBUSTNESS. A protocol achieves *robustness* if every correct player finally receives his outputs.

FAIRNESS. A protocol achieves *fairness* if the adversary gets no information about the correct players' inputs in case that robustness is not achieved.

More formally, MPC is modeled by an *ideal process* involving a mutually trusted party $\tau$ where the players secretly hand their inputs to $\tau$, followed by

$\tau$ computing the players' outputs and secretly handing them back to the corresponding players [Bea91, Can00, Gol01]. This model is referred to as the *ideal model* The goal of MPC is now to achieve the same functionality in the so-called *real model* where there is no such trusted party such that an adversary gets no advantage compared to an execution of the ideal protocol. An MPC protocol is defined to be secure if, for every adversary $\mathcal{A}$ in the protocol, there is an adversary $\mathcal{S}$ in the ideal model that, with similar costs, achieves (essentially) the same output distribution as the adversary in the protocol [Bea91, Can00, Gol01].

### 2.2 Broadcast

Broadcast is the special case of an MPC. In broadcast, one player $p_s$ is given an initial value which everybody is required to receive. The definition is as follows:

**Definition 1 (Broadcast).** *A protocol among $n$ players, where player $p_s \in P$ (called the* sender*) holds an input value $x_s$ and every player $p_i$ ($i \in \{1, \ldots, n\}$) computes an output value $y_i$, achieves* broadcast *if it satisfies:*

- Validity: *If the sender is correct then all correct players $p_i$ compute output $y_i = x_s$.*
- Consistency: *All correct players compute the same output value $y$.*

Often, it is also added to the definition that the protocol is always required to terminate. We will not mention this property explicitly since termination is obvious for our protocols.

Note that broadcast for any finite domain easily reduces to binary broadcast (where the sender sends a value from $\{0, 1\}$) as, e.g., shown by Turpin and Coan [TC84]. We will thus focus on binary broadcast.

During the simulation of broadcast, using signatures, it must be avoided that previous signatures can be reused by the adversary in a different context, i.e., an independent phase or another instance of the protocol. This fact was observed in [GLR95] and more profoundly treated in [LLR02]. To avoid such "replay attacks" values can be combined with unique sequence numbers before signing. The sequence numbers themselves do not have to be transferred since they can be generated in a predefined manner (encoding the protocol instance and the communication round). However, we will not explicitly state these details in the descriptions of the sequel.

### 2.3 Setting

We consider a set $P = \{p_1, \ldots, p_n\}$ of $n$ players that are connected via a complete synchronous network of pairwise secure channels. We assume a PKI to be set up among the players.

A given PKI is *consistent* if every player $p_i$ ($i \in \{1, \ldots, n\}$) has a secret-key/public-key pair ($SK_i$,$PK_i$) which was chosen by $p_i$ with respect to the key-generation algorithm of a digital signature scheme and, additionally, that each respective public key $PK_i$ is known to all players as $p_i$'s public key to be exclusively used for the verification of $p_i$'s signatures. Asserting that, with respect to every signer $p_i$, each player holds the same public key $PK_i$ guarantees that $p_i$'s signatures can be transferred between the players without losing conclusiveness. In our model, the PKI may or may not be consistent.[7]

We assume the existence of an active threshold adversary to corrupt some of the players. The adversary may or may not be able to forge signatures. Furthermore, for our protocols, the adversary is assumed to be adaptive (but non-mobile). In contrast, our proofs of optimality even hold with respect to a static adversary. As we have three bounds with respect to player corruption, $t_p$, $t_\sigma$, and $T$, we will make a stringent distinction between these bounds and the actual number of players corrupted at the end of the protocol, which we will denote by $f$.[8]

### 2.4 Protocol Notation

Protocols are specified with respect to a player set $P$ and stated with respect to the local view of player $p_i$, meaning that all players $p_i \in P$ execute this code in parallel with respect to their own identity $i$.

With respect to pairwise communication we also consider reflexive channels among the players for simplicity. Thus, when a player $p_i$ sends a value to each player then $p_i$ also receives a copy himself.

For simplicity, in our protocol notation, we do not explicitly state how to handle values received from corrupted players that are outside the specified domain. Such a value is always implicitly assumed to be replaced by a default value or by any arbitrary value inside the specified domain.

## 3 Generic Broadcast Simulation for $t < n/2$

Our broadcast simulations are based on the "phase king" protocol in [BGP89]. In [FM00], it was observed that any protocol for "weak broadcast" is sufficient in order to achieve broadcast secure against a faulty minority — as secure as the given protocol for weak broadcast. Since all of our tight bounds imply that strictly less than half of all players are corrupted we thus only need to give respective protocols for weak broadcast.

Weak broadcast (as called in [FM00]) was originally introduced in [Dol82] under the name *crusader agreement.* Weak broadcast is the same as broadcast for the case that the sender is correct but, if the sender is corrupted, then some players might end up with the "invalidity symbol" $\perp$ — but still, it guarantees that no two correct players end up with two different values in $\{0, 1\}$.

7 In case of unconditional pseudo-signatures the situation is slightly different since, instead of the same public key $PK_i$, each player $p_j$ holds a different "public key" $PK_{ij}$ (which is in fact secret).

8 As the adversary is adaptive, the number might increase during the execution of the protocol, and reach its maximum at the end.

**Definition 2 (Weak broadcast).** *A protocol where one player $p_s$ has an input $x_s \in \{0,1\}$ and every player $p_i$ computes an output $y_i \in \{0,1,\perp\}$ achieves weak broadcast if it satisfies the following conditions:*

- VALIDITY: *If $p_s$ is correct then every correct player $p_i$ computes output $y_i = x_s$.*
- CONSISTENCY: *If player $p_i$ is correct and computes $y_i \in \{0,1\}$ then every correct player $p_j$ computes $y_j \in \{y_i, \perp\}$.*

In Appendix A, we describe a reduction from broadcast to weak broadcast that is simpler and more efficient than the one in [FM00], yielding the following theorem.

**Theorem 1.** *If at most $t < n/2$ players are corrupted then efficient achievability of weak broadcast implies efficient achievability of broadcast.*

## 4 Tight Bounds

We now demonstrate the tightness of the bound given in Bound (1).

### 4.1 Efficient Protocol

We first give an efficient protocol for broadcast and then show how to plug it into the MPC protocol in [CDD$^+$99] in order to get out final protocol for efficient hybrid MPC.

**Broadcast.** For the constructive part, according to Theorem 1, it is sufficient to give a construction for weak broadcast. The following protocol is designed for any selection of thresholds $t_p$, $t_\sigma$, and $T$, ($t_p \leq t_\sigma \leq T$), satisfying Bound (1).

Let $x_s$ be $p_s$'s input value, and let $\sigma_s(x_s)$ be a signature by $p_s$ on the value $x_s$. Furthermore, let $V$ be the signature verification algorithm with respect to the underlying signature scheme computing $V(x, \sigma, \mathrm{PK}) = 1$ if $\sigma$ is a valid signature on $x$ with respect to public key PK, and $V(x, \sigma, \mathrm{PK}) = 0$ otherwise. Let $\mathrm{PK}_i^s$ be player $p_i$'s version of $p_s$'s public key. We use $V_i^s(x, \sigma)$ as a short cut for $V(x, \sigma, \mathrm{PK}_i^s)$. With respect to player $p_i$, we say that a given signature $\sigma$ is *valid* if it is valid with respect to $p_i$'s view, i.e., $V_i^s(x, \sigma) = 1$. In particular, a valid signature with respect to player $p_i$'s view might in fact not have been issued by the respective signer.

The protocol works as follows. The sender $p_s$ signs his input value and sends his input together with its signature to every other player: $(x_s, \sigma_s(x_s))$. Every player except for the sender now redistributes this information to everybody (but without signing this new message himself). Now, every player received $n$ values, one from every player. Each player $p_i$ now decides on the outcome of the protocol:

- Let $x_i^s$ be the bit he directly received from the sender. If the bit $x_i^s$ is received from at least $n - t_p$ different players overall then he computes output $y_i = x_i$.

- Otherwise, if he received the bit $x_i^s$ together with a valid signature by $p_s$ from the sender and at least $n - t_\sigma$ different players overall then he decides on $y_i = x_i$.
- Otherwise, if he received the bit $x_i$ together with a valid signature by $p_s$ from the sender and at least $n - T$ different players overall — but no single correct signature by $p_s$ for bit $1 - x_i$ — then he decides on $y_i = x_i$.
- Otherwise, he decides on $y_i = \bot$.

**Protocol 1** $\texttt{WeakBroadcast}_{p_s}(P, x_s)$

*1.* **if** $i = s$ **then** $\texttt{SendToAll}(x_s, \sigma_s(x_s))$ **fi**; $\qquad$ $\texttt{Receive}(x_i^s, \sigma_i^s)$;
*2.* **if** $i \neq s$ **then** $\texttt{SendToAll}(x_i^s, \sigma_i^s)$ **fi**; $\qquad$ $\forall j \neq s$: $\texttt{Receive}(x_i^j, \sigma_i^j)$;
*3.* $U_i^0 := \{p_j \in P | x_i^j = 0\}$; $U_i^1 := \{p_j \in P | x_i^j = 1\}$;
*4.* $S_i^0 := \{p_j \in P | x_i^j = 0 \ \wedge \ V_i^s(0, \sigma_i^j) = 1\}$;
$S_i^1 := \{p_j \in P | x_i^j = 1 \ \wedge \ V_i^s(1, \sigma_i^j) = 1\}$;
*5.* **if** $\left|U_i^{x_i^s}\right| \geq n - t_p$ **then** $y_i := x_i^s$ $\qquad$ *(A)*
**elseif** $p_s \in S_i^{x_i^s} \ \wedge \ \left|S_i^{x_i^s}\right| \geq n - t_\sigma$ **then** $y_i := x_i^s$ $\qquad$ *(B)*
**elseif** $p_s \in S_i^{x_i^s} \ \wedge \ \left|S_i^{x_i^s}\right| \geq n - T \ \wedge \ S_i^{1-x_i^s} = \emptyset$ **then** $y_i := x_i^s$ $\qquad$ *(C)*
**else** $y_i := \bot$ **fi**; $\qquad$ *(D)*
*6.* **return** $y_i$

**Lemma 1 (Weak Broadcast).** *Protocol 1 among the players $P = \{p_1, \ldots, p_n\}$ achieves efficient weak broadcast with sender $p_s \in P$ if $2T + t_p < n$ and $T + 2t_\sigma < n$.*

*Proof.* We show that the validity and consistency properties are satisfied. For this, let $f$ be the number of corrupted players at the end of the protocol. Efficiency is obvious.

VALIDITY: Suppose that the sender $p_s$ is correct. Hence, every correct player $p_i$ receives the sender's input $x_s$ during Step 1 of the protocol, $x_i^s = x_s$, and a signature $\sigma_i^s$.

If $f \leq t_p$ then every correct player $p_i$ receives the value $x_s$ from at least $n - t_p$ different players (including the sender) during Steps 1 and 2. Hence, $|U_i^{x_s}| \geq n - t_p$, and $p_i$ computes $y_i = x_s$ according to Condition (A) in Step 5.

If $t_p < f \leq t_\sigma$ then every correct player $p_i$ receives the value $x_s$ together with a valid signature by $p_s$ (note that the PKI is consistent in this case) from at least $n - t_\sigma$ different players (including the sender) during Steps 1 and 2. Hence, $|S_i^{x_s}| \geq n - t_\sigma$, and $p_i$ computes $y_i = x_s$ according to Conditions (A) or (B) in Step 5.

If $t_\sigma < f \leq T$ then every correct player $p_i$ receives the value $x_s$ together with a valid signature by $p_s$ from at least $n - T$ different players (including the sender) during Steps 1 and 2. Hence, $|S_i^{x_s}| \geq n - T$ and $S_i^{1-x_s} = \emptyset$ since the adversary cannot forge signatures in this case. Hence, $p_i$ computes $y_i = x_s$ according to Conditions (A), (B), or (C).

CONSISTENCY: Suppose that some correct player $p_i$ computes output $y_i \neq \bot$. We have to show that hence, every correct player $p_j$ computes an output $y_j \in \{y_i, \bot\}$.

Suppose first, that $p_i$ decides according to Condition (A) in Step 5, i.e., $|U_i^{y_i}| \geq n - t_p$. For $p_j$ this implies that $|U_j^{y_i}| \geq |U_i^{y_i}| - T \geq n - t_p - T > T$ and hence that $|S_j^{1-y_i}| \leq |U_j^{1-y_i}| < n - T$ and thus that $p_j$ cannot compute $y_j = 1 - y_i$, neither according to Conditions (A), (B), nor (C).

Second, suppose that $p_i$ decides according to Condition (B) in Step 5, i.e., $|S_i^{y_i}| \geq n - t_\sigma$. It remains to show that $p_j$ does not decide on $y_j = 1 - y_i$ according to Conditions (B) or (C) (the rest is out-ruled by the last paragraph). For $p_j$ the assumption implies that

$$|U_j^{y_i}| \geq |U_i^{y_i}| - f \geq n - t_\sigma - f \geq \begin{cases} n - 2t_\sigma > T & , \text{if } f \leq t_\sigma , \\ n - t_\sigma - T > t_\sigma & , \text{if } f \leq T . \end{cases}$$

Now, if $f \leq t_\sigma$ then $|S_j^{1-y_i}| \leq |U_j^{1-y_i}| < n - T$, and $p_j$ cannot compute $y_j = 1 - y_i$ according to Conditions (B) or (C). If $t_\sigma < f \leq T$ then $|U_j^{1-y_i}| < n - t_\sigma$, and $S_j^{y_i} \neq \emptyset$ (since the PKI is consistent and $p_i$ holds and redistributes a valid signature on $y_i$), and thus $p_j$ still cannot compute $y_j = 1 - y_i$ according to Conditions (B) or (C).

Third, suppose that $p_i$ decides according to Condition (C) in Step 5, i.e., $|S_i^{y_i}| \geq n - T$ and $S_i^{1-y_i} = \emptyset$. It remains to show that $p_j$ cannot decide on $y_j = 1 - y_i$ according to Condition (C). Now, $f \leq t_p$ implies $|U_j^{1-y_i}| < n - T$ (since $|U_j^{y_i}| \geq n - T - t_p > T$), and $f > t_p$ implies $S_j^{y_i} \neq \emptyset$ (since the PKI is consistent). Finally, both implications rule out that $p_j$ computes $y_j = 1 - y_i$ according to Condition (C). □

**Multi-party Computation.** The MPC protocol in [CDD+99] unconditionally tolerates an (adaptive) adversary that corrupts up to $t < n/2$ players — but assuming broadcast channels to be available.

**Theorem 2.** *Hybrid MPC is efficiently achievable if* $2T + t_p < n$ *and* $T + 2t_\sigma < n$.

*Proof.* Efficient achievability of hybrid broadcast for $2T + t_p < n$ and $T + 2t_\sigma < n$ follows from Lemma 1 and Theorem 1. We can now simulate each invocation of a broadcast channel in [CDD+99] with an instance of such a hybrid broadcast protocol. Since Bound (1) implies $2T < n$, we have that $t_p \leq t_\sigma \leq T < n/2$. Thus, an adversary that is tolerated in the broadcast protocol is automatically tolerated in the MPC protocol. □

### 4.2 Tightness

We now show that Bound (1) is tight. We do this in three steps. First, we show that hybrid broadcast is impossible if $t_p > 0$, $t_\sigma = 0$, and $2T + t_p \geq n$. Second, we show that hybrid broadcast is impossible if $t_p = 0$, $t_\sigma > 0$, and $T + 2t_\sigma \geq n$. Third, we use the fact that MPC is impossible whenever $2T \geq n$ [Cle86].

**Impossibility of Broadcast for $\mathbf{t_p + 2T \geq n}$ when $\mathbf{t_p > 0}$.** The proof proceeds along the lines of the proof in [FLM86] that unconditional broadcast for $t \geq n/3$ is impossible. The idea is to assume any protocol among $n$ players that (possibly) achieves broadcast for $n \leq 2T + t_p$ ($t_p > 0$, $t_\sigma = 0$) and to use it to build a different distributed system whose behavior demonstrates that the original protocol must be insecure. It is important to note that this new system is not required to achieve broadcast. It is simply a distributed system whose behavior is determined by the protocol or, more precisely, by the corresponding local programs of the involved players. Also, it is assumed that no adversary is present in this system. Rather, with respect to some of the players, the way the new system is composed simulates an admissible adversary with respect to these players in the original system. Thus, with respect to these players, all conditions of broadcast are required to be satisfied among them even in this new system. Finally, it is shown that all of these players' respective conditions cannot be satisfied simultaneously and thus that the protocol cannot achieve broadcast.

*Building the new system.* Assume any protocol $\Psi$ for a player set $P$ with sender $p_0$ and $|P| = n \geq 3$ that tolerates $2T + t_p \geq n$ (with $t_p > 0$).

Let $\Pi = \{\pi_0, \dots, \pi_{n-1}\}$ be the set of the players' corresponding processors with their local programs sharing a consistent PKI where player $p_i$'s secret-key/public-key pair is $(\mathrm{SK}_i, \mathrm{PK}_i)$ and player $p_j$'s copy of the respective public key is $\mathrm{PK}_{ij}$. Since $0 < t_p \leq T$, it is possible to partition the processors into three sets, $\Pi_0 \dot{\cup} \Pi_1 \dot{\cup} \Pi_2 = \Pi$, such that $1 \leq |\Pi_0| \leq t_p$, $1 \leq |\Pi_1| \leq T$, and $1 \leq |\Pi_2| \leq T$.

For each $\pi_i \in \Pi_0$, let $\pi_i'$ be an identical copy of processor $\pi_i$. Let the number $i$ denote the *type* of any processor $\pi_i$ (or $\pi_i'$, respectively). Furthermore, let $\Pi_0' = \{\pi_i' | \pi_i \in \Pi_0\}$ form an identical copy of set $\Pi_0$. For all $\pi_i \in \Pi_0'$, generate a new secret-key/public-key pair $(\mathrm{SK}_i', \mathrm{PK}_i')$ and overwrite $\pi_i$'s own secret key $\mathrm{SK}_i := \mathrm{SK}_i'$. Additionally, for all $\pi_j \in \Pi_2 \cup \Pi_0'$, overwrite $\pi_j$'s copy of $\pi_i$'s public key: $\mathrm{PK}_{ij} := \mathrm{PK}_i'$ (and $\mathrm{PK}_i := \mathrm{PK}_i'$). See Figure 3.

Instead of connecting the original processors as required for broadcast, we build a network involving all processors in $\Pi_0 \cup \Pi_1 \cup \Pi_2 \cup \Pi_0'$ with their pairwise communication channels connected in a way such that each processor $\pi_i$ (or $\pi_i'$) communicates with at most one processor of each type $j \in \{1, \dots, n\} \setminus \{i\}$.

Consider Figure 3. Exactly all pairs in $(\Pi_0 \cup \Pi_1) \times (\Pi_0 \cup \Pi_1)$, $(\Pi_1 \cup \Pi_2) \times (\Pi_1 \cup \Pi_2)$, and $(\Pi_2 \cup \Pi_0') \times (\Pi_2 \cup \Pi_0')$ are connected by pairwise channels. There are no connections between the sets $\Pi_0$ and $\Pi_2$, and no connections between the sets $\Pi_1$ and $\Pi_0'$. Messages that originally would have been sent from a processor in $\Pi_0$ to a processor in $\Pi_2$ are discarded. Messages that originally would have been sent from a processor in $\Pi_2$ to a processor in $\Pi_0$ are delivered to the corresponding processor in $\Pi_0'$. Messages sent from a processor in $\Pi_0'$ to a processor in $\Pi_1$ are discarded.

We now show that for the sets $\Pi_0 \cup \Pi_1$, $\Pi_1 \cup \Pi_2$, and $\Pi_2 \cup \Pi_0'$, and for inputs $x_0 = 0$ and $x_0' = 1$, each set's joint view is indistinguishable from its view in the original setting for an adversary corrupting the remaining processors in

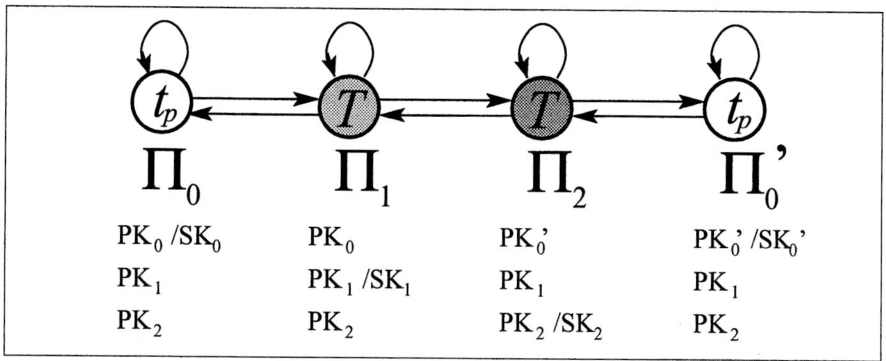


**Fig. 3.** *Rearrangement of processors in proof of Theorem 3.*

an admissible way, and possibly have made for a bad PKI if it corrupts at most $f \leq t_p$ processors.

**Lemma 2.** *If the input of $\pi_0$ is $x_0 = 0$ then the joint view of the processors in $\Pi_0 \cup \Pi_1$ is indistinguishable from their view in the original system when the adversary corrupts the processors in $\Pi_2$ in an admissible way.*

*Proof.* By corrupting all processors in $\Pi_2$ in the original system the adversary simulates all processors in $\Pi_2 \cup \Pi_0'$ of the new system. For all $\pi_i \in \Pi_0'$ it generates a new secret-key/public-key pair $(\mathrm{SK}_i', \mathrm{PK}_i')$ and overwrites $\pi_i$'s own secret key $\mathrm{SK}_i := \mathrm{SK}_i'$ and, for all $\pi_j \in \Pi_2 \cup \Pi_0'$, overwrites $\pi_j$'s copy of $\pi_i$'s public key: $\mathrm{PK}_{ij} := \mathrm{PK}_i'$ (and $\mathrm{PK}_i := \mathrm{PK}_i'$). Initially, the adversary overwrites input $x_0' := 1$. The PKI among the processors in $\Pi_0 \cup \Pi_1$ is still fully consistent and thus the joint view of the processors in $\Pi_0 \cup \Pi_1$ in the original system is exactly the same as their view in the new system. □

**Lemma 3.** *If the input of $\pi_0'$ is $x_0' = 1$ then the joint view of the processors in $\Pi_2 \cup \Pi_0'$ is indistinguishable from their view in the original system when the adversary corrupts the processors in $\Pi_1$ in an admissible way.*

*Proof.* By symmetry, this case follows from Lemma 2.[9] □

**Lemma 4.** *The joint view of the processors in $\Pi_1 \cup \Pi_2$ is indistinguishable from their view in the original system when the adversary corrupts the processors in $\Pi_0$ in an admissible way.*

[9] The only difference in this case is that $\Pi_0'$ takes the role of the original set and $\Pi_0$ the role of its copy. Accordingly, the initial key pairs are $(\mathrm{SK}_i', \mathrm{PK}_i')$, the pairs $(\mathrm{SK}_i, \mathrm{PK}_i)$ are newly generated by the adversary, and $x_0 := 0$ is overwritten.

*Proof.* Since $|\Pi_0| \leq t_p$ the adversary can have previously made the PKI inconsistent by generating and respectively distributing the key pairs $(\mathrm{SK}'_i, \mathrm{PK}'_i)$ for all $\pi_i \in \Pi'_0$ (according to Figure 3). By corrupting all processors in $\Pi_0$ in the original system the adversary can now simulate all processors in $\Pi_0 \cup \Pi'_0$ of the new system whereas, initially, it overwrites $x_0 := 0$ and $x'_0 := 1$. Thus the joint view of the processors in $\Pi_1 \cup \Pi_2$ in the original system is exactly the same as their view in the new system. □

**Theorem 3.** *If $2T + t_p \geq n$ and $t_p > 0$ then there exists no hybrid broadcast protocol. In particular, for every protocol there exists a sender input $x_0 \in \{0, 1\}$ such that a computationally bounded adversary can make the protocol fail with some non-negligible probability — by either corrupting $T$ players, or by corrupting $t_p$ players and additionally having made for an inconsistent PKI.*

*Proof.* Assume that $x_0 = 0$ and $x'_0 = 1$. Then, by Lemmas 2, 3, and 4, each mentioned set's joint view in the new system is indistinguishable from their view in the original system. However, for each run of the new system, either validity is violated among the processors in $S_{01} = \Pi_0 \cup \Pi_1$ or $S_{20'} = \Pi_2 \cup \Pi'_0$, or consistency is violated among the processors in $S_{12} = \Pi_1 \cup \Pi_2$.

Thus there is a sender input ($x_0 = 0$ or $x'_0 = 1$) such that the adversary can make the protocol fail with non-negligible probability by uniformly randomly choosing a processor set $\Pi_i$ and corrupting the respective processors correspondingly. □

**Impossibility of Broadcast for $\mathbf{2t_\sigma + T \geq n}$ when $\mathbf{t_\sigma > 0}$.** The proof of this case is very similar to the proof of Theorem 3.

**Theorem 4.** *If $T+2t_\sigma \geq n$ and $t_\sigma > 0$ then there exists no hybrid broadcast protocol. In particular, for every protocol there exists a sender input $x_0 \in \{0, 1\}$ such that the adversary can make the protocol fail with some non-negligible probability — by either corrupting $T$ players, or by corrupting $t_\sigma$ players and additionally being able to forge signatures with non-negligible probability.*

*Proof.* Assume any protocol $\Psi$ for a player set $P$ with sender $p_0$ and $|P| = n \geq 3$ that tolerates $2T + t_\sigma \geq n$ (with $t_\sigma > 0$).

Let $\Pi = \{\pi_0, \ldots, \pi_{n-1}\}$ be the set of the players' corresponding processors with their local programs. Since $0 < t_\sigma \leq T$, it is possible to partition the processors into three sets, $\Pi_0 \dot\cup \Pi_1 \dot\cup \Pi_2 = \Pi$, such that $1 \leq |\Pi_0| \leq T$, $1 \leq |\Pi_1| \leq t_\sigma$, and $1 \leq |\Pi_2| \leq t_\sigma$.

For each $\pi_i \in \Pi_0$, let $\pi'_i$ be an identical copy of processor $\pi_i$ and, as in the proof of Theorem 3, let $\Pi'_0 = \{\pi'_i \,|\, \pi_i \in \Pi_0\}$ form an identical copy of set $\Pi_0$.

Consider Figure 4. Exactly all pairs in $(\Pi_0 \cup \Pi_1) \times (\Pi_0 \cup \Pi_1)$, $(\Pi_1 \cup \Pi_2) \times (\Pi_1 \cup \Pi_2)$, and $(\Pi_2 \cup \Pi'_0) \times (\Pi_2 \cup \Pi'_0)$ are connected by pairwise channels.

Again, we show that for the sets $\Pi_0 \cup \Pi_1$, $\Pi_1 \cup \Pi_2$, and $\Pi_2 \cup \Pi'_0$, and for inputs $x_0 = 0$ and $x'_0 = 1$, each set's joint view is indistinguishable from its view in the original setting.

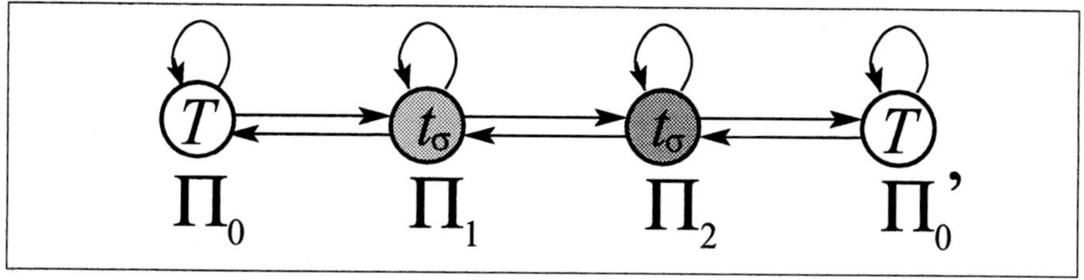


**Fig. 4.** *Rearrangement of processors in proof of Theorem 3.*

*Joint view of $\Pi_0 \cup \Pi_1$ with $x_0 = 0$.* By corrupting all processors in $\Pi_2$ in the original system the adversary simulates all processors in $\Pi_2 \cup \Pi_0'$ of the new system. Since $|\Pi_2| \leq t_\sigma$, the adversary can forge all signatures by processors in $\Pi_0'$ required for the simulation. Initially, the adversary overwrites input $x_0' := 1$. Thus the joint view of the processors in $\Pi_0 \cup \Pi_1$ in the original system is exactly the same as their view in the new system.

*Joint view of $\Pi_2 \cup \Pi_0'$ with $x_0' = 1$.* By symmetry, this case follows from the above paragraph.

*Joint view of $\Pi_1 \cup \Pi_2$.* By corrupting all processors in $\Pi_0$ in the original system the adversary can simulate all processors in $\Pi_0 \cup \Pi_0'$ of the new system whereas, initially, it overwrites $x_0 := 0$ and $x_0' := 1$. Note that, by corrupting the processors in $\Pi_0$, the adversary gains access to all corresponding secret keys and thus is not required to forge any signatures for the simulation. Thus the joint view of the processors in $\Pi_1 \cup \Pi_2$ in the original system is exactly the same as their view in the new system.

The theorem now follows along the lines of the proof of Theorem 3. □

**Multi-party Computation.** In order to complete our tightness argument, we require the following proposition.

**Proposition 1 ([Cle86]).** *There is no protocol for MPC secure against $T \geq n/2$ actively corrupted players. In particular, fairness cannot be guaranteed.*

**Theorem 5.** *Hybrid MPC is impossible if either $2T + t_p \geq n$ or $T + 2t_\sigma \geq n$.*

*Proof.* The theorem follows from Theorems 3 and 4, and Proposition 1. □

## 5 Conclusion and Open Problems

We can now conclude tight bounds for the achievability of hybrid MPC with respect to thresholds $t_p$, $t_\sigma$, and $T$.

**Theorem 6.** *Hybrid MPC is (efficiently) achievable if and only if* $2T + t_p < n$ *and* $T + 2t_\sigma < n$.

*Proof.* The theorem follows from Theorems 2 and 5. □

In particular, assuming the PKI to be consistent in any case (as in the alternative model for [GMW87] assuming a PKI instead of broadcast) we can drop parameter $t_p$ and immediately get the tight bound $2T < n \ \wedge \ T + 2t_\sigma < n$. This means that, with respect to this model, computational security against $f \leq T < n/2$ corrupted players can be combined with unconditional security against $f \leq t_\sigma$ corrupted players.

The characterization given in Theorem 6 is tight with respect to fully secure, robust MPC. However, as mentioned in the introduction, non-robust MPC is also possible in presence of a corrupted majority. Thus, for the case $t_p = 0$, it remains an open question whether hybrid non-robust MPC can be achieved for any $T + 2t_\sigma < n$.

## References


[Bea89] Donald Beaver. Multiparty protocols tolerating half faulty processors. In *Advances in Cryptology: CRYPTO '89*, volume 435 of *Lecture Notes in Computer Science*, pages 560–572. Springer-Verlag, 1989.

[Bea91] Donald Beaver. Foundations of secure interactive computation. In *Advances in Cryptology: CRYPTO '91*, volume 576 of *Lecture Notes in Computer Science*, pages 377–391. Springer-Verlag, 1991.

[BG89] Donald Beaver and Shafi Goldwasser. Multiparty computation with faulty majority. In *Proceedings of the 30th Annual IEEE Symposium on Foundations of Computer Science (FOCS '89)*, pages 468–473, 1989.

[BGP89] Piotr Berman, Juan A. Garay, and Kenneth J. Perry. Towards optimal distributed consensus (extended abstract). In *Proceedings of the 30th Annual IEEE Symposium on Foundations of Computer Science (FOCS '89)*, pages 410–415, 1989.

[BGW88] Michael Ben-Or, Shafi Goldwasser, and Avi Wigderson. Completeness theorems for non-cryptographic fault-tolerant distributed computation. In *Proceedings of the 20th Annual ACM Symposium on Theory of Computing (STOC '88)*, pages 1–10. Springer-Verlag, 1988.

[Can00] Ran Canetti. Security and composition of multiparty cryptographic protocols. *Journal of Cryptology*, 13(1):143–202, 2000.

[CCD88] David Chaum, Claude Crépeau, and Ivan Damgård. Multiparty unconditionally secure protocols (extended abstract). In *Proceedings of the 20th Annual ACM Symposium on Theory of Computing (STOC '88)*, pages 11–19. ACM Press, 1988.

[CDD+99] Ronald Cramer, Ivan Damgård, Stefan Dziembowski, Martin Hirt, and Tal Rabin. Efficient multiparty computations secure against an adaptive adversary. In *Advances in Cryptology: EUROCRYPT '99*, volume 1592 of *Lecture Notes in Computer Science*, 1999.

[Cha89] David Chaum. The spymasters double-agent problem. In *Advances in Cryptology: CRYPTO '89*, volume 435 of *Lecture Notes in Computer Science*, pages 591–602. Springer-Verlag, 1989.

[Cle86] Richard Cleve. Limits on the security of coin flips when half the processors are faulty. In *ACM Symposium on Theory of Computing (STOC '86)*, pages 364–369, Baltimore, USA, May 1986. ACM Press.

[DFF+82] Danny Dolev, Michael J. Fischer, Rob Fowler, Nancy A. Lynch, and H. Raymond Strong. An efficient algorithm for Byzantine agreement without authentication. *Information and Control*, 52(3):257–274, March 1982.

[Dol82] Danny Dolev. The Byzantine generals strike again. *Journal of Algorithms*, 3(1):14–30, 1982.

[DS82] Danny Dolev and H. Raymond Strong. Polynomial algorithms for multiple processor agreement. In *Proceedings of the 14th Annual ACM Symposium on Theory of Computing (STOC '82)*, pages 401–407, 1982.

[DS83] Danny Dolev and H. Raymond Strong. Authenticated algorithms for Byzantine agreement. *SIAM Journal on Computing*, 12(4):656–666, 1983.

[FGH+02] Matthias Fitzi, Daniel Gottesman, Martin Hirt, Thomas Holenstein, and Adam Smith. Detectable Byzantine agreement secure against faulty majorities. In *Proceedings of the 21st ACM Symposium on Principles of Distributed Computing (PODC '02)*, pages 118–126, 2002.

[FHHW03] Matthias Fitzi, Martin Hirt, Thomas Holenstein, and Jürg Wullschleger. Two-threshold broadcast and detectable multi-party computation. In Eli Biham, editor, *Advances in Cryptology — EUROCRYPT '03*, Lecture Notes in Computer Science. Springer-Verlag, May 2003.

[FLM86] Michael J. Fischer, Nancy A. Lynch, and Michael Merritt. Easy impossibility proofs for distributed consensus problems. *Distributed Computing*, 1:26–39, 1986.

[FM97] Pesech Feldman and Silvio Micali. An optimal probabilistic protocol for synchronous Byzantine agreement. *SIAM Journal on Computing*, 26(4):873–933, August 1997.

[FM00] Matthias Fitzi and Ueli Maurer. From partial consistency to global broadcast. In *Proceedings of the 32nd Annual ACM Symposium on Theory of Computing (STOC '00)*, pages 494–503, 2000.

[GHY87] Zvi Galil, Stuart Haber, and Moti Yung. Cryptographic computation: Secure fault-tolerant protocols and the public-key model. In *Advances in Cryptology: CRYPTO '87*, volume 293 of *Lecture Notes in Computer Science*, pages 135–155. Springer-Verlag, 1987.

[GL90] Shafi Goldwasser and Leonid Levin. Fair computation of general functions in presence of immoral majority. In *Advances in Cryptology: CRYPTO '90*, volume 537 of *Lecture Notes in Computer Science*, pages 11–15. Springer-Verlag, 1990.

[GL02] Shafi Goldwasser and Yehuda Lindell. Secure computation without agreement. 16th International Symposium on Distributed Computing (DISC '02). Preliminary version on http://www.research.ibm.com/people/l/lindell, 2002.

[GLR95] Li Gong, Patrick Lincoln, and John Rushby. Byzantine agreement with authentication: Observations and applications in tolerating hybrid and link faults. In *Proceedings of the 5th Conference on Dependable Computing for Critical Applications (DCCA-5)*, pages 79–90, 1995.

[GMW87] Oded Goldreich, Silvio Micali, and Avi Wigderson. How to play any mental game. In *Proceedings of the 19th Annual ACM Symposium on Theory of Computing (STOC '87)*, pages 218–229. ACM Press, 1987.

[Gol01] Oded Goldreich. Secure multi-party computation, working draft, version 1.3, June 2001.

[LLR02] Yehuda Lindell, Anna Lysyanskaya, and Tal Rabin. On the composition of authenticated Byzantine agreement. In *Proceedings of the 34th Annual ACM Symposium on Theory of Computing (STOC '02)*, pages 514–523. ACM Press, 2002.

[LSP82] Leslie Lamport, Robert Shostak, and Marshall Pease. The Byzantine generals problem. *ACM Transactions on Programming Languages and Systems*, 4(3):382–401, July 1982.

[PW96] Birgit Pfitzmann and Michael Waidner. Information-theoretic pseudosignatures and Byzantine agreement for $t >= n/3$. Technical Report RZ 2882 (#90830), IBM Research, 1996.

[RB89] Tal Rabin and Michael Ben-Or. Verifiable secret sharing and multiparty protocols with honest majority. In *Proceedings of the 21st Annual ACM Symposium on Theory of Computing (STOC '89)*, pages 73–85, 1989.

[TC84] Russell Turpin and Brian A. Coan. Extending binary Byzantine agreement to multivalued Byzantine agreement. *Information Processing Letters*, 18(2):73–76, February 1984.

[WP89] Michael Waidner and Birgit Pfitzmann. Unconditional sender and recipient untraceability in spite of active attacks — some remarks. Technical Report 5/89, Universität Karlsruhe, Institut für Rechnerentwurf und Fehlertoleranz, 1989.

[Yao82] Andrew C. Yao. Protocols for secure computations. In *Proceedings of the 23rd Annual IEEE Symposium on Foundations of Computer Science (FOCS '82)*, pages 160–164, 1982.

# A Reducing Broadcast to Weak Broadcast

In this section we describe how to efficiently reduce broadcast to weak broadcast in a way that is more direct than in [FM00]. Given that at most $t < n/2$ players are corrupted the resulting protocol for broadcast is as secure as the given protocol for weak broadcast.

In a first step, weak broadcast is transformed into a protocol for graded consensus (Section A.1), the "consensus variant" of graded broadcast introduced by Feldman and Micali in [FM97]; and finally, graded consensus is transformed into broadcast (Section A.2).

## A.1 Graded Consensus

In graded consensus, every player has an input $x$ and receives two outputs, a value $y \in \{0, 1\}$ and a grade $g \in \{0, 1\}$. If all correct players start with the same value $x$ then all players output $y = x$ and $g = 1$. Additionally, if any correct player ends up with grade $g = 1$ then all correct players output the same value $y$, i.e., computing $g = 1$ means "detecting agreement."

**Definition 3 (Graded Consensus).** *A protocol where every player $p_i$ has an input $x_i \in \{0, 1\}$ and computes two output values $y_i, g_i \in \{0, 1\}$ achieves graded consensus if it satisfies the following conditions:*

- Validity: *If all correct players have the same input value $x$ then every correct player $p_i$ computes $y_i = x$ and $g_i = 1$.*
- Consistency: *If any correct player $p_i$ computes $g_i = 1$ then every correct player $p_j$ computes $y_j = y_i$.*

The following protocol for graded consensus basically consists of two consecutive rounds wherein each player weak-broadcasts a value. Note that, in Step 4 of the protocol, the domain of weak broadcast is ternary, namely $\{0, 1, \perp\}$. Following the restriction to focus on protocols with binary domains we can simply interpret such a protocol as being simulated by two parallel invocations of binary weak broadcast.

**Protocol 2** `GradedConsensus`$(P, x_i)$

1. $\forall j \in \{1, \ldots, n\} : x_i^j :=$ `WeakBroadcast`$_{p_j}$ $(P, x_j)$;
2. $S_i^0 := \{j \in \{1, \ldots, n\} | x_i^j = 0\}$; $S_i^1 := \{j \in \{1, \ldots, n\} | x_i^j = 1\}$;
3. **if** $|S_i^{x_i}| \geq n - t$ **then** $z_i := x_i$ **else** $z_i := \perp$ **fi**;
4. $\forall j \in \{1, \ldots, n\} : z_i^j :=$ `WeakBroadcast`$_{p_j}$ $(P, z_j)$;
5. $T_i^0 := \{j \in \{1, \ldots, n\} | z_i^j = 0\}$; $T_i^1 := \{j \in \{1, \ldots, n\} | z_i^j = 1\}$;
6. **if** $|T_i^0| > |T_i^1|$ **then** $y_i := 0$ **else** $y_i := 1$ **fi**;
7. **if** $|T_i^{y_i}| \geq n - t$ **then** $g_i := 1$ **else** $g_i = 0$ **fi**;
8. **return** $(y_i, g_i)$

**Lemma 5 (Graded Consensus).** *If Protocol* `WeakBroadcast` *achieves weak broadcast then Protocol 2 achieves graded consensus secure against $t < n/2$ corrupted players.*

*Proof.*
Validity: If all correct players hold the same value $x$ at the beginning of the protocol then, by the validity property of weak broadcast, $|S_i^{x_i}| \geq n - t > t$ for every correct player $p_i$ and thus $z_i = x_i = x$. Finally, $|T_i^x| \geq n - t > t$, $|T_i^{1-x}| < n - t$, and $y_i = x$ and $g_i = 1$.

Consistency: Note that every correct player $p_i$ that does not compute $z_i = \perp$ (in Step 3) holds the same value $z_i = z$: By the validity property of weak broadcast, $|S_i^{x_i}| \geq n - t$ implies that $|S_j^{1-x_i}| \leq t < n - t$.

Now, let $p_i$ and $p_j$ be two correct players and suppose that $p_i$ decides on $y_i = y \in \{0, 1\}$ and $g_i = 1$. We have to show that $y_j = y$.

From $g_i = 1$ it follows that $|T_i^y| \geq n - t > t$ and thus that at least one correct player $p_k$ must have sent $z_k = y$ during Step 4, and with the above remark, that no correct player $p_k$ can have sent $z_k = 1 - y$ during Step 4.

Let $\ell$ be the number of corrupted players who distributed value $y$ during Step 4. Now, $|T_i^y| \geq n - t$ implies $|T_j^y| \geq n - t - \ell > t - \ell$ and $|T_j^{1-y}| \leq t - \ell$ since only the remaining $t - \ell$ corrupted players can have sent value $y$ during Step 4. Thus, $p_j$ computes $y_j := y = y_i$. □

### A.2 Broadcast

For simplicity, without loss of generality, assume $s = 1$, i.e., that $p_1$ is the sender of the broadcast.

**Protocol 3** `Broadcast`$_{p_1}$ $(P, x_1)$

1. $p_1$: `Send` $x_1$; `Receive`$(y_i)$
2. $p_i$: `for` $k = 2$ `to` $t + 1$ `do`
3. $\quad (y_i, g_i) :=$ `GradedConsensus` $(P, y_i)$;
4. $\quad p_k$: `Send` $y_k$; `Receive`$(y_i^k)$
5. $\quad$ `if` $g_i = 0$ `then` $y_i := y_i^k$ `fi`
6. `od`; `return` $y_i$

**Lemma 6.** *Suppose that Protocol* `GradedConsensus` *achieves graded consensus. If in Protocol 3, for some $k \in \{2, \ldots, t+1\}$, every correct player $p_i$ holds the same value $y_i = b$ at the beginning of Step 3 then $y_i = b$ holds at the end of the protocol.*

*Proof.* Suppose that $y_i = b \in \{0, 1\}$ for every correct player $p_i$ before Step 3 (for some $k$). Because of the validity property of graded consensus, after Step 3, (for $k$), every correct player $p_i$ holds $y_i = b$ and $g_i = 1$, and thus ignores Step 5, (for $k$). Thus, by induction, every correct player $p_i$ ends the protocol with $y_i = b$. □

**Lemma 7 (Broadcast).** *If Protocol* `GradedConsensus` *achieves graded consensus then Protocol 3 achieves broadcast with sender $p_1$ (for any $t < n$).*

*Proof.* We show that the validity and consistency properties of broadcast are satisfied.

VALIDITY: Suppose the sender $p_1$ to be correct with input $x_s = b$. Hence, every correct player $p_i$ holds value $y_i = b$ before Step 3 for $k = 2$. And by Lemma 6, every correct player $p_i$ ends the protocol with $y_i = b$.

CONSISTENCY: If the sender is correct then consistency is implied by the validity property. Assume now that $p_1$ is corrupted. Hence there is a correct player $p_k$ ($k \in \{2, \ldots, t+1\}$). We now argue that, for such a $k$ where $p_k$ is correct, every correct player $p_i$ holds the same value $y_i$ after Step 5. Then, together with Lemma 6, the consistency property follows.

First, suppose that every correct player $p_i$ holds $g_i = 0$ after Step 3. Then all of them adopt $p_k$'s value, $y_i = y_i^k$, and consistency follows from Lemma 6. Suppose now, that any correct player $p_i$ holds $g_i = 1$ after Step 3. Then, by the consistency property of graded consensus, $p_k$ and every other correct player $p_j$ hold $y_k = y_i = y_j$, and consistency follows from Lemma 6. □

**Theorem 1.** *If at most $t < n/2$ players are corrupted then efficient achievability of weak broadcast implies efficient achievability of broadcast.*

*Proof.* Since the given construction for broadcast involves a polynomial number of invocations of weak broadcast ($2n(t + 1)$), the theorem follows directly from Lemma 7. □

# On the Hardness of Information-Theoretic Multiparty Computation*

Yuval Ishai and Eyal Kushilevitz

Computer Science Department, Technion, Haifa 32000, Israel
{yuvali,eyalk}@cs.technion.ac.il

**Abstract.** We revisit the following open problem in information-theoretic cryptography: Does the communication complexity of unconditionally secure computation depend on the computational complexity of the function being computed? For instance, can computationally unbounded players compute an *arbitrary* function of their inputs with polynomial communication complexity and a linear threshold of unconditional privacy? Can this be done using a constant number of communication rounds?
We provide an explanation for the difficulty of resolving these questions by showing that they are closely related to the problem of obtaining efficient protocols for (information-theoretic) private information retrieval and hence also to the problem of constructing short locally-decodable error-correcting codes. The latter is currently considered to be among the most intriguing open problems in complexity theory.

**Keywords.** Information-theoretic cryptography, secure multiparty computation, private information retrieval, locally decodable codes.

## 1 Introduction

In STOC 1990, Beaver, Micali, and Rogaway [5] posed the following question:

> Is there a constant-round protocol that allows $k$ computationally unbounded players to defeat a computationally unbounded adversary, while using only a polynomial amount of communication in the total length of their inputs?

This question is still wide open today: it is not known whether all functions admit such a protocol, even in the simple case that the adversary can *passively* corrupt only a *single* player, and even without any restriction on the number of rounds.

A partial answer to the above question was given by Beaver, Feigenbaum, Kilian and Rogaway [4]. They showed that such a round- and communication-efficient protocol exists when the number of players is roughly as large as the

* Research supported in part by a grant from the Israel Science Foundation and by the Technion V.P.R. Fund.

C. Cachin and J. Camenisch (Eds.): EUROCRYPT 2004, LNCS 3027, pp. 439–455, 2004.

total input size. More precisely, every function $f$ of $n$ input bits can be $t$-securely computed by $k = O(tn/\log n)$ computationally unbounded players using poly$(n)$ communication complexity and a constant round complexity. Note that this result is meaningless when the number of players is fixed (even when $t = 1$), since it requires the number of players to grow with the input size . This should be contrasted with the fact that, ignoring complexity issues, the optimal security threshold is a constant fraction of the number of players, regardless of the input size. Again, the problem of resolving these difficulties was posed as an open question in [4].[1]

As noted above, if there is no limit on the resources used by the players, then any function $f$ can be computed by $k$ players with a linear threshold of information-theoretic security. This can also be done in a constant number of rounds. However, all general-purpose protocols achieving this have a somewhat unexpected common feature: their *communication complexity* depends on the computational complexity of $f$ (either its circuit complexity if there is no restriction on the number of rounds [7,10,12], or its formula- or branching program complexity in the constant-round case [2,19]). It seems quite unlikely that a purely information-theoretic complexity measure would be so closely linked with computational measures. However, so far there has been no significant negative evidence against this link nor a positive evidence to support it.

The main goal of this work is to establish a close connection between the above questions and other well-known open problems. These problems are discussed below.

**Private information retrieval (PIR).** A *private information retrieval* (PIR) protocol allows a user to retrieve an item $i$ from a database of size $N$ while hiding $i$ from the servers storing the database. The main cost-measure of such protocols is the *communication complexity* of retrieving one out of $N$ bits of data. There are two main settings for PIR. In the information-theoretic setting [11,1,6], there are $k \geq 2$ servers holding copies of the database and the default privacy requirement is that each *individual* server learn absolutely no information about $i$. In the computational setting for PIR [8,22,9] there is typically only a single server holding the database, and the privacy requirement is relaxed to *computational* privacy. While the complexity of PIR in the computational setting is pretty well understood (an "essentially optimal" protocol with polylogarithmic communication can be based on a reasonable cryptographic assumption [9]), the situation is very different in the information-theoretic setting. For any constant $k$, the best upper bound on the communication complexity of $k$-server PIR is some fixed polynomial in $N$, i.e., $O(N^{1/c_k})$ where $c_k$ is a constant depending on $k$. (The current best bound on $c_k$ is $\Omega(k \log k / \log\log k)$ [6].) On the other hand, the best known general lower bound on the communication complexity of $k$-server

[1] These questions should not be confused with another major open problem in information-theoretic MPC: does every *polynomial-time* computable function admit a constant-round protocol which is *computationally efficient*? Our results do not have direct relevance to this question. However, our results do have relevance to the variant of this question which allows the protocols to be computationally inefficient.

PIR is logarithmic in $N$ [23]. Hence, there is an exponential gap between known upper and lower bounds. From now on, the term PIR will refer by default to information-theoretic PIR.

**Symmetrically Private Information Retrieval (SPIR).** The original PIR model is not concerned with protecting the privacy of the data, and allows the user to learn arbitrary additional information (in addition to the selected bit). The stronger SPIR primitive [15] requires, on top of the PIR requirement, that the user learn no additional information about the database other than the selected bit. This may be viewed as an information-theoretic analogue of $\binom{N}{1}$-Oblivious Transfer. We use SPIR as an intermediate primitive for establishing the connection between PIR and multi-party computation. In doing so, we need to establish a tighter reduction from SPIR to PIR than the one shown in [15].

**Locally-decodable codes (LDC).** Standard error-correcting codes can provide high fault tolerance while only moderately expanding the encoded message. However, their decoding procedure requires to read the entire encoded message even if one is only interested in decoding a single bit of this message. LDC simultaneously provide high fault tolerance and a sublinear-time "local" decoding procedure. To make this possible, the decoding procedure must use randomness for selecting which bits to probe, and some error probability must be tolerated. More formally, a code $C : \{0,1\}^N \rightarrow \Sigma^M$ is said to be $(k, \delta, \epsilon)$-*locally decodable* if every bit $x_i$ of $x$ can be decoded from $y = C(x)$ with success probability$\geq 1/2+\epsilon$ by reading $k$ (randomly chosen) symbols of $y$, even if up to a $\delta$-fraction of the symbols in $y$ were adversarially corrupted. The main complexity question related to LDC is the following: Given a constant number of queries $k$, what is the minimal length $M(N)$ of a $(k, \delta, \epsilon)$-LDC? In studying this question, one typically requires $\delta, \epsilon$ to be bounded by some fixed constants (independently of $N$). However, the problem appears to be as difficult even if $\delta, \epsilon$ are sub-constant (say, $\delta, \epsilon = 2^{-\log^c n}$) as long as they are not exponentially small.

Katz and Trevisan [20] have shown an intimate connection between this question and information-theoretic PIR. In particular, a $k$-server PIR protocol in which the user sends $\alpha(N)$ bits to each server and receives $\beta(N)$ bits in return can be used to construct a $k$-query LDC of length $O(k2^{\alpha(N)})$ over $\Sigma = \{0,1\}^{\beta(N)}$. Accordingly, the best upper bound on the length of a $k$-query LDC is exponential in $N$ and the best general lower bound is polynomial in $N$ [20]. The question of obtaining stronger lower bounds for LDC has recently received a significant amount of attention [20,17,13,27,21], and progress on this question appears to be very difficult.

### 1.1 Our Results

We prove that the problem of obtaining communication-efficient constant-round protocols for arbitrary functions is closely related to the problem of obtaining communication-efficient PIR protocols. Relying on known connections between PIR and locally decodable codes [20], we obtain a similar connection between the communication complexity of unconditionally secure multiparty computation and the length of locally decodable codes. In particular, strong negative results for the former problem would imply strong negative results for the latter, which so far seem elusive.

The above high-level statements hide some subtleties. By default, we will view the number of players as constant, and measure the complexity of protocols in terms of their input size. Hence, by referring to the existence of communication-efficient protocols with a linear security threshold we mean the following: there exists a constant $c < 1/2$ such that for all $k$ there exists a polynomial $p(\cdot)$ (possibly depending on $k$) such that for all functions $f : \{0,1\}^n \rightarrow \{0,1\}$ there exists a $k$-player $\lfloor ck \rfloor$-private protocol that computes $f$ with $p(n)$ communication.[2]

Also, the term "security" refers here to security against honest-but-curious, computationally unbounded players (or equivalently a passive, unbounded external adversary).

With the above terminology in hand, we can now informally state our main results (which are actually special cases of more general theorems). The first of these results connect between the existence of very efficient PIR protocols and the existence of communication-efficient multiparty computation (MPC):

- (from PIR to MPC) If there exists a 1-round, polylog communication, PIR with a constant number of servers then there exist communication-efficient, statistically private, constant-round, multiparty protocols with a linear privacy threshold.
  Moreover, if the PIR protocol that we start with is so-called *linear* then this transformation yields perfect multiparty protocols.
- (from MPC to PIR) If there exist communication-efficient multiparty protocols with a linear privacy threshold then there exists polylog communication PIR with a constant number of servers. Moreover, this transformation maintains the number of rounds.

Using the above results, combined with the connections between PIR and locally decodable codes mentioned above, we get the following additional corollaries:

- (from LDC to MPC) If there exist constant-query LDCs of quasi-polynomial length and alphabet of quasi-polynomial size then there exist communication-efficient, statistically private, constant-round, multiparty protocols with a linear security threshold.
- (from MPC to LDC) If there exist communication-efficient multiparty protocols with a linear privacy threshold then there exists a constant-query LDC of quasi-polynomial length, quasi-polynomial size alphabet and parameters $\epsilon, \delta$ which are $1/\text{quasipoly}(N)$. (It should be noted that all currently known LDC with these parameters are of exponential size (i.e., $2^{N^{\Omega(1)}}$); therefore, codes with quasi-polynomial parameters, as those mentioned here, are considered non-trivial.)

[2] Here and in the following, the $n$ input bits of $f$ may be arbitrarily partitioned between the $k$ players.

To conclude, strong (upper or lower) bounds on the communication complexity of MPC should be roughly as difficult as strong bounds on LDCs, up to some loss in the achieved parameters.

### 1.2 Related Work

There is a vast literature on secure computation in the information-theoretic setting and on private information retrieval. However, most related to the current work are [4] and [24].

As noted above, [4] obtain communication-efficient protocols for arbitrary functions, whose security threshold decreases almost linearly with the input size. Their protocol was related to constructions of locally-random reductions [3], which in turn are related to PIR. However, the protocol of [4] made a heavy use of special "easiness" properties of the underlying locally-random reductions, and thus did not provide an indication that a more general relation exists.

Naor and Nissim [24] study the question of turning a communication-efficient two-party protocol into a secure one without incurring a significant communication overhead. In doing so, they make use of an idealized $\binom{N}{1}$-OT, which in turn (using reductions from [25,14]) can be based on single-server PIR with polylogarithmic communication. However, in the two-party setting considered in [24] our main result becomes trivial, as the secure computation of an arbitrary function reduces to a single table lookup.

**Organization:** In Section 2 we provide some necessary definitions and notation. Section 3 deals with transforming PIR protocols into SPIR protocols and Section 4 with transforming the latter into MPC protocols. Section 5 describes a construction of PIR protocols from multiparty protocols. Finally, in Section 6 we discuss the relation between LDC and PIR.

## 2 Preliminaries

In this section we sketch the definitions of the main primitives considered in this work. Since these are very basic and well known primitives, the purpose of this section is mainly to set up the notation and terminology used in this paper. For more detailed definitions the reader is referred to the relevant literature.

### 2.1 MPC

A secure multiparty computation (MPC) protocol allows a set of $k$ players $\mathcal{P}_1, \ldots, \mathcal{P}_k$ to compute some function $f$ of their local inputs while hiding the inputs from each other. By default, we consider functions $f : (\{0,1\}^n)^k \rightarrow \{0,1\}$. When computing such a function, each player $\mathcal{P}_i$ holds an $n$-bit input $a_i \in \{0,1\}^n$, and all players output $f(a_1, \ldots, a_k)$. Our results easily extend to more general types of functionalities (e.g., allowing non-boolean outputs and different outputs to different players).

In this work we consider MPC in the *pure information-theoretic setting*, where both the legitimate players running the protocol and the adversary attacking it have unlimited computational resources. We restrict our attention to security against a *passive* adversary (or honest-but-curious players), also referred to as *privacy*. In this setting, a $k$-party protocol is said to $t$-*privately* compute $f$ (where $1 \leq t \leq k$) if the following requirements are met:

- **Correctness.** The players always output the correct output $f(a_1, \ldots, a_k)$.
- ***t*-privacy.** The view of any set $B$ of at most $t$ players depends only on the inputs of the players in $B$ and the output of the function. That is, on any two input vectors $\boldsymbol{a}, \boldsymbol{a}'$ such that $\boldsymbol{a}_B, \boldsymbol{a}'_B$ and $f(\boldsymbol{a}) = f(\boldsymbol{a}')$, the view of players in $B$ is identically distributed.

The above perfect correctness requirement can be naturally relaxed to $\epsilon$-correctness, allowing the output to be incorrect with probability $\epsilon$. Similarly, the perfect privacy requirement can be relaxed to $(t, \epsilon)$-privacy, requiring that for any set $B$ of at most $t$ players the distributions of its view on any two inputs vectors $\boldsymbol{a}, \boldsymbol{a}'$ as above are in statistical distance of at most $\epsilon$. Moreover, it is convenient to assume that the above $\epsilon$-privacy requirement hold given *every* choice of the random inputs of players in $B$.[3]

While the case of perfect MPC is the more interesting one and is the one usually considered in the literature, some of our transformations will only yield non-perfect protocols. In all such cases, $\epsilon$ can be made negligible in $n$.

### 2.2 PIR

Private Information Retrieval (PIR) schemes are protocols for $k+1$ parties: *servers* $\mathcal{S}_1, \ldots, \mathcal{S}_k$, which are given an $N$-bit string $x \in \{0,1\}^N$ as input (sometimes referred to as a *database*), and a *user* $\mathcal{U}$, which is given as input an index $i \in [N]$. A PIR protocol allows communication between the user and the servers; we assume, without loss of generality, that the servers do not communicate with each other directly.[4] The goal of the protocol is for the user to learn the value $x_i$ while, at the same time, keeping $i$ private. This is captured by the following requirement.

**User-privacy:** Denote by $V_j[x, i]$ the random variable containing the *view* of server $\mathcal{S}_j$ in the protocol when the database is $x$ and the user wishes to retrieve $x_i$. *User-privacy* requires that, for any server $\mathcal{S}_j$, the view $V_j$ is independent of $i$ (i.e., for all $x, i, i'$ the views $V_j[x, i]$ and $V_j[x, i']$ are identically distributed). We will also consider a relaxed variant, termed $\epsilon$-*PIR*, in which we only require that

[3] This assumption is without loss of generality, since there is at most an $\sqrt{\epsilon}$-fraction of the random input choices given which the distance is larger than $\sqrt{\epsilon}$. For sufficiently small $\epsilon$, these bad choices can be eliminated without significantly altering the protocol's behavior.

[4] Since we are interested in the honest-but-curious setting, and since there is no privacy requirement with respect to the user, communication between the servers can always be done with the help of the user.

the statistical distance between $V_j[x,i]$ and $V_j[x,i']$ be bounded by $\epsilon$. The latter requirement will be referred to as $\epsilon$-user-privacy.

The complexity of PIR schemes is measured mainly by their *communication complexity*. We denote by $\alpha(N)$ the total number of bits sent in the protocol from the user to the servers, by $\beta(N)$ the total number of bits sent from the servers to the user, and by $m(N)$ the total communication (in either direction).

### 2.3 SPIR

Symmetrically Private Information Retrieval (SPIR) schemes are PIR schemes that satisfy an additional *data-privacy* requirement, guaranteeing that the only information obtained by the user in the protocol is the intended output $x_i$:

**Data-privacy:** Denote by $V_U[x,i]$ the random variable which is the *view* of the user in the protocol where the servers hold database $x$ the and user's input is $i$. We require that, for all $i$ and for all strings $x, x'$ such that $x_i = x'_i$, the views $V_U[x,i]$ and $V_U[x',i]$ are identically distributed. We will also consider a relaxed variant, termed $\epsilon$-*SPIR*, in which we require that the statistical distance between these two views be bounded by $\epsilon$ and, as in the case of PIR, also allow $\epsilon$-user-privacy.

It should be noted that in the literature (see [15]) information-theoretic SPIR is discussed in a setting where all servers share a common random string (CRS) which is unknown to the user. This assumption is necessary if no direct communication between the servers is allowed. In contrast, the use of SPIR in this paper cannot allow the servers to share a CRS. We therefore allow servers in a SPIR protocol to directly communicate with each other.

Note that SPIR in this setting can also be viewed as a special case of MPC: the MPC consists of $k+1$ players, the user and the $k$ servers, whose inputs are restricted to so that all servers hold an identical input $x$, and whose privacy constraints are those obtained by setting $t=1$ in the formal definitions of MPC. A similar view can be taken with respect to PIR, except that here the privacy constraint for the user is removed.

## 3 From PIR to SPIR

In this section we show how to transform a (perfect or non-perfect) PIR scheme with communication complexity $m(N)$ into an $\epsilon$-SPIR scheme with communication complexity poly$(m(N))$ (in the model where no CRS is available). This transformation maintains the number of rounds[5] but has a small penalty of increasing the number of servers from $k$ to $k+1$.

A good starting point for presenting our transformation is to recall the transformation of [15], obtaining SPIR with perfect data-privacy in the case where a

[5] In the context of PIR, a round is an exchange of messages from the user to the servers and back. In the context of SPIR we also allow, in parallel, a communication between the servers.

CRS is available. Its main disadvantage from our point of view is that the CRS in use is very long, and so modifying it to the setting with no CRS does not seem obvious. We will show, however, that such a modification can still be done. We therefore start with the solution from [15]; it assumes a CRS denoted $r$ of length $N$ that is available to $k+1$ servers $\mathcal{S}_1, \ldots, \mathcal{S}_k, \mathcal{S}_{k+1}$.

1. The user $\mathcal{U}$ picks a random shift $\Delta \in [N]$ and sends it to the servers $\mathcal{S}_1, \ldots, \mathcal{S}_k$.
   The user also sends the shifted index $i+\Delta$ to $\mathcal{S}_{k+1}$ (here and below, whenever an index is larger than $N$ it should be understood that $N$ is subtracted from it).
2. $\mathcal{U}, \mathcal{S}_1, \ldots, \mathcal{S}_k$ execute the assumed PIR scheme where $\mathcal{U}$ uses $i$ as its input and the servers use $y = x \oplus (r \ll \Delta)$ as their input. This scheme allows $\mathcal{U}$ to compute $y_i = x_i \oplus r_{i+\Delta}$ (but may potentially leak additional information about $y$).
   $\mathcal{U}$ also receives from $\mathcal{S}_{k+1}$ the bit $r_{i+\Delta}$. It xors this bit with $y_i$ to obtain $x_i$.

Intuitively, user-privacy follows from the fact that the view of each of $\mathcal{S}_1, \ldots, \mathcal{S}_k$ is exactly as in the PIR protocol and the view of $\mathcal{S}_{k+1}$ consists of a random index (independent of $i$). Data-privacy follows from the fact that $y$ is uniformly distributed in $\{0,1\}^N$ and that the only bit of $r$ which is available for $\mathcal{U}$ is $r_{i+\Delta}$. This intuition is formally proved in [15]. The communication complexity of the above SPIR protocol is dominated by the communication complexity of the PIR. The round complexity also remains unchanged (note that Step 1 can be executed in parallel to the first message of Step 2).

Next, we wish to modify the transformation to work in the setting where no CRS is available. A natural approach is to let the server $\mathcal{S}_{k+1}$ choose the string $r \in_R \{0,1\}^N$, distribute it among all other servers (but not the user) and then run the protocol above. While this modification still respects both user-privacy and data-privacy, the communication complexity grows by $k \cdot N$ (since the length of $r$ is $N$) and hence it makes the whole protocol useless for our purposes.

To overcome this, we will show the existence of a "small" set of strings $\mathcal{R} \subset \{0,1\}^N$ that "fools" the protocol; namely, the user's views obtained in the modified protocol in which $\mathcal{S}_{k+1}$ picks $r \in_R \mathcal{R}$ are statistically close to those obtained in the protocol above. The overhead of this transformation will only be $k \cdot \log |\mathcal{R}|$ (rather than $k \cdot N$), which will be small enough. However, the transformation will no longer obtain *perfect* data-privacy. The theorem that we prove is as follows:

**Theorem 1.** *Fix $k \geq 2$ and $\epsilon > 0$. Assume that there exists a $k$-server PIR protocol $\mathcal{P}$ with communication complexity $m(N)$ and round complexity $d(N)$ that satisfies $\epsilon_1$-user-privacy, for some $\epsilon_1 \geq 0$. Then, there exists a $(k+1)$-server SPIR protocol $\mathcal{P}'$ with communication complexity $O(m(N) + \log(1/\epsilon))$ and round complexity $d(N)$ that satisfies $\epsilon_1$-user-privacy and $\epsilon$-data-privacy.*

The rest of this section is organized as follows. We first formalize a technical lemma about the existence of a set $\mathcal{R}$ as needed. Then, based on this lemma, we present and analyze the modified transformation from PIR to SPIR. Finally, we prove the lemma (this is a fairly standard proof, in complexity theory contexts, that uses a probabilistic argument and is given here for the sake of completeness).

Let $\mathcal{R} \subseteq \{0,1\}^N$ and let $C : \{0,1\}^N \to [M]$ be a function. Denote by $C(\mathcal{R})$ the random variable obtained by applying $C$ to a random element of $\mathcal{R}$ and by $C(U)$ the random variable obtained by applying $C$ to a uniformly random $N$-bit string. We say that $\mathcal{R}$ $\epsilon$-fools the function $C$ if the statistical distance between $C(\mathcal{R})$ and $C(U)$ is bounded by $\epsilon$. Let $\mathcal{C}$ be a family of functions. We say that $\mathcal{R}$ $\epsilon$-fools $\mathcal{C}$ if it $\epsilon$-fools every function $C \in \mathcal{C}$.

**Lemma 1.** *Let $\mathcal{C}$ be a family of functions from $\{0,1\}^N$ to $[M]$ and let $\epsilon > 0$. Then, there exists a set $\mathcal{R}_\mathcal{C} \subset \{0,1\}^N$ of size* $\mathrm{poly}(1/\epsilon, M, \log|\mathcal{C}|)$ *that $\epsilon$-fools $\mathcal{C}$.*

It should be noted that we will apply the above claim with $\mathcal{C}$ which is significantly smaller than the set of all $M^{2^N}$ functions. Also note that $\mathcal{R}_\mathcal{C}$ may depend on $\mathcal{C}$; obviously, there can be no single $\mathcal{R}$ that is good for all families $\mathcal{C}$, even if $\mathcal{C}$ can only contain a single function.

We defer the proof of the lemma and now describe the modified transformation. We are given a PIR protocol $\mathcal{P}$, and assume for now that $\mathcal{P}$ is a perfect, one-round protocol (which is the case for all known PIR protocols; the multi-round case will be discussed in Remark 1 below and the non-perfect case in Remark 2 below). The protocol starts by server $\mathcal{S}_{k+1}$ picking $r \in_R \mathcal{R}$, from a carefully chosen $\mathcal{R}$ (specified below) and sending its index ($\log|\mathcal{R}|$ bits) to all other servers. The SPIR protocol then proceeds as the SPIR protocol described above.

User-privacy is easy to argue, independently of the choice of $\mathcal{R}$; indeed, user-privacy in the original transformation holds for *any* choice of $r$, in particular for all $r \in \mathcal{R}$. To argue the data-privacy, we first have to define the set $\mathcal{R}$. For this, we define a family of functions $\mathcal{C}$ that our set $\mathcal{R}$ will be able to fool. Fix some database $x \in \{0,1\}^N$ and a sequence of queries $q = (q_1, \ldots, q_k, q_{k+1})$ that may be sent in our protocol from the user to the $k+1$ servers. Let $C_{x,q}(r)$ be the function that returns the sequence of all answers that the user gets from the servers, as a function of $r$, when the database is $x$ and its queries were $q$. Let $\mathcal{C}$ be the family of all functions $C_{x,q}(r)$, parameterized by the choice of $x$ and $q$. Note that the length of the queries is bounded by $\alpha(N)$ and the length of the answers is bounded by $\beta(N)$ (it is therefore convenient to set $M \stackrel{\mathrm{def}}{=} 2^{\beta(N)}$). Also note that the size of $\mathcal{C}$ is $2^N \cdot 2^{\alpha(N)}$. For this family $\mathcal{C}$, we pick $\mathcal{R} = \mathcal{R}_\mathcal{C}$ as promised by Lemma 1. This choice of $\mathcal{R}$ guarantees that the view seen by the user (which is determined by $C_{x,q}(r)$) is $\epsilon$-close if $r$ is truly random or if $r \in_R \mathcal{R}$. Hence, by the perfect data-privacy of the original transformation, we get $\epsilon$-privacy of the modified transformation.

Finally the communication complexity consists of the communication complexity of the original PIR (which is $m(N) = \alpha(N) + \beta(N)$), the communication between the user and $\mathcal{S}_{k+1}$ (which is $\log N + 1$ bits) and the cost of sending $r$ from $\mathcal{S}_{k+1}$ to all other servers (which is $k \cdot \log|\mathcal{R}_\mathcal{C}| = O(\log 1/\epsilon + \log M + \log\log|\mathcal{C}|) =$

$O(\log 1/\epsilon + \beta(N) + \alpha(N) + \log N) = O(m(N) + \log 1/\epsilon))$. This implies that the communication overhead of the transformation is fairly small.

**Proof of Lemma 1:** We prove the lemma by picking at random a set $R \subset \{0,1\}^N$ of $w$ strings (each is chosen uniformly and they are all independent). To prove the lemma, it suffices to show that for all $C \in \mathcal{C}$ no (statistical) distinguisher can distinguish between the random variables $C(\mathcal{R})$ and $C(U)$ with more than an $\epsilon$-advantage, where a (statistical) distinguisher is just a subset $T \subset [M]$ of all possible outputs. For this, we first fix some $C$ and $T$ and bound from above the probability that, for a random $\mathcal{R}$, the distinguisher can tell apart $C(\mathcal{R})$ from $C(U)$. Namely, for some "small" $\delta$ we wish to prove that

$$\Pr_{\mathcal{R}}\left[|\Pr(C(U) \in T) - \Pr(C(\mathcal{R}) \in T)| > \epsilon\right] \leq \delta.$$

Let $p \stackrel{\text{def}}{=} \Pr(C(U) \in T)$. Therefore, we need to prove that when sampling $w$ times a binomial distribution that gives 1 with probability $p$, the probability that the average will deviate from $p$ by more than $\epsilon$ is bounded by $\delta$. This kind of bounds is given by Chernoff bounds. Specifically, it can be shown that if $w = \text{poly}(1/\epsilon, \log(1/\delta))$ then this probability is indeed bounded by $\delta$. Now, if we set $\delta < 1/(|\mathcal{C}| \cdot 2^{|M|})$ it follows by a union bound argument that there exists a choice of $\mathcal{R}$ such that for all $2^M$ distinguishers, and for each of the $|\mathcal{C}|$ functions $C \in \mathcal{C}$, we have $|\Pr(C(U) \in T) - \Pr(C(\mathcal{R}) \in T)| \leq \epsilon$, as needed. The size of this $\mathcal{R}$ is $w = \text{poly}(1/\epsilon, M, \log|\mathcal{C}|)$, as needed. ■

*Remark 1.* We dealt above with the case that the PIR scheme $\mathcal{P}$ is a one-round scheme. We outline here how a similar construction can be applied in the case where $\mathcal{P}$ is a multi-round PIR. Essentially, we apply the same transformation as above; we just need to re-define the set of functions $\mathcal{C}$ and as a result the set $\mathcal{R}_\mathcal{C}$ that fools these functions. The set $\mathcal{C}$ is defined by the collection of all functions $C_{x,q}$ as before, except that this time $q$ includes all the communication sent by the user in all rounds and $C_{x,q}(r)$ returns all the answers sent by the servers over all rounds. $\mathcal{R}_\mathcal{C}$ is now defined by applying the lemma to this $\mathcal{C}$ and with $\epsilon' = \epsilon/2^{\alpha(N)}$. We claim that the resulting SPIR protocol, $\mathcal{P}'$, is indeed $\epsilon$-private. Suppose to the contrary that there is a distinguisher $T$ that participates in $\mathcal{P}'$ and can tell, with advantage more than $\epsilon$, whether $r$ is chosen from $U$ or from $\mathcal{R}$. We argue that this allows us to construct a distinguisher $T'$ that can tell $C(R)$ from $C(U)$, for some $C$, with advantage better than $\epsilon'$, contradicting the choice of $\mathcal{R}$. The distinguisher $T'$ works by guessing $q$, i.e. guessing all the messages sent by the user over all rounds of the protocol (a total of $\alpha(N)$ bits), randomly picking the user's random input, and asking to see the value of $C_{x,q}(r)$. (In case where the servers in $\mathcal{P}$ are randomized, the latter should also depend on their uniformly chosen random inputs.) If the answers are consistent with the queries guessed by $T'$, it applies $T$ to guess whether $r$ comes from $U$ or from $\mathcal{R}$; otherwise, it just guesses at random. The advantage of $T'$ in this guess is $1/2^{\alpha(N)}$ (the probability of guessing $q$ correctly) times the advantage of $T$. Note that even though $\epsilon' \ll \epsilon$, since the communication grows by $\log |\mathcal{R}|$ the effect of using $\epsilon'$ rather than the original $\epsilon$ is just an additive factor of $\alpha(N)$.

*Remark 2.* The same transformation, as described above, can be applied to an $\epsilon_1$-PIR. The user-privacy of the SPIR that we obtain remains as in the PIR (i.e., $\epsilon_1$) and the data-privacy has a parameter $\epsilon$.

*Remark 3.* It is important to note that our transformation is inherently non-perfect. However, we point out that there is an important special case in which an alternative *perfect* transformation can be presented; this is the case of *linear* PIR (or LPIR, for short). LPIR is a variant of PIR discussed in the literature; it is a one-round protocol where the servers' answers are viewed as vectors in a space $F^{\beta}$, and the user computes its output $x_i$ by taking a linear combination of the $k$ answers, whose coefficients may depend on $i$ and on the user's random input. All information-theoretic PIR schemes from the literature are linear in this sense. The perfect transformation is now obtained as a combination of two facts: The first is that any $k$-server LPIR protocol with $m(N)$ communication can be transformed into a linear $2k$-server protocol with query length $m(N)$ in which the user outputs the sum of the $2k$ answers [16] (see [6] for details). The second is the existence of a simple MPC protocol to privately compute the sum of $k$ elements in $F$ with $O(k)$ communication and two rounds. Our transformation for this case will therefore work as follows: given the LPIR protocol, we construct the protocol with short answers but instead of the servers sending their answers to $\mathcal{U}$ they will invoke the private protocol for computing $x_j$ in a way that only $\mathcal{U}$ will learn the result. In fact, it is possible to avoid doubling the number of servers by replacing the second step above with a private multiparty protocol for the following function. The input of each server is its answer to the user's query in the LPIR protocol. The user's input is a vector $u$ representing the linear combination of the servers' answers which is needed to reconstruct $x_i$ (note that $u$ should remain private, as it may depend on $i$). The function should return the value of $x_i$, which is a degree-2 polynomial in the inputs. Note that, due to the easiness of the above function, it can be efficiently computed using standard MPC protocols (e.g., [7]).

## 4 From SPIR to MPC

In this section we show how to construct, based on a one-round $k$-server SPIR, constant-round, 1-private multiparty protocols for $k' = k^2 + 2$ players that can compute *any* function $f$. If the communication complexity of the SPIR is $m(N)$ then the communication complexity of the multiparty protocols will be poly$(m(N))$. If the SPIR protocol is only $\epsilon$-private then the MPC protocol is $O(\epsilon)$-private (where as usual, $k$ is viewed as a constant).

Let $\mathcal{P}$ be the given SPIR protocol. Denote the $k^2+2$ players of the multiparty protocol by $\mathcal{S}_{i,j}, i, j \in [k]$ and $\mathcal{P}_1, \mathcal{P}_2$. Also assume, without loss of generality, that in the given function $f$ only $\mathcal{P}_1, \mathcal{P}_2$ have inputs [6]. We therefore denote the input of these two players by $a_1, a_2$ and the desired output by $f(a_1, a_2)$. Intuitively, the protocol views the function as a table $F$ of size $N \times N$ where $N \stackrel{\text{def}}{=} 2^n$. The goal is for, say, $\mathcal{P}_1$ to retrieve the $(a_1, a_2)$ index of this table, which is just the desired $f(a_1, a_2)$. The MPC protocol proceeds as follows:

1. Player $\mathcal{P}_1$ applies the SPIR protocol with index $a_1 \in [N]$ to generate queries $q_1, \ldots, q_k$. It sends each query $q_i$ to all players $\mathcal{S}_{i,j}, j \in [k]$ (i.e., each query is sent to $k$ players; intuitively, this is done to create the replication needed in the next step of the protocol).[7]
   Each player $\mathcal{S}_{i,j}$, upon receiving the query $q_i$, computes (but does *not* send) the answer in the SPIR protocol to the query $q_i$ if the database was $F_{a_2}$, the $a_2$-th column of the table $F$; since the actual value of $a_2$ is not known to $\mathcal{S}_{i,j}$ it does so for all possible values $a_2 \in [N]$ hence obtaining a vector $A_i$ consisting of all $N$ answers to $q_i$ (each is a $\beta(N)$-bit string). In particular, note that each $A_i$ is therefore replicated among $k$ players.
2. Player $\mathcal{P}_2$ applies the SPIR protocol with index $a_2 \in [N]$ to generate queries $q'_1, \ldots, q'_k$. It sends each query $q'_j$ to all players $\mathcal{S}_{i,j}, i \in [k]$.
   Each player $\mathcal{S}_{i,j}$, upon receiving the query $q'_j$, computes the answer it would give in the SPIR protocol, when its database is $A_i$ (as computed in the previous step).[8] It sends this answer, $b_{i,j}$ to $\mathcal{P}_2$.
3. Upon receiving the answers $b_{i,j}$, the player $\mathcal{P}_2$ does the following: It uses, for each $i$, the $k$ answers $b_{i,j}, j \in [k]$ to obtain the $a_2$-th block of $A_i$ (for this it applies the reconstruction procedure as in the SPIR protocol). By definition of $A_i$, this block contains the answer given in the SPIR protocol to the query $q_i$ on database $F_{a_2}$. Denote this answer by $b_i$.
   $\mathcal{P}_2$ sends the reconstructed information $b_1, \ldots, b_k$ (total of $\beta(N)$ bits) to $\mathcal{P}_1$ who can now also apply the reconstruction procedure of the SPIR protocol to construct the $a_1$-th entry of $F_{a_2}$; $\mathcal{P}_1$ sends this value to all other players. This, by definition, is exactly $f(a_1, a_2)$, as needed.

The communication complexity of the above protocol is bounded by the communication complexity of applying the SPIR protocol $k+1$ times. Once, initiated by $\mathcal{P}_1$, on databases of length $N = 2^n$ but repeated $k$ times (hence its

[6] In the general case where all players have inputs we simply add a preliminary step where each player $\mathcal{S}_{i,j}$ shares its input between $\mathcal{P}_1, \mathcal{P}_2$. Then, we proceed as in the case where only these two players have an input, where the input for each of $\mathcal{P}_1, \mathcal{P}_2$ consists of its original input together with the shares received from other players

[7] If the SPIR protocol requires also communication among the servers then this is done in parallel to the described step.

[8] SPIR (as well as PIR) is defined above to allow the retrieval of a single bit. However, both primitives have a standard extension that deals with the retrieval of "blocks" [11]: the user sends one set of queries and the servers answer them by considering the blocks in a bitwise manner. If the blocks are of length $\ell$ then the query complexity of this solution, $\alpha(N)$, remains unchanged and the answer complexity, $\beta(N)$ grows by a factor of $\ell$. Since $A_i$ consists of blocks of $\ell = \beta(N)$ bits then this extension is needed here.

cost is $O(m(N))$) and the others, initiated by $\mathcal{P}_2$, for $k$ retrievals of $\beta(N)$-bit blocks (hence its cost is $O(m(N)\cdot\beta(N))$).[9] The total communication complexity is therefore poly$(m(N))$, as needed.

We turn to the 1-privacy of the protocol. Informally, we make the following observations: (1) player $\mathcal{P}_1$ has the same view as the user has in the first invocation of the SPIR protocol and hence from the $\epsilon$-data-privacy of the SPIR follows the $\epsilon$-privacy of the protocol for computing $f$, with respect to player $\mathcal{P}_1$. (2) for each $i \in [k]$, player $\mathcal{P}_2$ has the same view as the user has in a SPIR protocol for constructing the block $b_i$ from $A_i$. Also note that $b_1, \ldots, b_k$ may give information on $f(a_1, a_2)$ (and may even determine it completely); however, this information is legal since this is the output of the protocol (and nothing more). By the $\epsilon$-data-privacy of the SPIR it follows that the protocol for computing $f$ is $(k\cdot\epsilon)$-private with respect to player $\mathcal{P}_2$ (which, again, is $O(\epsilon)$ as $k$ is viewed as a constant). (3) each player $\mathcal{S}_{i,j}$ receives one query in each of two (independent) SPIR invocations; By the $\epsilon$-user-privacy of the SPIR protocol the view of such player in the multiparty protocol satisfies $\epsilon$-privacy.
To conclude, we have established the following:

**Theorem 2.** *Let $k \geq 2$ be a constant. Assume that there exist a $k$-server one-round SPIR protocol which satisfies $\epsilon$-privacy, for some $\epsilon \geq 0$, and has communication complexity $m(N)$. Then, for every function $f : (\{0,1\}^n)^{k'} \rightarrow \{0,1\}$, for $k' = k^2 + 2$, there exists a multiparty $(1, O(\epsilon))$-private protocol with communication complexity* poly$(m(2^n))$ *and round complexity* $O(1)$.

*Remark 4.* Similar results can also be proved for $t$-private MPC with $t > 1$ by applying the player simulation technique of Hirt and Maurer [18]. More specifically, $k$-party 1-private protocols can be composed with each other to obtain $k'$-party $\lfloor \frac{k'-1}{k} \rfloor$-private protocols, for any $k' > k$. However, this approach can be efficiently applied in our setting only for a constant number of players $k'$. It follows that the existence of communication-efficient 1-private protocols for a constant number of players implies the existence of communication-efficient protocols with a linear privacy threshold, in the sense defined in Section 1.1. It is interesting to note that in all other contexts we are aware of, the case of $t$-privacy can be handled directly without going through intermediate protocols for non-threshold structures as in [18]. We are not aware of a more direct way to obtain $t$-private protocols in our case, and leave open the question of obtaining protocols with a linear privacy threshold whose communication complexity is polynomial in both the number of players and the input length.

[9] In fact, a more careful examination of block retrieval shows that only the answer complexity grows to $O(\beta^2(N))$ while the query complexity remains at $2\cdot\alpha(N)$. Similarly, the analysis of the other invocations of the SPIR can also be optimized to take into account repeated messages etc.

## 5 From MPC to PIR

In this section we show that if *every* $k$-argument function $f$ admits a 1-private, $k$-party MPC protocol with communication complexity $c(n)$, then there exists a $k$-server PIR protocol with communication complexity $c(\log N) + O(\log N)$. [10] This transformation is *perfect* in the sense that if the MPC protocols are perfect then so is the PIR. The PIR protocol works as follows:

1. $\mathcal{U}$ picks at random $a, b$ subject to $a + b = i \bmod N$ (in other words $a, b$ form an additive secret-sharing of $i$). It also picks random bits $r_1, r_2$. It sends $a, r_1$ to server $\mathcal{S}_1$ and $b, r_2$ to server $\mathcal{S}_2$.
2. The $k$ servers execute the guaranteed MPC protocol for the function
$$f_x((a, r_1), (b, r_2)) \stackrel{\text{def}}{=} x_{a+b} \oplus r_1 \oplus r_2.$$
The output is sent to $\mathcal{U}$ who then masks it with $r_1 \oplus r_2$ to recover $x_i$.

Clearly, the communication complexity is as promised. To argue the the user-privacy, observe that the input to the MPC protocol provides 1-privacy (since it is a 1-private secret sharing of $i$), the output of the MPC also maintains the privacy since it is masked by random bits (and each server knows at most one of the two masking bits), and the last part in the view of each server is its view in the MPC protocol, which also maintains 1-privacy. It follows:

**Theorem 3.** *Let $k \geq 3$ be a constant. Assume that there exists a $k$-player $(1, \epsilon)$-private multiparty protocol for every function $f : (\{0,1\}^n)^k \to \{0,1\}$ with communication complexity $c(n)$ and round complexity $d(n)$. Then, there exists $\epsilon$-PIR with communication complexity $c(\log N) + O(\log N)$ and round complexity $d(\log N) + 1$.*

We note that any family of multiparty protocols with a linear privacy threshold can be easily turned into a 1-private protocol with a constant number of players by using a standard player partitioning argument.

## 6 Locally Decodable Codes Vs. PIR

Locally decodable codes (LDCs) were introduced in [20] where their close connection with PIR was pointed out. In this section we rely on this connection; most of the material in this section can be derived from explicit and implicit statements in [20].

Recall the relevant parameters for a LDC. We are given a string $x \in \{0,1\}^N$ and encode it into a codeword $y$ of length $M(N)$ over an alphabet $\Sigma$. The

[10] Note that if the complexity of every function $f$ can be bounded by some polynomial $c_f(n)$, then there must be a uniform polynomial bound $c(n)$ that is good for *all* functions $f$. Otherwise, for every $n$ let $f_n : (\{0,1\}^n)^k \to \{0,1\}$ be the "worst" function on $n$-bit inputs; the family of functions $f = \{f_n\}_n$ has superpolynomial complexity.

code is a $(k, \delta, \epsilon)$-LDC if after suffering an adversarial corruption of $\delta$ fraction of the symbols in the codeword $y$, it is still possible to reconstruct each bit $x_i$ with probability at least $0.5 + \epsilon$ by reading only $k$ symbols of the (corrupted) codeword.[11]

The first transformation (that follows from implicit statements in [20]) shows that given a $(k, \delta, \epsilon)$-LDC of length $M(N)$ over alphabet $\Sigma$ it is possible to construct 1-round $k$-server PIR with perfect privacy; its query complexity is $\alpha(N) = O(\log M(N))$, its answer complexity is $\beta(N) = \log |\Sigma|$ and its probability of success (i.e., the probability for correct reconstruction) is $0.5 + \epsilon^2\delta/(2q)$. This probability of success can be amplified to $1 - 2^{-\sigma}$ by repeating the protocol $O(\sigma)$ times.

In the opposite direction (again, using implicit statements in [20]) there is a transformation that takes a 1-round, $k$-server PIR protocol with success probability $0.5 + \epsilon$ and, for all $\delta > 0$, constructs $(k, \delta, \epsilon/2 - k\delta)$-LDC of length $M(N) = O(k \cdot 2^{\alpha(N)}/\epsilon)$ and alphabet $\Sigma = \{0,1\}^{\beta(N)}$. This already implies that a "standard" one-round PIR with polylog($N$) communication yields LDC with constant $\epsilon, \delta$ and length and alphabet size which are both quasi-polynomial in $N$.

We observe that a transformation similar to the one used to handle multi-round PIR protocols in Section 3 can be used to show that any multi-round PIR with query complexity $\alpha(N)$, answer complexity $\beta(N)$ and success probability $0.5 + \epsilon$ can be transformed into a one-round PIR with similar communication complexity and success probability of $0.5 + \epsilon/2^{\alpha(N)}$. Combining this observation with the transformation from one-round PIR to LDC, we get that if there exists a multi-round $k$-server PIR protocol with polylog($N$) communication then there exist LDC with length and alphabet size which are both quasi-polynomial in $N$ and $\delta, \epsilon$ which are both 1/quasi-poly($N$).

*Remark 5.* The above transformation from multi-round PIR to 1-round PIR applies also in the case where the servers in the multi-round PIR are randomized. However, the servers in the resulting 1-round PIR will also be randomized, in which case the transformation from PIR to LDC does not directly apply. It is possible to get around this difficulty by letting the user pick the servers' randomness and send it as part of its queries. Using Lemma 1, the amount of servers' randomness can be guaranteed to be of the same order of magnitude as the communication. Hence, this derandomization does not significantly increase the communication complexity of the original protocol.

## 7 Conclusions and Open Problems

Our results show close connections among several open problems in information-theoretic cryptography. Some of the techniques used in proving these connections

[11] This is a non-adaptive version of the definition. An adaptive version can also be considered.

may be of independent interest. In particular, the technique used in transforming PIR to SPIR can be used to reduce the amount of randomness used by more general information-theoretic protocols. Moreover, our transformation from PIR to MPC can be applied to get an information-theoretic analogue of the communication preserving secure protocol compiler from [24].

An interesting problem is to find an explicit construction of a set $\mathcal{R}$, whose existence is proved in Lemma 1, assuming that the functions it tries to fool are efficient. This requires an extension of the Nisan-Wigderson type pseudo-random generators [26] to ones that fool non-Boolean circuits. Good explicit generators of this type seem necessary for randomness reduction in *computationally-efficient* information-theoretic protocols.

**Acknowledgements.** We thanks Amos Beimel and Dieter van Melkebeek for helpful related discussions.

## References

1. A. Ambainis. Upper bound on the communication complexity of private information retrieval. In *Proc. of 24th ICALP*, pages 401–407, 1997.
2. J. Bar-Ilan and D. Beaver. Non-cryptographic fault-tolerant computing in a constant number of rounds. In *Proc. of 8th PODC*, pages 201–209, 1989.
3. D. Beaver and J. Feigenbaum. Hiding instances in multioracle queries. In *Proc. of 6th STACS*, pages 37–48, 1990.
4. D. Beaver, J. Feigenbaum, J. Kilian, and P. Rogaway. Security with low communication overhead. In *Proc. of CRYPTO '90*, pages 62–76, 1990.
5. D. Beaver, S. Micali, and P. Rogaway. The round complexity of secure protocols (extended abstract). In *Proc. of 22nd STOC*, pages 503–513, 1990.
6. A. Beimel, Y. Ishai, E. Kushilevitz, and J.-F. Raymond. Breaking the $O(n^{1/(2k-1)})$ Barrier for Information-Theoretic Private Information Retrieval. In *Proc. of 43rd FOCS*, pages 261-270, 2002.
7. M. Ben-Or, S. Goldwasser, and A. Wigderson. Completeness theorems for non-cryptographic fault-tolerant distributed computation. In *Proc. of 20th STOC*, pages 1–10, 1988.
8. B. Chor and N. Gilboa. Computationally private information retrieval. In *Proc. of the 29th STOC*, pages 304–313, 1997.
9. C. Cachin, S. Micali, and M. Stadler. Computationally private information retrieval with polylogarithmic communication. In *Proc. of EUROCRYPT '99*, pages 402–414, 1999.
10. D. Chaum, C. Crépeau, and I. Damgård. Multiparty unconditionally secure protocols (extended abstract). In *Proc. of 20th STOC*, pages 11–19, 1988.
11. B. Chor, O. Goldreich, E. Kushilevitz, and M. Sudan. Private information retrieval. In *Proc. of the 36th FOCS*, pages 41–51, 1995. Journal version: *J. of the ACM*, 45, pages 965–981, 1998.
12. R. Cramer, I. Damgård, and U. Maurer. General secure multi-party computation from any linear secret-sharing scheme. In *Proc. of EUROCRYPT 2000*, pages 316–334, 2000.
13. A. Deshpande, R. Jain, T Kavita, V. Lokam, and J. Radhakrishnan. Better lower bounds for locally decodable codes. In *Proc. of 16th CCC*, pages 184–193, 2002.

14. G. Di-Crescenzo, T. Malkin, and R. Ostrovsky. Single-database private information retrieval implies oblivious transfer. In *Proc. of EUROCRYPT 2000*, pages 122–138, 2000.
15. Y. Gertner, Y. Ishai, E. Kushilevitz, and T. Malkin. Protecting data privacy in private information retrieval schemes. In *Proc. of 30th STOC*, pages 151–160, 1998. Journal version: *J. of Computer and System Sciences*, 60(3), pages 592–629, 2000.
16. O. Goldreich. Personal communication, 2000 (cited in [6]).
17. O. Goldreich, H. Karloff, L. Schulman, and L. Trevisan. Lower bounds for linear locally decodable codes and PIR. In *Proc. of 16th CCC*, pp. 175 – 183, 2002.
18. M. Hirt and U. Maurer. Player Simulation and General Adversary Structures in Perfect Multiparty Computation. In *Journal of cryptology*, 13(1), pages 31–60, 2000.
19. Y. Ishai and E. Kushilevitz. Randomizing Polynomials: A New Representation with Applications to Round-Efficient Secure Computation. In *Proc. of 41st FOCS*, pages 294–304, 2000.
20. J. Katz and L. Trevisan. On the efficiency of local decoding procedures for error-correcting codes. In *Proc. of 32nd STOC*, pages 80–86, 2000.
21. I. Kerenidis and R. de Wolf. Exponential lower bound for 2-query locally decodable codes via a quantum argument. In *Proc. of 35th STOC*, pages 106–115, 2003.
22. E. Kushilevitz and R. Ostrovsky. Replication is not needed: Single database, computationally-private information retrieval. In *Proc. of 38th FOCS*, pages 364–373, 1997.
23. E. Mann. Private access to distributed information. Master's thesis, Technion – Israel Institute of Technology, Haifa, 1998.
24. M. Naor and K. Nissim. Communication Preserving Protocols for Secure Function Evaluation. In *Proc. of 33rd STOC*, pages 590–599, 2001.
25. M. Naor and B. Pinkas. Oblivious transfer and polynomial evaluation. *Proc. 31st STOC*, pages 245–254, 1999.
26. N. Nisan and A. Wigderson. Hardness vs Randomness. *J. Comput. Syst. Sci.* 49(2), pages 149-167, 1994.
27. K. Obata. Optimal Lower Bounds for 2-Query Locally Decodable Linear Codes. In *Proc. of 6th RANDOM*, pages 39–50, 2002.

# Dining Cryptographers Revisited

Philippe Golle[1] and Ari Juels[2]

[1] Palo Alto Research Center
3333 Coyote Hill Road
Palo Alto, CA 94304
pgolle@parc.com
[2] RSA Laboratories
Bedford, MA 01730, USA
ajuels@rsasecurity.com

**Abstract.** Dining cryptographers networks (or DC-nets) are a privacy-preserving primitive devised by Chaum for anonymous message publication. A very attractive feature of the basic DC-net is its *non-interactivity.* Subsequent to key establishment, players may publish their messages in a single broadcast round, with no player-to-player communication. This feature is not possible in other privacy-preserving tools like mixnets. A drawback to DC-nets, however, is that malicious players can easily jam them, i.e., corrupt or block the transmission of messages from honest parties, and may do so without being traced.
Several researchers have proposed valuable methods of detecting cheating players in DC-nets. This is usually at the cost, however, of multiple broadcast rounds, even in the optimistic case, and often of high computational and/or communications overhead, particularly for fault recovery.
We present new DC-net constructions that simultaneously achieve non-interactivity and high-probability detection and identification of cheating players. Our proposals are quite efficient, imposing a basic cost that is linear in the number of participating players. Moreover, even in the case of cheating in our proposed system, just one additional broadcast round suffices for full fault recovery. Among other tools, our constructions employ bilinear maps, a recently popular cryptographic technique for reducing communication complexity.

**Keywords:** anonymity, dining cryptographers, mix network, non-interactive, privacy.

## 1 Introduction

Anonymous message transmission is a fundamental privacy-preserving tool, both in the literature and in practice. Toward this aim, Chaum devised two seminal techniques: mixnets [10] and "dining-cryptographers" nets [11], also known as DC-nets. Mixnets have seen broad exploration in the literature, and serve as the basis for several fielded anonymity systems, e.g., [3,13,17,19]. (See [14] for a good bibliography.) DC-nets, by contrast, have remained relatively neglected,

C. Cachin and J. Camenisch (Eds.): EUROCRYPT 2004, LNCS 3027, pp. 456–473, 2004.

apart from a small scattering of papers, e.g., [1,2,11,21,22]. One reason for this is perhaps that DC-nets, unlike mixnets, cannot operate by proxy; in particular, the players operating a DC-net must be identical with those providing input. In many real-world cases, however, this is not necessarily a serious drawback, as in the Crowds system [19], where participants provide mutual protection of privacy. Moreover, as formulated by Chaum for the case involving honest players, DC-nets have one very compelling feature unavailable in mixnets:

> In a basic DC-net, anonymous message transmission may be accomplished by players in a *non-interactive* manner, i.e., in a single broadcast round.

Non-interactivity is of course very naturally attractive as a practical feature of system design. It also renders security definitions and proofs simpler than in the case of mixnets (for which formal definitions have been quite elusive).

There is a major drawback to DC-nets, however, and a large obstacle to their deployment: They are subject to straightforward jamming by malicious players. Such players can prevent the delivery of messages from honest participants, either by broadcasting invalid messages or even simply by dropping out of the protocol. Several valuable techniques have been proposed for addressing this problem, but to this point have had the limitation of requiring either unfeasibly intensive computation and/or multiple rounds of interaction among players.

Our first contribution in this paper is a set of techniques permitting the identification of cheating players with very high probability, *while retaining the property of non-interactivity.* The resulting DC-net constructions are computationally efficient: Assuming $n$ players, they require each participant to perform a number of modular exponentiations that is linear in $n$ during the broadcast phase. Any player, whether a participant or not, may perform a quadratic number of exponentiations for verification of the output. Indeed, the computational costs of our constructions are comparable to those of the most efficient mixnets (assuming $n$ players processing $n$ inputs). Our DC-net proposals are therefore reasonable for small sets of, say, some dozens of players.

Of equal importance, we propose techniques that permit recovery from lost or corrupted messages in a single, additional broadcast round, provided that there is a majority of honest players. Previous proposals have required multiple rounds for this purpose, or assumed a re-broadcast of messages. The computational costs for our recovery protocol are comparable to those for the basic message-transmission protocol.

Although it is possible to detect cheating by a player in a non-interactive mix network, we maintain that under any reasonable set of security assumptions, it is not possible for such a mix network to recover from failure (and thus from cheating) by even one player without an additional round of interaction. Our reasoning is as follows. Suppose that we could recover the inputs of all participating players regardless of who participated. Then if a given player $P_i$ did participate, and furnished message $m_i$ as input, an adversary could determine $m_i$ by taking the difference between the set $M$ of all messages submitted and the set $M'$ of all messages except that of $P_i$ (the adversary would obtain $M'$ by simulating the absence of $P_i$).

We describe two different DC-net constructions, which we characterize as *short* and *long*. In a short DC-net, the basic unit of message transmission is an algebraic group element. For such DC-nets, we propose techniques that detect cheating with overwhelming probability. A long DC-net, by contrast, permits efficient transmission of messages of arbitrary length essentially by means of a form of hybrid encryption. (It may be viewed as roughly analogous to a "hybrid" mixnet.) For long DC-nets, we describe techniques to detect cheating with high, but not overwhelming probability; an adversary in this case may feasibly perform some limited jamming of messages.

In both constructions, we make use of bilinear maps, cryptographic techniques that have achieved much recently popularity as tools for reducing protocol interactivity [5]. In consequence, the security of our constructions is predicated on the Decisional Bilinear Diffie-Hellman assumption (DBDH) (see, e.g., [6]), as well as the random oracle assumption [4].

**Organization** In section 2, we explain the basic concepts of DC-net construction, and describe previous results on the topic. We present our formal model and other preliminary material in section 3. In section 4, we describe our short DC-net construction, followed in section 5 by presentation of our long DC-net proposal. We conclude in section 6. In the paper appendix, we offer security definitions and proofs for the protocols presented in the body of the paper.

## 2 Background

The intuition behind DC-nets is best introduced with a simple two-player example. Suppose that Alice and Bob possess $k$-bit messages $m_A$ and $m_B$ respectively. They wish to publish these messages anonymously, that is, in such a way that an observer cannot determine which player published which message. Suppose further that Alice and Bob share $k$-bit secret keys $k_{AB}(0)$ and $k_{AB}(1)$, as well as a secret, random bit $b$. Alice and Bob publish message pairs as follows:

**if b = 0:** Alice: $M_{A,0} = k_{AB}(0) \oplus m_A$, $M_{A,1} = k_{AB}(1)$
Bob: $M_{B,0} = k_{AB}(0)$, $M_{B,1} = k_{AB}(1) \oplus m_B$

**if b = 1:** Alice: $M_{A,0} = k_{AB}(0)$, $M_{A,1} = k_{AB}(1) \oplus m_A$
Bob: $M_{B,0} = k_{AB}(0) \oplus m_B$, $M_{B,1} = k_{AB}(1)$

An observer can compute $M_{A,0} \oplus M_{B,0}$ and $M_{A,1} \oplus M_{B,1}$, yielding the (unordered) message pair $(m_A, m_B)$. The origin of these messages, however, remains unconditionally private: Without knowing the secrets shared by Alice and Bob, the observer cannot determine which player published which message. Observe that this protocol is non-interactive, in the sense that once their secrets are established, Alice and Bob need not communicate directly with one another.

This basic protocol may be extended to multiple players $P_1, P_2, \ldots, P_n$. Suppose that each pair of players $(P_i, P_j)$ shares a set of keys $k_{i,j}(w)$ for $i, j, w \in \{1, 2, \ldots, n\}$, where $k_{i,j}(w) = k_{j,i}(w)$.

Each player $P_i$ computes a vector of values as follows:

$$W_i = \{W_i(1) = \oplus_{j=1}^{n} k_{i,j}(1), W_i(2) = \oplus_{j=1}^{n} k_{i,j}(2), \ldots, W_i(n) = \oplus_{j=1}^{n} k_{i,j}(n)\}.$$

We refer to each message $W_i(w)$ as a *pad*, and refer to each value $k_{i,j}(w)$ as a *partial pad*. Observe that $\oplus_{i=1}^{n} W_i(w) = \mathbf{0}$, i.e., the pads in a given position $w$ cancel when XORed together.

To broadcast messages in this scheme, each player $P_i$ chooses a random position $c_i$ and XORs her message $m_i$ with the pad $W_i(c_i)$ in $W_i$. This yields a new vector $V_i = \{V_i(1), V_i(2) \ldots V_i(n)\}$ differing from $W_i$ in the position $c_i$. Provided that all players have selected different positions $c_i$, the vector $V = \oplus_{i=1}^{n} V_i$ (i.e., the vector formed by XORing all messages in a given position), will consist of the set of messages posted by all players. Provided that keys and position selections $\{c_i\}$ are secret, the privacy of messages, i.e., the hiding of their originators, is unconditional.

As noted in Chaum's original paper, shared secrets may be established non-interactively via Diffie-Hellman key exchange, yielding computationally secure privacy.

*A note on "collisions":* Even when all players are honest, a problem arises in multi-player DC-nets in the selection of message positions $\{c_i\}$. In particular, there is no good non-interactive means of enabling all players to select distinct message positions. Hence, with some probability, two (or more) players will attempt to transmit messages in the same slot. In other words, players $P_i$ and $P_j$ will select $c_i = c_j$, so that the message $m_i \oplus m_j$ appears in the final vector $V$, rather than the individual messages. Some multi-round DC-net protocols address this problem via *reservation* procedure, whereby players request "slots" in advance. In all cases, however, DC-nets involve collisions, whether of messages themselves or reservation requests. (The problem can be avoided through techniques like secure multiparty computation of a secretly distributed permutation of slots among players, but this is impractical.)

We do not treat the issue of collisions in this paper, but simply regard a DC-net as a primitive that provides only partial throughput, i.e., drops some fraction of messages. Better throughput may be achieved by high-layer protocols, e.g., protocol repetition, either serially or in parallel.

## 2.1 Previous Work

As already explained, a basic DC-net is subject to jamming by even a single dishonest player. Such a player $P_i$ may simply set the vector $V_i$ to a series of random pads. This effectively jams the DC-net: All elements in the final output $V$ will be random and thus no messages will be successfully delivered. Worse still, the very privacy guarantees of the DC-net render it impossible to trace the source of the jamming in this case. Alternatively, an attacker may corrupt messages by tampering with bits in a valid vector $V_i$. It is on this security problem that the literature on DC-nets mainly focuses.

In his original paper [11], Chaum proposes the detection of dishonest players via a system of "traps" in a multi-round protocol. Prior to message transmission, a reservation protocol takes place in which players reserve future message slots. At this time, each player commits to a declaration of "trap" or "non-trap" for her reserved slot. To jam the DC-net, a dishonest player must transmit a message in a slot she has not reserved. But if she tries to transmit a message in a slot that is a "trap," then the attack may be detected during a decommitment phase.

An important follow-up result is that of Waidner and Pfitzman [21], who identify a weakness in this original protocol, and show that an attacker can feasibly strip the anonymity of honest players. (Improved reservation techniques in [2] and [22] reduce this possibility to some extent.) They propose a multi-round solution to this problem, also based on the idea of setting "traps" during a reservation phase. Like Chaum's protocol, theirs is only guaranteed to identify one dishonest player for a given "trap." No obvious method for fault recovery is available, apart from re-broadcasting. That said, it should be noted that the goal of this work is a little different than ours. While these researchers have sought to achieve unconditional untraceability assuming only honest point-to-point communication, our aim is to achieve privacy under only computational hardness assumptions.

Most recently, in [1], von Ahn, Bortz, and Hopper consider a constant-round anonymous-broadcast protocol that is essentially a DC-net variant (with an initial partitioning of players into autonomous groups). They accomplish the distribution of secrets for each protocol invocation via a secret-sharing protocol. In their scheme, the correctness of pads is proven via a cut-and-choose protocol. In the optimistic case, their protocol requires three broadcast rounds, and has $O(n^2)$ communications complexity (assuming a constant number of cut-and-choose invocations). In the presence of cheating players, the communications complexity rises to $O(n^4)$.

One problem with these previous protocols is that the computational and communications costs of catching cheating players with overwhelming probability is very high, requiring either many "traps" or many cut-and-choose invocations. This may not be problematic in cases where players may be reliably identified and where cheating carries a high penalty. For Internet systems, however, in which identities are not trustworthy, and participation in anonymous systems may be short-lived, even a small amount of cheating in the form of, e.g., tampering with messages, may be highly problematic. There is the risk that a savvy attacker may simply create false identities and then discard them when cheating is detected.

Our work is similar to the approach of von Ahn *et al.* in that we employ cryptographic proofs of correctness rather than "traps" in order to detect cheating. We employ a different strategy for pad computation, however, that has the benefit of more efficient proofs of correct pad computation. In particular, for our short DC-net proposal, in which players perform only a linear number of modular exponentiations (in $n$) on furnishing inputs, we show how to detect cheating with overwhelming probability. Another critical feature of our proposal is, of course, its non-interactivity in the optimistic case. Additionally, even in the presence of faults, our protocols may be completed in just two broadcast rounds, and with $O(n^2)$ communications complexity.

## 3 Preliminaries

For the sake of simplicity, we assume throughout this paper the presence of a reliable broadcast channel. As is well known, such a channel may be simulated via Byzantine agreement in a network with reliable point-to-point delivery. (See [7] for recent results in this area.) Another possible instantiation would be a Web server that performs the function of receiving and publishing messages in an honest and reliable manner. (Our constructions may also be employed in the presence of an unreliable broadcast channel provided that a given message is seen either by all players or by none. In this case, a dropped message may be modelled as a faulty player.) We further assume that all messages are authenticated, i.e., securely bound to the identities of their originators. In practice, this may be accomplished via digital signatures.

We define next the component functions of DC-nets. We denote the set of participants in the DC-net by $P = P_1, P_2, \ldots, P_n$. In what follows, when we specify a value as public or published, we assume it is transmitted to all players in $P$ via an authenticated channel or entity. Setup is achieved by means of a parameter generation function paramgen and a key distribution function keydist. These functions are similar to those employed in standard discrete-log-based distributed cryptographic protocols. They are called once at the beginning to set up long-lived parameters shared by all players. A difference here, however, is that we employ admissible bilinear maps as a basic tool in our constructions, and must therefore make use of elliptic-curve based algebraic groups accordingly. We assume the appropriate background on the part of the reader, and refer to [5] for further details and notation.

- **Parameter generation:** Taking security parameter $l$ as input, the function paramgen outputs a quintuple $\rho = (p, G_1, G_2, \hat{e}, Q)$, where $G_1$ and $G_2$ are two groups of order $p$, $Q$ is a generator of $G_1$ and $\hat{e} : G_1 \times G_1 \rightarrow G_2$ is an admissible bilinear map [5]. We require furthermore that the Decisional Bilinear Diffie-Hellman (DBDH) assumption holds for $\hat{e}$. Using the terminology of [5], the function paramgen is a parameter generator that satisfies the DBDH assumption. (For our "long" DC-net construction, we may weaken our hardness assumption to the Bilinear Diffie-Hellman problem (BDH), i.e., the computational variant, rather than the decisional one.) In practice, the map $\hat{e}$ may be instantiated with the Weil pairing over a suitable elliptic curve. The function paramgen may be executed by a trusted entity, which is our working assumption here. (Alternatively, it may be accompanied by a non-interactive proof of correct execution.) The quintuple $\rho$ is published. We leave system parameters implicit in our notation where appropriate.
- **Key generation:** The function keydist takes as input the parameter specification $\rho$. It yields for each player $P_i$ a private key $x_i \in_U \mathbb{Z}_p$ and a corre-

sponding public key $y_i = x_i \cdot Q$. Each private key $x_i$ is additionally shared among other players in a $(k, n)$-threshold manner. In particular, let $f_i$ be a polynomial over $\mathbb{F}_p$ of degree $k - 1$ selected uniformly at random such that $f_i(0) = x_i$. Player $P_j \in P$ receives from player $P_i$ the private share $x_{i,j} = f_i(j)$, with a corresponding public share $y_{i,j} = x_{i,j} \cdot Q$. We assume that the function keygen is an algorithm executed by a trusted entity and a secure environment. (In practice, it may be instantiated by means of a distributed protocol; see [16] for an example of such a protocol and a discussion of the underlying security guarantees.)

We now describe the functions employed in the DC-net itself. We assume that players have access to a trustworthy global session counter $s$ and specification $\Pi_s \subset P$ of players participating in the session $s$. Note that the privacy properties of our construction (defined in appendix A) do not rely upon publication of $s$ or $\Pi_s$ in a trustworthy manner, but the robustness does.

**Posting:** $(V_{i,s}, \sigma_{i,s}, i, s) \leftarrow \mathsf{post}(s, m_i, x_i)$ ; $[\Pi_s, \{y_j\}_{j \in P}]$.
The function post is invoked in session $s$ by every player in $\Pi_s$. It returns to each player a set of outputs that hides that player's input, as well as auxiliary data that proves the correctness of the outputs. More precisely, the function post is a randomized function that takes as input the session counter $s$, a message $m_i$ and the private key $x_i$ of player $P_i$. Inputs to the function also include the set of players $\Pi_s$ participating in the sessions and all public keys. For visual clarity, we set off the latter parameters in square brackets. We define $\pi_s = |\Pi_s|$ to be the number of participants in session $s$. The function post outputs:

- An output vector $V_{i,s} = \big(V_{i,s}(1), \ldots, V_{i,s}(\pi_s)\big)$. Let us denote the vector of random pads used by player $P_i$ as $W_{i,s} = \big(W_{i,s}(1), \ldots, W_{i,s}(\pi_s)\big)$. The elements of the output vector and of the pad vector agree in all positions but one: the position $c_i$ where message $m_i$ is xored with the pad. In other words $V_{i,s}(w) = W_{i,s}(w)$ for all $w \neq c_i$ and $V_{i,s}(c_i) = m_i \oplus W_{i,s}(c_i)$
- Subsidiary data $\sigma_{i,s}$. The value $\sigma_{i,s}$ includes the identity of player $P_i$ and a proof of valid formatting of the vector $V_{i,s}$.

**Verification:** $\{0,1\} \leftarrow \mathsf{verify}((V, \sigma), s, i, \Pi_s)$ ; $[\{y_j\}_{j \in P}]$.
The function verify determines the correctness of the vector $V$ output by a given player $P_i$. When $V$ is deemed correct, verify outputs '1'; otherwise it outputs '0'. This function can be called non-interactively by any player who wishes to verify the correctness of an output vector produced by another player.

**Message extraction:** $M \leftarrow \mathsf{extract}(\{V'_{i,s}\}_{i \in \Pi}, \Pi_s)$ ; $[\{y_j\}_{j \in P}]$.
Once all players in $\Pi_s$ have posted their output vectors, it should be possible for any entity to extract the messages input to the mix procedure. We denote by extract the function that accomplishes this. The outputs of extract is a set $M$ of at most $\pi_s$ distinct messages.

**Pad reconstruction:** $W_{i,s} \leftarrow \mathsf{reconstruct}(i, \Pi_s, \{x_{i,j}\}_{j \in \Pi_s})$.
If a player $P_i \in \Pi_s$ fails to produce a correct output vector (or any output at all), a quorum of other players in $\Pi_s$ can reconstruct that missing output. We denote by reconstruct the function that accomplishes this.

We denote by $\mathsf{DC} = \{\mathsf{paramgen}, \mathsf{keydist}, \mathsf{post}, \mathsf{verify}, \mathsf{extract}, \mathsf{reconstruct}\}$ the complete set of functions constituting a DC-net.

# 4 Short DC-Net Protocol

## 4.1 Intuition and Tools

In our first construction, the basic message unit is an algebraic group element. We would like to enable players to prove correct behavior in this setting with overwhelming probability. This combination of features leads to two basic problems:

**Problem 1:** We would like any given player $P_i$ to be able to compute a partial pad $k_{i,j}(w)$ with any other player $P_j$ in a non-interactive way. In fact, $P_i$ must be able to compute many such partial pads non-interactively, namely one partial pad for every value $w$. Additionally, $P_i$ must be able to prove the correctness of any partial pad $k_{i,j}(w)$ (or more precisely, of any pad, which is composed of partial pads).

*The contradiction:* Suppose that $P_i$ computes partial pad $k_{i,j}(w)$ using a standard D-H protocol employing her own secret key $x_i$ and the public key $y_j$ of player $P_j$. (I.e., $k_{i,j}(w) = y_j^{x_i}$.) Since this computation is algebraic in form, $P_i$ can efficiently prove statements in zero knowledge about $k_{i,j}(w)$. On the other hand, it is only possible to perform this D-H computation once, and $P_i$ needs to do so for many different values of $w$! An alternative possibility is to hash $y_j^{x_i}$ with $w$ to generate partial pad $k_{i,j}(w)$. In this case, though, there is no way to prove that $k_{i,j}(w)$ was correctly constructed with overwhelming probability without inefficient techniques like cut-and-choose or general secure function evaluation.

*The solution:* It is in resolving this problem that bilinear mapping comes into play.[3] It is possible to think of a bilinear map as a way of effecting a D-H exchange non-interactively across many different algebraic bases. In particular, $P_i$ can compute the partial pad $k_{i,j}(w) = \hat{e}(y_j, x_i Q_w) = \hat{e}(Q, Q_w)^{x_i x_j}$, where $Q_w$ is a randomly selected elliptic-curve point specific to $w$. We may thus think of $P_i$ as performing a D-H exchange relative to a different algebraic base $\hat{e}(Q, Q_w)$ for every different value of $w$.

[3] There are other possible solutions to this problem without use of bilinear maps, e.g., changing keydist such that the sum $\sum_i x_i = 0 \bmod q$ becomes a special condition on private keys. This, however, would mean that in practice the protocol could never be efficiently realized by having players generate their own key pairs. Also, this type of solution would not work for the long DC-net construction.

**Problem 2:** When a player $P_i$ publishes a vector $V$ of pads, she must prove its correctness. This means proving that every element of $V$ is a correct pad – except the one element modified to contain the message $m_i$ that $P_i$ wants to publish. The problem here is that $P_i$ of course does not wish to reveal *which* element of $V$ contains the message $m_i$!

*The solution:* For each pad position $w$ in her published vector, player $P_i$ commits to a bit $b_w$. She lets $b_w = 0$ if the element in position $w$ represents a correct pad, and $b_w = 1$ otherwise. $P_i$ then proves two things:

1. For every position $w$, either the pad is correct OR the bit $b_w = 1$.
2. The sum $\sum_w b_w = 1$, i.e., the vector $V$ contains at most one message.

To prove both of these facts, we use standard techniques for non-interactive proofs regarding statements involving discrete logs. We do so over the groups $G_1$ and $G_2$. As explored in many papers, these techniques permit honest-verifier zero-knowledge proof of knowledge of discrete logs [20], proof of equivalence of discrete logs [12], and first-order logical statements on such statements [9]. The proof protocols may be made non-interactive through use of the Fiat-Shamir heuristic [15]; they may be perfectly simulated with application of the random oracle model to the hash function used to generate challenges. We draw on the notation of Camenisch and Stadler [8] for a unified treatment and formal specification of these proofs in our detailed protocol. (E.g., $\mathrm{PoK}\{x : e = g^x \bigwedge f = h^x\}$ means a proof of knowledge that $\log_g e = \log_h f$, and is NIZK for our purposes.)

### 4.2 Protocol Details

**Parameter and key generation.** The function $\mathsf{paramgen}$ outputs the set of parameters $\rho = (p, G_1, G_2, \hat{e}, Q)$. We also assume the existence of a hash functions $h : \{0,1\}^* \to G_1$ that is publicly known. The function $\mathsf{keydist}(\rho)$ then outputs a secret key $x_i \in \mathbb{Z}_p$ for each player $P_i$. Recall that shares of this secret key are distributed to other players and that all public keys are published.

**Message posting.** The pads $W_{i,s}(k)$ for player $P_i$ in session $s$ are computed as follows. We compute the point $Q_k = h(s||k)$ on $G_1$ and let

$$W_{i,s}(k) = \prod_{j \in \Pi_s; j \neq i} \hat{e}(Q_k, y_j)^{\delta_{i,j} x_i},$$

where $\delta_{i,j} = 1$ if $i < j$ and $\delta_{i,j} = -1$ if $j < i$. Player $P_i$ then chooses at random a value $c_i \in \Pi_s$ and multiplies the message $m_i \in G_2$ with pad $W_{i,s}(c_i) \in G_2$ to produce the output vector $V_{i,s}$. We turn now to the computation of the auxiliary verification data $\sigma_{i,s}$:

1. Let $g$ and $h$ be two fixed random generators in a group $G$ of order $q$ for which the discrete logarithm problem is hard. Player $P_i$ chooses independently at random $n$ values $r_1, \ldots, r_n \in \mathbb{Z}_q$. For $1 \leq k \leq n$, where $k \neq c_i$, $P_i$ computes $w_k = h^{r_k}$. $P_i$ computes $w_{c_i} = gh^{r_{c_i}}$.

2. The prover proves knowledge of $\log_h(g^{-1}\prod_{i=0}^{n} w_i)$, i.e., $\mathrm{PoK}\{r : \prod_{i=0}^{n} w_i/g = h^r\}$..
3. For $1 \leq k \leq \pi_s$, $P_i$ proves the following statement:

$$\left(W_{i,s}(k) = \hat{e}(\prod y_j, Q_k)^{x_i} \textbf{ and } P_i \text{ knows } \log_h(w_k)\right) \textbf{ or } \left(P_i \text{ knows} \log_h(w_k/g)\right)$$

i.e., $\mathrm{PoK}\{x, r : (W_{i,s}(k) = \hat{e}(\prod y_j, Q_k)^x \bigwedge w_k = h^r) \bigvee (w_k/g = h^r)\}$.

The string $\sigma_{i,s}$ consists of all the values computed in steps 1 and 2 above. Finally, the function post outputs $(V_{i,s}, \sigma_{i,s}, i, s)$.

**Verification.** Anyone can verify non-interactively that the values computed in $\sigma_{i,s}$ are correct.

**Message extraction.** Given the $\pi_s$ vectors $V_{1,s}, \ldots, V_{\pi_s,s}$ published by the players in $\Pi_s$, anyone can non-interactively compute $r_k = \prod_{i \in \Pi_s} V_{i,s}(k)$ for $k \in \Pi_s$. Recall that the definition of the pads is such that $\prod_{i \in \Pi_s} W_{i,s}(k) = 1$. We need now to introduce a notation for the subset of players who chose to publish their message in position $k$ for a given $k$. For $k \in \Pi_s$, we denote $c^{-1}(k) = \{i \in \Pi_s \mid c_i = k\}$. Note that the subset $c^{-1}(k)$ could be empty, or contain a single or multiple players. Now it is clear that in every position $k$ for which $c^{-1}(k)$ is a singleton $\{i\}$, we have $r_j = m_i$. All other messages $m_i$ for which $c^{-1}(c_i)$ is not a singleton are unrecoverable in the output. The output of the function extract is the set of messages $m_i$ which are recovered in the output.

**Pad reconstruction.** If a subset of players $\mathcal{P} \subseteq \Pi_s$ fail to publish their output vector, the remaining players can reconstruct the pads of missing players, and compute the output of the DC-net, as follows. Each player $P_i$ for $i \notin \mathcal{P}$ publishes $x_{j,i} \cdot Q_k$ for all $j \in \mathcal{P}$. Anyone can verify the correctness of these values by checking that $\hat{e}(Q, x_{j,i}Q_k) = \hat{e}(y_{j,i}, Q_k)$. Furthermore, these values enable any player to recompute the pads of missing player $P_j$ since $\hat{e}(Q_k, y_i)^{x_j}$ can be derived from the values $\hat{e}(x_{j,i}Q_k, y_i)$ by polynomial interpolation.

## 5 Long DC-Net Protocol

### 5.1 Intuition and Tools

In order to obtain a "stretched" pad of the desired length in our long DC-net, it is necessary to apply a PRNG to a secret seed $K$, i.e., to use symmetric-key techniques. In consequence, proofs based on the algebraic structure of pads are no longer possible, and there are no efficient techniques for effecting proofs with overwhelming probability. Our use of symmetric-key techniques thus engenders two basic problems:

**Problem 1:** We face the same basic problem as in the short DC-net: It is necessary to prove correct construction of vectors without revealing where the messages are positioned. But the use of symmetric-key primitives means that we cannot benefit from the same NIZK proof techniques as in the short DC-net.

*The solution:* We resolve this problem by employing proof techniques that detect cheating players with high, but not overwhelming probability. In particular, we use a technique very similar to that of "randomized partial checking" [18] for mixnets. The idea is for a player $P_i$ to prove correctness of her published vector $V$ by generating a random challenge $R$ non-interactively. This challenge $R$ specifies a subset of half of the elements in the vector $V$. $P_i$ reveals the underlying seeds for these as part of her proof. These seeds are derived essentially just like pads in the short DC-net. Thus, it is possible to provide a simple proof of correctness that may be non-interactively verified by other players.

One problem, of course, is that if $P_i$ transmits a message $m_i$, then with probability $1/2$, the challenge $R$ will lead to opening of the seed for the position containing that message. This problem may be resolved quite simply: $P_i$ chooses challenges until she finds one that does not lead to opening of the seed for the message position. Some tens of attempts will permit this with overwhelming probability.

Since only half of the seeds are revealed, some number of invalid pads can escape detection. In particular, for a given challenge, any seed will be revealed with probability $1/2$. Hence, given $u$ invalid pads, an adversary must perform work roughly $2^u$ to compute a challenge $R$ that does not reveal cheating. In practice, therefore, we would expect an adversarial player to be unable to insert more than, say, 80 invalid pads into a vector. Thus such a player can "jam" only a limited number of slots. Assuming large enough vectors and adversarial control of a small enough set of players, the throughput for the DC-net remains fairly high.

Thus, our proof protocol is as follows. Let $h$ be a hash function from $\{0,1\}^*$ to $\mathbb{Z}_n$ (modelled in our proof as a random oracle).

1. The player chooses a random seed r and computes $h(V||r||1), h(V||r||2), \ldots$ until all these values form a subset $S \subset \{1, \ldots, n\}$ of size $|S| = n/2$. Note that $i \neq j$ does not imply $h(V||r||i) \neq h(V||r||j)$ so that more than $n/2$ computations may be required to obtain the set $S$.
2. If $i_0 \in S$, the set $S$ is discarded. The prover returns to step 1 and chooses a new random seed. Step 1 is successful on average after 2 tries.
3. Otherwise, the protocol outputs the random seed $r$ and the set $S$. For all $j \in S$, the protocol also outputs the secret key $k_j$.
4. The verifier verifies that the set $S$ is correctly computed from randomness $r$. For all $j \in S$, the verifier uses the key $k_j$ to verify the correctness of $V_j$.

**Problem 2:** Since the seeds used to compute pads in our long DC-net assume the same form as those in the short DC-net, the reconstruction procedure is very similar. The only difference in the process is that once a seed is recovered, the

PRNG must be applied to obtain the corresponding pad. What we highlight, however, is that our use of bilinear maps is solving a fundamental problem in the long DC-net construction.

In the short DC-net, honest players could, in principle, make do without using bilinear maps. Indeed, they can reconstruct a pad in a verifiable way without revealing any long term secrets, by exploiting the algebraic structure of pads. (As explained in the footnote above, it is possible in principle to have, for example, secret keys $\{x_i\}$ that cancel, i.e., such that $\sum_i x_i = 0 \bmod q$, thereby engendering pads that "cancel." Note that this results in a very cumbersome key setup.) In the case of long DC-nets, however, there is no good way to do this. Briefly stated, the application of the PRNG eliminates algebraic structure on the pads.

The only way, therefore, to achieve "cancellation" of pads in a long DC-net, is for pairs of players to share secrets. But as already noted, in a standard setup without bilinear maps, it is possible for a pair of players $(P_i, P_j)$ to establish a shared secret $S$ non-interactively *only once* through application of D-H to their public keys. This secret $S$ can be used to generate new secrets for multiple sessions through application of symmetric-key primitives, e.g., secrets may be generated as $h(S, 1), h(S, 2), \ldots$. But without expensive general techniques, there is no way to reconstruct a given secret $h(S, w)$ without revealing $S$ itself and consequently compromising *all* shared secrets between $P_i$ and $P_j$.

*The solution:* This is where bilinear maps are helpful. As explained above, the intuition is that for a single pair of public keys, a bilinear map may be thought of as permitting non-interactive D-H key establishment across many different algebraic bases. Thus, each seed may be reconstructed individually by honest players holding shares of the private keys of $P_i$ and $P_j$. Under the (Bilinear) Diffie-Hellman assumption, this may be accomplished without compromising the privacy of other seeds. (In algebraic terms, one seed might assume the form $S_1 = {g_1}^{x_i x_j}$, while another assumes the form $S_2 = {g_2}^{x_i x_j}$. Provided that $g_1$ and $g_2$ are random, knowledge of $S_1$ does not permit computation of $S_2$.)

### 5.2 Protocol Details

In this section, we define our long DC-net protocol and highlight the differences with the short DC-net. The main differences between the long and short schemes lie in the definition of the auxiliary data $\sigma_{i,s}$ and the verification algorithm.

**Parameter and key generation.** This step is nearly identical to the short protocol. The function paramgen outputs parameters $\rho = (p, G_1, G_2, \hat{e}, Q)$. As in the short protocol, we use a hash functions $h : \{0,1\}^* \to G_1$. We also assume the existence of a publicly known pseudo-random number generator $f : G_2 \to \{0,1\}^l$, where $l$ is the length in bits of messages processed by the long DC-net. (For the purposes of our proofs, we model this as a random oracle.) The function keydist($\rho$) distributes keys to all players.

**Message posting.** Recall that we define the point $Q_k = h(s||k)$ on $G_1$. The pads $W_{i,s}(k)$ for player $P_i$ in session $s$ are computed as follows:

$$W_{i,s}(k) = \oplus_{j \in \Pi_s; j \neq i} \; f\left(\hat{e}(Q_k, y_j)^{x_i}\right)$$

Recall that player $P_i$ then chooses at random a value $c_i \in \Pi_s$ and XORs the message $m_i$ with pad $W_{i,s}(c_i)$ to produce the output vector $V_{i,s}$. We turn now to the computation of the auxiliary verification data $\sigma_{i,s}$:

1. Recall that the number of participants in session $s$ is denoted $\pi_s$. Let $\varphi$ be a hash function from $\{0,1\}^*$ to $\mathbb{Z}_{\pi_s}$. Using $\varphi$ and a random value $r$, the player $P_i$ computes a subset $S \subset \{1, \ldots, \pi_s\}$ of size $\pi_s/2$ such that $c_i \notin S$.
2. For all $j \in S$, $P_i$ proves that the value $V_{i,s}(j)$ is computed correctly by revealing $x_i Q_j$.

The string $\sigma_{i,s}$ consists of the values computed in steps 1 and 2 above. Finally, the function post outputs $(V_{i,s}, \sigma_{i,s}, i, s)$.

**Verification.** Anyone can verify non-interactively that the values computed in $\sigma_{i,s}$ are correct.

**Message extraction.** Given the $\pi_s$ vectors $V_{1,s}, \ldots, V_{\pi_s,s}$ published by the players in $\Pi_s$, anyone can non-interactively compute $r_k = \oplus_{i \in \Pi_s} V_{i,s}(k)$ for $k \in \Pi_s$. Recall that the definition of the pads is such that $\oplus_{i \in \Pi_s} W_{i,s}(k) = 0$. Using the same notations as in the short protocol, it is clear that $r_k = \oplus_{i \in c^{-1}(k)} m_i$. In other words, in every position $k$ for which $c^{-1}(k)$ is a singleton $\{i\}$, we have $r_j = m_i$. All other messages $m_i$ for which $c^{-1}(c_i)$ is not a singleton are unrecoverable in the output. The output of the function extract is the set of messages $m_i$ which are recovered in the output.

**Pad reconstruction.** If a subset of players $\mathcal{P} \subseteq \Pi_s$ fail to publish their output vector, the remaining players can reconstruct the pads of missing players, and compute the output of the DC-net, as follows. Each player $P_i$ for $i \notin \mathcal{P}$ publishes $x_{j,i} \cdot Q_k$ for all $j \in \mathcal{P}$. Anyone can verify the correctness of these values by checking that $\hat{e}(Q, x_{j,i} Q_k) = \hat{e}(y_{j,i}, Q_k)$. Furthermore, these values enable any player to recompute the seeds of missing player $P_j$ since the value $\hat{e}(Q_k, y_i)^{x_j}$ can be computed from the values $\hat{e}(x_{j,i} Q_k, y_i)$ by polynomial interpolation. The pads themselves may then be computed through application of $f$.

## 6 Conclusion

We have proposed two new DC-net constructions. Unlike previous DC-net proposals, our constructions allow for efficient detection and identification of cheating players with high probability. When cheating is detected, a single additional broadcast round enables full fault recovery. Our DC-net protocols are thus resilient to the jamming attacks that negated the simplicity and non-interactivity of earlier DC-net proposals.

In the appendix, we define a formal model in which we prove the privacy and correctness of our constructions. We observe that our comparatively simple definitions and proofs are made possible by the non-interactivity of DC-nets.

## References


1. L. von Ahn, A. Bortz and N. J. Hopper. $k$-anonymous message transmission. In *Proc. of ACM CCS '03*, pp. 122-130. ACM Press, 2003.
2. J. Bos and B. den Boer. Detection of disrupters in the DC protocol. In *Proc. of Eurocrypt '89*, pp. 320-327. LNCS 434.
3. P. Boucher, A. Shostack, and I. Goldberg. Freedom Systems 2.0 architecture. Zero Knowledge Systems, Inc. White Paper, December 2000. Available at http://freehaven.net/anonbib/
4. M. Bellare and P. Rogaway. Random oracles are practical: a paradigm for designing efficient protocols. In *Proc. of ACM CCS '93*, pp. 62-73. ACM Press, 1993.
5. D. Boneh and M. Franklin. Identity based encryption from the Weil pairing. In *SIAM J. of Computing*, 32(3), pp. 586-615, 2003.
6. R. Canetti, S. Halevi, and J. Katz. A forward-secure public-key encryption scheme. In *Proc. of Eurocrypt '03*, pp. 255-271. LNCS 2656.
7. C. Cachin, K. Kursawe, F. Petzold, and V. Shoup. Secure and efficient asynchronous broadcast protocols. In *Proc. of Crypto '01*, pp. 524-541. LNCS 2139.
8. J. Camenisch and M. Stadler. Efficient group signature schemes for large groups. In *Proc. of Crypto '97*, pp. 410-424. LNCS 740.
9. R. Cramer, I. Damgaard, and B. Schoenmakers. Proofs of partial knowledge and simplified design of witness hiding protocols. In *Proc. of Crypto '94*, pp. 174-187. LNCS 839.
10. D. Chaum. Untraceable electronic mail, return addresses, and digital pseudonyms. In *Communications of the ACM*, 24(2), pp. 84-88, 1981.
11. D. Chaum. The dining cryptographers problem: unconditional sender and recipient untraceability. In *Journal of Cryptology*, 1(1), pp. 65-75, 1988.
12. D. Chaum and H. van Antwerpen. Undeniable signatures. In *Proc. of Crypto '89*, pp. 212-216. LNCS 435.
13. G. Danezis, R. Dingledine and N. Mathewson. Mixminion: design of a type III anonymous remailer protocol. In *IEEE Symposium on Security and Privacy 2003*, pp. 2-15.
14. R. Dingledine. Anonymity bibliography. Available on the web at http://freehaven.net/anonbib/
15. A. Fiat and A. Shamir. How to prove yourself: practical solutions to identification and signature problems. In *Proc. of Crypto '86*, pp. 186-194. LNCS 263.
16. R. Gennaro, S. Jarecki, H. Krawczyk and T. Rabin. Secure distributed key generation for discrete-log based cryptosystems. In *Proc. of Eurocrypt '99*, pp. 295-310. LNCS 1592.
17. D. Goldschlag, M. Reed and P. Syverson. Onion routing. In *Communications of the ACM*, 42(2), pp. 39-41. 1999.
18. M. Jakobsson, A. Juels and R. Rivest. Making mix nets robust for electronic voting by randomized partial checking. In *Proc of USENIX'02*.
19. M. Reiter and A. Rubin. Anonymous web transactions with Crowds. In *Communications of the ACM*, 42(2), pp. 32-38. 1999.

20. C. Schnorr. Efficient signature generation by smart cards. In *Journal of Cryptology*, 4(3), pp. 161-174. 1991.
21. M. Waidner and B. Pfitzmann. The Dining Cryptographers in the disco: unconditional sender and recipient untraceability with computationally secure serviceability. In *Proc. of Eurocrypt '89*, p. 690. LNCS 434.
22. M. Waidner. Unconditional sender and recipient untraceability in spite of active attacks. In *Proc. of Eurocrypt'89*, pp. 302-319. LNCS 434.

# A Security Definitions

## A.1 Privacy

We consider a static adversary $\mathcal{A}$ capable of actively corrupting a set $P_{\mathcal{A}}$ of fewer than $n/2$ players in $P$. We regard a mix network DC as private if $\mathcal{A}$ is unable to determine the origin of any message input by an honest player with probability significantly better than a random guess. We capture this concept by way of an experiment in which $\mathcal{A}$ selects just two honest players $p_0$ and $p_1$ as targets. The adversary may also choose a pair of plaintexts $(m_0, m_1)$ to serve as inputs for these two target players. The plaintexts are randomly assigned to $p_0$ and $p_1$; the task of $\mathcal{A}$ is to guess this assignment.

We let $\tilde{\mathsf{post}}_i(\cdot,\cdot,\cdot)$ denote an oracle that posts a message on behalf of player $P_i$. The adversary may specify $s, m$ and $\Pi_s$. The oracle is assumed to have access to the private keys of $P_i$. The adversary may not invoke a given oracle twice on the same session identifier $s$. (In a real-world protocol, this restriction is easily enforced through use of a local counter.)

The oracle $\tilde{\mathsf{post}}_i(\cdot,\cdot,\cdot)$ also produces auxiliary data $\sigma_{i,s}$. A small difficulty arises in the long protocol, where $\sigma_{i,s}$ reveals half the pads of player $P_i$. If the pads of all honest players are revealed in the positions where $p_0$ and $p_1$ posted $m_0$ and $m_1$, then $\mathcal{A}$ can trivially determine which player posted which message. This happens with low probability if the number of honest players is large. In our privacy experiment, we assume that the auxiliary data does not reveal the pads used by $p_0$ and $p_1$ in the positions where they posted $m_0$ and $m_1$.

The oracle $\tilde{\mathsf{post}}^*(\cdot,\cdot,\cdot,\cdot)$ is a special oracle call that causes the two targeted players to post the chosen messages $(m_0, m_1)$. In particular, this oracle call is equivalent to two successive oracle calls: $\tilde{\mathsf{post}}_{p_0}(m_b, s, \Pi_s, \cdot)$ and $\tilde{\mathsf{post}}_{p_1}(m_{1-b}, s, \Pi_s, \cdot)$, where $p_0, p_1 \in \Pi_s$.

We let $\tilde{\mathsf{reconstruct}}_i(\cdot,\cdot)$ denote an oracle that returns the reconstructed pad of player $P_i$. The adversary may specify the session $s$ and $\Pi_s$. The oracle is assumed to have access to the private key held by $P_i$. The oracle $\tilde{\mathsf{reconstruct}}_i$ may be called by $\mathcal{A}$ at any point during the experiment, with the following restriction: $\mathcal{A}$ may not call $\tilde{\mathsf{reconstruct}}_{p_0}$ or $\tilde{\mathsf{reconstruct}}_{p_1}$ for the session $s$ in which $\mathcal{A}$ chose to call the special oracle $\tilde{\mathsf{post}}^*$. This restriction is natural: it simply states that $\mathcal{A}$ is not allowed to ask for the pads of players $p_0$ and $p_1$ in the session in which it must guess the assignment of messages $m_0, m_1$ to $p_0, p_1$.

We let $\in_U$ denote uniform, random selection from a set. Security parameters are left implicit.

Experiment $\mathbf{Exp}^{priv}_{\mathcal{A}}(\mathsf{DC}); [k, n, l]$
  $\mathsf{paramgen}(l)$; $\mathsf{keydist}$;
  $P_{\mathcal{A}} \leftarrow \mathcal{A}(\{PK_i\})$;
  $(m_0, m_1, p_0, p_1) \leftarrow \mathcal{A}^{\{\tilde{\mathsf{post}}_i(\cdot,\cdot,\cdot,\cdot)\}_{i \in P - P_{\mathcal{A}}}, \tilde{\mathsf{reconstruct}}_i(\cdot,\cdot)}$;
  $b \in_U \{0, 1\}$;
  $b' \leftarrow \mathcal{A}^{\{\tilde{\mathsf{post}}_i(\cdot,\cdot,\cdot,\cdot)\}_{i \in P - P_{\mathcal{A}}}, \tilde{\mathsf{reconstruct}}_i(\cdot,\cdot), \tilde{\mathsf{post}}^*(\cdot,\cdot,\cdot,\cdot)}$;
  if $b' = b$ output '1' else output '0';

We define the advantage of $\mathcal{A}$ in this experiment as

$$\mathbf{Adv}^{priv}_{\mathcal{A}}(\mathsf{DC}); [k, n, l] = \mathsf{pr}[\mathbf{Exp}^{priv}_{\mathcal{A}}(\mathsf{DC}); [k, n, l] = \text{'1'}] - 1/2 \,.$$

We say that our scheme is private if this advantage is negligible for all adversaries $\mathcal{A}$ with polynomial running time (where the quantities are defined asymptotically with respect to $l$ in the usual manner). The following propositions show that our short and long DC-nets are private. (The proofs are in appendix B.)

**Proposition 1.** *The short DC-net protocol of section 4 is private if the Decisional Bilinear Diffie-Hellman (DBDH) problem is hard in the group $G_1$.*

**Proposition 2.** *The long DC-net protocol of section 5 is private if the Bilinear Diffie-Hellman (BDH) problem is hard in the group $G_1$.*

**Remark:** the non-interactivity of the mix network DC makes possible this relatively simple definition of privacy. In a mix network involving interaction among players, an adversary can change the behavior of honest players by inducing errors or failures in the outputs of corrupted players. The resulting broad scope of adversarial behavior induces considerably more complex privacy definitions.

### A.2 Correctness

We define correctness in terms of the ability of a corrupted player $P_i$ to post a vector $V$ that has an invalid format, but is accepted by the function $\mathsf{verify}$. Invalid formatting may mean that $V$ includes incorrectly computed pads or, alternatively, that $V$ contains an inadmissibly large number of messages. More formally, we deem a vector $V$ as correct if it constitutes a valid output of $\mathsf{post}$ for the private key of the correct player. (Other definitions are possible.) We use the triangular brackets '$\langle\ \rangle$' to denote the set of possible function outputs.

Experiment $\mathbf{Exp}^{corr}_{\mathcal{A}}(\mathsf{DC}); [k, n, l]$
  $\mathsf{paramgen}(l)$; $\mathsf{keydist}$;
  $P_{\mathcal{A}} \leftarrow \mathcal{A}(\{PK_i\})$; $((V, \sigma, i, s), \Pi_s) \leftarrow \mathcal{A}$;
  if $(V, \sigma, i, s) \notin \langle \mathsf{post}(s, m, x_i); [\Pi_s, \{y_j\}_{j \in P}] \rangle$ for any $m$ and
    $\mathsf{verify}((V, \sigma), s, i, \Pi_s); [\{y_j\}_{j \in P}] = \text{'1'}$ then output '1';
  else output '0';

We define the advantage of $\mathcal{A}$ in this experiment as $\mathbf{Adv}^{priv}_{\mathcal{A}}(\mathsf{DC});[k,n,l] =$ $\mathsf{pr}[\mathbf{Exp}^{corr}_{\mathcal{A}}(\mathsf{DC});[k,n,l] = \text{'1'}]$. We regard our scheme as providing correctness if for all adversaries $\mathcal{A}$ with polynomial running time, this advantage is negligible.

**Proposition 3.** *The short DC-net protocol of section 4 is correct.*

**Proposition 4.** *The long DC-net protocol of section 5 satisfies a weaker property. If an adversary submits an output in which k pads out of n are incorrectly computed, the probability that* verify *accepts this output is* $2^{-k}$.

## B Proofs of Privacy

**Proposition 1.** *The short DC-net protocol of section 4 is private if the Decisional Bilinear Diffie-Hellman (DBDH) problem is hard in the group* $G_1$.

*Proof.* Let $\mathcal{A}$ be a polynomial-time adversary who wins $\mathbf{Exp}^{priv}_{\mathcal{A}}(\mathsf{DC})$ with non-negligible advantage $\epsilon$. We use $\mathcal{A}$ to solve DBDH challenges with non-negligible advantage as follows. We first call $\mathsf{paramgen}(l)$ to get parameters $(p, G_1, G_2, \hat{e}, Q)$ where $G_1$ and $G_2$ are groups of order $p$, $Q$ is a generator of $G_1$ and $\hat{e} : G_1 \times G_1 \rightarrow G_2$ is an admissible bilinear map. Let $(aQ, bQ, cQ, dQ)$ be a DBDH challenge in $G_1$ (the challenge is to determine whether $d = abc$ or $d$ is random).

We give $\mathcal{A}$ the output of $\mathsf{paramgen}(l)$. Next we simulate $\mathsf{keydist}$ for $\mathcal{A}$. We let the public keys of two players (say $P_1$ and $P_2$) be $y_1 = aQ$ and $y_2 = bQ$. For every other player, we choose a private key $x_i \in_U \mathbb{Z}_p$ and compute the corresponding public key $y_i = x_i \cdot Q$. Given all these public keys, $\mathcal{A}$ returns the set $P_{\mathcal{A}}$ of players it controls. If $P_1 \in P_{\mathcal{A}}$ or $P_2 \in P_{\mathcal{A}}$, we abort. Otherwise, we give $\mathcal{A}$ the private keys of all the players in $P_{\mathcal{A}}$. We also give $\mathcal{A}$ the shares of the private keys held by all the players in $P_{\mathcal{A}}$. For the private key of player $P_1$ and $P_2$, which we do not know, we generate random shares.

$\mathcal{A}$ can then call the oracle $\tilde{\mathsf{post}}_i(\cdot,\cdot,\cdot,\cdot)$ any number of times for $i \in P - P_{\mathcal{A}}$. For all but one session for which $\mathcal{A}$ calls $\tilde{\mathsf{post}}$, we let $h(s||k) = r_{s,k}Q$, where the values $r_{s,k} \in_U \mathbb{Z}_p$. For one session $s_0$, we define $h(s_0||k)$ differently. We choose 2 "special" positions $k_0, k_1 \in_U \{1, \ldots, n\}$ as well as $R \in_U \mathbb{Z}_p$. We define $h(s_0||k_0) = cQ$, $h(s_0||k_1) = RcQ$ and for $k \not\in \{k_0, k_1\}$, we let $h(s_0||k) = r_{s_0,k}Q$ for values $r_{s_0,k}$ chosen at random in $\mathbb{Z}_p$.

To simulate $\tilde{\mathsf{post}}_i(\cdot,\cdot,\cdot,\cdot)$ for $\mathcal{A}$ in session $s$, we need the pads $W_{i,s}(k) = \prod_{j \in \Pi_s; j \neq i} \hat{e}(Q_k, y_j)^{\delta_{i,j} x_i}$, where $Q_k = h(s||k)$. For all session $s \neq s_0$, we have $Q_k = h(s||k) = r_{s,k}Q$ and therefore we can compute the pad $W_{i,s}(k)$ for all players $P_i$ (even for $P_1, P_2$) using the equality $\hat{e}(Q_k, y_j)^{x_i} = \hat{e}(y_i, y_j)^{r_{s,k}}$. For session $s_0$, we can compute the pads of all players except $P_1$ and $P_2$ whose private key we do not know. If $\mathcal{A}$ calls $\tilde{\mathsf{post}}$ for $P_1$ or $P_2$ in session $s_0$, we abort.

Note that knowledge of the pads also enables us to simulate the auxiliary data $\sigma_{i,s}$ in both the short and the long protocol, as well as the oracle $\tilde{\mathsf{reconstruct}}_i$.

$\mathcal{A}$ then chooses two messages $m_0, m_1$ to be posted by two players $p_0, p_1$ of the adversary's choice. If $(p_0, p_1) \neq (P_1, P_2)$, we abort the simulation. $\mathcal{A}$ may again call $\tilde{\mathsf{post}}_i$ and we simulate that oracle as before.

Finally, $\mathcal{A}$ calls $\tilde{\mathsf{post}}^*$ for a particular session. If that session is not $s_0$, we abort. Otherwise, we simulate $\tilde{\mathsf{post}}^*$ as follows. For $P_1$, we define the pads:

$$W_{1,s_0}(k_0) = \hat{e}(Q, dQ)^{\delta_{1,2}} \prod_{3 \leq j \leq n} \hat{e}(cQ, y_1)^{\delta_{1,j} x_j},$$

$$W_{1,s_0}(k_1) = \hat{e}(Q, dQ)^{\delta_{1,2}} \prod_{3 \leq j \leq n} \hat{e}(RcQ, y_1)^{\delta_{1,j} x_j},$$

$$W_{1,s_0}(k) = \prod_{2 \leq j \leq n} \hat{e}(y_1, y_j)^{\delta_{1,j} r_{s_0,k}} \quad \text{for } k \notin \{k_0, k_1\}$$

We define the pads for $P_2$ similarly. We choose a bit $b$ at random and let $P_1$ post $m_b$ in position $k_1$ and $P_2$ post $m_{1-b}$ in position $k_2$. We simulate the corresponding NIZK proofs for the auxiliary data using standard techniques by allowing the simulator to set random oracle responses before making commitments.

$\mathcal{A}$ outputs a guess $b'$. If $b' = b$, we guess that $(aQ, bQ, cQ, dQ)$ is a DBDH tuple, and otherwise that it is not. It remains to show that our guess is correct with non-negligible advantage:

- When $d = abc$, by definition of $\mathcal{A}$, we have $b' = b$ with advantage $\epsilon$.
- When $d \neq abc$, our simulation of the pads $W_{1,s_0}(k_0), W_{1,s_0}(k_1), W_{2,s_0}(k_0)$ and $W_{2,s_0}(k_1)$ was incorrect. There is consequently no way for $\mathcal{A}$ to distinguish between respective partial pads for $P_1$ and $P_2$ of the form $(V_1, V_2) = (\mathit{Rand} \oplus m_1, \mathit{Rand})$ and $(V_1, V_2) = (\mathit{Rand}, \mathit{Rand} \oplus m_2)$, because they are identically distributed (here, *Rand* denotes random values). In other words, $\mathcal{A}$ can't possibly guess the bit $b$ with non-negligible advantage.

This shows that when the simulation does not abort, $\mathcal{A}$ solves DBDH challenges with advantage $\epsilon/2$. The probability that the simulation does not abort is greater than a value that is polynomial in the security parameter. Overall, we have used $\mathcal{A}$ to solve DBDH challenges with non-negligible advantage. □

**Proposition 2.** *The long DC-net protocol of section 5 is private if the Bilinear Diffie-Hellman (BDH) problem is hard in the group $G_1$.*

*Proof.* The proof is similar to that of Proposition 1. Let $\mathcal{A}$ be a polynomial-time adversary who wins $\mathbf{Exp}_{\mathcal{A}}^{priv}(\mathsf{DC})$ with non-negligible advantage $\epsilon$ and let $(aQ, bQ, cQ)$ be a BDH challenge (the challenge is to compute $dQ$, where $d = abc$). We embed the BDH challenge as before. The difference worth noting is that the output of the bilinear function, in the long protocol, is expanded with a PRNG $f$. We model $f$ as a random oracle. There are two possible distributions for the simulator: distribution $D$, where the simulator calls $f(dQ)$ (for the correct BDH value $d$), and distribution $\tilde{D}$, where the simulator uses a random value. $\mathcal{A}$ cannot distinguish between $D$ and $\tilde{D}$ unless it calls $f$ on input $dQ$.

If $\mathcal{A}$ cannot distinguish $D$ from $D'$, it cannot distinguish a real-world protocol invocation from one in which random pads are used and therefore cannot learn anything about which player posted which message. $\mathcal{A}$ then must be able to distinguish $D$ from $D'$ and so must call the random oracle on input $dQ$ occasionally. We answer the BDH challenge with one of $\mathcal{A}$'s calls to the random oracle and win with non-negligible probability since $\mathcal{A}$ is polynomially bounded. □

# Algebraic Attacks and Decomposition of Boolean Functions

Willi Meier[1], Enes Pasalic[2], and Claude Carlet[2]

[1] FH Aargau, CH-5210 Windisch, Switzerland
meierw@fh-aargau.ch

[2] INRIA, projet CODES, Domaine de Voluceau, Rocquencourt
BP 105, 78153 Le Chesnay Cedex, France
{Enes.Pasalic,Claude.Carlet}@inria.fr.

**Abstract.** Algebraic attacks on LFSR-based stream ciphers recover the secret key by solving an overdefined system of multivariate algebraic equations. They exploit multivariate relations involving key bits and output bits and become very efficient if such relations of low degrees may be found. Low degree relations have been shown to exist for several well known constructions of stream ciphers immune to all previously known attacks. Such relations may be derived by multiplying the output function of a stream cipher by a well chosen low degree function such that the product function is again of low degree. In view of algebraic attacks, low degree multiples of Boolean functions are a basic concern in the design of stream ciphers as well as of block ciphers.

This paper investigates the existence of low degree multiples of Boolean functions in several directions: The known scenarios under which low degree multiples exist are reduced and simplified to two scenarios, that are treated differently in algebraic attacks. A new algorithm is proposed that allows to successfully decide whether a Boolean function has low degree multiples. This represents a significant step towards provable security against algebraic attacks. Furthermore, it is shown that a recently introduced class of degree optimized Maiorana-McFarland functions immanently has low degree multiples. Finally, the probability that a random Boolean function has a low degree multiple is estimated.

**Keywords :** Algebraic attacks, Stream ciphers, Boolean functions, Algebraic degree, Annihilator, Low degree multiple, Resiliency.

## 1 Introduction

Algebraic attacks on stream ciphers based on linear feedback shift registers (LFSR's) have been proposed in [8]. Many stream ciphers consist of a linear part, producing a sequence with a large period, usually composed of one or several LFSR's, and a nonlinear combining function $f$ that produces the output, given the state of the linear part. Algebraic attacks recover the secret key by solving an overdefined system of multivariate algebraic equations. These attacks

C. Cachin and J. Camenisch (Eds.): EUROCRYPT 2004, LNCS 3027, pp. 474–491, 2004.

exploit multivariate relations involving key/state bits and output bits of $f$. If one such relation is found that is of low degree in the key/state bits, algebraic attacks are very efficient, [6].

In [8] it is demonstrated that low degree relations and thus successful algebraic attacks exist for several well known constructions of stream ciphers that are immune to all previously known attacks. In particular, low degree relations are proven to exist for ciphers using a combining function $f$ with a small number of inputs. These low degree relations are obtained by producing low degree polynomial multiples of $f$, i.e., by multiplying the Boolean function $f$ by a well chosen low degree function $g$ such that the product function $f * g$ is again of low degree.

There have become known alternative methods to attack stream ciphers by solving overdefined systems of equations using Gröbner bases, [11]. In order to be efficient, these methods rest on the existence of low degree multiples as well.

To counter algebraic attacks, it is recommended in [8], that the combining function $f$ should have at least 32 inputs. But even then, by now it cannot be excluded for certain, that $f$ has low degree multiples that would then make a fielded or a new design vulnerable to algebraic attacks. This is in strong contrast to other attacks on stream ciphers: A variety of proposed stream ciphers have been shown to be provably resistant, e.g., against the Berlekamp-Massey shift register synthesis algorithm.

In a different direction, in view of algebraic attacks on block ciphers, [7], it may be desirable to know for certain, e.g., that there are no low degree equations, relating output bits of a (reduced round) block cipher, plaintext bits and key bits. We mention also that recently the framework of algebraic attacks has been extended to combiners with memory [6, 1].

As a consequence, investigation of Boolean functions with regard to existence of low degree multiples is of both, theoretical and practical interest.

The results of this paper contribute to this problem in four directions: We reduce and simplify the scenarios found in [8], under which low degree multiples may exist. As a significant step towards provable resistance against algebraic attacks we propose an algorithm that allows to successfully decide whether a Boolean function has low degree multiples. This new algorithm can be efficient for input sizes of $f$ of 32 bits or larger. Furthermore, we show that for a recently proposed class of Boolean functions, the degree optimized Maiorana-McFarland class [18], relatively low degree multiples are immanent. Finally we derive upper bounds on the probability that a random Boolean function has a low degree multiple. This is partly done by using results from coding theory. These bounds are shown to give strong estimates for input sizes of practical interest.

To further explain some of our results, recall that the main cryptographic criteria for Boolean functions $f$ used for stream cipher applications had previously been a high algebraic degree, to counter linear synthesis by Berlekamp-Massey algorithm, some order of correlation immunity (resiliency), and large distance to affine functions (high nonlinearity), to withstand different types of correlation and linear attacks [17, 13, 3]. There are some known tradeoffs between the criteria, e.g., there is the bound by Siegenthaler [19], that the algebraic degree of $f$ is upper bounded by $n - t - 1$, where $n$ is the number of inputs of $f$ and $t < n - 1$ is its order of resiliency.

The more recent algebraic attacks impose a new restriction on the combining function $f$ chosen: $f$ shouldn't have low degree multiples. In [8], essentially three different scenarios are described which lead to low degree multiples of a Boolean function which can be exploited in algebraic attacks. We show that these scenarios can be reduced to two, to be treated differently in algebraic attacks. This simplified description of scenarios leads to a precise measure of algebraic immunity of a Boolean function $f$: *The algebraic immunity $AI(f)$ is the minimum value of $d$ such that $f$ or $f+1$ admits an annihilating function of degree $d$.* Recall that an annihilator of $f$ is a non-zero function $g$ such that $f * g = 0$.

The new criterion that $f$ shouldn't have a low algebraic immunity, may be in conflict with some established criteria. This is exemplified for the Maiorana-McFarland class. These functions can have high resiliency, high nonlinearity, and optimum algebraic degree [10, 2, 4, 18]. Nevertheless it is shown in this paper that such functions can have relatively low algebraic immunity (Example 1). This is done by deriving a useful representation for the complete set of annihilators for a given function $f$. Any annihilator can be viewed as a concatenation of annihilators from some smaller variable space. This method when applied to a function in the standard Maiorana-McFarland class [10, 2] only yields annihilators of degree larger than the degree of the function itself. However, this method may be successfully applied to the degree optimized Maiorana-McFarland class [18], showing that relatively low degree annihilators are immanent for this class.

In the design of stream ciphers, this property needs to be avoided. Therefore, it is desirable to have an efficient algorithm for deciding whether a given Boolean function has no low degree annihilator. Such an algorithm is derived in this paper (Algorithm 2). A refined version allows to decide whether a Boolean function with $n$ inputs has no annihilator of degree $d$ at most 5, in about $\binom{n}{d-1}^3$ operations, which e.g. for $n = 32$ is certainly feasible. If for a stream cipher a degree $d$ annihilator with $d = 4$ (say) of its combining function $f$ (or $f + 1$) is found by our algorithm, we can break this cipher. On the other hand, if $f$ and $f + 1$ are shown to have no annihilator of degree $d \leq 5$, this cipher has some amount of immunity against algebraic attacks, as for $d = 6$ and for a size of the initial state of 128 bits, the computational complexity of the basic algebraic attack in [8] is already about $2^{96}$.

The paper is organized as follows. In Section 2 the basic definitions and notions regarding Boolean functions are introduced. Section 3 recalls and simplifies the various scenarios of algebraic attacks. Algebraic properties of annihilators for an arbitrary function $f$ are addressed in Section 4, where an alternative representation of annihilators is given which is useful for the analysis of some well known classes of Boolean functions. Section 5 deals with the fundamental problem of efficiently deciding whether the combining function in a stream cipher has annihilators of low degrees. In Section 6 we estimate an upper bound on the probability that a random function has annihilators of certain degree.

## 2 Preliminaries

A Boolean function on $n$ variables may be viewed as a mapping from $\{0,1\}^n$ into $\{0,1\}$. A Boolean function $f(x_1,\ldots,x_n)$ is also interpreted as the output column of its *truth table* $f$, i.e., a binary string of length $2^n$,

$$\overline{f} = [f(0,0,\cdots,0), f(1,0,\cdots,0), f(0,1,\cdots,0),\ldots, f(1,1,\cdots,1)].$$

The *Hamming distance* between $n$-variable functions $f, g$, denoted by $d(f,g)$, is

$$d(f,g) = \#\{x \in \mathbb{F}_2^n \mid f(x) \neq g(x)\}.$$

Also the *Hamming weight* or simply the weight of $f$ is the number of ones in $\overline{f}$. This is denoted by $wt(f)$. An $n$-variable function $f$ is said to be *balanced* if its output column in the truth table contains equal number of 0's and 1's (i.e., $wt(f) = 2^{n-1}$).

The Galois field of order $2^n$ will be denoted by $\mathbb{F}_{2^n}$ and the corresponding vector space by $\mathbb{F}_2^n$. Addition operator over $\mathbb{F}_2$ is denoted by $\oplus$, and if no confusion is to arise we use the usual addition operator $+$. An $n$-variable Boolean function $f(x_1,\ldots,x_n)$ can be considered to be a multivariate polynomial over $\mathbb{F}_2$. This polynomial can be expressed as a sum of products representation of all distinct $r$-th order products $(0 \leq r \leq n)$ of the variables. More precisely, $f(x_1,\ldots,x_n)$ can be written as

$$f(x_1,\ldots,x_n) = \sum_{u\in\mathbb{F}_2^n} \lambda_u \left(\prod_{i=1}^{n} x_i^{u_i}\right) \,, \quad \lambda_u \in \mathbb{F}_2 \,, u = (u_1,\ldots,u_n). \tag{1}$$

This representation of $f$ is called the *algebraic normal form* (ANF) of $f$. The *algebraic degree* of $f$, denoted by $deg(f)$ or sometimes simply $d$, is the maximal value of the Hamming weight of $u$ such that $\lambda_u \neq 0$. There is a one-to-one correspondence between the truth table and the ANF via so called inversion formulae. The set of $x$ values for which $f(x) = 1$ respectively $f(x) = 0$ is called the on-set respectively the off-set, denoted by $S_1(f)$ and $S_0(f)$. The ANF of $f$ is fully specified by its on-set using the following expansion,

$$f(x_1,\ldots,x_n) = \sum_{\tau\in S_1(f)} \left(\prod_{i=1}^{n}(x_i+\tau_i+1)\right), \quad \tau = (\tau_1,\ldots,\tau_n). \tag{2}$$

The set of all Boolean functions in $n$ variables is denoted by $\mathcal{R}_n$. For any $0 \leq b \leq n$ an $n$-variable function is called non degenerate on $b$ variables if its ANF contains exactly $b$ distinct input variables. Functions of degree at most one are called *affine* functions. An affine function with constant term equal to zero is called a *linear* function. The set of all $n$-variable affine (respectively linear) functions is denoted by $\mathcal{A}_n$ (respectively $\mathcal{L}_n$). The *concatenation*, denoted by $''||''$ simply means that the truth tables of the functions are merged. For instance, for $f_1, f_2 \in \mathcal{R}_{n-1}$ one may construct $\overline{f} = \overline{f_1}||\overline{f_2}$ (where $f \in \mathcal{R}_n$), meaning that the upper half part of the truth table of $f$ correspond to $f_1$ and the lower part to $f_2$. The ANF of $f$ is then given by $f(x_1,\ldots,x_n) = (1+x_n)f_1(x_1,\ldots,x_{n-1}) + x_n f_2(x_1,\ldots,x_{n-1})$.

## 3 Algebraic Attacks: Scenarios Revisited

In [8], three different scenarios (S3a, S3b, S3c) are described under which low degree relations (that hold with probability 1) may exist and how they can be exploited in algebraic attacks. The aim of this section is to show that these can be reduced to essentially two scenarios, and to clarify how to use them in an attack.
To recall the scenarios in [8], let the Boolean function $f$ have high degree.

- S3a Assume that there exists a function $g$ of *low* degree such that the product function is of low degree, *i.e.*, $f * g = h$, where $h$ is a nonzero function of low degree.
- S3b Assume there exists a function $g$ of low degree such that $f * g = 0$.
- S3c Assume there exists a function $g$ of *high* degree such that $f * g = h$ where $h$ is nonzero and of low degree.

Consider scenario S3c. Then $f*g = h \neq 0$. Multiply this equation by $f$. As $f^2 = f$ does hold over $\mathbb{F}_2$, we get $f^2*g = f*h = f*g = h$. Hence $f*h = h$. As $h$ is of low degree, we are in scenario S3a. Therefore, scenario S3c is redundant. Further, one might consider another scenario (not contained in [8]): Factorizations of the form $f = g * h$, where $g$ and/or $h$ are of low degree. However, $g * (1 + g) = 0$ over $\mathbb{F}_2$. Hence by multiplying $f = g * h$ by $1 + g$, we get $f * (1 + g) = 0$, i.e., we are back in scenario S3b. These considerations suggest that in algebraic attacks one can always restrict to scenarios S3a and S3b. There is an interesting relation between the two:

**Proposition 1** *Assume that $f * g = h \neq 0$, does hold for some functions $g$ and $h$ of degrees at most $d$ (scenario S3a). Suppose in addition that $g \neq h$. Then there is a function $g'$ of degree at most $d$ such that $f * g' = 0$ (scenario S3b).*

*Proof.* As above, we have $f^2 * g = f * g = f * h = h$. Hence $f * (g + h) = 0$. □

The argument just given shows that we can reduce ourselves to scenario S3a in case where $g = h$, and scenario S3b. However, S3a with $g = h$ is equivalent to scenario S3b for the function $f + 1$.

The existence of algebraic attacks will impose that neither $f$ nor $f + 1$ does admit an annihilating function of low degree. This motivates the notion "algebraic immunity" of $f$, denoted by $AI(f)$, which is the minimum value of $d$ such that $f$ or $f + 1$ admits an annihilating function of degree $d$.

In [8], low degree relations according to scenarios S3a or S3b are proven to exist for any Boolean function $f$ with a small number of inputs:

**Theorem 6.0.1** [8, 9] *Let $f$ be any Boolean function with $n$ inputs. Then there is a Boolean function $g \neq 0$ of degree at most $\lceil n/2 \rceil$ such that $f * g$ is of degree at most $\lceil n/2 \rceil$.*

**Remark 1** *Without restricting the form of the function, the upper bound given above cannot be improved for the case of annihilators, i.e.* $f * g = 0$. *For instance one example of a function not admitting annihilators of degree lower than* $n/2$ *is given in [11]. Namely the function in* 6 *variables, denoted there CanFil* 8, *has annihilators of degree* $d \geq 3$ *only. Moreover, [9, Table 3] gives experimental evidence that a random function with* 10 *variables is not likely to have an annhilator of degree lower than* 5.

To exploit low degree relations as in scenarios S3a and S3b, assume that $N_d$ linearly independent functions $h$ with $f * g = h$ have been found, where $h$ and $g$ have low degree $d$. Similarly, assume that $N_d'$ linearly independent functions $g$ of low degree $d$ have been found such that $f * g = 0$.

In an algebraic attack on an LFSR-based stream cipher, it is assumed that the feedback connections are known. Let $(s_0, ..., s_{k-1})$ be the initial state of the driving LFSR's. Then the output of the cipher is given by:

$$\begin{cases} b_0 &= f(s_0, ..., s_{k-1}) \\ b_1 &= f(L(s_0, ..., s_{k-1})) \\ b_2 &= f(L^2(s_0, ... s_{k-1})) \\ ... &= ... \end{cases}$$

Here $L$ denotes the linear update function to the next state of the LFSR's involved. The problem is to recover the $k$-bit key $(s_0, ..., s_{k-1})$. Let $x$ equal $L^i(s_0, ..., s_{k-1})$.

If the output bit $b_i = 1$, we use scenario S3b, i.e., $f * g = 0$, and get an equation $g(x) = 0$. Alternatively, we can use scenario S3a, $f * g = h$, and take $g(x) = h(x)$. However, either $g = h$, which gives nothing, or $g \neq h$, which gives $g + h = 0$, i.e. we are back in scenario S3b.

If $b_i = 0$, use scenario S3a: $h(x) = 0$. Hence for any known output bit $b_i$ we get $N_d$ equations, if $b_i = 0$, and $N_d'$ equations, if $b_i = 1$.

If we get at least one such equation for each of sufficiently many output bits, we obtain a very overdefined system of multivariate equations of low degree $d$, that can be solved efficiently: There are about $T \approx \binom{n}{d}$ monomials of degree at most $d$ in the $k$ variables $s_i$, $i = 0, \ldots, k-1$ (assuming $d \ll n/2$). Consider each of these monomials as a new variable $V_j$. Given $R \geq \binom{n}{d}$ equations, we get a system of $R \geq T$ linear equations in the $V_j$'s that can be solved by Gaussian elimination. If more than one equation holds per output bit, the output stream needed reduces accordingly.

## 4 Properties of the Annihilator Set

As set out in the introductory part, in the realm of algebraic attacks there is one major concern: Given a Boolean function $f$ used in a stream cipher, the task is to determine whether this function has low algebraic immunity, i.e., whether $f$ or $f + 1$ has a low degree annihilator. In this section we specify the structure of the set of annihilators for a given $f$, and also give an alternative representation of their ANF. Let $An(f) = \{g \mid f * g = 0\}$ denotes the annihilator set for the function $f$ in the Boolean ring $\mathcal{R}_n = \mathbb{F}_2[x_1, \ldots, x_n]/I$, $I$ being an ideal generated by the polynomials $x_i^2 - x_i$, $i = 1, \ldots, n$. Since in this ring $f(1+f) = 0$ for any $f \in \mathcal{R}_n$ the set $An(f)$ is nonempty.

**Theorem 1.** *Let $f$ be any Boolean function in $\mathcal{R}_n$. Then $An(f)$ is a principal ideal in $\mathcal{R}_n$ generated by $(1+f)$, i.e. $An(f) = \{(1+f)r \mid r \in \mathcal{R}_n\} =< 1+f >$. Its cardinality equals to $|An(f)| = 2^{2^n - |S_1(f)|}$. In particular when $f$ is balanced $|An(f)| = 2^{2^{n-1}}$.*

*Proof.* In order to show that $An(f)$ is a principal ideal in the Boolean ring $\mathcal{R}_n$ generated by $(1+f)$, we prove firstly that $An(f)$ is a subring of $\mathcal{R}_n$, then an ideal which is principal.

To prove that $An(f)$ is a subring of $\mathcal{R}_n$ it is enough to demonstrate that $An(f)$ is closed under the operations $'+'$ and $'*'$. Clearly $An(f)$ is nonempty since $(1+f) \in An(f)$. Let $g, h \in An(f)$. Then $f * (g+h) = f * g + f * h = 0$, and $f * (g * h) = (f * g) * h = 0$. Hence $An(f)$ is closed under $'+'$ and $'*'$ and therefore a subring of $\mathcal{R}_n$.

Obviously for any $r \in \mathcal{R}_n$, $g \in An(f)$, we have $r * g \in An(f)$. Thus $An(f)$ is an ideal. Let us prove that $An(f)$ is a principal ideal. For if $h \in An(f)$ and $h \notin < 1+f >$, then $f * h = 0$ implying $h * (1+f) = h$, so $h \in < 1+f >$.

Next we prove the assertion on the cardinality of $An(f)$. Note that the condition $f(x) * g(x) = 0$ implies that

$$f(x) = 1 \Rightarrow g(x) = 0 \;\; \forall x \in \mathbb{F}_2^n.$$

Then at any position $\tau \in \mathbb{F}_2^n$ for which $f(\tau) = 0$, $g(\tau)$ may be selected arbitrary, i.e. there are $2^{2^n - |S_1(f)|}$ possibilities for $g$. Hence $|An(f)| = 2^{2^n - |S_1(f)|}$. In particular if $f$ is balanced then $|An(f)| = 2^{2^{n-1}}$. □

Henceforth we restrict our discussion to balanced functions having much wider cryptographic applications (at least in the case of stream ciphers). For a balanced function $f$ the quotient ring $\mathcal{R}_n/An(f)$ has $2^{2^{n-1}}$ elements. As noticed, there is a strong symmetry between the two different attacks based on the annihilators $f * g = 0$ and the multiples of low degree $f * r = h$. Indeed, the cardinality of nonzero annihilators $\#\{An(f) \setminus 0\} = 2^{2^{n-1}} - 1$ is the same as the number of distinct $h$ when considering $f * r = h$. This is confirmed by noting that any function $r$ in the coset $a + An(f)$ gives $f * r = f * a = h$, and there are $2^{2^{n-1}} - 1$ such cosets for $a \neq 0$. In other words, finding low degree annihilators is equivalent to designing a low degree function $g$ defined on some subset of $S_0(f)$. Similarly, as any $g$ defined on the subset of $S_0(f)$ gives $f * g = 0$, the existence of low degree multiples of the form $f * r = h$ may always be viewed as design of the low degree $h$ on the subset of $S_1(f)$ due to the deccomposition of the form $r = g + h$. We attempt to deduce some properties of the cosets of $An(f)$ regarding its minimum degree.

**Proposition 2** *Let $f \in \mathcal{R}_n$ be a nonaffine balanced function. Then $An(f)$ contains exactly one balanced function, namely the function $1 + f$.*
*In particular, there are no nonzero affine functions in $An(f)$.*

*Proof.* In order that $f * g = 0$ the function $g$ must satisfy $g(x) = 0$ whenever $f(x) = 1$. Since $f$ is balanced $S_0(g) \geq 2^{n-1}$. Then if $g$ is to be balanced it must be that $g = 1 + f$. In particular, since any affine function is balanced and $1 + f$ is nonlinear by assumption, there are no nonzero affine functions in $An(f)$. □

**Corollary 1** *There is exactly one nonzero annihilator of degree one for any affine function $a \in \mathcal{A}_n$ given by $(1 + a)$.*

*Proof.* Since $a$ is affine the only balanced annihilator is of the form $1 + a$ which is an affine function. □

### 4.1 Concatenating Annihilators with Application to Maiorana-McFarland Class

We know that $M = \{1, x_1, \ldots, x_n, x_1x_2, \ldots, x_{n-1}x_n, \ldots, x_1x_2 \cdots x_n\}$, the set of $2^n$ monomials, constitutes the basis of $\mathcal{R}_n$ which we call the *monomial basis.* An alternative basis may be derived by considering all the products of the form $\prod_{i=1}^{n}(x_i + \tau_i + 1)$ when $\tau$ runs through $\mathbb{F}_2^n$. It is clear that any such product $\prod_{i=1}^{n}(x_i+\tau_i+1)$ specifies the function defined to be nonzero exactly for $x = \tau$ and zero otherwise. Hence the set $E_n = \{\prod_{i=1}^{n}(x_i + \tau_i + 1) \mid \tau \in \mathbb{F}_2^n\}$ constitutes the basis of $\mathcal{R}_n$ which will be called *polynomial basis.* In fact distinct basis elements are orthogonal to each other, that is $e * e' = 0$ for $e \neq e' \in E_n$ with exception that for any $e \in E_n$ we have $e * e = e$ which is in accordance to the property that any element in the Boolean ring $\mathcal{R}_n$ is idempotent.

An important application of these ideas is a general result on the set of annihilators.

**Theorem 2.** *Let $f$ be a balanced Boolean function in $\mathcal{R}_n$. In general, for a positive integer $m$, $1 \leq m \leq n-1$, write $f$ as*

$$f(y, x) = \bigoplus_{\tau \in \mathbb{F}_2^{n-m}} \Big(\prod_{i=1}^{n-m} (y_i + \tau_i + 1)\Big) r_\tau(x),$$

*for $(y, x) \in \mathbb{F}_2^{n-m} \times \mathbb{F}_2^m$, and not necessarily distinct functions $r_\tau$ in $\mathcal{R}_m$. Then any annihilator of $f$ can be written in the form,*

$$g(y, x) = \bigoplus_{\tau \in \mathbb{F}_2^{n-m}} \Big(\prod_{i=1}^{n-m} (y_i + \tau_i + 1)\Big) g_\tau(x), \tag{3}$$

*where $g_\tau$ is any annihilator of $r_\tau$.*

*Proof.* Due to the orthogonality of distinct products $\prod_{i=1}^{n-m}(y_i + \tau_i + 1)$ and the fact that $g_\tau$ is annihilator of $r_\tau$ for any $\tau \in \mathbb{F}_2^{n-m}$, it is easily verified that $fg = 0$. By Theorem 1 for a function $r_\tau \in \mathcal{R}_m$ there are $2^{2^m - S_1(r_\tau)}$ distinct annihilators. Let $G = \{g \mid g_\tau \in A(r_\tau), \tau \in \mathbb{F}_2^{n-m}\}$ and denote by $r_0, \ldots, r_{2^{n-m}-1}$ the subfunctions of $f$ when $\tau$ runs through $\mathbb{F}_2^{n-m}$. Then,

$$|G| = 2^{2^m - |S_1(r_0)|} \ldots 2^{2^m - |S_1(r_{2^{n-m}-1})|} = 2^{2^{n-m} \cdot 2^m - \sum_{i=0}^{2^{n-m}-1} |S_1(r_i)|} = 2^{2^{n-1}},$$

which is in accordance with Theorem 1, that is $|G| = |An(f)|$. It is obvious that the functions in $G$ are two-by-two distinct, hence all annihilators are in $G$. □

This approach is a very efficient method for annihilating the functions which have a subfunction of low degree on some $(n-m)$-dimensional flat.

**Example 1** *The functions in the standard Maiorana-McFarland class may be viewed as a concatenation of affine functions from some smaller variable space. That is* $f(y, x) = \bigoplus_{\tau \in \mathbb{F}_2^{n-m}} \left( \prod_{i=1}^{n-m}(y_i + \tau_i + 1) \right) a_\tau(x)$, *where* $a_\tau(x) \in \mathcal{A}_m$ *are affine functions in* $m$ *variables for all* $\tau$. *Then the annihilators of degree* $n-m+1$ *are for instance obtained by choosing* $g_{\tau^c}(x) = 1 + a_{\tau^c}(x)$ *in (3) for a fixed* $\tau^c \in \mathbb{F}_2^{n-m}$ *and otherwise* $g_\tau(x) = 0$. *But the degree of such an annihilator is* $n-m+1$ *which equals to the maximum degree of the Maiorana-McFarland class of functions and therefore not of practical use.* □

The result above is more successfully applied to the degree optimized Maiorana-McFarland class that has been introduced in [18]. Here some affine functions in $\mathcal{A}_m$ (at least one) are replaced by suitably chosen nonlinear function(s) $h_i$ of degree $m - t - 1$, $t$ being the order of resiliency. Then the degree of $f$ is optimized, i.e. $deg(f) = n - t - 1$. Still, multiplying this function by $g(y, x) = (\prod_{i=1}^{n-m}(y_i + \tau_i + 1))(1 + a_\tau(x))$ (for $\tau \in \mathbb{F}_2^{n-m}$ chosen such that $f$ is affine on that $m$-dimensional flat) the degree of $f$ is decreased from $n-t-1$ to $n-m+1$. As $m > n/2$ when $t > 0$ for this class, in many cases one obtains annihilators of degree $< n/2$.

## 5 How to Decide the (Non-) Existence of Annihilators

In this section we derive an efficient algorithm to decide whether a given boolean function $f$ in $n$ variables $x = (x_1, ..., x_n)$ has low algebraic immunity, i.e., whether $f$ or $f+1$ has an an annihilator of low degree. From ([9], proof of Theorem C.0.1) one deduces the following algorithm for determining annihilating functions for $f$, i.e., functions $g$ such that $f(x) * g(x) = 0$ for all $x$:

A necessary and sufficient condition for $f * g = 0$ is that the function $g$ vanishes for all arguments $x$ for which $f(x) = 1$. The algebraic normal form ANF of a function $g$ in $n$ variables of degree $d$ is a sum of a constant and monomials $a_{i_1,i_2,\ldots,i_m} x^{i_1} x^{i_2} \cdots x^{i_m}$, $1 \le m \le d$, determined by its coefficients $a_{i_1,i_2,\ldots,i_m}$, whose number equals to $\sum_{i=0}^{d} \binom{n}{i}$. In some complexity estimates, we approximate this number by the summand $\binom{n}{d}$, which is dominant for $d < n/2$. In order to determine the unknown coefficients of an annihilating function $g$, substitute all arguments $x$ in $g(x)$ with $f(x) = 1$. For balanced $f$ these are $2^{n-1}$ arguments. We thus get $2^{n-1}$ linear equations for the coefficients of $g$, which can be solved by Gaussian elimination. This method immediately allows to decide whether there is an annihilator $g$ of degree at most $d$, and if so, to determine a set of linearly independent annihilators (of degree at most $d$). In view of Theorem C.0.1 in [9] we assume $d \leq \lceil n/2 \rceil$.

**Algorithm 1**

1. Substitute all $N$ arguments $x$ with $f(x) = 1$ in the ANF of a general boolean function $g(x)$ of degree $d$. This gives a system of $N$ linear equations for the coefficients of $g(x)$.
2. Solve this linear system.
3. If there is no (nontrivial) solution, output `no annihilator of degree` $d$, else determine sets of coefficients for linearly independent annihilators.

For $n$ not much larger than about 10, solving this system of linear equations is quite easy. However, in [12] it is recommended that the combining function $f$ in a stream cipher should have more than 10 (e.g. 32) arguments, to prevent algebraic attacks.

For such numbers of inputs, Algorithm 1 becomes infeasible, as the number of equations is on the order of $2^{n-1}$, and the complexity of Gaussian elimination already for $n = 20$ inputs is about $2^{57}$. In [11] there are given two alternative algorithms for determining low degree annihilators and low degree multiples of functions, both of which are based on Gröbner bases. The examples of functions given in [11] have at most $n = 10$ variables. No complexity estimates are given in [11] for determining the necessary Gröbner bases for general $n$, however it seems that these methods become infeasible as well for larger numbers of variables.

Here we propose an accelerated method for deciding whether a Boolean function has an annihilator of low degree $d$. As in Algorithm 1, let the (candidate) annihilators $g$ of degree $d$ of $f$ be described as ANF with unknown coefficients.

We assume that $f$ behaves roughly like a random function, i.e., the coefficients in the ANF of $f$ are roughly chosen at random. If this is not the case, e.g., if the nonzero coefficients are sparse, the algorithm may be adapted to be even more efficient. (However, for cipher design, we do not advocate sparse functions.) Suppose $f$ is (close to) balanced. Then the number of arguments $x$ with weight $w \leq d$ and $f(x) = 1$ is about half the number of coefficients of $g(x)$.

The idea is to exploit some specific structure of the system of equations occurring in Algorithm 1. To see this, start with arguments $x$ with Hamming weight $w = 1$. Suppose the only value 1 in $x$ is at position $i$. Then substituting this $x$ in $g(x) = 0$ gives $a_i + a_0 = 0$. Thus $a_i = a_0$. There are about $n/2$ arguments $x$ of weight 1 with $f(x) = 1$. Assume $d \geq 2$. Consider all arguments $x$ of weight 2 with $f(x) = 1$, and with value 1 in positions $i$ and $j$. Then one gets $a_{ij} + a_i + a_j + a_0 = 0$. Hence $a_{ij}$ for these indices can be expressed by coefficients of monomials of degrees 0 and 1. In general, for any argument $x$ of weight $w$, $1 \leq w \leq d$, the resulting linear equation in the coefficients of $g(x)$ has a similar structure: There is exactly one coefficient of a monomial of degree $w$, (we term this a *coefficient of weight* $w$) which can immediately be expressed by coefficients of lower weight. By iterating this process for increasing weight $w$, until $w = d$, we can eliminate roughly half of the coefficients in $g(x)$ almost for free. We describe a basic version of an algorithm which for low degree $d$ will later be considerably improved.

**Algorithm 2**

1. Let weight $w = 1$.
2. For all $x$ of weight $w$ with $f(x) = 1$ substitute $x$ in $g(x) = 0$ to derive a linear equation in the coefficients of $g$, with a single coefficient of weight $w$. Use this equation to express this coefficient iteratively by coefficients of lower weight.
3. If $w < d$, increment $w$ by 1 and go to step 2.
4. Choose random arguments $x$ of arbitrary weight such that $f(x) = 1$ and substitute in $g(x) = 0$, until there are the same number of equations as unknowns.
5. Solve the linear system. If there is no solution, output `no annihilator of degree` $d$.

Algorithm 2 is aimed at showing that $f$ has *no* annihilator of given degree $d$. However, if the system turns out to be solvable, one may try another set of arguments $x$ in step 5. If the new system is again solvable, one checks whether the solutions found are consistent. In case the number of variables $n$ of $f$ is not too large, one may directly verify whether one has found an annihilator, by formally expanding $f(x)*g(x)$ and by checking whether the result is identically 0.

We estimate the computational and data complexity of Algorithm 2. The expressions of those coefficients that in step 2 have been replaced by linear combinations of coefficients of lower weight, need to be memorized for step 4. As the number of coefficients involved in these expressions is of order $\frac{1}{2}\binom{n}{d-1}$, and we have a number of $\frac{1}{2}\binom{n}{d}$ memorized coefficients in step 2, the number of memory bits is of order $M = \frac{1}{4}\binom{n}{d} \cdot \binom{n}{d-1}$. In the evaluation of $g(x)$ in step 4, one has to substitute the linear expressions found in step 2. The complexity of substituting $x$ depends on its weight, and is at most of order $M$ elementary operations. This needs to be done for about $\frac{1}{2}\binom{n}{d}$ values of $x$, as we have about this number of remaining unknowns. Hence we get a computational complexity in step 4 of order $\frac{1}{2}\binom{n}{d} * M = \frac{1}{8}\binom{n}{d}^2 \cdot \binom{n}{d-1}$. The computational complexity of step 5, and hence of Algorithm 2, is of order $\frac{1}{8}\binom{n}{d}^3$, if the exponent for Gaussian elimination $\omega = 3$. Thus Algorithm 2 does run roughly 8 times faster than Algorithm 1, when modified for low degree $d$ (i.e., by taking a number of linear equations equal to the number of unknown coefficients in $g(x)$). To summarize, Algorithm 2 has the complexities as shown:

| Memory | $\frac{1}{4}\binom{n}{d} \cdot \binom{n}{d-1}$ |
|---|---|
| Complexity | $\frac{1}{8}\binom{n}{d}^3$ |

Note that the memory requirement is not stringent when compared to Algorithm 1, where a linear system of equations with about $\binom{n}{d}^2$ coefficients needs to be memorized.

In order to improve efficiency over Algorithm 2, we use arguments $x$ of higher weight than $d$: Consider all arguments $x$ with weight $d+1$ such that $f(x) = 1$. For each such $x$, a linear equation arises where $\binom{d+1}{d} = d+1$ coefficients of weight $d$ (and coefficients of lower weight) are involved. In some fraction of arguments $x$, exactly $d$ coefficients of weight $d$ were already expressed by coefficients of lower weight. Thus the remaining coefficient can be expressed as well by coefficients of lower weight. This procedure can be iterated for $w = d+2$, and so on, with higher number of coefficients of weight $d$ involved, but with higher probability that a coefficient has already been replaced in an earlier step. The gain of efficiency for increasing weight is dependent on $n$ and $d$. The necessary estimates are given in a Lemma.

**Lemma 2.** *Let $f$ be a random Boolean function with $n$ variables, and let $d$ be the degree of an annihilator $g$ of $f$. Then the following statements hold:*

*a) A fraction*

$$p = \frac{1}{2} + (n-d) \cdot 2^{-(d+2)} \tag{4}$$

*of weight $d$ coefficients can be replaced by lower weight coefficients by substituting all weight $w$ arguments $x$ with $f(x) = 1$, and with $w \leq d+1$.*

*b) Suppose that according to a) a fraction $p$ of coefficients of weight $d$ have been replaced. Then an additional number $A$ of coefficients can be replaced by substituting arguments of weight $w = d+2$, where*

$$A = \frac{1}{2}\binom{n}{d+2} \cdot \binom{d+2}{2}(1-p)p^{\binom{d+2}{2}-1} \tag{5}$$

*Proof.* a): By following steps 1 to 3 of Algorithm 2, about $\frac{1}{2}\binom{n}{d}$ coefficients of weight $d$ have already been replaced by lower weight coefficients. There are about $\frac{1}{2}\binom{n}{d+1}$ arguments $x$ of weight $w = d+1$ with $f(x) = 1$. Substitute these in $g(x)$. Then in the average, for $\frac{1}{2}\binom{n}{d+1} * (d+1) * 2^{-(d+1)}$ of arguments, we have that amongst the $d+1$ weight $d$ coefficients involved, exactly $d$ coefficients have previously been expressed by coefficients of lower weight. Thus the remaining coefficient can be expressed by coefficients of lower weight. The average fraction of coefficients of weight $d$ replaced by now is got by dividing by $\binom{n}{d}$ and is as claimed.

b) is similar, and is omitted. □

The improved algorithm is illustrated for degrees $d = 4$ and $d = 5$.

**Case** $d = 4$**:** Let the number of variables of $f$ be $n \geq 20$. Search for potential annihilators of degree $d = 4$. First assume $n = 20$. Formula (4) shows that by using all arguments of weight up to $w \leq 5$, a fraction $p = 0.75$ of the $\binom{20}{4}$

coefficients of weight 4 can be replaced. Thus with $n = 20$, there remain 1211 coefficients to be replaced. According to Formula (5), an average number $A$ of new coefficients of weight $d$ can be replaced by using arguments of weight $d+2$. With $n = 20$, $d = 4$, and $p = 0.75$, one gets 1294. Thus with high probability (almost) all coefficients of weight $d = 4$ can be replaced. Using formulas (4) and (5) one can show that this probability quickly increases for increasing $n$. Hence the number of remaining unknowns (and equations) is of order $\frac{1}{2}\binom{n}{d-1}$. Thus we are able to reduce deciding the existence of annihilators of degree at most 4 from $\binom{n}{d}^3$, when using Algorithm 1, to $\frac{1}{8}\binom{n}{d-1}^3$, when using our refinement of Algorithm 2.

If $n = 32$, i.e., one of our target values, this complexity is about $\frac{1}{8}\binom{32}{3}^3 \approx 2^{34}$, compared to about $\binom{32}{4}^3 \approx 2^{45}$, when Algorithm 1 (modified to $d = 4$) would be directly applied.

Recall that the final system of linear equations to be solved, is found by substituting linear relations for coefficients of $g(x)$, for various arguments $x$. This should be done in a way such that it doesn't exceed the cost for solving this system. To get a linear system of largest possible rank, one should take arguments with arbitrary weight, so that all monomials in $f$ contribute to the evaluation of $f$. A majority of arguments $x$ have weight about $n/2$. Hence only about $\binom{n/2}{d}$ monomials in $g(x)$ are nonzero. Thus in this case the complexity of substituting linear expressions in $g(x)$ to get a linear equation in unknowns has complexity about $\binom{n}{d-1} \cdot \binom{n/2}{d}$. Doing this for $\binom{n}{d-1}$ equations, for values $n$ and $d$ under consideration, the average complexity is not larger than $\binom{n}{d-1}^3$. When taking arguments with weight close to $n$, one better computes the linear equation got from the weight $n$ argument $x$, and then modifies this equation by setting some components in $x$ to 0.

**Case $d = 5$:** Let $n \geq 32$. Assume $n = 32$, (the case $n > 32$ works even better). Then according to formula (4), $p = 0.7109375$. The number of coefficients of weight $d = 5$ after using all arguments of weight up to $d + 1 = 6$ is 58210. After using weight $d + 2 = 7$ arguments, we can replace another 22229 coefficients of weight 5. Hence there remain 35981, which is of the same order as $\binom{32}{4} = 35960$. As half of coefficients of weight at most 4 have already been replaced by basic step 4 of Algorithm 2, and as the case $n > 32$ is more favorable, we conclude that the remaining number of unknowns is of order $\binom{n}{4}$. Hence the complexity of deciding existence of an annihilator of degree at most 5 is of order $\binom{n}{4}^3$, e.g., for $n = 32$, it is of order $2^{45}$ (compared to $2^{53}$, when modified Algorithm 1 would be directly applied).

The cases $d < 4$ work similar as the cases $d = 4$ and $d = 5$ just given. However, for $d = 6$, and $n < 50$, formula (4) shows that the probability $p$ is already close to 0.5, so that in this case by using arguments with weight larger than 6 only weak refinements over the basic Algorithm 2 may be expected.

## 6 Bounds on the Probability of Annihilators' Existence

In the last section we have proposed an algorithm for deciding whether a given function $f$ admits annihilators of degree $\leq d$. However the complexity of the algorithm is strongly related with the inputs $n, d$ and it turns out that this task becomes infeasible for $n \geq 32$ and $d \geq 6$. Hence using more inputs to the function might be an obvious solution to protect from algebraic attacks. It cannot be precluded however that finding annihilators for larger $n$ and $d$ may still be feasible by using methods related to Gröbner basis, although this seems open. In such a setting it is important to derive bounds on the probability that a function admits annihilators.

An easy upper bound for the probability that an $n$-variable balanced function admits an annihilator of degree at most $d$, is deduced from the minimum weight of any nonzero function of degree less or equal to $d$. As $f$ is assumed to be balanced, this extends to a statement on the algebraic immunity of $f$:

**Proposition 3** *The probability that a random $n$-variable balanced function $f$ has algebraic immunity at most $d$ is upper bounded by the number:*

$$Pb\{AI(f) \leq d\} \leq \frac{2(2^{1+n+\cdots+\binom{n}{d}} - 1)\binom{2^n-2^{n-d}}{2^{n-1}-2^{n-d}}}{\binom{2^n}{2^{n-1}}}. \tag{6}$$

*Proof.* The size of the set $A$ of nonzero functions of degrees at most $d$ equals $2^{1+n+\cdots+\binom{n}{d}} - 1$. For every such function $g$, the number of balanced functions $f$ such that the support of $g$ is included in $S_0(f)$ equals $N_g = \binom{2^n-wt(g)}{2^{n-1}-wt(g)}$, where $wt(g)$ denotes the Hamming weight of $g$. Since every such function $g$ has weight at least $2^{n-d}$, we have $\binom{2^n-wt(g)}{2^{n-1}-wt(g)} \leq \binom{2^n-2^{n-d}}{2^{n-1}-2^{n-d}}$. Thus, the number of balanced functions admitting an annihilator of degree at most $d$ is smaller than or equal to $\sum_{g\in A} N_g \leq (2^{1+n+\cdots+\binom{n}{d}} - 1)\binom{2^n-2^{n-d}}{2^{n-1}-2^{n-d}}$; indeed, the size of a union of sets is smaller than or equal to the sum of the sizes of the sets. Since $\binom{2^n}{2^{n-1}}$ is the number of balanced functions, this completes the proof. □

Even though this bound is not tight, it helps us to determine the asymptotic behavior of the probability of annihilator's existence.

**Theorem 3.** *Let $d_n$ be a sequence of positive integers such that $d_n \leq \mu n$ where $\mu = \frac{1}{2}(1 + \frac{\ln 2}{2} - \sqrt{(1 + \frac{\ln 2}{2})^2 - 1}) \approx 0.22$. Then*

$$Pb\{AI(f) \leq d_n\} \to 0, \; n \to \infty. \tag{7}$$

*Proof.* We know that, for every positive integer $N$ and every $0 < \lambda < 1/2$:

$$\sum_{0 \leq i \leq \lambda N} \binom{N}{i} \leq 2^N e^{-2N(1/2-\lambda)^2},$$

(e.g., see C. Carlet [5]). We deduce that for every $n$ and every $d_n < n/2$:

$$1+n+\cdots+\binom{n}{d_n} \leq 2^n e^{-2n(1/2-d_n/n)^2},$$

and denoting the number $\frac{1/2-2^{-d_n}}{1-2^{-d_n}}$ by $\lambda_n$ we have:

$$\binom{2^n-2^{n-d_n}}{2^{n-1}-2^{n-d_n}} \leq 2^{2^n-2^{n-d_n}} e^{-2(2^n-2^{n-d_n})(1/2-\lambda_n)^2}.$$

Thus

$$(2^{1+n+\cdots+\binom{n}{d_n}}-1)\binom{2^n-2^{n-d_n}}{2^{n-1}-2^{n-d_n}} \leq$$

$$2^{2^n e^{-2n(1/2-d_n/n)^2}+2^n-2^{n-d_n}} e^{-2(2^n-2^{n-d_n})(1/2-\lambda_n)^2},$$

and therefore

$$\log_2\left[(2^{1+n+\cdots+\binom{n}{d_n}}-1)\binom{2^n-2^{n-d_n}}{2^{n-1}-2^{n-d_n}}\right] \leq$$

$$2^n e^{-2n(1/2-d_n/n)^2}+2^n-2^{n-d_n}-2(\log_2 e)(2^n-2^{n-d_n})(1/2-\lambda_n)^2.$$

We have also $\binom{2^n}{2^{n-1}} \sim k2^{2^n-n/2}$, where $k$ is a constant, according to Stirling formula. Hence, if $n/2$ is negligible with respect to

$$2^{n-d_n}-2^n e^{-2n(1/2-d_n/n)^2}+2(\log_2 e)(2^n-2^{n-d_n})(1/2-\lambda_n)^2 =$$

$$2^n\left[2^{-d_n}-e^{-2n(1/2-d_n/n)^2}+2(\log_2 e)(1-2^{-d_n})(1/2-\lambda_n)^2\right]$$

then $\frac{(2^{1+n+\cdots+\binom{n}{d_n}}-1)\binom{2^n-2^{n-d_n}}{2^{n-1}-2^{n-d_n}}}{\binom{2^n}{2^{n-1}}}$ tends to zero.

A sufficient condition is that $2^{-d_n} \geq e^{-2n(1/2-d_n/n)^2}$ and that $n/2$ is negligible with respect to $2^n\left[2(\log_2 e)(1-2^{-d_n})(1/2-\lambda_n)^2\right]$. We have

$$2^{-d_n} \geq e^{-2n(1/2-d_n/n)^2} \Leftrightarrow d_n \leq 2n(\log_2 e)(1/2-d_n/n)^2,$$

that is,

$$d_n/n \leq 2(\log_2 e)(1/2-d_n/n)^2.$$

The equation $x = 2(\log_2 e)(1/2-x)^2$ is equivalent to $x\ln 2/2 = (1/2-x)^2$, that is, $x^2 - x(1+\frac{\ln 2}{2})+\frac{1}{4} = 0$, which roots are both positive.

Its smallest root is $\mu$. Thus $d_n \leq \mu n$ implies $2^{-d_n} \geq e^{-2n(1/2-d_n/n)^2}$. If $d_n \leq \mu n$, then

$$(1-2^{-d_n})(\frac{1}{2}-\lambda_n)^2 = (1-2^{-d_n})\left(\frac{1}{2}-\frac{1/2-2^{-d_n}}{1-2^{-d_n}}\right)^2 = \frac{2^{-2d_n}}{4(1-2^{-d_n})} \geq \frac{2^{-2\mu n}}{4(1-2^{-\mu n})}.$$

Hence, since $2\mu$ is strictly smaller than 1, then $n/2$ is negligible with respect to

$$2^n\left[2(\log_2 e)(1-2^{-d_n})(1/2-\lambda_n)^2\right].$$

$\square$

For practical applications we are interested in concrete values of this bound for moderate $n$ rather than the asymptotical values. For instance, we can compute the probability that a random balanced function $f$ in $n = 32$ variables admits annihilators of degree $d \leq 6$. In view of Theorem 3, $d = 6$ satisfies the inequality $d \leq 0.22n$ for $n = 32$. Then computing (6) for $n = 32$, $d = 6$, gives a probability of order $10^{-300}$ which is negligibly small. Notice that in this case, due to the complexity reasons, we cannot confirm the (non)existence of annihilators through Algorithm 2.

However the upper bound as derived above is based on the property that all annihilators have weights at least $2^{n-d}$. This bound can be sharpened by using some known results on the weight distribution and enumeration of the codewords in the Reed-Muller code $\mathcal{R}(d,n)$. Let us denote by $A_w$ the number of codewords of weight $w$ in $\mathcal{R}(d,n)$; then $A_{2^{n-d}}$ equals $2^d \prod_{i=0}^{n-d-1} \frac{2^{n-i}-1}{2^{n-d-i}-1}$ due to McWilliams-Sloane [16]. Furthermore, Kasami and Tokura [14] have done the weight enumeration of codewords of weight $w$ in $\mathcal{R}(d,n)$ for all $2^{n-d} < w < 2^{n-d+1}$. These results are found in [16, pg. 446] and can be used to derive a tighter upper bound from the following easy improvement of Proposition 3:

**Theorem 4.** *For a random balanced function $f \in \mathcal{B}_n$ the upper bound on the probability, denoted $Pb^d$, that $AI(f) \leq d$ is given by*

$$Pb^d \leq \sum_{w=2^{n-d}}^{w<2^{n-d+1}} A_w \cdot \frac{\binom{2^n-w}{2^{n-1}-w}}{\binom{2^n}{2^{n-1}}} + \left(2^{\sum_{i=0}^{d}\binom{n}{i}} - \sum_{w=2^{n-d}}^{w<2^{n-d+1}} A_w\right) \cdot \frac{\binom{2^n-2^{n-d}}{2^{n-1}-2^{n-d}}}{\binom{2^n}{2^{n-1}}}. \quad (8)$$

Note that, for every $w$, we have $\frac{\binom{2^n-w}{2^{n-1}-w}}{\binom{2^n}{2^{n-1}}} \leq \left(\frac{1}{2}\right)^w$.

**Remark 2** *This upper bound can be further tightened by using more values of $w$ for which the exact number of codewords is known. This has been done in [15] for the weights $w$ in the range $2^{n-d} \leq w < 2.5 \cdot 2^{n-d}$.*

For the bound of Theorem 4, it seems to be much harder to estimate the value of $\mu$ as it has been done in Theorem 3. By computations one can deduce the same behavior of this bound but with slightly shifted limit value, that is $\mu' \approx 0.27$. This gives a better value than Theorem 3 as for increasing $n$ the sequence $d_n \leq \mu' n$ has a larger range.

The upper bounds above are important tools for estimating the security of a stream cipher. For instance assuming that the computational complexity of breaking a cipher whose multiples are of degrees strictly greater than say $d = 5$, then Theorem 4 gives $n = 18$ which is the lowest value of $n$ such that the probability that there exists annihilators of degree $d \leq 5$ is close to zero. Hence assuming that $f$ has no particular structure that might be exploited, the value of $n = 18$ and the key length of $k = 128$ should guarantee that the known attacks are infeasible. Assuming the existence of multiples/annihilators of degree $d = 6$ this would give a computational complexity of order $\approx \binom{128}{6}^{\omega} = (2^{32})^{\omega}$, which for $\omega = 3$ yields $2^{96}$. If a more secure cipher is preferred then the obvious method is to increase $n$. In Table 1 below we list some other interesting cases. Each entry relates a given degree of annihilators $d$ to the minimum value of $n$ for which $Pb\{AI(f) \leq d\} \approx 0$. We apply the results above to the stream cipher LILI-128,

| $n; P_b$ | $d = 5$ | $d = 6$ | $d = 7$ | $d = 8$ |
|---|---|---|---|---|
| | $18; 10^{-1134}$ | $22; 10^{-6326}$ | $26; 10^{-23138}$ | $31; 10^{-10^7}$ |

**Table 1.** Upper bound on the probability for the annihilators

for which 14 linearly independent annihilators of degree $d = 4$ have been found in [8].

**Example 2** *In [8], Courtois and Meier (see also [11]) have investigated the algebraic properties of LILI-128. They have found that the function $f$ in $n =$ 10 variables used in LILI-128 is rather weak, since one could find* 14 *linearly independent annihilators of degree* 4.

*Note that the probability $Pb\{AI(f) \leq 5\}$ is equal to* 1 *due to the Theorem 6.0.1 in [8]. The upper bound is not tight for* $d = 4, 5$ *giving a probability greater than* 1. *However, applying the upper bound for the case* $d = 3$ *one deduces that*

$$Pb\{AI(f) \leq 3\} \leq 0.30 \cdot 10^{-24}.$$

□

Example 2 shows that the upper bound in particular for low values of $n$ is not tight. However, Table 1 illustrates that this bound gives very strong estimates for larger $n$ of interest.

**Acknowledgment** We are indebted to Jean-Pierre Tillich for hepful discussions.

## References

[1] F. Armknecht and Krause M. Algebraic attacks on stream combiners with memory. In *Advances in Cryptology—CRYPTO 2003*, volume LNCS 2729, pages 162–176. Springer-Verlag, 2003.

[2] P. Camion, C. Carlet, P. Charpin, and N. Sendrier. On correlation-immune functions. In *Advances in Cryptology—EUROCRYPT'91*, volume LNCS 547, pages 86–100. Springer-Verlag, 1991.

[3] A. Canteaut and M. Trabbia. Improved fast correlation attacks using parity-check equations of weight 4 and 5. In *Advances in Cryptology—EUROCRYPT 2000*, volume LNCS 1807, pages 573–588. Springer-Verlag, 2000.

[4] C. Carlet. A larger class of cryptographic Boolean functions via a study of the Maiorana-McFarland constructions. In *Advances in Cryptology—CRYPTO 2002*, volume LNCS 2442, pages 549–564. Springer-Verlag, 2002.

[5] C. Carlet. On the algebraic thickness and non-normality of Boolean functions. In *Proceedings of 2003 IEEE Information Theory Workshop, Paris, France*, 2003.

[6] N. Courtois. Fast algebraic attacks on stream ciphers with linear feedback. In *Advances in Cryptology—CRYPTO 2003*, volume LNCS 2729, pages 176–194. Springer-Verlag, 2003.

[7] N. Courtois and Pieprzyk J. Cryptanalysis of block ciphers with overdefined systems of equations. In *Advances in Cryptology—ASIACRYPT 2002*, volume LNCS 2501. Springer-Verlag, 2002.

[8] N. Courtois and W. Meier. Algebraic attacks on stream ciphers with linear feedback. In *Advances in Cryptology—EUROCRYPT 2003*, volume LNCS 2656, pages 346–359. Springer-Verlag, 2003.

[9] N. Courtois and W. Meier. Algebraic attacks on stream ciphers with linear feedback. Extended version of [8], available at http://www.cryptosystem.net/stream/, 2003.

[10] J. F. Dillon. Elementary Haddamard Difference Sets. Ph. D. thesis, University of Maryland, U.S.A., 1974.

[11] J.-Ch. Faugère and G. Ars. An algebraic cryptanalysis of nonlinear filter generators using Gröbner bases. Available on the web, 2003. http://www.inria.fr/rrrt/rr-4739.html.

[12] J. Dj. Golić. On the security of nonlinear filter generators. In *Fast Software Encryption'96*, volume LNCS 1039, pages 173–188. Springer-Verlag, 1996.

[13] T. Johansson and F. Jönsson. Fast correlation attacks through reconstruction of linear polynomials. In *Advances in Cryptology—CRYPTO 2000*, volume LNCS 1880, pages 300–315. Springer-Verlag, 2000.

[14] T. Kasami and N. Tokura. On the weight structure of Reed-Muller codes. *IEEE Trans. on Inform. Theory*, IT-16(6):pages 752–759, 1970.

[15] T. Kasami, N. Tokura, and S. Asumi. On the weight enumeration of weights less than 2.5$d$ of Reed-Muller codes. *Information and Control*, vol. 30(4):pages 380–395, 1974.

[16] F. J. MacWilliams and N. J. A. Sloane. *The Theory of Error-Correcting Codes.* North-Holland, Amsterdam, 1977.

[17] W. Meier and O. Staffelbach. Fast correlation attacks on stream ciphers. In *Advances in Cryptology—EUROCRYPT'88*, volume LNCS 330, pages 301–314. Springer-Verlag, 1988.

[18] E. Pasalic. Degree optimized resilient Boolean functions from Maiorana-McFarland class. In *9-th IMA Conference on Cryptography and Coding*, 2003.

[19] T. Siegenthaler. Correlation-immunity of nonlinear combining functions for cryptographic applications. *IEEE Trans. on Inform. Theory*, IT-30:pages 776–780, 1984.

# Finding Small Roots of Bivariate Integer Polynomial Equations Revisited

Jean-Sébastien Coron

Gemplus Card International
34 rue Guynemer, 92447 Issy-les-Moulineaux, France
jean-sebastien.coron@gemplus.com

**Abstract.** At Eurocrypt '96, Coppersmith proposed an algorithm for finding small roots of bivariate integer polynomial equations, based on lattice reduction techniques. But the approach is difficult to understand. In this paper, we present a much simpler algorithm for solving the same problem. Our simplification is analogous to the simplification brought by Howgrave-Graham to Coppersmith's algorithm for finding small roots of univariate modular polynomial equations. As an application, we illustrate the new algorithm with the problem of finding the factors of $n = pq$ if we are given the high order $1/4 \log_2 n$ bits of $p$.

## 1 Introduction

An important application of lattice reduction found by Coppersmith in 1996 is finding small roots of low-degree polynomial equations [3,4,5]. This includes modular univariate polynomial equations, and bivariate integer equations.

The problem of solving univariate polynomial equations modulo an integer $N$ of unknown factorization seems to be hard, as for some polynomials it is equivalent to the knowledge of the factorization of $N$. Moreover, the problem of inverting RSA, *i.e.* extracting $e$-th root modulo $N$, is a particular case of this problem. However, at Eurocrypt '96, Coppersmith showed that the problem of finding *small* roots is easy [3,5], using the LLL lattice reduction algorithm [9]:

**Theorem 1 (Coppersmith).** *Given a monic polynomial $P(x)$ of degree $\delta$, modulo an integer $N$ of unknown factorization, one can find in time polynomial in $(\log N, 2^\delta)$ all integers $x_0$ such that $P(x_0) = 0 \mod N$ and $|x_0| \leq N^{1/\delta}$.*

The algorithm can be extended to handle multivariate modular polynomial equations, but the extension is heuristic only. Coppersmith's algorithm has many applications in cryptography: cryptanalysis of RSA with small public exponent when some part of the message is known [5], cryptanalysis of RSA with private exponent $d$ smaller than $N^{0.29}$ [1], polynomial-time factorization of $N = p^r q$ for large $r$ [2], and even an improved security proof for OAEP with small public exponent [13] (see [12] for a nice survey).

Coppersmith's algorithm for solving univariate modular polynomial equations was further simplified by Howgrave-Graham in [7]. Apart from being simpler

C. Cachin and J. Camenisch (Eds.): EUROCRYPT 2004, LNCS 3027, pp. 492–505, 2004.

to understand and implement, a significant advantage of Howgrave-Graham's approach is the heuristic extension to multivariate modular polynomial: indeed, depending on the shape of the polynomial, there is much flexibility in selecting the parameters of the algorithm, and Howgrave-Graham's approach enables to easily derive the corresponding bound for the roots. This approach is actually used in all previously cited variants of Coppersmith's technique [1,2].

Similarly, the problem of solving bivariate integer polynomial equations seems to be hard. Letting $p(x, y)$ be a polynomial in two variables with integer coefficients,

$$p(x, y) = \sum_{i,j} p_{i,j} \cdot x^i y^j$$

it consists in finding all integer pairs $(x_0, y_0)$ such that $p(x_0, y_0) = 0$. We see that integer factorization is a special case as one can take $p(x, y) = N - x \cdot y$. However, at Eurocrypt '96, Coppersmith showed [4,5] that using LLL, the problem of finding *small* roots of bivariate polynomial equations is easy:

**Theorem 2.** *Let $p(x, y)$ be an irreducible polynomial in two variables over $\mathbb{Z}$, of maximum degree $\delta$ in each variable separately. Let $X$ and $Y$ be upper bounds on the desired integer solution $(x_0, y_0)$, and let $W = \max_{i,j} |p_{ij}| X^i Y^j$. If $XY < W^{2/(3\delta)}$, then in time polynomial in $(\log W, 2^\delta)$, one can find all integer pairs $(x_0, y_0)$ such that $p(x_0, y_0) = 0$, $|x_0| \leq X$, and $|y_0| \leq Y$.*

Moreover, there can be improved bounds depending on the shape of the polynomial $p(x, y)$. For example, for a polynomial $p(x, y)$ of total degree $\delta$ in $x$ and $y$, the bound is $XY < W^{1/\delta}$. As for the univariate modular case, the technique can be heuristically extended to more than two variables. An application of Coppersmith's algorithm for the bivariate integer polynomial case is to factor in polynomial-time an RSA-modulus $n = pq$ such that half of the least significant or most significants bits of $p$ are known [5].

However, as noted in [6], the approach for the bivariate integer case is rather difficult to understand. This means that the algorithm is difficult to implement in practice, and that improved bounds depending on the shape of the polynomial are more difficult to derive. In particular, what makes the analysis harder is that one has to derive the determinant of lattices which are not full rank. The particular case of factoring $n = pq$ when half of the least significant or most significants bits of $p$ are known, was further simplified by Howgrave-Graham in [7], but as noted in [6], this particular simplification does not seem to extend to the general case of bivariate polynomial equations. As suggested in [12], a simplification analogue to what has been obtained by Howgrave-Graham for the univariate modular case would be useful.

In this paper, we present a simple and efficient algorithm for finding small roots of bivariate integer polynomials. Our simplification is analogous to the simplification obtained by Howgrave-Graham for the univariate modular case. We apply lattice reduction to a full rank lattice that admits a natural triangular basis. It is then straightforward to derive the determinant and improved bounds depending on the shape of the polynomial; the heuristic extension to more than two variables is also simpler. However, our algorithm is slightly less efficient than Coppersmith's algorithm, because our algorithm has a polynomial-time complexity only if $XY < W^{2/3\delta-\varepsilon}$ for any fixed $\varepsilon > 0$, whereas Coppersmith's algorithm requires $XY < W^{2/3\delta}$, a slightly weaker condition. In section 7, we illustrate our algorithm with the problem of finding the factors of $n = pq$ if we are given the high order $1/4 \log_2 n$ bits of $p$, and show that our algorithm is rather efficient in practice.

## 2 Solving Bivariate Integer Equations: an Illustration

In this section, we first illustrate our technique with a bivariate integer polynomial of the form

$$p(x, y) = a + bx + cy + dxy,$$

with $a \neq 0$ and $d \neq 0$. We assume that $p(x, y)$ is irreducible and has a small root $(x_0, y_0)$. Our goal is to recover $(x_0, y_0)$. As in theorem 2, we let $X, Y$ be some bound on $x_0, y_0$, that is we have $|x_0| \leq X$ and $|y_0| \leq Y$, and let $W = \max\{|a|, |b|X, |c|Y, |d|XY\}$. Moreover, given a polynomial $h(x, y) = \sum_{i,j} h_{ij} x^i y^j$, we define $\|h(x, y)\|^2 := \sum_{i,j} |h_{ij}|^2$ and $\|h(x, y)\|_\infty := \max_{i,j} |h_{ij}|$. Note that we have:

$$W = \|p(xX, yY)\|_\infty \tag{1}$$

First, we generate an integer $n$ such that :

$$W \leq n < 2 \cdot W \tag{2}$$

and $\gcd(n, a) = 1$. One can take $n = W + ((1 - W) \mod |a|)$. Then we define the polynomial:

$$\begin{aligned} q_{00}(x, y) &= a^{-1} p(x, y) \mod n \\ &= 1 + b'x + c'y + d'xy \end{aligned}$$

We also consider the polynomials $q_{10}(x, y) = nx$, $q_{01}(x, y) = ny$ and $q_{11}(x, y) = nxy$. Note that for all four polynomials $q_{ij}(x, y)$, we have that $q_{ij}(x_0, y_0) = 0 \mod n$.

We consider the four polynomials $\tilde{q}_{ij}(x, y) = q_{ij}(xX, yY)$; we are interested in finding a small linear integer combination of the polynomials $\tilde{q}_{ij}(x, y)$. Therefore, we consider the lattice generated by all linear integer combinations of the coefficient vectors of the $\tilde{q}_{ij}(x, y)$. A basis of the lattice is given by the following matrix $L$ of row vectors:

$$L = \begin{bmatrix} 1 & b'X & c'Y & d'XY \\ & nX & & \\ & & nY & \\ & & & nXY \end{bmatrix}$$

We know that the LLL algorithm [9], given a lattice spanned by $(u_1, \ldots, u_\omega)$, finds in polynomial time a lattice vector $b_1$ such that $\|b_1\| \le 2^{(\omega-1)/4} \det(L)^{1/\omega}$. More background on lattice reduction techniques will be given in the next section. With $\omega = 4$ and $\det L = n^3 (XY)^2$ we obtain in polynomial time a non-zero polynomial $h(x, y)$ such that

$$\|h(xX, yY)\| \le 2 \cdot n^{3/4} (XY)^{1/2} \tag{3}$$

Note that we have $h(x_0, y_0) = 0 \mod n$. The following lemma, due to Howgrave-Graham, shows that if the coefficients of $h(x, y)$ are sufficiently small, then the equality $h(x_0, y_0) = 0$ holds not only modulo $n$, but also over $\mathbb{Z}$.

**Lemma 1 (Howgrave-Graham).** *Let $h(x, y) \in \mathbb{Z}[x, y]$ which is a sum of at most $\omega$ monomials. Suppose that $h(x_0, y_0) = 0 \mod n$ where $|x_0| \le X$ and $|y_0| \le Y$ and $\|h(xX, yY)\| < n/\sqrt{\omega}$. Then $h(x_0, y_0) = 0$ holds over the integers.*

*Proof.* We have:

$$\begin{aligned} |h(x_0, y_0)| &= \left|\sum h_{ij} x_0^i y_0^i\right| = \left|\sum h_{ij} X^i Y^j \left(\frac{x_0}{X}\right)^i \left(\frac{y_0}{Y}\right)^j\right| \\ &\le \sum \left|h_{ij} X^i Y^j \left(\frac{x_0}{X}\right)^i \left(\frac{y_0}{Y}\right)^j\right| \le \sum |h_{ij} X^i Y^j| \\ &\le \sqrt{\omega} \|h(xX, yY)\| < n \end{aligned}$$

Since $h(x_0, y_0) = 0 \mod n$, this gives $h(x_0, y_0) = 0$. □

Assume now that:

$$XY < n^{1/2}/16 \tag{4}$$

Then inequality (3) gives:

$$\|h(xX, yY)\| < n/2 \tag{5}$$

which implies that $h(x_0, y_0) = 0$. Moreover, from (1), (2) and (5) we get:

$$\|h(xX, yY)\| < n/2 < W \le \|p(xX, yY)\|_\infty \le \|p(xX, yY)\|$$

This shows that $h(x, y)$ cannot be a multiple of $p(x, y)$. Namely, if $h(x, y)$ is a multiple of $p(x, y)$, then it follows from the definition of $p$ and $h$ that we must have $h(x, y) = \lambda \cdot p(x, y)$ with $\lambda \in \mathbb{Z}^*$. This would give $\|h(xX, yY)\| = |\lambda| \cdot \|p(xX, yY)\| \ge \|p(xX, yY)\|$, a contradiction.

Eventually, since $p(x, y)$ is irreducible and $h(x, y)$ is not a multiple of $p(x, y)$,

$$Q(x) = \text{Resultant}_y(h(x, y), p(x, y))$$

gives a non-zero integer polynomial such that $Q(x_0) = 0$. Using any standard root-finding algorithm, we can recover $x_0$, and finally $y_0$ by solving $p(x_0, y) = 0$. Using inequality (4) and $n \ge W$, this shows that if :

$$XY < \frac{W^{1/2}}{16}$$

one can find in time polynomial in $\log W$ all integer pairs $(x_0, y_0)$ such that $p(x_0, y_0) = 0$, $|x_0| \le X$, and $|y_0| \le Y$.

This bound is weaker than the bound $XY < W^{2/3}$ given by theorem 2 for $\delta = 1$. We will see in section 4 that by adding more multiples of $p(x, y)$ into the lattice, we recover the desired bound.

# 3 Background on Lattices and Polynomials

## 3.1 The LLL Algorithm

Let $u_1, \dots, u_\omega \in \mathbb{Z}^n$ be linearly independent vectors with $\omega \le n$. The lattice $L$ spanned by $< u_1, \dots, u_\omega >$ consists of all integral linear combinations of $u_1, \dots, u_\omega$, that is:

$$L = \{ \sum_{i=1}^{\omega} n_i \cdot u_i | \ n_i \in \mathbb{Z} \}$$

Such a set of vectors $u_i$'s is called a lattice *basis*. All the bases have the same number of elements, called the *dimension* or *rank* of the lattice. We say that the lattice is full rank if $\omega = n$. Any two bases of the same lattice $L$ are related by some integral matrix of determinant $\pm 1$. Therefore, all the bases have the same Gramian determinant $\det_{1 \le i,j \le d} < u_i, u_j >$. One defines the *determinant* of the lattice as the square root of the Gramian determinant. If the lattice is full rank, then the determinant of $L$ is equal to the absolute value of the determinant of the $\omega \times \omega$ matrix whose rows are the basis vectors $u_1, \dots, u_\omega$.

The LLL algorithm [9] computes a short vector in a lattice :

**Theorem 3 (LLL).** *Let $L$ be a lattice spanned by $(u_1, \dots, u_\omega)$. The LLL algorithm, given $(u_1, \dots, u_\omega)$, finds in polynomial time a vector $b_1$ such that:*

$$\|b_1\| \le 2^{(\omega-1)/4} \det(L)^{1/\omega}$$

## 3.2 Bound on the Factors of Polynomials

We use the following notation: given a polynomial $h(x) = \sum_i h_i x^i$, we define $\|h\|^2 := \sum_i |h_i|^2$ and $\|h\|_\infty := \max_i |h_i|$. We use the same notations for bivariate polynomials, as defined in section 2. The following two lemmata will be useful in the next section:

**Lemma 2.** *Let $a(x, y)$ and $b(x, y)$ be two non-zero polynomials over $\mathbb{Z}$ of maximum degree $d$ separately in $x$ and $y$, such that $b(x, y)$ is a multiple of $a(x, y)$ in $\mathbb{Z}[x, y]$. Then:*

$$\|b\| \ge 2^{-(d+1)^2} \cdot \|a\|_\infty$$

*Proof.* The proof is based on the following result of Mignotte [11]: let $f(x)$ and $g(x)$ be two non-zero polynomials over the integers, such that $\deg f \le k$ and $f$ divides $g$ in $\mathbb{Z}[X]$; then :

$$\|g\| \ge 2^{-k} \cdot \|f\|_\infty$$

Let $f(x) = a(x, x^{d+1})$. Then we have $\deg f \le (d+1)^2$ and the polynomials $a(x,y)$ and $f(x)$ have the same list of non-zero coefficients, which gives $\|f\|_\infty = \|a\|_\infty$. Similarly, letting $g(x) = b(x, x^{d+1})$, we have $\|g\| = \|b\|$. Moreover $f(x)$ divides $g(x)$ in $\mathbb{Z}[x]$. Using the previous result of Mignotte, this proves lemma 2. □

**Lemma 3.** *Let $a(x,y)$ and $b(x,y)$ be as in lemma 2. Assume that $a(0,0) \ne 0$ and $b(x,y)$ is divisible by a non-zero integer $r$ such that $\gcd(r, a(0,0)) = 1$. Then $b(x,y)$ is divisible by $r \cdot a(x,y)$ and:*

$$\|b\| \ge 2^{-(d+1)^2} \cdot |r| \cdot \|a\|_\infty$$

*Proof.* Let $\lambda(x,y)$ be the polynomial such that $a(x,y) \cdot \lambda(x,y) = b(x,y)$. We show that $r$ divides $\lambda(x,y)$. Assume that this is not the case, and let $\lambda_{ij}$ be a coefficient of $x^i y^j$ in $\lambda(x,y)$ not divisible by $r$. Take the smallest $(i,j)$ for the lexicographic ordering. Then we have that $b_{ij} = \lambda_{ij} \cdot a(0,0) \mod r$, where $b_{ij}$ is the coefficient of $x^i y^j$ in $b(x,y)$. Since $a(0,0)$ is invertible modulo $r$ and $b_{ij} = 0 \mod r$, this gives a contradiction. This shows that $r \cdot a(x,y)$ divides $b(x,y)$. Applying the previous lemma to $r \cdot a(x,y)$ and $b(x,y)$, this terminates the proof. □

## 4 Finding Small Roots of Bivariate Integer Polynomials

We prove the following theorem:

**Theorem 4.** *Let $p(x,y)$ be an irreducible polynomial in two variables over $\mathbb{Z}$, of maximum degree $\delta$ in each variable separately. Let $X$ and $Y$ be upper bounds on the desired integer solution $(x_0, y_0)$, and let $W = \max_{i,j} |p_{ij}| X^i Y^j$. If for some $\varepsilon > 0$,*

$$XY < W^{2/(3\delta) - \varepsilon} \tag{6}$$

*then in time polynomial in $(\log W, 2^\delta)$, one can find all integer pairs $(x_0, y_0)$ such that $p(x_0, y_0) = 0$, $|x_0| \le X$, and $|y_0| \le Y$.*

*Proof.* We write:

$$p(x,y) = \sum_{0 \le i,j \le \delta} p_{ij} x^i y^j$$

and let $(x_0, y_0)$ be an integer root of $p(x,y)$. As previously we let

$$W = \|p(xX, yY)\|_\infty$$

First we assume that $p_{00} \ne 0$ and $\gcd(p_{00}, XY) = 1$. We will see in appendix A how to handle the general case.

We select an integer $k \geq 0$ and let $\omega = (\delta + k + 1)^2$. We generate an integer $u$ such that $\sqrt{\omega} \cdot 2^{-\omega} \cdot W \leq u < 2W$ and $\gcd(p_{00}, u) = 1$. As in section 2, one can take

$$u = W + ((1 - W) \mod |p_{00}|) .$$

We let $n = u \cdot (XY)^k$. We have that $\gcd(p_{00}, n) = 1$ and:

$$\sqrt{\omega} \cdot 2^{-\omega} \cdot (XY)^k \cdot W \leq n < 2 \cdot (XY)^k \cdot W \tag{7}$$

As in section 2, we must find a polynomial $h(x, y)$ such that $h(x_0, y_0) = 0$ and $h(x, y)$ is not a multiple of $p(x, y)$. We let $q(x, y)$ be the polynomial:

$$\begin{aligned} q(x, y) &= p_{00}^{-1} \cdot p(x, y) \mod n \\ &= 1 + \sum_{(i,j) \neq (0,0)} a_{ij} x^i y^j \end{aligned}$$

For all $0 \leq i, j \leq k$, we form the polynomials:

$$q_{ij}(x, y) = x^i y^j X^{k-i} Y^{k-j} q(x, y)$$

For all $(i, j) \in [0, \delta + k]^2 \setminus [0, k]^2$, we also form the polynomials:

$$q_{ij}(x, y) = x^i y^j n$$

We consider the corresponding polynomials $\tilde{q}_{ij}(x, y) = q_{ij}(xX, yY)$. Note that for all $(i, j) \in [0, \delta + k]^2$, we have that $q_{ij}(x_0, y_0) = 0 \mod n$, and the polynomial $\tilde{q}_{ij}(x, y)$ is divisible by $(XY)^k$.

Let $h(x, y)$ be a linear integer combination of the polynomials $q_{ij}(x, y)$; the polynomial $\tilde{h}(x, y) = h(xX, yX)$ is also a linear combination of the $\tilde{q}_{ij}(x, y)$ with the same integer coefficients. We have that $h(x_0, y_0) = 0 \mod n$ and $(XY)^k$ divides $h(xX, yY)$. Moreover $h(x, y)$ has maximum degree $\delta + k$ independently in $x$ and $y$, therefore it is the sum of at most $\omega$ monomials. As in section 2, we are interested in finding a polynomial $h(x, y)$ such that the coefficients of $h(xX, yY)$ are small enough, for the following two reasons:

1) if the coefficients of $h(xX, yY)$ are sufficiently small, then the equality $h(x_0, y_0) = 0$ holds not only modulo $n$, but also over $\mathbb{Z}$. From lemma 1, the condition is:

$$\|h(xX, yY)\| < \frac{n}{\sqrt{\omega}} \tag{8}$$

2) if the coefficients of $h(xX, yY)$ are sufficiently small, then $h(x, y)$ cannot be a multiple of $p(x, y)$. Using lemma 3, the condition is:

$$\|h(xX, yY)\| < 2^{-\omega} \cdot (XY)^k \cdot W \tag{9}$$

This condition is obtained by applying lemma 3 with $a(x, y) = p(xX, yY)$, $b(x, y) = h(xX, yY)$ and $r = (XY)^k$. Then we have $a(0, 0) = p_{00} \neq 0$ and

$\gcd(a(0,0),(XY)^k) = 1$. Under condition (9), $h(xX, yY)$ cannot be a multiple of $p(xX, yY)$ and therefore $h(x, y)$ cannot be a multiple of $p(x, y)$.

Using inequality (7), we obtain that the first condition (8) is satisfied whenever the second condition (9) is satisfied.

We form the lattice $L$ spanned by the coefficients of the polynomials $\tilde{q}_{ij}(x, y)$. The polynomials $q_{ij}(x, y)$ have a maximum degree of $\delta + k$ separately in $x$ and $y$; therefore, there are $(\delta + k + 1)^2$ such coefficients. Moreover, there is a total of $(\delta+k+1)^2$ polynomials. This gives a full rank lattice of dimension $\omega = (\delta+k+1)^2$. In figure 1, we illustrate the lattice for $\delta = 1$ and $k = 1$.

| | $1$ | $x$ | $y$ | $xy$ | $x^2$ | $x^2y$ | $y^2$ | $xy^2$ | $x^2y^2$ |
|---|---|---|---|---|---|---|---|---|---|
| $XYq$ | $XY$ | $a_{10}X^2Y$ | $a_{01}XY^2$ | $a_{11}X^2Y^2$ | | | | | |
| $Yxq$ | | $XY$ | | $a_{01}XY^2$ | $a_{10}X^2Y$ | $a_{11}X^2Y^2$ | | | |
| $Xyq$ | | | $XY$ | $a_{10}X^2Y$ | | | $a_{01}XY^2$ | $a_{11}X^2Y^2$ | |
| $xyq$ | | | | $XY$ | | $a_{10}X^2Y$ | | $a_{01}XY^2$ | $a_{11}X^2Y^2$ |
| $x^2n$ | | | | | $X^2n$ | | | | |
| $x^2yn$ | | | | | | $X^2Yn$ | | | |
| $y^2n$ | | | | | | | $Y^2n$ | | |
| $xy^2n$ | | | | | | | | $XY^2n$ | |
| $x^2y^2n$ | | | | | | | | | $X^2Y^2n$ |

**Fig. 1.** The lattice $L$ for $\delta = 1$ and $k = 1$

It is easy to see that the coefficient vectors of the polynomials $\tilde{q}_{ij}(x, y)$ form a triangular basis of $L$. The determinant is then the product of the diagonal entries. For $0 \leq i, j \leq k$, the contribution of the polynomials $\tilde{q}_{ij}(x, y)$ to the determinant is given by:

$$\prod_{0 \leq i,j \leq k} (XY)^k = (XY)^{k(k+1)^2}$$

The contribution of the other polynomials $\tilde{q}_{ij}(x, y)$ is then:

$$\prod_{(i,j)\in[0,\delta+k]^2\setminus[0,k]^2} X^i Y^j n = (XY)^{\frac{(\delta+k)(\delta+k+1)^2}{2} - \frac{k(k+1)^2}{2}} n^{\delta(\delta+2k+2)}$$

Therefore, the determinant of $L$ is given by:

$$\det(L) = (XY)^{\frac{(\delta+k)(\delta+k+1)^2 + k(k+1)^2}{2}} n^{\delta(\delta+2k+2)} \tag{10}$$

Using LLL (see theorem 3), we obtain in time polynomial in $(\log W, \omega)$ a non-zero polynomial $h(x, y)$ such that:

$$\|h(xX, yY)\| \leq 2^{(\omega-1)/4} \cdot \det(L)^{1/\omega} \tag{11}$$

Note that any vector in the lattice $L$ has integer coefficients divisible by $(XY)^k$; this means that in practice, it is more efficient to apply LLL to the lattice $(XY)^{-k}L$.

From inequality (11) we obtain that the conditions (8) and (9) are satisfied when:

$$2^{(\omega-1)/4} \cdot \det(L)^{1/\omega} < 2^{-\omega} \cdot (XY)^k \cdot W \tag{12}$$

In this case, we have that $h(x_0, y_0) = 0$ and $h(x, y)$ is not a multiple of $p(x, y)$. Since $p(x, y)$ is irreducible,

$$Q(x) = \mathrm{Resultant}_y(h(x, y), p(x, y))$$

gives a non-zero integer polynomial such that $Q(x_0) = 0$. Using any standard root-finding algorithm, we can recover $x_0$, and finally $y_0$ by solving $p(x_0, y) = 0$.

Using inequality (7), we obtain that inequality (12) is satisfied when:

$$XY < 2^{-\beta} W^{\alpha} \tag{13}$$

where

$$\alpha = \frac{2(k+1)^2}{(\delta+k)(\delta+k+1)^2 - k(k+1)^2} \tag{14}$$

$$\beta = \frac{10}{4} \cdot \frac{(\delta+k+1)^4 + (\delta+k+1)^2}{(\delta+k)(\delta+k+1)^2 - k(k+1)^2} \tag{15}$$

We have that for all $\delta \geq 1$ and $k \geq 0$ :

$$\alpha \geq \frac{2}{3\delta} - \frac{2}{3 \cdot (k+1)} \tag{16}$$

and:

$$\beta \leq \frac{4k^2}{\delta} + 13 \cdot \delta \tag{17}$$

Then, taking $k = \lfloor 1/\varepsilon \rfloor$, we obtain from (13), (16) and (17) the following condition for $XY$:

$$XY < W^{2/(3\delta)-\varepsilon} \cdot 2^{-4/(\delta \cdot \varepsilon^2) - 13\delta} \tag{18}$$

For an $XY$ satisfying (18), we obtain a bivariate integer polynomial root-finding algorithm running in time polynomial in $(\log W, \delta, 1/\varepsilon)$.

For an $XY$ satisfying the slightly weaker condition (6), we exhaustively search the high order $4/(\delta \cdot \varepsilon^2) + 13\delta$ bits of $x_0$, so that condition (18) applies, and for each possible value we use the algorithm described previously. For a fixed $\varepsilon > 0$, the running time is polynomial in $(\log W, 2^{\delta})$. This terminates the proof of theorem 4. □

As in [5], the efficiency of our algorithm depends on the shape of the polynomial $p(x, y)$. The previous theorem applies when $p(x, y)$ has maximum degree $\delta$ separately in $x$ and $y$. If we assume that $p(x, y)$ has a total degree $\delta$ in $x$ and $y$, we obtain the following theorem, analogous to theorem 3 in [5] (the proof is given in appendix B).

**Theorem 5.** *Under the hypothesis of theorem 4, except that $p(x,y)$ has total degree $\delta$, the appropriate bound is:*

$$XY < W^{1/\delta-\varepsilon}$$

## 5 Comparison with Coppersmith's Algorithm

We note that under the following condition, stronger than (6) :

$$XY < W^{2/(3\delta)-\varepsilon} \cdot 2^{-13\delta}$$

Coppersmith's algorithm is polynomial-time in $(\log W, \delta, 1/\varepsilon)$ (see [5], theorem 2), whereas our algorithm is polynomial-time in $(\log W, \delta)$ but exponential-time in $1/\varepsilon$. Coppersmith's algorithm is therefore more efficient than ours for small values of $\varepsilon$. This implies that under the following condition, weaker than (6) :

$$XY < W^{2/(3\delta)}$$

Coppersmith's algorithm is still polynomial in $(\log W, 2^{\delta})$ (see [5], corollary 2), which is no longer the case for our algorithm.

## 6 Extension to More Variables

Our algorithm can be extended to solve integer polynomial equations with more than two variables. As for Coppersmith's algorithm, the extension is heuristic only.

Let $p(x,y,z)$ be a polynomial in three variables over the integers, of degree $\delta$ independently in $x, y$ and $z$. Let $(x_0, y_0, z_0)$ be an integer root of $p(x,y,z)$, with $|x_0| \le X$, $|y_0| \le Y$ and $|z_0| \le Z$. Let $\ell$ be an integer $\ge 0$. As for the bivariate case, we generate an integer $n$ such that $n = 0 \mod (XYZ)^{\ell}$, and a polynomial $q(x,y,z)$ such that $q(x_0,y_0,z_0) = 0 \mod n$ and $q(0,0,0) = 1 \mod n$. Then we consider the lattice $L$ generated by all linear integer combinations of the polynomials $x^i y^j z^k X^{\ell-i} Y^{\ell-j} Z^{\ell-k} q(xX, yY, zZ)$ for $0 \le i,j,k \le \ell$ and the polynomials $(xX)^i (yY)^j (zZ)^k \cdot n$ for $(i,j,k) \in [0, \delta+\ell]^3 \setminus [0,\ell]^3$. If the ranges $X, Y, Z$ are small enough, then by using LLL we are guaranteed to find a polynomial $h_1(x,y,z)$ such that $h_1(x_0,y_0,z_0) = 0$ over $\mathbb{Z}$ and $h_1(x,y,z)$ is not a multiple of $p(x,y,z)$. Unfortunately, this is not enough. For small enough ranges $X, Y, Z$, we can also obtain a second polynomial $h_2(x,y,z)$ satisfying the same property. This can be done by bounding the norm of the second vector produced by LLL, as in [1,8]. Then we could take the resultant between the three polynomials $p(x,y,z)$, $h_1(x,y,z)$ and $h_2(x,y,z)$ in order to obtain a polynomial $f(x)$ such that $f(x_0) = 0$. But we have no guarantee that the polynomials $h_1(x,y,z)$ and $h_2(x,y,z)$ will be algebraically independent, for example we might have $h_2(x,y,z) = x \cdot h_1(x,y,z)$. This makes the method heuristic only.

## 7 Practical Experiments

An application of solving bivariate equations described in [5] is factoring an RSA modulus $n = pq$ when the high-order bits of $p$ are known. Using our algorithm from theorem 4, we obtain the following theorem, whose proof is given in appendix C.

**Theorem 6.** *For any $\varepsilon > 0$, given $n = pq$ and the high-order $(1/4 + \varepsilon) \log_2 n$ bits of $p$, we can recover the factorization of $n$ in time polynomial in $\log n$.*

By comparison, Coppersmith's algorithm provides a slightly better result since only the high-order $1/4 \log_2 n$ bits of $p$ are required (see theorem 4, [5]). The result of practical experiments are summarized in table 2, using Shoup's NTL library [14]. It shows that our bivariate polynomial root-finding algorithm works well in practice.

| $N$ | bits of $p$ given | lattice dimension | running time |
|---|---|---|---|
| 512 bits | 144 bits | 25 | 35 sec |
| 512 bits | 141 bits | 36 | 3 min |
| 1024 bits | 282 bits | 36 | 20 min |

**Fig. 2.** Running times for factoring $N = pq$ given the high-order bits of $p$, using our bivariate integer polynomial root finding algorithm on a 733 Mhz PC running under Linux.

We have also implemented the factorization of $n = pq$ with high-order bits known using the simplification of Howgrave-Graham [7]. Results are given in table 3. It shows that the simplification of Howgrave-Graham is much more efficient in practice. Namely, the factorization of a 1024-bit RSA modulus knowing the high-order 282 bits of $p$ takes roughly 20 minutes using our bivariate polynomial root finding algorithm, and only one second using Howgrave-Graham's simplification. This is due to the fact that the Howgrave-Graham simplification enables to obtain a lattice with a lower dimension (but it applies only to the particular case of factoring with high-bits known, not to the general case of finding small roots of bivariate integer polynomials).

| $N$ | bits of $p$ given | lattice dimension | running time |
|---|---|---|---|
| 1024 bits | 282 bits | 11 | 1 sec |
| 1024 bits | 266 bits | 25 | 1 min |
| 1536 bits | 396 bits | 33 | 19 min |

**Fig. 3.** Running times for factoring $N = pq$ given the high-order bits of $p$, using Howgrave-Graham's algorithm on a 733 Mhz PC running under Linux.

## 8 Conclusion

We have presented an algorithm for finding small roots of bivariate integer polynomials, simpler than Coppersmith's algorithm. The bivariate integer case is now as simple to analyze and implement as the univariate modular case. Our algorithm is asymptotically less efficient than Coppersmith's algorithm, but experiments show that it works well in practice; however, for the particular case of integer factorization with high-bits known, the Howgrave-Graham simplification appears to be more efficient.

## References


1. D. Boneh and G. Durfee, *Crypanalysis of RSA with private key d less than* $N^{0.292}$, proceedings of Eurocrypt '99, vol. 1592, Lecture Notes in Computer Science.
2. D. Boneh, G. Durfee and N.A. Howgrave-Graham, *Factoring* $n = p^r q$ *for large* $r$, proceedings of Crypto '99, vol. 1666, Lecture Notes in Computer Science.
3. D. Coppersmith, *Finding a Small Root of a Univariate Modular Equation*, proceedings of Eurocrypt '96, vol. 1070, Lecture Notes in Computer Science.
4. D. Coppersmith, *Finding a Small Root of a Bivariate Integer Equation; Factoring with High Bits Known*, proceedings of Eurocrypt' 96, vol. 1070, Lecture Notes in Computer Science.
5. D. Coppersmith, *Small solutions to polynomial equations, and low exponent vulnerabilities*. J. of Cryptology, 10(4)233-260, 1997. Revised version of two articles of Eurocrypt '96.
6. D. Coppersmith, *Finding small solutions to small degree polynomials*. In Proc. of CALC '01, LNCS, Sptinger-Verlag, 2001.
7. N.A. Howgrave-Graham, *Finding small roots of univariate modular equations revisited*. In Cryptography and Coding, volume 1355 of LNCS, pp. 131-142. Springer Verlag, 1997.
8. C.S. Jutla, *On finding small solutions of modular multivariate polynomial equations*. Proceedings of Eurocrypt '98, Lecture Notes in Computer Science. vol. 1402.
9. A.K. Lenstra, H.W. Lenstra, Jr., and L. Lovász, *Factoring polynomials with rational coefficients*. Mathematische Ann., 261:513-534, 1982.
10. U. Maurer, *Fast Generation of Prime Numbers and Secure Public-Key Cryptographic Parameters*, Journal of Cryptology, vol. 8, no. 3, pp. 123-155, 1995.
11. M. Mignotte, *An inequality about factors of polynomials*. Math Comp. 28, 1153-1157, 1974.
12. P.Q. Nguyen and J. Stern, *The two faces of lattices in cryptology*. Proceedings of CALC '01, LNCS vol. 2146.
13. V. Shoup, *OAEP reconsidered*. Proceedings of Crypto '01, vol. 2139, Lecture Notes in Computer Science.
14. V. Shoup, *Number Theory C++ Library (NTL) version 5.3.1*. Available at `www.shoup.net`.

# A Finding Small Roots in the General Case

The algorithm described in section 4 assumes that $p(0,0) \neq 0$ and also that $\gcd(p(0,0), XY) = 1$. Here we show how to handle the general case.

If $p(0,0) = 0$, we use a simple change of variable to derive a polynomial $p^*(x,y)$ such that $p^*(0,0) \neq 0$. This is done as follows: we write $p$ as $p(x,y) = x \cdot a_0(x) + y \cdot c(x,y)$, where $a_0$ is a polynomial of maximum degree $\delta - 1$. Since $p(x,y)$ is irreducible, we must have $a_0 \neq 0$. Since $\deg a_0 \leq \delta - 1$, there exists $0 < i \leq \delta$ such that $a_0(i) \neq 0$. Then $p(i,0) \neq 0$ and letting $p^*(x,y) = p(x+i,y)$, we obtain that $p^*(0,0) \neq 0$ and use $p^*(x,y)$ instead of $p(x,y)$.

If $\gcd(p(0,0), XY) \neq 1$, we generate two random primes $X'$ and $Y'$ such that $X < X' < 2X$ and $Y < Y' < 2Y$, and $X'$ and $Y'$ do not divide $p(0,0)$. This can be done in polynomial-time using the recursive prime generation algorithm described in [10]. We then use $X', Y'$ instead of $X, Y$.

# B Proof of Theorem 5

We use the same $n$ and the same $q(x,y)$ as in section 4. We use the same polynomials $q_{ij}(x,y) = x^i y^j X^{k-i} Y^{k-j} q(x,y)$, but only for $0 \leq i+j \leq k$ (instead of $0 \leq i,j \leq k$). We also use the polynomials $q_{ij}(x,y) = x^i y^j n$ for $k < i+j \leq k+\delta$.

We obtain a full-rank lattice $L$ of dimension $\omega = (k+\delta+1)(k+\delta+2)/2$, where the coefficient vectors of the polynomials $\tilde{q}_{ij}(x,y)$ form a triangular basis. The contribution of the polynomials $\tilde{q}_{ij}(x,y)$ for $0 \leq i+j \leq k$ to the determinant is given by:

$$\prod_{0 \leq i+j \leq k} (XY)^k = (XY)^{\frac{k(k+1)(k+2)}{2}}$$

and the contribution of the remaining polynomial is:

$$\prod_{k < i+j \leq k+\delta} X^i Y^j n = (XY)^{d \cdot (2+d^2+6k+3k^2+3d(1+k))/6} \cdot n^{d(3+d+2k)/2}$$

which gives:

$$\det L = (XY)^{\frac{3k(1+k)(2+k)+d(2+d^2+6k+3k^2+3d(1+k))}{6}} \cdot n^{d(3+d+2k)/2}$$

As before, the condition is:

$$2^{(\omega-1)/4} \det(L)^{1/\omega} < 2^{-(k+\delta+1)^2} \cdot (XY)^k \cdot W$$

from which we derive the following condition on $XY$:

$$XY < W^{(1/\delta)-\varepsilon} \cdot 2^{-4/(\delta\varepsilon^2)-13\delta}$$

for $\varepsilon = \mathcal{O}(1/k)$. As previously, we exhaustive search on the high-order $4/(\delta\varepsilon^2) + 13\delta$ bits of $x_0$, to obtain the bound:

$$XY < W^{(1/\delta)-\varepsilon}$$

while remaining polynomial-time in $(\log W, 2^\delta)$.

## C Factoring with High-Bits Known: Proof of Theorem 6

Let $N = pq$ be an RSA-modulus and assume that we know that high-order $(1/4+\varepsilon)\log_2 N$ bits of $p$, for $\varepsilon > 0$. By division we also know the high-order $(1/4+\varepsilon)\log_2 N$ bits of $q$. We write:

$$p = p_0 + x_0 \quad q = q_0 + y_0$$

$$|x_0| < p_0 N^{-1/4-\varepsilon} = X \quad |y_0| < q_0 N^{-1/4-\varepsilon} = Y$$

where $p_0$ and $q_0$ are known and $x_0$ and $y_0$ are unknown. We define the polynomial:

$$p(x,y) = (p_0 + x)\cdot(q_0 + y) - N = (p_0 q_0 - N) + q_0 x + p_0 y + xy$$

We have that $p(x_0, y_0) = 0$ and:

$$W = \max(|p_0 q_0 - n|, q_0 X, p_0 Y, XY) > q_0 X > \frac{1}{2} N^{3/4-\varepsilon}$$

We have:

$$XY = p_0 q_0 N^{-1/2-2\varepsilon} < N^{1/2-2\varepsilon}$$

which gives:

$$XY < 2W^{2/3-\varepsilon}$$

so that by guessing one additional bit of $x_0$ we are under the conditions of theorem 4.

# Public Key Encryption with Keyword Search

Dan Boneh[1]*, Giovanni Di Crescenzo[2], Rafail Ostrovsky[3]**, and Giuseppe Persiano[4]***

[1] Stanford University
dabo@cs.stanford.edu
[2] Telcordia
giovanni@research.telcordia.com
[3] UCLA
rafail@cs.ucla.edu
[4] Università di Salerno
giuper@dia.unisa.it

**Abstract.** We study the problem of searching on data that is encrypted using a public key system. Consider user Bob who sends email to user Alice encrypted under Alice's public key. An email gateway wants to test whether the email contains the keyword "urgent" so that it could route the email accordingly. Alice, on the other hand does not wish to give the gateway the ability to decrypt all her messages. We define and construct a mechanism that enables Alice to provide a key to the gateway that enables the gateway to test whether the word "urgent" is a keyword in the email without learning anything else about the email. We refer to this mechanism as *Public Key Encryption with keyword Search.* As another example, consider a mail server that stores various messages publicly encrypted for Alice by others. Using our mechanism Alice can send the mail server a key that will enable the server to identify all messages containing some specific keyword, but learn nothing else. We define the concept of public key encryption with keyword search and give several constructions.

## 1 Introduction

Suppose user Alice wishes to read her email on a number of devices: laptop, desktop, pager, etc. Alice's mail gateway is supposed to route email to the appropriate device based on the keywords in the email. For example, when Bob sends email with the keyword "urgent" the mail is routed to Alice's pager. When Bob sends email with the keyword "lunch" the mail is routed to Alice's desktop for reading later. One expects each email to contain a small number of keywords. For example, all words on the subject line as well as the sender's email address

* Supported by NSF and the Packard foundation.
** Partially supported by a gift from Teradata. Preliminary work done while visiting Stanford and while at Telcordia.
*** Part of this work done while visiting DIMACS. Work supported by NoE ECRYPT.

C. Cachin and J. Camenisch (Eds.): EUROCRYPT 2004, LNCS 3027, pp. 506–522, 2004.

could be used as keywords. The mobile people project [24] provides this email processing capability.

Now, suppose Bob sends encrypted email to Alice using Alice's public key. Both the contents of the email and the keywords are encrypted. In this case the mail gateway cannot see the keywords and hence cannot make routing decisions. As a result, the mobile people project is unable to process secure email without violating user privacy. Our goal is to enable Alice to give the gateway the ability to test whether "urgent" is a keyword in the email, but the gateway should learn nothing else about the email. More generally, Alice should be able to specify a few keywords that the mail gateway can search for, but learn nothing else about incoming mail. We give precise definitions in section 2.

To do so, Bob encrypts his email using a standard public key system. He then appends to the resulting ciphertext a *Public-Key Encryption with keyword Search* (PEKS) of each keyword. To send a message $M$ with keywords $W_1, \ldots, W_m$ Bob sends

$$E_{A_{pub}}(M) \parallel \mathsf{PEKS}(A_{pub}, W_1) \parallel \cdots \parallel \mathsf{PEKS}(A_{pub}, W_m)$$

Where $A_{pub}$ is Alice's public key. The point of this form of encryption is that Alice can give the gateway a certain trapdoor $T_W$ that enables the gateway to test whether one of the keywords associated with the message is equal to the word $W$ of Alice's choice. Given $\mathsf{PEKS}(A_{pub}, W')$ and $T_W$ the gateway can test whether $W = W'$. If $W \neq W'$ the gateway learns nothing more about $W'$. Note that Alice and Bob do not communicate in this entire process. Bob generates the searchable encryption for $W'$ just given Alice's public key.

In some cases, it is instructive to view the email gateway as an IMAP or POP email server. The server stores many emails and each email contains a small number of keywords. As before, all these emails are created by various people sending mail to Alice encrypted using her public key. We want to enable Alice to ask queries of the form: do any of the messages on the server contain the keyword "urgent"? Alice would do this by giving the server a trapdoor $T_W$, thus enabling the server to retrieve emails containing the keyword $W$. The server learns nothing else about the emails.

**Related work.** A related issue deals with privacy of database data. There are two different scenarios: *public* databases and *private* databases, and the solutions for each are different.

*Private databases:* In this settings a user wishes to upload its private data to a remote database and wishes to keep the data private from the remote database administrator. Later, the user must be able to retrieve from the remote database all records that contain a particular keyword. Solutions to this problem were presented in the early 1990's by Ostrovsky [26] and Ostrovsky and Goldreich [17] and more recently by Song at al. [28]. The solution of Song. at al [28] requires very little communication between the user and the database (proportional to the security parameter) and only one round of interaction. The database performs work that is linear in its size per query. The solution of [26,17] requires poly-logarithmic rounds (in the size of the database) between the user and the database, but allows the database to do only poly-logarithmic work per query. An additional privacy requirement that might be appealing in some scenarios is to hide from the database administrator any information regarding the *access pattern*, i.e. if some item was retrieved more then once, some item was not retrieved at all, etc. The work of [26,17] achieves this property as well, with the same poly-logarithmic cost[5] per query both for the database-user interaction and the actual database work. We stress that both the constructions of [26,17] and the more recent work of [10,28,16] apply only to the private-key setting for users who own their data and wish to upload it to a third-party database that they do not trust.

*Public Databases* Here the database data is public (such as stock quotes) but the user is unaware of it and wishes to retrieve some data-item or search for some data-item, without revealing to the database administrator which item it is. The naive solution is that the user can download the entire database. Public Information Retrieval (PIR) protocols allow user to retrieve data from a public database with far smaller communication then just downloading the entire database. PIR was first shown to be possible only in the setting where there are many copies of the same database and none of the copies can talk to each other [5]. PIR was shown to be possible for a single database by Kushilevitz and Ostrovsky [22] (using homomorphic encryption scheme of [19]). The communication complexity of [22] solution (i.e. the number of bits transmitted between the user and the database) is $O(n^\epsilon)$, where $n$ is the size of the database and $\epsilon > 0$. This was reduced to poly-logarithmic overhead by Cachin, Micali, and Stadler [4]. As pointed out in [22], the model of PIR can be extended to one-out-of-$n$ Oblivious Transfer and keyword searching on public data, and received a lot of additional attention in the literature (see, for example, [22,8,20,9,23,25,27]. We stress though that in all these settings the database is public, and the user is trying to retrieve or find certain items without revealing to the database administrator what it is searching for. In the setting of a single public database, it can be shown that the database must always perform work which is at least linear in the size of the database.

Our problem does not fit either of the two models mentioned above. Unlike the private-key setting, data collected by the mail-server is from third parties, and can not be "organized" by the user in any convenient way. Unlike the publicly available database, the data is not public, and hence the PIR solutions do not apply.

We point out that in practical applications, due to the computation cost of public key encryption, our constructions are applicable to searching on a small number of keywords rather than an entire file. Recently, Waters et al. [30] showed that public key encryption with keyword search can be used to build an encrypted and searchable audit log. Other methods for searching on encrypted data are described in [16,12].

[5] The poly-logarithmic construction of [26,17] requires large constants, which makes it impractical; however their basic $O(\sqrt{n})$ solution was recently shown to be applicable for some practical applications [10].

## 2 Public Key Encryption with Searching: Definitions

Throughout the paper we use the term *negligible function* to refer to a function $f : \mathbb{R} \to [0,1]$ where $f(s) < 1/g(s)$ for any polynomial $g$ and sufficiently large $s$.

We start by precisely defining what is a secure Public Key Encryption with keyword Search (PEKS) scheme. Here "public-key" refers to the fact that ciphertexts are created by various people using Alice's public key. Suppose user Bob is about to send an encrypted email to Alice with keywords $W_1, \ldots, W_k$ (e.g., words in the subject line and the sender's address could be used as keywords, so that $k$ is relatively small). Bob sends the following message:

$$\left[E_{A_{pub}}[msg],\ \mathsf{PEKS}(A_{pub}, W_1), \ldots, \mathsf{PEKS}(A_{pub}, W_k)\right] \quad (1)$$

where $A_{pub}$ is Alice's public key, $msg$ is the email body, and PEKS is an algorithm with properties discussed below. The PEKS values do not reveal any information about the message, but enable searching for specific keywords. For the rest of the paper, we use as our sample application a mail server that stores all incoming email.

Our goal is to enable Alice to send a short secret key $T_W$ to the mail server that will enable the server to locate all messages containing the keyword $W$, but learn nothing else. Alice produces this trapdoor $T_W$ using her private key. The server simply sends the relevant emails back to Alice. We call such a system *non-interactive public key encryption with keyword search*, or as a shorthand "searchable public-key encryption".

**Definition 1.** A non-interactive public key encryption with keyword search (we sometimes abbreviate it as "searchable encryption") scheme consists of the following polynomial time randomized algorithms:

1. KeyGen(s): Takes a security parameter, $s$, and generates a public/private key pair $A_{pub}, A_{priv}$.
2. $\mathsf{PEKS}(A_{pub}, W)$: for a public key $A_{pub}$ and a word $W$, produces a searchable encryption of $W$.
3. $\mathsf{Trapdoor}(A_{priv}, W)$: given Alice's private key and a word $W$ produces a trapdoor $T_W$.
4. $\mathsf{Test}(A_{pub}, S, T_W)$: given Alice's public key, a searchable encryption $S = \mathsf{PEKS}(A_{pub}, W')$, and a trapdoor $T_W = \mathsf{Trapdoor}(A_{priv}, W)$, outputs 'yes' if $W = W'$ and 'no' otherwise.

Alice runs the KeyGen algorithm to generate her public/private key pair. She uses Trapdoor to generate trapdoors $T_W$ for any keywords $W$ that she wants the mail server or mail gateway to search for. The mail server uses the given trapdoors as input to the Test() algorithm to determine whether a given email contains one of the keywords $W$ specified by Alice.

Next, we define security for a PEKS in the sense of semantic-security. We need to ensure that an $\mathsf{PEKS}(A_{pub}, W)$ does not reveal any information about $W$ unless $T_W$ is available. We define security against an active attacker who is able to obtain trapdoors $T_W$ for any $W$ of his choice. Even under such attack the attacker should not be able to distinguish an encryption of a keyword $W_0$ from an encryption of a keyword $W_1$ for which he did not obtain the trapdoor. Formally, we define security against an active attacker $\mathcal{A}$ using the following game between a challenger and the attacker (the security parameter $s$ is given to both players as input).

PEKS Security game:

1. The challenger runs the KeyGen($s$) algorithm to generate $A_{pub}$ and $A_{priv}$. It gives $A_{pub}$ to the attacker.
2. The attacker can adaptively ask the challenger for the trapdoor $T_W$ for any keyword $W \in \{0,1\}^*$ of his choice.
3. At some point, the attacker $\mathcal{A}$ sends the challenger two words $W_0, W_1$ on which it wishes to be challenged. The only restriction is that the attacker did not previously ask for the trapdoors $T_{W_0}$ or $T_{W_1}$. The challenger picks a random $b \in \{0,1\}$ and gives the attacker $C = \mathsf{PEKS}(A_{pub}, W_b)$. We refer to $C$ as the challenge PEKS.
4. The attacker can continue to ask for trapdoors $T_W$ for any keyword $W$ of his choice as long as $W \neq W_0, W_1$.
5. Eventually, the attacker $\mathcal{A}$ outputs $b' \in \{0,1\}$ and wins the game if $b = b'$.

In other words, the attacker wins the game if he can correctly guess whether he was given the PEKS for $W_0$ or $W_1$. We define $\mathcal{A}$'s advantage in breaking the PEKS as

$$\mathrm{Adv}_{\mathcal{A}}(s) = |\Pr[b = b'] - \frac{1}{2}|$$

**Definition 2.** We say that a PEKS is semantically secure against an adaptive chosen keyword attack if for any polynomial time attacker $\mathcal{A}$ we have that $\mathrm{Adv}_{\mathcal{A}}(s)$ is a negligible function.

*Chosen Ciphertext Security.* We note that Definition 2 ensures that the construction given in Eq. (1) is semantically secure whenever the public key encryption system $E_{A_{pub}}$ is semantically secure. However, as is, the construction is not chosen ciphertext secure. Indeed, a chosen ciphertext attacker can break semantic security by reordering the keywords in Eq. (1) and submitting the resulting ciphertext for decryption. A standard technique can make this construction chosen ciphertext secure using the methods of [7]. We defer this to the full version of the paper.

## 2.1 PEKS Implies Identity Based Encryption

Public key encryption with keyword search is related to Identity Based Encryption (IBE) [29,2]. Constructing a secure PEKS appears to be a harder problem than constructing an IBE. Indeed, the following lemma shows that PEKS implies Identity Based Encryption. The converse is probably false. Security notions for IBE, and in particular chosen ciphertext secure IBE (IND-ID-CCA), are defined in [2].

**Lemma 1.** *A non-interactive searchable encryption scheme (*PEKS*) that is semantically secure against an adaptive chosen keyword attack gives rise to a chosen ciphertext secure IBE system (IND-ID-CCA).*

**Proof sketch:** Given a PEKS (KeyGen, PEKS, Trapdoor, Test) the IBE system is as follows:

1. Setup: Run the PEKS KeyGen algorithm to generate $A_{pub}/A_{priv}$. The IBE system parameters are $A_{pub}$. The master-key is $A_{priv}$.
2. KeyGen: The IBE private key associated with a public key $X \in \{0,1\}^*$ is

$$d_X = [\mathsf{Trapdoor}(A_{priv}, X\|0),\ \ \mathsf{Trapdoor}(A_{priv}, X\|1)],$$

   where $\|$ denotes concatenation.
3. Encrypt: Encrypt a bit $b \in \{0,1\}$ using a public key $X \in \{0,1\}^*$ as: $CT = \mathsf{PEKS}(A_{pub}, X\|b)$.
4. Decrypt: To decrypt $CT = \mathsf{PEKS}(A_{pub}, X\|b)$ using the private key $d_X = (d_0, d_1)$ output '0' if $\mathsf{Test}(A_{pub}, CT, d_0)$ = 'yes' and output '1' if $\mathsf{Test}(A_{pub}, CT, d_1)$ = 'yes'

One can show that the resulting system is IND-ID-CCA assuming the PEKS is semantically secure against an adaptive chosen message attack. □

This shows that building non-interactive public-key searchable encryption is at least as hard as building an IBE system. One might be tempted to prove the converse (i.e., IBE implies PEKS) by defining

$$\mathsf{PEKS}(A_{pub}, W) = E_W[0^k] \tag{2}$$

i.e. encrypt a string of $k$ zeros with the IBE public key $W \in \{0,1\}^*$. The Test algorithm attempts to decrypt $E_W[0]$ and checks that the resulting plaintext is $0^k$. Unfortunately, this does not necessarily give a secure searchable encryption scheme. The problem is that the ciphertext $CT$ could expose the public key ($W$) used to create $CT$. Generally, an encryption scheme need not hide the public key that was used to create a given ciphertext. But this property is essential for the PEKS construction given in (2). We note that public key privacy was previously studied by Bellare et al. [1].

Generally, it appears that constructing a searchable public-key encryption is a harder problem than constructing an IBE scheme. Nevertheless, our first PEKS construction is based on a recent construction for an IBE system. We are able to prove security by exploiting extra properties of this system.

## 3 Constructions

We give two constructions for public-key searchable encryption: (1) an efficient system based on a variant of the Decision Diffie-Hellman assumption (assuming a random oracle) and (2) a limited system based on general trapdoor permutations (without assuming the random oracle), but less efficient.

### 3.1 Construction Using Bilinear Maps

Our first construction is based on a variant of the Computational Diffie-Hellman problem. Boneh and Franklin [2] recently used bilinear maps on elliptic curves to build an efficient IBE system. Abstractly, they use two groups $G_1, G_2$ of prime order $p$ and a bilinear map $e : G_1 \times G_1 \to G_2$ between them. The map satisfies the following properties:

1. Computable: given $g, h \in G_1$ there is a polynomial time algorithms to compute $e(g, h) \in G_2$.
2. Bilinear: for any integers $x, y \in [1, p]$ we have $e(g^x, g^y) = e(g, g)^{xy}$
3. Non-degenerate: if $g$ is a generator of $G_1$ then $e(g, g)$ is a generator of $G_2$.

The size of $G_1, G_2$ is determined by the security parameter.

We build a non-interactive searchable encryption scheme from such a bilinear map. The construction is based on [2]. We will need hash functions $H_1 : \{0,1\}^* \to G_1$ and $H_2 : G_2 \to \{0,1\}^{\log p}$. Our PEKS works as follows:

- KeyGen: The input security parameter determines the size, $p$, of the groups $G_1$ and $G_2$. The algorithm picks a random $\alpha \in \mathbb{Z}_p^*$ and a generator $g$ of $G_1$. It outputs $A_{pub} = [g, h = g^\alpha]$ and $A_{priv} = \alpha$.
- PEKS($A_{pub}, W$): First compute $t = e(H_1(W), h^r) \in G_2$ for a random $r \in \mathbb{Z}_p^*$. Output PEKS($A_{pub}, W$) $= [g^r,\ H_2(t)]$.
- Trapdoor($A_{priv}, W$): output $T_W = H_1(W)^\alpha \in G_1$.
- Test($A_{pub}, S, T_W$): let $S = [A, B]$. Test if $H_2(e(T_W, A)) = B$. If so, output 'yes'; if not, output 'no'.

We prove that this system is a non-interactive searchable encryption scheme semantically secure against a chosen keyword attack in the random oracle model. The proof of security relies on the difficulty of the Bilinear Diffie-Hellman problem (BDH) [2,21].

*Bilinear Diffie-Hellman Problem (BDH):* Fix a generator $g$ of $G_1$. The BDH problem is as follows: given $g, g^a, g^b, g^c \in G_1$ as input, compute $e(g, g)^{abc} \in G_2$. We say that BDH is intractable if all polynomial time algorithms have a negligible advantage in solving BDH.

We note that the Boneh-Franklin IBE system [2] relies on the same intractability assumption for security. The security of our PEKS is proved in the following theorem. The proof is set in the random oracle model. Indeed, it is currently an open problem to build a secure IBE, and hence a PEKS, without the random oracle model.

**Theorem 1.** *The non-interactive searchable encryption scheme (PEKS) above is semantically secure against a chosen keyword attack in the random oracle model assuming BDH is intractable.*

**Proof :** Suppose $\mathcal{A}$ is an attack algorithm that has advantage $\epsilon$ in breaking the PEKS. Suppose $\mathcal{A}$ makes at most $q_{H_2}$ hash function queries to $H_2$ and at most $q_T$ trapdoor queries (we assume $q_T$ and $q_{H_2}$ are positive). We construct an algorithm $\mathcal{B}$ that solves the BDH problem with probability at least $\epsilon' = \epsilon/(eq_Tq_{H_2})$, where $e$ is the base of the natural logarithm. Algorithm $\mathcal{B}$'s running time is approximately the same as $\mathcal{A}$'s. Hence, if the BDH assumption holds in $G_1$ then $\epsilon'$ is a negligible function and consequently $\epsilon$ must be a negligible function in the security parameter.

Let $g$ be a generator of $G_1$. Algorithm $\mathcal{B}$ is given $g, u_1 = g^\alpha, u_2 = g^\beta, u_3 = g^\gamma \in G_1$. Its goal is to output $v = e(g,g)^{\alpha\beta\gamma} \in G_2$. Algorithm $\mathcal{B}$ simulates the challenger and interacts with forger $\mathcal{A}$ as follows:

KeyGen. Algorithm $\mathcal{B}$ starts by giving $\mathcal{A}$ the public key $A_{pub} = [g, u_1]$.

**$H_1, H_2$-queries.** At any time algorithm $\mathcal{A}$ can query the random oracles $H_1$ or $H_2$. To respond to $H_1$ queries algorithm $\mathcal{B}$ maintains a list of tuples $\langle W_j, h_j, a_j, c_j\rangle$ called the $H_1$-list. The list is initially empty. When $\mathcal{A}$ queries the random oracle $H_1$ at a point $W_i \in \{0,1\}^*$, algorithm $\mathcal{B}$ responds as follows:

1. If the query $W_i$ already appears on the $H_1$-list in a tuple $\langle W_i, h_i, a_i, c_i\rangle$ then algorithm $\mathcal{B}$ responds with $H_1(W_i) = h_i \in G_1$.
2. Otherwise, $\mathcal{B}$ generates a random coin $c_i \in \{0,1\}$ so that $\Pr[c_i = 0] = 1/(q_T + 1)$.
3. Algorithm $\mathcal{B}$ picks a random $a_i \in \mathbb{Z}_p$.
If $c_i = 0$, $\mathcal{B}$ computes $h_i \leftarrow u_2 \cdot g^{a_i} \in G_1$.
If $c_i = 1$, $\mathcal{B}$ computes $h_i \leftarrow g^{a_i} \in G_1$.
4. Algorithm $\mathcal{B}$ adds the tuple $\langle W_i, h_i, a_i, c_i\rangle$ to the $H_1$-list and responds to $\mathcal{A}$ by setting $H_1(W_i) = h_i$. Note that either way $h_i$ is uniform in $G_1$ and is independent of $\mathcal{A}$'s current view as required.

Similarly, at any time $\mathcal{A}$ can issue a query to $H_2$. Algorithm $\mathcal{B}$ responds to a query for $H_2(t)$ by picking a new random value $V \in \{0,1\}^{\log p}$ for each new $t$ and setting $H_2(t) = V$. In addition, $\mathcal{B}$ keeps track of all $H_2$ queries by adding the pair $(t, V)$ to an $H_2$-list. The $H_2$-list is initially empty.

**Trapdoor queries.** When $\mathcal{A}$ issues a query for the trapdoor corresponding to the word $W_i$ algorithm $\mathcal{B}$ responds as follows:

1. Algorithm $\mathcal{B}$ runs the above algorithm for responding to $H_1$-queries to obtain an $h_i \in G_1$ such that $H_1(W_i) = h_i$. Let $\langle W_i, h_i, a_i, c_i\rangle$ be the corresponding tuple on the $H_1$-list. If $c_i = 0$ then $\mathcal{B}$ reports failure and terminates.
2. Otherwise, we know $c_i = 1$ and hence $h_i = g^{a_i} \in G_1$. Define $T_i = u_1^{a_i}$. Observe that $T_i = H(W_i)^\alpha$ and therefore $T_i$ is the correct trapdoor for the keyword $W_i$ under the public key $A_{pub} = [g, u_1]$. Algorithm $\mathcal{B}$ gives $T_i$ to algorithm $\mathcal{A}$.

**Challenge.** Eventually algorithm $\mathcal{A}$ produces a pair of keywords $W_0$ and $W_1$ that it wishes to be challenged on. Algorithm $\mathcal{B}$ generates the challenge PEKS as follows:

1. Algorithm $\mathcal{B}$ runs the above algorithm for responding to $H_1$-queries twice to obtain a $h_0, h_1 \in G_1$ such that $H_1(W_0) = h_0$ and $H_1(W_1) = h_1$. For $i = 0, 1$ let $\langle W_i, h_i, a_i, c_i\rangle$ be the corresponding tuples on the $H_1$-list. If both $c_0 = 1$ and $c_1 = 1$ then $\mathcal{B}$ reports failure and terminates.

2. We know that at least one of $c_0, c_1$ is equal to 0. Algorithm $\mathcal{B}$ randomly picks a $b \in \{0,1\}$ such that $c_b = 0$ (if only one $c_b$ is equal to 0 then no randomness is needed since there is only one choice).
3. Algorithm $\mathcal{B}$ responds with the challenge PEKS $C = [u_3,\ J]$ for a random $J \in \{0,1\}^{\log p}$.

Note that this challenge implicitly defines $H_2(e(H_1(W_b), u_1^{\gamma})) = J$. In other words,

$$J = H_2(e(H_1(W_b), u_1^{\gamma})) = H_2(e(u_2 g^{a_b}, g^{\alpha\gamma})) = H_2(e(g,g)^{\alpha\gamma(\beta+a_b)})$$

With this definition, $C$ is a valid PEKS for $W_b$ as required.

**More trapdoor queries.** $\mathcal{A}$ can continue to issue trapdoor queries for keywords $W_i$ where the only restriction is that $W_i \neq W_0, W_1$. Algorithm $\mathcal{B}$ responds to these queries as before.

**Output.** Eventually, $\mathcal{A}$ outputs its guess $b' \in \{0,1\}$ indicating whether the challenge $C$ is the result of $\mathsf{PEKS}(A_{pub}, W_0)$ or $\mathsf{PEKS}(A_{pub}, W_1)$. At this point, algorithm $\mathcal{B}$ picks a random pair $(t, V)$ from the $H_2$-list and outputs $t/e(u_1,u_3)^{a_b}$ as its guess for $e(g,g)^{\alpha\beta\gamma}$, where $a_b$ is the value used in the Challenge step. The reason this works is that, as we will show, $\mathcal{A}$ must have issued a query for either $H_2(e(H_1(W_0), u_1^{\gamma}))$ or $H_2(e(H_1(W_1), u_1^{\gamma}))$. Therefore, with probability 1/2 the $H_2$-list contains a pair whose left hand side is $t = e(H_1(W_b), u_1^{\gamma}) = e(g,g)^{\alpha\gamma(\beta+a_b)}$. If $\mathcal{B}$ picks this pair $(t, V)$ from the $H_2$-list then $t/e(u_1,u_3)^{a_b} = e(g,g)^{\alpha\beta\gamma}$ as required.

This completes the description of algorithm $\mathcal{B}$. It remains to show that $\mathcal{B}$ correctly outputs $e(g,g)^{\alpha\beta\gamma}$ with probability at least $\epsilon'$. To do so, we first analyze the probability that $\mathcal{B}$ does not abort during the simulation. We define two events:

$\mathcal{E}_1$**:** $\mathcal{B}$ does not abort as a result of any of $\mathcal{A}$'s trapdoor queries.
$\mathcal{E}_2$**:** $\mathcal{B}$ does not abort during the challenge phase.

We first argue as in [6] that both events $\mathcal{E}_1$ and $\mathcal{E}_2$ occur with sufficiently high probability.

**Claim 1:** The probability that algorithm $\mathcal{B}$ does not abort as a result of $\mathcal{A}$'s trapdoor queries is at least $1/e$. Hence, $\Pr[\mathcal{E}_1] \geq 1/e$.

*Proof.* Without loss of generality we assume that $\mathcal{A}$ does not ask for the trapdoor of the same keyword twice. The probability that a trapdoor query causes $\mathcal{B}$ to abort is $1/(q_T+1)$. To see this, let $W_i$ be $\mathcal{A}$'s $i$'th trapdoor query and let $\langle W_i, h_i, a_i, c_i \rangle$ be the corresponding tuple on the $H_1$-list. Prior to issuing the query, the bit $c_i$ is independent of $\mathcal{A}$'s view — the only value that could be given to $\mathcal{A}$ that depends on $c_i$ is $H(W_i)$, but the distribution on $H(W_i)$ is the same whether $c_i = 0$ or $c_i = 1$. Therefore, the probability that this query causes $\mathcal{B}$ to abort is at most $1/(q_T+1)$. Since $\mathcal{A}$ makes at most $q_T$ trapdoor queries the probability that $\mathcal{B}$ does not abort as a result of all trapdoor queries is at least $(1-1/(q_T+1))^{q_T} \geq 1/e$. □

**Claim 2:** The probability that algorithm $\mathcal{B}$ does not abort during the challenge phase is at least $1/q_T$. Hence, $\Pr[\mathcal{E}_2] \geq 1/q_T$.

*Proof.* Algorithm $\mathcal{B}$ will abort during the challenge phase if $\mathcal{A}$ is able to produce $W_0, W_1$ with the following property: $c_0 = c_1 = 1$ where for $i = 0, 1$ the tuple $\langle W_i, h_i, a_i, c_i \rangle$ is the tuple on the $H_1$-list corresponding to $W_i$. Since $\mathcal{A}$ has not queried for the trapdoor for $W_0, W_1$ we have that both $c_0, c_1$ are independent of $\mathcal{A}$'s current view. Therefore, since $\Pr[c_i = 0] = 1/(q_T + 1)$ for $i = 0, 1$, and the two values are independent of one another, we have that $\Pr[c_0 = c_1 = 1] = (1 - 1/(q_T + 1))^2 \leq 1 - 1/q_T$. Hence, the probability that $\mathcal{B}$ does not abort is at least $1/q_T$. □

Observe that since $\mathcal{A}$ can never issue a trapdoor query for the challenge keywords $W_0, W_1$ the two events $\mathcal{E}_1$ and $\mathcal{E}_2$ are independent. Therefore, $\Pr[\mathcal{E}_1 \wedge \mathcal{E}_2] \geq 1/(eq_T)$.

To complete the proof of Theorem 1 it remains to show that $\mathcal{B}$ outputs the solution to the given BDH instance with probability at least $\epsilon/q_{H_2}$. To do we show that during the simulation $\mathcal{A}$ issues a query for $H_2(e(H_1(W_b), u_1^\gamma))$ with probability at least $\epsilon$.

**Claim 3:** Suppose that in a real attack game $\mathcal{A}$ is given the public key $[g, u_1]$ and $\mathcal{A}$ asks to be challenged on words $W_0$ and $W_1$. In response, $\mathcal{A}$ is given a challenge $C = [g^r, J]$. Then, in the real attack game $\mathcal{A}$ issues an $H_2$ query for either $H_2(e(H_1(W_0), u_1^r))$ or $H_2(e(H_1(W_1), u_1^r))$ with probability at least $2\epsilon$.

*Proof.* Let $\mathcal{E}_3$ be the event that in the real attack $\mathcal{A}$ does not issue a query for either one of $H_2(e(H_1(W_0), u_1^r))$ and $H_2(e(H_1(W_1), u_1^r))$. Then, when $\mathcal{E}_3$ occurs we know that the bit $b \in \{0, 1\}$ indicating whether $C$ is a PEKS of $W_0$ or $W_1$ is independent of $\mathcal{A}$'s view. Therefore, $\mathcal{A}$'s output $b'$ will satisfy $b = b'$ with probability at most $\frac{1}{2}$. By definition of $\mathcal{A}$, we know that in the real attack $|\Pr[b = b'] - 1/2| \geq \epsilon$. We show that these two facts imply that $\Pr[\neg\mathcal{E}_3] \geq 2\epsilon$. To do so, we first derive simple upper and lower bounds on $\Pr[b = b']$:

$$\begin{aligned} \Pr[b = b'] &= \Pr[b = b' | \mathcal{E}_3] \Pr[\mathcal{E}_3] + \Pr[b = b' | \neg\mathcal{E}_3] \Pr[\neg\mathcal{E}_3] \\ &\leq \Pr[b = b' | \mathcal{E}_3] \Pr[\mathcal{E}_3] + \Pr[\neg\mathcal{E}_3] \\ &= \frac{1}{2} \Pr[\mathcal{E}_3] + \Pr[\neg\mathcal{E}_3] \\ &= \frac{1}{2} + \frac{1}{2} \Pr[\neg\mathcal{E}_3], \end{aligned}$$

$$\Pr[b = b'] \geq \Pr[b = b' | \mathcal{E}_3] \Pr[\mathcal{E}_3] = \frac{1}{2} \Pr[\mathcal{E}_3] = \frac{1}{2} - \frac{1}{2} \Pr[\neg\mathcal{E}_3].$$

It follows that $\epsilon \leq |\Pr[b = b'] - 1/2| \leq \frac{1}{2} \Pr[\neg\mathcal{E}_3]$. Therefore, in the real attack, $\Pr[\neg\mathcal{E}_3] \geq 2\epsilon$ as required. □

Now, assuming $\mathcal{B}$ does not abort, we know that $\mathcal{B}$ simulates a real attack game perfectly up to the moment when $\mathcal{A}$ issues a query for either $H_2(e(H_1(W_0), u_1^\gamma))$ or $H_2(e(H_1(W_1), u_1^\gamma))$. Therefore, by Claim 3, by the end of the simulation $\mathcal{A}$ will have issued a query for either $H_2(e(H_1(W_0), u_1^\gamma))$ or $H_2(e(H_1(W_1), u_1^\gamma))$ with probability at least $2\epsilon$. It follows that $\mathcal{A}$ issues a query for $H_2(e(H_1(W_b), u_1^\gamma))$ with probability at least $\epsilon$. Consequently, the value $e(H_1(W_b), u_1^\gamma) = e(g^{\beta + a_b}, g)^{\alpha\gamma}$ will appear on the left hand side of some pair in the $H_2$-list. Algorithm $\mathcal{B}$ will choose the correct pair with probability at least $1/q_{H_2}$ and therefore, assuming $\mathcal{B}$ does not abort during the simulation, it will produce the correct answer with probability at least $\epsilon/q_{H_2}$. Since $\mathcal{B}$ does not abort with probability at least $1/(eq_T)$ we see that $\mathcal{B}$'s success probability overall is at least $\epsilon/(eq_T q_{H_2})$ as required. □

### 3.2 A Limited Construction Using Any Trapdoor Permutation

Our second PEKS construction is based on general trapdoor permutations, assuming that the total number of keywords that the user wishes to search for is bounded by some polynomial function in the security parameter. (As a first step in our construction, we will make an even stronger assumption that the total number of words $\Sigma \subset \{0,1\}^*$ in the dictionary is also bounded by a polynomial function, we will later show how to remove this additional assumption.) We will also need a family of semantically-secure encryptions where given a ciphertext it is computationally hard to say which public-key this ciphertext is associated with. This notion was formalized by Bellare et al. [1]. We say that a public-key system that has this property is **source-indistinguishable**. More precisely, source-indistinguishability for an encryption scheme $(G, E, D)$ is defined using the following game between a challenger and an attacker $\mathcal{A}$ (here $G$ is the key generation algorithm, and $E/D$ are encryption/decryption algorithms). The security parameter $s$ is given to both players.
Source Indistinguishability security game:

1. The challenger runs algorithm $G(s)$ two times to generate two public/private key pairs $(PK_0, Priv_0)$ and $(PK_1, Priv_1)$.
2. The challenger picks a random $M \in \{0,1\}^s$ and a random $b \in \{0,1\}$ and computes an encryption $C = PK_b(M)$. The challenger gives $(M, C)$ to the attacker.
3. The attacker outputs $b'$ and wins the game if $b = b'$.

In other words, the attacker wins if he correctly guesses whether he was given the encryption of $M$ under $PK_0$ or under $PK_1$. We define $\mathcal{A}$'s advantage in winning the game as:

$$\mathrm{AdvSI}_{\mathcal{A}}(s) = |\Pr[b = b'] - \frac{1}{2}|$$

**Definition 3.** We say that a public-key encryption scheme is source indistinguishable if for any polynomial time attacker $\mathcal{A}$ we have that $\mathrm{AdvSI}_{\mathcal{A}}(s)$ is a negligible function.

We note that Bellare et al. [1] define a stronger notion of source indistinguishability than the one above by allowing the adversary to choose the challenge message $M$. For our purposes, giving the adversary an encryption of a random message is sufficient.

It is easy to check that source indistinguishability can be attained from any trapdoor permutation family, where for a given security parameter all permutations in the family are defined over the same domain. Such a family can be constructed from any family of trapdoor permutations as described in [1]. Then to encrypt a bit $b$ we pick a random $x$, and output $[f(x), GL(x) \oplus b]$ where GL is the Goldreich-Levin hard-core bit [19]. We therefore obtain the following lemma:

**Lemma 2.** *Given any trapdoor permutation family we can construct a semantically secure source indistinguishable encryption scheme.*

We note that source indistinguishability is an orthogonal property to semantic security. One can build a semantically secure system that is not source indistinguishable (by embedding the public key in every ciphertext). Conversely, one can build a source indistinguishable system that is not semantically secure (by embedding the plaintext in every ciphertext).

*A simple* PEKS *from trapdoor permutations.* When the keyword family $\Sigma$ is of polynomial size (in the security parameter) it is easy to construct searchable encryption from any source-indistinguishable public-key system $(G, E, D)$. We let $s$ be the security parameter for the scheme.

- KeyGen: For each $W \in \Sigma$ run $G(s)$ to generate a new public/private key pair $PK_W/Priv_W$ for the source-indistinguishable encryption scheme. The PEKS public key is $A_{pub} = \{PK_W \mid W \in \Sigma\}$. The private key is $A_{priv} = \{Priv_W \mid W \in \Sigma\}$.
- PEKS($A_{pub}, W$): Pick a random $M \in \{0,1\}^s$ and output PEKS($A_{pub}, W$) = $(M,\ E[PK_W, M])$, i.e. encrypt $M$ using the public key $PK_W$.
- Trapdoor($A_{priv}, W$): The trapdoor for word $W$ is simply $T_W = Priv_W$.
- Test($A_{pub}, S, T_W$): Test if the decryption $D[T_W, S] = 0^s$. Output 'yes' if so and 'no' otherwise.

Note that the dictionary must be of polynomial size (in $s$) so that the public and private keys are of polynomial size (in $s$).

This construction gives a semantically secure PEKS as stated in the following simple theorem. Semantically secure PEKS is defined as in Definition 2 except that the adversary is not allowed to make chosen keyword queries.

**Theorem 2.** *The* PEKS *scheme above is semantically secure assuming the underlying public key encryption scheme $(G, E, D)$ is source-indistinguishable.*

**Proof sketch:** Let $\Sigma = \{W_1, \ldots, W_k\}$ be the keyword dictionary. Suppose we have a PEKS attacker $\mathcal{A}$ for which $\mathrm{Adv}_{\mathcal{A}}(s) > \epsilon(s)$. We build an attacker $\mathcal{B}$ that breaks the source indistinguishability of $(G, E, D)$ where $\mathrm{AdvSI}_{\mathcal{B}}(s) > \epsilon(s)/k^2$.

The reduction is immediate: $\mathcal{B}$ is given two public keys $PK_0, PK_1$ and a pair $(M, C)$ where $M$ is random in $\{0,1\}^s$ and $C = PK_b(M)$ for $b \in \{0,1\}$. Algorithm $\mathcal{B}$ generates $k-2$ additional public/private keys using $G(s)$. It creates $A_{pub}$ as a list of all $k$ public keys with $PK_0, PK_1$ embedded in a random location in the list. Let $W_i, W_j$ be the words associated with the public keys $PK_0, PK_1$. $\mathcal{B}$ sends $A_{pub}$ to $\mathcal{A}$ who then responds with two words $W_k, W_\ell \in \Sigma$ on which $\mathcal{A}$ wishes to be challenged. If $\{i, j\} \neq \{k, \ell\}$ algorithm $\mathcal{B}$ reports failure and aborts. Otherwise, $\mathcal{B}$ sends the challenge $(M, C)$ to $\mathcal{A}$ who then responds with a $b' \in$

$\{0,1\}$. Algorithm $\mathcal{B}$ outputs $b'$ as its response to the source indistinguishability challenge. We have that $b = b'$ if algorithm $\mathcal{B}$ did not abort and $\mathcal{A}$'s response was correct. This happens with probability at least $\frac{1}{2} + \epsilon/k^2$. Hence, $\text{AdvSI}_{\mathcal{B}}(s) > \epsilon(s)/k^2$ as required. □

We note that this PEKS can be viewed as derived from an IBE system with a limited number of identities. For each identity there is a pre-specified public key. Such an IBE system is implied in the work of Dodis et al. [13]. They propose reducing the size of the public-key using cover-free set systems. We apply the same idea below to reduce the size of the public key in the PEKS above.

*Reducing the public key size.* The drawback of the above scheme is that the public key length grows linearly with the total dictionary size. If we have an upper-bound on the total number of keyword trapdoors that the user will release to the email gateway (though we do not need to know these keywords a-priori) we can do much better using cover-free families [15] and can allow keyword dictionary to be of exponential size. Since typically a user will only allow a third party (such as e-mail server) to search for a limited number of keywords so that assuming an upper bound on the number of released trapdoors is within reason. We begin by recalling the definition of cover-free families.

**Definition 4. Cover-free families.** *Let $d, t, k$ be positive integers, let $G$ be a ground set of size $d$, and let $F = \{S_1, \ldots, S_k\}$ be a family of subsets of $G$. We say that subset $S_j$ does not* cover *$S_i$ if it holds that $S_i \not\subseteq S_j$. We say that family $F$ is $t$-*cover free *over $G$ if each subset in $F$ is not covered by the union of $t$ subsets in $F$. Moreover, we say that a family of subsets is $q$-*uniform *if all subsets in the family have size $q$.*

We will use the following fact from [14].

**Lemma 3.** [14] There exists a deterministic algorithm that, for any fixed $t, k$, constructs a $q$-uniform $t$-cover free family $F$ over a ground set of size $d$, for $q = \lceil d/4t \rceil$ and $d \leq 16t^2(1 + \log(k/2)/\log 3)$.

*The* PEKS. Given the previous PEKS construction as a starting point, we can significantly reduce the size of public file $A_{pub}$ by allowing user to re-use individual public keys for different keywords. We associate to each keyword a subset of public keys chosen from a cover free family. Let $k$ be the size of the dictionary $\Sigma = \{W_1, \ldots, W_k\}$ and let $t$ be an upper bound on the number of keyword trapdoors released to the mail gateway by user Alice. Let $d, q$ satisfy the bounds of Lemma 3. The PEKS$(d, t, k, q)$ construction is as follows:

- KeyGen: For $i = 1, \ldots, d$ run algorithm $G(s)$ to generate a new public/private key pair $PK_i/Priv_i$ for the source-indistinguishable encryption scheme. The PEKS public key is $A_{pub} = \{PK_1, \ldots, PK_d\}$. The private key is $A_{priv} = \{Priv_1, \ldots, Priv_d\}$. We will be using a $q$-uniform $t$-cover free family of subsets $F = \{S_1, \ldots, S_k\}$ of $\{PK_1, \ldots, PK_d\}$. Hence, each $S_i$ is a subset of public keys.

- $\mathsf{PEKS}(A_{pub}, W_i)$: Let $S_i \in F$ be the subset associated with the word $W_i \in \Sigma$. Let $S_i = \{PK^{(1)}, \ldots, PK^{(q)}\}$. Pick random messages $M_1, \ldots, M_q \in \{0,1\}^s$ and let $M = M_1 \oplus \cdots \oplus M_q$. Output the tuple:

$$\mathsf{PEKS}(A_{pub}, W_i) = \Big(M,\ E[PK^{(1)}, M_1],\ \ldots,\ E[PK^{(q)}, M_q]\Big)$$

- $\mathsf{Trapdoor}(A_{priv}, W_i)$: Let $S_i \in F$ be the subset associated with word $W_i \in \Sigma$. The trapdoor for word $W_i$ is simply the set of private keys that correspond to the public keys in the set $S_i$.
- $\mathsf{Test}(A_{pub}, R, T_W)$:
Let $T_W = \{Priv^{(1)}, \ldots, Priv^{(q)}\}$ and let $R = (M,\ C_1,\ \ldots,\ C_q)$ be a PEKS. For $i = 1, \ldots, q$ decrypt each $C_i$ using private key $Priv^{(i)}$ to obtain $M_i$. Output 'yes' if $M = M_1 \oplus \cdots \oplus M_q$, and output 'no' otherwise.

The size of the public key file $A_{pub}$ is much smaller now: logarithmic in the size of the dictionary. The downside is that Alice can only release $t$ keywords to the email gateway. Once $t$ trapdoors are released privacy is no longer guaranteed. Also, notice that the size of the PEKS is larger now (logarithmic in the dictionary size and linear in $t$). The following corollary of Theorem 2 shows that the resulting PEKS is secure.

**Corollary 1.** *Let $d, t, k, q$ satisfy the bounds of Lemma 3. The* $\mathsf{PEKS}(d, t, k, q)$ *scheme above is semantically secure under a chosen keyword attack assuming the underlying public key encryption scheme $(G, E, D)$ is source-indistinguishable and semantically secure, and that the adversary makes no more than $t$ trapdoors queries.*

**Proof sketch:** Let $\Sigma = \{W_1, \ldots, W_k\}$ be the keyword dictionary. Suppose we have a PEKS attacker $\mathcal{A}$ for which $\mathrm{Adv}_{\mathcal{A}}(s) > \epsilon(s)$. We build an attacker $\mathcal{B}$ that breaks the source indistinguishability of $(G, E, D)$.

Algorithm $\mathcal{B}$ is given two public keys $PK_0, PK_1$ and a pair $(M, C)$ where $M$ is random in $\{0,1\}^s$ and $C = PK_b(M)$ for $b \in \{0,1\}$. Its goal is to output a guess for $b$ which it does by interacting with $\mathcal{A}$. Algorithm $\mathcal{B}$ generates $d-2$ additional public/private keys using $G(s)$. It creates $A_{pub}$ as a list of all $d$ public keys with $PK_0, PK_1$ embedded in a random location in the list. Let $W_i, W_j$ be the words associated with the public keys $PK_0, PK_1$.

$\mathcal{B}$ sends $A_{pub}$ to $\mathcal{A}$. Algorithm $\mathcal{A}$ issues up to $t$ trapdoor queries. $\mathcal{B}$ responds to a trapdoor query for $W \in \Sigma$ as follows: let $S \in F$ be the subset corresponding to the word $W$. If $PK_0 \in S$ or $PK_1 \in S$ algorithm $\mathcal{B}$ reports failure and aborts. Otherwise, $\mathcal{B}$ gives $\mathcal{A}$ the set of private keys $\{Priv_i \,|\, i \in S\}$.

At some point, Algorithm $\mathcal{A}$ outputs two words $W_0', W_1' \in \Sigma$ on which it wishes to be challenged. Let $S_0', S_1' \in F$ be the subsets corresponding to $W_0', W_1'$ respectively. Let $\mathcal{E}$ be the event that $PK_0 \in S_0'$ and $PK_1 \in S_1'$. If event $\mathcal{E}$ did not happen then $\mathcal{B}$ reports failure and aborts.

We now know that $PK_0 \in S_0'$ and $PK_1 \in S_1'$. For $j = 0,1$ let $S_j' = \{PK_j^{(1)}, \ldots, PK_j^{(q)}\}$. We arrange things so that $PK_0 = PK_0^{(c)}$ and $PK_1 =$

$PK_1^{(c)}$ for some random $1 \leq c \leq q$. Next, $\mathcal{B}$ picks random $M_1, \ldots, M_{c-1}, M_{c+1}, \ldots, M_q \in \{0,1\}^s$ and sets $M_c = M$. Let $M' = M_1 \oplus \cdots \oplus M_q$. Algorithm $\mathcal{B}$ defines the following hybrid tuple:

$$R = \Big(M',\ E[PK_0^{(1)}, M_1],\ \ldots,\ E[PK_0^{(c-1)}, M_{c-1}],\ C,$$
$$E[PK_1^{(c+1)}, M_{c+1}],\ \ldots,\ E[PK_1^{(q)}, M_q]\Big)$$

It gives $R$ as the challenge PEKS to algorithm $\mathcal{A}$. Algorithm $\mathcal{A}$ eventually responds with some $b' \in \{0,1\}$ indicating whether $R$ is PEKS$(A_{pub}, W_0')$ or PEKS$(A_{pub}, W_1')$. Algorithm $\mathcal{B}$ outputs $b'$ as its guess for $b$. One can show using a standard hybrid argument that if $\mathcal{B}$ does not abort then $|\Pr[b = b'] - \frac{1}{2}| > \epsilon/q^2$. The probability that $\mathcal{B}$ does not abort at a result of a trapdoor query is at least $1 - (tq/d)$. The probability that $\mathcal{B}$ does not abort as a result of the choice of words $W_0', W_1'$ is at least $(q/d)^2$. Hence, $\mathcal{B}$ does not abort with probability at least $1/poly(t, q, d)$. Repeatedly running $\mathcal{B}$ until it does not abort shows that we can get advantage $\epsilon/q^2$ in breaking the source indistinguishability of $(G, E, D)$ in expected polynomial time in the running time of $\mathcal{A}$. □

## 4 Construction Using Jacobi Symbols

Given the relation between Identity Based Encryption and PEKS it is tempting to construct a PEKS from an IBE system due to Cocks [3]. The security of Cocks' IBE system is based on the difficulty of distinguishing quadratic residues from non-residues modulo $N = pq$ where $p = q = 3(\text{mod} 4)$.

Unfortunately, Galbraith [11] shows that the Cocks system as described in [3] is not public-key private in the sense of Bellare et al. [1]. Therefore it appears that the Cocks system cannot be directly used to construct a PEKS. It provides a good example that constructing a PEKS is a harder problem than constructing an IBE.

## 5 Conclusions

We defined the concept of a public key encryption with keyword search (PEKS) and gave two constructions. Constructing a PEKS is related to Identity Based Encryption (IBE), though PEKS seems to be harder to construct. We showed that PEKS implies Identity Based Encryption, but the converse is currently an open problem. Our constructions for PEKS are based on recent IBE constructions. We are able to prove security by exploiting extra properties of these schemes.

## Acknowledgments

We thank Glenn Durfee for suggesting the use of $H_2$ in the construction of Section 3.1. We thank Yevgeniy Dodis, David Molnar, and Steven Galbraith for helpful comments on this work.

## References


1. M. Bellare, A. Boldyreva, A. Desai, and D. Pointcheval "Key-Privacy in Public-Key Encryption," in Advances in Cryptology - Asiacrypt 2001 Proceedings, LNCS Vol. 2248, Springer-Verlag, 2001.
2. D. Boneh and M. Franklin, *Identity-based Encryption from the Weil Pairing,* SIAM J. of Computing, Vol. 32, No. 3, pp. 586-615, 2003, Extended abstract in Crypto 2001.
3. C. Cocks, *An identity based encryption scheme based on quadratic residues*, Eighth IMA International Conference on Cryptography and Coding, Dec. 2001, Royal Agricultural College, Cirencester, UK.
4. C. Cachin, S. Micali, M. Stadler *Computationally Private Information Retrieval with Polylogarithmic Communication* Eurcrypt 1999.
5. B. Chor, O. Goldreich, E. Kushilevitz and M. Sudan, *Private Information Retrieval,* in FOCS 95 (also Journal of ACM).
6. J. Coron, "On the exact security of Full-Domain-Hash", in *Advances in Cryptology – Crypto 2000*, Lecture Notes in Computer Science, Vol. 1880, Springer-Verlag, pp. 229–235, 2000.
7. D. Dolev, C. Dwork, and M. Naor, "Non-Malleable Cryptography," in SIAM Journal on Computing, 2000. Early version in proceedings of STOC '91.
8. G. Di Crescenzo, Y. Ishai, and R. Ostrovsky. Universal service-providers for database private information retrieval. In *Proc. of the 17th Annu. ACM Symp. on Principles of Distributed Computing*, pages 91-100, 1998.
9. G. Di Crescenzo, T. Malkin, and R. Ostrovsky. Single-database private information retrieval implies oblivious transfer. In *Advances in Cryptology - EUROCRYPT 2000*, 2000.
10. A. Iliev, S. Smith Privacy-enhanced credential services. Second annual PKI workshop. (see also Darthmoth Technical Report TR-2003-442; http://www.cs.dartmouth.edu/ sws/papers/ilsm03.pdf.
11. S. Galbraith, private communications.
12. Y. Desmedt, "Computer security by redefining what a computer is," in Proceedings New Security Paradigms II Workshop, pp. 160–166, 1992.
13. Y. Dodis, J. Katz, S. Xu, and M. Yung. "Key-insulated public key cryptosystems," in Advances in Cryptology – Eurocrypt 2002, LNCS, Springer-Verlag, pp. 65–82, 2002.
14. D. Z. Du and F. K. Hwang, *Combinatorial Group Testing and its Applications,* World Scientific, Singapore, 1993.
15. P. Erdos, P. Frankl and Z. Furedi, *Families of finite sets in which no set is covered by the union of r others,* in Israeli Journal of Mathematics, 51: 79–89, 1985.
16. E. Goh, "Building Secure Indexes for Searching Efficiently on Encrypted Compressed Data," http://eprint.iacr.org/2003/216/
17. O. Goldreich and R. Ostrovsky. Software protection and simulation by oblivious RAMs. *JACM*, 1996.
18. Goldreich, O., S. Goldwasser, and S. Micali, "How To Construct Random Functions," *Journal of the Association for Computing Machinery*, Vol. 33, No. 4 (October 1986), 792-807.
19. S. Goldwasser and S. Micali, *Probabilistic Encryption,* in Journal of Computer and System Sciences. vol. 28 (1984), n. 2, pp. 270–299.
20. Y. Gertner, Y. Ishai, E. Kushilevitz, and T. Malkin Protecting data privacy in private information retrieval schemes. In Proc. of the *30th Annual ACM Symposium on the Theory of Computing*, pp. 151-160, 1998.

21. A. Joux, "The Weil and Tate Pairings as Building Blocks for Public Key Cryptosystems", in *Proc. Fifth Algorithmic Number Theory Symposium*, Lecture Notes in Computer Science, Springer-Verlag, 2002.
22. E. Kushilevitz and R. Ostrovsky, *Replication is not needed: Single Database, Computationally-Private Information Retrieval,* in FOCS 97.
23. E. Kushilevitz and R. Ostrovsky. One-way Trapdoor Permutations are Sufficient for Non-Trivial Single-Database Computationally-Private Information Retrieval. In *Proc. of EUROCRYPT '00*, 2000.
24. P. Maniatis, M. Roussopoulos, E. Swierk, K. Lai, G. Appenzeller, X. Zhao, and M. Baker, *The Mobile People Architecture.* ACM Mobile Computing and Communications Review (MC2R), Volume 3, Number 3, July 1999.
25. M. Naor and B. Pinkas. Oblivious transfer and polynomial evaluation. In *Proc. of the 31th Annu. ACM Symp. on the Theory of Computing*, pages 245–254, 1999.
26. R. Ostrovsky. Software protection and simulation on oblivious RAMs. MIT Ph.D. Thesis, 1992. Preliminary version in *Proc. 22nd Annual ACM Symp. Theory Comp.*, 1990.
27. W. Ogata and K. Kurosawa, "Oblivious keyword search," to appear in J. of Complexity.
28. D. Song, D. Wagner, and A. Perrig, *Practical Techniques for Searches on Encrypted Data,* in Proc. of the 2000 IEEE symposium on Security and Privacy (S&P 2000).
29. A. Shamir, *Identity-based Cryptosystems and Signature Schemes,* in CRYPTO 84.
30. B. Waters, D. Balfanz, G. Durfee, D. Smetters, "Building an encrypted and searchable audit log", to appear in NDSS '04.

# Fuzzy Extractors: How to Generate Strong Keys from Biometrics and Other Noisy Data


Yevgeniy Dodis[1], Leonid Reyzin[2], and Adam Smith[3]

[1] New York University, dodis@cs.nyu.edu
[2] Boston University, reyzin@cs.bu.edu
[3] MIT, asmith@csail.mit.edu



**Abstract.** We provide formal definitions and efficient secure techniques for

- turning biometric information into keys usable for *any* cryptographic application, and
- reliably and securely authenticating biometric data.

Our techniques apply not just to biometric information, but to any keying material that, unlike traditional cryptographic keys, is (1) not reproducible precisely and (2) not distributed uniformly. We propose two primitives: a *fuzzy extractor* extracts nearly uniform randomness $R$ from its biometric input; the extraction is error-tolerant in the sense that $R$ will be the same even if the input changes, as long as it remains reasonably close to the original. Thus, $R$ can be used as a key in any cryptographic application. A *secure sketch* produces public information about its biometric input $w$ that does not reveal $w$, and yet allows exact recovery of $w$ given another value that is close to $w$. Thus, it can be used to reliably reproduce error-prone biometric inputs without incurring the security risk inherent in storing them.

In addition to formally introducing our new primitives, we provide nearly optimal constructions of both primitives for various measures of "closeness" of input data, such as Hamming distance, edit distance, and set difference.


## 1 Introduction

Cryptography traditionally relies on uniformly distributed random strings for its secrets. Reality, however, makes it difficult to create, store, and reliably retrieve such strings. Strings that are neither uniformly random nor reliably reproducible seem to be more plentiful. For example, a random person's fingerprint or iris scan is clearly not a uniform random string, nor does it get reproduced precisely each time it is measured. Similarly, a long pass-phrase (or answers to 15 questions [12] or a list of favorite movies [16]) is not uniformly random and is difficult to remember for a human user. This work is about using such nonuniform and unreliable secrets in cryptographic applications. Our approach is rigorous and general, and our results have both theoretical and practical value.


C. Cachin and J. Camenisch (Eds.): EUROCRYPT 2004, LNCS 3027, pp. 523–540, 2004.

To illustrate the use of random strings on a simple example, let us consider the task of password authentication. A user Alice has a password $w$ and wants to gain access to her account. A trusted server stores some information $y = f(w)$ about the password. When Alice enters $w$, the server lets Alice in only if $f(w) = y$. In this simple application, we assume that it is safe for Alice to enter the password for the verification. However, the server's long-term storage is not assumed to be secure (e.g., $y$ is stored in a publicly readable `/etc/passwd` file in UNIX). The goal, then, is to design an efficient $f$ that is hard to invert (i.e., given $y$ it is hard to find $w'$ s.t. $f(w') = y$), so that no one can figure out Alice's password from $y$. Recall that such functions $f$ are called *one-way functions.*

Unfortunately, the solution above has several problems when used with passwords $w$ available in real life. First, the definition of a one-way function assumes that $w$ is *truly uniform*, and guarantees nothing if this is not the case. However, human-generated and biometric passwords are far from uniform, although they do have some unpredictability in them. Second, Alice has to reproduce her password *exactly* each time she authenticates herself. This restriction severely limits the kinds of passwords that can be used. Indeed, a human can precisely memorize and reliably type in only relatively short passwords, which do not provide an adequate level of security. Greater levels of security are achieved by longer human-generated and biometric passwords, such as pass-phrases, answers to questionnaires, handwritten signatures, fingerprints, retina scans, voice commands, and other values selected by humans or provided by nature, possibly in combination (see [11] for a survey). However, two biometric readings are rarely identical, even though they are likely to be close; similarly, humans are unlikely to precisely remember their answers to multiple question from time to time, though such answers will likely be similar. In other words, the ability to tolerate a (limited) number of errors in the password while retaining security is crucial if we are to obtain greater security than provided by typical user-chosen short passwords.

The password authentication described above is just one example of a cryptographic application where the issues of nonuniformity and error tolerance naturally come up. Other examples include any cryptographic application, such as encryption, signatures, or identification, where the secret key comes in the form of "biometric" data.

OUR DEFINITIONS. We propose two primitives, termed *secure sketch* and *fuzzy extractor.*

A secure sketch addresses the problem of error tolerance. It is a (probabilistic) function outputting a public value $v$ about its biometric input $w$, that, while revealing little about $w$, allows its exact reconstruction from any other input $w'$ that is sufficiently close. The price for this error tolerance is that the application will have to work with a lower level of entropy of the input, since publishing $v$ effectively reduces the entropy of $w$. However, in a good secure sketch, this reduction will be small, and $w$ will still have enough entropy to be useful, even if the adversary knows $v$. A secure sketch, however, does not address nonuniformity of inputs.

A fuzzy extractor addresses both error tolerance and nonuniformity. It reliably extracts a uniformly random string $R$ from its biometric input $w$ in an error-tolerant way. If the input changes but remains close, the extracted $R$ remains the same. To assist in recovering $R$ from $w'$, a fuzzy extractor outputs a public string $P$ (much like a secure sketch outputs $v$ to assist in recovering $w$). However, $R$ remains uniformly random even given $P$.

Our approach is general: our primitives can be naturally combined with *any* cryptographic system. Indeed, $R$ extracted from $w$ by a fuzzy extractor can be used as a key in any cryptographic application, but, unlike traditional keys, need not be stored (because it can be recovered from any $w'$ that is close to $w$). We define our primitives to be *information-theoretically* secure, thus allowing them to be used in combination with any cryptographic system without additional assumptions (however, the cryptographic application itself will typically have computational, rather than information-theoretic, security).

For a concrete example of how to use fuzzy extractors, in the password authentication case, the server can store $\langle P, f(R) \rangle$. When the user inputs $w'$ close to $w$, the server recovers the actual $R$ and checks if $f(R)$ matches what it stores. Similarly, $R$ can be used for symmetric encryption, for generating a public-secret key pair, or any other application. Secure sketches and extractors can thus be viewed as providing fuzzy key storage: they allow recovery of the secret key ($w$ or $R$) from a faulty reading $w'$ of the password $w$, by using some public information ($v$ or $P$). In particular, fuzzy extractors can be viewed as error- and nonuniformity-tolerant secret key *key-encapsulation mechanisms* [27].

Because different biometric information has different error patterns, we do not assume any particular notion of closeness between $w'$ and $w$. Rather, in defining our primitives, we simply assume that $w$ comes from some metric space, and that $w'$ is no more that a certain distance from $w$ in that space. We only consider particular metrics when building concrete constructions.

General Results. Before proceeding to construct our primitives for concrete metrics, we make some observations about our definitions. We demonstrate that fuzzy extractors can be built out of secure sketches by utilizing (the usual) strong randomness extractors [24], such as, for example, pairwise-independent hash functions. We also demonstrate that the existence of secure sketches and fuzzy extractors over a particular metric space implies the existence of certain error-correcting codes in that space, thus producing lower bounds on the best parameters a secure fingerprint and fuzzy extractor can achieve. Finally, we define a notion of a *biometric embedding* of one metric space into another, and show that the existence of a fuzzy extractor in the target space implies, combined with a biometric embedding of the source into the target, the existence of a fuzzy extractor in the source space.

These general results help us in building and analyzing our constructions.

Our Constructions. We provide constructions of secure sketches and extractors in three metrics: Hamming distance, set difference, and edit distance.

Hamming distance (i.e., the number of bit positions that differ between $w$ and $w'$) is perhaps the most natural metric to consider. We observe that the "fuzzy-commitment" construction of Juels and Wattenberg [15] based on error-correcting codes can be viewed as a (nearly optimal) secure sketch. We then apply our general result to convert it into a nearly optimal fuzzy extractor. While our results on the Hamming distance essentially use previously known constructions, they serve as an important stepping stone for the rest of the work.

The set difference metric (i.e., size of the symmetric difference of two input sets $w$ and $w'$) comes up naturally whenever the biometric input is represented as a subset of features from a universe of possible features.[4] We demonstrate the existence of optimal (with respect to entropy loss) secure sketches (and therefore also fuzzy extractors) for this metric. However, this result is mainly of theoretical interest, because (1) it relies on optimal constant-weight codes, which we do not know how construct and (2) it produces sketches of length proportional to the universe size. We then turn our attention to more efficient constructions for this metric, and provide two of them.

First, we observe that the "fuzzy vault" construction of Juels and Sudan [16] can be viewed as a secure sketch in this metric (and then converted to a fuzzy extractor using our general result). We provide a new, simpler analysis for this construction, which bounds the entropy lost from $w$ given $v$. Our bound on the loss is quite high unless one makes the size of the output $v$ very large. We then provide an improvement to the Juels-Sudan construction to reduce the entropy loss to near optimal, while keeping $v$ short (essentially as long as $w$).

Second, we note that in the case of a small universe, a set can be simply encoded as its characteristic vector (1 if an element is in the set, 0 if it is not), and set difference becomes Hamming distance. However, the length of such a vector becomes unmanageable as the universe size grows. Nonetheless, we demonstrate that this approach can be made to work efficiently even for exponentially large universes. This involves a result that may be of independent interest: we show that BCH codes can be decoded in time polynomial in the *weight* of the received corrupted word (i.e., in *sublinear* time if the weight is small). The resulting secure sketch scheme compares favorably to the modified Juels-Sudan construction: it has the same near-optimal entropy loss, while the public output $v$ is even shorter (proportional to the number of errors tolerated, rather than the input length).

Finally, edit distance (i.e., the number of insertions and deletions needed to convert one string into the other) naturally comes up, for example, when the password is entered as a string, due to typing errors or mistakes made in hand-writing recognition. We construct a biometric embedding from the edit metric into the set difference metric, and then apply our general result to show such an embedding yields a fuzzy extractor for edit distance, because we already have fuzzy extractors for set difference. We note that the edit metric is quite difficult

[4] A perhaps unexpected application of the set difference metric was explored in [16]: a user would like to encrypt a file (e.g., her phone number) using a small subset of values from a large universe (e.g., her favorite movies) in such a way that those and only those with a similar subset (e.g., similar taste in movies) can decrypt it.

to work with, and the existence of such an embedding is not a priori obvious: for example, low-distortion embeddings of the edit distance into the Hamming distance are unknown and seem hard [2]. It is the particular properties of biometric embeddings, as we define them, that help us construct this embedding.

RELATION TO PREVIOUS WORK. Since our work combines elements of error correction, randomness extraction and password authentication, there has been a lot of related work.

The need to deal with nonuniform and low-entropy passwords has long been realized in the security community, and many approaches have been proposed. For example, Ellison et al. [10] propose asking the user a series of $n$ personalized questions, and use these answers to encrypt the "actual" truly random secret $R$. A similar approach using user's keyboard dynamics (and, subsequently, voice [21,22]) was proposed by Monrose et al [20]. Of course, this technique reduces the question to that of designing a secure "fuzzy encryption". While heuristic approaches were suggested in the above works (using various forms of Shamir's secret sharing), no formal analysis was given. Additionally, error tolerance was addressed only by brute force search.

A formal approach to error tolerance in biometrics was taken by Juels and Wattenberg [15] (for less formal solutions, see [8,20,10]), who provided a simple way to tolerate errors in *uniformly distributed* passwords. Frykholm and Juels [12] extended this solution; our analysis is quite similar to theirs in the Hamming distance case. Almost the same construction appeared implicitly in earlier, seemingly unrelated, literature on information reconciliation and privacy amplification (see, e.g., [3,4,7]). We discuss the connections between these works and our work further in Section 4.

Juels and Sudan [16] provided the first construction for a metric other than Hamming: they construct a "fuzzy vault" scheme for the set difference metric. The main difference is that [16] lacks a cryptographically strong definition of the object constructed. In particular, their construction leaks a significant amount of information about their analog of $R$, even though it leaves the adversary with provably "many valid choices" for $R$. In retrospect, their notion can be viewed as an (information-theoretically) one-way function, rather than a semantically-secure key encapsulation mechanism, like the one considered in this work. Nonetheless, their informal notion is very closely related to our secure sketches, and we improve their construction in Section 5.

Linnartz and Tuyls [18] define and construct a primitive very similar to a fuzzy extractor (that line of work was continued in [28].) The definition of [18] focuses on the continuous space $\mathbb{R}^n$, and assumes a particular input distribution (typically a known, multivariate Gaussian). Thus, our definition of a fuzzy extractor can be viewed as a generalization of the notion of a "shielding function" from [18]. However, our constructions focus on discrete metric spaces.

Work on privacy amplification [3,4], as well as work on de-randomization and hardness amplification [14,24], also addressed the need to extract uniform randomness from a random variable about which some information has been leaked. A major focus of research in that literature has been the development of (ordinary, not fuzzy) extractors with short seeds (see [26] for a survey). We use extractors in this work (though for our purposes, pairwise independent hashing [3,14] is sufficient). Conversely, our work has been applied recently to privacy amplification: Ding [9] uses fuzzy extractors for noise tolerance in Maurer's bounded storage model.

EXTENSIONS. We can relax the error correction properties of sketches and fuzzy extractors to allow *list decoding*: instead of outputting one correct secret, we can output a short list of secrets, one of which is correct. For many applications (e.g., password authentication), this is sufficient, while the advantage is that we can possibly tolerate many more errors in the password. Not surprisingly, by using list-decodable codes (see [13] and the references therein) in our constructions, we can achieve this relaxation and considerably improve our error tolerance. Other similar extensions would be to allow small error probability in error-correction, to ensure correction of only *average-case* errors, or to consider nonbinary alphabets. Again, many of our results will extend to these settings. Finally, an interesting new direction is to consider other metrics not considered in this work.

## 2 Preliminaries

Unless explicitly stated otherwise, all logarithms below are base 2. We use $U_\ell$ to denote the uniform distribution on $\ell$-bit binary strings.

ENTROPY. The *min-entropy* $\mathbf{H}_\infty(A)$ of a random variable $A$ is $-\log(\max_a \Pr(A = a))$. For a pair of (possibly correlated) random variables $A, B$, a conventional notion of "average min-entropy" of $A$ given $B$ would be $\mathbb{E}_{b\leftarrow B}\left[\mathbf{H}_\infty(A \mid B = b)\right]$. However, for the purposes of this paper, the following slightly modified notion will be more robust: we let $\tilde{\mathbf{H}}_\infty(A \mid B) = -\log\left(\mathbb{E}_{b\leftarrow B}\left[2^{-\mathbf{H}_\infty(A|B=b)}\right]\right)$. Namely, we define *average min-entropy* of $A$ given $B$ to be the logarithm of the average probability of the most likely value of $A$ given $B$. One can easily verify that if $B$ is an $\ell$-bit string, then $\tilde{\mathbf{H}}_\infty(A \mid B) \geq \mathbf{H}_\infty(A) - \ell$.

STRONG EXTRACTORS. The *statistical distance between* two probability distributions $A$ and $B$ is $\mathbf{SD}(A, B) = \frac{1}{2}\sum_v |\Pr(A = v) - \Pr(B = v)|$. We can now define *strong randomness extractors* [24].

**Definition 1.** *An efficient* $(n, m', \ell, \epsilon)$-strong extractor *is a polynomial time probabilistic function* $\mathsf{Ext} : \{0,1\}^n \to \{0,1\}^\ell$ *such that for all min-entropy* $m'$ *distributions* $W$*, we have* $\mathbf{SD}(\langle \mathsf{Ext}(W; X), X\rangle, \langle U_\ell, X\rangle) \leq \epsilon$*, where* $\mathsf{Ext}(W; X)$ *stands for applying* $\mathsf{Ext}$ *to* $W$ *using (uniformly distributed) randomness* $X$.

Strong extractors can extract at most $\ell = m' - 2\log(1/\epsilon) + O(1)$ nearly random bits [25]. Many constructions match this bound (see Shaltiels' survey [26] for references). Extractor constructions are often complex since they seek to minimize the length of the seed $X$. For our purposes, the length of $X$ will be less important, so 2-wise independent hash functions will already give us optimal $\ell = m' - 2\log(1/\epsilon)$ [3,14].

Metric Spaces. A metric space is a set $\mathcal{M}$ with a distance function $\mathsf{dis} : \mathcal{M} \times \mathcal{M} \to \mathbb{R}^+ = [0, \infty)$ which obeys various natural properties. In this work, $\mathcal{M}$ will always be a finite set, and the distance function will only take on integer values. The size of the $\mathcal{M}$ will always be denoted $N = |\mathcal{M}|$. We will assume that any point in $\mathcal{M}$ can be naturally represented as a binary string of appropriate length $O(\log N)$.

We will concentrate on the following metrics. (1) *Hamming metric.* Here $\mathcal{M} = \mathcal{F}^n$ over some alphabet $\mathcal{F}$ (we will mainly use $\mathcal{F} = \{0,1\}$), and $\mathsf{dis}(w, w')$ is the number of positions in which they differ. (2) *Set Difference metric.* Here $\mathcal{M}$ consists of all $s$-element subsets in a universe $\mathcal{U} = [n] = \{1, ..., n\}$. The distance between two sets $A, B$ is the number of points in $A$ that are not in $B$. Since $A$ and $B$ have the same size, the distance is half of the size of their symmetric difference: $\mathsf{dis}(A, B) = \frac{1}{2}|A \triangle B|$. (3) *Edit metric.* Here again $\mathcal{M} = \mathcal{F}^n$, but the distance between $w$ and $w'$ is defined to be one half of the smallest number of character insertions and deletions needed to transform $w$ into $w'$.

As already mentioned, all three metrics seem natural for biometric data.

Coding. Since we want to achieve error tolerance in various metric spaces, we will use *error-correcting codes* in the corresponding metric space $\mathcal{M}$. A code $C$ is a subset $\{w_1, \ldots, w_K\}$ of $K$ elements of $\mathcal{M}$ (for efficiency purposes, we want the map from $i$ to $w_i$ to be polynomial-time). The *minimum distance* of $C$ is the smallest $d > 0$ such that for all $i \neq j$ we have $\mathsf{dis}(w_i, w_j) \geq d$. In our case of integer metrics, this means that one can detect up to $(d-1)$ "errors" in any codeword. The *error-correcting distance* of $C$ is the largest number $t > 0$ such that for every $w \in \mathcal{M}$ there exists at most one codeword $w_i$ in the ball of radius $t$ around $w$: $\mathsf{dis}(w, w_i) \leq t$ for at most one $i$. Clearly, for integer metrics we have $t = \lfloor (d-1)/2 \rfloor$. Since error correction will be more important in our applications, we denote the corresponding codes by $(\mathcal{M}, K, t)$-codes. For the Hamming and the edit metrics on strings of length $n$ over some alphabet $\mathcal{F}$, we will sometimes call $k = \log_{|F|} K$ the *dimension* on the code, and denote the code itself as an $[n, k, d = 2t+1]$-code, following the standard notation in the literature.

## 3 Definitions and General Lemmas

Let $\mathcal{M}$ be a metric space on $N$ points with distance function $\mathsf{dis}$.

**Definition 2.** *An* $(\mathcal{M}, m, m', t)$-secure sketch *is a randomized map* $\mathsf{SS} : \mathcal{M} \to \{0,1\}^*$ *with the following properties.*

1. *There exists a deterministic recovery function* $\mathsf{Rec}$ *allowing to recover* $w$ *from its sketch* $\mathsf{SS}(w)$ *and any vector* $w'$ *close to* $w$*: for all* $w, w' \in \mathcal{M}$ *satisfying* $\mathsf{dis}(w, w') \leq t$*, we have* $\mathsf{Rec}(w', \mathsf{SS}(w)) = w$*.*
2. *For all random variables* $W$ *over* $\mathcal{M}$ *with min-entropy* $m$*, the average min-entropy of* $W$ *given* $\mathsf{SS}(W)$ *is at least* $m'$*. That is,* $\tilde{\mathbf{H}}_\infty(W \mid \mathsf{SS}(W)) \geq m'$*.*

*The secure sketch is* efficient *if* SS *and* Rec *run in time polynomial in the representation size of a point in* $\mathcal{M}$. *We denote the random output of* SS *by* $\mathsf{SS}(W)$, *or by* $\mathsf{SS}(W;X)$ *when we wish to make the randomness explicit.*

We will have several examples of secure sketches when we discuss specific metrics. The quantity $m - m'$ is called the *entropy loss* of a secure sketch. Our proofs in fact bound $m - m'$, and the same bound holds for all values of $m$.

**Definition 3.** *An* $(\mathcal{M}, m, \ell, t, \epsilon)$ fuzzy extractor *is a given by two procedures* (Gen, Rep).

1. Gen *is a probabilistic generation procedure, which on input* $w \in \mathcal{M}$ *outputs an "extracted" string* $R \in \{0,1\}^\ell$ *and a public string* $P$. *We require that for any distribution* $W$ *on* $\mathcal{M}$ *of min-entropy* $m$, *if* $\langle R, P\rangle \leftarrow \mathsf{Gen}(W)$, *then we have* $\mathbf{SD}(\langle R, P\rangle, \langle U_\ell, P\rangle) \leq \epsilon$.
2. Rep *is a deterministic reproduction procedure allowing to recover* $R$ *from the corresponding public string* $P$ *and any vector* $w'$ *close to* $w$: *for all* $w, w' \in \mathcal{M}$ *satisfying* $\mathsf{dis}(w, w') \leq t$, *if* $\langle R, P\rangle \leftarrow \mathsf{Gen}(w)$, *then we have* $\mathsf{Rep}(w', P) = R$.

*The fuzzy extractor is* efficient *if* Gen *and* Rep *run in time polynomial in the representation size of a point in* $\mathcal{M}$.

In other words, fuzzy extractors allow one to extract some randomness $R$ from $w$ and then successfully reproduce $R$ from any string $w'$ that is close to $w$. The reproduction is done with the help of the public string $P$ produced during the initial extraction; yet $R$ looks truly random even given $P$. To justify our terminology, notice that strong extractors (as defined in Section 2) can indeed be seen as "nonfuzzy" analogs of fuzzy extractors, corresponding to $t = 0$, $P = X$ (and $\mathcal{M} = \{0,1\}^n$).

Construction of Fuzzy Extractors from Secure Sketches. Not surprisingly, secure sketches come up very handy in constructing fuzzy extractors. Specifically, we construct fuzzy extractors from secure sketches and strong extractors. For that, we assume that one can naturally represent a point $w$ in $\mathcal{M}$ using $n$ bits. The strong extractor we use is the standard pairwise-independent hashing construction, which has (optimal) entropy loss $2\log\left(\frac{1}{\epsilon}\right)$. The proof of the following lemma uses the "left-over hash" (a.k.a. "privacy amplification") lemma of [14,4], and can be found in the full version of our paper.

**Lemma 1 (Fuzzy Extractors from Sketches).** *Assume* SS *is a* $(\mathcal{M}, m, m', t)$*-secure sketch with recovery procedure* Rec, *and let* Ext *be the* $(n, m', \ell, \epsilon)$*-strong extractor based on pairwise-independent hashing (in particular,* $\ell = m' - 2\log\left(\frac{1}{\epsilon}\right)$*). Then the following* (Gen, Rep) *is a* $(\mathcal{M}, m, \ell, t, \epsilon)$*-fuzzy extractor:*

- $\mathsf{Gen}(W; X_1, X_2)$: *set* $P = \langle \mathsf{SS}(W; X_1), X_2\rangle$, $R = \mathsf{Ext}(W; X_2)$, *output* $\langle R, P\rangle$.
- $\mathsf{Rep}(W', \langle V, X_2\rangle)$: *recover* $W = \mathsf{Rec}(W', V)$ *and output* $R = \mathsf{Ext}(W; X_2)$.

*Remark 1.* One can prove an analogous form of Lemma 1 using any strong extractor. However, in general, the resulting reduction leads to fuzzy extractors with min-entropy loss $3\log\left(\frac{1}{\epsilon}\right)$ instead of $2\log\left(\frac{1}{\epsilon}\right)$. This may happen in the case when the extractor does not have a convex tradeoff between the input entropy and the distance from uniform of the output. Then one can instead use a high-probability bound on the min-entropy of the input (that is, if $\tilde{\mathbf{H}}_\infty(X|Y) \geq m'$ then the event $\mathbf{H}_\infty(X|Y=y) \geq m' - \log\left(\frac{1}{\epsilon}\right)$ happens with probability $1-\epsilon$).

SKETCHES FOR TRANSITIVE METRIC SPACES. We give a general technique for building secure sketches in *transitive* metric spaces, which we now define. A permutation $\pi$ on a metric space $\mathcal{M}$ is an *isometry* if it preserves distances, i.e. $\mathsf{dis}(a,b) = \mathsf{dis}(\pi(a),\pi(b))$. A family of permutations $\Pi = \{\pi_i\}_{i\in\mathcal{I}}$ acts *transitively* on $\mathcal{M}$ if for any two elements $a, b \in \mathcal{M}$, there exists $\pi_i \in \Pi$ such that $\pi_i(a) = b$. Suppose we have a family $\Pi$ of transitive isometries for $\mathcal{M}$ (we will call such $\mathcal{M}$ *transtive*). For example, in the Hamming space, the set of all shifts $\pi_x(w) = w \oplus x$ is such a family (see Section 4 for more details on this example).

Let $C$ be an $(\mathcal{M}, K, t)$-code. Then the general sketching scheme is the following: given a input $w \in \mathcal{M}$, pick a random codeword $b \in C$, pick a random permutation $\pi \in \Pi$ such that $\pi(w) = b$, and output $\mathsf{SS}(w) = \pi$. To recover $w$ given $w'$ and the sketch $\pi$, find the closest codeword $b'$ to $\pi(w')$, and output $\pi^{-1}(b')$. This works when $\mathsf{dis}((,w),w') \leq t$, because then $\mathsf{dis}((,b),\pi(w')) \leq t$, so decoding $\pi(w')$ will result in $b' = b$, which in turn means that $\pi^{-1}(b') = w$.

A bound on the entropy loss of this scheme, which follows simply from "counting" entropies, is $|``\pi"| - \log K$, where $|``\pi"|$ is the size, in bits, of a canonical description of $\pi$. (We omit the proof, as it is a simple generalization of the proof of Lemma 3.) Clearly, this quantity will be small if the family $\Pi$ of transifitive isometries is small and the code $C$ is dense. (For the scheme to be usable, we also need the operations on the code, as well as $\pi$ and $\pi^{-1}$, to be implementable reasonably efficiently.)

CONSTRUCTIONS FROM BIOMETRIC EMBEDDINGS. We now introduce a general technique that allows one to build good fuzzy extractors in some metric space $\mathcal{M}_1$ from good fuzzy extractors in some other metric space $\mathcal{M}_2$. Below, we let $\mathsf{dis}(\cdot,\cdot)_i$ denote the distance function in $\mathcal{M}_i$. The technique is to *embed* $\mathcal{M}_1$ into $\mathcal{M}_2$ so as to "preserve" relevant parameters for fuzzy extraction.

**Definition 4.** *A function $f : \mathcal{M}_1 \to \mathcal{M}_2$ is called a $(t_1, t_2, m_1, m_2)$-biometric embedding if the following two conditions hold:*

- $\forall\, w_1, w_1' \in \mathcal{M}_1$ *such that* $\mathsf{dis}(w_1, w_1')_1 \leq t_1$, *we have* $\mathsf{dis}(f(w_1), f(w_2))_2 \leq t_2$.
- $\forall\, W_1$ *on* $\mathcal{M}_1$ *such that* $\mathbf{H}_\infty(W_1) \geq m_1$, *we have* $\mathbf{H}_\infty(f(W_1)) \geq m_2$.

The following lemma is immediate:

**Lemma 2.** *If $f$ is $(t_1, t_2, m_1, m_2)$-biometric embedding of $\mathcal{M}_1$ into $\mathcal{M}_2$ and* $(\mathsf{Gen}_1(\cdot), \mathsf{Rep}_1(\cdot,\cdot))$ *is a $(\mathcal{M}_2, m_2, \ell, t_2, \epsilon)$-fuzzy extractor, then* $(\mathsf{Gen}_1(f(\cdot)), \mathsf{Rep}_1(f(\cdot),\cdot))$ *is a $(\mathcal{M}_1, m_1, \ell, t_1, \epsilon)$-fuzzy extractor.*

Notice that a similar result does not hold for secure sketches, unless $f$ is injective (and efficiently invertible).

We will see the utility of this particular notion of embedding (as opposed to previously defined notions) in Section 6.

## 4 Constructions for Hamming Distance

In this section we consider constructions for the space $\mathcal{M} = \{0,1\}^n$ under the Hamming distance metric.

THE CODE-OFFSET CONSTRUCTION. Juels and Wattenberg [15] considered a notion of "fuzzy commitment." [5] Given a binary $[n, k, 2t+1]$ error-correcting code $C$ (not necessarily linear), they fuzzy-commit to $X$ by publishing $W \oplus C(X)$. Their construction can be rephrased in our language to give a very simple construction of secure sketches: for random $X \leftarrow \{0,1\}^k$, set

$$\mathsf{SS}(W; X) = W \oplus C(X)\,.$$

(Note that if $W$ is uniform, this secure sketch direcly yields a fuzzy extractor with $R = X$).

When the code $C$ is linear, this is equivalent to revealing the syndrome of the input $w$, and so we do not need the randomness $X$. Namely, in this case we could have set $\mathsf{SS}(w) = \mathsf{syn}_C(w)$ (as mentioned in the introduction, this construction also appears implicitly in the information reconciliation literature, e.g. [3,4,7]: when Alice and Bob hold secret values which are very close in Hamming distance, one way to correct the differences with few bits of communication is for Alice to send to Bob the *syndrome* of her word $w$ with respect to a good linear code.)

Since the syndrome of a $k$-dimensional linear code is $n-k$ bits long, it is clear that $\mathsf{SS}(w)$ leaks only $n-k$ bits about $w$. In fact, we show the same is true even for nonlinear codes.

**Lemma 3.** *For any $[n, k, 2t+1]$ code $C$ and any $m$, $\mathsf{SS}$ above is a $(\mathcal{M}, m, m+k-n, t)$ secure sketch. It is efficient if the code $C$ allows decoding errors in polynomial time.*

*Proof.* Let $D$ be the decoding procedure of our code $C$. Since $D$ can correct up to $t$ errors, if $v = w \oplus C(x)$ and $\mathsf{dis}(w, w') \le t$, then $D(w' \oplus v) = x$. Thus, we can set $\mathsf{Rec}(w', v) = v \oplus C(D(w' \oplus v))$.

Let $A$ be the joint variable $(X, W)$. Together, these have min-entropy $m+k$ when $\mathbf{H}_\infty(W) = m$. Since $\mathsf{SS}(W) \in \{0,1\}^n$, we have $\tilde{\mathbf{H}}_\infty(W, X \mid \mathsf{SS}(W)) \ge m + k - n$. Now given $\mathsf{SS}(W)$, $W$ and $X$ determine each other uniquely, and so $\tilde{\mathbf{H}}_\infty(W \mid \mathsf{SS}(W)) \ge m + k - n$ as well. □

In the full version, we present some generic lower bounds on secure sketches and extractors. Let $A(n, d)$ denote the maximum number of codewords possible

[5] In their interpretation, one commits to $X$ by picking a random $W$ and publishing $\mathsf{SS}(W; X)$.

in a code of distance $d$ in $\{0,1\}^n$. Then the entropy loss of a secure sketch for the Hamming metric is at least $n - \log A(n, 2t+1)$, when the input is uniform (that is, when $m = n$). This means that the code-offset construction above is optimal for the case of uniform inputs. Of course, we do not know the exact value of $A(n, d)$, never mind of efficiently decodable codes which meet the bound, for most settings of $n$ and $d$. Nonetheless, the code-offset scheme gets as close to optimality as is possible in coding.

GETTING FUZZY EXTRACTORS. As a warm-up, consider the case when $W$ is uniform ($m = n$) and look at the code-offset sketch construction: $V = W \oplus C(X)$. Setting $R = X$, $P = V$ and $\mathsf{Rep}(W', V) = D(V \oplus W')$, we clearly get an $(\mathcal{M}, n, k, t, 0)$ fuzzy extractor, since $V$ is truly random when $W$ is random, and therefore independent of $X$. In fact, this is exactly the usage proposed by Juels-Wattenberg, except they viewed the above fuzzy extractor as a way to use $W$ to "fuzzy commit" to $X$, without revealing information about $X$.

Unfortunately, the above construction setting $R = X$ only works for uniform $W$, since otherwise $V$ would leak information about $X$. However, by using the construction in Lemma 1, we get

**Lemma 4.** *Given any $[n, k, 2t+1]$ code $C$ and any $m, \epsilon$, we can get an $(\mathcal{M}, m, \ell, t, \epsilon)$ fuzzy extractor, where $\ell = m+k-n-2\log(1/\epsilon)$. The recovery* Rep *is efficient if $C$ allows decoding errors in polynomial time.*

## 5 Constructions for Set Difference

Consider the collection of all sets of a particular size $s$ in a universe $\mathcal{U} = [n] = \{1, ..., n\}$. The distance between two sets $A, B$ is the number of points in $A$ that are not in $B$. Since $A$ and $B$ have the same size, the distance is half of the size of their symmetric difference: $\frac{1}{2}\mathsf{dis}(A, B) = |A \triangle B|$. If $A$ and $B$ are viewed as $n$-bit characteristic vectors over $[n]$, this metric is the same as the Hamming metric (scaled by 1/2). Thus, the set difference metric can be viewed as a restriction of the binary Hamming metric to all the strings with exactly $s$ nonzero components. However, one typically assumes that $n$ is much larger than $s$, so that representing a set by $n$ bits is much less efficient than, say writing down a list of elements, which requires $(s \log n)$ bits.

LARGE VERSUS SMALL UNIVERSES. Most of this section studies situations where the universe size $n$ is super-polynomial in the set size $s$. We call this the large universe setting. By contrast, the small universe setting refers to situations in which $n = \mathit{poly}(s)$. We want our various constructions to run in polynomial time and use polynomial storage space. Thus, the large universe setting is exactly the setting in which the $n$-bit string representation of a set becomes too large to be usable. We consider the small-universe setting first, since it appears simpler (Section 5.1). The remaining subsections consider large universes.

### 5.1 Small Universes

When the universe size is polynomial in $s$, there are a number of natural constructions. Perhaps the most direct one, given previous work, is the construction of Juels and Sudan [16]. Unfortunately, that scheme achieves relatively poor parameters (see Section 5.2). We suggest two possible constructions. The first one represents sets as $n$-bit strings and uses the constructions of the previous section (with the caveat that Hamming distance is off by a factor of 2 from set difference).

The second construction goes directly through codes for set difference, also called "constant-weight" codes. A constant-weight code is a ordinary error-correcting code in $\{0,1\}^n$ in which all of the codewords have the same Hamming weight $s$. The set difference metric is transitive—the metric is invariant under permutations of the underlying universe $\mathcal{U}$, and for any two sets of the same size $A, B \subseteq \mathcal{U}$, there is a permutation of $\mathcal{U}$ that maps $A$ to $B$. Thus, one can use the general scheme for secure sketches in transitive metrics (Section 3) to get a secure sketch for set difference with output length about $n \log n$.

The full version of the paper contains a more detailed comparison of the two constructions. Briefly: The second construction achieves better parameters since, according to currently proved bounds, it seems that constant-weight codes can be more dense than ordinary codes. On the other hand, explicit codes which highlight this difference are not known, and much more is known about efficient implementations of decoding for ordinary codes. In practice, the Hamming-based scheme is likely to be more useful.

### 5.2 Modifying the Construction of Juels and Sudan

We now turn to the large universe setting, where $n$ is super-polynomial in $s$. Juels and Sudan [16] proposed a secure sketch for the set difference metric (called a "fuzzy vault" in that paper). They assume for simplicity that $n = |\mathcal{U}|$ is a prime power and work over the field $\mathcal{F} = GF(n)$. On input set $A$, the sketch they produce is a set of $r$ pairs of points $(x_i, y_i)$ in $\mathcal{F}$, with $s < r \leq n$. Of the $x_i$ values, $s$ are the elements of $A$, and their corresponding $y_i$ value are evaluations of a random degree-$(s-2t-1)$ polynomial $p$ at $x_i$; the remaining $r-s$ of the $(x_i, y_i)$ values are chosen at random but not on $p$. The original analysis [16] does not extend to the case of a nonuniform password in a large universe. However, we give a simpler analysis which does cover that range of parameters. Their actual scheme, as well as our new analysis, can be found in the full version of the paper. We summarize here:

**Lemma 5.** *The entropy loss of the Juels-Sudan scheme is at most* $m - m' = 2t \log n + \log \binom{n}{r} - \log \binom{n-s}{r-s}$.

Their scheme requires storage $2r \log n$. In the large universe setting, we will have $r \ll n$ (since we wish to have storage polynomial in $s$). In that setting, the bound on the entropy loss of the Juels-Sudan scheme is in fact very large. We can rewrite the entropy loss as $2t \log n - \log \binom{r}{s} + \log \binom{n}{s}$, using the identity

$\binom{n}{r}\binom{r}{s} = \binom{n}{s}\binom{n-s}{r-s}$. Now the entropy of $A$ is at most $\binom{n}{s}$, and so our lower bound on the remaining entropy is $(\log \binom{r}{s} - 2t\log n)$. To make this quantity large requires making $r$ very large.

MODIFIED JS SKETCHES. We suggest a modification of the Juels-Sudan scheme with entropy loss at most $2t\log n$ and storage $s\log n$. Our scheme has the advantage of being even simpler to analyze. As before, we assume $n$ is a prime power and work over $\mathcal{F} = GF(n)$. An intuition for the scheme is that the numbers $y_{s+1}, ..., y_r$ from the JS scheme need not be chosen at random. One can instead evaluate them as $y_i = p'(x_i)$ for some polynomial $p'$. One can then represent the entire list of pairs $(x_i, y_i)$ using only the coefficients of $p'$.

**Algorithm 1 (Modified JS Secure Sketch).** Input: a set $A \subseteq \mathcal{U}$.

1. Choose $p()$ at random from the set of polynomials of degree at most $k = s - 2t - 1$ over $\mathcal{F}$.
2. Let $p'()$ be the unique monic polynomial of degree exactly $s$ such that $p'(x) = p(x)$ for all $x \in A$.
   (Write $p'(x) = x^s + \sum_{i=0}^{s-1} a_i x^i$. Solve for $a_0, ..., a_{s-1}$ using the $s$ linear constraints $p'(x) = p(x)$, $x \in A$.)
3. Output the list of coefficients of $p'()$, that is $\mathsf{SS}(A) = (a_0, ..., a_{s-1})$.

First, observe that solving for $p'()$ in Step 2 is always possible, since the $s$ constraints $\sum_{i=0}^{s-1} a_i x^i = p(x) - x^s$ are in fact linearly independent (this is just polynomial interpolation).

Second, this sketch scheme can tolerate $t$ set difference errors. Suppose we are given a set $B \subseteq \mathcal{U}$ which agrees with $A$ in at least $s - t$ positions. Given $p' = \mathsf{SS}(A)$, one can evaluate $p'$ on all the points in the set $B$. The resulting vector agrees with $p$ on at least $s-t$ positions, and using the decoding algorithm for Reed-Solomon codes, one can thus reconstruct $p$ exactly (since $k = s-2t-1$). Finally, the set $A$ can be recovered by finding the roots of the polynomial $p' - p$: since $p' - p$ is not identically zero and has degree exactly $s$, it can have at most $s$ roots and so $p' - p$ is zero only on $A$.

We now turn to the entropy loss of the scheme. The sketching scheme invests $(s - 2t)\log n$ bits of randomness to choose the polynomial $p$. The number of possible outputs $p'$ is $n^s$. If $X$ is the invested randomness, then the (average) min-entropy $(A, X)$ given $\mathsf{SS}(A)$ is at least $\tilde{\mathbf{H}}_\infty(A) - 2t\log n$. The randomness $X$ can be recovered from $A$ and $\mathsf{SS}(A)$, and so we have $\tilde{\mathbf{H}}_\infty(A \mid \mathsf{SS}(A)) \geq \tilde{\mathbf{H}}_\infty(A) - 2t\log n$. We have proved:

**Lemma 6 (Analysis of Modified JS).** *The entropy loss of the modified JS scheme is at most* $2t\log n$. *The scheme has storage* $(s+1)\log n$ *for sets of size* $s$ *in* $[n]$, *and both the sketch generation* $\mathsf{SS}()$ *and the recovery procedure* $\mathsf{Rec}()$ *run in polynomial time.*

The short length of the sketch makes this scheme feasible for essentially any ratio of set size to universe size (we only need $\log n$ to be polynomial in $s$). Moreover, for large universes the entropy loss $2t\log n$ is essentially optimal for the uniform case $m = \log \binom{n}{s}$. Our lower bound (in the full version) shows that for a uniformly distributed input, the best possible entropy loss is $m - m' \geq \log \binom{n}{s} - \log A(n, s, 4t + 1)$, where $A(n, s, d)$ is the maximum size of a code of constant weight $s$ and minimum Hamming distance $d$. Using a bound of Agrell *et al* ([1], Theorem 12), the entropy loss is at least:

$$m - m' \geq \log \binom{n}{s} - \log A(n, s, 4t + 1) \geq \log \binom{n - s + 2t}{2t}$$

When $n \geq s$, this last quantity is roughly $2t \log n$, as desired.

### 5.3 Large Universes via the Hamming Metric: Sublinear-Time Decoding

In this section, we show that code-offset construction can in fact be adapted for small sets in large universe, using specific properties of algebraic codes. We will show that BCH codes, which contain Hamming and Reed-Solomon codes as special cases, have these properties.

SYNDROMES OF LINEAR CODES. For a $[n, k, d]$ linear code $C$ with parity check matrix $H$, recall that the syndrome of a word $w \in \{0, 1\}^n$ is $\mathsf{syn}(w) = Hw$. The syndrome has length $n - k$, and the code is exactly the set of words $c$ such that $\mathsf{syn}(c) = 0^{n-k}$. The syndrome captures all the information necessary for decoding. That is, suppose a codeword $c$ is sent through a channel and the word $w = c \oplus e$ is received. First, the syndrome of $w$ is the syndrome of $e$: $\mathsf{syn}(w) = \mathsf{syn}(c) \oplus \mathsf{syn}(e) = 0 \oplus \mathsf{syn}(e) = \mathsf{syn}(e)$. Moreover, for any value $u$, there is at most one word $e$ of weight less than $d/2$ such that $\mathsf{syn}(e) = u$ (the existence of a pair of distinct words $e_1, e_2$ would mean that $e_1 + e_2$ is a codeword of weight less than $d$). Thus, knowing syndrome $\mathsf{syn}(w)$ is enough to determine the error pattern $e$ if not too many errors occurred.

As mentioned before, we can reformulate the code-offset construction in terms of syndrome: $\mathsf{SS}(w) = \mathsf{syn}(w)$. The two schemes are equivalent: given $\mathsf{syn}(w)$ one can sample from $w \oplus C(X)$ by choosing a random string $v$ with $\mathsf{syn}(v) = \mathsf{syn}(w)$; conversely, $\mathsf{syn}(w \oplus C(X)) = \mathsf{syn}(w)$. This reformulation gives us no special advantage when the universe is small: storing $w + C(X)$ is not a problem. However, it's a substantial improvement when $n \gg n - k$.

SYNDROME MANIPULATION FOR SMALL-WEIGHT WORDS. Suppose now that we have a small set $A \subseteq [n]$ of size $s$, where $n \gg s$. Let $x_A \in \{0, 1\}^n$ denote the characteristic vector of $A$. If we want to use $\mathsf{syn}(x_A)$ as the sketch of $A$, then we must choose a code with $n - k \leq \log \binom{n}{s} \approx s \log n$, since the sketch has entropy loss $(n - k)$ and the maximum entropy of $A$ is $\log \binom{n}{s}$.

Binary BCH codes are a family of $[n, k, d]$ linear codes with $d = 4t + 1$ and $k = n - 2t \log n$ (assuming $n + 1$ is a power of 2) (see, e.g. [19]). These codes are optimal for $t \ll n$ by the Hamming bound, which implies that $k \leq n - \log \binom{n}{2t}$ [19]. Using the code-offset sketch with a BCH code $C$, we get entropy loss $n - k = 2t \log n$, just as we did for the modified Juels-Sudan scheme (recall that $d \geq 4t+1$ allows us to correct $t$ set difference errors).

The only problem is that the scheme appears to require computation time $\Omega(n)$, since we must compute $\mathsf{syn}(x_A) = Hx_A$ and, later, run a decoding algorithm to recover $x_A$. For BCH codes, this difficulty can be overcome. A word of small weight $x$ can be described by listing the positions on which it is nonzero. We call this description the *support* of $x$ and write $\mathsf{supp}(x)$ (that is $\mathsf{supp}(x_A) = A$)).

**Lemma 7.** *For a $[n, k, d]$ binary BCH code $C$ one can compute:*

1. $\mathsf{syn}(x)$, *given* $\mathsf{supp}(x)$, *and*
2. $\mathsf{supp}(x)$, *given* $\mathsf{syn}(x)$ *(when $x$ has weight at most $(d-1)/2$),*

*in time polynomial in $|\mathsf{supp}(x)| = weight(x) \cdot \log(n)$ and $|\mathsf{syn}(x)| = n - k$.*

The proof of Lemma 7 mainly requires a careful reworking of the standard BCH decoding algorithm. The details are presented in the full version of the paper. For now, we present the resulting sketching scheme for set difference. The algorithm works in the field $GF(2^m) = GF(n+1)$, and assumes a generator $\alpha$ for $GF(2^m)$ has been chosen ahead of time.

**Algorithm 2 (BCH-based Secure Sketch).** Input: a set $A \in [n]$ of size $s$, where $n = 2^m - 1$. (Here $\alpha$ is a generator for $GF(2^m)$, fixed ahead of time.)

1. Let $p(x) = \sum_{i \in A} x^i$.
2. Output $\mathsf{SS}(A) = (p(\alpha), p(\alpha^3), p(\alpha^5), \ldots, p(\alpha^{4t+1}))$ (computations in $GF(2^m)$).

Lemma 7 yields the algorithm $\mathsf{Rec}()$ which recovers $A$ from $\mathsf{SS}(A)$ and any set which intersects $A$ in at least $s - t$ points. However, the bound on entropy loss is easy to see: the output is $2t \log n$ bits long, and hence the entropy loss is at most $2t \log n$. We obtain:

**Theorem 1.** *The BCH scheme above is a $[m, m-2t \log n, t]$ secure sketch scheme for set difference with storage $2t \log n$. The algorithms $\mathsf{SS}$ and $\mathsf{Rec}$ both run in polynomial time.*

## 6 Constructions for Edit Distance

First we note that simply applying the same approach as we took for the transitive metric spaces before (the Hamming space and the set difference space for small universe sizes) does not work here, because the edit metric does not seem to be transitive. Indeed, it is unclear how to build a permutation $\pi$ such that for any $w'$ close to $w$, we also have $\pi(w')$ close to $x = \pi(w)$. For example, setting $\pi(y) = y \oplus (x \oplus w)$ is easily seen not to work with insertions and deletions. Similarly, if $I$ is some sequence of insertions and deletions mapping $w$ to $x$, it is not true that applying $I$ to $w'$ (which is close to $w$) will necessarily result in some $x'$ close to $x$. In fact, then we could even get $\mathsf{dis}(w', x') = 2\mathsf{dis}(w, x) + \mathsf{dis}(w, w')$.

Perhaps one could try to simply embed the edit metric into the Hamming metric using known embeddings, such as conventionally used low-distorion embeddings, which provide that all distances are preserved up to some small "distortion" factor. However, there are no known nontrivial low-distortion embeddings from the edit metric to the Hamming metric. Moreover, it was recently proved by Andoni et al [2] that no such embedding can have distortion less than 3/2, and it was conjectured that a much stronger lower bound should hold.

Thus, as the previous approaches don't work, we turn to the embeddings we defined specifically for fuzzy extractors: biometric embeddings. Unlike low-distortion embeddings, biometric embeddings do not care about relative distances, as long as points that were "close" (closer than $t_1$) do not become "distant" (farther apart than $t_2$). The only additional requirement of biometric embeddings is that they preserve some min-entropy: we do not want too many points to collide together, although collisions are allowed, even collisions of distant points. We will build a biometric embedding from the edit distance to the set difference.

A *c-shingle* [5], which is a length-$c$ consecutive substring of a given string $w$. A *c-shingling* [5] of a string $w$ of length $n$ is the set (ignoring order or repetition) of all $(n-c+1)$ $c$-shingles of $w$. Thus, the range of the $c$-shingling operation consists of all nonempty subsets of size at most $n-c+1$ of $\{0,1\}^c$. To simplify our future computations, we will always arbitrarily pad the $c$-shingling of any string $w$ to contain precisely $n$ distinct shingles (say, by adding the first $n-|c\text{-shingling}|$ elements of $\{0,1\}^c$ not present in the given $c$-shingling). Thus, we can define a deterministic map $\mathsf{SH}_c(w)$ which maps $w$ into $n$ substrings of $\{0,1\}^c$, where we assume that $c \geq \log_2 n$. Let $\mathsf{Edit}(n)$ stand for the edit metric over $\{0,1\}^n$, and $\mathsf{SDif}(N,s)$ stand for the set difference metric over $[N]$ where the set sizes are $s$. We now show that $c$-shingling yields pretty good biometric embeddings for our purposes.

**Lemma 8.** *For any $c > \log_2 n$, $\mathsf{SH}_c$ is a $(t_1, t_2 = ct_1, m_1, m_2 = m_1 - \frac{n\log_2 n}{c})$-biometric embedding of $\mathsf{Edit}(n)$ into $\mathsf{SDif}(2^c, n)$.*

*Proof.* Assume $\mathsf{dis}(w_1, w_1')_{ed} \leq t_1$ and that $I$ is the smallest set of $2t_1$ insertions and deletions which transforms $w$ into $w'$. It is easy to see that each character deletion or insertion affects at most $c$ shingles, and thus the symmetric difference between $\mathsf{SH}_c(w_1)$ and $\mathsf{SH}_c(w_2) \leq 2ct_1$, which implies that $\mathsf{dis}(\mathsf{SH}_c(w_1), \mathsf{SH}_c(w_2))_{sd} \leq ct_1$, as needed.

Now, assume $w_1$ is any string. Define $g_c(w_1)$ as follows. One computes $\mathsf{SH}_c(w_1)$, and stores $n$ resulting shingles in lexicographic order $h_1 \ldots h_n$. Next, one naturally partitions $w_1$ into $n/c$ disjoint shingles of length $c$, call them $k_1 \ldots k_{n/c}$. Next, for $1 \leq j \leq n/c$, one sets $p_c(j)$ to be the index $i \in \{1 \ldots n\}$ such that $k_j = h_i$. Namely, it tells the index of the $j$-th disjoint shingle of $w_1$ in the ordered $n$-set $\mathsf{SH}_c(w_1)$. Finally, one sets $g_c(w_1) = (p_c(1) \ldots p_c(n/c))$. Notice, the length of $g_c(w_1)$ is $\frac{n}{c} \cdot \log_2 n$, and also that $w_1$ can be completely recovered from $\mathsf{SH}_c(w_1)$ and $g_c(w_1)$.

Now, assume $W_1$ is any distribution of min-entropy at least $m_1$ on $\mathsf{Edit}(n)$. Since $g_c(W)$ has length $(n \log_2 n/c)$, its min-entropy is at most this much as well. But since min-entropy of $W_1$ drops to 0 when given $\mathsf{SH}_c(W_1)$ and $g_c(W_1)$, it means that the min-entropy of $\mathsf{SH}_c(W_1)$ must be at least $m_2 \geq m_1 - (n \log_2 n)/c$, as claimed.

We can now optimize the value $c$. By either Lemma 6 or Theorem 1, for arbitrary universe size (in our case $2^c$) and distance threshold $t_2 = ct_1$, we can construct a secure sketch for the set difference metric with min-entropy loss $2t_2 \log_2(2^c) = 2t_1c^2$, which leaves us total min-entropy $m_2' = m_2 - 2t_1c^2 \geq m_1 - \frac{n \log n}{c} - 2t_1c^2$. Applying further Lemma 1, we can convert it into a fuzzy extractor over $\mathsf{SDif}(2^c, n)$ for the min-entropy level $m_2$ with error $\epsilon$, which can extract at least $\ell = m_2' - 2\log\left(\frac{1}{\epsilon}\right) \geq m_1 - \frac{n \log n}{c} - 2t_1c^2 - 2\log\left(\frac{1}{\epsilon}\right)$ bits, while still correcting $t_2 = ct_1$ of errors in $\mathsf{SDif}(2^c, n)$. We can now apply Lemma 2 to get an $(\mathsf{Edit}(n), m_1, m_1 - \frac{n \log n}{c} - 2t_1c^2 - 2\log\left(\frac{1}{\epsilon}\right), t_1, \epsilon)$-fuzzy extractor. Let us now optimize for the value of $c \geq \log_2 n$. We can set $\frac{n \log n}{c} = 2t_1c^2$, which gives $c = (\frac{n \log n}{2t_1})^{1/3}$. We get $\ell = m_1 - (2t_1n^2 \log^2 n)^{1/3} - 2\log\left(\frac{1}{\epsilon}\right)$ and therefore

**Theorem 2.** *There is an efficient* $(\mathsf{Edit}(n), m_1, m_1 - (2t_1n^2 \log^2 n)^{1/3} - 2\log\left(\frac{1}{\epsilon}\right),$ $t_1, \epsilon)$ *fuzzy extractor. Setting* $t_1 = m_1^3/(16n^2 \log^2 n)$*, we get an efficient* $(\mathsf{Edit}(n),$ $m_1, \frac{m_1}{2} - 2\log\left(\frac{1}{\epsilon}\right), \frac{m_1^3}{16n^2 \log^2 n}, \epsilon)$ *fuzzy extractor. In particular, if* $m_1 = \Omega(n)$*, one can extract* $\Omega(n)$ *bits while tolerating* $\Omega(n/\log^2 n)$ *insertions and deletions.*

## Acknowledgements

We thank Piotr Indyk for discussions about embeddings and for his help in the proof of Lemma 8. We are also thankful to Madhu Sudan for helpful discussions about the construction of [16] and the uses of error-correcting codes. Finally, we thank Rafi Ostrovsky for discussions in the initial phases of this work and Pim Tuyls for pointing out relevant previous work.

The work of the first author was partly funded by the National Science Foundation under CAREER Award No. CCR-0133806 and Trusted Computing Grant No. CCR-0311095, and by the New York University Research Challenge Fund 25-74100-N5237. The work of the second author was partly funded by the National Science Foundation under Grant No. CCR-0311485. The work of the third author was partly funded by US A.R.O. grant DAAD19-00-1-0177 and by a Microsoft Fellowship.

## References

1. E. Agrell, A. Vardy, and K. Zeger. Upper bounds for constant-weight codes. *IEEE Transactions on Information Theory*, **46**(7), pp. 2373–2395, 2000.
2. A. Andoni, M. Deza, A. Gupta, P. Indyk, S. Raskhodnikova. Lower bounds for embedding edit distance into normed spaces. In *Proc. ACM Symp. on Discrete Algorithms, 2003*, pp. 523–526.
3. C. Bennett, G. Brassard, and J. Robert. Privacy Amplification by Public Discussion. *SIAM J. on Computing*, **17**(2), pp. 210–229, 1988.
4. C. Bennett, G. Brassard, C. Crépeau, and U. Maurer. Generalized Privacy Amplification. *IEEE Transactions on Information Theory*, **41**(6), pp. 1915-1923, 1995.
5. A. Broder. On the resemblence and containment of documents. In *Compression and Complexity of Sequences*, 1997.

6. A. E. Brouwer, J. B. Shearer, N. J. A. Sloane, and W. D. Smith, "A new table of constant weight codes," *IEEE Transactions on Information Theory*, **36**, p. 1334-1380, 1990.
7. C. Crépeau. Efficient Cryptographic Protocols Based on Noisy Channels. In *Advances in Cryptology — EUROCRYPT 1997*, pp. 306–317.
8. G. Davida, Y. Frankel, B. Matt. On enabling secure applications through off-line biometric identification. In *Proc. IEEE Symp. on Security and Privacy*, pp. 148–157, 1998.
9. Y.Z. Ding. Manuscript.
10. C. Ellison, C. Hall, R. Milbert, B. Schneier. Protecting Keys with Personal Entropy. *Future Generation Computer Systems*, **16**, pp. 311–318, 2000.
11. N. Frykholm. Passwords: Beyond the Terminal Interaction Model. *Master's Thesis*, Umea University.
12. N. Frykholm, A. Juels. Error-Tolerant Password Recovery. In *Proc. ACM Conf. Computer and Communications Security, 2001*, pp. 1–8.
13. V. Guruswami, M. Sudan. Improved Decoding of Reed-Solomon and Algebraic-Geometric Codes. In *Proc. 39th IEEE Symp. on Foundations of Computer Science*, 1998, pp. 28–39.
14. J. Håstad, R. Impagliazzo, L. Levin, M. Luby. A Pseudorandom generator from any one-way function. In *Proc. 21st ACM Symp. on Theory of Computing*, 1989.
15. A. Juels, M. Wattenberg. A Fuzzy Commitment Scheme. In *Proc. ACM Conf. Computer and Communications Security, 1999*, pp. 28–36.
16. A. Juels and M. Sudan. A Fuzzy Vault Scheme. In *IEEE International Symposium on Information Theory*, 2002.
17. J. Kelsey, B. Schneier, C. Hall, D. Wagner. Secure Applications of Low-Entropy Keys. In *Proc. of Information Security Workshop*, pp. 121–134, 1997.
18. J.-P. M. G. Linnartz, P. Tuyls. New Shielding Functions to Enhance Privacy and Prevent Misuse of Biometric Templates. In *AVBPA 2003*, p. 393–402.
19. J.H. van Lint. *Introduction to Coding Theory.* Springer-Verlag, 1992, 183 pp.
20. F. Monrose, M. Reiter, S. Wetzel. Password Hardening Based on Keystroke Dynamics. In *Proc. ACM Conf. Computer and Communications Security, 1999*, p. 73–82.
21. F. Monrose, M. Reiter, Q. Li, S. Wetzel. Cryptographic key generation from voice. In *Proc. IEEE Symp. on Security and Privacy*, 2001.
22. F. Monrose, M. Reiter, Q. Li, S. Wetzel. Using voice to generate cryptographic keys. In *Proc. of Odyssey 2001, The Speaker Verification Workshop*, 2001.
23. N. Nisan, A. Ta-Shma. Extracting Randomness: a survey and new constructions. In *JCSS*, **58**(1), pp. 148–173, 1999.
24. N. Nisan, D. Zuckerman. Randomness is Linear in Space. In *JCSS*, **52**(1), pp. 43–52, 1996.
25. J. Radhakrishnan and A. Ta-Shma. Tight bounds for depth-two superconcentrators. In *Proc. 38th IEEE Symp. on Foundations of Computer Science*, 1997, pp. 585–594.
26. R. Shaltiel. Recent developments in Explicit Constructions of Extractors. *Bulletin of the EATCS*, **77**, pp. 67–95, 2002.
27. V. Shoup. A Proposal for an ISO Standard for Public Key Encryption. Available at `http://eprint.iacr.org/2001/112`, 2001.
28. E. Verbitskiy, P. Tyls, D. Denteneer, J.-P. Linnartz. Reliable Biometric Authentication with Privacy Protection. In *Proc. 24th Benelux Symposium on Information theory*, 2003.

# Merkle Tree Traversal in Log Space and Time

Michael Szydlo

RSA Laboratories, Bedford, MA 01730.
mszydlo@rsasecurity.com

**Abstract.** We present a technique for Merkle tree traversal which requires only logarithmic space and time. For a tree with $N$ leaves, our algorithm computes sequential tree leaves and authentication path data in time $2\log_2(N)$ and space less than $3\log_2(N)$, where the units of computation are hash function evaluations or leaf value computations, and the units of space are the number of node values stored. This result is an asymptotic improvement over all other previous results (for example, measuring $cost = space * time$). We also prove that the complexity of our algorithm is optimal: There can exist no Merkle tree traversal algorithm which consumes both less than $O(\log_2(N))$ space and less than $O(\log_2(N))$ time. Our algorithm is especially of practical interest when space efficiency is required.

**Keywords:** amortization, authentication path, Merkle tree, tail zipping, binary tree, fractal traversal, pebbling

## 1 Introduction

Twenty years ago, Merkle suggested the use of complete binary trees for producing multiple *one-time signatures* [4] associated to a single public key. Since this introduction, a *Merkle tree* [8] has been defined to be a complete binary tree with a $k$ bit value associated to each node such that each interior node value is a one-way function of the node values of its children.

Merkle trees have found many uses in theoretical cryptographic constructions, having been specifically designed so that a leaf value can be verified with respect to a publicly known root value and the *authentication data* of the leaf. This authentication data consists of one node value at each height, where these nodes are the siblings of the nodes on the path connecting the leaf to the root. The *Merkle tree traversal problem* is the task of finding an efficient algorithm to output this authentication data for successive leaves. The trivial solution of storing every node value in memory requires too much space. On the other hand, the approach of computing the authentication nodes on the round they are required will be very expensive for some nodes. The challenge is to conserve both space and computation by amortizing the cost of computing such expensive nodes. Thus, this goal is different from other, more well known, tree traversal problems found in the literature.

In practice, Merkle trees have not been appealing due to the large amount of computation or storage required. However, with more efficient traversal techniques, Merkle trees may once again become more compelling, especially given

C. Cachin and J. Camenisch (Eds.): EUROCRYPT 2004, LNCS 3027, pp. 541–554, 2004.

the advantage that cryptographic constructions based on Merkle trees do not require any number theoretic assumptions.

**Our Contribution.** We present a Merkle tree-traversal algorithm which has a better space and time complexity than the previously known algorithms. Specifically, to traverse a tree with $N$ leaves, our algorithm requires computation of at most $2\log_2(N)$ elementary operations per round and requires storage of less than 3 $\log_2(N)$ node values. In this analysis, a hash function computation, and a leaf value computation are each counted as a single elementary operation[1]. The improvement over previous traversal algorithms is achieved as a result of a new approach to scheduling the node computations. We also prove that this complexity is optimal in the sense that there can be no Merkle Tree traversal algorithm which requires both less than $O(log(N))$ space and less than $O(log(N))$ space.

**History and Related Work.** In his original presentation [7], Merkle proposed a straightforward technique to amortize the costs associated with tree traversal. His method requires storage of up to $\log^2(N)/2$ hash values, and computation of about 2 $\log(N)$ hash evaluations per round. This complexity had been conjectured to be optimal.

In [6], an algorithm is presented which allows various time-space trade-offs. A parameter choice which minimizes space requires a maximum storage of about $1.5 \log^2(N)/\log(\log(N))$ hash values, and requires $2\log(N)/\log(\log(N))$ hash evaluations per round. The basic logarithmic space and time algorithm of our paper does not provide for any time-space trade-offs, but our scheduling techniques can be used to enhance the methods of [6].

Other work on tree traversal in the cryptographic literature (e.g. [5]) considers a different type of traversal problem. Related work includes efficient hash chain traversal (e.g [1,2]). Finally, we remark that because the verifier is indifferent to the technique used to produce the authentication path data, these new traversal techniques apply to many existing constructions.

**Applications.** The standard application of Merkle trees is to digital signatures [4,8]. The leaves of such a binary tree may also be used individually for authentication purposes. For example, see TESLA [11]. Other applications include certificate refreshal [9], and micro-payments [3,12]. Because this algorithm just deals with traversing a binary tree, it's applications need not be restricted to cryptography.

**Outline.** We begin by presenting the background and standard algorithms of Merkle trees (Section 2). We then introduce some notation and review the classic Merkle traversal algorithm (Section 3). After providing some intuition (Section 4), we present the new algorithm (Section 5). We prove the time and space

[1] This differs from the measurement of *total* computational cost, which includes, e.g., the scheduling algorithm itself.

bounds in (Section 6), and discuss the optimal asymptotic nature of this result in (Section 7). We conclude with some comments on efficiency enhancements and future work. (Section 8). In the appendix we sketch the proof of the theorem stating that our complexity result is asymptotically optimal.

## 2 Merkle Trees and Background

The definitions and algorithms in this section are well known, but are useful to precisely explain our traversal algorithm.

**Binary Trees.** A complete binary tree $T$ is said to have *height* $H$ if it has $2^H$ leaves, and $2^H - 1$ interior nodes. By labeling each left child node with a "0" and each right child node with a "1", the digits along the path from the root identify each node. Interpreting the string as a binary number, the leaves are naturally indexed by the integers in the range $\{0, 1, \ldots 2^H - 1\}$. The higher the leaf index, the further to the right that leaf is. Leaves are said to have *height* 0, while the *height* of an interior node is the length of the path to a leaf below it. Thus, the root has height $H$, and below each node at height $h$, there are $2^h$ leaves.

**Merkle Trees.** A Merkle tree is a complete binary tree equipped with a function $hash$ and an assignment, $\Phi$, which maps the set of nodes to the set of $k$-length strings: $n \mapsto \Phi(n) \in \{0,1\}^k$. For the two child nodes, $n_{left}$ and $n_{right}$, of any interior node, $n_{parent}$, the assignment $\Phi$ is required to satisfy

$$\Phi(n_{parent}) = hash(\Phi(n_{left})||\Phi(n_{right})). \tag{1}$$

The function $hash$ is a candidate one-way function such as SHA-1 [13].

For each leaf $l$, the value $\Phi(l)$ may be chosen arbitrarily, and then equation (1) determines the values of all the interior nodes. While choosing arbitrary leaf values $\Phi(l)$ might be feasible for a small tree, a better way is to generate them with a keyed pseudo-random number generator. When the leaf value is the hash of the random number, this number is called a *leaf-preimage*. An application might calculate the leaf values in a more complex way, but we focus on the traversal itself and model a leaf calculation with an oracle *LEAFCALC*, which will produces $\Phi(l)$ at the cost of single computational unit[2].

**Authentication Paths.** The goal of Merkle tree traversal is the sequential output of the leaf values, with the associated authentication data. For each height $h < H$, we define $Auth_h$ to be the value of the sibling of the height $h$ node on the path from the leaf to the root. The authentication data is then the set $\{Auth_i \mid 0 \leq i < H\}$.

The correctness of a leaf value may be verified as follows: It is first hashed together with its sibling $Auth_0$, which, in turn, is hashed together with $Auth_1$, etc., all the way up to the root. If the calculated root value is equal to the published root value, then the leaf value is accepted as authentic. Fortunately, when

[2] It is straightforward to adapt the analysis to more expensive leaf value calculations.

the leaves are naturally ordered from left to right, consecutive leaves typically share a large portion of the authentication data.

**Efficiently Computing Nodes.** By construction, each interior node value $\Phi(n)$ (also abbreviated $\Phi_n$) is determined from the leaf values below it. The following well known algorithm, which we call *TREEHASH*, conserves space. During the required $2^{h+1}-1$ steps, it stores a maximum of $h+1$ hash values at once. The *TREEHASH* algorithm consolidates node values at the same height before calculating a new leaf, and it is commonly implemented with a stack.

**Algorithm 1:** TREEHASH (start, maxheight)

1. Set $leaf = start$ and create empty stack.
2. **Consolidate** If top 2 nodes on the stack are equal height:
- Pop node value $\Phi_{right}$ from stack.
- Pop node value $\Phi_{left}$ from stack.
- Compute $\Phi_{parent} = hash(\Phi_{left}||\Phi_{right})$.
- If height of $\Phi_{parent}$ = maxheight, output $\Phi_{parent}$ and stop.
- Push $\Phi_{parent}$ onto the stack.
3. **New Leaf** Otherwise:
- Compute $\Phi_l = LEAFCALC(leaf)$.
- Push $\Phi_l$ onto stack.
- Increment $leaf$ .
4. Loop to step 2.

Often, multiple instances of *TREEHASH* are integrated into a larger algorithm. To do this, one might define an object with two methods, *initialize*, and *update*. The initialization step simply sets the starting leaf index, and height of the desired output. The update method executes either step 2 or step 3, and modifies the contents of the stack. When it is done, the sole remaining value on the stack is $\Phi(n)$. We call the intermediate values stored in the stack *tail node* values.

## 3 The Classic Traversal

The challenge of Merkle tree traversal is to ensure that all node values are ready when needed, but are computed in a manner which conserves space and time. To motivate our own algorithm, we first discuss what the average per-round computation is expected to be, and review the classic Merkle tree traversal.

**Average Costs.** Each node in the tree is eventually part of an authentication path, so one useful measure is the total cost of computing each node value exactly once. There are $2^{H-h}$ right (respectively, left) nodes at height $h$, and if computed independently, each costs $2^{h+1}-1$ operations. Rounding up, this is $2^{H+1} = 2N$ operations, or two per round. Adding together the costs for each height $h$ $(0 \leq h < H)$, we expect, on average, $2H = 2\log(N)$ operations per round to be required.

**Three Components.** As with a digital signature scheme, the tree-traversal algorithms consists of three components: *key generation*, *output*, and *verification.* During key generation, the root of the tree, the first authentication path, and some upcoming authentication node values are computed. The root node value plays the role of a public key, and the leaf values play the role of one-time private keys.

The *output* phase consists of $N$ rounds, one for each leaf. During round $leaf$, the leaf's value, $\Phi(leaf)$ (or leaf pre-image) is output. The authentication path, $\{Auth_i\}$, is also output. Additionally, the algorithm's state is modified in order to prepare for future outputs.

As mentioned above, the *verification* phase is identical to the traditional verification phase for Merkle trees.

**Notation.** In addition to denoting the current authentication nodes $Auth_h$, we need some notation to describe the stacks used to compute upcoming needed nodes. Define $Stack_h$ to be an object which contains a stack of node values as in the description of $TREEHASH$ above. $Stack_h.initialize$ and $Stack_h.update$ will be methods to setup and incrementally compute $TREEHASH$. Additionally, define $Stack_h.low$ to be the height of the lowest node in $Stack_h$, except in two cases: if the stack is empty $Stack_h.low$ is defined to be $h$, and if the $TREEHASH$ algorithm has completed $Stack_h.low$ is defined to be $\infty$.

## 3.1 Key Generation and Setup

The main task of key generation is to compute and publish the root value. This is a direct application of $TREEHASH$. In the process of this computation, every node value is computed, and, it is important to record the initial values $\{Auth_i\}$, as well as the upcoming values for each of the $\{Auth_i\}$.

If we denote the $j$'th node at height $h$ by $n_{h,j}$, we have $Auth_h = \Phi(n_{h,1})$ (these are right nodes). The "upcoming" authentication node at height $h$ is $\Phi(n_{h,0})$ (these are left nodes). These node values are used to initialize $Stack_h$ to be in the state of having completed $TREEHASH$.

**Algorithm 2:** Key-Gen and Setup

1. **Initial Authentication Nodes** For each $i \in \{0, 1, \ldots H-1\}$: Calculate $Auth_i = \Phi(n_{i,1})$.
2. **Initial Next Nodes** For each $i \in \{0, 1, \ldots H-1\}$: Setup $Stack_k$ with the single node value $Auth_i = \Phi(n_{i,1})$.
3. **Public Key** Calculate and publish tree root, $\Phi(root)$.

## 3.2 Output and Update (Classic)

Merkle's tree traversal algorithm runs one instance of $TREEHASH$ for each height $h$ to compute the next authentication node value for that level. Every $2^h$

rounds, the authentication path will shift to the right at level $h$, thus requiring a new node (its sibling) as the height $h$ authentication node.

At each round the $TREEHASH$ state is updated with two units of computation. After $2^h$ rounds this node value computation will be completed, and a new instance of $TREEHASH$ begins for the next authentication node at that level.

To specify how to refresh the $Auth$ nodes, we observe how to easily determine which heights need updating: height $h$ needs updating if and only if $2^h$ divides $leaf + 1$ evenly. Furthermore, we note that at round $leaf + 1 + 2^h$, the authentication *path* will pass though the $(leaf + 1 + 2^h)/2^h$'th node at height $h$. Thus, its *sibling's* value, (the new required upcoming $Auth_h$) is determined from the $2^h$ leaf values starting from leaf number $(leaf + 1 + 2^h) \oplus 2^h$, where $\oplus$ denotes bitwise XOR.

In this language, we summarize Merkle's classic traversal algorithm.

**Algorithm 3:** Classic Merkle Tree Traversal

1. Set $leaf = 0$.
2. **Output:**
- Compute and output $\Phi(leaf)$ with $LEAFCALC(leaf)$.
- For each $h \in [0, H-1]$ output $\{Auth_h\}$.
3. **Refresh Auth Nodes:**
For all $h$ such that $2^h$ divides $leaf + 1$:
- Set $Auth_h$ be the sole node value in $Stack_h$.
- Set startnode $= (leaf + 1 + 2^h) \oplus 2^h$.
- $Stack_k.initialize(startnode, h)$.
4. **Build Stacks:**
For all $h \in [0, H-1]$:
- $Stack_h.update(2)$.
5. **Loop:**
- Set $leaf = leaf + 1$.
- If $leaf < 2^H$ go to Step 2.

## 4 Intuition for an Improvement

Let us make some observations about the classic traversal algorithm. We see that with the classic algorithm above, up to $H$ instances of $TREEHASH$ may be concurrently active, one for each height less than $H$. One can conceptualize them as $H$ processes running in parallel, each requiring also a certain amount of space for the "tail nodes" of the $TREEHASH$ algorithm, and receiving a budget of two hash value computations per round, clearly enough to complete the $2^{h+1} - 1$ hash computations required over the $2^h$ available rounds.

Because the stack employed by $TREEHASH$ may contain up to $h+1$ node values, we are only guaranteed a space bound of $1+2+\cdots+N$. The possibility of so many tail nodes is the source of the $\Omega(N^2/2)$ space complexity in the classic algorithm.

Considering that for the larger $h$, the $TREEHASH$ calculations have many rounds to complete, it appears that it might be wasteful to save so many intermediate nodes at once. Our idea is to schedule the concurrent $TREEHASH$ calculations differently, so that at any given round, the associated stacks are mostly empty. We chose a schedule which generally favors computation of upcoming $Auth_h$ for lower $h$, (because they are required sooner), but delays beginning of a new $TREEHASH$ instance slightly, waiting until all stacks $\{Stack_i\}$ are partially completed, containing no tail nodes of height less than $h$.

This delay, was motivated by the observation that in general, if the computation of two nodes at the same height in different $TREEHASH$ stacks are computed serially, rather than in parallel, less space will be used. Informally, we call the delay in starting new stack computations "zipping up the tails". We will need to prove the fact, which is no longer obvious, that the upcoming needed nodes will always be ready in time.

## 5 The New Traversal Algorithm

In this section we describe the new scheduling algorithm. Comparing to the classic traversal algorithm, the only difference will be in how the budget of $2H$ hash function evaluations will be allocated among the potentially $H$ concurrent $TREEHASH$ processes.

Using the idea of zipping up the tails, there is more than one way to invent a scheduling algorithm which will take advantage of this savings. The one we present here is not optimal, but it is simple to describe. For example, an earlier version of this work presented a more efficient, but more difficult algorithm.

**Algorithm 4:** Logarithmic Merkle Tree Traversal

1. Set $leaf = 0$.
2. **Output:**
   - Compute and output $\Phi(leaf)$ with $LEAFCALC(leaf)$.
   - For each $h \in [0, H-1]$ output $\{Auth_h\}$.
3. **Refresh Auth Nodes:**
   For all $h$ such that $2^h$ divides $leaf + 1$:
   - Set $Auth_h$ be the sole node value in $Stack_h$.
   - Set startnode $= (leaf + 1 + 2^h) \oplus 2^h$.
   - $Stack_k.initialize(startnode, h)$.
4. **Build Stacks:**
   Repeat the following $2H - 1$ times:
   - Let $l_{min}$ be the minimum of $\{Stack_h.low\}$.
   - Let $focus$ be the least $h$ so $Stack_h.low = l_{min}$.
   - $Stack_{focus}.update(1)$.
5. **Loop:**
   - Set $leaf = leaf + 1$.
   - If $leaf < 2^H$ go to Step 2.

This version can be concisely described as follows. The upcoming needed authentication nodes are computed as in the classic traversal, but the various stacks do not all receive equal attention. Each *TREEHASH* instance can be characterized as being either not started, partially completed, or completed. Our schedule to prefers to complete $Stack_h$ for the lowest $h$ values first, *unless another stack has a lower tail node.* We express this preference by defining $l_{min}$ be the minimum of the $h$ values $\{Stack_h.low\}$, then choosing to focus our attention on the smallest level $h$ attaining this minimum. (setting $Stack_h.low = \infty$ for completed stacks effectively skips them over).

In other words, all stacks must be completed to a stage where there are no tail nodes at height $h$ or less before we start a new $Stack_h$ *TREEHASH* computation. The final algorithm is summarized in the box above.

# 6 Correctness and Analysis

In this section we show that our computational budget of $2H$ is indeed sufficient to complete every $Stack_h$ computation before it is required as an authentication node. We also show that the space required for hash values is less than $3H$.

## 6.1 Nodes Are Computed on Time

As presented above, our algorithm allocates exactly a budget of $2H$ computational units per round to spend updating the $h$ stacks. Here, a computational unit is defined to be either a call to *LEAFCALC*, or the computation of a hash value. We do not model any extra expense due to complex leaf calculations.

To prove this, we focus on a given height $h$, and consider the period starting from the time $Stack_h$ is created and ending at the time when the upcoming authentication node (denoted $Need_h$ here) is required to be completed. This is not immediately clear, due to the complicated scheduling algorithm. Our approach to prove that $Need_h$ is completed on time is to showing that the total budget over this period exceeds the cost of *all* nodes computed within this period which can be computed before $Need_h$.

**Node Costs.** The node $Need_h$ itself costs only $2^{h+1} - 1$ units, a tractable amount given that there are $2^h$ rounds between the time $Stack_h$ is created, and the time by which $Need_h$ must be completed. However, a non trivial calculation is required, since in addition to the resources required by $Need_h$, many other nodes compete for the total budget of $2H2^h$ computational units available in this period. These nodes include all the future needed nodes $Need_i$, $(i < h)$, for lower levels, and the $2^N$ output nodes of Algorithm 4, Step 2. Finally there may be a partial contribution to a node $Need_i\, i > h$, so that its stack contains no low nodes by the time $Need_h$ is computed.

It is easy to count the number of such needed nodes in the interval, and we know the cost of each one. As for the contributions to higher stacks, we at least know that the cost to raise any low node to height $h$ must be less than $2^{h+1} - 1$

(the total cost of a height $h$ node). We summarize these quantities and costs in the following figure.

**Nodes built during $2^h$ rounds for $Need_h$.**

| Node Type | Quantity | Cost Each |
|---|---|---|
| $Need_h$ | 1 | $2^{h+1}-1$ |
| $Need_{h-1}$ | 2 | $2^h-1$ |
| ... | ... | ... |
| $Need_k$ | $2^{h-k}$ | $2^{k+1}-1$ |
| ... | ... | ... |
| $Need_0$ | $2^h$ | 1 |
| $Output$ | $2^h$ | 1 |
| $Tail$ | 1 | $\leq 2^{h+1}-2$ |

We proceed to tally up the total cost incurred during the interval. Notice that the rows beginning $Need_0$ and $Output$ require a total of $2^{h+1}$ computational units. For ever other row in the node chart, the number of nodes of a given type multiplied by the cost per node is less than $2^{h+1}$. There are $h+1$ such rows, so the total cost of all nodes represented in the chart is

$$TotalCost_h < (h+2)2^h.$$

For heights $h \leq H-2$, it is clear that this total cost is less than $2H2^H$. It is also true for the remaining case of $h = H-1$, because there are no tail nodes in this case.

We conclude that, as claimed, the budget of $2H$ units per round is indeed always sufficient to prepare $Need_h$ on time, for any $0 \leq h < H$.

## 6.2 Space Is Bounded by 3H

Our motivation leading to this relatively complex scheduling is to use as little space as possible. To prove this, we simply add up the quantities of each kind of node. We know there are always $H$ nodes $Auth_h$. Let $C < H$ be the number completed nodes $Next_h$.

$$\#Auth_i + \#Next_i = H + C. \tag{2}$$

We must finally consider the number of tail nodes in the $\{Stack_h\}$. As for these, we observe that since a $Stack_h$ never becomes active until all nodes in "higher" stacks are of height at least $h$, there can never be two distinct stacks, each containing a node of the same height. Furthermore, recalling algorithm *TREEHASH*, we know there is at most one height for which a stack has two node values. In all, there is at most one tail node at each height ($0 \leq h \leq H-3$), plus up to one additional tail node per non-completed stack. Thus

$$\#Tail \leq H - 2 + (H - C). \tag{3}$$

Adding all types of nodes we obtain:

$$\#Auth_i + \#Next_i + \#Tail \leq 3H - 2. \tag{4}$$

This proves the assertion. There are at most $3H - 2$ stored nodes.

## 7 Asymptotic Optimality Result

An interesting optimality result states that a traversal algorithm can never beat both *time* $= O(\log(N))$ and *space* $= O(\log(N))$. It is clear that at least $H - 2$ nodes are required for the TREEHASH algorithm, so our task is essentially to show that if space is limited by any constant multiple of $\log(N)$, then the computational complexity must be $\Omega(\log(N))$. Let us be clear that this theorem does not quantify the constants. Clearly, with greater space, computation time can be reduced.

**Theorem 1.** *Suppose that there is a Merkle tree traversal algorithm for which the space is bounded by $\alpha \, log(N)$. Then there exists some constant $\beta$ so that the time required is at least $\beta \, log(N)$.*

The theorem simply states that it is not possible to reduce space complexity below logarithmic without increasing the time complexity beyond logarithmic!

The proof of this technical statement is found in the appendix, but we will briefly describe the approach here. We consider only right nodes for the proof. We divide all right nodes into two groups: those which must be computed (at a cost of $2^{h+1} - 1$), and those which have been saved from some earlier calculation. The proof assumes a sub-logarithmic time complexity and derives a contradiction.

The more nodes in the second category, the faster the traversal can go. However, such a large quantity of nodes would be required to be saved in order to reduce the time complexity to sub-logarithmic, that the average number of saved node values would have to exceed a linear amount! The rather technical proof in the appendix uses a certain sequence of subtrees to formulate the contradiction.

## 8 Efficiency Improvements and Future Work

**Halving the required time.** The scheduling algorithm we presented above is not optimally efficient. In an earlier version of this paper, we described a technique to half the number of required hash computations per round. The trick was to notice that all of the *left* nodes in the tree could be calculated nearly for free. Unfortunately, this resulted in a more complicated algorithm which is less appealing for a transparent exposition.

**Other Variants.** A space-time trade-off is the subject of [6]. For our algorithm, clearly a few extra node values stored near the top of the tree will reduce total computation, but there are also other strategies to exploit extra space and save time. For Merkle tree traversal all such approaches are based on the idea that during a node computation (such as that of $Need_i$) saving some wisely chosen set of intermediate node values will avoid their duplicate future recomputation, and thus save time.

**Future work.** It might be interesting to explicitly combine the idea in this paper with the work in [6]. One might ask the question, for any size tree, what is the least number of hash computations per round which will suffice, if only $S$ hash nodes may be stored at one time.

Perhaps a more interesting direction will be to look for applications for which an efficient Merkle tree traversal would be useful. Because the traversal algorithms are a relatively general construction, applications outside of cryptography might be discovered.

Within cryptography, there is some interest in understanding which constructions would still be possible if no public-key functionality turned out to exist. (For example due to quantum computers). Then the efficiency of a signature scheme based on Merkle tree's would be of practical interest. Finally, in any practical implementation designed to conserve space, it is important to take into consideration the size of the algorithm's code itself.

## 9 Acknowledgments

The author wishes to thank Markus Jakobsson, and Silvio Micali, and the anonymous reviewers for encouraging discussions and useful comments. The author especially appreciates the reviewer's advice to simplify the algorithm, even for a small cost in efficiency, as it may have helped to improve the presentation of this result, making the paper easier to read.

## References

1. D. Coppersmith and M. Jakobsson, "Almost Optimal Hash Sequence Traversal," Financial Crypto '02. Available at `www.markus-jakobsson.com`.
2. M. Jakobsson, "Fractal Hash Sequence Representation and Traversal," ISIT '02, p. 437. Available at `www.markus-jakobsson.com`.
3. C. Jutla and M. Yung, "PayTree: Amortized-Signature for Flexible Micropayments," 2nd USENIX Workshop on Electronic Commerce, pp. 213–221, 1996.
4. L. Lamport, "Constructing Digital Signatures from a One Way Function," SRI International Technical Report CSL-98 (October 1979).
5. H. Lipmaa, "On Optimal Hash Tree Traversal for Interval Time-Stamping," In Proceedings of Information Security Conference 2002, volume 2433 of Lecture Notes in Computer Science, pp. 357–371. Available at `www.tcs.hut.fi/~helger/papers/lip02a/`.
6. M. Jakobsson, T. Leighton, S. Micali, M. Szydlo, "Fractal Merkle Tree Representation and Traversal," In RSA Cryptographers Track, RSA Security Conference 2003.
7. R. Merkle, "Secrecy, Authentication, and Public Key Systems," UMI Research Press, 1982. Also appears as a Stanford Ph.D. thesis in 1979.

8. R. Merkle, "A Digital Signature Based on a Conventional Encryption Function," Proceedings of Crypto '87, pp. 369–378.
9. S. Micali, "Efficient Certificate Revocation," In RSA Cryptographers Track, RSA Security Conference 1997, and U.S. Patent No. 5,666,416.
10. T. Malkin, D. Micciancio, and S. Miner, "Efficient Generic Forward-Secure Signatures With An Unbounded Number Of Time Periods" Proceedings of Eurocrypt '02, pp. 400-417.
11. A. Perrig, R. Canetti, D. Tygar, and D. Song, "The TESLA Broadcast Authentication Protocol," Cryptobytes, Volume 5, No. 2 (RSA Laboratories, Summer/Fall 2002), pp. 2–13. Available at `www.rsasecurity.com/rsalabs/cryptobytes/`.
12. R. Rivest and A. Shamir, "PayWord and MicroMint–Two Simple Micropayment Schemes," CryptoBytes, Volume 2, No. 1 (RSA Laboratories, Spring 1996), pp. 7–11. Available at `www.rsasecurity.com/rsalabs/cryptobytes/`.
13. FIPS PUB 180-1, "Secure Hash Standard, SHA-1". Available at `www.itl.nist.gov/fipspubs/fip180-1.htm`.

# A Complexity Proof

We now begin the technical proof of Theorem 1. This will be a proof by contradiction. We assume that the time complexity is sub logarithmic, and show that this is incompatible with the assumption that the space complexity is $O(log(N))$.

Our strategy to produce a contradiction is to find a bound on some linear combination of the average time and the average amount of space consumed.

**Notation** The theorem is an asymptotic statement, so we will be considering trees of height $H = log(N)$, for large $H$. We need to consider $L$ *levels* of subtrees of height $k$, where $kL = H$. Within the main tree, the roots of these subtrees will be at heights $k, 2*k, 3*k \ldots H$. We say that the subtree is at *level* $i$ if its root is at height $(i+1)k$. This subtree notation is similar to that used in [6].

Note that we will only need to consider right nodes to complete our argument. Recall that during a complete tree traversal every single right node is eventually output as part of the authentication data. This prompts us to categorize the right nodes in three classes.

1. Those already present after the key generation: *free nodes.*
2. Those explicitly calculated (e.g. with $TREEHASH$): *computed nodes.*
3. Those retained from another node's calculation (e.g from another node's $TREEHASH$): *saved nodes.*

Notice how type 2 nodes require computational effort, whereas type 1 and type 3 nodes require some period of storage. We need further notation to conveniently reason about these nodes. Let $a_i$ denote the number of level $i$ subtrees which contain *at least 1* non-root computed (right) node. Similarly, let $b_i$ denote the number of level $i$ subtrees which contain *zero* computed nodes. Just by counting the total number of level $i$ subtrees we have the relation.

$$a_i + b_i = N/2^{(i+1)k}. \tag{5}$$

**Computational costs** Let us tally the cost of some of the computed nodes. There are $a_i$ subtrees containing a node of type 2, which must be of height at least $ik$. Each such node will cost at least $2^{ik+1} - 1$ operations to compute. Rounding down, we find a simple lower bound for the cost of the nodes at level $i$.

$$Cost > \Sigma_0^{L-1}(a_i 2^{ik}). \tag{6}$$

**Storage costs** Let us tally the lifespans of some of the retained nodes. Measuring units of *space* × *rounds* is natural when considering average space consumed. In general, a saved node, $S$, results from a calculation of some computed node $C$, say, located at height $h$. We know that $S$ has been produced before $C$ is even needed, and $S$ will never become an authentication node before $C$ is discarded. We conclude that such a node $S$ must be therefore be stored in memory for at least $2^h$ rounds.

Even (most of) the free nodes at height $h$ remain in memory for at least $2^{h+1}$ rounds. In fact, there can be at most one exception: the first right node at level $h$.

Now consider one of the $b_i$ subtrees at level $i$ containing only free or stored nodes. Except for the leftmost subtree at each level, which may contain a free node waiting in memory less than $2^{(i+1)k}$ rounds, every other node in this subtree takes up space for at least $2^{(i+1)k}$ rounds. There are $2^k - 1$ nodes in a subtree and thus we find a simple lower bound on the *space* × *rounds*.

$$Space * Rounds \geq \Sigma_0^{L-1}(b_i - 1)(2^k - 1)2^{(i+1)k}. \tag{7}$$

Note that the $(b_i - 1)$ term reflects the possible omission of the leftmost level $i$ subtree.

**Mixed Bounds** We can now use simple algebra with Equations (5), (6), and (7) to yield combined bounds. First the cost is related to the $b_i$, which is then related to a space bound.

$$2^k Cost > \Sigma_0^{L-1} a_i 2^{(i+1)k} = \Sigma_0^{L-1} N - 2^{(i+1)k} b_i. \tag{8}$$

As series of similar algebraic manipulations finally yield (somewhat weaker) very useful bounds.

$$2^k Cost + \Sigma_0^{L-1} 2^{(i+1)k} b_i > NL. \tag{9}$$

$$2^k Cost + \Sigma_0^{L-1} 2^{(i+1)k}/(2^{k-1}) + Space * Rounds./(2^{k-1}) > NL \tag{10}$$

$$2^k Cost + 2N + Space * Rounds/(2^{k-1}) > NL. \tag{11}$$

$$2^k Average\,Cost + Average\,Space/(2^{k-1}) > (L-2) \geq L/2. \tag{12}$$

$$(k\,2^{k+1}) Average\,Cost + (k/2^{k-2}) Average\,Space > L/2 * 2k = H. \tag{13}$$

This last bound on the sum of average cost and space requirements will allow us to find a contradiction.

**Proof by Contradiction** Let us assume the opposite of the statement of Theorem 1. Then there is some $\alpha$ such that the space is bounded above by $\alpha\, log(N)$. Secondly, the time complexity is supposed to be sub-logarithmic, so for every small $\beta$ the time required is less than $\beta\, log(N)$ for sufficiently large $N$.

With these assumptions we are now able to choose a useful value of $k$. We pick $k$ to be large enough so that $\alpha > 1/k2^{k+3}$. We also choose $\beta$ to be less than $1/k2^{k+2}$. With these choices we obtain two relations.

$$(k\, 2^{k+1}) Average\, Cost < H/2. \tag{14}$$

$$(k/2^{k-2}) Average\, Space < H/2. \tag{15}$$

By adding these two last equations, we contradict Equation (13).
**QED.**

# Can We Trust Cryptographic Software? Cryptographic Flaws in GNU Privacy Guard v1.2.3

Phong Q. Nguyen

CNRS/École normale supérieure
Département d'informatique
45 rue d'Ulm, 75230 Paris Cedex 05, France.
Phong.Nguyen@ens.fr
http://www.di.ens.fr/~pnguyen

**Abstract.** More and more software use cryptography. But how can one know if what is implemented is good cryptography? For proprietary software, one cannot say much unless one proceeds to reverse-engineering, and history tends to show that bad cryptography is much more frequent than good cryptography there. Open source software thus sounds like a good solution, but the fact that a source code can be read does not imply that it is actually read, especially by cryptography experts. In this paper, we illustrate this point by examining the case of a basic Internet application of cryptography: secure email. We analyze parts of the source code of the latest version of GNU Privacy Guard (GnuPG or GPG), a free open source alternative to the famous PGP software, compliant with the OpenPGP standard, and included in most GNU/Linux distributions such as Debian, MandrakeSoft, Red Hat and SuSE. We observe several cryptographic flaws in GPG v1.2.3. The most serious flaw has been present in GPG for almost four years: we show that as soon as one (GPG-generated) ElGamal signature of an arbitrary message is released, one can recover the signer's private key in less than a second on a PC. As a consequence, ElGamal signatures and the so-called ElGamal sign+encrypt keys have recently been removed from GPG. Fortunately, ElGamal was not GPG's default option for signing keys.

**Keywords:** Public-key cryptography, GnuPG, GPG, OpenPGP, Cryptanalysis, RSA, ElGamal, Implementation.

## 1 Introduction

With the advent of standardization in the cryptography world (RSA PKCS [20], IEEE P1363 [14], CRYPTREC [15], NESSIE [8], *etc.*), one may think that there is more and more good cryptography. But as cryptography becomes "global", how can one be sure that what is implemented in the real world is actually good cryptography? Numerous examples (such as [4,2,21]) have shown that the frontier between good cryptography and bad cryptography is very thin. For proprietary software, it seems difficult to make any statement unless one proceeds

C. Cachin and J. Camenisch (Eds.): EUROCRYPT 2004, LNCS 3027, pp. 555–570, 2004.

to the tedious task of reverse-engineering. If a proprietary software claims to implement 2048-bit RSA and 128-bit AES, it does not say much about the actual cryptographic security: which RSA is being used? Could it be textbook RSA [5] (with zero-padding) encrypting a 128-bit AES key with public exponent 3? Are secret keys generated by a weak pseudo-random number generator like old versions of Netscape [9]? Who knows if it is really RSA–OAEP which is implemented [21]? With proprietary software, it is ultimately a matter of trust: unfortunately, history has shown that there is a lot of bad cryptography in proprietary software (see for instance [28,12] for explanations). Open source software thus sounds like a good solution. However, the fact that a source code can be read does not necessarily imply that it is actually read, especially by cryptography experts.

The present paper illustrates this point by examining the case of "perhaps the most mature cryptographic technology in use on the Internet" (according to [1]): secure email, which enables Internet users to authentify and/or encrypt emails. Secure email became popular in the early 90s with the appearance of the now famous Pretty Good Privacy (PGP) [27] software developed by Phil Zimmermann in the US. Not so long ago, because of strict export restrictions and other US laws, PGP was unsuitable for many areas outside the US. Although the source code of PGP has been published, it is unknown whether future versions of PGP will be shipped with access to the source code.

GNU Privacy Guard [10] (GnuPG, or GPG in short) was developed in the late 90s as an answer to those PGP issues. GPG is a full implementation of OpenPGP [26], the Internet standard that extends PGP. GPG has been released as free software under the GNU General Public License (GNU GPL): As such, full access to the source code is provided at [10], and GPG can be viewed as a free replacement for PGP. The German Federal Ministry of Economics and Technology granted funds for the further development of GPG. GPG has a fairly significant user base: it is included in most GNU/Linux distributions, such as Debian, MandrakeSoft, Red Hat and SuSE. The first stable version of GPG was released on September 7th, 1999. Here, we review the main public-key aspects of the source code of v1.2.3, which was the current stable version (released on August 22nd, 2003) when this paper was submitted to Eurocrypt '04. Our comments seem to also apply to several previous versions of GPG. However, we stress that our analysis is not claimed to be complete, even for the public-key aspects of GPG.

We observe several cryptographic flaws in GPG v1.2.3. The most serious flaw (which turns out to have been present in GPG for almost four years) is related to ElGamal signatures: we present a lattice-based attack which recovers the signer's private key in less than a second on a PC, given any (GPG-generated) ElGamal signature of a (known) arbitrary message and the corresponding public key. This is because both a short private exponent and a short nonce are used for the generation of ElGamal signatures, when the GPG version in use is between 1.0.2 (January 2000) and 1.2.3 (August 2003). As a result, GPG–ElGamal signing keys have been considered compromised [19], especially the so-called primary ElGamal sign+encrypt keys: with such keys, one signature is always readily available, because such keys automatically come up with a signature to bind the user identity to the public key, thus leaking the private key used for both encryption and signature. Hence, ElGamal signatures and ElGamal sign+encrypt keys have recently been removed from GPG (see [10] for more information).

We notice that GPG encryption provides no chosen-ciphertext security, due to its compliance with OpenPGP [26], which uses the old PKCS #1 v1.5 standard [17]: Bleichenbacher's chosen-ciphertext attack [4] applies to OpenPGP, either when RSA or ElGamal is used. Although the relevance of chosen-ciphertext attacks to the context of email communications is debatable, we hope that OpenPGP will replace PKCS #1 v1.5 to achieve chosen-ciphertext security. The other flaws do not seem to lead to any reasonable attack, they only underline the lack of state-of-the-art cryptography in GPG and sometimes OpenPGP. It is worth noting that the OpenPGP standard is fairly loose: it gives non-negligible freedom over the implementation of cryptographic functions, especially regarding key generation. Perhaps stricter and less ambiguous guidelines should be given in order to decrease security risks.

The only published research on the cryptographic strength of GPG we are aware of is [18,16], which presented chosen-ciphertext attacks with respect to the symmetric encryption used in PGP and GPG. The rest of the paper is organized as follows. In Section 2, we give an overview of the GPG software v1.2.3. In Section 3, we review the GPG implementation of ElGamal and present the attack on ElGamal signatures. In Section 4, we review the GPG implementation of RSA. There is no section devoted to the GPG implementation of DSA, since we have not found any noteworthy weakness in it. In Appendix A, we give a brief introduction to lattice theory, because the attack on ElGamal signatures uses lattices. In Appendix B, we provide a proof (in an idealized model) of the lattice-based attack on ElGamal signatures.

## 2 An Overview of GPG v1.2.3

GPG v1.2.3 [10] supports ElGamal (signature and encryption), DSA, RSA, AES, 3DES, Blowfish, Twofish, CAST5, MD5, SHA-1, RIPE-MD-160 and TIGER. GPG decrypts and verifies PGP 5, 6 and 7 messages: It is compliant with the OpenPGP standard [26], which is described in RFC 2440 [6].

GPG provides secrecy and/or authentication to emails: it enables users to encrypt/decrypt and/or sign/verify emails using public-key cryptography. The public-key infrastructure is the famous web of trust: users certify public key of other users.

GPG v.1.2.3 allows the user to generate several types of public/private keys, with the command `gpg --gen-key`:

- Choices available in the standard mode:
  - (1) DSA and ElGamal: this is the default option. The DSA keys are signing keys, while the ElGamal keys are encryption keys (type 16 in the OpenPGP terminology).

  - (2) DSA: only for signatures.
  - (5) RSA: only for signatures.
- Additional choices available in the expert mode:
  - (4) ElGamal for both signature and encryption. In the OpenPGP terminology, these are keys of type 20.
  - (7) RSA for both signature and encryption.

In particular, an ElGamal signing key is also an encryption key, but an ElGamal encryption key may be restricted to encryption. In GPG v1.2.3, ElGamal signing keys cannot be created unless one runs the expert mode: however, this was not always the case in previous versions. For instance, the standard mode of GPG v1.0.7 (which was released in April 2002) proposes the choices (1), (2), (4) and (5).

## 2.1 Encryption

GPG uses hybrid encryption to encrypt emails. A session key (of a symmetric encryption scheme) is encrypted by a public-key encryption scheme: either RSA or ElGamal (in a group $\mathbb{Z}_p^*$, where $p$ is a prime number). The session key is formatted as specified by OpenPGP (see Figure 1): First, the session key is prefixed with a one-octet algorithm identifier that specifies the symmetric encryption algorithm to be used; Then a two-octet checksum is appended which is equal to the sum of the preceding session key octets, not including the algorithm identifier, modulo 65536.

| One-octet identifier of the symmetric encryption algorithm | Key of the symmetric encryption algorithm | Two-octet checksum over the key bytes |
|---|---|---|

**Fig. 1.** Session key format in OpenPGP.

This value is then padded as described in PKCS#1 v1.5 block type 02 (see [17] and Figure 2): a zero byte is added to the left, as well as as many non-zero random bytes as necessary in such a way that the first two bytes of the final value are 00 02 followed by as many nonzero random bytes as necessary, and the rest. Note that this formatting is applied to both RSA and ElGamal encryption, even though PKCS#1 v1.5 was only designed for RSA encryption. The randomness required to generate nonzero random bytes is obtained by a process following the principles of [11].

| 00 | 02 | Non-zero random bytes | 00 | Message |
|---|---|---|---|---|

**Fig. 2.** PKCS#1 v1.5 encryption padding, block type 02.

### 2.2 Signature

GPG supports the following signature schemes: RSA, DSA and ElGamal. The current GPG FAQ includes the following comment: *As for the key algorithms, you should stick with the default (i.e., DSA signature and ElGamal encryption). An ElGamal signing key has the following disadvantages: the signature is larger, it is hard to create such a key useful for signatures which can withstand some real world attacks, you don't get any extra security compared to DSA, and there might be compatibility problems with certain PGP versions. It has only been introduced because at the time it was not clear whether there was a patent on DSA.* The `README` file of GPG includes the following comment: *ElGamal for signing is available, but because of the larger size of such signatures it is strongly deprecated (Please note that the GnuPG implementation of ElGamal signatures is *not* insecure).* Thus, ElGamal signatures are not really recommended (mainly for efficiency reasons), but they are nevertheless supported by GPG, and they were not supposed to be insecure.

When RSA or ElGamal is used, the message is first hashed (using the hash function selected by the user), and the hash value is encoded as described in PKCS#1 v1.5 (see [17] and Figure 3): a certain constant (depending on the hash function) is added to the left, then a zero byte is added to the left, as well as as many FF bytes as necessary in such a way that the first two bytes of the final value are 00 01 followed by the FF bytes and the rest. With DSA, there is

| 00 | 01 | FF bytes | 00 | Constant | Hashed message |
|---|---|---|---|---|---|

**Fig. 3.** PKCS#1 v1.5 signature padding, block type 01.

no need to apply a signature padding, as the DSS standard completely specifies how a message is signed.

The randomness required by ElGamal and DSA is obtained by a process following the principles of [11].

## 3 The Implementation of ElGamal

GPG uses the same key generation for signature and encryption. It implements ElGamal in a multiplicative group $\mathbb{Z}_p^*$ (where $p$ is a large prime) with generator $g$. The private key is denoted by $x$, and the corresponding public key is $y = g^x \pmod{p}$.

### 3.1 Key Generation

The large prime number $p$ is chosen in such a way that the factorization of $p-1$ is completely known and all the prime factors of $(p-1)/2$ have bit-length larger than a threshold $q_{bit}$ depending on the requested bit-length of $p$. The correspondance between the size of $p$ and the threshold is given by the so-called Wiener table (see Figure 4): notice that $4q_{bit}$ is always less than the bit length of $p$.

| Bit-length of $p$ | 512 | 768 | 1024 | 1280 | 1536 | 1792 | 2048 | 2304 | 2560 | 2816 | 3072 | 3328 | 3584 | 3840 |
|---|---|---|---|---|---|---|---|---|---|---|---|---|---|---|
| $q_{bit}$ | 119 | 145 | 165 | 183 | 198 | 212 | 225 | 237 | 249 | 259 | 269 | 279 | 288 | 296 |

**Fig. 4.** The Wiener table used to generate ElGamal primes.

Once $p$ is selected, a generator $g$ of $\mathbb{Z}_p^*$ is found by testing successive potential generators (thanks to the known factorization of $p-1$), starting with the number 3: If 3 turns out not to be a generator, then one tries with 4, and so on. The generation of the pair $(p, g)$ is specified in the procedure `generate_elg_prime` of the file `cipher/primegen.c`. The generation of the pair of public and private keys is done in the procedure `generate` of the file `cipher/elgamal.c`. Although the generator $g$ is likely to be small, we note that because all the factors of $(p-1)/2$ have at least $q_{bit} \geq 119$ bits, and $g > 2$, Bleichenbacher's forgery [3] of ElGamal signatures does not seem to apply here.

The private exponent $x$ is not chosen as a pseudo-random number modulo $p-1$, although GPG makes the following comment: *select a random number which has these properties: $0 < x < p-1$. This must be a very good random number because this is the secret part.* Instead, $x$ is chosen as a pseudo-random number of bit-length $3q_{bit}/2$, which is explained by the following comment (and which somehow contradicts the previous one): *I don't see a reason to have a $x$ of about the same size as the $p$. It should be sufficient to have one about the size of $q$ or the later used $k$ plus a large safety margin. Decryption will be much faster with such an $x$.* Thus, one chooses an $x$ much smaller than $p$ to speed-up the private operations. Unfortunately, we will see that this has implications on the security of GPG–ElGamal signatures.

### 3.2 Signature

**Description.** The signature of a message already formatted as an integer $m$ modulo $p$ (as decribed in Section 2.2), is the pair $(a, b)$ where: $a = g^k \bmod p$ and $b = (m - ax)k^{-1} \pmod{p-1}$. The integer $k$ is a "random" number coprime with $p-1$, which must be generated at each signature. GPG verifies a signature $(a, b)$ by checking that $0 < a < p$ and $y^a a^b \equiv g^m \pmod{p}$. We note that such a signature verification does not prevent malleability (see [30] for a discussion on malleability): if $(a, b)$ is a valid signature of $m$, then $(a, b + u(p-1))$ is another valid signature of $m$ for all integer $u$, because there is no range check over $b$. This is a minor problem, but there is worse.

Theoretically, $k$ should be a cryptographically secure random number modulo $p-1$ such that $k$ is coprime to $p-1$. Recent attacks on discrete-log signature schemes (see [22,2,13]) have shown that any leakage of information (or any peculiar property) on $k$ may enable an attacker to recover the signer's private key in a much shorter time than what is usually required to solve discrete logarithms, provided that sufficiently many message/signature pairs are available. Intuitively, if partial information on $k$ is available, each message/signature pair discloses information on the signer's private key, even though the signatures use different $k$'s: when enough such pairs are gathered, it might become possible to extract the private key. Unfortunately, the GPG generation of $k$ falls short of such recommendations.

The generation of $k$ is described in the procedure **gen_k** of the file `cipher/elgamal.c`. It turns out that $k$ is first chosen with $3q_{bit}/2$ pseudo-random bits (as in the generation of the private exponent $x$, except that $k$ may have less than $3q_{bit}/2$ bits). Next, as while as $k$ is not coprime with $p-1$, $k$ is incremented. Obviously, the final $k$ is much smaller than $p$, and therefore far from being uniformly distributed modulo $p-1$: the bit-length of $k$ should still be around $3q_{bit}/2$, while that of $p$ is at least $4q_{bit}$. This is explained in the following comment: *IMO using a k much lesser than p is sufficient and it greatly improves the encryption performance. We use Wiener's table and add a large safety margin.* One should bear in mind that the same generation of $k$ is used for both encryption and signature. However, the choice of a small $k$ turns out to be dramatic for signature, rather than for encryption.

**Attacking GPG–ElGamal Signatures.** Independently of the choice of the private exponent $x$, because $k$ is much smaller than $p-1$, one could apply the lattice-based attack of Nguyen-Shparlinski [22] with very slight modifications, provided that a few signatures are known. However, because $x$ is so small, there is even a simpler attack, using only a single signature! Indeed, we have the following congruence:

$$bk + ax \equiv m \pmod{p-1}. \tag{1}$$

In this congruence, only $k$ and $x$ are unknowns, and they are unusually small: From Wiener's table (Figure 4), we know that $k$ and $x$ are much smaller than $\sqrt{p}$.

Linear congruences with small unknowns occur frequently in public-key cryptanalysis (see for instance the survey [24]), and they are typically solved with lattice techniques. We assume that the reader is familiar with lattice theory (see Appendix A and the references of [24]). Following the classical strategy described in [24], we view the problem as a closest vector problem in a two-dimensional lattice, using lattices defined by a single linear congruence. The following lemma introduces the kind of two-dimensional lattices we need:

**Lemma 1.** *Let $(\alpha, \beta) \in \mathbb{Z}^2$ and $n$ be a positive integer. Let $d$ be the greatest common divisor of $\alpha$ and $n$. Let $e$ be the greatest common divisor of $\alpha$, $\beta$ and $n$. Let $L$ be the set of all $(u, v) \in \mathbb{Z}^2$ such that $\alpha u + \beta v \equiv 0 \pmod{n}$. Then:*

1. *$L$ is a two-dimensional lattice in $\mathbb{Z}^2$.*
2. *The determinant of $L$ is equal to $n/e$.*
3. *There exists $u \in \mathbb{Z}$ such that $\alpha u + (\beta/e)d \equiv 0 \pmod{n}$.*
4. *The vectors $(n/d, 0)$ and $(u, d/e)$ form a basis of $L$.*

*Proof.* By definition, $L$ is a subgroup of $\mathbb{Z}^2$, hence a lattice. Besides, $L$ contains the two linearly independent vectors $(n, 0)$ and $(0, n)$, which proves statement 1. Let $f$ be the function that maps $(u, v) \in \mathbb{Z}^2$ to $\alpha u + \beta v$ modulo $n$. $f$ is a group morphism between $\mathbb{Z}^2$ and the additive group $\mathbb{Z}_n$. The image of $f$ is the subgroup (of $\mathbb{Z}_n$) spanned by the greatest common divisor of $\alpha$ and $\beta$: it follows that the image of $f$ has exactly $n/e$ elements. Since $L$ is the kernel of $f$, we deduce that the index of $L$ in $\mathbb{Z}^2$ is equal to $n/e$, which proves statement 2. Statement 3 holds because the greatest common divisor of $\alpha$ and $n$ is $d$, which divides the integer $(\beta/e)d$. By definition of $u$, the vector $(u, d/e)$ belongs to $L$. Obviously, the vector $(n/d, 0)$ belongs to $L$. But the determinant of those two vectors is $n/e$, that is, the determinant of $L$. This proves statement 4. □

We use the following lattice:

$$L = \{(u, v) \in \mathbb{Z}^2 : bu + av \equiv 0 \pmod{p-1}\}. \tag{2}$$

By Lemma 1, a basis of $L$ can easily be found. We then compute an arbitrary pair $(k', x') \in \mathbb{Z}^2$ such that $bk' + ax' \equiv m \pmod{p-1}$. To do so, we can apply the extended Euclidean algorithm to express the greatest common divisor of $a, b$ and $p-1$ as a linear combination of $a$, $b$ and $p-1$. This gcd must divide $m$ by (1), and therefore, a suitable multiplication of the coefficients of the linear combination gives an appropriate $(k', x')$.

The vector $\boldsymbol{l} = (k' - k, x' - x)$ belongs to $L$ and is quite close to the vector $\boldsymbol{t} = (k' - 2^{3q_{bit}/2-1}, x' - 2^{3q_{bit}/2-1})$. Indeed, $k$ has about $3q_{bit}/2$ bits and $x$ has exactly $3q_{bit}/2$ bits, therefore the distance between $\boldsymbol{l}$ and $\boldsymbol{t}$ is about $2^{(3q_{bit}-1)/2}$, which is much smaller than $\det(L)^{1/2}$, because from Lemma 1:

$$\det(L) = \frac{p-1}{\gcd(a, b, p-1)}.$$

From the structure of $p-1$ and the way $a$ and $b$ are defined, we thus expect $\det(L)$ to be around $p$. Hence, we can hope that $\boldsymbol{l}$ is the closest vector of $\boldsymbol{t}$ in the lattice $L$, due to the huge size difference between $2^{(3q_{bit}-1)/2}$ and $\sqrt{p}$, for all the values of $q_{bit}$ given by Figure 4: this heuristic reasoning is frequent in lattice-based cryptanalysis, here it can however be proved if we assume that the distribution of $a$ and $b$ is uniform modulo $p-1$ (see Appendix B). If $\boldsymbol{l}$ is the closest vector of $\boldsymbol{t}$, $\boldsymbol{l}$ and therefore the private exponent $x$ can be recovered from a two-dimensional closest vector computation (we know $t$ and a basis of $L$). And such a closest vector computation can be done in quadratic time (see for instance [23]), using the classical Gaussian algorithm for two-dimensional lattice reduction. Figure 5 sums up the attack, which clearly runs in polynomial time.

Alternatively, if one wants to program as less as possible, one can mount another lattice-based attack, by simply computing a shortest vector of the 4-dimensional lattice $L'$ spanned by the following row vectors, where $K$ is a large constant:

$$\begin{pmatrix} (p-1)K & 0 & 0 & 0 \\ -mK & 2^{3q_{bit}/2} & 0 & 0 \\ bK & 0 & 1 & 0 \\ aK & 0 & 0 & 1 \end{pmatrix}$$

**Input:** The public parameters and a GPG–ElGamal signature $(a, b)$ of $m$.
**Expected output:** The signer's private exponent $x$.

1. Compute a basis of the lattice $L$ of (2), using statement 4 of Lemma 1.
2. Compute $(k', x') \in \mathbb{Z}^2$ such that $bk' + ax' \equiv m \pmod{p-1}$, using the Euclidean algorithm.
3. Compute the target vector $\boldsymbol{t} = (k' - 2^{3q_{bit}/2-1}, x' - 2^{3q_{bit}/2-1})$.
4. Compute the lattice vector $\boldsymbol{l}$ closest to $\boldsymbol{t}$ in the two-dimensional lattice $L$.
5. Return $x'$ minus the second coordinate of $\boldsymbol{l}$.

**Fig. 5.** An attack using a single GPG–ElGamal signature.

This shortest vector computation can be done in quadratic time using the lattice reduction algorithm of [23]. The lattice $L'$ contains the short vector $(0, 2^{3q_{bit}/2}, k, x)$ because of (1). This vector is expected to be a shortest lattice vector under roughly the same condition on $q_{bit}$ and $p$ as in the previous lattice attack (we omit the details). Thus, for all values of Wiener's table (see Figure 4), one can hope to recover the private exponent $x$ as the absolute value of the last coordinate of any shortest nonzero vector of $L'$.

We implemented the last attack with Shoup's NTL library [29], using the integer LLL algorithm to obtain short vectors. In our experiments, the attack worked for all the values of Wiener's table, and the total running time was negligible (less than a second).

**Practical Impact.** We have shown that GPG's implementation of the ElGamal signature is totally insecure: an attacker can recover the signer's private key from the public key and a single message/signature pair in less than a second. Thus, GPG–ElGamal signing keys should be be considered compromised, as announced by the GPG development team [19]. There are two types of ElGamal signing keys in GPG:

- The primary ElGamal sign+encrypt keys. When running the command `gpg --list-keys`, such keys can be spotted by a prefix of the form `pub 2048G/` where 2048 can be replaced by any possible keylength. The prefix `pub` specifies a primary key, while the capital letter `G` indicates an ElGamal sign+encrypt key.
- The ElGamal sign+encrypt subkeys. When running the command `gpg --list-keys`, such keys can be spotted by a prefix of the form `sub 2048G/` where 2048 can be replaced by any possible keylength. The prefix `sub` indicates a subkey.

The primary keys are definitely compromised because such keys automatically come up with a signature to bind the user identity to the public key, thus disclosing the private key immediately. The subkeys may not be compromised if no signature has ever been generated. In both cases, it is worth noting that the signing key is also an encryption key, so the damage is not limited to authentication: a compromised ElGamal signing key would also disclose all communications encrypted with the corresponding public key.

The mistake of using a small $k$ and a small $x$ dates back to GPG v1.0.2 (which was released in January 2000), when the generation of $k$ and $x$ was changed to improve performances: the flaw has therefore been present in GPG for almost four years. A signing key created prior to GPG v1.0.2 may still be compromised if a signature using that key has been generated with GPG v1.0.2 or later.

Nobody knows how many ElGamal sign+encrypt keys there are. What one knows is the number of ElGamal sign+encrypt keys that have been registered on keyservers. According to keyserver statistics (see [19]), there are 848 registered primary ElGamal sign+encrypt keys (which is a mere 0.04% percent of all primary keys on keyservers) and 324 registered ElGamal sign+encrypt subkeys: of course, GPG advised all the owners of such keys to revoke their keys. These (fortunately) small numbers can be explained by the fact that ElGamal signing keys were never GPG's default option for signing, and their use was not really advocated.

As a consequence, ElGamal signatures and ElGamal sign+encrypt keys have recently been removed from GPG, and the GNU/Linux distributions which include GPG have been updated accordingly.

### 3.3 Encryption

Let $m$ be the message to be encrypted. The message $m$ is formatted in the way described in Section 2.1. The ciphertext is the pair $(a, b)$ where: $a = g^k \bmod p$ and $b = my^k \bmod p$. The integer $k$ is a "random" number coprime with $p - 1$. Theoretically, $k$ should be a cryptographically secure random number modulo $p-1$ such that $k$ is coprime to $p-1$. But the generation of $k$ is performed using the same procedure **gen_k** called by the ElGamal signature generation process. Thus, $k$ is first selected with $3q_{bit}/2$ pseudo-random bits. Next, as while as $k$ is not coprime with $p-1$, $k$ is incremented. Hence, $k$ is much smaller than $p-1$.

The security assumption for the hardness of decryption is no longer the standard Diffie-Hellman problem: instead, this is the Diffie-Hellman problem with short exponents (see [25]). Because the key generation makes sure that all the factors of $(p-1)/2$ have bit-length $\geq q_{bit}$, the best attack known to recover the plaintext requires at least $2^{q_{bit}/2}$ time, which is not a real threat.

However, the session key is formatted according to a specific padding, PKCS#1 v1.5 block type 02, which does not provide chosen-ciphertext security (see [4]). If we had access to a validity-checking oracle (which is weaker than a decryption oracle) that tells whether or not a given ciphertext is the ElGamal encryption of a message formatted with PKCS#1 v1.5 block type 02, we could apply Bleichenbacher's attack [4] to decrypt any ciphertext. Indeed, even though Bleichenbacher's attack was originally described with RSA, it also applies to ElGamal due to its homomorphic property: if $(a, b)$ and $(a', b')$ are ElGamal ciphertexts of respectively $m$ and $m'$, then $(aa' \bmod p, bb' \bmod p)$ is an ElGamal ciphertext of $mm' \bmod p$. One could argue that a validity-checking oracle is feasible in the situation where a user has configured his software to automatically decrypt any encrypted emails he receives: if an encrypted email turns out not to be valid, the user would inform the sender. However, Bleichenbacher's attack require a large number of oracle calls, which makes the attack debatable in an email context. Nevertheless, it would be better if OpenPGP recommended a provably secure variant of ElGamal encryption such as ACE-KEM selected by NESSIE [8].

# 4 The Implementation of RSA

## 4.1 Key Generation

To generate the parameters $p, q, n, e, d$, GPG implements the process described in Figure 6. Although the process does not lead to any realistic attack, it is

**Input:** Bit-length $k$ of the RSA modulus.

Repeat<br>
    Generate a pseudo-random prime $p$ of $k/2$ bits.<br>
    Generate a pseudo-random prime $q$ of $k/2$ bits.<br>
    If $p > q$, swap $p$ and $q$.<br>
    $n \longleftarrow pq$.<br>
Until the bit-length of $n$ is equal to $k$.<br>
If 41 is coprime with $\varphi(n)$, then $e \longleftarrow 41$<br>
Else if 257 is coprime with $\varphi(n)$, then $e \longleftarrow 257$<br>
Else<br>
    $e \longleftarrow 65537$<br>
    While $e$ is not coprime with $\varphi(n)$, $e \longleftarrow e+2$<br>
Let $d$ be the inverse of $e$ modulo $\varphi(n)$.

**Fig. 6.** The RSA key generation in GnuPG.

worth noting that the process leaks information on the private key. Indeed, the value of the RSA public exponent $e$ discloses additional information on $\varphi(n)$. For instance, if we see a GPG–RSA public key with $e \geq 65539$, we know that $\varphi(n)$ is divisible by the prime numbers 41, 257 and 65537: we learn a 30-bit factor of $\varphi(n)$, namely $41 \times 257 \times 65537$. However, the probability of getting $e \geq 65539$ after the process is very small. To our knowledge, efficient attacks to factor $n$ from partial knowledge of $\varphi(n)$ require a factor of $\varphi(n)$ larger than approximately $n^{1/4}$. Thus, this flaw does not lead to a serious attack, since the probability of getting a factor $\geq n^{1/4}$ after the process is way too small.

Nevertheless, any leakage on $\varphi(n)$ (apart from the fact that $e$ is coprime with $\varphi(n)$) is not recommended: if one really wants a small public exponent, one should rather select $e$ first, and then generate the primes $p$ and $q$ until both $p-1$ and $q-1$ are coprime with $e$.

### 4.2 Encryption

As already mentioned in Section 2, GPG implements RSA encryption as defined by PKCS#1 v1.5. This is not state-of-the-art cryptography: like with ElGamal, Bleichenbacher's chosen-ciphertext attack [4] can decrypt any ciphertext. But, as mentioned in 3.3, the relevance of such attacks to the email world is debatable, in part because of the high number of oracle calls. We hope that future versions of the OpenPGP standard, will recommend better RSA encryption standards (see for instance PKCS#1 v2.1 [20] or NESSIE [8]).

### 4.3 Signature

GPG implements RSA signatures as defined by PKCS#1 v1.5. Again, this is not state-of-the-art cryptography (no security proof is known for this padding), but we are unaware of any realistic attack with the GPG setting, as opposed to some other paddings (see [7]). The RSA verification does not seem to check the range of the signature with respect to the modulus, which gives (marginal) malleability (see [30]): given a signature $s$ of $m$, one can forge another signature $s'$ of $m$. As with encryption, we hope that future versions of the OpenPGP standard will recommend a better RSA signature standard (see for instance PKCS#1 v2.1 [20] or NESSIE [8]).

## References

1. D. M. Bellovin. Cryptography and the Internet. In *Proc. of Crypto '98*, volume 1462 of *LNCS*. IACR, Springer-Verlag, 1998.
2. D. Bleichenbacher. On the generation of one-time keys in DSS. Manuscript, February 2001. Result presented at the Monteverita workshop of March 2001.
3. D. Bleichenbacher. Generating ElGamal signatures without knowing the secret key. In *Proc. of Eurocrypt '96*, volume 1070 of *LNCS*, pages 10–18. IACR, Springer-Verlag, 1996.
4. D. Bleichenbacher. Chosen ciphertext attacks against protocols based on the RSA encryption standard PKCS #1. In *Proc. of Crypto '98*, volume 1462 of *LNCS*, pages 1–12. IACR, Springer-Verlag, 1998.
5. D. Boneh, A. Joux, and P. Q. Nguyen. Why textbook ElGamal and RSA encryption are insecure. In *Proc. of Asiacrypt '00*, volume 1976 of *LNCS*, pages 30–43. IACR, Springer-Verlag, 2000.
6. J. Callas, L. Donnerhacke, H. Finney, and R. Thayer. OpenPGP message format: Request for Comments 2440. Available as `http://www.ietf.org/rfc/rfc2440.txt`.
7. J.-S. Coron, D. Naccache, and J. P. Stern. On the security of RSA padding. In *Proc. of Crypto '99*, volume 1666 of *LNCS*, pages 1–18. IACR, Springer-Verlag, 1999.
8. European Union. European project IST-1999-12324: New European Schemes for Signatures, Integrity, and Encryption (NESSIE). `http://www.cryptonessie.org`.
9. I. Goldberg and D. Wagner. Randomness and the Netscape browser. *Dr Dobb's*, January 1996.

10. GPG. The GNU privacy guard. http://www.gnupg.org.
11. P. Gutmann. Software generation of practically strong random numbers. In *Proc. of the 7th Usenix Security Symposium*, 1998.
12. P. Gutmann. Lessons learned in implementing and deploying crypto software. In *Proc. of the 11th Usenix Security Symposium*, 2002.
13. N. A. Howgrave-Graham and N. P. Smart. Lattice attacks on digital signature schemes. *Design, Codes and Cryptography*, 23:283–290, 2001.
14. IEEE. P1363: Standard specifications for public-key cryptography. Available at http://grouper.ieee.org/groups/1363/.
15. IPA. Cryptrec: Evaluation of cryptographic techniques. Available at http://www.ipa.go.jp/security/enc/CRYPTREC/index-e.html.
16. K. Jallad, J. Katz, and B. Schneier. Implementation of chosen-ciphertext attacks against PGP and GnuPG. In *Proc. of ISC '02*, volume 2433 of *LNCS*. Springer-Verlag, 2002.
17. B. Kaliski. PKCS #1: RSA encryption version 1.5: Request for Comments 2313. Available as http://www.ietf.org/rfc/rfc2313.txt.
18. J. Katz and B. Schneier. A chosen ciphertext attack against several E-Mail encryption protocols. In *Proc. of the 9th Usenix Security Symposium*, 2000.
19. W. Koch. GnuPG's ElGamal signing keys compromised. Internet public announcement on November 27th, 2003.
20. RSA Labs. PKCS #1: RSA cryptography standard. Available at http://www.rsasecurity.com/rsalabs/pkcs/pkcs-1/index.html.
21. J. Manger. A chosen ciphertext attack on RSA Optimal Asymmetric Encryption Padding (OAEP) as standardized in PKCS #1 v2.0. In *Proc. of Crypto '01*, volume 2139 of *LNCS*, pages 230–231. IACR, Springer-Verlag, 2001.
22. P. Q. Nguyen and I. E. Shparlinski. The insecurity of the Digital Signature Algorithm with partially known nonces. *Journal of Cryptology*, 15(3), 2002.
23. P. Q. Nguyen and D. Stehlé. Low-dimensional lattice basis reduction revisited. In *Algorithmic Number Theory – Proc. of ANTS-VI*, LNCS. Springer-Verlag, 2004.
24. P. Q. Nguyen and J. Stern. The two faces of lattices in cryptology. In *Proc. Workshop on Cryptography and Lattices (CALC '01)*, volume 2146 of *LNCS*, pages 146–180. Springer-Verlag, 2001.
25. P. C. van Oorschot and M. J. Wiener. On Diffie-Hellman key agreement with short exponents. In *Proc. of Eurocrypt '96*, volume 1070 of *LNCS*, pages 332–343. IACR, Springer-Verlag, 1996.
26. OpenPGP. http://www.openpgp.org.
27. PGP. Pretty good privacy. http://www.pgp.com.
28. B. Schneier. Security in the real world: How to evaluate security technology. *Computer Security Journal*, XV(4), 1999.
29. V. Shoup. Number Theory C++ Library (NTL) version 5.3.1. Available at http://www.shoup.net/ntl/.
30. J. Stern, D. Pointcheval, J. Malone-Lee, and N. P. Smart. Flaws in applying proof methodologies to signature schemes. In *Proc. of Crypto '02*, volume 2442 of *LNCS*, pages 93–110. IACR, Springer-Verlag, 2002.

## A Lattices in a Nutshell

We recall basic facts about lattices. To learn more about lattices, see [24] for a list of references. Informally speaking, a lattice is a regular arrangement of points in $n$-dimensional space. In this paper, by the term lattice, we actually mean an integral lattice.

An integral lattice is a subgroup of $(\mathbb{Z}^n, +)$, that is, a non-empty subset $L$ of $\mathbb{Z}^n$ which is stable by subtraction: $\boldsymbol{x} - \boldsymbol{y} \in L$ whenever $(\boldsymbol{x}, \boldsymbol{y}) \in L^2$. The simplest lattice is $\mathbb{Z}^n$. It turns out that in any lattice $L$, not just $\mathbb{Z}^n$, there must exist linearly independent vectors $\boldsymbol{b}_1, \ldots, \boldsymbol{b}_d \in L$ such that:

$$L = \left\{ \sum_{i=1}^{d} n_i \boldsymbol{b}_i \mid n_i \in \mathbb{Z} \right\}.$$

Any such $d$-uple of vectors $\boldsymbol{b}_1, \ldots, \boldsymbol{b}_d$ is called a basis of $L$: a lattice can be represented by a basis, that is, a matrix. Reciprocally, if one considers $d$ integral vectors $\boldsymbol{b}_1, \ldots, \boldsymbol{b}_d \in \mathbb{Z}^n$, the previous set of all integral linear combinations of the $\boldsymbol{b}_i$'s is a subgroup of $\mathbb{Z}^n$, and therefore a lattice.

The *dimension* of a lattice $L$ is the dimension $d$ of the linear span of $L$: any basis of $L$ has exactly $d$ elements. It turns out that the $d$-dimensional volume of the parallelepiped spanned by an arbitrary basis of $L$ only depends on $L$, not on the basis itself: this volume is called the *determinant* (or *volume*) of $L$. When the lattice is full-rank, that is, when the lattice dimension $d$ equals the space dimension $n$, the determinant of $L$ is simply the absolute value of the determinant of any basis. Thus, the volume of $\mathbb{Z}^n$ is 1.

Since our lattices are subsets of $\mathbb{Z}^n$, they must have a shortest nonzero vector: In any lattice $L \subseteq \mathbb{Z}^n$, there is at least one nonzero vector $\boldsymbol{v} \in L$ such that no other nonzero lattice vector has a Euclidean norm strictly smaller than that of $\boldsymbol{v}$. Finding such a vector $\boldsymbol{v}$ from a basis of $L$ is called the shortest vector problem. When the lattice dimension is fixed, it is possible to solve the shortest vector problem in polynomial time (with respect to the size of the basis), using lattice reduction techniques. But the problem becomes much more difficult if the lattice dimension varies. In this article, we only deal with low-dimensional lattices, so the shortest vector problem is really not a problem.

The lattice determinant is often used to estimate the size of short lattice vectors. In a typical $d$-dimensional lattice $L$, if one knows a nonzero vector $\boldsymbol{v} \in L$ whose Euclidean norm is much smaller than $\det(L)^{1/d}$, then this vector is likely to be the shortest vector, in which case it can be found by solving the shortest vector problem, because any shortest vector would be expected to be equal to $\pm\boldsymbol{v}$. Although this can sometimes be proved, this is not a theorem: there are counter-examples, but it is often true with the lattices one is faced with in practice, which is what we mean by a typical lattice.

Another problem which causes no troubles when the lattice dimension is fixed is the closest vector problem: given a basis of $L \subseteq \mathbb{Z}^n$ and a point $\boldsymbol{t} \in \mathbb{Q}^n$, find a lattice vector $\boldsymbol{l} \in L$ minimizing the Euclidean norm of $\boldsymbol{l} - \boldsymbol{t}$. Again, in a typical $d$-dimensional lattice $L$, if one knows a vector $\boldsymbol{t}$ and a lattice vector $\boldsymbol{l}$ such that the norm of $\boldsymbol{t} - \boldsymbol{l}$ is much smaller than $\det(L)^{1/d}$, then $\boldsymbol{l}$ is likely to be the closest lattice vector of $\boldsymbol{t}$ in $L$, in which case $\boldsymbol{l}$ can be found by solving the closest vector problem. Indeed, if there was another lattice vector $\boldsymbol{l}'$ close to $\boldsymbol{t}$, then $\boldsymbol{l} - \boldsymbol{l}'$ would be a lattice vector of norm much smaller than $\det(L)^{1/d}$: it should be zero.

## B Proving the GPG–ElGamal Attack

We use the same notation as in Section 3.2. Let $(a, b)$ be an GPG–ElGamal signature of $m$. If we make the simplifying assumption that both $a$ and $b$ are uniformly distributed modulo $p-1$, then the attack of Figure 5 can be proved, using the following lemma (which is not meant to be optimal, but is sufficient for our purpose):

**Lemma 2.** *Let $\varepsilon > 0$. Let $p$ be a prime number such that all the prime factors of $(p-1)/2$ are $\geq 2^{q_{bit}}$. Let $a$ and $b$ be chosen uniformly at random over $\{0, \ldots, p-2\}$. Let $L$ be the lattice defined by (2). Then the probability (over the choice of $a$ and $b$) that there exists a non-zero $(u, v) \in L$ such that both $|u|$ and $|v|$ are $< 2^{3q_{bit}/2+\varepsilon}$ is less than:*

$$\frac{2^{7q_{bit}/2+5+3\varepsilon} \log_2 p}{(p-1)q_{bit}}.$$

*Proof.* This probability $P$ is less than the sum of all the probabilities $P_{u,v}$, where the sum is over all the $(u, v) \neq (0, 0)$ such that both $|u|$ and $|v|$ are $< 2^{3q_{bit}/2+\varepsilon}$, and $P_{u,v}$ denotes the probability (over the choice of $a$ and $b$) that $(u, v) \in L$. Let $(u, v) \in \mathbb{Z}^2$ be fixed and nonzero. If $v = 0$, there are at most $2\gcd(u, (p-1)/2)$ values of $b$ in the set $\{0, \ldots, p-2\}$ such that:

$$bu + av \equiv 0 \pmod{(p-1)/2} \tag{3}$$

It follows that:

$$P_{u,0} \leq \frac{2\gcd(u, (p-1)/2)}{p-1}.$$

If $v \neq 0$: for any $b$, there are at most $2\gcd(v, (p-1)/2)$ values of $a$ in the set $\{0, \ldots, p-2\}$ which satisfy (3), therefore:

$$P_{u,v} \leq \frac{2\gcd(v, (p-1)/2)}{p-1}.$$

Hence:

$$P \leq S + 2^{3q_{bit}/2+1+\varepsilon} S \leq 2^{3q_{bit}/2+2+\varepsilon} S,$$

where

$$S = \sum_{0<|u|<2^{3q_{bit}/2+\varepsilon}} \frac{2\gcd(u, (p-1)/2)}{p-1} = \sum_{0<|v|<2^{3q_{bit}/2+\varepsilon}} \frac{2\gcd(v, (p-1)/2)}{p-1}.$$

To bound $S$, we split the sum in two parts, depending on whether or not $\gcd(u, (p-1)/2) > 1$. If $\gcd(u, (p-1)/2) > 1$, then $\gcd(u, (p-1)/2) \leq |u| < 2^{3q_{bit}/2+\varepsilon}$ and $u$ must be divisible by a prime factor of $(p-1)/2$ which is necessarily $\geq 2^{q_{bit}}$: the number of such $u$'s is less than $2^{q_{bit}/2+1+\varepsilon}(\log_2 p)/q_{bit}$ because the number of prime factors of $(p-1)/2$ is less than $(\log_2 p)/q_{bit}$. We obtain:

$$S \leq 2^{3q_{bit}/2+1+\varepsilon} \times \frac{2}{p-1} + 2^{q_{bit}/2+1+\varepsilon}(\log_2 p)/q_{bit} \times \frac{2^{3q_{bit}/2+1+\varepsilon}}{p-1}$$
$$\leq \frac{2^{2q_{bit}+3+2\varepsilon}\log_2 p}{(p-1)q_{bit}}$$

This completes the proof since $P \leq 2^{3q_{bit}/2+2+\varepsilon}S$. □

Because $p$ is always much larger than $2^{4q_{bit}}$, the lemma shows that if $\varepsilon$ is not too big, then with overwhelming probability, there is no non-zero $(u, v) \in L$ such that both $|u|$ and $|v|$ are $< 2^{3q_{bit}/2+\varepsilon}$. If $\boldsymbol{l}$ was not the closest vector of $\boldsymbol{t}$ in $L$, there would be another lattice vector $\boldsymbol{l}' \in L$ closer to $\boldsymbol{t}$: the distance between $\boldsymbol{l}'$ and $\boldsymbol{t}$ would be less than $2^{(3q_{bit}-1)/2}$. But then, the lattice vector $(u, v) = \boldsymbol{l} - \boldsymbol{l}'$ would contradict the lemma, for some small $\varepsilon$. Hence, $\boldsymbol{l}$ is the closest vector of $\boldsymbol{t}$ in $L$ with overwhelming probability, which proves the attack. However, the initial assumption that both $a$ and $b$ are uniformly distributed modulo $p-1$ is an idealized model, compared to the actual way $a$ and $b$ are generated by GPG. In this sense, the lemma explains why the attack works, but it does not provide a complete proof.

# Traceable Signatures

Aggelos Kiayias[1], Yiannis Tsiounis[2], and Moti Yung[3]

[1] Computer Science and Engineering, University of Connecticut
Storrs, CT, USA
aggelos@cse.uconn.edu

[2] Etolian Capital Management, LP.
New York, NY, USA
yiannist@etolian.com*

[3] Computer Science, Columbia University
New York, NY, USA
moti@cs.columbia.edu

**Abstract.** This work presents a new privacy primitive called "Traceable Signatures", together with an efficient provably secure implementation. To this end, we develop the underlying mathematical and protocol tools, present the concepts and the underlying security model, and then realize the scheme and its security proof. Traceable signatures support an extended set of fairness mechanisms (mechanisms for anonymity management and revocation) when compared with the traditional group signature mechanism. The extended functionality of traceable signatures is needed for proper operation and adequate level of privacy in various settings and applications. For example, the new notion allows (distributed) tracing of all signatures of a single (misbehaving) party without opening signatures and revealing identities of any other user in the system. In contrast, if such tracing is implemented by a state of the art group signature system, such wide opening of all signatures of a single user is a (centralized) operation that requires the opening of *all* anonymous signatures and revealing the users associated with them, an act that violates the privacy of all users.

To allow efficient implementation of our scheme we develop a number of basic tools, zero-knowledge proofs, protocols, and primitives that we use extensively throughout. These novel mechanisms work directly over a group of unknown order, contributing to the efficiency and modularity of our design, and may be of independent interest. The interactive version of our signature scheme yields the notion of "traceable (anonymous) identification."

## 1 Introduction

A number of basic primitives have been suggested in cryptographic research to deal with the issue of privacy. The most flexible private authentication tool to date is "group-signatures," a primitive where each group member is equipped with a signing algorithm that incorporates a proof of group-membership. Group-signatures were introduced by Chaum and Van Heyst in [14] and were further studied and improved in many ways in [15, 13, 7, 12, 4, 2, 25]. Each signature value is anonymous, in the sense that it only

* Research supported in part by NIST, under grant SB1341-02-W-1113

C. Cachin and J. Camenisch (Eds.): EUROCRYPT 2004, LNCS 3027, pp. 571–589, 2004.

reveals that the issuer is a member of the group, without even linking signatures by the same signer.

Privacy comes at a price. Unconditional privacy seems to be an attractive notion from the user's viewpoint, nevertheless it can potentially be a very dangerous tool against public safety (and can even be abused against the user herself). Undoubtedly everybody understands that privacy is a right of law-abiding citizens, while at the same time a community must be capable of revoking such privacy when illegal behavior (performed under the "mask of privacy") is detected; this balancing act is thus called "fairness". Group-signatures were designed with one embedded fairness mechanism which, in fact, allows for the "opening" of an atomic signature value, revealing the identity of its signer.

We observe that while group signatures are a very general "private credentials" tool, their opening capability is not a sufficient mechanism to ensure safety and/or privacy in a number of settings. What we need is additional mechanisms for lifting of privacy conditions. It may sound paradoxical that offering more mechanisms for revoking privacy actually contributes to privacy; still, consider the following scenario: a certain member of the group is suspected of illegal activity (potentially, its identity was revealed by opening a signature value). It is then crucial to detect which signatures were issued by this particular member so that his/her transactions are traced. The only solution with the existing group signature schemes is to have the Group Manager (GM) open all signatures, thus violating the privacy of all (including law-abiding) group members. Furthermore, this operation is also scalability impairing, since the GM would have to open all signatures in the system and these signatures may be distributed in various locations. What would be desirable, instead, is to have a mechanism that allows the selective linking of the existing signatures of a misbehaving user without violating the privacy of law-abiding group members; this mechanism should be efficient (e.g. done in parallel by numerous agents when required). This capability, in fact, implements an "oblivious data mining" operation where only signature values of a selected misbehaving user are traced. Such traceability property should be offered in conjunction with the standard opening capability of group signatures.

Another type of traceability, "self-traceability," is helpful to the user and is important in our setting. It suggests that a user should also be capable of claiming that he is the originator of a certain signature value if he wishes (or when a certain application protocol requires this). In other words, a group-member should be capable of stepping out and *claiming* a certain group-signature value as his own, *without* compromising the privacy of the remaining past or future group-signatures that he/she issues. Adding self-traceability to the existing efficient solutions in group-signatures is also far from ideal: the user will be required to remember her private random coin-tosses for all the signatures she signed, which is an unreasonable user storage overhead in many settings.

**Our Notion:** Motivated by the above, in this work we introduce a new basic primitive which we call *Traceable Signatures.* It incorporates the following three different types of traceability: (i) user tracing: check whether a signature was issued by a given user; it can be applied to all signatures by agents running in parallel; (ii) signature opening: reveal the signer of a given signature (as in group signature); and (iii) signature claiming: the signer of a signature provably claims a given signature that it has signed (in a stateless fashion). When recovering all transactions by performing user tracing it may be useful to avoid collecting all signatures to a central location and in order to reduce the burden of the GM (which may be a distributed entity), we divide user tracing into two steps: the first is executed by the GM and reveals some secret information about the user; this is given to a set of designated agents (clerks) that scan all signatures in parallel and reveal those signed by the suspected user. Note that the secret information revealed should not allow the agents to impersonate the user or violate the anonymity of law-abiding users.

**Modeling:** We model our concepts of traceable signatures and their interactive version (as traceable identification) and define their correctness and security.

We introduce a novel general way of modeling privacy systems. The model includes the definition of correctness and of security properties of the system. In a security system, like encryption, it is obvious who is the attacker and who tries to defend the encryption device, so adversary modeling is relatively easy. In a privacy system, on the other hand, a protocol between many parties may involve mutually distrusting, malicious users attacking each other from many sides and in various coalitions: e.g., a server (perhaps collaborating with a subset of some users) trying to violate the user's privacy interacting with a user trying to impersonate a group member. Since in privacy systems we deal with mutually adversarial parties, we develop a model that copes with this situation. The adversaries are described in the spirit that adversaries against a signature scheme or an encryption scheme have been dealt with in the past (i.e., by describing attack capabilities and goals for an adversary), while the model is constructed with simulation-based security proofs in mind.

To this effect, we introduce a set of queries (basic capabilities) by which adversaries can manipulate the system (and the simulator during the security proof). Then we present an "array of security definitions," where each definition is modeled as an adversary with partial access to the queries, representing a capability that the attack captures. This allows us to deal with various notions of simultaneous adversarial behavior within one system, modeling them as an "array of attacks" and proving security against each of them. Specifically in our setting, we classify three general security requirements that cover all perceived adversarial activities: misidentification attacks, anonymity attacks and framing attacks. We note that previous intuitive security notions that have appeared in the group signature literature such as unforgeability, coalition-resistance and exculpability are subsumed by our classification. We also compare our model to other models.

**Constructions:** Our construction is motivated by the state of the art and in particular by the mathematical assumptions that allow a group of users to generate a multitude of keys modulo a composite number that are private, namely are (partially) unknown even to the group manager who owns a trapdoor (prime factorization of the composite); such an ingenious mathematical setting was presented in [2]. Due to the refined notions of fairness of our model and its extended functionality, we need to introduce a number of new tools as well as employ a number of new cryptographic constructs that enable the various mechanisms that our model and scheme employ. We also note that our scheme is consistent with the present state-of-the-art revocation method for group signatures presented in [9], thus member revocation can be added modularly to our construction.

We remark that the user tracing (combined with the GM publishing the user's "tracing trapdoor") can be used to implement a type of "CRL-based revocation" that nullifies all signatures by a private key. This type of revocation has been considered recently in [3] (also [21] has been brought to our attention).

In order to implement the scheme efficiently, we design a number of basic protocols and primitives that we use extensively throughout (as useful subroutines). A pleasing feature of these novel notions and protocols is that they work directly over a group of unknown order. We show useful properties of such groups of quadratic residues that are required for the security proofs. We then introduce the notion of "discrete-log relation sets" which is a generic way of designing zero-knowledge proof systems that allows an entity to prove efficiently the knowledge of a number of witnesses for any such relation set that involves various discrete-logarithms and satisfies a condition that we call "triangularity." Discrete-log relation sets are employed extensively in our protocols but, in fact, they are a useful as an abstraction that can be used elsewhere and are therefore of independent interest. We then define a notion called "discrete-log representations of arbitrary powers," as well as a mechanism we call "drawing random powers" which is a two party protocol wherein one party gets a secret discrete logarithm whose value she does not control, while at the same time the other party gets the public key version, i.e., the exponentiated value.

Based on the above primitives we present traceable signatures and prove their correctness and security. We remark that our traceable signature scheme adds only a constant overhead to the complexity measures of the state of the art group signature scheme of [2].

**Applications:** One generic application of traceable signatures is transforming an anonymous system to one with "fair privacy" (by combing traceable signature with the original system). Membership revocation of the CRL-type is also an immediate application.

Due to lack of space proofs and many details are omitted. We refer to [24] for an extended version.

**Notations:** The notation $S(a,b)$ (called a sphere of radius $b$ centered at $a$) where $a, b \in \mathbb{Z}$ denotes the set $\{a-b+1,\ldots,a+b-1\}$. A function in $w$ will be called negligible if it holds that it is smaller than any fraction of the form $\frac{1}{w^c}$ for any $c$ and sufficiently large $w$; we use the notation $\mathsf{negl}(w)$ for such functions. The concatenation of two strings $a, b$ will be denoted by $a||b$. If $a$ is a bitstring we denote by $(a)_{l,\ldots,j}$ the substring $(a)_l||\ldots||(a)_j$ where $(a)_i$ denotes the $i$-th bit of $a$. The cardinality of a set $A$, will be denoted by $\#A$. If $X$ and $Y$ are parameterized probability distributions with the same support, we will write $X \approx Y$ if the statistical distance between $X, Y$ is a negligible function in the parameter. Furthermore, if $f$ and $g$ are functions over a variable, we will write $f \approx g$ if their absolute distance is a negligible function in the same variable. Finally note that log denotes the logarithm base 2, PPT stands for "probabilistic polynomial-time," and $=_{\mathsf{df}}$ means "equal by definition."

## 2 Preliminaries

Throughout the paper we work (unless noted otherwise) in the group of quadratic residues modulo $n$, denoted by $QR(n)$, with $n = pq$ and $p = 2p' + 1$ and $q = 2q' + 1$.

All operations are to be interpreted as modulo $n$ (unless noted otherwise). We will employ various related security parameters (as introduced in the sequel); with respect to $QR(n)$ the relevant security parameter is the number of bits needed to represent the order of the group, denoted by $\nu =_{\mathsf{df}} \lfloor \log p'q' \rfloor + 1$. Next we define the Cryptographic Intractability Assumptions that will be relevant in proving the security properties of our constructions.

The first assumption is the so called Strong-RSA assumption. It is similar in nature to the assumption of the difficulty of finding $e$-th roots of arbitrary elements in $\mathbb{Z}_n^*$ with the difference that the exponent $e$ is not fixed (part of the instance).

**Definition 1. Strong-RSA.** *Given a composite $n$ and $z \in QR(n)$, it is infeasible to find $u \in \mathbb{Z}_n^*$ and $e > 1$ such that $u^e = z(\mathrm{mod} n)$, in time polynomial in $\nu$.*

The second assumption that we will employ is the Decisional Diffie-Hellman Assumption over the quadratic residues modulo $n$; in stating this assumption we also take into account the fact that the exponents may belong to pre-specified integer spheres $\mathcal{B} \subseteq \{1, \ldots, p'q'\}$.

**Definition 2. Decisional Diffie-Hellman** *(over $\mathcal{B}_1, \mathcal{B}_2, \mathcal{B}_3$) Given a generator $g$ of a cyclic group $QR(n)$ where $n$ is as above, a DDH distinguisher $\mathcal{A}$ is a polynomial in $\nu$ time* PPT *that distinguishes the family of triples of the form $\langle g^x, g^y, g^z \rangle$ from the family of triples of the form $\langle g^x, g^y, g^{xy} \rangle$, where $x \in_R \mathcal{B}_1$, $y \in_R \mathcal{B}_2$, and $z \in_R \mathcal{B}_3$.*

*The maximum distance of these two distributions of triples as quantified over all possible* PPT *distinguishers will be denoted by $\mathsf{Adv}^{DDH}_{\mathcal{B}_1,\mathcal{B}_2,\mathcal{B}_3}(\nu)$; if $\mathcal{B}_1 = \mathcal{B}_2 = \mathcal{B}_3 = \{1, \ldots, p'q'\}$ we will write simply $\mathsf{Adv}^{DDH}(\nu)$ instead. The* DDH *assumption suggests that this advantage is a negligible function in $\nu$.*

We remark that when the size of the spheres $\mathcal{B}_1, \mathcal{B}_2, \mathcal{B}_3$ are sufficiently close to the order of $QR(n)$ it will hold that $\mathsf{Adv}^{DDH}_{\mathcal{B}_1,\mathcal{B}_2,\mathcal{B}_3}(\nu) \approx \mathsf{Adv}^{DDH}(\nu)$. Nevertheless we discover that the spheres can be selected to be much smaller than that without any degradation in security (see the remark at the end of section 3).

Finally, we will employ the discrete-logarithm assumption over the quadratic residues modulo $n$ and a pre-specified sphere $\mathcal{B}$, when the factorization of $n$ is known:

**Definition 3. Discrete-Logarithm.** *Given two values $a, b$ that belong to the set of quadratic residues modulo $n$ with known factorization, so that $\exists x \in \mathcal{B} : a^x = b$, find in time polynomial in $\nu$ the integer $x$ so that $a^x = b$. Again $\mathcal{B}$ is an integer sphere into the set $\{1, \ldots, p'q'\}$.*

**Conventions.** our proofs of knowledge will only be proven to work properly in the honest-verifier setting. On the one hand, the honest-verifier setting is sufficient for producing signatures. On the other hand, even in the general interactive setting the honest-verifier scenario can be enforced by assuming the existence, e.g., of a beacon, or some other mechanism that can produce trusted randomness; alternatively the participants may execute a distributed coin flipping algorithm (which are by now standard tools for converting random coin honest verifier scenario to a general proof). Such protocols where the randomness that is used to select the challenge is trusted will be called "canonical."

## 3 Sphere Truncations of Quadratic Residues

Let $n$ be a composite so that $n = pq$ and $p = 2p' + 1$ and $q = 2q' + 1$ with $p, q, p', q'$ all prime. Let $a$ be a generator of the cyclic group of quadratic residues modulo $n$. Recall that the order of $QR(n)$ is $p'q'$. Let $S(2^\ell, 2^\mu) = \{2^\ell - 2^\mu + 1, \ldots, 2^\ell + 2^\mu - 1\}$ be a sphere for two parameters $\ell, \mu \in \mathbb{N}$. Observe that $\#S(2^\ell, 2^\mu) = 2^{\mu+1} - 1$.

In this section we will prove a basic result that will be helpful later in the analysis of our scheme. In particular we will show that, assuming factoring is hard and the fact the sphere $S(2^\ell, 2^\mu)$ is sufficiently large (but still not very large) the random variable $a^x$ with $x \in_R S(2^\ell, 2^\mu)$ is indistinguishable from the uniform distribution over $QR(n)$; note that the result becomes trivial if the size of the sphere is very close to the order of $QR(n)$; we will be interested in cases where the size of the sphere is exponentially smaller (but still sufficiently large). Intuitively, this means that a truncation of the $QR(n)$ as defined by the sphere $S(2^\ell, 2^\mu)$ is indistinguishable to any probabilistic polynomial-time observer.

Consider the function $f_{g,n}(x) = g^x(\text{mod} n)$ defined for all $x < n$. The inverse of this function $f_{g,n}^{-1}$ is defined for any element in $QR(n)$ so that $f_{a,n}^{-1}(y) = x$ where $x \leq p'q'$ and it holds that $a^x = y(\text{mod} n)$. Observe that $x$ can be written as a $\nu$-bitstring. Note that if $y$ is uniformly distributed over $\mathbb{Z}_n^*$ it holds that every bit $(x)_i$ of $x$ with $i = 1, \ldots, \nu$ follows a probability distribution $\mathcal{D}_i^\nu$ with support the set $\{0, 1\}$. Note that for the $\mathcal{O}(\log \nu)$ most significant bits $i$ it holds that the distribution $\mathcal{D}_i^\nu$ is biased towards 0, whereas for the remaining bits the distribution $\mathcal{D}_i^\nu$ is uniform; this bias is due to the distance between $2^\nu$ and $p'q'$. Below we define the simultaneous hardness of the bits of the discrete-logarithm function, (cf. [22]):

**Definition 4.** *The bits $[l, \ldots, j]$, $l > j$, of $f_{g,n}^{-1}$ are* simultaneously hard *if the following two distributions are* PPT*-indistinguishable:*

- *the $\mathcal{SD}_i^j$ distribution: $\langle (f_{g,n}^{-1}(y))_{i,\ldots,j}, y\rangle$ where $y \in_R QR(n)$.*
- *the $\mathcal{SR}_i^j$ distribution: $\langle r_l || \ldots || r_j, y\rangle$ where $y \in_R QR(n)$ and $r_i \leftarrow \mathcal{D}_i^\nu$ for $i = l, \ldots, j$.*

Håstad et al. [22] studied the simultaneous hardness of of the discrete-logarithm over composite groups and one of their results imply the following theorem:

**Theorem 1.** *The bits $[\nu, \ldots, j]$ of $f_{g,n}^{-1}$ are simultaneously hard under the assumption that factoring $n$ is hard, provided that $j = \lceil \frac{\nu}{2} \rceil - \mathcal{O}(\log \nu)$.*

Now let us return to the study of the subset of $QR(n)$ defined by the sphere $S(2^\ell, 2^\mu)$. Consider the uniform probability distribution $\mathcal{U}$ over $QR(n)$ and the probability distribution $\mathcal{D}_a^{S(2^\ell, 2^\mu)}$ with support $QR(n)$ that assigns the probability $1/(2^{\mu+1} - 1)$ to all elements $a^x$ with $x \in S(2^\ell, 2^\mu)$ and probability 0 to all remaining elements of the support. The main result of this section is the following theorem:

**Theorem 2.** *The probability distributions $\mathcal{D}_a^{S(2^\ell, 2^\mu)}$ and $\mathcal{U}$ with support $QR(n)$ are* PPT*-indistinguishable under the assumption that factoring $n$ is hard, provided that $\#S(2^\ell, 2^\mu) = 2^{\lceil \frac{\nu}{2} \rceil - \mathcal{O}(\log \nu)}$.*

**Remark.** The results of this section suggest that we may truncate the range of a random variable $a^x$, $x \in_R \{1, \ldots, p'q'\}$, into a subset of $QR(n)$ that is of size approximately $\sqrt{p'q'}$; this truncation will not affect the behavior of any polynomial-time bounded observer. In particular, for the case of the Decisional Diffie Hellman assumption in $QR(n)$ over the spheres $\mathcal{B}_1, \mathcal{B}_2, \mathcal{B}_3$, we may use spheres of size approximately $\sqrt{p'q'}$; under the assumption that factoring is hard, we will still maintain that $\mathsf{Adv}^{DDH}_{\mathcal{B}_1,\mathcal{B}_2,\mathcal{B}_3}(\nu) \approx \mathsf{Adv}^{DDH}(\nu)$. In some few cases we may need to employ the DDH over spheres that are smaller in size than $\sqrt{p'q'}$ (in particular we will employ the sphere $\mathcal{B}_2$ to be of size approximately $\sqrt[4]{p'q'}$). While the DDH over such sphere selection does not appear to be easier it could be possible that this version of DDH is a stronger intractability assumption. Nevertheless we remark that if we assume that factoring remains hard even if $\lceil \nu/4 \rceil$ of bits of the prime factors of $n$ are known[4] then as stated in [22] approximately 3/4 of the bits of $f^{-1}_{g,n}$ are simultaneously hard and thus, using the methodology developed in this section, we can still argue that $\mathsf{Adv}^{DDH}_{\mathcal{B}_1,\mathcal{B}_2,\mathcal{B}_3}(\nu) \approx \mathsf{Adv}^{DDH}(\nu)$, even if $\mathcal{B}_2$ is of size approximately $\sqrt[4]{p'q'}$.

## 4 Discrete-Log Relation Sets

Discrete-log relation sets are quite useful in planning complex proofs of knowledge for protocols operating over groups of unknown order in general. We note that special instances of such proofs have been investigated individually in the literature, see e.g. [12, 11](also, various discrete-log based protocols over known and unknown order subgroups have been utilized extensively in the literature, [16, 19, 17]). Our approach, that builds on this previous work, homogenizes previous instantiations in the context of signatures into a more generic framework. Below, let $G$ be the unknown order group of quadratic residues modulo $n$, denoted also by $QR(n)$.

**Definition 5.** *A discrete-log relation set $R$ with $z$ relations over $r$ variables and $m$ objects is a set of relations defined over the objects $A_1, \ldots, A_m \in G$ and the free variables $\alpha_1, \ldots, \alpha_r$ with the following specifications: (1) The $i$-th relation in the set $R$ is specified by a tuple $\langle a^i_1, \ldots, a^i_m \rangle$ so that each $a^i_j$ is selected to be one of the free variables $\{\alpha_1, \ldots, \alpha_r\}$ or an element of $\mathbb{Z}$. The relation is to be interpreted as $\prod_{j=1}^m A_j^{a^i_j} = 1$. (2) Every free variable $\alpha_j$ is assumed to take values in a finite integer range $S(2^{\ell_j}, 2^{\mu_j})$ where $\ell_j, \mu_j \geq 0$.*

*We will write $R(\alpha_1, \ldots, \alpha_r)$ to denote the conjunction of all relations $\prod_{j=1}^m A_j^{a^i_j} = 1$ that are included in $R$.*

Below we will design a 3-move honest verifier zero-knowledge proof (see e.g. [16]) that allows to a prover that knows witnesses $x_1, \ldots, x_r$ such that $R(x_1, \ldots, x_r) = 1$ to prove knowledge of these values. We will concentrate on a discrete-log relation sets that have a specific structure that is sufficient for our setting: a discrete-log relation set $R$ is said to be *triangular*, if for each relation $i$ involving the free variables $\alpha_w, \alpha_{w_1}, \ldots, \alpha_{w_b}$ it holds that the free-variables $\alpha_{w_1}, \ldots, \alpha_{w_b}$ are contained in relations $1, \ldots, i-1$.

[4] Efficient factorization techniques are known when at least $\lceil \nu/3 \rceil$ bits of the prime factors of $n$ are known, [22].

**Proof of knowledge for a Discrete-Log Relation Set** $R$

objects $A_1, \ldots, A_m$, $r$ free-variables $\alpha_1, \ldots, \alpha_r$, parameters: $\epsilon > 1, k \in \mathbb{N}$,

Each variable $\alpha_j$ takes values in the range $S(2^{\ell_j}, 2^{\mu_j})$

$\mathcal{P}$ proves knowledge of the witnesses $x_j \in S(2^{\ell_j}, 2^{\epsilon(\mu_j+k)+2})$ s.t. $R(x_1, \ldots, x_r) = 1$

$\mathcal{P}$ $\qquad\qquad\qquad\qquad\qquad\qquad\qquad\qquad$ $\mathcal{V}$

for $w \in \{1, \ldots, r\}$ select $t_w \in_R \pm\{0,1\}^{\epsilon(\mu_w+k)}$

for $i \in \{1, \ldots, z\}$ set $B_i = \prod_{j:\exists w, a^i_j = \alpha_w} A_j^{t_w}$ $\quad \xrightarrow{B_1,\ldots,B_z}$ $\quad c \in_R \{0,1\}^k$

$\xleftarrow{c}$

for $w \in \{1, \ldots, r\}$ set $s_w = t_w - c \cdot (x_w - 2^{\ell_w})$ $\quad \xrightarrow{s_1,\ldots,s_r}$ $\quad$ Verify:

for $w \in \{1, \ldots, r\}$

$s_w \in_? \pm\{0,1\}^{\epsilon(\mu_w+k)+1}$

for $i \in \{1, \ldots, z\}$

$$\prod_{j:\exists w, a^i_j = \alpha_w} A_j^{s_w} \stackrel{?}{=} B_i \left(\prod_{j:a^i_j \in \mathbb{Z}} A_j^{a^i_j} \prod_{j:\exists w, a^i_j = \alpha_w} A_j^{2^{\ell_w}}\right)^c$$

**Fig. 1.** *Proof of Knowledge for a Discrete-Log relation set R.*

**Theorem 3.** *For any triangular discrete-log relation set R the 3-move protocol of figure 1 is a honest verifier zero-knowledge proof that can be used by a party (prover) knowing a witness for R to prove knowledge of the witness to a second party (verifier).*

*We remark that the proof assumes that the prover is incapable of solving the Strong-RSA problem; under this assumption the cheating probability of the prover is* $1/2^k$. *Regarding the length of the proof we note that the proof requires the first communication flow from the prover to the verifier to be of size* $z$ $QR(n)$ *elements (where* $z$ *is the number of relations in R) and the second communication flow from the prover to the verifier to be of total bit-length* $\sum_{w=1}^{r}(\epsilon(\mu_w + k) + 1)$.

Below, for a sphere $S(2^\ell, 2^\mu)$, the notation $S^k_\epsilon(2^\ell, 2^\mu) =_{\mathrm{df}} S(2^\ell, 2^{\frac{\mu-2}{\epsilon}-k})$ will be called the innersphere of $S(2^\ell, 2^\mu)$ for parameters $\epsilon, k$.

## 5 Discrete Log Representations of Arbitrary Powers

In this section we introduce and present some basic facts about "discrete log representations of arbitrary powers" inside the set of Quadratic Residues $QR(n)$ where $n$. We will define three spheres $\Lambda, \Gamma, \mathrm{M}$ inside the set $\{0, \ldots, 2^\nu - 1\}$ so that the following conditions are satisfied:
[S1.] $(\min \Gamma)^2 > \max \Gamma$. [S2.] M has size approximately equal to $2^{\lceil \nu/2 \rceil}$. [S3.] $\min \Gamma > \max \mathrm{M} \max \Lambda + \max \Lambda + \max \mathrm{M}$. This set of conditions is attainable as shown by the following possible selection: for simplicity, we assume that $\nu$ is divisible by 4: $\Lambda = S(2^{\frac{\nu}{4}-1}, 2^{\frac{\nu}{4}-1})$, note that $\#\Lambda = 2^{\frac{\nu}{4}} - 1$ and $\max \Lambda = 2^{\frac{\nu}{4}} - 1$. $\mathrm{M} = S(2^{\frac{\nu}{2}-1}, 2^{\frac{\nu}{2}-1})$, note that $\#\mathrm{M} = 2^{\frac{\nu}{2}} - 1$ and $\max \mathrm{M} = 2^{\frac{\nu}{2}} - 1$. $\Gamma = S(2^{\frac{3\nu}{4}} + 2^{\frac{\nu}{4}-1}, 2^{\frac{\nu}{4}-1})$, note that $\#\Gamma = 2^{\frac{\nu}{4}} - 1$, $\min \Gamma = 2^{\frac{3\nu}{4}} + 1 > \max \Lambda \max \mathrm{M} + \max \Lambda + \max \mathrm{M} = 2^{\frac{3\nu}{4}} - 1$.

In the exposition below we use some fixed values $a_0, a, b \in QR(n)$.

**Definition 6.** *A discrete-log representation of an arbitrary power is a tuple* $\langle A, e : x, x' \rangle$ *so that it holds* $A^e = a_0 a^x b^{x'}$ *with* $x, x' \in \Lambda$ *and* $e \in \Gamma$.

In this work we will be interested in the following computational problem:

◇ *The One-more Representation Problem.* Given $n, a_0, a, b$ and $K$ discrete-log representations of arbitrary powers find "one-more" discrete-log representation of an arbitrary power inside $QR(n)$.

The theorem below establishes that solving the One-more representation problem cannot be substantially easier than solving the Strong-RSA problem. We remark that a variant of this problem and of the theorem below has been proposed and proved in a recent work of Camenisch and Lysyanskaya [10] (without the sphere constraints). Note that the sphere constraints that we employ will allow shorter membership certificates later on, thus contributing in the efficiency of the general design.

**Theorem 4.** *Fix* $a_0, a, b \in QR(n)$ *and spheres* $\Lambda, \mathrm{M}, \Gamma$ *satisfying the above properties. Let* $\mathcal{M}$ *be a* PPT *algorithm that given* $K$ *discrete-log representations of arbitrary powers inside* $QR(n)$ *it outputs a different discrete-log representation of an arbitrary power inside* $QR(n)$ *with non-negligible probability* $\alpha$. *Then, the Strong-RSA problem can be solved with non-negligible probability at least* $\alpha/2K$.

## 6 Non-adaptive Drawings of Random Powers

Consider the following game between two players A and B: player A wishes to select a random power $a^x$ so that $x \in_R S(2^\ell, 2^\mu)$ where $a \in QR(n)$. Player B wants to ensure that the value $x$ is selected "non-adaptively" from its respective domain. The output specifications of the game is that player A returns $x$ and that player B returns $a^x$. Player B is assumed to know the factorization of $n$. In this section we will carefully model and implement a protocol for achieving this two-player functionality. The reader is referred to [20] for a general discussion of modeling secure two-party computations.

In the ideal world the above game is played by two Interactive TM's (ITM's) $A_0, B_0$ and the help of a trusted third party ITM $T$ following the specifications below. We note that we use a special symbol $\perp$ to denote failure (or unwillingness to participate); if an ITM terminates with any other output other than $\perp$ we say that it accepts; in the other case we say it rejects. From all the possible ways to implement $A_0, B_0$ one is considered to be the honest one; this will be marked as $A_0^H, B_0^H$ and is also specified below.

0. The modulus $n$ is available to all parties and its factorization is known to $B_0$. The sphere $S(2^\ell, 2^\mu)$ is also public and fixed.
1. $A_0$ sends a message in $\{\text{go}, \perp\}$ to $T$. $A_0^H$ transmits go.
2. $B_0$ sends a message in $\{\text{go}, \perp\}$ to $T$. $B_0^H$ transmits go.
3. If $T$ receives go from both parties, it selects $x \in_R S(2^\ell, 2^\mu)$ and returns $x$ to $A_0$; otherwise $T$ transmits $\perp$ to both parties.
4. $A_0$ selects a value $C \in \mathbb{Z}_n^*$ and transmits either $C$ or $\perp$ to $T$. $A_0^H$ transmits $C = a^x \bmod n$.

5. $T$ verifies that $a^x \equiv C(\text{mod} n)$ and if this is the case it transmits $C$ to both players. Otherwise, (or in the case $A_0$ transmitted $\perp$ in step 4), $T$ transmits $\perp$ to both players. $B_0^H$ terminates by returning $C$ or $\perp$ in the case of receiving $\perp$ from $T$. Similarly $A_0^H$ terminates by returning $x$, or $\perp$ in the case of receiving $\perp$ from $T$.

Let $\mathsf{Im}_T =_{\mathsf{df}} \langle A_0, B_0 \rangle$ be two ITM's that implement the above protocol with the help of the ITM $T$. We define by $\mathsf{OUT}_{A_0}^{\mathsf{Im}_T}(\mathsf{init}_A(\nu))$ and $\mathsf{OUT}_{B_0}^{\mathsf{Im}_T}(\mathsf{init}_B(\nu))$ be the output probability distributions of the two players. Note that $\mathsf{init}_A(\nu)$ contains the initialization string of player A which contains the modulus $n$, and the description of the sphere $S(2^\ell, 2^\mu)$; similarly $\mathsf{init}_B(\nu)$ is defined as $\mathsf{init}_A(\nu)$ with the addition of the factorization of $n$. Below we will use the notation $\mathsf{IDEAL}^{\mathsf{Im}_T}(in_A, in_B)$ to denote the pair $\langle \mathsf{OUT}_{A_0}^{\mathsf{Im}_T}(in_A), \mathsf{OUT}_{B_0}^{\mathsf{Im}_T}(in_B) \rangle$. Finally, we denote by $\mathsf{Im}_T^H$ the pair $\langle A_0^H, B_0^H \rangle$.

The goal of a protocol for non-adaptive drawing of random powers is the simulation of the trusted third party by the two players. Let $\mathsf{Im} = \langle A_1, B_1 \rangle$ be a two-player system of interactive TM's that implement the above game without interacting with the trusted third party $T$. As above we will denote by $\mathsf{OUT}_{A_1}^{\mathsf{Im}}(in_A)$ the output probability distribution $A_1$, and likewise for $\mathsf{OUT}_{B_1}^{\mathsf{Im}}(in_B)$. Also we denote by $\mathsf{REAL}^{\mathsf{Im}}(in_A, in_B)$ the concatenation of these two distributions.

**Definition 7.** *(Correctness) An implementation* $\mathsf{Im} = \langle A_1, B_1 \rangle$ *for non-adaptive drawings of random powers is* correct *if the following is true:*

$$\mathsf{REAL}^{\mathsf{Im}}(in_A, in_B) \approx \mathsf{IDEAL}^{\mathsf{Im}_T^H}(in_A, in_B)$$

*where* $in_A \leftarrow \mathsf{init}_A(\nu)$ *and* $in_B \leftarrow \mathsf{init}_B(\nu)$. *Intuitively the above definition means that the implementation* $\mathsf{Im}$ *should achieve essentially the same output functionality for the two players as the ideal honest implementation.*

Defining security is naturally a bit trickier as the two players may misbehave arbitrarily when executing the prescribed protocol implementation $\mathsf{Im} = \langle A_1, B_1 \rangle$.

**Definition 8.** *(Security) An implementation* $\mathsf{Im} = \langle A_1, B_1 \rangle$ *for non-adaptive drawings of random powers is* secure *if the following is true:*

$$\forall A_1^* \ \exists A_0^* \ \mathsf{REAL}^{\langle A_1^*, B_1 \rangle}(in_A, in_B) \approx \mathsf{IDEAL}^{\langle A_0^*, B_0^H \rangle}(in_A, in_B)$$

$$\forall B_1^* \ \exists B_0^* \ \mathsf{REAL}^{\langle A_1, B_1^* \rangle}(in_A, in_B) \approx \mathsf{IDEAL}^{\langle A_0^H, B_0^* \rangle}(in_A, in_B)$$

*where* $in_A \leftarrow \mathsf{init}_A(\nu)$ *and* $in_B \leftarrow \mathsf{init}_B(\nu)$. *Intuitively the above definition means that no matter what adversarial strategy is followed by either player it holds that it can be transformed to the ideal world setting without affecting the output distribution.*

Having defined the goals, we now take on the task of designing an implementation $\mathsf{Im}$ without a trusted third party; below we denote by $\tilde{m} =_{\mathsf{df}} \#S(2^\ell, 2^\mu) = 2^{\mu+1} - 1$.

1. The two players read their inputs and initiate a protocol dialog.
2. Player A selects $\tilde{x} \in_R \mathbb{Z}_{\tilde{m}}, \tilde{r} \in_R \{0, \ldots, n^2 - 1\}$ and transmits to player B the value $C_1 = g^{\tilde{x}} h^{\tilde{r}}(\text{mod} n)$ and $C_2 = y^{\tilde{r}}(\text{mod} n)$.

3. Player A engages player B in a proof of knowledge for the discrete-log relation set $\langle -1, 0, \tilde{x}, \tilde{r}, 0\rangle$ and $\langle 0, -1, 0, 0, \tilde{r}\rangle$ over the objects $C_1, C_2, g, h, y$. Observe that the relation set is triangular.
4. Player B selects $\tilde{y} \in_R \mathbb{Z}_{\tilde{m}}$ and transmits $\tilde{y}$ to A.
5. Player A computes $x' = \tilde{x} + \tilde{y}(\mathrm{mod}\tilde{m})$ and transmits to player B the value $C_3 = a^{x'}$.
6. Player A engages player B in a proof of knowledge for the discrete-log relation set $\langle -1, 0, \alpha, \beta, \gamma, 0, 0\rangle, \langle 0, -1, 0, 0, 0, 0, 0, \gamma\rangle, \langle 0, 0, -1, 0, 0, 0, \alpha, 0\rangle$ over the objects $C_1 g^{\tilde{y}}, C_2, C_3, g, g^{\tilde{m}}, h, a, y$ (observe again, that the relation set is triangular).
7. Player A engages with player B to a tight interval proof for $C_3$ ensuring that $\log_a C_3 \in \mathbb{Z}_{\tilde{m}}$ (treating $\mathbb{Z}_{\tilde{m}}$ as an integer range); this is done as described in [6].
8. Player A outputs $x := x' + 2^{\ell} - 2^{\mu} + 1$ and Player B outputs $C := C_3 a^{2^{\ell} - 2^{\mu} + 1}$.

**Theorem 5.** *The above protocol implementation for non-adaptive drawing of random powers is correct and secure (as in definitions 7 and 8) under the Strong-RSA and DDH assumptions.*

## 7 Traceable Signatures and Identification

In this section we describe the traceable signature syntax and model, focusing first on the interactive version, called a traceable identification scheme. Traceable identification employs seven sub-protocols Setup, Join, Identify, Open, Reveal, Trace, Claim that are executed by the active participants of the system, which are identified by the Group Manager (GM), a set of users and other non-trusted third parties called tracers.

Setup (executed by the GM). For a given security parameter $\nu$, the GM produces a publicly-known string $\mathsf{pk}_{\mathcal{GM}}$ and some private string $\mathsf{sk}_{\mathcal{GM}}$ to be used for user key generation.

Join (a protocol between a new user and the GM). In the course of the protocol the GM employs the secret-key string $\mathsf{sk}_{\mathcal{GM}}$. The outcome of the protocol results in a membership certificate $\mathsf{cert}_i$ that becomes known to the new user. The entire Join protocol transcript is stored by the GM in a database that will be denoted by Jtrans. This is a private database and each Join transcript contains also all the coin tosses that were used by the GM during the execution.

Identify (traceable identification) It is a proof system between a prover and a verifier with the user playing the role of the prover and the verifier played by any non-trusted third party. The Identify protocol is a proof of knowledge of a membership certificate $\mathsf{cert}_i$. In our setting, we will restrict the protocol to operate in 3 rounds, with the verifier selecting honestly a random challenge of appropriate length in the second round.

Open (invoked by the Trustee) A PPT TM which, given an Identify protocol transcript, the secret-key $\mathsf{sk}_{\mathcal{GM}}$ and access to the database Jtrans it outputs the identity of the signer.

Reveal (invoked by the GM) A PPT TM which, given the Join transcript for a user $i$, it outputs the "tracing trapdoor" for the user $i$ denoted by $\mathsf{trace}_i$.

Trace (invoked by designated parties, called tracers). A PPT TM which, given an Identify protocol transcript $\pi$ and the tracing trapdoor of a certain user $\mathsf{trace}_i$, checks if $\pi$ was produced by user $i$.

Claim. It is a proof system between a prover and a verifier where the role of the prover is played by the user and the role of the verifier is played by any claim recipient. In our setting, the Claim protocol is a proof of knowledge that binds to a given Identify protocol transcript and employs the membership certificate $\mathsf{cert}_i$ of the user. As in the case of Identify protocol we restrict Claim to be a 3-round protocol so that in round 2 the verifier selects honestly a random challenge of appropriate length.

**Definition 9. (Correctness for traceable identification)** *A traceable identification scheme with security parameter $\nu$ is* **correct** *if the following four conditions are satisfied (with overwhelming probability in $\nu$). Let* $\mathsf{Identify}_{\mathcal{U}}(\mathsf{pk}_{\mathcal{GM}})$ *be the distribution of* Identify *protocol transcripts generated by user $\mathcal{U}$ and* $\mathsf{Claim}_{\mathcal{U}}(\pi)$ *the distribution of* Claim *protocol transcripts generated by user $\mathcal{U}$ for an* Identify *protocol transcript $\pi$.*

*(1)* **Identify-Correctness:** *The* Identify *protocol is a proof of knowledge of a membership certificate for the public-key* $\mathsf{pk}_{\mathcal{U}}$ *that satisfies completeness.*

*(2)* **Open-Correctness:** $\mathsf{Open}(\mathsf{sk}_{\mathcal{GM}}, \mathsf{Jtrans}, \mathsf{Identify}_{\mathcal{U}}) = \mathcal{U}$.

*(3)* **Trace-Correctness:** $\mathsf{Trace}(\mathsf{Reveal}(\mathcal{U}, \mathsf{Jtrans}), \mathsf{Identify}_{\mathcal{U}}) = \mathsf{true}$ *and for any* $\mathcal{U}' \neq \mathcal{U}$, $\mathsf{Trace}(\mathsf{Reveal}\ (\mathcal{U}, \mathsf{Jtrans}), \mathsf{Identify}_{\mathcal{U}'}) = \mathsf{false}$.

*(4)* **Claim-Correctness:** *The* Claim *protocol over the* Identify *transcript $\pi$, is a proof of knowledge of the membership certificate embedded into $\pi$ that satisfies completeness.*

Given an traceable identification scheme as described above, we will derive a traceable signature by employing the Fiat-Shamir transformation [18].

### 7.1 Security Model for Traceable Schemes

In this section we formalize the security model for traceable schemes. To claim security we will define the notion of an interface $\mathcal{I}$ for a traceable scheme which is a PTM that simulates the operation of the system. The purpose behind the definition of $\mathcal{I}$ is to capture all possible adversarial activities against a traceable scheme in an intuitive way. As in the previous section, we will focus first on traceable identification. We model the security of a traceable identification scheme as an interaction between the adversary $\mathcal{A}$ and an entity called the *interface*. The interface maintains a (private) state denoted by $\mathsf{state}_{\mathcal{I}}$ (or simply state) and communicates with the adversary over a handful of pre-specified *query actions* that allow the adversary to learn information about $\mathsf{state}_{\mathcal{I}}$; these queries are specified below. The initial state of the interface is set to $\mathsf{state}_{\mathcal{I}} = \langle \mathsf{sk}_{\mathcal{GM}}, \mathsf{pk}_{\mathcal{GM}} \rangle$. The interface also employs an "internal user counter" denoted by n which is initialized to 0. Moreover three sets are initialized $U^p, U^a, U^b, U^r$ to $\emptyset$. Note that $\mathsf{state}_{\mathcal{I}}$ is also assumed to contain $U^p, U^a, U^b, U^r$ and n. Finally the interface employs two other strings denoted and initialized as follows: $\mathsf{Jtrans} = \epsilon$ and $\mathsf{Itrans} = \epsilon$. The various query action specifications are listed below:

- $\langle \mathcal{Q}_{\mathsf{pub}} \rangle$. The interface returns the string $\langle \mathsf{n}, \mathsf{pk}_{\mathcal{GM}} \rangle$. This allows to an adversary to learn the public-information of the system, i.e., the number of users and the public-key information.
- $\langle \mathcal{Q}_{\mathsf{key}} \rangle$. The interface returns $\mathsf{sk}_{\mathcal{GM}}$; this query action allows to the adversary to corrupt the group-manager.
- $\langle \mathcal{Q}_{\mathsf{p-join}} \rangle$. The interface simulates the Join protocol in *private*, increases the user count n by 1, and sets $\mathsf{state} := \mathsf{state}_{\mathcal{I}} || \langle \mathsf{n}, \mathsf{transcript}_{\mathsf{n}}, \mathsf{cert}_{\mathsf{n}} \rangle$. It also adds n into $U^p$ and sets $\mathsf{Jtrans} := \mathsf{Jtrans} || \langle \mathsf{n}, \mathsf{transcript}_{\mathsf{n}} \rangle$.
  This query action allows to the adversary to introduce a new user to the system (that is not adversarially controlled).
- $\langle \mathcal{Q}_{\mathsf{a-join}} \rangle$. The interface initiates an active Join dialog with the adversary; the interface increases the user count n by 1, and assumes the role of the GM where the adversary assumes the role of the prospective user. If the dialog terminates successfully, the interface sets $\mathsf{state}_{\mathcal{I}} := \mathsf{state}_{\mathcal{I}} || \langle \mathsf{n}, \mathsf{transcript}_{\mathsf{n}}, \perp \rangle$. It finally adds n into the set $U^a$ and $\mathsf{Jtrans} := \mathsf{Jtrans} || \langle \mathsf{n}, \mathsf{transcript}_{\mathsf{n}} \rangle$.
  This query action allows to the adversary to introduce an adversarially controlled user to the system. The adversary has the chance to interact with the GM through the Join dialog.
- $\langle \mathcal{Q}_{\mathsf{b-join}} \rangle$. The interface initiates an active Join dialog with the adversary; the interface increases the user count n by 1 and assumes the role of the prospective user and the adversary assumes the role of the GM. If the dialog terminates successfully the interface sets $\mathsf{state}_{\mathcal{I}} := \mathsf{state}_{\mathcal{I}} || \langle \mathsf{n}, \perp, \mathsf{cert}_{\mathsf{n}} \rangle$. It also adds n into $U^b$.
  This query allows the adversary to introduce users to the system acting as a GM.
- $\langle \mathcal{Q}_{\mathsf{id}}, i \rangle$. The interface parses $\mathsf{state}_{\mathcal{I}}$ and to recover an entry of the form $\langle i, \cdot, \mathsf{cert}_i \rangle$; then it produces an Identify protocol transcript using the certificate $\mathsf{cert}_i$ and selecting the verifier challenge at random; if no such entry is discovered or if $i \in U^a$ the interface returns $\perp$. Finally, if $\pi$ is the protocol transcript the interface sets $\mathsf{Itrans} = \mathsf{Itrans} || \langle i, \pi \rangle$.
- $\langle \mathcal{Q}_{\mathsf{reveal}}, i \rangle$. The interface returns the output of $\mathsf{Reveal}(i, \mathsf{Jtrans})$ and places $i \in U^r$. Sometimes we will write $\mathcal{Q}_{\mathsf{reveal}}^{\neg A}$ to restrict the interface from revealing users in $A$. Note that this query returns $\perp$ in case user $i$ does not exist or $i \in U^b$.

Given the above definition of an interface we proceed to characterize the various security properties that a traceable scheme should satisfy. We will use the notation $\mathcal{I}[a, \mathcal{Q}_1, \ldots, \mathcal{Q}_r]$ to denote the operation of the interface with (initial) state $a$ that responds to the query actions $\mathcal{Q}_1, \ldots, \mathcal{Q}_r$ (a subset of the query actions defined above). In general we assume that the interface serves one query at a time: this applies to the queries $\mathcal{Q}_{\mathsf{a-join}}$ and $\mathcal{Q}_{\mathsf{b-join}}$ that require interaction with the adversary (i.e., the interface does not allow the adversary to cascade such queries). For a traceable identification scheme we will denote by iV the verifier algorithm for the canonical Identify 3-move protocol as well as by cV the verifier algorithm of the canonical Claim 3-move protocol.

Our definition of security, stated below, is based on the definitions of the three named security properties in the coming subsections.

**Definition 10.** *A traceable scheme is said to be* **secure** *provided that it satisfies security against misidentification, anonymity and framing attacks.*

Regarding traceable signatures, we note that we model security using canonical 3-move proofs of knowledge and passive impersonation-type of attacks; we remark that identification security in this type of model facilitates the employment of the Fiat-Shamir transform for proving signature security; thus, proving security for the interactive version will be sufficient for ensuring security of the traceable signature in the random oracle model following the proof techniques of [1].

**Misidentification Attacks.** In a misidentification attack against a traceable scheme, the adversary is allowed to control a number of users of the system (in an adaptive fashion). The adversary is also allowed to observe and control the operation of the system in the way that users are added and produce identification transcripts. In addition, the adversary is allowed to invoke $\mathcal{Q}_{\mathsf{reveal}}$, i.e., participate in the system as a tracer. The objective of the adversary can take either of the following forms: (i) produce an identification transcript that satisfies either one of the following properties: (ia): the adversarial identification transcript does not open to any of the users controlled by the adversary, or (ib): the adversarial identification transcript does not trace to any of the users controlled by the adversary. Alternatively, (ii) produce a claim for an Identify transcript of one of the users that he does not control (in the set $U^p$). We will formalize this attack using the experiment presented in figure 2.

$$
\mathsf{Exp}^{\mathcal{A}}_{\mathsf{mis}}(\nu) : \left|
\begin{array}{l}
\mathsf{state}_{\mathcal{I}} = \langle \mathsf{pk}_{\mathcal{GM}}, \mathsf{sk}_{\mathcal{GM}} \rangle \leftarrow \mathsf{Setup}(1^\nu);\\
\langle \mathsf{s}, d, \rho_1 \rangle \leftarrow \mathcal{A}^{\mathcal{I}[\mathsf{state}_{\mathcal{I}}, \mathcal{Q}_{\mathsf{pub}}, \mathcal{Q}_{\mathsf{p-join}}, \mathcal{Q}_{\mathsf{a-join}}, \mathcal{Q}_{\mathsf{id}}, \mathcal{Q}_{\mathsf{reveal}}]}(\mathsf{first}, 1^\nu);\\
c \stackrel{r}{\longleftarrow} \{0,1\}^k;\\
\rho_2 \leftarrow \mathcal{A}(\mathsf{second}, d, \rho_1, c);\\
\mathtt{if}\ \mathsf{iV}(\mathsf{pk}_{\mathcal{GM}}, \rho_1, c, \rho_2) = \mathtt{true\ and}\\
\qquad \mathtt{if}\ \mathsf{Open}(\mathsf{sk}_{\mathcal{GM}}, \mathsf{Jtrans}, \rho_1) \notin U^a\\
\qquad\quad \mathtt{or}\ \wedge_{i \in U^a} \mathsf{Trace}(\mathsf{Reveal}(i, \mathsf{Jtrans}), \rho_1, c, \rho_2) = \mathtt{false}\\
\qquad \mathtt{then\ output\ 1}\\
\mathtt{else\ if}\ \mathsf{s}\ \mathtt{is\ such\ that}\ \langle i, \mathsf{s} \rangle \in \mathsf{Itrans}\ \mathsf{and}\ i \in U^p \cup U^r\\
\qquad \mathtt{and}\ \mathsf{cV}(\mathsf{s}, \rho_1, c, \rho_2) = \mathsf{true}\ \mathtt{then\ output\ 1}\\
\mathtt{else\ output\ 0}
\end{array}
\right.
$$

**Fig. 2.** The misidentification experiment

We will say that a traceable identification scheme satisfies security against misidentification if for any PPT $\mathcal{A}$, it holds that $\mathbf{Prob}[\mathsf{Exp}^{\mathcal{A}}_{\mathsf{mis}}(\nu) = 1] = \mathsf{negl}(\nu)$.

**Anonymity Attacks** An anonymity attack is best understood in terms of the following experiment that is played with the adversary $\mathcal{A}$ who is assumed to operate in two phases called play and guess. In the play phase, the adversary interacts with the interface, introduces users in the system, and selects two target users he does not control; then receives an identification transcript that corresponds to one of the two at random; in the guess stage the adversary tries to guess which of the two produced the identification transcript (while accessing the system but without revealing the challenge transcripts). We remark that we allow the adversary to participate in the system also as a tracer (i.e., one of the agents that assist in the tracing functionality). The experiment is presented in figure 3. A traceability scheme is said to satisfy anonymity if for any attacker $\mathcal{A}$ it holds that $|\mathbf{Prob}[\mathsf{Exp}^{\mathcal{A}}_{\mathsf{anon}}(\nu) = 1] - \frac{1}{2}| = \mathsf{negl}(\nu)$.

$$\mathsf{Exp}^{\mathcal{A}}_{\mathsf{anon}}(\nu) : \left| \begin{array}{l} \mathsf{state}_{\mathcal{I}} = \langle \mathsf{pk}_{\mathcal{GM}}, \mathsf{sk}_{\mathcal{GM}} \rangle \leftarrow \mathsf{Setup}(1^\nu); \\ \langle d, i_0, i_1 \rangle \leftarrow \mathcal{A}^{\mathcal{I}[\mathsf{state}_{\mathcal{I}}, \mathcal{Q}_{\mathsf{pub}}, \mathcal{Q}_{\mathsf{p-join}}, \mathcal{Q}_{\mathsf{a-join}}, \mathcal{Q}_{\mathsf{id}}, \mathcal{Q}_{\mathsf{reveal}}]}(\mathsf{play}, 1^\nu); \\ \texttt{if } i_0 \texttt{ or } i_1 \texttt{ belong to } U^a \cup U^r \texttt{ output } \perp. \\ b \xleftarrow{r} \{0,1\}. \\ \texttt{parse } \mathsf{state}_{\mathcal{I}} \texttt{ and find the entry } \langle i_b, \mathsf{transcript}_{i_b}, \mathsf{cert}_{i_b} \rangle. \\ \texttt{execute the } \textbf{Identify} \texttt{ protocol for } \mathsf{cert}_{i_b} \texttt{ to obtain } \langle \rho_1, c, \rho_2 \rangle. \\ b_* \leftarrow \mathcal{A}^{\mathcal{I}[\mathsf{state}_{\mathcal{I}}, \mathcal{Q}_{\mathsf{pub}}, \mathcal{Q}_{\mathsf{p-join}}, \mathcal{Q}_{\mathsf{a-join}}, \mathcal{Q}_{\mathsf{id}}, \mathcal{Q}^{\neg(i_0,i_1)}_{\mathsf{reveal}}]}(\mathsf{guess}, 1^\nu, d, \langle \rho_1, c, \rho_2 \rangle); \\ \texttt{if } b = b_* \texttt{ then output } 1 \texttt{ else output } 0. \end{array} \right.$$

**Fig. 3.** The anonymity attack experiment

**Framing Attacks** A user may be framed by the system in two different ways: the GM may construct a signature that opens or trace to an innocent user, or it may claim a signature that was generated by the user. We capture these two framing notions with the experiment described in figure 4 (we remark that "exculpability" of group signatures [2] is integrated in this experiment).

$$\mathsf{Exp}^{\mathcal{A}}_{\mathsf{fra}}(\nu) : \left| \begin{array}{l} \mathsf{state}_{\mathcal{I}} = \langle \mathsf{pk}_{\mathcal{GM}}, \mathsf{sk}_{\mathcal{GM}} \rangle \leftarrow \mathsf{Setup}(1^\nu); \\ \langle \mathsf{s}, d, \rho_1 \rangle \leftarrow \mathcal{A}^{\mathcal{I}[\mathsf{state}_{\mathcal{I}}, \mathcal{Q}_{\mathsf{pub}}, \mathcal{Q}_{\mathsf{key}}, \mathcal{Q}_{\mathsf{b-join}}, \mathcal{Q}_{\mathsf{id}}]}(\mathsf{first}, 1^\nu); \\ c \xleftarrow{r} \{0,1\}^k; \\ \rho_2 \leftarrow \mathcal{A}(\mathsf{second}, d, \rho_1, c); \\ \texttt{if } \mathsf{iV}(\mathsf{pk}_{\mathcal{GM}}, \rho_1, c, \rho_2) = \texttt{true and} \\ \qquad \texttt{if } \mathsf{Open}(\mathsf{sk}_{\mathcal{GM}}, \mathsf{Jtrans}, \rho_1) \in U^b \\ \qquad\quad \texttt{or } \exists i \in U^b : \mathsf{Trace}(\mathsf{Reveal}(i, \mathsf{Jtrans}), \rho_1, c, \rho_2) = \texttt{true} \\ \qquad \texttt{then output 1} \\ \texttt{else if } \mathsf{s} \texttt{ is such that } \langle i, \mathsf{s} \rangle \in \mathsf{Itrans} \texttt{ and } i \in U^b \\ \qquad \texttt{and } \mathsf{cV}(\mathsf{s}, \rho_1, c, \rho_2) = \mathsf{true} \texttt{ then output 1} \\ \texttt{else output 0} \end{array} \right.$$

**Fig. 4.** The framing attack experiment

A traceable scheme satisfies security against framing provided that for any probabilistic polynomial-time $\mathcal{A}$ it holds that $\mathbf{Prob}[\mathsf{Exp}^{\mathcal{A}}_{\mathsf{fra}}(\nu) = 1] = \mathsf{negl}(\nu)$.

**Comments** (i) In modeling misidentification and anonymity attacks we do not allow the adversary to submit "open signature" queries to the interface. This models the fact

that opening a signature is an internal operation performed by the GM. On the contrary, this is not assumed for the tracing operation, since we model it as a distributed operation whose results are made available to distributed agents (and thus the $\mathcal{Q}_{\mathsf{reveal}}$ oracle query is available to the adversary). Allowing opening oracles to be part of the adversarial control is possible, but will require our encryptions and commitments to be of the chosen ciphertext secure type.
(ii) Misidentification and Framing in traceable schemes capture two perspectives of adversarial behavior: in the first case the adversary does not corrupt the GM (and thus does not have at its disposal the GM's keys) and attempts to subvert the system. In the second case, the adversary is essentially the system itself (controls the GM) and attempts to frame innocent users. We find that the distinction of these two perspectives is important in the terms of our modeling of traceable signatures and as we see they rely on different intractability assumptions.
(iii) It is worth noting here the comparison of our model to previous approaches to formal modeling of primitives related to traceable signatures, in particular identity escrow and group signatures. Camenisch and Lysyanskaya [8] formalize security in identity escrow schemes based on a real vs. ideal model formulation, whereas our approach is more along the lines of security against adversaries of signature schemes with adversarial system access capabilities and adversarial goals in mind. Bellare et al. [5] provide a formal model for a relaxed group signature scenario where a dealer is supposed to run the user key-generation mechanism (rather than the user itself interactively with the group manager via the Join protocol). Our approach, employing active interaction between the adversary and the interface that represents the system (and simulates it in a security proof), is more suitable for the traceable schemes setting, which, in turn, follows the setting and attacks considered in [2] (where the group manager enters users into the system and, at the same time, he lacks full knowledge of the joining users' keys).

## 8 Design of a Traceable Scheme

**Parameters**. The parameters of the scheme are $\epsilon \in \mathbb{R}$ with $\epsilon > 1$, $k \in \mathbb{N}$ as well as three spheres $\Lambda, \mathrm{M}, \Gamma$ satisfying the properties presented in 5; Below we will denote by $\Lambda_\epsilon^k$, and $\Gamma_\epsilon^k$ the inner spheres of $\Lambda, \mathrm{M}$ and $\Gamma$ w.r.t. the parameters $\epsilon, k$ .
**Setup** The GM generates two primes $p', q'$ with $p = 2p' + 1, q = 2q' + 1$ also primes. The modulus is set to $n = pq$. The spheres $\Lambda, \mathrm{M}, \Gamma$ are embedded into $\{0, \ldots, p'q'-1\}$. Also the GM selects $a, a_0, b, g, h \in_R QR(n)$ of order $p'q'$. The secret-key $\mathsf{sk}_{\mathcal{GM}}$ of the GM is set to $p, q$. The public-key of the system is subsequently set to $\mathsf{pk}_{\mathcal{GM}} := \langle n, a, a_0, b, y, g, h\rangle$.
**Join** (a protocol executed by a new user and the GM). The prospective user and the GM execute the protocol for non-adaptive drawing a random power $x' \in \Lambda_\epsilon^k$ over $b$ (see section 6) with the user playing the role of player A and the GM playing the role of player B; upon successful completion of the protocol the user obtains $x_i'$ and the GM obtains the value $C_i = b^{x_i'}$.

Subsequently the GM selects a random prime $e_i \in \Gamma_\epsilon^k$ and $x_i \in \Lambda_\epsilon^k$ and then computes $A_i = (C_i a^{x_i} a_0)^{e_i^{-1}} (\bmod n)$ and sends to the user the values $\langle A_i, e_i, x_i\rangle$.

The user forms the membership certificate as $\mathsf{cert}_i := \langle A_i, e_i, x_i, x_i' \rangle$. Observe that $\langle A_i, e_i : x_i, x_i' \rangle$ is a discrete-log representation of an arbitrary power in $QR(n)$ (see section 5); furthermore observe that the portion of the certificate $x_i$ is known to the GM and will be used as the user's tracing trapdoor.

**Identify**. To identify herself a user first computes the values,

$$T_1 = A_i y^r,\ T_2 = g^r,\ T_3 = g^{e_i} h^r,\ T_4 = g^{x_i k},\ T_5 = g^k,\ T_6 = g^{x_i' k'},\ T_7 = g^{k'}$$

where $r, k, k' \in_R$ M. Subsequently the user proceeds to execute the proof of knowledge of the following triangular discrete-log relation set defined over the objects $g, h, y, a_0, a, b, T_1^{-1}, T_2^{-1}, T_3, T_4, T_5, T_6, T_7$ and the free variables are $x, x' \in \Lambda_\epsilon^k, e \in \Gamma_\epsilon^k, r, h'$.

$$\left[\begin{array}{r@{\;:\;}ccccccccccccc}
\multicolumn{1}{c}{} & g & h & (T_2)^{-1} & T_5 & T_7 & y & (T_1)^{-1} & a & b & a_0 & T_3 & T_4 & T_6 \\
T_2 = g^r & r & 0 & 1 & 0 & 0 & 0 & 0 & 0 & 0 & 0 & 0 & 0 & 0 \\
T_3 = g^e h^r & e & r & 0 & 0 & 0 & 0 & 0 & 0 & 0 & 0 & -1 & 0 & 0 \\
T_2^e = g^{h'} & h' & 0 & e & 0 & 0 & 0 & 0 & 0 & 0 & 0 & 0 & 0 & 0 \\
T_5^x = T_4 & 0 & 0 & 0 & x & 0 & 0 & 0 & 0 & 0 & 0 & 0 & -1 & 0 \\
T_7^{x'} = T_6 & 0 & 0 & 0 & 0 & x' & 0 & 0 & 0 & 0 & 0 & 0 & 0 & -1 \\
a_0 a^x b^{x'} y^{h'} = T_1^e & 0 & 0 & 0 & 0 & 0 & h' & e & x & x' & 1 & 0 & 0 & 0
\end{array}\right]$$

Observe that the above proof of knowledge ensures that the values $T_1, \ldots, T_7$ are properly formed and "contain" a valid certificate. In particular the above proof not only enforces the certificate condition $A_i^{e_i} = a_0 a^{x_i} b^{x_i'}$ but also the fact that $e_i \in \Gamma$ and $x_i, x_i' \in \Lambda$.

**Open**. (invoked by the GM) Given a Identify transcript $\langle \rho_1, c, \rho_2 \rangle$ and all Join transcripts the GM does the following: it parses $\rho_1$ for the sequence $\langle T_1, \ldots, T_7 \rangle$ and computes the value $A = (T_2)^{-x} T_1$. Then it searches the membership certificates $\langle A_i, e_i \rangle$ (available from the Join transcripts) to discover the index $i$ such that $A = A_i$; the index $i$ identifies the signer of the message.

**Reveal**. (invoked by the GM) Given the Join transcript of the $i$-th user the GM parses the Join transcript to recover the tracing trapdoor $\mathsf{trace}_i := x_i$.

**Trace**. (invoked by any agent/clerk) Given the value $\mathsf{trace}_i$ and an Identify protocol transcript $\langle \rho_1, c, \rho_2 \rangle$ the agent parses the sequence $\langle T_1, T_2, T_3, T_4, T_5, T_6, T_7 \rangle$ from $\rho_1$; subsequently it checks whether $T_5^{x_i} = T_4$; if this is the case the agent concludes that user $i$ is the originator of the given Identify protocol transcript.

**Claim**. (invoked by the user) Given an Identify protocol transcript that was generated by user $i$ and contains the sequence $\langle T_1, T_2, T_3, T_4, T_5, T_6, T_7 \rangle$, the user $i$ can claim that he is the originator as follows: he initiates a proof of knowledge of the discrete-log of $T_6$ base $T_7$ (which is a discrete-log relation set, see section 4). As a side-note, we remark here that if the proof is directed to a specific entity the proof can be targeted to the receiver using a designated verifier proof, see [23]; such proofs can be easily coupled with our proofs of knowledge for discrete-log relation sets.

**Theorem 6.** *The traceable identification scheme above is correct according to definition 9 and secure according to definition 10. In particular it satisfies (i) security against misidentification attacks based on the Strong-RSA and the DDH assumptions; (ii) security against anonymity attacks based on the DDH assumption; (iii) security against* *framing attacks based on the discrete-logarithm problem over $QR(n)$ when the factorization of $n$ is known.*

## 9 Applications

One immediate application of traceable signatures is membership revocation of the CRL-type. Another motivation for traceable signatures is the development of a generic way to transform any system $S$ that provides anonymity into a system that provides "fair" or conditional anonymity taking advantage of the various traceability procedures we developed. An anonymity system is comprised of a population of units which, depending on the system's function, exchange messages using anonymous channels. An anonymity system with *fairness* allows the identification of the origin of messages, as well as the tracing of all messages of a suspect unit, if this is mandated by the authorities. A sketch of the idea of using traceable signatures to transform any such an anonymous system into a system with fair anonymity is as follows: each unit of the anonymous system becomes a member of a traceable signature system; any message that is sent by a unit must be signed using the traceable signature mechanism. Messages that are not accompanied by a valid traceable signature are rejected by the recipients. This simple transformation is powerful and generic enough to add "fair" anonymity to a large class of anonymous systems (for example mix-networks).

## References


[1] M. Abdalla, J. H. An, M. Bellare, and C. Namprempre. From identification to signatures via the Fiat-Shamir transform: Minimizing assumptions for security and forward-security. In L. Knudsen, editor, *Advances in Cryptology – EUROCRYPT 2002*, volume 2332 of *LNCS*, pages 418–433. Springer-Verlag, 2002.

[2] G. Ateniese, J. Camenisch, M. Joye, and G. Tsudik. A practical and provably secure coalition-resistant group signature scheme. In M. Bellare, editor, *Advances in Cryptology - CRYPTO 2000*, volume 1880 of *LNCS*, pages 255–270, 2000.

[3] G. Ateniese, G. Song, and G. Tsudik. Quasi-efficient revocation of group signatures. In M. Blaze, editor, *Financial Cryptography 2002*, volume 2357 of *LNCS*, pages 183–197, 2002.

[4] G. Ateniese and G. Tsudik. Some open issues and new directions in group signatures. In M. Franklin, editor, *Financial Cryptography 1999*, volume 1648 of *LNCS*, pages 196–211. Springer-Verlag, 1999.

[5] M. Bellare, D. Micciancio, and B. Warinschi. Foundations of group signatures: Formal definitions, simplified requirements, and a construction based on general assumptions. In E. Biham, editor, *Advances in Cryptology – EUROCRYPT 2003*, volume 2656 of *LNCS*, pages 614–629. Springer, 2003.

[6] F. Boudot. Efficient proofs that a committed number lies in an interval. In B. Preneel, editor, *Advances in Cryptology – EUROCRYPT 2000*, volume 1807 of *LNCS*, pages 431–444. Springer-Verlag, 2000.

[7] J. Camenisch. Efficient and generalized group signatures. In W. Fumy, editor, *Advances in Cryptology - EUROCRYPT 1997*, volume 1233 of *LNCS*, pages 465–479. Springer, 1997.

[8] J. Camenisch and A. Lysyanskaya. An identity escrow scheme with appointed verifiers. In J. Kilian, editor, *Advances in Cryptology - CRYPTO 2001*, volume 2139 of *LNCS*, pages 388–407. Springer, 2001.

[9] J. Camenisch and A. Lysyanskaya. Dynamic accumulators and application to efficient revocation of anonymous credentials. In M. Yung, editor, *Advances in Cryptology - CRYPTO 2002*, volume 2442 of *LNCS*, pages 61–76. Springer, 2002.

[10] J. Camenisch and A. Lysyanskaya. A signature scheme with efficient protocols. In S. Cimato, C. Galdi, and G. Persiano, editors, *International Conference on Security in Communication Networks – SCN 2002*, volume 2576 of *LNCS*, pages 268–289. Springer Verlag, 2002.

[11] J. Camenisch and M. Michels. A group signature scheme based on an RSA-variant. *BRICS Technical Report*, RS98-27, 1998.

[12] J. Camenisch and M. Michels. A group signature scheme with improved efficiency. In K. Ohta and D. Pei, editors, *Advances in Cryptology - ASIACRYPT 1998*, volume 1514 of *LNCS*, pages 160–174. Springer, 1998.

[13] J. Camenisch and M. Stadler. Efficient group signature schemes for large groups. In B. Kaliski, editor, *Advances in Cryptology — CRYPTO 1997*, LNCS, pages 410–424. Springer-Verlag, 1997.

[14] D. Chaum and E. van Heyst. Group signatures. In D. W. Davies, editor, *Advances in Cryptology – EUROCRYPT 1991*, volume 547 of *LNCS*, pages 257–265. Springer-Verlag, 1991.

[15] L. Chen and T. P. Pedersen. New group signature schemes (extended abstract). In A. De Santis, editor, *Advances in Cryptology—EUROCRYPT 1994*, volume 950 of *LNCS*, pages 171–181. Springer-Verlag, 1994.

[16] R. Cramer, I. Damgård, and B. Schoenmakers. Proofs of partial knowledge and simplified design of witness hiding protocols. In Y. G. Desmedt, editor, *Advances in Cryptology—CRYPTO 1994*, volume 839 of *LNCS*, pages 174–187. Springer-Verlag, 1994.

[17] I. Damgård and E. Fujisaki. A statistically-hiding integer commitment scheme based on groups with hidden order. In Y. Zheng, editor, *Advances in Cryptology – ASIACRYPT 2002*, volume 2501 of *LNCS*, pages 125–142. Springer-Verlag, 2002.

[18] A. Fiat and A. Shamir. How to prove yourself: Practical solutions to identification and signature problems. In A. Odlyzko, editor, *Advances in Cryptology — CRYPTO 1986*, volume 263 of *LNCS*, pages 186–194. Springer-Verlag, 1987.

[19] E. Fujisaki and T. Okamoto. Statistical zero knowledge protocols to prove modular polynomial relations. In B. S. Kaliski, editor, *Advances in Cryptology - CRYPTO 1997*, volume 1294 of *LNCS*, pages 16–30, 1997.

[20] O. Goldreich. Secure multi-party computation, manuscript available from the web. http://www.wisdom.weizmann.ac.il/~oded/ , 1998.

[21] J. Groth. Group signatures: revisiting definitions, assumptions and revocation, 2004. manuscript.

[22] J. Håstad, A. W. Schrift, and A. Shamir. The discrete logarithm modulo a composite hides O(n) bits. *JCSS: Journal of Computer and System Sciences*, 47(3):376–404, 1993.

[23] M. Jakobsson, K. Sako, and R. Impagliazzo. Designated verifier proofs and their applications. In U. M. Maurer, editor, *Advances in Cryptology - EUROCRYPT 1996*, volume 1070 of *LNCS*, pages 143–154, 1996.

[24] A. Kiayias, Y. Tsiounis, and M. Yung. Traceable signatures. E-Print Cryptology Archive, Report 2004/007, 2004. http://eprint.iacr.org/.

[25] A. Kiayias and M. Yung. Extracting group signatures from traitor tracing schemes. In E. Biham, editor, *Advances in Cryptology – EUROCRYPT 2003*, volume 2656 of *LNCS*, pages 630–648. Springer, 2003.

# Handcuffing Big Brother: an Abuse-Resilient Transaction Escrow Scheme
## (Extended Abstract)

Stanislaw Jarecki[1] and Vitaly Shmatikov[2*]

[1] University of California, Irvine, CA**
stasio@ics.uci.edu
[2] SRI International, Menlo Park, CA
shmat@csl.sri.com

**Abstract.** We propose a practical abuse-resilient transaction escrow scheme with applications to privacy-preserving audit and monitoring of electronic transactions. Our scheme ensures correctness of escrows as long as at least one of the participating parties is honest, and it ensures privacy and anonymity of transactions even if the escrow agent is corrupt or malicious. The escrowed information is secret and anonymous, but the escrow agent can efficiently find transactions involving some user in response to a subpoena or a search warrant. Moreover, for applications such as abuse-resilient monitoring of unusually high levels of certain transactions, the escrow agent can identify escrows with particular common characteristics and automatically (*i.e.*, without a subpoena) open them once their number has reached a pre-specified threshold.
Our solution for transaction escrow is based on the use of Verifiable Random Functions. We show that by *tagging* the entries in the escrow database using VRFs indexed by users' private keys, we can protect users' anonymity while enabling efficient and, optionally, automatic de-escrow of these entries. We give a practical instantiation of a transaction escrow scheme utilizing a simple and efficient VRF family secure under the DDH assumption in the Random Oracle Model.

## 1 Introduction

Massive collection of personal and business data is increasingly seen as a necessary measure to detect and thwart crime, fraud, and terrorism. For example, all U.S. banks must report transactions over $10,000. Regulations of the U.S. Securities and Exchange Commission effectively require financial firms to store all emails in case they are subpoenaed in some future investigation. Government authorities often demand that financial transactions, internal corporate communications, and so on be escrowed with law enforcement or regulatory agencies in

* Supported in part by ONR grants N00014-01-1-0837 and N00014-03-1-0961.
** Part of this work was done while visiting the Applied Cryptography Group at Stanford University.

C. Cachin and J. Camenisch (Eds.): EUROCRYPT 2004, LNCS 3027, pp. 590–608, 2004.

such a way that the escrow agency can open the data pertaining to some user within the time period for which a subpoena or search warrant has been issued, or mine the collected data without a warrant for evidence of suspicious activity.

**Existing techniques.** Information stored in the escrow agency's database must be protected both from abuse by the escrow agency's employees and from external attacks. Unfortunately, existing escrow schemes sacrifice either user privacy, or efficiency of the escrow operation. Moreover, existing techniques allow mining of the escrowed data for evidence of suspicious activity only by letting the escrow agency de-escrow any entry at will.

Key escrow techniques [Mic92, KL95] implicitly assume that escrowed data are tagged by the key owner's identity or address. This enables efficient de-escrow of a subset of records pertaining to some user (*e.g.*, in response to a subpoena), but fails to protect anonymity of records against malicious employees of the escrow agency who can learn the number and timing of transactions performed by a given person, find correlations between transactions of different people, and so on. On the other hand, if escrows are *not* tagged, then there is no efficient procedure for opening the relevant escrows in response to a subpoena. Each entry in the escrow database must be decrypted to determine whether it involves the subpoenaed user. This is prohibitively inefficient, especially if the decryption key of the escrow agency is shared, as it should be, among a group of trustees.

**Our contribution.** We propose a *verifiable transaction escrow* (VTE) scheme which offers strong privacy protection *and* enables efficient operation of the escrow agent. Our scheme furnishes transaction participants with a provably secure privacy guarantee which we call *category-preserving anonymity.* We say that two transactions belong to the same category if and only if they were performed by the same user and are of the same *type* (*e.g.*, both are money transfers). An escrow scheme is *category-preserving anonymous* if the only information about any two transactions that the (malicious) escrow agent can learn from the corresponding escrow entries is whether the transactions fall into the same category or not. The agent cannot learn *which* category either transaction belongs to.

Of course, a malicious participant may reveal the transaction to the escrow agent. However, regardless of the user's transactions with dishonest parties who leak information to the escrow agent, all of his transactions with honest parties remain private in the sense of category-preserving anonymity — even if they belong to the same category as compromised transactions. While it does not provide perfect anonymity, category-preserving anonymity seems to give out no useful information, especially if transaction volume is high. (If volume is low, there may be undesirable information leaks, *e.g.*, the escrow agent may observe that only one category is ever used, and deduce that only one user is active.)

We present a VTE scheme with two variants. The first variant has an inexpensive escrow protocol, but does not achieve full category-preserving anonymity. The privacy guarantees it does offer might be acceptable in practice, however. The second variant achieves category-preserving anonymity at the cost of adding an expensive cut-and-choose zero-knowledge proof to the escrow protocol.

Our VTE scheme supports both (1) efficient identification and opening of escrows in response to a subpoena, and (2) efficient *automatic* opening of escrows that fall into the same category once their number reaches some pre-specified threshold. The scheme is also *tamper-resistant* in the sense that a malicious escrow agent cannot add any valid-looking escrows to the database. Finally, our scheme ensures correctness of the escrow entry as long as at least one participant in the escrowed transaction is honest. Note that there is no way to ensure escrow of transactions between parties who cooperate in concealing the transaction.

Our scheme employs Verifiable Random Functions. We show that by *tagging* entries in the escrow database using VRFs indexed by users' private keys, we enable efficient and, if necessary, automatic *de-escrow* (*disclosure*) of these entries, while providing category-preserving anonymity for the users. We instantiate our scheme with a practical construction based on a simple and efficient (shareable) VRF family secure under the DDH assumption in the Random Oracle Model.

**Applications.** A VTE scheme can be used in any scenario where transaction data must be escrowed but should remain private and anonymous. For example, a financial regulatory agency may collect escrows of all money transfers to ensure availability of evidence for future investigations of money laundering. Unless a court warrant is obtained, the agency should not be able to extract any useful information from the escrows, not even participants' identities. At the same time, the automatic opening capability of our VTE scheme can also support a scenario where the agency needs to identify all transfers which are made from the same account and share the same type, *e.g.*, all involve a certain organization or country, or more than a certain amount. These transactions should be secret and anonymous until their number reaches a pre-specified threshold, in which case the authority gains the ability to extract all corresponding plaintexts.

**Related work.** The problem of efficient classification and opening of escrows is related to the problem of search on encrypted data [SWP00, BCOP03]. In the latter problem, however, there is no notion of a malicious user who submits incorrect ciphertexts or interferes with record retrieval. Moreover, their techniques require the user to generate search-specific trapdoors, while we are also interested in scenarios where the escrow agent is able to open all escrows in a given category not because he received some category-specific trapdoor but because the *number* of escrows within a category reached a pre-specific threshold.

**Paper organization.** In section 2, we define verifiable transaction escrow and describe its security properties. In section 3, we present the simpler variant of our VTE construction, which is practical but does not achieve full category-preserving anonymity. In section 4, we present another variant which does achieve category-preserving anonymity, but employs an expensive cut-and-choose zero-knowledge protocol. In section 5, we show how to extend either construction to support automatic de-escrow capability. For lack of space, we omit all proofs from these proceedings. The full version of the paper, including all proofs, will be made available on eprint [JS04].

# 2 Definition of a Verifiable Transaction Escrow Scheme

A *Verifiable Transaction Escrow* (VTE) system involves an escrow *Agent* and any number of users. We assume that each transaction occurs between a *User* and a *Counterparty*. The two roles are naturally symmetric (users may act as counterparties for each other), but in some applications the escrow agent may only be interested in monitoring users (*e.g.*, bank clients), but not the counterparties (banks).

We assume that each transaction is adequately described by some bitstring $m$, and that there is a public and easily computable function $Type$, where $Type(m)$ of transaction $m$ is application-specific, *e.g.*, "this transaction is a money transfer," or "this transaction is a money transfer between \$1,000 and \$10,000." The *category* of a transaction is the ⟨user identity,type⟩ pair.

## 2.1 Basic Properties of a Verifiable Transaction Escrow Scheme

A VTE scheme is a tuple $(AKG, UKG, U_1, A, U_2, C, U_3, J)$ of the following probabilistic polynomial-time (PPT) algorithms:

- $AKG$ and $UKG$ are *key generation* algorithms, which on input of a security parameter $\tau$ generate, respectively, Agent's key pair $(k_A, pk_A)$ and, for each User, key pair $(k_U, pk_U)$.

- $(U_1, A)$ are interactive algorithms which define an *escrow protocol*. Its aim is to add an escrow of a transaction to the Agent's database in exchange for a *receipt* which will be later verified by the transaction Counterparty. The protocol runs between User $(U_1)$ and Agent $(A)$, on public input of Agent's public key $pk_A$. User's private input is $(k_U, m)$, where $m$ is the transaction description. Agent's private input is $(k_A, D)$ where $D$ is the state of Agent's escrow database. User's output is a receipt *rcpt*, and Agent's output is an *escrow* item $e$, which defines a new state of Agent's database as $D' = D \cup \{e\}$.

- $(U_2, C)$ are interactive algorithms which define a *verification protocol*. Its aim is for the Counterparty to verify the receipt certifying that the transaction was properly escrowed with the Agent. The protocol runs between User $(U_2)$ and Counterparty $(C)$, on public input $(pk_U, m, pk_A)$. User's private input is $k_U$, *rcpt*. Counterparty outputs decision $d = \mathsf{accept}/\mathsf{reject}$.

- $(U_3, J)$ is a pair of interactive algorithms which defines a *subpoena protocol*. Its aim is to identify all transactions of a given type in which the user participated, and *only* those transactions. The protocol runs between User $(U_3)$ and a public *Judge* $(J)$, on public inputs $(pk_U, T, D)$, where $pk_U, T$ identify the ⟨user,type⟩ category to be subpoenaed, and $D$ is Agent's database. User's private input is $k_U$. Judge has no private inputs. Algorithm $J$ outputs $M$, which is either a symbol $\mathsf{contempt}$ if the User refuses to cooperate, or a (possibly empty) list $(m_1, m_2, ...)$ of transactions of type $T$ involving user $pk_U$.

**Completeness.** If parties follow the protocol, then every escrowed transaction can be de-escrowed in the subpoena. In other words, for all keys $(k_A, pk_A)$ and

$(k_U, pk_U)$ generated by $AKG$ and $UKG$, and for every $m, D, D'$, if $\langle U_1(k_U, m), A(k_A, D)\rangle(pk_A)$ outputs $(rcpt, e)$ then $\langle U_2(k_U, rcpt), C\rangle(pk_U, m, pk_A)$ outputs $d = \mathsf{accept}$ and $\langle U_3(k_U), J\rangle(pk_U, Type(m), D' \cup \{e\})$ outputs $M$ s.t. $m \in M$.

For notational convenience, we define predicate $Prop(e, m, pk_U)$ to be *true* if and only if $\langle U_3(k_U), J\rangle(pk_U, Type(m), D' \cup \{e\})$ outputs $M$ s.t. $m \in M$.

**Verifiability.** The escrow agent receives a correct escrow of the transaction as long as at least one party in the transaction is honest. In particular, a malicious User has only negligible probability[3] of getting an honest Counterparty to *accept* in an escrow protocol unless the User gives to the Agent a proper escrow. Formally, for every PPT algorithms $U_1^*, U_2^*$, for every $D, m$,

$$\begin{aligned}\Pr[\ Prop(e, m, pk_U) \ \mid \ (k_A, pk_A) \leftarrow AKG(1^\tau);\ (k_U, pk_U) \leftarrow UKG(1^\tau);\\ (rcpt^*, e) \leftarrow \langle U_1^*(k_U, m), A(k_A, D)\rangle(pk_A);\\ \mathsf{accept} \leftarrow \langle U_2^*(rcpt^*), C\rangle(pk_U, m, pk_A)\ ] \ \geq 1 - \mathsf{negl}(\tau)\end{aligned}$$

**Efficient and unavoidable subpoena.** The subpoena procedure is *unavoidable* in the sense that the user is either publicly identified as refusing to cooperate, or all entries in the escrow database which involve the user and the specified type are publicly revealed. Namely, for every PPT algorithm $U_3^*$, for every $D'$, $m$, $e$, for $T = Type(m)$,

$$\begin{aligned}\Pr[M = \mathsf{contempt} \vee m \in M \mid (k_A, pk_A) \leftarrow AKG(1^\tau);\ (k_U, pk_U) \leftarrow UKG(1^\tau);\\ M \leftarrow \langle U_3^*(k_U), J\rangle(pk_U, T, D' \cup \{e\});\ Prop(e, m, pk_U)] \geq 1 - \mathsf{negl}(\tau)\end{aligned}$$

Moreover, the subpoena protocol is *efficient* in the sense that its running time is linear in the number of escrows of the subpoenaed ⟨user,type⟩ category in the database $D$, rather than in the size of the whole escrow database $D$.

**Tamper resistance.** A malicious Agent can't add entries to the escrow database which would be identified as transactions involving some user during the public subpoena process, unless that user created these escrows himself. Namely, for every PPT algorithm $A^*$, for random keys $k_U, pk_U$ generated by $UKG$, if $A^*$ has access to user oracles $O_{U_1}(\cdot,\cdot)$, $O_{U_2}(\cdot,\cdot,\cdot)$, and $O_{U_3}(\cdot,\cdot)$, where $O_{U_1}(m, pk_A)$ follows the $U_1$ protocol on $(k_U, m)$ and $pk_A$, $O_{U_2}(m, rcpt, pk_A)$ follows the $U_2$ protocol on $(k_U, rcpt)$ and $pk_A$, and $O_{U_3}(T, D)$ follows the $U_3$ protocol on $k_U$ and $(pk_U, T, D)$, then there is only negligible probability that $A^{*O_{U_1,U_2,U_3}(\cdot,\cdot,\cdot)}(pk_U)$ produces $T^*, D^*$ s.t. $M \leftarrow \langle U_3(k_U), J\rangle(pk_U, T^*, D^*)$ where $M$ contains some message $m^*$ s.t. $A^*$ did *not* run oracle $O_{U_1}(\cdot,\cdot)$ on $m^*$ and some $pk_A$.

**Category-preserving anonymity.** By default, the only information learned by a malicious Agent about any two instances of the escrow protocol is whether the two transactions fall into the same *category*, *i.e.*, correspond to the same

[3] We say that a function $f(\tau)$ is *negligible* if for any polynomial $p(\cdot)$, there exists $\tau_0$ s.t. for every $\tau \geq \tau_0$, $f(\tau) < 1/p(\tau)$. We denote a negligible function by $\mathsf{negl}(\cdot)$.

⟨user,type⟩ pair or not. Moreover, neither the transactions opened in the subpoena protocol, nor transactions reported to the Agent by some malicious Counterparties, should help the malicious Agent to crack the privacy of transactions done with honest Counterparties and which were not subpoenaed.

Formally, consider the following game between any PPT algorithms $A^*, C^*$ and the VTE system. First, polynomially many user keys $\{(k_i, pk_i)\}$ are generated by the $UKG$ algorithm. Then, if $A^*$ has access to *flexible* user oracles $O_{U_1}(\cdot,\cdot,\cdot)$, $O_{U_2}(\cdot,\cdot,\cdot,\cdot)$, and $O_{U_3}(\cdot,\cdot,\cdot)$, where $O_{U_1}(i, m, pk_A)$ follows the $U_1$ protocol on $(k_i, m)$ and $pk_A$, $O_{U_2}(i, m, rcpt, pk_A)$ follows the $U_2$ protocol on $(k_i, rcpt)$ and $(pk_i, m, pk_A)$, and $O_{U_3}(i, T, D)$ follows the $U_3$ protocol on $k_i$ and $(pk_i, T, D)$, the following holds:

$$\begin{aligned} \Pr[\ b = b' \mid (i_0, i_1, m_0, m_1, st, pk_A) \leftarrow A^{*O_{U_1,U_2,U_3}(\cdot,\cdot,\cdot,\cdot)}(pk_1, ..., pk_{p(\tau)}); \\ b \leftarrow \{0,1\};\ (rcpt_b, st') \leftarrow \langle U_1(k_{i_b}, m_b), A^*(st)\rangle(pk_A); \\ \bar{b} = \neg b;\ (rcpt_{\bar{b}}, st'') \leftarrow \langle U_1(k_{i_{\bar{b}}}, m_{\bar{b}}), A^*(st')\rangle(pk_A); \\ (st''') \leftarrow \langle U_2(k_{i_0}, rcpt_0), C^*(st'')\rangle(pk_{i_0}, m_0, pk_A); \\ (st'''') \leftarrow \langle U_2(k_{i_1}, rcpt_1), C^*(st''')\rangle(pk_{i_1}, m_1, pk_A); \\ b' \leftarrow A^{*O_{U_1,U_2,U_3}(\cdot,\cdot,\cdot,\cdot)}(st'''');\ ] \ \leq\ \frac{1}{2} + \mathsf{negl}(\tau) \end{aligned}$$

where the *test transactions* $(i_0, m_0)$ and $(i_1, m_1)$ and the queries of $A^*$ to $O_{U_1}$ and $O_{U_3}$ oracles are restricted as follows:
(1) The test transactions are not subpoenaed, *i.e.*, $O_{U_3}$ is not queried on either $(i_0, Type(m_0))$ or $(i_1, Type(m_1))$.
(2) If any of the ⟨user,type⟩ pairs involved in the test transactions are seen by the Agent in some query to $O_{U_1}$ or $O_{U_3}$, then the two test transactions must have the same ⟨user,type⟩ pairs, *i.e.*, if for any $\beta = 0, 1$, either $O_{U_3}$ was queried on $(i_\beta, Type(m_\beta))$ or $O_{U_1}$ was queried on $(i_\beta, m'_\beta)$ s.t. $Type(m'_\beta) = Type(m_\beta)$, then $i_0 = i_1$ and $Type(m_0) = Type(m_1)$.

## 2.2 Additional Desirable Properties of a VTE Scheme

**Automatic threshold disclosure.** A VTE scheme may support automatic opening of escrows involving transactions with the same ⟨user,type⟩ once their number reaches some threshold value, pre-set for transactions of this type. We show an example of such extension in Section 5.

**Key management.** In practice, a VTE scheme requires a *Key Certification Authority* serving as strong PKI. If a user's key is lost or compromised, the CA must not only revoke that key and certify a new one, but also reconstruct the old key to facilitate the subpoena of transactions which were escrowed under it. To avoid a single point of failure, the CA should implement this *key escrow* functionality via a group of trustees using standard threshold techniques. We stress that although majority of the CA trustees must be trusted, this is not a severe limitation of the proposed scheme because CA is invoked only when a new user enrolls in the system, or when the key of some user is subpoenaed and he refuses to cooperate. Moreover, the *secret keys* of the CA trustees need only be used during reconstruction of some user's key in the case of key loss and/or user's refusal to cooperate with a subpoena, both of which should be relatively infrequent events. Interestingly, while PKI is often viewed as a threat to privacy, in our scheme it actually *helps* privacy. Without PKI, escrow can only be implemented via a public-key scheme that cannot guarantee both user anonymity *and* efficient operation of the escrow scheme.

## 3 Basic Construction of a VTE Scheme

We present the simpler variant of our VTE scheme. As we explain in section 3.1, this scheme does not achieve full category-preserving anonymity, but its privacy protection can be good enough in practice. In section 4, we show a variant of the same VTE scheme which does achieve full category-preserving anonymity. Both variants use cryptographic primitives of *verifiable anonymous encryption*, *verifiable anonymous tagging*, and *anonymous signatures*, which we define and implement in section 3.2. In section 3.3, we discuss key management issues.

**VTE construction overview.** In our VTE construction, an escrow consists of (1) an encryption of the transaction plaintext, (2) a signature, and (3) a deterministically computed *tag* which is an output of a pseudorandom function indexed by the user's private key and applied to the *type* of the transaction. The tags enable the Agent to group entries in the escrow database into "bins" corresponding to tag values. Because a pseudorandom function assigns outputs to inputs deterministically, escrows corresponding to the same ⟨user,type⟩ category are always placed in the same bin, enabling efficient identification of the escrowed entries of a given category during the subpoena. However, the pseudorandomness helps to ensure that the tags reveal no more information than permitted by *category-preserving anonymity*, *i.e.*, the only information learned by the escrow agent about any two escrows is whether they belong to the same category.

The signature is included to disable Agent's tampering with the escrowed entries. The encryption and the tag must preserve *secrecy* of the transaction plaintext against chosen-plaintext attack, because a malicious Agent can cause a user to participate in transactions of Agent's choice and see the corresponding escrow entries (see the definition of category-preserving anonymity). The whole escrow must also protect user's *key privacy* against the same chosen-plaintext attack. To enable verification that an escrow is correctly formed, both the tag, the ciphertext, and the signature must be *verifiable* by the transaction counterparty, *i.e.*, given the transaction plaintext and the user's public key.

**Initialization:** Every user is initialized with a public/private key pair implemented as in section 3.2. The escrow agent is initialized with a key pair of any CMA-secure signature scheme.

**Escrow protocol:** We assume that before the escrow protocol starts, the user and the counterparty agree on transaction description $m$ of type $T = Type(m)$.

1. The user sends to the escrow agent an *escrow* $e = (c, t, s)$ s.t.:
   (a) $c = \mathsf{Enc}_k\{m\}$ is a *verifiable anonymous symmetric encryption* of $m$.
   (b) $t = \mathsf{Tag}_k\{T\}$ is an output of a *verifiable anonymous tagging function.*
   (c) $s = \mathsf{sig}_k\{c, t\}$ is an *anonymous signature* on the (ciphertext,tag) pair.
2. The agent places escrow $e$ in the escrow database in the bin indexed by the tag $t$, and sends his signature *rcpt* on $e$ to the user.

**Verification protocol:**

1. The user forwards the escrow $e$ and the agent's signature *rcpt* to the counterparty, together with a proof that:
   (a) $c$ is a ciphertext of $m$ under a key $k$ corresponding to the public key $pk$.
   (b) $t$ is a tag computed on type $T$ under key $k$ corresponding to $pk$.
   (c) $s$ is an anonymous signature computed on $(c, t)$ under the public key $pk$.
2. The transaction counterparty accepts if he verifies the agent's signature on $e$ and the correctness of the above three proofs.

**Subpoena protocol:** The protocol proceeds on a public input of any subset $D$ of the escrow database, the type $T$ of the subpoenaed transactions, and the identity $pk$ of the subpoenaed user:

1. The user computes tag $t = \mathsf{Tag}_k\{T\}$ and proves its correctness under $pk$.
2. Entries $(e_1, e_2, ...)$ in $D$ which are indexed by tag $t$ are publicly identified, and for each $e_i = (c_i, t, s_i)$, the user verifies the signature $s_i$ on $(c_i, t)$.
   (a) If the signature does not match, the user *provably denies* that the signature is valid under $pk$, and if the proof is correct the entry is skipped.
   (b) If the signature matches, the user publishes the transaction plaintext $m_i$ by decrypting the ciphertext $c_i$ under $k$, and proving correctness of the decryption under key $k$ corresponding to $pk$.
3. If the user cooperates, the output includes all (and only) transactions of the subpoenaed type for that user. If any of the above proofs fails, the public output is the special symbol $\mathsf{contempt}$.

From the properties of the cryptographic primitives used in this VTE construction, the following theorem follows:

**Theorem 1.** *The basic VTE scheme satisfies (1) verifiability, (2) efficient and unavoidable subpoena, and (3) tamper resistance.*

### 3.1 Privacy Leakage of the Basic VTE Scheme

In the above scheme, the user presents the (ciphertext, tag, signature) tuple to both the agent and the counterparty. This allows a malicious counterparty and a malicious agent to *link* their views of the escrow and verification protocols, and since the counterparty knows the user identity and the message plaintext, a malicious agent can learn an association between a tag and a ⟨user,type⟩ pair. This would violate category-preserving anonymity, because with this knowledge the escrow agent can learn the type and user identity of *all* transactions with the same tag, even those conducted with other, *honest* counterparties.

In practice, privacy protection can be increased by allowing the *type* of the transaction to range over some small set, for example of a hundred constants. If the index of the constant used for a given transaction is chosen by hashing the counterparty's identity, then there is only 1% chance that a dishonest counterparty can endanger the anonymity of transactions of the same type with any other honest counterparty. On the other hand, when a user is subpoenaed on a given type, he has to identify a hundred categories instead of one. Such privacy/efficiency trade-off may be acceptable in some applications.

### 3.2 Definitions and Constructions for Cryptographic Primitives

Let $p, q$ be large primes s.t. $p = 2q + 1$, and let $g$ be a generator of $\mathbb{Z}_p^*$. The security of our constructions relies on the hardness of the Decisional Diffie-Hellman (DDH) problem in subgroup $QR_p$ of quadratic residues in $\mathbb{Z}_p^*$, which says that tuples $(h, h^a, h^b, h^{ab})$ are indistinguishable from tuples $(h, h^a, h^b, h^c)$ for $h \in QR_p$ and random $a, b, c$ in $\mathbb{Z}_q$ (see, *e.g.*, [Bon98]). Our security arguments follow the so-called "Random Oracle Model" methodology of [BR93]. Namely, we assume an "ideal hash function" $H : \{0,1\}^* \rightarrow \mathbb{Z}_p^*$ which can be treated as a random function in the context of our constructions.

**Verifiable random functions.** A VRF family [MRV99] is defined by three algorithms: a key generation algorithm $\mathsf{KGen}$ outputing private key $k$ and public key $pk$, an evaluation algorithm $\mathsf{Eval}(k, x) = (y, \pi)$ which on input $x$ outputs the value of the function $y = f_k(x)$ and a proof $\pi$ that the value is computed correctly, and a verification algorithm $\mathsf{Ver}$ which can verify $\pi$ on inputs $(pk, x, y, \pi)$. The VRF is secure if it is infeasible to distinguish an interaction with function $f_k$, for a randomly chosen key $k$, from an interaction with a purely random function which outputs uniformly distributed values in the same range. Moreover, the VRF needs to be verifiable, in the sense that any proof will be rejected unless the returned value $y$ is indeed $f_k(x)$. The VRF concept and constructions were originally proposed for the standard model [MRV99, Lys02, Dod03], *i.e.*, without assuming ideal hash functions, but evaluation/verification cost for these constructions involves $\Omega(\tau)$ cryptographic operations. In contrast, in the Random Oracle Model, a simple VRF family can be constructed based on the DDH assumption, with evaluation and verification cost of 1-3 exponentiations. Similar or identical constructions were used before [CP92, NPR99, CKS00], without explicitly noting that the result is a VRF family.

We relax (slightly) the standard definition of VRF [MRV99] by replacing the uniqueness requirement with a computational soundness requirement.

**Definition 1.** *A VRF family (for a group family $\{G_i\}_{i=1,2,\ldots}$) is given by a tuple of polynomial-time algorithms (*$\mathsf{KGen}$, $\mathsf{Eval}$, $\mathsf{Ver}$*) where* $\mathsf{KGen}(1^\tau)$ *outputs a pair of keys* $(k, pk)$, $\mathsf{Eval}$ *is a deterministic algorithm which, on any $x$, outputs* $(y, \pi) \leftarrow \mathsf{Eval}(k, x)$ *s.t. $y \in G_n$, and* $\mathsf{Ver}(pk, x, y, \pi)$ *outputs* 0 *or* 1, *which satisfy the following requirements:*

1. Completeness: *For every* $\tau$ *and* $x$, *if* $(k, pk) \leftarrow \mathsf{KGen}(1^\tau)$ *and* $(y, \pi) = \mathsf{Eval}(k, x)$ *then* $\mathsf{Ver}(pk, x, y, \pi) = 1$.
2. Soundness: *For any probabilistic polynomial-time algorithm* $A$, *for any values* $pk$ *and* $x$, *the following probability is negligible:*

$$\Pr[\mathsf{Ver}(pk, x, y, \pi) = \mathsf{Ver}(pk, x, y', \pi') = 1 \wedge y \neq y' \mid (y, y', \pi, \pi') \leftarrow A(pk, x)]$$

3. Pseudorandomness: *For all probabilistic polynomial-time algorithms* $A_1, A_2$,

$$\Pr[\ b = b' \mid (k, pk) \leftarrow \mathsf{KGen}(1^\tau); (x, st) \leftarrow A_1^{O_{\mathsf{Eval}}(k,\cdot)}(pk); y_0 \leftarrow \mathsf{Eval}(k, x); \\ y_1 \leftarrow G_n; b \leftarrow \{0,1\}; b' \leftarrow A_2^{O_{\mathsf{Eval}}(k,\cdot)}(st, y_b)\ ] \ \leq\ \frac{1}{2} + \mathsf{negl}(\tau)$$

*where* $A_1$ *and* $A_2$ *are restricted from querying oracle* $O_{\mathsf{Eval}}(k, \cdot)$ *on the challenge input* $x$ *chosen by* $A_1$.

**Construction:** Let $H : \{0,1\}^* \rightarrow \mathbb{Z}_p^*$ be an ideal hash function (modeled as a random oracle). Formally, the key generation picks a triple $(p, q, g)$ as above s.t. the hardness of the DDH problem in $QR_p$ is good enough for the security parameter. For ease of discussion, we treat $(p, q, g)$ as chosen once and for all. We will construct a $VRF$ function family indexed by such triples, whose range is the group of quadratic residues $QR_p$. The key generation algorithm picks a secret key $k \in \mathbb{Z}_q^*$ and the public key $pk = g^{2k} \bmod p$. The evaluation algorithm $\mathsf{Eval}(k, x)$ returns $y = h^{2k} \bmod p$ where $h = H(x)$, and a non-interactive zero-knowledge proof $\pi$ of equality of discrete logarithm $x = \mathsf{DL}_h(y) = \mathsf{DL}_g(pk)$. This is a standard ZKPK proof of discrete-log equality which can be made non-interactive in the ROM model, *e.g.*, [CS97].

**Theorem 2.** *Algorithms* $(\mathsf{KGen}, \mathsf{Eval}, \mathsf{Ver})$ *define a Verifiable Random Function family, under the DDH assumption in the Random Oracle Model.*

**Verifiable anonymous tagging function.** We define a verifiable anonymous *tagging* function simply as a VRF, and we implement it as $\mathsf{Tag}_k\{x\} = f_k(x)$. It is easy to see that tags $\mathsf{Tag}_k\{T\}$ give no information about the category they represent, *i.e.*, user's identity $pk$ and the transaction type $T$, except that, whatever category this is, it is identified with tag $\mathsf{Tag}_k\{T\}$. It is also easy to see that a VRF has good enough collision-resistance so that escrows of two categories go to different bins. In fact, a much stronger property holds:

**Theorem 3.** *Under the discrete log assumption, in the Random Oracle Model, the VRF family* $(\mathsf{KGen}, \mathsf{Eval}, \mathsf{Ver})$ *has a strong collision resistance property in the sense that it is infeasible to find pair* $(k, x) \neq (k', x')$ *s.t.* $\mathsf{Eval}(k, x) = \mathsf{Eval}(k', x')$.

**Verifiable anonymous symmetric encryption.** For escrows to be anonymous, the symmetric encryption $\mathsf{Enc}$ used by the user must be not only chosen-plaintext secure, but also *key-hiding*. Following [Fis99, BBDP01], we combine these in one definition that implies several natural anonymity properties. Even an adversary who decides who encrypts what, cannot tell, for ciphertexts created outside of his control, whether the messages and keys satisfy any non-trivial relation this adversary is interested in. For example, the adversary cannot tell if a ciphertext is an encryption under any given key, if two ciphertexts are encryptions under the same key, if two ciphertexts encrypt related messages, *etc.*

Let $(\mathsf{KGen}, \mathsf{Enc}, \mathsf{Dec})$ be a symmetric encryption scheme. In our experiment, first the key generation algorithm is executed $p(\tau)$ times where $p(\cdot)$ is some polynomial and $\tau$ is the security parameter. Denote the keys as $k_i$, for $i \in \{1, p(\tau)\}$. Adversary can query the following *flexible* encryption oracle $O_{\mathsf{Enc}}(\cdot,\cdot)$: on input $(i, m)$, $i \in \{1, p(\tau)\}$ and $m \in \{0,1\}^*$, $O_{\mathsf{Enc}}(i, m)$ outputs $\mathsf{Enc}(k_i, m)$.

**Definition 2.** *We say that a symmetric encryption scheme* $(\mathsf{KGen}, \mathsf{Enc}, \mathsf{Dec})$ *is (chosen-plaintext-secure)* anonymous *if, for any polynomial* $p(\cdot)$ *and probabilistic polynomial-time adversary* $A_1, A_2$,

$$\Pr[\ b = b' \mid (k_1, ..., k_{p(\tau)}) \leftarrow (\mathsf{KGen}(1^\tau))^{p(\tau)}; (i_0, i_1, m_0, m_1, st) \leftarrow A_1^{O_{\mathsf{Enc}}(\cdot,\cdot)}(1^\tau); \\ b \leftarrow \{0,1\}; c \leftarrow \mathsf{Enc}(k_b, m_b); b' \leftarrow A_2^{O_{\mathsf{Enc}}(\cdot,\cdot)}(st, c)] \ \leq \ \frac{1}{2} + \mathsf{negl}(\tau)$$

We also extend the notion of (CPA-secure and anonymous) symmetric encryption by a *verifiability* property. We stress that this property is different from what is referred to as verifiable encryption in the context of *asymmetric* encryption schemes [ASW98, CD00]. We require that the secret key $k$ of an anonymous encryption be generated together with a *commitment* to this secret key, which we will call a *public key pk*. This public key, however, is used not to encrypt but to enable efficient verification that a given ciphertext is a correct encryption of a given plaintext. In fact, our verifiability property for symmetric encryption is very similar to the verifiability property of VRFs. Namely, we require that the encryption procedure $\mathsf{Enc}$ is augmented so that along with output $c = \mathsf{Enc}_k\{m\}$ it produces a *proof* $\pi$ of correct encryption evaluation. We also require an efficient procedure $\mathsf{Ver}$ which takes as inputs message $m$, ciphertext $c$, and a proof $\pi$. The algorithms $(\mathsf{KGen}, \mathsf{Enc}, \mathsf{Dec}, \mathsf{Ver})$ must then satisfy an obvious *completeness* property, *i.e.*, that a correctly computed proof always verifies, and a *soundness* property, which says that it is intractable, for any $(k, pk)$, to find a tuple $(m, m', c, \pi, \pi')$ s.t. $m \neq m'$ but $\mathsf{Ver}(pk, m, c, \pi) = \mathsf{Ver}(pk, m', c, \pi') = 1$.
**Construction:** Instead of using our VRF family to encrypt directly, we replace the hash function in our VRF construction with a Feistel-like padding scheme $pad^H(m|r)$ similar to the OAEP padding [BR94, Sho01]. Assume message length is $|m| = \tau_1 = |p| - 2\tau - 2$ where $\tau$ is the security parameter. We define our padding scheme as $\mathsf{pad}^H(m|r) = (h_1|h_2)$ for $h_1 = H_1(r) \oplus m$ and $h2 = H_2(h_1) \oplus r$, where hash functions $H_1, H_2$ output bit strings of length $\tau_1$ and $2\tau$, respectively, and $r$ is a random string of length $2\tau$. Note that $(m|r)$ can be recovered from $(h_1|h_2)$. This padding is simpler than the OAEP padding and its variants because our (symmetric, anonymous) encryption needs only chosen plaintext security rather than chosen ciphertext security.

Using such padding we can encrypt as follows. $\mathsf{KGen}$ is the same as in the VRF scheme. $\mathsf{Enc}_k(m) = o^{2k} \bmod p$ where $o = \mathsf{pad}^H(m|r)$ is treated as an element in

$\mathbb{Z}_p^*$. The decryption $\mathsf{Dec}_k(c)$ computes candidates $o'$ and $-o' \bmod p$ for $o$, where $o' = c^{k'} \bmod p$, and $k' = \alpha * k^{-1} \bmod q$ where $\alpha = (q+1)/2$ (in integers). To decrypt we take as $o$ either $o'$ or $-o' \bmod p$, depending on which one is smaller than $2^{|p|-2}$. We then recover $m|r$ by inverting the padding scheme $\mathsf{pad}^H$ on $o$. The proof of correct encryption consists of the randomness $r$ and a proof $\pi$ of discrete-log equality $\mathsf{DL}_o(c) = \mathsf{DL}_g(pk)$.

**Theorem 4.** *The above scheme is a verifiable anonymous symmetric encryption scheme secure under the DDH assumption in ROM.*

**Anonymous signatures.** An *anonymous signature* is an undeniable signature scheme [CP92] with an additional property of *key-privacy*. Recall that an undeniable signature scheme requires that the recipient of a signature $s$ produced under public key $pk$ on message $m$ cannot prove to a third party that this is a valid signature under $pk$. Instead, the third party must ask $U$ to verify the signature validity or invalidity via an interactive proof protocol. Here we additionally require *key privacy* in the sense corresponding to the CPA-security of the anonymous symmetric encryption, *i.e.*, that it is infeasible to tell from a (message,signature) pair what public key was used in computing it.
**Construction:** Any VRF family immediately yields an anonymous signature scheme. In fact, the undeniable signature construction of [CP92] already has the required properties, because it is implicitly constructed from the same VRF construction as here. For better concrete security, we slightly modify the [CP92] construction. The signature on $m$ is a pair $s = (r, \tilde{s})$ where $r$ is a random string of length $2\tau$, and $\tilde{s} = f_k(m|r) = H(m|r)^{2k} \bmod p$. The proof of (in)correctness of a signature under public key $pk$ is a zero-knowledge proof of (in)equality of discrete logarithm (*e.g.*, [CS03]) between tuples $(g, pk)$ and $(H(m|r), \tilde{s})$.

### 3.3 Key Management for Discrete-Log Based VTE Schemes

The discrete-log based keys used in our scheme can be efficiently secret-shared by the user with the CA trustees using Feldman's verifiable secret sharing (see, *e.g.*, [GJKR99] for an exposition). Using recent techniques of [CS03], the user can deliver a secret-share to each trustee encrypted under the trustee's public key, and the trustee can verify the share's correctness without the use of the trustee's private key. The resulting shares can then be efficiently used by the trustees in the subpoena process. For example, if the user refuses to cooperate, the CA trustees can efficiently compute the tag $t = (H(Type))^{2k} \bmod p$ for the subpoenaed user and type via threshold exponentiation protocol such as [GJKR99]. The trustees can also use the same protocol to verify signatures on and decrypt the escrows.

## 4 VTE Scheme with Unlinkable Receipts

As explained in section 3.1, category-preserving anonymity is hard to achieve unless the escrow agent and the transaction counterparty are somehow prevented from linking their views of the escrow and the verification protocols. We show how to achieve such separation of agent's and counterparty's views by replacing the standard signature scheme used by our basic VTE scheme with the *CL signature* scheme of [CL01, CL02], which enables the user to prove his possession of the agent's receipt to the counterparty in zero-knowledge. To integrate CL signatures into our VTE scheme, in section 4.1 we introduce a novel zero-knowledge proof of knowledge of *committed key and plaintext* (CKP-ZKPK).

**Diophantine commitments.** To use the CL signature scheme, we need a commitment scheme of [FO98, DF01] which allows a commitment to *integers* rather than to elements in a finite field. Consider a special RSA modulus $n = p'q'$, where $p', q', (p'-1)/2, (q'-1)/2$ are all prime and $|p'|, |q'|$ are polynomial in the security parameter $\tau$. Consider also a random element $b$ of group $QR_n$ of quadratic residues modulo $n$, and a random element $a$ of the subgroup generated by $b$ in $\mathbb{Z}_n^*$. The commitment to an *integer* value $m$ is $C = a^m b^{m'} \bmod n$ where $m'$ is chosen uniformly in $\mathbb{Z}_n$. This commitment scheme is statistically hiding, and it is binding if strong RSA assumption holds for $n$ [FO98, DF01].

**CL signatures.** The public key in CL signature consists of a special RSA modulus $n$ as above, and three uniformly chosen elements $a, b, d$ in $QR_n$. Let $l_m$ be a parameter upper-bounding the length of messages that need to be signed. The public key is $(n, a, b, d)$. The signature on $m$ is a triple $(v, e, s)$ where $v^e = a^m b^s d \bmod n$ and $2^{l_e} > e > 2^{l_e+1}$ where $l_e \geq l_m + 2$. This signature scheme is CMA-secure under the strong RSA assumption [CL02].

The CL signature comes with two protocols: (1) the *CL signing protocol*, in which the signer can issue signature $(v, e, s)$ on $m \in \{0,1\}^{l_m}$ given only a commitment $C_m$ to $m$; and (2) the *CL verification protocol* which is a zero-knowledge proof in which the prover can prove the knowledge of a signature on $m$ to the verifier who knows only a commitment to $m$.

The commitments to $m$ used in protocols (1) and (2) can be independent of the CL signature public key. However, for simplicity, in our application the instance of the Diophantine commitment scheme used in the CL signing protocol will be formed by values $(n, a, b)$ which are parts of the CL signature public key.

Before we show how to use them, we need to make two modifications to the CL signatures as shown above. First, we use the [CL02] extension of the above scheme to signing a *block* of three messages $(m_1, m_2, m_3)$. This is done simply by including three random elements $a_1, a_2, a_3$ in $QR_n$ instead of one $a$ in the public key of the CL signature scheme. The signature is a triple $(v, e, s)$ where $v^e = a_1^{m_1} a_2^{m_2} a_3^{m_3} b^s d \bmod n$. In the CL signing and verification protocols adapted to a block of three messages, both the signer and the verifier know three separate commitments on these messages.

Second, we note that if in the CL signature verification protocol the verifier knows the message $m$ itself instead of a commitment to it, the protocol still works and even gets easier. Similarly, if the verifier knows not the above Diophantine commitment to $m$, but $g^m \bmod p$ (also a commitment to $m$), the protocol still works, but the prover only shows knowledge of a signature on *some* integer $m'$ s.t. $m' = m \bmod 2q$ (recall that $p = 2q+1$, $p, q$ are primes, and $g$ is a generator of $\mathbb{Z}_p^*$).

The same holds for the CL verification protocol extended to a block of messages $(m_1, m_2, m_3)$. In our case, the verifier will know messages $m_1$ and $m_2$, and a commitment $g^{m_3} \bmod p$ to message $m_3$, and the prover will show possession of CL signature on block of messages $(m_1, m_2, m_3')$ s.t. $m_3' = m_3 \bmod q$.

**VTE scheme with unlinkable receipts.** We recall the VTE construction of section 3, where $k$ is the user's secret key, $pk = g^{2k} \bmod p$ is the public key, and $m$ is the transaction plaintext. The escrow is a triple $e = (c, t, s)$ where $c = o^{2k} \bmod p$, $t = h^{2k} \bmod p$, $s = (r, \tilde{s})$, $\tilde{s} = H((c,t)|r)^{2k} \bmod p$, $h = H(Type(m))$, $o = \mathsf{pad}^H(m|r')$, and $r, r'$ are random strings of length $2\tau$.

Let $l_m$, the maximum message length, be $|p|$, enough to represent elements in either $\mathbb{Z}_p^*$ or $\mathbb{Z}_q^*$. The public key of the escrow agent is the public key $(n, a, b, c)$ of the CL signature scheme, except that $a$ is chosen at random from the subgroup generated by $b$ in $\mathbb{Z}_n^*$. If the escrow agent generates his key himself, he must prove knowledge of $i$ s.t. $a = b^i \bmod n$.

The user sends $e = (c, t, s) = (c, t, (r, \tilde{s}))$ to the escrow agent as in the basic VTE scheme, but here he also includes three diophantine commitments $C_o, C_h, C_k$ on *integer* values $o, h, k$ using $(n, a, b)$ as the instance of the commitment scheme. Using the zero-knowledge proof CKP-ZKPK of *committed key and plaintext* (see section 4.1), the user then proves his knowledge of integer values $(o', h', k')$ s.t. $o', h', k'$ are committed to in $C_o, C_h, C_k$, and $c = (o')^{2k'} \bmod p$, $t = (h')^{2k'} \bmod p$, and $\tilde{s} = H((c,t)|r)^{2k'} \bmod p$. If the proof succeeds, the user and the escrow agent run the CL signing protocol on the commitments $C_o, C_h, C_k$ at the end of which the user holds a CL signature on the block $(o', h', k')$ of the committed messages.

In the verification phase, the user sends to the transaction counterparty values $(o, r')$, together with the transaction plaintext $m$ and his public key $pk = g^{2k} \bmod p$. The counterparty computes $h = H(Type(m))$ and verifies if $o = \mathsf{pad}^H(m|r')$. The user and the counterparty then run the CL verification protocol in which the user proves possession of a CL signature on integer values $o, h, k'$ where the verifier knows $o$ and $h$ and $pk = g^{2k'} \bmod p$.

If the user passes both proofs, the first with the escrow agent as the verifier and the second with the transaction counterparty as the verifier, then under the strong RSA assumption needed for the diophantine commitment to be binding, $o' = o$, $h' = h$, and $k' = k \bmod q$, thus the escrow entry $e = (c, t, s)$ is computed correctly. Furthermore, the escrow agent learns only the (ciphertext,tag) pair $(c, t) = (o^{2k} \bmod p, h^{2k} \bmod p)$ and the signature $s$, while the counterparty learns only the values $o, h$ associated with the plaintext $m$ and the public key $pk = g^{2k} \bmod p$.

From the properties of the basic VTE scheme and the CKP-ZKPK proof system (see section 4.1), the following theorem follows:

**Theorem 5.** *The VTE scheme with unlinkable receipts satisfies (1) verifiability, (2) efficient and unavoidable subpoena, (3) tamper resistance, and (4) category-preserving anonymity, under the DDH and strong RSA assumptions in ROM.*

## 4.1 Zero-Knowledge Proof of Committed Key and Plaintext

We present the ZK proof protocol required by the unlinkable-receipt VTE construction of the previous section. Recall that the user needs to prove in zero-knowledge to the escrow agent his knowledge of integer values $o, h, k$ s.t. $o, h, k$ are committed to in $C_o, C_h, C_k$, and $c = o^{2k} \bmod p$, $t = h^{2k} \bmod p$, and $\tilde{s} = H((c,t)|r)^{2k} \bmod p$. The public inputs in this proof are values $(p, q, g)$, $(n, a, b)$, $(C_o, C_h, C_k)$, and $(c, t, r, \tilde{s})$. The prover's inputs are $o, h \in \mathbb{Z}_p^*$, $k \in \mathbb{Z}_q^*$, and the decommitment values $o', h', k'$ in $\mathbb{Z}_n$.

**ZKPK of Committed Key and Plaintext**

Prover's Input: $k \in \mathbb{Z}_q^*$, the secret key
$o \in \mathbb{Z}_p^*$, the "plaintext"
$o', k' \in \mathbb{Z}_n$, the decommitment values

Common Input: $(p, q, g)$, the discrete-log group setting
$(n, a, b)$, the instance of a diophantine commitment scheme
$C_k = a^k b^{k'} \bmod n$, commitment to $k$
$C_o = a^o b^{o'} \bmod n$, commitment to $o$
$c = o^{2k} \bmod p$, the ciphertext

1. Prover P picks $\tilde{o} \leftarrow \mathbb{Z}_p^*$ and $\tilde{o}' \leftarrow \mathbb{Z}_n$, and sends $C_{\tilde{o}} = a^{\tilde{o}} b^{\tilde{o}'} \bmod n$, and $\tilde{c} = (\tilde{o})^{2k} \bmod p$ to the Verifier V
2. Verifier V sends to P a random binary challenge $b = 0$ or $1$
3. P responds as follows:

$b = 0$: (a) P sends $(s, s') = (\tilde{o}, \tilde{o}')$ to V
(b) P performs a standard ZKPK proof of knowledge (*e.g.*, [CM99]) of $(k, k')$ s.t. $a^k b^{k'} = C_k \bmod n$ and $s^{2k} = \tilde{c} \bmod p$

$b = 1$: (a) P sends $s = o * \tilde{o} \bmod p$ to V
(b) P performs a standard ZKPK proof of knowledge (*e.g.*, [CM99]) of $(k, k')$ s.t. $a^k b^{k'} = C_k \bmod n$ and $s^{2k} = c * \tilde{c} \bmod p$
(c) P performs a ZKPK given by [CM99], of knowledge of values $(o, o', \tilde{o}, \tilde{o}')$ s.t. $a^o b^{o'} = C_o \bmod n$, $a^{\tilde{o}} b^{\tilde{o}'} = C_{\tilde{o}} \bmod n$, and $o * \tilde{o} = s \bmod p$

4. In both cases V accepts only if the appropriate ZKPK proofs verify. Additionally, if $b = 0$, V checks also if $a^s b^{s'} = C_{\tilde{o}} \bmod n$.

**Fig. 1.** Binary challenge proof system CKP-ZKPK

To simplify the presentation, we will show a ZKPK system for a slightly simpler problem, namely the ZK proof of knowledge *of committed key and plaintext* (CKP-ZKPK). Namely, the public values are $(p, q, g)$, $(n, a, b)$, $(C_o, C_k, c)$ and the prover proves knowledge of *integer* values $o, k$ s.t. (1) they are committed to in $C_o, C_k$ under commitment instance $(n, a, b)$, and (2) $c = o^{2k} \bmod p$. One can see that the required ZKPK system is created by running three proofs in parallel: (i) one CKP-ZKPK proof for secrets $o, k$ and public $(C_o, C_k, c)$, (ii) another CKP-ZKPK proof for secrets $h, k$ and public $(C_h, C_k, t)$, and (iii) a standard ZKPK proof of knowledge (*e.g.* [CM99]) of $k$ s.t. $H((c,t)|r)^{2k} = \tilde{s} \bmod p$ and $k$ is committed to in $C_k$, where the public inputs are $(C_k, c, t, r, \tilde{s})$.

We present the CKP-ZKPK proof protocol in Figure 1. We note that this is a binary challenge protocol with $1/2$ soundness error, so to get security parameter $\tau$ this proof should be repeated $\tau$ times, or in the Random Oracle Model it can be made non-interactive by preparing the $\tau$ instances of it in independently in parallel, except that the challenge bits are computed by hashing together the first prover's messages of all these $\tau$ instances. The resulting protocol involves $O(\tau)$ exponentiations for both the prover and the verifier, which unfortunately makes this protocol quite expensive in practice.

Note that both ZKPK proofs referred to in the CKP-ZKPK protocol can be non-interactive in the Random Oracle Model considered here, and that they involve a small constant amount of exponentiations. We remark that the protocol proof system of [CM99] used in step (c) of case $b = 1$ for proving modular multiplication on committed values, can be simplified in our case, because here the multiplicative factor $s = o * \tilde{o}$ and the modulus $p$ are publicly known, in contrast to the general case considered by [CM99], where the verifier knows $s$ and $p$ only in a committed form.

**Theorem 6.** *CKP-ZKPK proof system is computational zero-knowledge if the DDH problem for group $QR_p$ is hard.*

**Theorem 7.** *CKP-ZKPK proof system is a proof of knowledge with soundness error $1/2$ if the strong RSA problem in group $\mathbb{Z}_n$ is hard.*

## 5 VTE Scheme with Automatic Threshold Disclosure

We describe an extension of the VTE scheme which enables the escrow agent to automatically open escrows that (1) fall into the same *bin*, *i.e.*, share the same $\langle$user,type$\rangle$ category, and (2) their number is no less than some fixed threshold, pre-specified for transactions of this type. This can be used, for example, to implement oversight of financial transactions which the following disclosure condition: if some user requests more than 10 transfers, via any set of banks, to some pre-specified "offshore haven," the plaintexts of the corresponding escrows must be automatically disclosed to the overseeing authority.

Using Feldman's non-interactive verifiable secret sharing scheme [Fel87], we modify the VTE scheme of section 3 as follows. To create an escrow of plaintext $m$ under key $k$, the user computes the tag $t = \mathsf{Tag}_k\{T\}$ where $T = \mathit{Type}(m)$ as in section 3, but the ciphertext is computed differently. Let $d$ be the publicly pre-specified threshold disclosure value that corresponds to this $T$. The user picks a unique $d$-degree *secret-sharing polynomial* $f(\cdot)$ by applying $d+1$ times a pseudorandom function indexed by the secret $k$, *i.e.*, $k_i = H(k, T, i)$ for $i = 0, \ldots, d$, where $H : \{0,1\}^* \rightarrow \mathbb{Z}_q$, and setting $f(x) = k_0 + k_1 x + \ldots k_d x^d \bmod q$.

A set of values $\{C_0, \ldots, C_d\}$ where $C_i = g^{2k_i} \bmod p$ serves as *public verification information* for this secret-sharing polynomial, The ciphertext is now $c' = (c, \{C_i\}_{i=0..d}, x, f(x), d)$, where $c = \mathsf{Enc}_{k_0}\{m\}$, $\pi$ is the proof that $c$ is a correct encryption of $m$ under the "quasi-one-time" private key $k_0$ (and its public counterpart $C_0 = g^{2k_0} \bmod p$), and $x$ is some unique value corresponding to this transaction, *e.g.*, $x = H(c)$. The user computes the private signature $s = (r, \tilde{s})$ on $(c', t)$, and hands the escrow $e = (c', t, s)$ to the escrow agent.

The escrow agent checks that $(x, f(x))$ is a true data point on the polynomial committed to in set $\{C_i\}_{i=0..d}$ by verifying that $g^{2f(x)} = (C_0) * (C_1)^x * \ldots * (C_d)^{x^d} \bmod p$. Moreover, if the bin tagged with tag $t$ in the escrow database has other entries, the agent checks that the argument $x$ has not been used before with the tag $t$, and the values $\{C_0, \ldots, C_d\}$ are the same for this $t$ as before. The agent then releases his signature on the escrow $e$ to the user. The user presents it to the counterparty, who verifies it as before, except that correctness of the ciphertext $c = \mathsf{Enc}_{k_0}\{m\}$ is verified on $(C_0, m, c, \pi)$ instead of $(pk, m, c, \pi)$, and it is checked that $d$ is the threshold value corresponding to type $T$.

To prevent the counterparty and the escrow agent from linking their views, the same mechanism as in section 4 may be deployed. The user sends commitments $C_o, C_h, C_k$ on values $o, h, k$ to the escrow agent (note the difference between $C_o$ and $C_0$), proving his knowledge of $o, h, k, k_0$ s.t. $c = o^{2k_0} \bmod p$, $C_0 = g^{2k_0} \bmod p$, $t = h^{2k} \bmod p$, and $\tilde{s} = H((c', t)|r)^{2k} \bmod p$. The same zero-knowledge protocol as in section 4 may be used, and is even slightly simpler since $C_0$ is a simpler commitment to $k_0$ than the Diophantine commitment. After checking the proofs, the user and the escrow agent perform the CL signing protocol to give the user a CL signature on the block of messages $(o, h, k, d)$. The user then sends to the counterparty values $(o, r')$ as in section 4, together with $d$. The counterparty checks that $o$ is properly formed and $d$ is the proper threshold value for the given transaction type, and they run the CL verification protocol to prove the user's knowledge of a CL signature on values $(o, h, k, d)$ where the verifier knows $o, h, d$ and $pk = g^{2k} \bmod p$.

**Acknowledgments.** We want to thank Dan Boneh, Pat Lincoln, and Anna Lysyanskaya for helpful discussions and for proposing extensions and improvements. We also thank the anonymous referees for their suggestions.

## References

[ASW98] N. Asokan, V. Shoup, and M. Waidner. Asynchronous protocols for optimistic fair exchange. In *Proc. IEEE Symposium on Security and Privacy*, pages 86–99, 1998.

[BBDP01] M. Bellare, A. Boldyreva, A. Desai, and D. Pointcheval. Key-privacy in public-key encryption. In *Proc. Asiacrypt '01*, pages 566–582, 2001.

[BCOP03] D. Boneh, G. Di Crescenzo, R. Ostrovsky, and G. Persiano. Searchable public key encryption. IACR 2003/195, 2003.

[Bon98] D. Boneh. The decisional Diffie-Hellman problem. In *Proc. 3rd Algorithmic Number Theory Symposium*, pages 48–63, 1998.

[BR93] M. Bellare and P. Rogaway. Random oracles are practical: a paradigm for designing efficient protocols. In *Proc. ACM Conference on Computer and Communications Security*, pages 62–73, 1993.

[BR94] M. Bellare and P. Rogaway. Optimal asymmetric encryption. In *Proc. Eurocrypt '94*, pages 92–111, 1994.

[CD00] J. Camenisch and I. Damgard. Verifiable encryption, group encryption, and their applications to group signatures and signature sharing schemes. In *Proc. of Asiacrypt '00*, pages 331–345, 2000.

[CKS00] C. Cachin, K. Kursawe, and V. Shoup. Random oracles in Constantinople. In *Proc. of PODC '00*, pages 123–132, 2000.

[CL01] J. Camenisch and A. Lysyanskaya. An efficient system for non-transferable anonymous credentials with optional anonymity revocation. In *Proc. Eurocrypt '01*, pages 93–118, 2001.

[CL02] J. Camenish and A. Lysyanskaya. A signature scheme with efficient protocols. In *Proc. SCN '02*, pages 268–289, 2002.

[CM99] J. Camenisch and M. Michels. Proving in zero-knowledge that a number is the product of two safe primes. In *Proc. Eurocrypt '99*, pages 107–122, 1999.

[CP92] D. Chaum and T. Pedersen. Wallet databases with observers. In *Proc. Crypto '92*, pages 89–105, 1992.

[CS97] J. Camenisch and M. Stadler. Proof systems for general statements about discrete logarithms. Technical Report TR 260, ETH Zürich, March 1997.

[CS03] J. Camenisch and V. Shoup. Verifiable encryption and decryption of discrete logarithms. In *Proc. of Crypto '03*, pages 126–144, 2003.

[DF01] I. Damgard and E. Fujisaki. An integer commitment scheme based on groups with hidden order. IACR 2001/064, 2001.

[Dod03] Y. Dodis. Efficient construction of (distributed) verifiable random functions. In *Proc. Public Key Cryptography '03*, pages 1–17, 2003.

[Fel87] P. Feldman. A practical scheme for non-interactive verifiable secret sharing. In *Proc. FOCS '87*, pages 427–438, 1987.

[Fis99] M. Fischlin. Pseudorandom function tribe ensembles based on one-way permutations: improvements and applications. In *Proc. Eurocrypt '99*, pages 432–445, 1999.

[FO98] E. Fujisaki and T. Okamoto. A practical and porvably secure scheme for publicly verifiable secret sharing and its applications. In *Proc. Eurocrypt '98*, pages 32–46, 1998.

[GJKR99] R. Gennaro, S. Jarecki, H. Krawczyk, and T. Rabin. Secure distributed key generation for discrete-log based cryptosystems. In *Proc. of Eurocrypt '99*, pages 295–310, 1999.

[JS04] S. Jarecki and V. Shmatikov. Handcuffing Big Brother: An abuse-resilient transaction escrow scheme. IACR eprint, 2004.

[KL95] J. Kilian and F.T. Leighton. Fair cryptosystems revisited. In *Proc. Crypto '95*, pages 208–221, 1995.

[Lys02] A. Lysyanskaya. Unique signatures and verifiable random functions from the DH-DDH separation. In *Proc. Crypto '02*, pages 597–612, 2002.

[Mic92] S. Micali. Fair public-key cryptosystems. In *Proc. Crypto '92*, pages 113–138, 1992.

[MRV99] S. Micali, M. Rabin, and S. Vadhan. Verifiable random functions. In *Proc. FOCS '99*, pages 120–130, 1999.

[NPR99] M. Naor, B. Pinkas, and O. Reingold. Distributed pseudo-random functions and KDCs. In *Proc. of Eurocrypt '99*, pages 327–346, 1999.

[Sho01] V. Shoup. OAEP reconsidered. In *Proc. of Crypto '01*, pages 239–259, 2001.

[SWP00] D.X. Song, D. Wagner, and A. Perrig. Practical techniques for searches on encrypted data. In *Proc. IEEE Symposium on Security and Privacy*, pages 44–55, 2000.

# Anonymous Identification in *Ad Hoc* Groups

Yevgeniy Dodis[1], Aggelos Kiayias[2], Antonio Nicolosi[1], and Victor Shoup[1]

[1] Courant Institute of Mathematical Sciences, New York University, NY, USA
{dodis,nicolosi,shoup}@cs.nyu.edu
[2] Department of Computer Science and Eng., University of Connecticut, CT, USA
aggelos@cse.uconn.edu

**Abstract.** We introduce *Ad hoc* Anonymous Identification schemes, a new multi-user cryptographic primitive that allows participants from a user population to form *ad-hoc* groups, and then prove membership anonymously in such groups. Our schemes are based on the notion of *accumulator with one-way domain*, a natural extension of cryptographic accumulators we introduce in this work. We provide a formal model for *Ad hoc* Anonymous Identification schemes and design secure such schemes both generically (based on any accumulator with one-way domain) and for a specific efficient implementation of such an accumulator based on the Strong RSA Assumption. A salient feature of our approach is that all the identification protocols take time independent of the size of the ad-hoc group. All our schemes and notions can be generally and efficiently amended so that they allow the recovery of the signer's identity by an authority, if the latter is desired.
Using the Fiat-Shamir transform, we also obtain *constant-size*, signer-ambiguous group and ring signatures (provably secure in the Random Oracle Model). For ring signatures, this is the first such constant-size scheme, as all the previous proposals had signature size proportional to the size of the ring. For group signatures, we obtain schemes comparable in performance with state-of-the-art schemes, with the additional feature that the role of the group manager during key registration is extremely simple and essentially passive: all it does is accept the public key of the new member (and update the constant-size public key of the group).

## 1 Introduction

Anonymous identification is an oxymoron with many useful applications. Consider the setting, for a known user population and a known set of resources, where a user wants to gain access to a certain resource. In many cases, accessing the resource is an action that *does not* mandate positive identification of the user. Instead, it would be sufficient for the user to prove that he belongs to the subset of the population that is supposed to have access to the resource. This would allow the user to lawfully access the resource while protect his real identity and thus "anonymously identify" himself.

Given the close relationships between identification schemes and digital signatures, one can easily extend the above reasoning to settings where a user

C. Cachin and J. Camenisch (Eds.): EUROCRYPT 2004, LNCS 3027, pp. 609–626, 2004.

produces a signature that is "signer-ambiguous" i.e., such that the verifier is not capable of distinguishing the actual signer among a subgroup of potential signers. In fact, it was in the digital signature setting that such an anonymous scheme was presented for the first time, with the introduction of the group signature model [19], which additionally mandates the presence of a designated party able to reveal the identity of the signer, were the need to arise.

Subsequent work on group signatures and on anonymous identification in general [20, 24, 13, 18, 16, 23, 3, 1, 11, 14, 6, 2] allowed for more efficient designs and formal modelling of the primitive, with the current state of the art being the scheme by Ateniese et al. [1]. In general, existing group signature schemes are derived from their interactive counterpart (*ID Escrow* schemes [32]) via the Fiat-Shamir transform [28].

A related notion, but of slightly different nature, is that of ring signatures, introduced by Rivest, Shamir and Tauman in [34] and further studied in [12, 33]. Ring signatures differ from group signatures in that they allow group formation to happen in an ad-hoc fashion: group must be formed without the help of a group manager; in fact, a user might not even know that he has been included in a certain group. This is in sharp contrast to the group signature setting where the user must execute a Join protocol with the group manager and obtain a group-membership certificate that cannot be constructed without the help of the group manager. Note that ad-hoc group formation in the context of ring signatures is always understood within the context of a user population and an associated PKI. Based on the PKI, ad-hoc subsets of the user population can be formed without the help of a "subset manager"—but it is assumed that every user has a registered public key.

While ring signatures are attractive because they have simple group formation procedures that can be executed by any user individually, they have the shortcoming that the length of the signature is proportional to the group size. For large groups, the length of a ring signature (growing linearly with the group size) will become impractical. To the contrary, schemes with *constant-size* signatures have been successfully designed in the group signature setting [1]. We remark that in the setting of anonymous identification, the counterpart of "signature size" is the bandwidth consumed by the protocol, which is thus an important complexity measure to minimize.

Based on the above discussion, an important open question in the context of anonymous identification and signature schemes, recently posed by Naor in [33], is the following:

> Is it possible to design secure anonymous identification schemes that enable ad-hoc group formation in the sense of ring signatures and at the same time possess constant-size signature (or proof) length?

**Our contribution.** In this work we provide an affirmative answer to the above question. Specifically, we introduce a new primitive called *Ad hoc* Anonymous Identification schemes; this is a family of schemes where participants from a user population can form groups in ad-hoc fashion (without the help of a group manager) and then get anonymously identified as members of such groups.

Our main tool in the construction of *Ad hoc* Anonymous Identification schemes is a new cryptographic primitive, *accumulator with one-way domain*, which extends the notion of a collision-resistant accumulator [7, 4, 15]. In simple terms, in an accumulator with one-way domain, the set of values that can be accumulated are associated with a "witness space" such that it is computationally intractable to find witnesses for random values in the accumulator's domain.

First, we demonstrate the relationship between such accumulators and *Ad hoc* Anonymous Identification schemes by presenting a generic construction based on *any* accumulator with one-way domain. Second, we design an efficient implementation of accumulator with a one-way domain based on the Strong RSA Assumption, from which we obtain a more efficient construction of *Ad hoc* Anonymous Identification scheme whose security rests upon the Strong RSA Assumption.

We remark that previous work on anonymous identification that allowed subset queries was done by Boneh and Franklin [8]. They define a more limited security model, and show a protocol which imposes on both parties a computational load proportional to the subset size at each run. Moreover, their scheme is susceptible to collusion attacks (both against the soundness and against the anonymity of the scheme) that do not apply to our setting.

In our Strong-RSA-based *Ad hoc* Anonymous Identification scheme, the computational and communication complexity on both ends is constant in the size of the group. Thus, the signature version of our ad-hoc anonymous identification scheme yields a ring signature with constant size signatures (over a dedicated PKI). Other applications of our scheme include "ad-hoc" group signatures (group signature schemes where the group manager can be offline during the group formation) and identity escrow over ad-hoc groups.

Recently, work by Tsudik and Xu [35], building on the work by Camenisch and Lysyanskaya [15], investigated techniques to obtain more flexible dynamic accumulators, on which to base group signature schemes (which is one of our applications). The specific method used by [35] bears many similarities with our Strong-RSA-based instantiation, with some important differences. Namely, in their solution anonymity revocation takes time proportional to the user population, due to subtle problems concerning the accumulation of composite values inside the accumulator. Our work resolves this technical problem. Moreover, we present a new notion of *Ad hoc* Anonymous Identification scheme, which has more applications than those specific to group signature schemes: for example, they allow us to build the first constant-size ring signature schemes. We present a general construction for our primitives from any accumulator and not just the one of [15]. Last, our formal definitional framework is of independent interest.

## 2 Preliminaries

### 2.1 NP-Relations and $\Sigma$-Protocols

Throughout the paper, we assume familiarity with the GMR notation [30].

An **NP**-relation $R$ is a relation over bitstrings for which there is an efficient algorithm to decide whether $(x, w) \in R$ in time polynomial in the length of $x$. The **NP**-language $L_R$ associated to $R$ is defined as $L_R \doteq \{x \mid (\exists w)[(x, w) \in R]\}$

A $\Sigma$-protocol [22, 21] for an **NP**-relation $R$ is an efficient 3-round two-party protocol, such that for every input $(x, w)$ to $P$ and $x$ to $V$, the first $P$-round yields a *commitment* message, the subsequent $V$-round replies with a random *challenge* message, and the last $P$-round concludes by sending a *response* message. At the end of a run, $V$ outputs a 0/1 value, functionally dependent on $x$ and the transcript $\pi$ only. Additionally, a $\Sigma$-protocol satisfies *Special Soundness*, meaning that for any $(x, w) \notin R$ and any commitment message, there is at most one pair of challenge/response messages for which $V$ would output 1; and *Special Honest-Verifier Zero-Knowledge*, meaning that there is an efficient algorithm (called a *Simulator*) that on input $x \in L_R$ and any challenge message, outputs a pair of commitment/response messages for which $V$ would output 1.

The main result we will need about $\Sigma$-protocols is the following:

**Theorem 1 ([29, 27]).** *A $\Sigma$-protocol for any* **NP**-*relation can be efficiently constructed if one-way functions exist.*

### 2.2 Accumulators

An *accumulator family* is a pair $(\{F_\lambda\}_{\lambda\in\mathbb{N}}, \{X_\lambda\}_{\lambda\in\mathbb{N}})$, where $\{F_\lambda\}_{\lambda\in\mathbb{N}}$ is a sequence of families of functions such that each $f \in F_\lambda$ is defined as $f : U_f \times X_f^{\text{ext}} \to U_f$ for some $X_f^{\text{ext}} \supseteq X_\lambda$ and additionally the following properties are satisfied:

- (efficient generation) There exists an efficient algorithm $G$ that on input a security parameter $1^\lambda$ outputs a random element $f$ of $F_\lambda$, possibly together with some auxiliary information $a_f$.
- (efficient evaluation) Any $f \in F_\lambda$ is computable in time polynomial in $\lambda$.
- (quasi-commutativity) For all $\lambda \in \mathbb{N}$, $f \in F_\lambda$, $u \in U_f$, $x_1, x_2 \in X_\lambda$,

$$f(f(u, x_1), x_2) = f(f(u, x_2), x_1)$$

We will refer to $\{X_\lambda\}_{\lambda\in\mathbb{N}}$ as the *value domain* of the accumulator. For any $\lambda \in \mathbb{N}, f \in F_\lambda$ and $X = \{x_1, \ldots, x_s\} \subset X_\lambda$, we will refer to $f(\ldots f(u, x_1) \ldots, x_s)$ as the *accumulated value* of the set $X$ over $u$: due to quasi-commutativity, such value is independent of the order of the $x_i$'s and will be denoted by $f(u, X)$.

**Definition 1.** *An accumulator is said to be* collision resistant *if for any $\lambda \in \mathbb{N}$ and any adversary $\mathcal{A}$:*

$$\begin{aligned} &Pr[f \stackrel{R}{\leftarrow} F_\lambda; u \stackrel{R}{\leftarrow} U_f; (x, w, X) \stackrel{R}{\leftarrow} \mathcal{A}(f, U_f, u) \quad | \\ &\quad (X \subseteq X_\lambda) \wedge (w \in U_f) \wedge (x \in X_f^{\text{ext}} \setminus X) \wedge (f(w, x) = f(u, X))] = \nu(\lambda) \end{aligned}$$

For $\lambda \in \mathbb{N}$ and $f \in F_\lambda$, we say that $w \in U_f$ is a *witness* for the fact that $x \in X_\lambda$ has been accumulated within $v \in U_f$ (or simply that $w$ is a witness for $x$ in $v$) whenever $f(w, x) = v$. We extend the notion of witness for a set of values $X = \{x_1, \ldots, x_s\}$ in a straightforward manner.

**Accumulators with One-Way Domain.** An *accumulator with one-way domain* is a quadruple $(\{F_\lambda\}_{\lambda\in\mathbb{N}}, \{X_\lambda\}_{\lambda\in\mathbb{N}}, \{Z_\lambda\}_{\lambda\in\mathbb{N}}, \{R_\lambda\}_{\lambda\in\mathbb{N}})$, such that the pair $(\{F_\lambda\}_{\lambda\in\mathbb{N}}, \{X_\lambda\}_{\lambda\in\mathbb{N}})$ is a collision-resistant accumulator, and each $R_\lambda$ is a relation over $X_\lambda \times Z_\lambda$ with the following properties:

- (efficient verification) There exists an efficient algorithm $D$ that on input $(x, z) \in X_\lambda \times Z_\lambda$, returns 1 if and only if $(x, z) \in R_\lambda$.
- (efficient sampling) There exists a probabilistic algorithm $W$ that on input $1^\lambda$ returns a pair $(x, z) \in X_\lambda \times Z_\lambda$ such that $(x, z) \in R_\lambda$. We refer to $z$ as a *pre-image* of $x$.
- (one-wayness) It is computationally hard to compute any pre-image $z'$ of an $x$ that was sampled with $W$. Formally, for any adversary $\mathcal{A}$:

$$Pr[(x, z) \stackrel{\mathrm{R}}{\leftarrow} W(1^\lambda); z' \stackrel{\mathrm{R}}{\leftarrow} \mathcal{A}(1^\lambda, x) \mid (x, z') \in R_\lambda] = \nu(\lambda)$$

### 2.3 The Strong RSA Assumption

We briefly review some definitions [7, 4] regarding the computational assumption underlying our efficient construction in Section 5.

A number $n$ is an *RSA integer* if $n = pq$ for distinct primes $p$ and $q$ such that $|p| = |q|$. For $\lambda \in \mathbf{N}$, let $\mathsf{RSA}_\lambda$ be the set of RSA integers of size $\lambda$. A number $p$ is a *safe prime* if $p = 2p' + 1$ and both $p$ and $p'$ are odd primes. A number $n$ is a *rigid integer* if $n = pq$ for distinct safe primes $p$ and $q$ such that $|p| = |q|$. For $\lambda \in \mathbf{N}$, let $\mathsf{Rig}_\lambda$ be the set of $\lambda$-bit rigid integers.

**Definition 2 (Strong RSA Assumption, [4]).**
*For any integer $\lambda$ and for any adversary $\mathcal{A}$:*

$$Pr[n \stackrel{R}{\leftarrow} \mathit{Rig}_\lambda; z \stackrel{R}{\leftarrow} \mathbb{Z}_n^*; (x', y') \stackrel{R}{\leftarrow} \mathcal{A}(1^\lambda, n, z) \mid (y' > 1) \wedge ((x')^{y'} \equiv z(n))] < \nu(\lambda)$$

*the probability being over the random choice of $n$ and $z$, and $\mathcal{A}$'s random coins.*

## 3 *Ad Hoc* Anonymous Identification Scheme

### 3.1 Syntax

An *Ad hoc* Anonymous Identification scheme is a six-tuple of efficient algorithms ($\mathsf{Setup}$, $\mathsf{Register}$, $\mathsf{Make\text{-}GPK}$, $\mathsf{Make\text{-}GSK}$, $\mathsf{Anon\text{-}ID}^{\mathsf{P}}$, $\mathsf{Anon\text{-}ID}^{\mathsf{V}}$), where:

- $\mathsf{Setup}$ initializes the state of the system: on input a security parameter $1^\lambda$, $\mathsf{Setup}$ creates a public database $\mathsf{DB}$ (that will be used to store information about the users' public keys), and then generates the system's parameters $\mathsf{param}$; its output implicitly defines a domain of possible global parameters.
- $\mathsf{Register}$, the *registration algorithm*, allows users to initially register with the system. On input the system's parameters $\mathsf{param}$ and the identity of the new user $u$ (from a suitable universe of users' identity $\mathcal{U}$), $\mathsf{Register}$ returns a secret key/public key pair $(sk, pk)$. To complete the subscription process,

the user then sends his public key to a bulletin board for inclusion in a public database DB.
The Register algorithm implicitly defines a domain $\mathcal{SK}$ of possible user secret keys and a domain $\mathcal{PK}$ of possible user public keys; its output induces a relation over user secret key/public key pairs, that we will denote by $\rightleftharpoons$. We also require a superset $\mathcal{PK}' \supseteq \mathcal{PK}$ to be specified, such that membership to $\mathcal{PK}'$ can be tested in polynomial time.

- Make-GPK, the *group public key construction algorithm*, is a deterministic algorithm used to combine a set of user public keys $S$ into a single group public key $gpk_S$, suitable for use in the Anon-ID protocol described below. Syntactically, Make-GPK takes as input param and a set $S \subseteq \mathcal{PK}'$; its output implicitly defines a domain $\mathcal{GPK}$ of possible group public keys. We also require a superset $\mathcal{GPK}' \supseteq \mathcal{GPK}$ to be specified, such that membership to $\mathcal{GPK}'$ can be tested in polynomial time.
  The Make-GPK algorithm shall run in time linear in the number of public keys being aggregated; we also remark here that our definition forces Make-GPK to be *order-independent* i.e., the order in which the public keys to be aggregated are provided shall not matter.
- Make-GSK, the *group secret key construction algorithm*, is a deterministic algorithm used to combine a set of user public keys $S'$, along with a secret key/public key pair $(sk_u, pk_u)$, into a single group secret key $gsk_u$, suitable for use in the Anon-ID protocol described below.
  Make-GSK takes as input param, a set $S' \subseteq \mathcal{PK}'$ and a key pair $(sk_u, pk_u)$ satisfying $sk_u \rightleftharpoons pk_u$, and it shall run in time proportional to the size of $S'$. Its output implicitly defines a domain $\mathcal{GSK}$ of possible group secret keys.
  The Make-GPK and Make-GSK algorithms can be used to extend the $\rightleftharpoons$-relation to $\mathcal{GSK} \times \mathcal{GPK}$, as follows: A group secret key $gsk \doteq \mathsf{Make\text{-}GSK}(\mathsf{param}, S', (sk, pk))$ is in $\rightleftharpoons$-relation with a group public key $gpk \doteq \mathsf{Make\text{-}GPK}(\mathsf{param}, S)$ if and only if $S = S' \cup \{pk\}$. Observe that even in the case that the $\rightleftharpoons$-relation is one-to-one over $\mathcal{SK} \times \mathcal{PK}$, it is usually many-to-one over $\mathcal{GSK} \times \mathcal{GPK}$, as more than one group secret key correspond to the same group public key.
- $\mathsf{Anon\text{-}ID} \doteq (\mathsf{Anon\text{-}ID}^{\mathsf{P}}, \mathsf{Anon\text{-}ID}^{\mathsf{V}})$, the *Anonymous Identification Protocol*, is an efficient two-party protocol, in which both $\mathsf{Anon\text{-}ID}^{\mathsf{P}}$ (the *prover*) and $\mathsf{Anon\text{-}ID}^{\mathsf{V}}$ (the *verifier*) get in input the system's parameters param and a group public key $gpk$ (corresponding to some set $S$ of user public keys i.e., $gpk \doteq \mathsf{Make\text{-}GPK}(\mathsf{param}, S)$); $\mathsf{Anon\text{-}ID}^{\mathsf{P}}$ is also given a group secret key $gsk$ as an additional input.
  Any execution of the Anon-ID protocol shall complete in time independent from the number of public keys that were aggregated when constructing $gpk$ and/or $gsk$; at the end of each protocol run, $\mathsf{Anon\text{-}ID}^{\mathsf{V}}$ outputs a 0/1-valued answer.

**Correctness.** For correctness, we require that any execution of the Anon-ID protocol in which the additional input to $\mathsf{Anon\text{-}ID}^{\mathsf{P}}$ is a group secret key $gsk$ $\rightleftharpoons$-related to the common input $gpk$, shall terminate with $\mathsf{Anon\text{-}ID}^{\mathsf{V}}$ outputting a 1 answer, with overwhelming probability.

| Honest user registration oracle $\mathcal{O}_{\mathsf{HReg}}$ | User corruption oracle $\mathcal{O}_{\mathsf{Cor}}$ |
|---|---|
| IN: $u \in \mathcal{U}$ | IN: $pk_u \in \mathcal{PK}'$ |
| RUN: 1. $(sk_u, pk_u) \overset{\mathrm{R}}{\leftarrow} \mathsf{Register}(\mathsf{param}, u)$<br>2. $\mathsf{DB.Store}(sk_u, pk_u)$ | RUN: 1. $sk_u \leftarrow \mathsf{DB.Lookup}(pk_u)$<br>/* $sk_u \leftarrow \bot$ if no match found */ |
| OUT: $pk_u$ | OUT: $sk_u$ |

| Transcript oracle $\mathcal{O}_{\mathsf{Scr}}$ |
|---|
| IN: $S' \subseteq \mathcal{PK}', pk_u \in \mathcal{PK}'$ |
| RUN: 1. $sk_u \leftarrow \mathsf{DB.Lookup}(pk_u)$<br>2. **if** $sk_u = \bot$<br>3. **then** $\pi \leftarrow \bot$<br>4. **else** $gpk \leftarrow \mathsf{Make\text{-}GPK}(\mathsf{param}, S' \cup \{pk_u\})$<br>5. $gsk \leftarrow \mathsf{Make\text{-}GSK}(\mathsf{param}, S', (sk_u, pk_u))$<br>6. $\pi \overset{\mathrm{R}}{\leftarrow} \mathsf{Anon\text{-}ID^P}(\mathsf{param}, gpk, gsk) \leftrightarrow \mathsf{Anon\text{-}ID^V}(\mathsf{param}, gpk)$ |
| OUT: $\pi$ |

**Fig. 1.** Oracles for the soundness attack game. DB denotes a database storing user secret key/public key pairs, indexed by public key.

### 3.2 Soundness

**The Attack Game.** We formalize the soundness guarantees that we require from an *Ad hoc* Anonymous Identification scheme in terms of a game being played between an honest dealer and an adversary $\mathcal{A}$. In this game, the adversary is allowed to interact with three oracles $\mathcal{O}_{\mathsf{HReg}}$ (the *honest user registration oracle*), $\mathcal{O}_{\mathsf{Cor}}$ (the *user corruption oracle*), and $\mathcal{O}_{\mathsf{Scr}}$ (the *transcript oracle*) (see Fig. 1).

The game begins with the honest dealer running the Setup algorithm for the security parameter $1^\lambda$, and handing the resulting global parameters param to the adversary. Then, $\mathcal{A}$ arbitrarily interleaves queries to the three oracles, according to any adaptive strategy she wishes: eventually, she outputs a *target group* $S^* \subseteq \mathcal{PK}'$. At this point, $\mathcal{A}$ starts executing, in the role of the prover, a run of the Anon-ID protocol with the honest dealer, on common inputs param and $gpk^* \doteq \mathsf{Make\text{-}GPK}(\mathsf{param}, S^*)$. Notice that during such interaction, the adversary is still allowed to query the three oracles $\mathcal{O}_{\mathsf{Reg}}, \mathcal{O}_{\mathsf{Scr}}$ and $\mathcal{O}_{\mathsf{Cor}}$. Let $\tilde{\pi}$ be the transcript resulting from such run of the Anon-ID protocol. $\mathcal{A}$ wins the game if the following conditions hold:

1. for all $pk^* \in \mathcal{S}^*$, there is an entry indexed by $pk^*$ in the SK-DB Database, **and**
2. $\tilde{\pi}$ is a valid transcript i.e., the run completed with the honest dealer outputting 1, **and**
3. for all $pk^* \in \mathcal{S}^*$, $\mathcal{A}$ never queried $\mathcal{O}_{\mathsf{Cor}}$ on input $pk^*$;

Define $\mathsf{Succ}^{\mathsf{Imp}}_{\mathcal{A}}(\lambda)$ to be the probability that $\mathcal{A}$ wins the above game.

**Definition 3.** *An* Ad hoc *Anonymous Identification scheme is sound against passive chosen-group attacks if any adversary $\mathcal{A}$ has negligible advantage to win the above game:*

$$(\forall \lambda \in \mathbf{N})(\forall PPT \mathcal{A})[\mathsf{Succ}^{\mathit{Snd}}_{\mathcal{A}}(\lambda) \leq \nu(\lambda)]$$

| Challenge oracle $\mathcal{O}_{\mathsf{Ch}}$ |
|---|
| IN: $S' \subseteq \mathcal{PK}', (sk_0, pk_0), (sk_1, pk_1)$ |
| RUN: 1. $b^* \stackrel{R}{\leftarrow} \{0,1\}$ |
| 2. **if** $sk_0 \not\Leftarrow pk_0$ **or** $sk_1 \not\Leftarrow pk_1$ **then abort** |
| 3. $gpk \leftarrow \mathsf{Make\text{-}GSK}(\mathsf{param}, S' \cup \{pk_0, pk_1\})$ |
| 4. $gsk^* \leftarrow \mathsf{Make\text{-}GSK}(\mathsf{param}, S' \cup \{pk_{1-b^*}\}, (sk_{b^*}, pk_{b^*}))$ |
| 5. $\pi^* \stackrel{R}{\leftarrow} \mathsf{Anon\text{-}ID^P}(\mathsf{param}, gpk, gsk^*) \leftrightarrow \mathsf{Anon\text{-}ID^V}(\mathsf{param}, gpk)$ |
| OUT: $\pi^*$ |

**Fig. 2.** The oracle for the anonymity attack game.

**A Note on Active Security.** Our definition of soundness models an adversary that, in her attempt to fool an honest verifier into accepting a "fake" run of the Anon-ID protocol, can actively (and, in fact, adaptively) corrupt users, but can only passively eavesdrop the communication between honest provers and verifiers. One could, of course, define stronger notions of security by considering active, concurrent or even reset attacks, along the lines of previous work on Identification Schemes [26, 5]; however, we refrain from doing so, both to keep the presentation simpler, and because the main application of our *Ad hoc* Anonymous Identification schemes is to obtain new ring and group signatures scheme by means of the Fiat-Shamir Heuristic (see Section 6.3), for which security against a passive adversary suffices.

### 3.3 Anonymity

**The Attack Game.** We formalize the anonymity guarantees that we require from an *Ad hoc* Anonymous Identification scheme in terms of a game being played between an honest dealer and an adversary $\mathcal{A}$. In this game, the adversary is allowed to interact only once with a "challenge" oracle $\mathcal{O}_{\mathsf{Ch}}$, described in Fig. 2.

The game begins with the honest dealer running the Setup algorithm for the security parameter $1^\lambda$, and handing the resulting global parameters param to the adversary. Then, the adversary $\mathcal{A}$ creates as many user secret key/public key pairs as she wishes, and experiments with the Make-GPK, Make-GSK, $\mathsf{Anon\text{-}ID^P}$ and $\mathsf{Anon\text{-}ID^V}$ algorithms as long as she deems necessary; eventually, she queries the $\mathcal{O}_{\mathsf{Ch}}$ oracle, getting back a "challenge" transcript $\pi^*$. The adversary then continues experimenting with the algorithms of the system, trying to infer the random bit $b^*$ used by the oracle $\mathcal{O}_{\mathsf{Ch}}$ to construct the challenge $\pi^*$; finally, $\mathcal{A}$ outputs a single bit $\tilde{b}$, her best guess to the "challenge" bit $b^*$.

Define $\mathsf{Succ}_{\mathcal{A}}^{\mathsf{Anon}}(\lambda)$ to be the probability that the bit $\tilde{b}$ output by $\mathcal{A}$ at the end of the above game is equal to the random bit $b^*$ used by the $\mathcal{O}_{\mathsf{Ch}}$ oracle.

**Definition 4.** *An* Ad hoc *Anonymous Identification scheme is* fully anonymizing *if any probabilistic, polynomial-time adversary $\mathcal{A}$ has success probability at most negligibly greater than one half:*

$$(\forall \lambda \in \mathbf{N})(\forall PPT \mathcal{A})\left[\left|\mathsf{Succ}_{\mathcal{A}}^{\mathsf{Anon}}(\lambda) - \frac{1}{2}\right| \leq \nu(\lambda)\right]$$

### 3.4 Extensions

**Identity Escrow.** In some scenarios, complete anonymity might create more security concerns than what it actually solves. Instead, some degree of "limited anonymity", not hindering user accountability, would be preferable. In our context, this can be achieved with the help of a trusted *Identity Escrow Authority*, or IEA (also called *Anonymity Revocation Manager* elsewhere [15]), endowed with the capability of "reading" the identity of the prover "between the lines" of any transcript produced by a run of the Anon-ID protocol.

To enable such escrow capability, the definition of *Ad hoc* Anonymous Identification scheme from Section 3.1 is modified as follows:

- The Setup algorithm is run by the IEA, and it additionally outputs an identity escrow key $sk_{\mathsf{IE}}$ (from some domain $\mathcal{SK}_{\mathsf{IE}}$), which the IEA keeps for himself.
- Register is replaced by an efficient two-party protocol ($\mathsf{Register}^{\mathsf{user}}$, $\mathsf{Register}^{\mathsf{IEA}}$), meant to be run between the prospective user and the IEA, at the end of which the IEA learns the user's newly generated public key $pk_u$ (possibly along with some other information $aux_u$ about $u$ that the IEA stores in a public registry database DB), but he doesn't learn anything about the corresponding secret key $sk_u$.
- An additional (deterministic) Extract algorithm is defined, which takes as input a transcript $\pi$ (for the Anon-ID protocol), along with the Identity Escrow secret key $sk_{\mathsf{IE}}$ and the registry database DB, and returns a public key $pk \in \mathcal{PK}'$ or one of the special symbols $\perp$ and ?. Intuitively, the algorithm should be able to recover the identity of the user who participated as the prover in the run of the Anon-ID protocol that produced $\pi$ as transcript; the symbol $\perp$ should be output when $\pi$ is ill-formed (e.g., when $\pi$ comes from a ZK simulator), whereas ? indicates failure to trace the correct identity.

Our definitions of the security properties of the system have to be adjusted, since we now have an additional functionality that the adversary may try to attack; moreover, the presence of the IEA may open new attack possibilities to the adversary.

The security requirements for the new Extract algorithm are formalized by augmenting the attack scenario defining the soundness property (Section 3.2). In this new, combined game, the adversary initially gets the IEA's secret key $sk_{\mathsf{IE}}$, along with the public parameters param of the system. Then, the game proceeds as described in Section 3.2, except that we loosen the conditions under which the adversary is considered to win the game, substituting the last two caveats with the following:

2′. $\tilde{\pi}$ is a valid transcript i.e., $\mathsf{Extract}(\tilde{\pi}, sk_{\mathsf{IE}}, \mathsf{DB}) \neq \perp$ **and**
3′. for all $pk^* \in \mathcal{S}^*$, either $\mathsf{Extract}(\tilde{\pi}, sk_{\mathsf{IE}}, \mathsf{DB}) \neq pk^*$, **or** $\mathcal{A}$ never queried $\mathcal{O}_{\mathsf{Cor}}$ on input $pk^*$;

As for the anonymity property, the definition from Section 3.3 is changed in that the adversary is now given access to two more oracles (beside the challenge oracle $\mathcal{O}_{\mathsf{Ch}}$): a *corrupted-user registration oracle* $\mathcal{O}_{\mathsf{CReg}}() \doteq$

$\mathsf{Register}^{\mathsf{IEA}}(sk_{\mathsf{IE}}, \mathsf{param}, \mathsf{DB})$, and a *user identity extraction oracle* $\mathcal{O}_{\mathsf{Xtr}}(\cdot) \doteq \mathsf{Extract}(\cdot, sk_{\mathsf{IE}}, \mathsf{DB})$. The adversary wins the game if she successfully guesses the random bit chosen by the challenge oracle $\mathcal{O}_{\mathsf{Ch}}$, without ever submitting the challenge transcript $\pi^*$ to the extraction oracle $\mathcal{O}_{\mathsf{Xtr}}$.

**Supporting Multiple Large *Ad Hoc* Groups.** In many applications where *Ad hoc* Anonymous Identification schemes could be useful, new *ad hoc* groups are often created as supersets of existing ones: for example, if *ad hoc* groups are used to enforce access control, new users may be added to the group of principals authorized to access a given resource. In such cases, the ability to "augment" a group public key with the a new user's public key can be very handy, especially if coupled with algorithms to efficiently create the corresponding group secret key for the new user, and to update the group secret keys for the existing users. Our model can be easily amended to capture this *incremental* functionality; we refer the reader to the full version of this paper [25] for the details.

## 4 Generic Construction

In this section, we will establish the fact that the existence of accumulators with one way domain implies the existence of *Ad hoc* Anonymous Identification schemes. Below we describe how the algorithms (Setup, Register, Make-GPK, Make-GSK, $\mathsf{Anon\text{-}ID}^{\mathsf{P}}$, $\mathsf{Anon\text{-}ID}^{\mathsf{V}}$) can be implemented given an accumulator with one-way domain $(\{F_\lambda\}_{\lambda\in\mathbb{N}}, \{X_\lambda\}_{\lambda\in\mathbb{N}}, \{Z_\lambda\}_{\lambda\in\mathbb{N}}, \{R_\lambda\}_{\lambda\in\mathbb{N}},)$.

- Setup executes the accumulator generation algorithm $G$ on $1^\lambda$ to obtain $f \in F_\lambda$. Then it samples $U_f$ to obtain $u \in_R U_f$. Setup terminates by setting $\mathsf{param} := (\lambda, u, f, D, W)$, where $D$ and $W$ are polynomial-time algorithms respectively to decide and to sample the relation $R_\lambda$.
- Register first samples a pair $(x, z) \in X_\lambda \times Z_\lambda$ such that $(x, z) \in R_\lambda$ using the sampling algorithm $W$ of the relation $R_\lambda$ on input $1^\lambda$. Then, Register outputs $sk \doteq z$ (the user secret key) and $pk \doteq x$ (the user public key). Observe that $\mathcal{SK}' = \mathcal{SK} \doteq Z_\lambda$, $\mathcal{PK}' = X_f^{\text{ext}}$ and $\mathcal{PK} \doteq X_\lambda$.
- Make-GPK operates as follows: given a set of user public keys $S = \{x_1, \ldots, x_t\}$ and the parameters $(\lambda, u, f, D)$, it sets the group public key of $S$ to be the (unique) accumulated value of $S$ over $u$ i.e., $gpk_S \doteq f(u, S)$. Note that thanks to the quasi-commutativity property of $f$, Make-GPK is indeed order-independent.
- Make-GSK operates as follows: given the set of user public keys $S' \doteq \{x_1, \ldots, x_t\}$, a user secret key/public key pair $(z, x)$ and the system parameters $\mathsf{param} = (\lambda, u, f, D, W)$, it first computes the accumulated value $w \doteq f(u, S')$, and then sets the group secret key $gsk$ to be the tuple $(x, z, w)$ (where $S \doteq S' \cup \{x\}$). Observe that $w$ is a witness for $x$ in $f(u, S')$, and that $\mathcal{GSK} \doteq X_\lambda \times Z_\lambda \times U_f$ and $\mathcal{GPK} \doteq U_f$.
- $\mathsf{Anon\text{-}ID}^{\mathsf{P}}$ and $\mathsf{Anon\text{-}ID}^{\mathsf{V}}$ are obtained generically as the $\Sigma$-protocol corresponding to the following **NP**-relation $\mathcal{R}_{\mathsf{param}} \subset \mathcal{GPK} \times \mathcal{GSK}$:

$$\mathcal{R}_{\mathsf{param}} \doteq \{(v, (x, z, w)) \mid ((x, z) \in R_\lambda) \wedge (f(w, x) = v)\}$$

It is easy to see that the above relation is polynomial-time verifiable: indeed, given $v$ and $(x, z, w)$, one can check in time polynomial in $|v|$ whether $(x, z) \in R_\lambda$ (by verifying that $D(x, z) = 1$), and whether $w$ is indeed a witness for $x$ in $v$ (by verifying that $f(w, x) = v$). Thus, by Theorem 1, we can construct a $\Sigma$-protocol $(P, V)$ for the **NP**-relation $\mathcal{R}_{\mathsf{param}}$. In the resulting protocol, the common input to the prover and the verifier is the accumulated value $v$ (i.e. a group public key) and the additional input to the prover is a tuple of the form $(x, z, w)$ (i.e., a group secret key). Hence, the protocol $(P, V)$ meets the specification of the Anon-ID protocol.

As for the correctness of the above construction, observe that relation $\mathcal{R}_{\mathsf{param}}$ is essentially equivalent to the $\rightleftharpoons$ relation. Consequently, a prover holding a group secret key $gsk \doteq (x, z, w)$ $\rightleftharpoons$-related to the group public key $gpk \doteq v$ given as input to the verifier, possesses a tuple belonging to the relation $\mathcal{R}_{\mathsf{param}}$, so that the execution of the Anon-ID protocol will terminate with the verifier outputting 1, with overwhelming probability.

**Soundness.** Intuitively, the soundness of the above generic construction stems from the following considerations. The Special Honest-Verifier Zero-Knowledge property of the $\Sigma$-protocol Anon-ID guarantees that the Transcript Oracle doesn't leak any information to the adversary that she could not compute herself. By the Special Soundness property, in order to make the honest dealer accept (with non-negligible probability) a run of the Anon-ID protocol in which the group public key $gpk \doteq v$ consists solely of the aggregation of public keys of non-corrupted users, $\mathcal{A}$ should posses a tuple $gsk \doteq (x, z, w)$ such that $(x, z) \in R_\lambda$ and $w$ is a witness of $x$ in $v$. Now, the collision resistance of the accumulator implies that the user public key $x$ must indeed have been accumulated within $v$, which means (by the third caveat of the soundness attack game in Section 3.2) that $x$ belongs to a non-corrupted user. Hence, the adversary didn't obtain the pre-image $z$ via the user corruption oracle, which implies that $\mathcal{A}$ was able to find it by herself, contradicting the one-wayness of the accumulator's domain.

The above intuition could be turned into a rigorous reduction argument: we refer the reader to the full version [25] for a formal proof.

**Anonymity.** In attacking the anonymity of the proposed scheme, the adversary basically chooses a group public key $gpk \doteq v$ and two group secret keys $gsk_1 \doteq (x_1, z_1, w_1)$ and $gsk_2 \doteq (x_2, z_2, w_2)$, both $\rightleftharpoons$-related to $gpk$. To subvert anonymity, the adversary should then be able (cfr. Section 3.3) to tell whether $gsk_1$ or $gsk_2$ was used in producing the (honest) "challenge" transcript. Since in the generic construction above the Anon-ID protocol is implemented as a $\Sigma$-protocol, this would mean that the adversary is able to tell which "witness" ($gsk_1$ or $gsk_2$) was used by the prover to show that $v$ belongs to the **NP**-language $\mathcal{L}_{\mathsf{param}}$ associated to the **NP**-relation $\mathcal{R}_{\mathsf{param}}$. In other words, a successful adversary would break the Witness Indistinguishability of the Anon-ID protocol, which contradicts the fact that Anon-ID enjoys Special Honest-Verifier Zero-Knowledge.

The reader is referred to [25] for a formalization of the above argument.

### 4.1 Adding ID Escrow

The generic construction described above can be extended to deal with Identity Escrow as follows. During the initialization, the Setup algorithm additionally runs the key generation algorithm $\mathcal{K}$ of some CCA2-secure encryption scheme $(\mathcal{K}, \mathcal{E}, \mathcal{D})$. The resulting public key $pk_{\mathsf{IE}}$ is included in the system parameters param, and the secret key $sk_{\mathsf{IE}}$ is given to the Identity Escrow Authority (IEA).

As for the user registration phase, each new user, after choosing his user secret key/public key pair $(sk, pk) \doteq (z, x)$, registers his public key with the IEA, which simply stores his identity and public key in a publicly-available database DB.

The Anon-ID protocol is also changed to be the $\Sigma$-protocol corresponding to the following **NP**-relation $\mathcal{R}^{\mathsf{IE}}_{\mathsf{param}}$:

$$\mathcal{R}^{\mathsf{IE}}_{\mathsf{param}} \doteq \left\{ \big( (v, \psi), (x, z, w) \big) \mid \big( (x, z) \in R_\lambda \big) \wedge \big( f(w, x) = v \big) \wedge \big( \psi \text{ decrypts to } x \big) \right\}$$

In other words, the prover now additionally encrypts his public key $x$ under the IEA's public key $pk_{\mathsf{IE}}$, and proves to the verifier that he did so correctly.

Finally, the Extract algorithm, on input a transcript $\pi$, recovers the ciphertext $\psi$ from $\pi$ and decrypts $\psi$, thus obtaining the identity of the user that played the role of the prover.

It is not hard to check that the above changes do not affect the soundness and anonymity properties of the generic construction: in particular, the CCA2-security of the encryption scheme (which is needed since a malicious party could trick the IEA into acting as a decryption oracle) guarantees that honest transcripts cannot be modified so as to alter the prover identity hidden under the ciphertext $\psi$. See [25] for a security analysis of the extended scheme.

## 5 Efficient Implementation

### 5.1 Construction of an Accumulator with One-Way Domain

An efficient construction of a collision-resistant accumulator was presented in [15], based on earlier work by [4] and [7]. Based on this construction, we present an efficient implementation of an accumulator with one-way domain.

For $\lambda \in \mathbb{N}$, the family $F_\lambda$ consists of the exponentiation functions modulo $\lambda$-bit rigid integers:

$$f : (\mathbb{Z}_n^*)^2 \times \mathbb{Z}_{n/4} \to (\mathbb{Z}_n^*)^2$$
$$f : (u, x) \mapsto u^x \bmod n$$

where $n \in \mathsf{Rig}_\lambda$ and $(\mathbb{Z}_n^*)^2$ denotes the set of quadratic residues modulo $n$.

The accumulator domain $\{X_\lambda\}_{\lambda \in \mathbb{N}}$ is defined by:

$$X_\lambda \doteq \left\{ e \text{ prime} \mid \left( \frac{e-1}{2} \in \mathsf{RSA}_\ell \right) \wedge \left( e \in S(2^\ell, 2^\mu) \right) \right\}$$

where $S(2^\ell, 2^\mu)$ is the integer range $(2^\ell - 2^\mu, 2^\ell + 2^\mu)$ that is embedded within $(0, 2^\lambda)$ with $\lambda - 2 > \ell$ and $\ell/2 > \mu + 1$. The pre-image domain $\{Z_\lambda\}_{\lambda \in \mathbb{N}}$ and the one-way relation $\{R_\lambda\}_{\lambda \in \mathbb{N}}$ are defined as follows:

$$Z_\lambda \doteq \{(e_1, e_2) \mid e_1, e_2 \text{ are distinct } \ell/2\text{-bit primes and } e_2 \in S(2^{\ell/2}, 2^\mu)\}$$
$$R_\lambda \doteq \{(x, (e_1, e_2)) \in X_\lambda \times Z_\lambda \mid (x = 2e_1e_2 + 1)\}$$

The collision resistance of the above construction can be based on the Strong RSA Assumption, as showed in [15]. Regarding the added one-wayness of the domain, assuming the hardness of factoring RSA integers, it is easy to see that the **NP**-relation $R_\lambda$ satisfies our one-wayness requirement (cfr. Section 2.2): hence, the above construction yields a secure accumulator with one-way domain.

### 5.2 Efficient Proof of Witnesses for the Accumulator

The generic construction described in Section 4 derives algorithms $\mathsf{Anon\text{-}ID}^\mathsf{P}$ and $\mathsf{Anon\text{-}ID}^\mathsf{V}$ from the $\Sigma$-protocol corresponding to some **NP**-relation $\mathcal{R}_{\mathsf{param}}$: for our RSA-based accumulator with one-way domain, the relation is defined as:

$$\mathcal{R}^{RSA}_{\mathsf{param}} \doteq \{(v, (x, (e_1, e_2), w)) \mid (w^x \equiv v \bmod n) \wedge (x \in S(2^\ell, 2^\mu)) \\ \wedge (x - 1 = 2e_1e_2) \wedge (e_2 \in S(2^{\ell/2}, 2^\mu))\}$$

However, the protocol generically obtained in virtue of Theorem 1, though polynomial time, is not efficient enough to be useful in practice; thus, below we describe how a practical $\Sigma$-protocol for relation $\mathcal{R}^{RSA}_{\mathsf{param}}$ could be constructed, exploiting the framework of discrete-log relation sets [31], which provides a simple method to construct complex proofs of knowledge over groups of unknown order. A discrete-log relation set $R$ is a set of vectors of length $m$ defined over $\mathbb{Z} \cup \{\alpha_1, \ldots, \alpha_r\}$ (where the $\alpha_j$'s are called the free variables of the relation) and involves a sequence of base elements $A_1, \ldots, A_m \in (\mathbb{Z}_n^*)^2$. For any vector $\langle a_1^i, \ldots, a_m^i \rangle$ the corresponding relation is defined as $\prod_{j=1}^m A_i^{a_j^i} = 1$. The conjunction of all the relations is denoted as $R(\alpha_1, \ldots, \alpha_r)$. In [31], an efficient $\Sigma$-protocol is presented for any discrete-log relation set $R$, by which the prover can prove of knowledge of a sequence of witnesses $x_1, \ldots, x_r$, with $x_i \in S(2^{\ell_i}, 2^{\mu_i})$ that satisfy $R(x_1, \ldots, x_r) \wedge \left( \wedge_{i=1}^r (x_i \in S(2^{\ell_i}, 2^{\epsilon(\mu_i + k) + 2})\right)$, where $\epsilon > 1, k \in \mathbb{N}$ are security parameters. Note that the tightness of the integer ranges can be increased by employing the range proofs of [10], nevertheless the tightness achieved above is sufficient for our purposes, and incurs a lower overhead.

In order to prove the relation $\mathcal{R}^{RSA}_{\mathsf{param}}$, we assume that the public parameters $\mathsf{param}$ include the elements $g, h, y, t, s \in (\mathbb{Z}_n^*)^2$ with unknown relative discrete-logarithms. In order to construct the proof, the prover provides a sequence of public values $T_1, T_2, T_3, T_4, T_5$ such that $T_1 = g^r, T_2 = h^r g^x, T_3 = s^r g^{e_2}, T_4 = wy^r, T_5 = t^r g^{2e_1}$, where $r \overset{\mathrm{R}}{\leftarrow} [0, \lfloor n/4 \rfloor - 1]$.

The proof is constructed as a discrete-log relation set that corresponds to the equations $T_1 = g^r$, $T_2 = h^r g^x$, $(T_1)^x = y^{a_1}$, $(T_1)^{e_2} = y^{a_2}$, $T_3 = s^r g^{e_2}$ $(T_4)^x = vy^{a_1}$, $(T_5)^{e_2} g = t^{a_2} g^x$, for the free variables $r, x, e_2, a_1, a_2$ such that $x \in S(2^\ell, 2^\mu), e_2 \in S(2^{\ell/2}, 2^\mu)$, $a_1 = rx$ and $a_2 = re_2$. The matrix of the discrete-log relation set is shown below:

$$\left[\begin{array}{rcccccccccccc}
 & g & h & y & t & s & v & T_1^{-1} & T_2^{-1} & T_3^{-1} & T_4^{-1} & T_5^{-1} & g^{-1} \\
T_1 = g^r : & r & 0 & 0 & 0 & 0 & 0 & 1 & 0 & 0 & 0 & 0 & 0 \\
T_2 = h^r g^x : & x & r & 0 & 0 & 0 & 0 & 0 & 1 & 0 & 0 & 0 & 0 \\
(T_1)^x = g^{a_1} : & a_1 & 0 & 0 & 0 & 0 & 0 & x & 0 & 0 & 0 & 0 & 0 \\
T_3 = s^r g^{e_2} : & e_2 & 0 & 0 & 0 & r & 0 & 0 & 0 & 1 & 0 & 0 & 0 \\
(T_1)^{e_2} = g^{a_2} : & a_2 & 0 & 0 & 0 & 0 & 0 & e_2 & 0 & 0 & 0 & 0 & 0 \\
(T_4)^x = vy^{a_1} : & 0 & 0 & a_1 & 0 & 0 & 1 & 0 & 0 & 0 & x & 0 & 0 \\
(T_5)^{e_2} g = t^{a_2} g^x : & x & 0 & 0 & a_2 & 0 & 0 & 0 & 0 & 0 & 0 & e_2 & 1
\end{array}\right]$$

Observe that a proof of the above discrete-log relation set ensures that (i) the prover knows a witness $w$ for some value $x$ in the ad-hoc group accumulated value $v$, and (ii) for the same $x$, the value $x - 1$ can be split by the prover into two integers one of which belongs to $S(2^{\ell/2}, 2^\mu)$. This latter range-property guarantees the non-triviality of the splitting i.e., that the prover knows a non-trivial factor of $x - 1$ (i.e., different than $-1, 1, 2$). Note that this will require that the parameters $\ell, \mu, \epsilon, k$ should be selected such that $\ell/2 > \epsilon(\mu + k) + 2$.

### 5.3 ID Escrow

In Section 4.1, we discussed a generic transformation to add Identity Escrow to an *Ad hoc* Anonymous Identification scheme. Most of the required changes do not affect the system's efficiency, except for the need to resort to a generic derivation of the Anon-ID protocol.

This performance penalty is not unavoidable, however: in fact, escrow capabilities can be directly supported by the $\Sigma$-protocol for Anonymous Identification described in Section 5.2. using protocols for verifiable encryption and decryption of discrete logarithms from [17].

With notation as in Section 5.2, the Anon-ID protocol is augmented as follows: after sending the commitment $T_2$ to the verifier, the prover verifiably encrypts an opening of $T_2$ (namely, $x$ and $r$) under the IEA public key. By checking that the encryption was carried out correctly, the verifier can be assured that, should the need arise, the IEA would be able to identify the prover by decrypting such opening, which would yield the prover's public key $x$. Moreover, by using verifiable decryption in the Extract algorithm, we can prevent the IEA from untruthfully opening the identity of the prover for a given transcript, or falsely claiming that the prover's identity cannot be properly recovered.

Alternatively, if only honest users are assumed to have access to the Escrow functionality (so that malicious parties cannot exploit the IEA as a "decryption oracle"), then a more efficient solution is possible, by having the IEA knowing the value $\log_g(h)$ in the proof of knowledge from Section 5. Then, given a transcript of the protocol (which includes the values $T_1, T_2, T_3, T_4, T_5$) the IEA can recover the value $g^x = T_2 T_1^{-\log_g(h)}$, from which the prover's identity can be recovered by comparing $g^x$ to the public keys published in the public DB database.

# 6 Applications

## 6.1 *Ad Hoc* Identification Schemes

This is the most direct application. Imagine a large universe of users, where each user has a public certificate, but otherwise there is no central authority in the system. Now, some party "from the street" has a resources which he is willing to share with some subset of the users. For example, an Internet provider $P$ may want to enable internet access to all its subscribers. However, privacy considerations may lead a user to refuse to positively identify himself; in fact, this is not strictly necessary, as long as he proves he belongs to the group of current subscribers. Our ad-hoc identification schemes are ideally suited for this application, bringing several very convenient feautures. First, $P$ can simply take all the public keys of the users (call this set $S$) and combine them into one short group public key $gpk_S$. Notice, this initial setup is the only operation $P$ performs which requires time proportional to the group size. As for each user $u \in S$, once again he will use his secret key and the public keys of other user to prepare one short group secret key $gsk_u$. After that, all identifications that $u$ makes to $P$ require computation and communication independent of the size of the group. Now, another provider $P'$ can do the same for a totally different subgroup, and so on, allowing truly ad-hoc groups with no trusted authority needed in the system. Additionally, with incremental *Ad hoc* Anonymous Identification schemes (defined in the full version of this paper [25]), one can preserve efficiency even when the ad-hoc group is built gradually, as each new member addition only requires constant computation by $P$ and by every pre-existing user in the system.

## 6.2 Constant Size Ring Signatures

This is one of our main applications, since it dramatically improves the efficiency of all known ring signature schemes (e.g. [34, 12, 9]). Recall, in a ring signature scheme there again is a universe of registered users, but no trusted authority. Any user $u$ can then form a ring $S$, and sign a message $m$ in such a way that any verifier (who knows $S$) can confidently conclude that "the message $m$ was signed by some member $u$ of $S$", but gets no information about $u$ beyond $u \in S$. Previous papers on the subject suggested that linear dependence of the ring signature size on the size of the ring $S$ was inevitable, since the group is ad-hoc, so the verifier needs to know at least the description of the ring. While the latter is true, in practical situations the ring often stays the same for a long period of time (in fact, there could be many "popular" rings that are used very often by various members of the ring), or has an implicit short decryption (e.g., the ring of public keys of all members of the President's Cabinet). Thus, we feel that the right measure of "signature size" in this situation is that of an "actual signature"—the string one needs in addition to the group description. Indeed, when the ring stays the same for a long time or has a short description, this actual signature is all that the verifier needs in order to verify its correctness. With this in mind, there is no reason why the signature size must be linear in the size of the ring.

In fact, our result shows that it does not have to be. Specifically, by applying the Fiat-Shamir heuristics to our ad-hoc identification scheme, we immediately get ring signatures of constant size. Moreover, our ring signatures enjoy several desirable features not generally required by ring signatures (even those of constant size). For example, both the signer and the verifier need to perform a one-time computation proportional to the size of the ring, and get some constant-size information ($gsk_S$ and $gpk_S$, respectively) which allows them to produce/verify many subsequent signatures in constant time.

### 6.3 *Ad Hoc* ID Escrow and Group Signatures

As mentioned in Section 3.4, in some situations complete anonymity might not be desirable. In this case, one wishes to introduce a trusted *Identity Escrow Authority* (IEA), who can reveal the true identity of the user given the transcript of the identification procedure (presumably, when some "anonymity abuse" happens). Such schemes are called ID Escrow schemes [32] and have traditionally been considered for fixed groups. ID Escrow schemes are duals of group signature schemes [19, 1], which again maintain a single group of signers, and where a similar concern is an issue when signing a document anonymously. As argued in Section 4.1 and Section 5.3, our *Ad hoc* Anonymous Identification schemes and the corresponding signer-ambiguous signature schemes can efficiently support identity escrow capabilities. As a result, we get an ID Escrow and a group signature scheme with the following nice features. (For concreteness, we concentrate on group signatures below.) First, just like in current state-of-the-art group signature schemes, the signature is of constant size. Second, a user can join any group by simply telling the group manager about its public key: no expensive interactive protocols, where the user will "get a special certificate" have to be run. Thus, the group manager only needs to decide if the user can join the group, and periodically certify the "current" public key of the group. In other words, we can imagine a simple bulletin board, where the group manager periodically publishes the (certified) group public key the group, the description of the group, and potentially the history of how the public key evolved (which is very handy for incremental *Ad hoc* Anonymous Identification schemes; see [25]). From this information, each member of the group can figure out its group secret key and sign arbitrary many messages efficiently. (Of course, when signing a message the signer should also include the certified version of the current group key, so that "old" signatures do not get invalidated when the group key changes.)

## References


[1] G. Ateniese, J. Camenisch, M. Joye, and G. Tsudik. A practical and provably secure coalition-resistant group signature scheme. In *Advances in Cryptology—CRYPTO '00*, volume 1880 of *LNCS*, pages 255–270. Springer, 2000.

[2] G. Ateniese and B. de Medeiros. Efficient group signatures without trapdoors. In *Advances in Cryptology—ASIACRYPT '03*, volume 2894 of *LNCS*, pages 246–268. Springer, 2002.

[3] G. Ateniese and G. Tsudik. Some open issues and new directions in group signatures. In *Financial Cryptography (FC '99)*, volume 1648 of *LNCS*, pages 196–211. Springer, 1999.

[4] N. Barić and B. Pfitzmann. Collision-free accumulators and fail-stop signature schemes without trees. In *Advances in Cryptology—EUROCRYPT '97*, volume 1233 of *LNCS*, pages 480–494. Springer, 1997.

[5] M. Bellare, M. Fischlin, S. Goldwasser, and S. Micali. Identification protocols secure against reset attacks. In *Advances in Cryptology—EUROCRYPT '01*, volume 2045 of *LNCS*, pages 495–511. Springer, 2001.

[6] M. Bellare, D. Micciancio, and B. Warinschi. Foundations of group signatures: Formal definitions, simplified requirements, and a construction based on general assumptions. In *Advances in Cryptology—EUROCRYPT '03*, volume 2656, pages 630–648. Springer, 2003.

[7] J. Benaloh and M. de Mare. One-way accumulators: a decentralized alternative to digital signatures. In *Advances in Cryptology—EUROCRYPT '93*, volume 765 of *LNCS*, pages 274–285. Springer, 1993.

[8] D. Boneh and M. Franklin. Anonymous authentication with subset queries. In *ACM Conference on Computer and Communications Security—CCS '99*, pages 113–119. ACM Press, 1999.

[9] D. Boneh, C. Gentry, B. Lynn, and H. Shacham. Aggregate and verifiably encrypted signatures from bilinear maps. In *Advances in Cryptology—EUROCRYPT '03*, volume 2656 of *LNCS*, pages 416–432. Springer, 2003.

[10] F. Boudot. Efficient proofs that a commited number lies in an interval. In *Advances in Cryptology—EUROCRYPT '00*, volume 1807 of *LNCS*, pages 431–444. Springer, 2000.

[11] E. Bresson and J. Stern. Efficient revocation in group signatures. In *Public Key Cryptography (PKC '01)*, volume 1992 of *LNCS*, pages 190–206. Springer, 2001.

[12] E. Bresson, J. Stern, and M. Szydlo. Threshold ring signatures and applications to ad-hoc groups. In *CRYPTO 2002*, volume 2442 of *LNCS*, pages 465–480. Springer, 2002.

[13] J. Camenisch. Efficient and generalized group signatures. In *Advances in Cryptology—EUROCRYPT '97*, volume 1233 of *LNCS*, pages 465–479. Springer, 1997.

[14] J. Camenisch and A. Lysyanskaya. An identity escrow scheme with appointed verifiers. In *Advances in Cryptology—CRYPTO '01*, volume 2139 of *LNCS*, pages 388–407. Springer, 2001.

[15] J. Camenisch and A. Lysyanskaya. Dynamic accumulators and applications to efficient revocation of anonymous credentials. In *Advances in Cryptology—CRYPTO '02*, volume 2442 of *LNCS*, pages 61–76. Springer, 2002.

[16] J. Camenisch and M. Michels. A group signature scheme with improved efficiency. In *Advances in Cryptology—ASIACRYPT '98*, volume 1514 of *LNCS*, pages 160–174. Springer, 1998.

[17] J. Camenisch and V. Shoup. Practical verifiable encryption and decryption of discrete logarithms. Full length version of extended abstract in CRYPTO'03, available at: `http://shoup.net/papers/`, 2003.

[18] J. Camenisch and M. Stadler. Efficient group signature schemes for large groups. In *Advances in Cryptology—CRYPTO '97*, volume 1294 of *LNCS*, pages 410–424. Springer, 1997.

[19] D. Chaum and E. van Heyst. Group signatures. In *Advances in Cryptology—EUROCRYPT '91*, volume 547 of *LNCS*, pages 257–265. Springer, 1991.

[20] L. Chen and T. P. Pedersen. New group signature schemes (extended abstract). In *Advances in Cryptology—EUROCRYPT 94*, volume 950 of *LNCS*, pages 171–181. Springer, 1994.

[21] R. Cramer. *Modular Design of Secure yet Practical Cryptographic Protocols.* PhD thesis, CWI and University of Amsterdam, November 1996.

[22] R. Cramer, I. B. Damgård, and B. Schoenmakers. Proofs of partial knowledge and simplified design of witness hiding protocols. In *Advances in Cryptology—CRYPTO '94*, volume 839 of *LNCS*, pages 174–187. Springer, 1994.

[23] A. De Santis, G. Di Crescenzo, and G. Persiano. Communication-efficient anonymous group identification. In *5th ACM Conference on Computer and Communications Security*, pages 73–82. ACM Press, 1998.

[24] A. De Santis, G. Di Crescenzo, G. Persiano, and M. Yung. On monotone formula closure of SZK. In *35th Annual Symposium on Foundations of Computer Science*, pages 454–465. IEEE Computer Society Press, 1994.

[25] Y. Dodis, A. Kiayias, A. Nicolosi, and V. Shoup. Anonymous identification in *Ad Hoc* groups. Full version: `http://www.cs.nyu.edu/~nicolosi/papers/`, 2004.

[26] U. Feige, A. Fiat, and A. Shamir. Zero-knowledge proof of identity. *Journal of Cryptology*, 1(2):77–94, 1988.

[27] U. Feige and A. Shamir. Zero knowledge proofs of knowledge in two rounds. In *Advances in Cryptology—CRYPTO '89*, volume 435 of *LNCS*, pages 526–544. Springer, 1989.

[28] A. Fiat and A. Shamir. How to Prove Yourself: Practical Solutions to Identification and Signature Problems. In *Advances in Cryptology—CRYPTO '86*, volume 263 of *LNCS*, pages 186–194. Springer, 1986.

[29] O. Goldreich, S. Micali, and A. Wigderson. Proof that yield nothing bur their validity or all languages in np have zero-knowledge proof systems. *Journal of the ACM*, 38(3):691–729, 1991.

[30] S. Goldwasser, S. Micali, and C. Rackoff. The knowledge complexity of interactive proof systems. *SIAM Journal on computing*, 18(1):186–208, 1989.

[31] A. Kiayias, Y. Tsiounis, and M. Yung. Traceable signatures. In *Advances in Cryptology—EUROCRYPT 04*, LNCS. Springer, 2004.

[32] J. Kilian and E. Petrank. Identity escrow. In *Advances in Cryptology—CRYPTO '98*, volume 1462 of *LNCS*, pages 169–185. Springer, 1998.

[33] M. Naor. Deniable ring authentication. In *Advances in Cryptology—CRYPTO '02*, volume 2442 of *LNCS*, pages 481–498. Springer, 2002.

[34] R. Rivest, A. Shamir, and Y. Tauman. How to leak a secret. In *Advances in Cryptology—ASIACRYPT '01*, volume 2248 of *LNCS*, pages 552–565. Springer, 2001.

[35] G. Tsudik and S. Xu. Accumulating composites and improved group signing. In *Advances in Cryptology—ASIACRYPT '03*, volume 2894 of *LNCS*, pages 269–286. Springer, 2003.

# Author Index

# Lecture Notes in Computer Science

For information about Vols. 1–2898

please contact your bookseller or Springer-Verlag

Vol. 3027: C. Cachin, J. Camenisch (Eds.), Advances in Cryptology - EUROCRYPT 2004. XII, 628 pages. 2004.

Vol. 3025: G.A. Vouros, T. Panayiotopoulos (Eds.), Methods and Applications of Artificial Intelligence. XV, 546 pages. 2004. (Subseries LNAI).

Vol. 3015: C. Barakat, I. Pratt (Eds.), Passive and Active Network Measurement. XI, 300 pages. 2004.

Vol. 3011: J.-C. Régin, M. Rueher (Eds.), Integration of AI and OR Techniques in Constraint Programming for Combinatorial Optimization Problems. XI, 415 pages. 2004.

Vol. 3010: K.R. Apt, F. Fages, F. Rossi, P. Szeredi, J. Váncza (Eds.), Recent Advances in Constraints. VIII, 285 pages. 2004. (Subseries LNAI).

Vol. 3009: F. Bomarius, H. Iida (Eds.), Product Focused Software Process Improvement. XIV, 584 pages. 2004.

Vol. 3007: J.X. Yu, X. Lin, H. Lu, Y. Zhang (Eds.), Advanced Web Technologies and Applications. XXII, 936 pages. 2004.

Vol. 3006: M. Matsui, R. Zuccherato (Eds.), Selected Areas in Cryptography. XI, 361 pages. 2004.

Vol. 3005: G.R. Raidl, S. Cagnoni, J. Branke, D.W. Corne, R. Drechsler, Y. Jin, C.G. Johnson, P. Machado, E. Marchiori, F. Rothlauf, G.D. Smith, G. Squillero (Eds.), Applications of Evolutionary Computing. XVII, 562 pages. 2004.

Vol. 3004: J. Gottlieb, G.R. Raidl (Eds.), Evolutionary Computation in Combinatorial Optimization. X, 241 pages. 2004.

Vol. 3003: M. Keijzer, U.-M. O'Reilly, S.M. Lucas, E. Costa, T. Soule (Eds.), Genetic Programming. XI, 410 pages. 2004.

Vol. 3001: A. Ferscha, F. Mattern (Eds.), Pervasive Computing. XIII, 358 pages. 2004.

Vol. 2999: E.A. Boiten, J. Derrick, G. Smith (Eds.), Integrated Formal Methods. XI, 541 pages. 2004.

Vol. 2998: Y. Kameyama, P.J. Stuckey (Eds.), Functional and Logic Programming. X, 307 pages. 2004.

Vol. 2997: S. McDonald, J. Tait (Eds.), Advances in Information Retrieval. XIII, 427 pages. 2004.

Vol. 2996: V. Diekert, M. Habib (Eds.), STACS 2004. XVI, 658 pages. 2004.

Vol. 2995: C. Jensen, S. Poslad, T. Dimitrakos (Eds.), Trust Management. XIII, 377 pages. 2004.

Vol. 2994: E. Rahm (Ed.), Data Integration in the Life Sciences. X, 221 pages. 2004. (Subseries LNBI).

Vol. 2993: R. Alur, G.J. Pappas (Eds.), Hybrid Systems: Computation and Control. XII, 674 pages. 2004.

Vol. 2992: E. Bertino, S. Christodoulakis, D. Plexousakis, V. Christophides, M. Koubarakis, K. Böhm, E. Ferrari (Eds.), Advances in Database Technology - EDBT 2004. XVIII, 877 pages. 2004.

Vol. 2991: R. Alt, A. Frommer, R.B. Kearfott, W. Luther (Eds.), Numerical Software with Result Verification. X, 315 pages. 2004.

Vol. 2989: S. Graf, L. Mounier (Eds.), Model Checking Software. X, 309 pages. 2004.

Vol. 2988: K. Jensen, A. Podelski (Eds.), Tools and Algorithms for the Construction and Analysis of Systems. XIV, 608 pages. 2004.

Vol. 2987: I. Walukiewicz (Ed.), Foundations of Software Science and Computation Structures. XIII, 529 pages. 2004.

Vol. 2986: D. Schmidt (Ed.), Programming Languages and Systems. XII, 417 pages. 2004.

Vol. 2985: E. Duesterwald (Ed.), Compiler Construction. X, 313 pages. 2004.

Vol. 2984: M. Wermelinger, T. Margaria-Steffen (Eds.), Fundamental Approaches to Software Engineering. XII, 389 pages. 2004.

Vol. 2983: S. Istrail, M.S. Waterman, A. Clark (Eds.), Computational Methods for SNPs and Haplotype Inference. IX, 153 pages. 2004. (Subseries LNBI).

Vol. 2982: N. Wakamiya, M. Solarski, J. Sterbenz (Eds.), Active Networks. XI, 308 pages. 2004.

Vol. 2981: C. Müller-Schloer, T. Ungerer, B. Bauer (Eds.), Organic and Pervasive Computing – ARCS 2004. XI, 339 pages. 2004.

Vol. 2980: A. Blackwell, K. Marriott, A. Shimojima (Eds.), Diagrammatic Representation and Inference. XV, 448 pages. 2004. (Subseries LNAI).

Vol. 2978: R. Groz, R.M. Hierons (Eds.), Testing of Communicating Systems. XII, 225 pages. 2004.

Vol. 2977: G. Di Marzo Serugendo, A. Karageorgos, O.F. Rana, F. Zambonelli (Eds.), Engineering Self-Organising Systems. X, 299 pages. 2004. (Subseries LNAI).

Vol. 2976: M. Farach-Colton (Ed.), LATIN 2004: Theoretical Informatics. XV, 626 pages. 2004.

Vol. 2973: Y. Lee, J. Li, K.-Y. Whang, D. Lee (Eds.), Database Systems for Advanced Applications. XXIV, 925 pages. 2004.

Vol. 2972: R. Monroy, G. Arroyo-Figueroa, L.E. Sucar, H. Sossa (Eds.), MICAI 2004: Advances in Artificial Intelligence. XVII, 923 pages. 2004. (Subseries LNAI).

Vol. 2971: J.I. Lim, D.H. Lee (Eds.), Information Security and Cryptology -ICISC 2003. XI, 458 pages. 2004.

Vol. 2970: F. Fernández Rivera, M. Bubak, A. Gómez Tato, R. Doallo (Eds.), Grid Computing. XI, 328 pages. 2004.

Vol. 2964: T. Okamoto (Ed.), Topics in Cryptology – CT-RSA 2004. XI, 387 pages. 2004.

Vol. 2963: R. Sharp, Higher Level Hardware Synthesis. XVI, 195 pages. 2004.

Vol. 2962: S. Bistarelli, Semirings for Soft Constraint Solving and Programming. XII, 279 pages. 2004.

Vol. 2961: P. Eklund (Ed.), Concept Lattices. IX, 411 pages. 2004. (Subseries LNAI).

Vol. 2960: P.D. Mosses (Ed.), CASL Reference Manual. XVII, 528 pages. 2004.

Vol. 2958: L. Rauchwerger (Ed.), Languages and Compilers for Parallel Computing. XI, 556 pages. 2004.

Vol. 2957: P. Langendoerfer, M. Liu, I. Matta, V. Tsaoussidis (Eds.), Wired/Wireless Internet Communications. XI, 307 pages. 2004.

Vol. 2954: F. Crestani, M. Dunlop, S. Mizzaro (Eds.), Mobile and Ubiquitous Information Access. X, 299 pages. 2004.

Vol. 2953: K. Konrad, Model Generation for Natural Language Interpretation and Analysis. XIII, 166 pages. 2004. (Subseries LNAI).

Vol. 2952: N. Guelfi, E. Astesiano, G. Reggio (Eds.), Scientific Engineering of Distributed Java Applications. X, 157 pages. 2004.

Vol. 2951: M. Naor (Ed.), Theory of Cryptography. XI, 523 pages. 2004.

Vol. 2949: R. De Nicola, G. Ferrari, G. Meredith (Eds.), Coordination Models and Languages. X, 323 pages. 2004.

Vol. 2948: G.L. Mullen, A. Poli, H. Stichtenoth (Eds.), Finite Fields and Applications. VIII, 263 pages. 2004.

Vol. 2947: F. Bao, R. Deng, J. Zhou (Eds.), Public Key Cryptography – PKC 2004. XI, 455 pages. 2004.

Vol. 2946: R. Focardi, R. Gorrieri (Eds.), Foundations of Security Analysis and Design II. VII, 267 pages. 2004.

Vol. 2943: J. Chen, J. Reif (Eds.), DNA Computing. X, 225 pages. 2004.

Vol. 2941: M. Wirsing, A. Knapp, S. Balsamo (Eds.), Radical Innovations of Software and Systems Engineering in the Future. X, 359 pages. 2004.

Vol. 2940: C. Lucena, A. Garcia, A. Romanovsky, J. Castro, P.S. Alencar (Eds.), Software Engineering for Multi-Agent Systems II. XII, 279 pages. 2004.

Vol. 2939: T. Kalker, I.J. Cox, Y.M. Ro (Eds.), Digital Watermarking. XII, 602 pages. 2004.

Vol. 2937: B. Steffen, G. Levi (Eds.), Verification, Model Checking, and Abstract Interpretation. XI, 325 pages. 2004.

Vol. 2936: P. Liardet, P. Collet, C. Fonlupt, E. Lutton, M. Schoenauer (Eds.), Artificial Evolution. XIV, 410 pages. 2004.

Vol. 2934: G. Lindemann, D. Moldt, M. Paolucci (Eds.), Regulated Agent-Based Social Systems. X, 301 pages. 2004. (Subseries LNAI).

Vol. 2930: F. Winkler (Ed.), Automated Deduction in Geometry. VII, 231 pages. 2004. (Subseries LNAI).

Vol. 2929: H. de Swart, E. Orlowska, G. Schmidt, M. Roubens (Eds.), Theory and Applications of Relational Structures as Knowledge Instruments. VII, 273 pages. 2003.

Vol. 2926: L. van Elst, V. Dignum, A. Abecker (Eds.), Agent-Mediated Knowledge Management. XI, 428 pages. 2004. (Subseries LNAI).

Vol. 2923: V. Lifschitz, I. Niemelä (Eds.), Logic Programming and Nonmonotonic Reasoning. IX, 365 pages. 2004. (Subseries LNAI).

Vol. 2919: E. Giunchiglia, A. Tacchella (Eds.), Theory and Applications of Satisfiability Testing. XI, 530 pages. 2004.

Vol. 2917: E. Quintarelli, Model-Checking Based Data Retrieval. XVI, 134 pages. 2004.

Vol. 2916: C. Palamidessi (Ed.), Logic Programming. XII, 520 pages. 2003.

Vol. 2915: A. Camurri, G. Volpe (Eds.), Gesture-Based Communication in Human-Computer Interaction. XIII, 558 pages. 2004. (Subseries LNAI).

Vol. 2914: P.K. Pandya, J. Radhakrishnan (Eds.), FST TCS 2003: Foundations of Software Technology and Theoretical Computer Science. XIII, 446 pages. 2003.

Vol. 2913: T.M. Pinkston, V.K. Prasanna (Eds.), High Performance Computing - HiPC 2003. XX, 512 pages. 2003. (Subseries LNAI).

Vol. 2911: T.M.T. Sembok, H.B. Zaman, H. Chen, S.R. Urs, S.H. Myaeng (Eds.), Digital Libraries: Technology and Management of Indigenous Knowledge for Global Access. XX, 703 pages. 2003.

Vol. 2910: M.E. Orlowska, S. Weerawarana, M.M.P. Papazoglou, J. Yang (Eds.), Service-Oriented Computing - ICSOC 2003. XIV, 576 pages. 2003.

Vol. 2909: R. Solis-Oba, K. Jansen (Eds.), Approximation and Online Algorithms. VIII, 269 pages. 2004.

Vol. 2908: K. Chae, M. Yung (Eds.), Information Security Applications. XII, 506 pages. 2004.

Vol. 2907: I. Lirkov, S. Margenov, J. Wasniewski, P. Yalamov (Eds.), Large-Scale Scientific Computing. XI, 490 pages. 2004.

Vol. 2906: T. Ibaraki, N. Katoh, H. Ono (Eds.), Algorithms and Computation. XVII, 748 pages. 2003.

Vol. 2905: A. Sanfeliu, J. Ruiz-Shulcloper (Eds.), Progress in Pattern Recognition, Speech and Image Analysis. XVII, 693 pages. 2003.

Vol. 2904: T. Johansson, S. Maitra (Eds.), Progress in Cryptology - INDOCRYPT 2003. XI, 431 pages. 2003.

Vol. 2903: T.D. Gedeon, L.C.C. Fung (Eds.), AI 2003: Advances in Artificial Intelligence. XVI, 1075 pages. 2003. (Subseries LNAI).

Vol. 2902: F.M. Pires, S.P. Abreu (Eds.), Progress in Artificial Intelligence. XV, 504 pages. 2003. (Subseries LNAI).

Vol. 2901: F. Bry, N. Henze, J. Ma luszyński (Eds.), Principles and Practice of Semantic Web Reasoning. X, 209 pages. 2003.

Vol. 2900: M. Bidoit, P.D. Mosses (Eds.), Casl User Manual. XIII, 240 pages. 2004.

Vol. 2899: G. Ventre, R. Canonico (Eds.), Interactive Multimedia on Next Generation Networks. XIV, 420 pages. 2003.